Variscan-Appalachian dynamics: The building of the late Paleozoic basement

Edited by

José R. Martínez Catalán
Departamento de Geología
Universidad de Salamanca
37008 Salamanca
Spain

Robert D. Hatcher Jr.
Department of Geological Sciences
University of Tennessee
Knoxville, Tennessee 37996-1410
USA

Ricardo Arenas
Departamento de Petrología y Geoquímica
Universidad Complutense
28040 Madrid
Spain

and

Florentino Díaz García
Departamento de Geología
Universidad de Oviedo
33005 Oviedo
Spain

THE
GEOLOGICAL
SOCIETY
OF AMERICA

Special Paper 364

3300 Penrose Place, P.O. Box 9140 ▪ Boulder, Colorado 80301-9140, USA

2002

Published by The Geological Society of America, Inc.
3300 Penrose Place, P.O. Box 9140, Boulder, Colorado 80301
www.geosociety.org

Printed in U.S.A.

GSA Books Science Editor: Abhijit Basu
Cover design: Heather L. Sutphin

Library of Congress Cataloging-in-Publication Data

Variscan-Appalachian dynamics : the building of the Late Paleozoic basement / edited by José R. Martínez Catalán . . . [et al.].
p. cm.—(Special paper ; SPE 364)
Includes bibliographical references and index.
ISBN 0-8137-2364-7
1. Geology, Stratigraphic—Paleozoic—Congresses. 2. Geology, Structural—Europe—Congresses. 3. Geology, Structural—East (U.S.)—Congresses. I. Martínez Catalán, José Ramón. II. International Conference on Basement Tectonics (15th : 2000 : La Coruña, Spain) III. Special papers (Geological Society of America) ; 364.

QE654 .V365 2002
551.7′2—dc21

2002029789

Cover: The cliffs of Uzal, in the northern part of the Cabo Ortegal Complex (northwestern Spain). High-pressure and high-temperature mafic granulites of the upper tectonic unit (light gray, center) crop out in the core of a recumbent fold, surrounded by ultramafic rocks (yellowish-brown) of the same unit. Photograph by José R. Martínez Catalán, 2002.

10 9 8 7 6 5 4 3 2 1

Contents

Preface

This book contains relevant contributions from among the keynote, invited, and volunteered papers presented at the Fifteenth International Conference on Basement Tectonics, held July 4–8, 2000, at the Laboratorio Xeolóxico de Laxe near A Coruña, Galicia, in northwestern Spain, and several other independently submitted contributions. This conference is one of a series of international gatherings sponsored by the International Basement Tectonics Association (IBTA) that began with a meeting convened in Salt Lake City, Utah, more than 25 years ago. The Galicia conference, entitled "Variscan-Appalachian Dynamics: The Building of the Upper Paleozoic Basement," was based on the themes of the construction of the Paleozoic basement of Europe and eastern North America, and the evolution of the Gondwanan and Laurentian continental margins and their intervening ocean basins. This remains the general theme of this volume.

About half of the contributions composing this volume refer directly to terranes in the Iberian Massif, or include this massif in a broad context. The remaining contributions concern aspects of other Variscan-Appalachian domains, including the Urals. The weight of Iberian contribution partly reflects the fact that many of the original contributions were from Spanish scientists; moreover, there are relatively few recent compendia of Iberian Variscan tectonics. Late Paleozoic reconstruction of the continental masses (i.e., at the end of the Variscan cycle) places Iberia adjacent to the Grand Banks of Newfoundland and close to the junction of three important Paleozoic belts: the Appalachians, the Caledonides, and the Variscides. This makes the Iberian Peninsula a key site for understanding the late Paleozoic geodynamics in the circum-Atlantic. The Iberian allochthonous terranes contain much information on past plate dynamics and the evolution of the peri-Gondwanan terranes.

The chapters cover a wide range, from individual massifs to syntheses of the entire Appalachian and Variscan orogens, but all contribute to improving our knowledge of late Paleozoic plate dynamics. The tectonic syntheses in this compendium are based in part on sedimentological or petrologic identification of past geodynamic realms, such as passive continental margins, volcanic arcs, oceanic lithosphere, and subduction zones (and their polarity). Criteria derived from structural studies are also widely used. The relative positions and interplay of the major and minor plates, mainly peri-Gondwanan terranes, are described in several of the chapters according to different criteria, and a few of them pay special attention to the controversial existence of the Armorica plate.

The book begins with the contributions interpreting a given area or unit in terms of a paleotectonic environment, according to paleobiogeographical, stratigraphic, or petrologic data. Metamorphic petrology is used in several chapters to estimate pressure and temperature conditions of subducted portions of the Variscan belt, and the processes involved in their exhumation. Structural geology assumes progressively greater importance as the emplacement of allochthonous units is discussed in the following chapters, and the large-scale tectonics and plate interactions is dealt with in the last part of the book, which culminates with a review of seismic signatures in the Caledonides, Variscides, and Uralides and their tectonic implications.

Several specialists in the different fields reviewed the original manuscripts, and their suggestions and criticisms resulted in a great improvement of the quality of the definitive versions. We are grateful to the following colleagues who kindly reviewed the manuscripts: James Alcock, James R. Andrews, Denis Brown, Jean Pierre Burg, Ramón Carbonell, Rodolfo Carosi, Evan N.K. Clarkson, Peter Clift, L. Robin M. Cocks, Helmut Echtler, Jean Bernard Edel, J. Fernández Suárez, Peter Floor, Wolfgang Franke, David G. Gee, M. Charles Gilbert, Gaston Godard, Reinhard O. Greiling, G. Randy Keller, Andrew C. Kerr, Vicente López Sánchez-Vizcaíno, Jacques Malavieille, Walter V. Maresch, Stephen Marshak, Miren Mendia, J. Brendan Murphy, Franz Neubauer, Patrick J. O'Brien, Andrés Pérez-Estaún, Cecilio Quesada, Nicholas Rast, Dietrich Roeder, James Skehan, Alan G. Smith, Christian Teyssier, William A. Thomas, Arlo Weil, and Joseph B. Whalen. We especially acknowledge Nicholas Rast, whose review may have been one of the last things he accomplished before he passed away in the summer of 2001.

José R. Martínez Catalán
Robert D. Hatcher Jr.
Ricardo Arenas
Florentino Díaz García

Printed in the U.S.A.

Geological Society of America
Special Paper 364
2002

Alternative approach to the Variscan Belt in southwestern Europe: Preorogenic paleobiogeographical constraints

Michel Robardet
*Géosciences-Rennes (UMR 6118 CNRS), Université de Rennes I,
Campus de Beaulieu, F-35042 Rennes cedex, France*

ABSTRACT

Most of the models proposed for the Variscan Belt of southwestern Europe have been based on structural, metamorphic, and magmatic data. An alternative approach is proposed here, based on sedimentological and faunal data that allows the definition and delimitation of different domains and evaluation of their relative positions and relations within the preorogenic paleogeography. Several previous models, which have not taken account of pre-Variscan paleogeographical data, can no longer be sustained. In particular, models that propose two distinct oceanic sutures within the French part of the belt should be abandoned. A model with a single Variscan ocean would accord much more satisfactorily with paleobiogeographical limitations. If it is accepted that, during pre-Variscan time, the Mid-North Armorican regions were not in their present positions, but formed an extension of the Central Iberian region, a hypothesis based on faunal and sedimentary evidence, then the single-ocean model proves more appropriate for the French section of the Variscan Belt.

INTRODUCTION

Present-day geography gives only a partial and distorted picture of the Variscan Belt, which now appears as more or less isolated Variscan massifs (Fig. 1) separated by Mesozoic-Cenozoic deposits (Aquitaine and Paris Basins), younger mountain belts (Alps and Pyrenees), and oceanic areas (Bay of Biscay). Several late Variscan and post-Variscan events have strongly modified the late Paleozoic configuration, and the different parts of the Variscan Belt now constitute a puzzle (cf. Bard, 1997).

The final late Paleozoic configuration of the Variscan Belt cannot be restored easily, and the connection between the different massifs and their main structural lines has led to various interpretations. There are also conflicting views concerning geodynamic evolution, the number and the limits of the continental units, and the number and location of the oceanic sutures.

Most of the proposed reconstructions have been based on structural, metamorphic, and magmatic characteristics of the different massifs (i.e., synorogenic to late orogenic data), and have generally paid little attention to the sedimentary rocks and faunas.

An alternative approach to interpreting the Variscan Belt is to reconstitute its preorogenic paleogeography, identify and delimit the main paleogeographical units, and evaluate their affinities and relations. This approach is based on the sedimentological and faunal characteristics of the different regions; therefore, these data can be obtained only from areas that have not been strongly metamorphosed and that offer the best-preserved sedimentary and fossil records. This approach is based on the following.

1. Some lithofacies are indicative of paleoclimate and thus of paleolatitude (see references in Witzke, 1990; Scotese and Barrett, 1990); more generally, the type of succession can also be characteristic of a paleogeographic unit.

2. Benthic faunas living on epicontinental marine shelves have limited dispersive abilities (only during larval life) and generally cannot cross wide deep oceans. Such benthic faunas are therefore especially important in characterizing the different

Robardet, M., 2002, Alternative approach to the Variscan Belt in southwestern Europe: Preorogenic paleobiogeographical constraints, *in* Martínez Catalán, J.R., Hatcher, R.D., Jr., Arenas, R., and Díaz García, F., eds., Variscan-Appalachian dynamics: The building of the late Paleozoic basement: Boulder, Colorado, Geological Society of America Special Paper 364, p. 1–15.

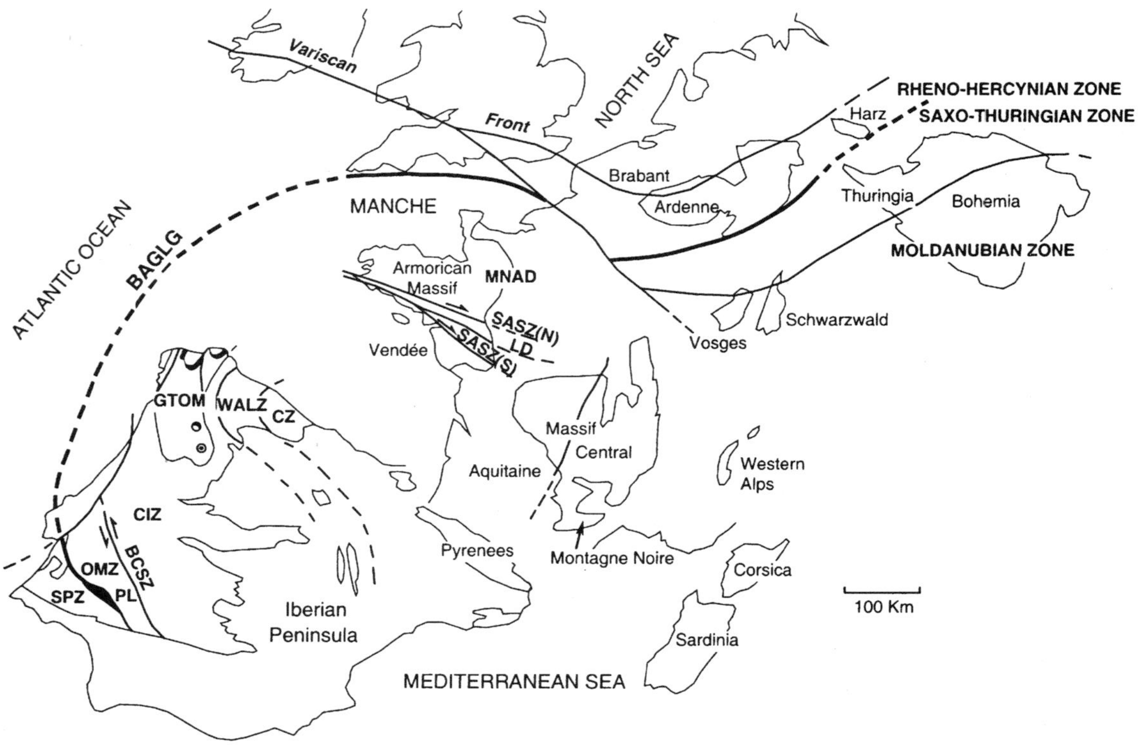

Figure 1. Variscan massifs in southwestern and central Europe. Western France—MAM, Manche magnetic anomaly; MABP, magnetic anomaly of Paris Basin; MNAD, Mid-North Armorican domain; LD, Ligerian domain; SASZ (N) and SASZ (S), South Armorican shear zone, northern and southern branches, respectively. Iberian Peninsula—CZ, Cantabrian zone; WALZ, West Asturian Leonese zone; GTOM, Galicia-Trás os Montes zone; CIZ, Central Iberian zone; OMZ, Ossa Morena zone; SPZ, South Portuguese zone; BCSZ, Badajoz-Cordoba shear zone; PL, Pulo de Lobo oceanic units.

shelf areas and evaluating their affinities, their connection or separation, and the existence of intervening oceanic areas acting as faunal barriers and controlling the geographical distribution of the faunas.

3. Marine faunas with a pelagic mode of life and a large ability for dispersion are especially important for precise biostratigraphy and for time control of the paleogeographical evolution. Moreover, their geographical distribution is controlled by latitude; therefore, they can also, in some cases (when the assemblages are strongly different), indicate the existence of wide oceanic areas (see Paris, 1993; Servais et al., 2000).

The results of this approach, based on preorogenic data, can be compared with the independent geodynamical reconstructions mainly based on synorogenic to late orogenic data. They can help to reconstitute the continuity of the different parts of the belt and can impose limitations on the models proposed to explain its formation.

Herein we consider the regions of the Variscan Belt of southwestern Europe and related areas, in order to define (1) on a large scale, the main paleogeographical units concerned and involved in the formation of the orogen; (2) on a smaller scale, the paleogeographical arrangement within the North Gondwanan province, and (3) the main paleogeographical limitations of any model proposing to explain the formation of the Variscan Belt.

GLOBAL PALEOGEOGRAPHY

Spjeldnaes (1961), on the basis of lithofacies and benthic fauna distribution, recognized and delineated the main faunal provinces and climatic zones that existed in the present day peri-Atlantic regions during the Ordovician Period.

Numerous studies subsequently corroborated and refined these views, and a paleogeographical organization comprising, during the Ordovician and the Early Silurian (Fig. 2), three major continental blocks with marine shelf areas (Laurentia, Baltica, Gondwana) separated by the Iapetus and Rheic (mid-European) oceans has been proposed (see references in Cocks and Fortey, 1982; Paris and Robardet, 1990, 1994; Robardet et al., 1990, 1993, 1994b; Scotese and Barrett, 1990; Scotese and McKerrow, 1990; Witzke, 1990).

Gondwana–North Gondwana (Armorica) relations

During the Ordovician, North Africa and southern-central Europe were parts of the North Gondwanan province, which was characterized by cold waters and high latitudes. This is attested to by the poorly diversified faunas, the almost entirely terrigenous successions, and the traces of a continental glaciation in Africa at the end of the Ordovician Period (glaciomarine de-

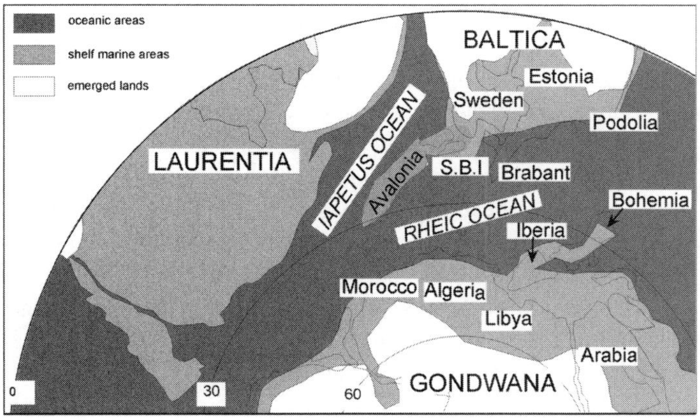

Figure 2. Early Silurian paleogeography of present-day peri-Atlantic regions (after Paris, 1993). SBI—southern British Isles; white—emergent land areas; light gray—marine shelf domains; dark gray—oceanic areas.

posits in the North Gondwanan regions), and confirmed by paleomagnetic data (see references in Kent and Van der Voo, 1990; Van der Voo, 1993; Bachtadse et al., 1995; Tait et al., 1997).

Paleontological studies have shown that the close benthic faunal affinities in the Ordovician Period between North Gondwanan Europe and Gondwana persisted until the Devonian (see references in Paris and Robardet, 1990, 1994; Mélou et al., 1999).

Recent studies of the Devonian benthic faunas (e.g., brachiopods, corals, crinoids, trilobites) from the Ougarta area (western Algerian Sahara) fully confirm strong affinities, at the species level, with the faunas of the Armorican Massif and the Iberian Peninsula (e.g., Plusquellec et al., 1997; Le Menn, 1997; Paris, 2000; Robardet et al., 2001).

Considering the modalities of dispersion of benthic species, it seems impossible that any wide ocean might have separated North Gondwanan Europe from stable Gondwana during the Paleozoic (see Fig. 2), and, as noted by Martínez-Catalán (1990), there is no evidence that Armorica (North Gondwana) was a microcontinent or a microplate detached from stable Gondwana.

This conclusion contrasts sharply with various paleogeographical scenarios suggesting that at some period of the Paleozoic a wide ocean separated North Gondwana (Armorica) from Gondwana. Such interpretations appeared in different models, and were based on paleomagnetic data poorly defined in age; most of the data were younger Variscan remagnetizations (see references in Paris and Robardet, 1990; Robardet et al., 1990, 1993, 2001; Van der Voo, 1993).

North Gondwana–Baltica relations: The Rheic ocean

However, it is also firmly established that North Gondwanan Europe and northern Europe (Baltica plus Avalonia) were separated by the mid-European Rheic ocean during the Ordovician and the Silurian (Fig. 2). This is attested to by the distribution of lithofacies (Webby, 1984; Scotese and Barrett, 1990; Witzke, 1990), benthic faunas (see references in Paris and Robardet, 1990, 1994; Robardet et al., 1994a), and the differences in latitude-controlled microplankton assemblages (Paris, 1993).

In the latest Silurian (Pridoli) and the earliest Devonian (Lochkovian), the faunal affinities between North Gondwana and the marine shelf extending along the southern margin of the Old Red Sandstone continent (Laurentia + Baltica + Avalonia = Laurussia) were very weak or nonexistent. The Rheic ocean was thus still acting as a faunal barrier for the benthic faunas. However, this was no longer the case for the pelagic microfaunas, because the same species of chitinozoans are known from both sides of the ocean (Paris, 1993). During the Pragian and the Emsian, the benthic faunal affinities between all these regions increased progressively (Racheboeuf, 1990; Robardet et al., 1990, 1993), indicating a narrowing of the Rheic ocean that heralded its closure and the general assembly of continents in the late Paleozoic.

Most of the paleogeographic models, based on paleomagnetic data, that suppose a wide ocean between Armorica (i.e., North Gondwanan Europe) and Gondwana generally consider that Armorica had joined Laurussia during the Silurian or the earliest Devonian. The evolution of the faunal affinities between southern and northern Europe shows that the Rheic ocean still existed during these periods; accordingly, such conclusions are questionable.

PALEOGEOGRAPHY OF NORTH GONDWANAN EUROPE

Considering more precisely the Variscan Belt of southwestern Europe (Fig. 3) and related areas, it appears that, during pre-Variscan time, the following was true.

1. Northern and northeastern France (Artois and Ardenne) were parts of Avalonia, and later of Laurussia (see Paris and Robardet, 1990; Robardet et al., 1990, and references therein), which also comprised the southern part of the British Isles, Belgium, and the Rheno-Hercynian zone of Germany (except the allochthonous units).
2. The rest of France (including the Armorican Massif), the entire Iberian Peninsula (except the South Portuguese zone), and the Saxo-Thuringian and Moldanubian zones of central Europe were parts of the North Gondwanan province.
3. These two paleogeographical units were separated by the mid-European Rheic ocean.

Mid-North Armorican–Central Iberian domain

The similarities between the Armorican Massif and the Iberian Peninsula (Fig. 3) led to the concept of the Ibero-Armorican arc, although the proposed correspondences between zones and domains of the two regions varied depending on the author.

The strongest similarities between the pre-Variscan Paleozoic successions and benthic faunas are between the Mid-North Armorican domain (north of the northern branch of the South Armorican shear zone) in France and the southern part (Luso-

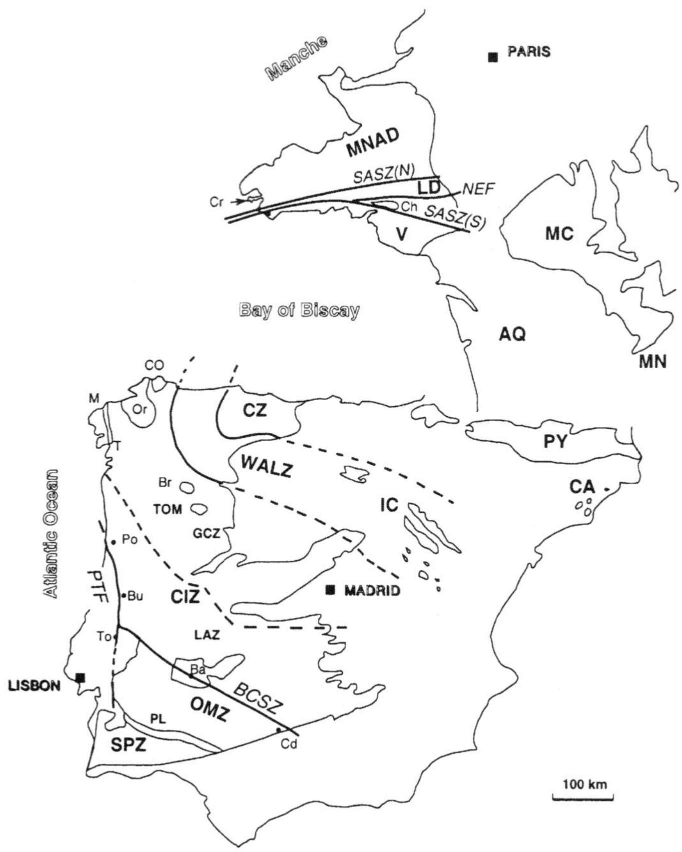

Figure 3. Variscan Belt in southwestern Europe. Abbreviations as in Figure 1. France—V, Vendée; MC, Massif Central; MN, Montagne Noire; AQ, Aquitaine; PY, Pyrénées; NEF, Nort-sur-Erdre fault; Ch, Champtoceaux; Cr, Crozon. Iberian Peninsula—GCZ, Galician-Castillan zone and LAZ, Luso-Alcudian zone (respectively northern and southern parts of Central Iberian zone, CIZ); CA, Catalonian Coastal Ranges; PTF, Porto-Tomar fault; CO, Cabo Ortegal; Or, Ordenes; M, Malpica; T, Tuy; Br, Bragança; Po, Porto; Bu, Buçaco; To, Tomar; Ba, Badajoz; Cd, Cordoba.

alcudian) of the Central Iberian zone (Central Iberian domain) in Spain and Portugal (Paris and Robardet, 1977; Robardet et al., 1990). The so-called North Armorican shear zone (Jegouzo, 1980) did not correspond to a paleogeographical boundary; the same lithological successions and faunas occur in the North and Middle Armorican regions (see Paris and Robardet, 1977; Lardeux et al., 1977). Sedimentary and benthic faunal similarities between the Mid-North Amorican domain and the Central Iberian domain are especially well documented for the Ordovician, but similar affinities also persisted during the Silurian and the Devonian. Discrete sedimentary and faunal differences allow distinction of three subdomains that existed both in the Mid-North Amorican and Central Iberian domains during the Middle Ordovician (Fig. 4), with a similar but symmetrical bathymetric polarity in the Armorican Massif (deepening from north to south) and the Iberian Peninsula (from south to north). There is a striking similarity in the entire Ordovician succession (Fig. 5) and benthic faunas (at the species level) between Crozon (western Brittany) and Buçaco (central Portugal) (Henry et al., 1974; Robardet et al., 1990; Young, 1989, 1990).

This shows that both regions were parts of the same Mid-North Armorican–Central Iberian domain and that western Brittany and central Portugal were contiguous within this domain; some parts of the Ordovician succession that occur both at Buçaco and Crozon cannot be followed in detail over a long distance in either of the two regions.

The Mid-North Armorican–Central Iberian domain thus appears to be a key element for the pre-Variscan paleogeographical reconstructions of southwestern Europe.

The Ordovician-Devonian successions of the whole Mid-North Armorican–Central Iberian domain are dominated by proximal shallow shelf lithofacies and benthic faunas, frequently corresponding to tidal or littoral environments, as are the arenaceous sediments deposited during the Arenig, the Caradoc, and the Pridoli-Lochkovian. The Mid-North Amorican and the Central Iberian domains are clearly the regions of southern Europe that show the shallowest depositional environments

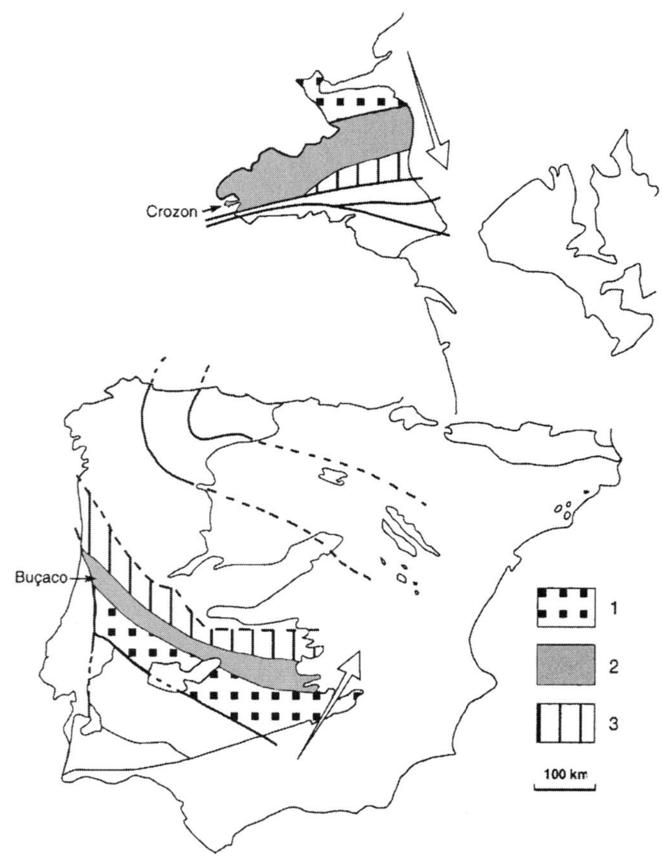

Figure 4. Iberian-Armorican relations in Middle Ordovician: sedimentary and faunal subdomains in Mid-North Armorican and Central Iberian domains (after Robardet et al., 1990). 1—dominant sandstone lithofacies; 2—shale, siltstone, and sandstone lithofacies; 3—shale-siltstone lithofacies; arrows indicate deepening trend in proximal part of marine shelf.

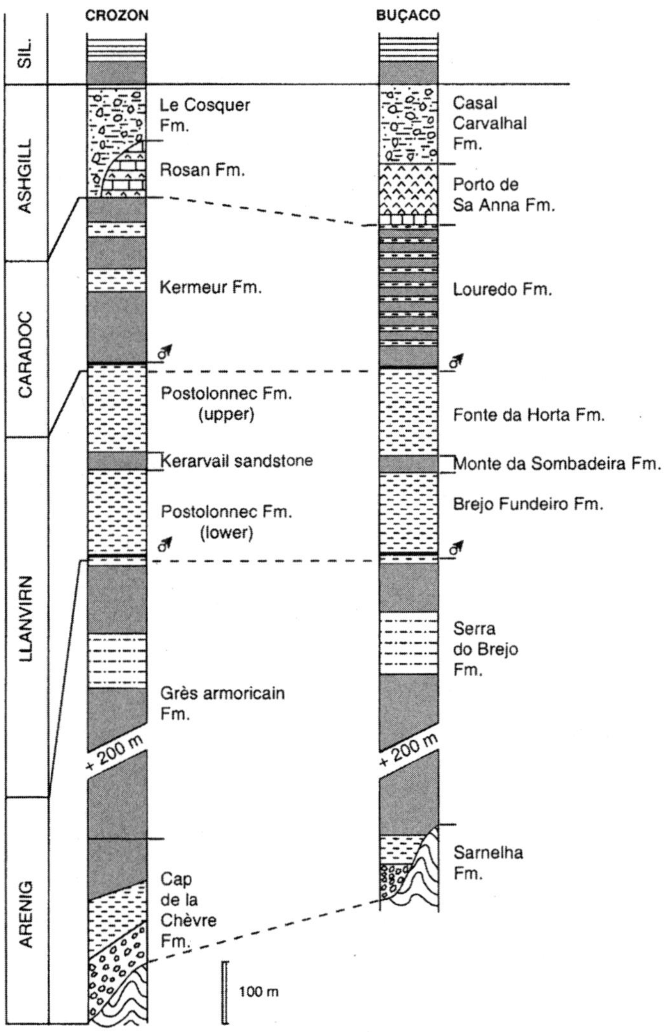

Figure 5. Ordovician successions of Crozon, western Brittany, and Buçaco, central Portugal (after Robardet et al., 1990). Nomenclature of formations (Fm.) for Buçaco is after Young (1990). Ordovician series are after Fortey et al. (1995).

during pre-Variscan time. The northern position, within North Gondwanan Europe, of the Mid-North Amorican domain in present-day geography thus appears a paradox. Deep distal shelf or continental slope environmental characteristics should have been expected in the northernmost external part of the North Gondwanan province, just at the southern border of the Rheic ocean.

The Mid-North Amorican domain disappears under the Mesozoic-Cenozoic deposits of the Paris Basin, which prevents observation of its eastern prolongation. However, in many reconstructions of the Variscan Belt, the Mid-North Amorican domain is supposed to extend into the Saxo-Thuringian zone of Germany, although some (see Franke and Engel, 1982) have serious doubts about whether this is realistic. When the pre-Variscan Paleozoic successions of the Mid-North Amorican domain and the Saxo-Thuringian zone are considered, it appears that, although both regions are North Gondwanan, they do not show close similarities. The Early-Middle Ordovician proximal shelf deposits (e.g., the Grès Armoricain Formation) that are so typical of the Mid-North Amorican domain (and the Central Iberian domain) do not exist in the Saxo-Thuringian zone, which shows more distal and deeper lithofacies (see Heuse, 2000). The Silurian–earliest Devonian successions are also very different; there are shallow proximal shelf formations in the Mid-North Amorican domain and a condensed, totally euxinic, deeper outer shelf succession in the Saxo-Thuringian zone. There is therefore no sedimentary or faunal argument that might suggest that the Mid-North Amorican domain and the Saxo-Thuringian zone were closely connected during pre-Variscan Paleozoic time.

However, although the two regions are currently juxtaposed along the South Armorican shear zone (Jegouzo, 1980), the pre-Variscan Paleozoic succession of the Mid-North Amorican domain does not show any trace of the tectonothermal eo-Variscan events (Silurian-Devonian) that characterize the geodynamical evolution of southern Brittany (see following). In the Mid-North Amorican domain sedimentation was continuous, without any break other than those related to the Late Ordovician glaciation, and the first indications of tectonic events only appeared in the latest Devonian or earliest Carboniferous with the so-called Bretonic phase. This is a supplementary argument to demonstrate that, during pre-Variscan time, the Mid-North Amorican domain was not juxtaposed or close to southern Brittany.

It has therefore been considered that the present-day position of the Mid-North Amorican domain does not correspond to its original pre-Variscan location, and a southern location has been recently proposed (Fig. 6) as a direct prolongation of the Central Iberian zone of the Iberian Peninsula (Hammann, 1992; Robardet et al., 1994b).

Although not totally understood, the important north-northwest–south-southeast geophysical magnetic anomaly detected under the Paris Basin (Fig. 1), between Rouen and Bourges (see Chantraine et al., 1992; Autran et al., 1994), might correspond to the eastern limit of the Mid-North Amorican domain block or terrane, and it can be imagined that the basic-ultrabasic rocks supposed to be the cause of the magnetic anomaly could correspond to part of the oceanic crust of the Rheic ocean.

When a southern location and a reverse north-south orientation are accepted for the Mid-North Amorican domain during pre-Variscan time, the Late Proterozoic Cadomian Belt and the old Icartian relicts of Early Proterozoic age (2000 Ma) of northern Brittany and Normandy can be more easily integrated in the global context of the Pan-African orogeny of Gondwana (see Nance and Murphy, 1996).

An original southern position of the Mid-North Amorican domain has important consequences for understanding the Variscan Belt. In this paleogeographical arrangement (Fig. 6), (1) there was no marine shelf area between the southern border of Baltica-Avalonia (and later Laurussia) and North Gondwana, (2) the northernmost units of North Gondwana corresponded to the southern regions of the Armorican Massif and to the French Mas-

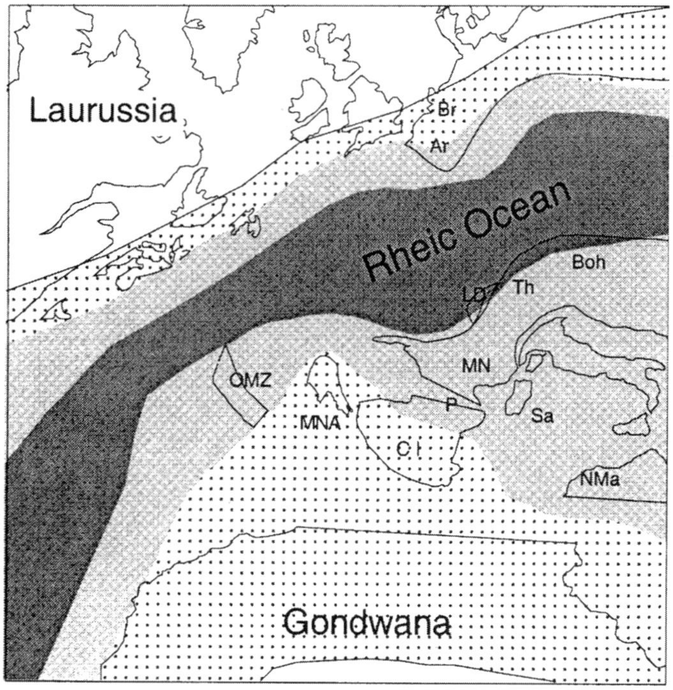

Figure 6. Pre-Variscan Late Silurian paleogeographic reconstruction (after Robardet et al., 1994b, slightly modified). White—emergent land areas; dotted—marine continental shelf proximal parts; light gray—marine continental shelf distal parts; dark gray—oceanic areas; Br—Brabant; Ar—Ardenne; Boh—Bohemia; Th—Thuringia; LD—Ligerian domain; MN—Montagne Noire; P—Pyrénées; Sa—Sardinia; OMZ—Ossa Morena zone; MNA—Mid-North Armorican domain; CI—Central Iberian domain; N Ma—northern Maghreb.

sif Central, and (3) there was no South Armorican ocean separating the Mid-North Amorican domain and southern Brittany.

Other domains in the Iberian Peninsula

From the southern Central Iberian zone, in a northward direction, the rest of the Iberian Peninsula, except the Cantabrian zone and the Western Pyrenees, i.e., the northern part of the Central Iberian zone, the West Asturian Leonese zone, Catalonia and the Central and Eastern Pyrenees (Fig. 3), show a global deepening trend toward deeper outer shelf conditions. This is suggested by the Ordovician successions (Robardet and Gutiérrez-Marco, 1990b) and even more clear for the Silurian–Early Devonian sequences (Gutiérrez-Marco et al., 1998; Sarmiento et al., 1999).

In the southern part of the Iberian Peninsula (Fig. 3), the Ossa Morena zone is now juxtaposed to the Central Iberian zone, along the Tomar-Badajoz-Cordoba shear zone. Although both regions are clearly North Gondwanan, the pre-Variscan Paleozoic of the Ossa Morena zone distinctly differs from that of the Central Iberian zone and comprises more distal lithofacies and faunas, typical of outer shelf environments (see Robardet, 1976; Robardet and Gutiérrez-Marco, 1990a, 1990b). This is especially clear for the Silurian and Early Devonian characteristics that closely resemble those of Thuringia (Jaeger and Robardet, 1979). The juxtaposition of the two zones does not reflect their original geographical relations (Fig. 6), and took place during the Variscan orogeny, due to sinistral strike-slip movements along the Tomar-Badajoz-Cordoba shear zone.

The South Portuguese zone (Fig. 3), which only shows Upper Devonian and Carboniferous rocks, is separated from the Ossa Morena zone by the South Iberian Pulo de Lobo oceanic suture (see following), and was probably part of Avalonia during pre-Variscan time (see Oliveira and Quesada, 1998).

Other domains in France

Southern Armorican Massif. To the south of the northern branch of the South Armorican shear zone (Jegouzo, 1980), the southern part of the Armorican Massif (Fig. 3) comprises the metamorphic units of southern Brittany and the Paleozoic units of the Ligerian and the Vendean domains. These South Armorican domains are separated by a series of anastomosed west-northwest–east-southeast or west-east thrusts and strike-slip faults. These strike-slip faults were active for a long time during the Paleozoic and the distinct units they separate have probably undergone large displacements; they now constitute a heterogeneous assemblage of allochthonous units, the original pre-Variscan relations of which can hardly be recognized (see Martínez-Catalán, 1990; Shelley and Bossière, 2000). It is thus difficult to have a clear idea of the pre-Variscan configuration of these domains or to reconstitute their geodynamic history.

Southern Brittany. Southern Brittany (south of the southern branch of the South Amorican shear zone) comprises several plutonic intrusions and various metamorphic complexes with isolated occurrences of high-pressure–low-temperature blueschists, fragmented oceanic relicts (in eclogite facies), and a high-temperature–low-pressure migmatite belt with associated micaschists and orthogneiss bodies. Southern Brittany and its southeastern prolongation into the Massif Central (Fig. 3) are generally considered to represent the active zone of the western part of the Variscan Belt (Ligerian Cordillera of Cogné, 1977; Ligerian Mobile zone of Jones, 1991), and are key areas for reconstructing the tectonic-metamorphic history of the Variscan regions in southwestern Europe. However, there are contrasting views regarding the course of evolution as proposed by different authors. Some consider that the Ligerian Cordillera resulted from a continuous continental convergence from the Late Silurian to the early Carboniferous. Others propose a polycyclic discontinuous evolution, with (1) a Late Silurian–Early Devonian eo-Variscan stage of subduction and collision (Ballèvre et al., 1994) and (2) an early Carboniferous Variscan sensu stricto postcollisional stage with intracontinental deformation (see references in Keppie, 1994; Faure et al., 1997).

Other conflicting views between the different interpretations and models concern the number of continental blocks and oceans involved in these processes, the location of the oceanic suture (or sutures), and the direction of subduction.

Ligerian units. The Ligerian domain (Fig. 3) extends between the northern and the southern branches of the South Armorican shear zone and comprises the southern part of the Saint-Julien-de-Vouvantes unit and the Saint-Georges-sur-Loire and Ancenis units. In these units, the pre-Carboniferous Paleozoic rocks occur as dispersed and discontinuous outcrops that have been interpreted as olistoliths and olistostromes redeposited within early Carboniferous basins (see Dubreuil, 1980, 1987; Lardeux and Cavet, 1994). These pre-Carboniferous rocks clearly differ from those of the Mid-North Amorican domain and correspond to deeper and distal North Gondwanan deposits (mainly shales, siltstones, and carbonates) with abundant pelagic fossils and benthic assemblages of Bohemian affinities. The present relations of the Ligerian units and the Mid-North Amorican domain, abruptly juxtaposed along the northern branch of the South Amorican shear zone, without any gradual transition, and the probably olistolithic character of the Ligerian rocks make difficult the reconstitution of the original relations between these two domains. However, it seems that the pre-Carboniferous Ligerian sediments were deposited originally in some area of the outer part of the North Gondwanan shelf that had simple connections with the Barrandian area of Bohemia (and was thus possibly situated in a more eastern position).

Considering the Paleozoic succession reconstructed (from isolated outcrops) for the Ligerian domain (Le Corre et al., 1991; Lardeux and Cavet, 1994), where no hiatus can be detected before the chaotic sedimentation of the early Carboniferous, leads to the conclusion that the domain where these pre-Variscan sediments were originally deposited had never registered any echo of eo-Variscan events that characterize southern Brittany.

Vendean domain. The Vendean domain (Fig. 3; see Colchen and Poncet, 1994) comprises several stacked lithostructural units, the geometry of which is not completely understood. Within the Paleozoic rocks, the fossil record is poor and the tentative stratigraphical assignments imprecise. However, the latest Silurian and/or Early Devonian volcaniclastic rocks of the Mareuil-sur-Lay Formation and the post Middle Devonian basaltic rocks of the La Meilleraie Formation (with oceanic and arc tholeiites suggesting a backarc basin) could possibly be related to eo-Variscan events.

Southern France. In the southern parts of France (Fig. 3), Paleozoic rocks are known from the basement of the Aquitaine Basin, the Montagne Noire (southern Massif Central), and the Pyrenees (see references in Paris and Le Pochat, 1994; Feist et al., 1994; Robardet et al., 1994b). The pre-Variscan successions of these regions consist mainly of Ordovician shaly and silty deposits, Silurian black shales with black cherts and limestones, and Devonian limestones. These sediments show outer shelf (or even slope) characteristics, confirmed by the abundance of pelagic faunas. These regions are thus considered to have been part of the deeper northern part of the North Gondwanan shelf, on the southern border of the Rheic ocean. Such a location is corroborated by the sporadic occurrence, within typical North Gondwanan faunal assemblages, of invader faunal elements of Baltic or Avalonian origin (e.g., Mélou, 1982). This immigration of Baltic taxa was repeated on several occasions during the early Paleozoic. It was probably due to favorable oceanic currents that allowed these northern elements to cross the mid-European Rheic ocean and to reach the northernmost part of the North Gondwana shelf. This phenomenon is more marked in the Barrandian area of central Europe (cf. Perunica microcontinent of Havlicek et al., 1994), whereas it was much more exceptional, almost nonexistent in the Mid-North Amorican and Central Iberian domains.

DISCUSSION: CHECKING GEODYNAMICAL MODELS

Considering the preceding discussion of pre-Variscan Paleozoic successions of the different regions of southwestern Europe, and the resulting pre-Variscan paleobiogeographical limitations, I check various geodynamical models proposed for the Variscan Belt of southwestern Europe against these data and see to what extent these models are in accordance with these limitations.

Main paleogeographical constraints

It seems clear that no wide ocean ever separated North Gondwanan southern Europe from stable Gondwana during the Paleozoic, and no wide ocean ever existed within the North Gondwanan regions (Figs. 2 and 6). A small and narrow ocean would probably not be detected by comparison of the sedimentary successions and by the distribution of benthic faunas, but closure of such a small oceanic area could not have been responsible for the formation of an orogenic belt.

This means that the Variscan ocean, the closure of which led to the Variscan orogeny, must have been located outside the Gondwanan and North Gondwanan regions; the best candidate is the mid-European Rheic ocean that separated North Gondwana from Baltica-Avalonia and later Laurussia, the existence of which is clear from the geographical distribution of lithofacies and benthic faunas.

Considering more precisely southwestern Europe, there is no doubt that the Mid-North Armorican domain and all other Armorican regions were parts of North Gondwana, and it is thus impossible to accept any model where the Mid-North Amorican domain would be considered as part of (or annexed to) a northern continental block that collided with Gondwana after closure of an intervening wide ocean.

Moreover, the Mid-North Armorican and Central Iberian regions constituted a continuous inner shelf sedimentary domain within which western Brittany (Crozon) and central Portugal (Buçaco) were contiguous. The original pre-Variscan location of the Mid-North Amorican domain was different from the present one, as a direct prolongation of the Central Iberian regions.

This paleogeographical arrangement (Figs. 2 and 6) implies that a South Armorican Ligerian ocean distinct from the Rheic ocean did not exist, and that these two supposed oceans were one unique mid-European Rheic ocean, and the south Armorican regions and southern France were situated in the northern outer part of the shelf of North Gondwana.

Models that have partly taken into account these constraints

A few authors have paid some attention to the sedimentary and faunal characteristics of the different regions of the Variscan Belt and have more or less taken into account some of the paleogeographical constraints summarized here.

In an early attempt to connect the different parts of the Variscan Belt of southwestern Europe, Matte (1976) made a review of the pretectonic markers that included the pre-Paleozoic basement and the Paleozoic successions. However, at that time the paleobiogeographical analysis was not yet very precise, and several of the similarities proposed by Matte for the Paleozoic now appear unfounded (e.g., the connection between the northern part of the Armorican Massif and the Ossa Morena zone of the Iberian Peninsula).

Julivert and Martinez (1983, 1987) also paid attention to faunal and sedimentological data, especially the strong similarities between the Mid-North Armorican and the Central Iberian regions. They discussed the different geodynamical models proposed at that time and, making an attempt at integrating all available data, considered that the Variscan Belt was probably a complex mosaic of microplates separated by small oceanic areas.

The Iapetus model of Neugebauer (1989) is explicitly based on the fact that, during the Paleozoic, all the Variscan regions were interconnected and formed part of the North Gondwanan margin, which implies that no wide ocean existed within these areas and that the Variscan ocean must have been outside the Variscan regions. In this model (Fig. 7), the oceanic area between North Gondwana and Avalonia is regarded as a marginal basin of the Iapetus ocean and Variscan evolution is considered to have resulted from the subduction of this marginal basin and from the compression and shortening of the North Gondwanan margin, which consisted of ridges and ensialic basins (probably with patches of oceanic crust). Some of these ideas are similar to those propounded herein, although the oceanic area separating North Gondwana and the southern part of Laurussia is regarded here as a true major ocean and not a marginal basin of the Iapetus. However, there are clear differences in detail; Neugebauer (1989, Figs. 13–16) considered the Mid-North Armorican domain as part of Avalonia and unrelated to the Iberian Peninsula located within the North Gondwanan margin, on the other side of an oceanic area.

Martínez Catalán (1990) also gave an important place to the sedimentological and faunal data, both on a large scale (Gondwana-Laurussia relations) and on a smaller scale (Iberian-Armorican relations), and dismissed the idea that Armorica was a microplate independent from Gondwana. He also stressed the fact that it was impossible to propose a cylindrical model for the Variscan Belt in southwestern Europe, because the present configuration of the South Armorican regions is the result of transcurrent movements (mostly dextral) that occurred during the convergence between Gondwana and Laurussia. This paper (Martínez Catalán, 1990) has probably been more critical than any other about previous schemes and currently accepted ideas.

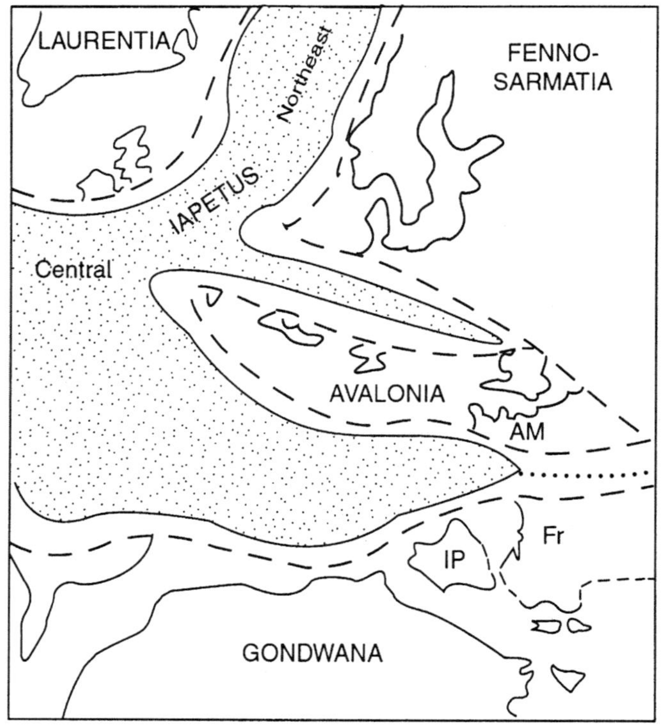

Figure 7. Iapetus model (after Neugebauer, 1989, simplified). AM—Armorican Massif; Fr—France; IP—Iberian Peninsula; dotted—oceanic areas.

Ballèvre et al. (1992) proposed (Fig. 8) Iberian-Armorican correlations based on (1) the strong affinities between the Mid-North Americal and Central Iberian domains, and (2) the similar lithologies and metamorphic histories of the Champtoceaux nappe in the Armorican Massif and the Malpica-Tuy unit in the northwestern Iberian Peninsula. These correlations suppose that a South Armorican suture (cut by the southern branch of the South Amorican shear zone) would prolongate along the western limit of the Malpica-Tuy unit and, within the Central Iberian zone, along the limit between the "domain of vertical folds" (cf. Luso-Alcudian zone) and the "domain of recumbent folds" (cf. Galician-Castillan zone) of the Central Iberian zone. In this interpretation, the paleobiogeographical relations between the Mid-North Americal domain and the southern Central Iberian zone are fully respected. However, there are two difficulties: (1) the supposed prolongation of the suture does not appear clearly within the Central Iberian zone, and (2) within the Iberian Peninsula, as well as within the Armorican Massif, the proposed suture would separate Paleozoic units that were all located within North Gondwana. The correspondences proposed by Ballèvre et al. (1992) for (1) the Mid-North Americal and the south-

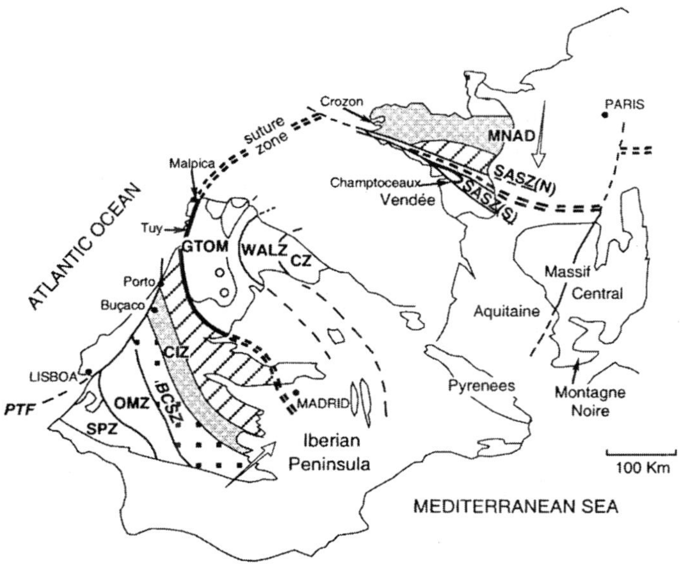

Figure 8. Iberian-Armorican correlations (after Ballèvre et al., 1992). Same abbreviations as in Figure 1. PTF—Porto-Tomar fault; subdomains within Mid-North Armorican and southern Central Iberian domains, see Figure 4.

ern Central Iberian regions and (2) the Champtoceaux nappe and the Malpica-Tuy unit are still valid and must be maintained. However, the connection between the Armorican Massif and the Iberian Peninsula cannot be regarded as an oceanic suture.

The presence of such a suture within the Central Iberian zone can also be found in Burg (1999, Fig. 18) and in Bresser and Walter (1999, Fig. 8); in these models, the Mid-North Armorican–Central Iberian domain appears as an intermediate domain between Laurussia and North Gondwana (that includes northern Iberia and southern France). This would suggest that the Mid-North Armorican–Central Iberian domain was detached from Gondwana and separated from it by an ocean, a situation that does not match the paleontological and biogeographical data.

Other models

Models without oceanic closure. Considering the converging data provided by lithological, paleontological, and paleomagnetic studies, it seems clear that the Variscan Belt resulted from an oceanic closure that led to the collision of Laurussia and Gondwana. It is thus impossible to accept any model that tries to explain the Variscan Belt without reference to such plate tectonic processes. Some authors proposed an ensialic evolution for the Variscan Belt of Europe, but these views seem now abandoned (see references and discussion in Bard et al., 1980; Julivert and Martinez, 1987).

The model proposed by Vai (1991) goes back to more or less similar "fixist" concepts and considers (1) that there is no support for a wide ocean separating northern Europe from southern Europe and (2) that the Variscan Belt resulted from fragmentation and displacement of microblocks generated by strike-slip, oblique rift pulses within a complex large-scale transform system connecting two oceans extending, respectively, to the east and west of Gondwana. This interpretation does not match satisfactorily the paleogeographical data.

Models where the Armorican Massif is not North Gondwanan. It is also impossible to accept the various models that consider (explicitly or implicitly) that the Armorican regions were not part of North Gondwana and were separated from it by an oceanic area. In such models, either the whole Armorican Massif, or the Mid-North Amorican domain, are regarded as subsidiary blocks appended to northern Europe (cf. Riding, 1974; Badham and Halls, 1975; Lefort et al., 1988), as part of Avalonia (Lefort, 1983, 1990; Cogné and Lefort, 1985; Neugebauer, 1989), or as a microplate (Piqué et al., 1994) that had detached and drifted away from Gondwana. As argued here, such paleogeographical assignments cannot be accepted.

This is the case of the model (Lefort, 1983, 1990; Cogné and Lefort, 1985; Lefort et al., 1988) based on geophysical criteria (Fig. 9). According to this model, (1) the South Armorican suture is currently situated offshore, south of the Armorican Massif, and can be followed westward as far as the Petite Sole Bank; (2) this suture is interrupted by the Petite Sole–Porto–Badajoz–Cordoba transform fault, and (3) finds its prolongation in the southern part of the Iberian Peninsula, between the Ossa Morena and South Portuguese zones. This South Iberian suture extends westward as far as southern New England (Collector Anomaly suture). This model considers that the proposed oceanic suture represents the northern boundary of Gondwana, which means that (1) the entire Armorican Massif was not North Gondwanan, but detached from Gondwana and part of Laurussia, and (2) the Armorican Massif and the Iberian Peninsula were separated by the South Armorican ocean. Both propositions are in complete contradiction with what is now known about global pre-Variscan paleogeography and the Iberian-Armorican relations.

Models with two Variscan oceans. Many models do not take into account the paleogeographical affinities of the different regions and focus only on their tectonic-metamorphic characteristics and on the rocks of supposed oceanic origin. These models differ in their evaluation of the number of oceanic sutures and with respect to the connections they propose between the Armorican and the Iberian parts of the belt. Despite their diversity, most of them do not satisfactorily accord with the known paleogeographical data.

Several authors, whose models have much in common, and have progressively integrated new data (see Bard et al., 1980; Matte 1986a, 1986b, 1991; Bard, 1997), suppose, in the Armorican Massif, the existence of two distinct oceanic sutures corresponding to two distinct Paleozoic oceans, the South Armorican ocean (or Ligerian or Massif Central ocean) and the Rheic ocean (Figs. 10 and 11). The Rheic ocean, separating the Mid-North Amorican domain and the southern border of Laurussia, is at-

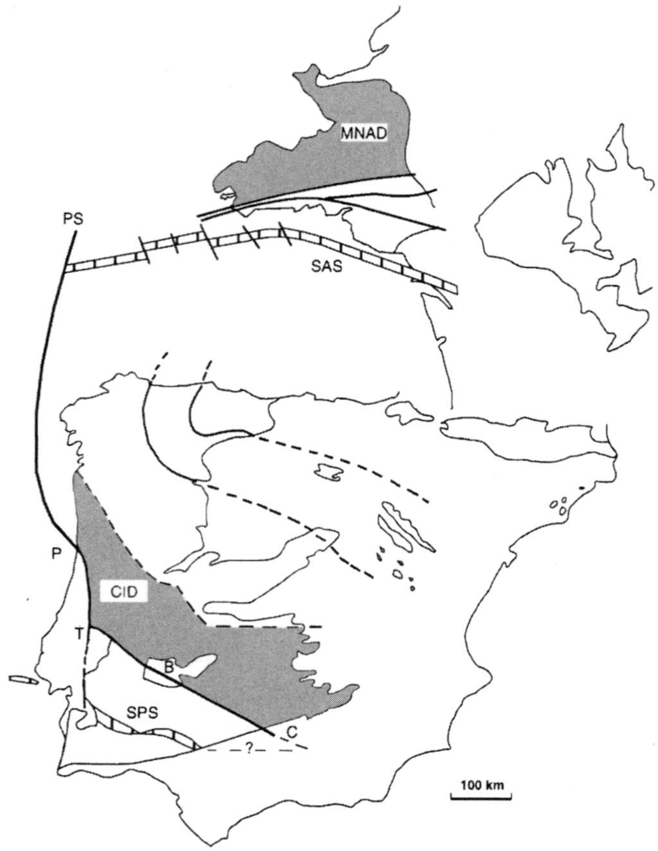

Figure 9. Model based on geophysical criteria (after Lefort, 1983, 1990). SAS—South Armorican suture; SPS—South Portuguese suture; MNAD—Mid-North Armorican domain; CID—Central Iberian domain; PS—Petite Sole Bank; P—Porto; T—Tomar; B—Badajoz; C—Cordoba.

tested by the contrast in the distribution of benthic fauna, whereas the South Armorican ocean, between the Mid-North Amorican regions and southern Brittany, is deduced from the tectonic-metamorphic evolution of southern Brittany.

Moreover, most of these authors consider that the South Armorican suture is extended, in the Iberian Peninsula, into the Badajoz-Cordoba shear zone, which would represent (Bard et al., 1980; Burg et al., 1981) a cryptic early Paleozoic oceanic suture later obliterated by a major Variscan shear zone. This interpretation was maintained (Matte, 1991), even when the South Iberian Beja–Acebuches–Pulo de Lobo units were recognized as oceanic, which would bring the number of Variscan oceans in southwestern Europe to three (Fig. 11).

It is generally considered (Matte, 1986a, 1986b, 1991; Jones, 1991) that the Rheic ocean subducted toward the south and the South Armorican–Massif Central ocean subducted toward the north. However, Faure et al. (1997) considered that the latter hypothesis is not supported by any conclusive argument and that a southward subduction can equally well be proposed.

In these models, (1) the South Armorican oceanic suture crosscuts the Mid-North Amorican–Central Iberian domain and

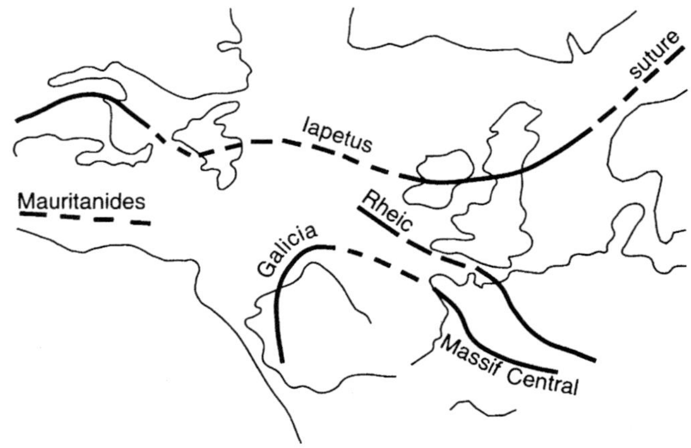

Figure 10. Oceanic sutures in Variscan Belt of southwestern Europe and related areas (after Matte, 1986a, 1986b).

places the Mid-North Amorican regions and the Central Iberian regions in two distinct continental blocks of the pre-Variscan paleogeography, which is in contradiction with the Iberian-Armorican relations described herein, and (2) the Iberian part of this supposed suture would separate the Central Iberian regions from the Ossa Morena zone, when both were parts of the same North Gondwanan shelf (see Fig. 12).

Such models with two oceanic sutures have been proposed by French geologists on the basis of the study of the French part of the Variscan Belt, whereas most Spanish and Portuguese geologists consider that the Iberian Peninsula only shows a single suture, in its southern part, between the Ossa Morena and the South Portuguese zones. It thus appears that the necessity of two oceanic sutures is due to the fact that these models have been

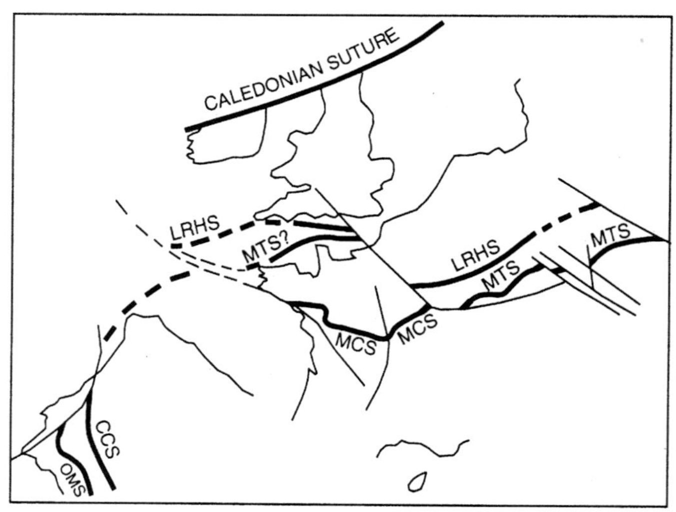

Figure 11. Oceanic sutures in Variscan Belt of southwestern and Central Europe (after Matte, 1991). LRHS—Lizard-Rhenish suture; MTS—Munchberg-Tepla suture; MCS—Massif Central suture (South Armorican suture); CCS—Coimbra-Cordoba suture; OMS—Ossa Morena suture (Pulo de Lobo suture).

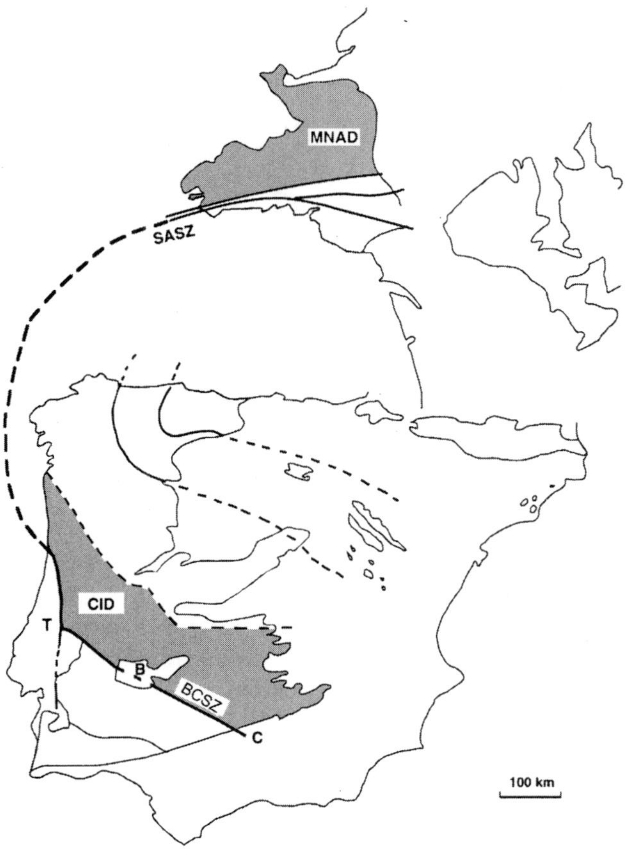

Figure 12. Schematic map illustrating disagreement between (1) geodynamical models with oceanic suture connecting South Armorican shear zone (SASZ) and Badajoz-Cordoba shear zone (BCSZ) and (2) preorogenic Iberian-Armorican relations between MNAD (Mid-North Amorican) and CID (Central Iberian) domains. T—Tomar; B—Badajoz; C—Cordoba.

elaborated on the basis of a paleogeographical configuration almost similar to the present-day geography, and infer that the Mid-North Amorican regions were already in their present northern position during pre-Variscan time.

When an original pre-Variscan southern position of the Mid-North Amorican domain is adopted (Hammann, 1992; Robardet et al., 1994b), it appears that (1) a separate South Armorican ocean did not actually exist and corresponded to the southern part of a unique Rheic ocean separating northern Europe (Laurussia) from southern Europe (North Gondwana), and (2) the regions located within the northernmost part of the North Gondwana shelf were the Montagne Noire, the Pyrénées, the Aquitaine, probably the Vendée, and the regions of the Massif Central and southern Brittany that now appear as metamorphic areas.

Models with a single Variscan ocean. A number of Spanish and Portuguese geologists consider (1) that the Iberian Peninsula (see Figs. 3 and 13) only shows a single suture, in its southern part, between the Ossa Morena and the South Portuguese zones, and (2) that the Badajoz-Cordoba shear zone does not correspond to the suture of a Paleozoic ocean but only to a major strike-slip fault (see Ribeiro et al., 1990; Quesada, 1991; Quesada et al., 1994).

The South Iberian suture comprises the Beja-Acebuches Ophiolitic Complex (probably corresponding to a marginal backarc basin) and the Pulo de Lobo accretionary terrane (mostly metasediments with ophiolitic remnants). This oceanic suture (Pulo de Lobo ocean) is considered to extend west of Galicia (Galicia ocean), under the present-day Atlantic Ocean (Fig. 13).

The connection of this unique Iberian suture with one of the two sutures described in France then becomes a serious problem. Quesada et al. (1994, p. 43–44) noted that the simplest model would be to consider that the Pulo de Lobo–Galicia ocean was connected with the Rheic ocean and that the South Armorican–Massif Central ocean was only a minor bifurcation of this main ocean. The problem would become simpler if it is accepted, as argued and proposed here, that the South Armorican–Massif Central ocean and the Rheic were in fact the same unique ocean.

When a simpler model, with a single Rheic ocean, is adopted, it appears that this ocean (Fig. 13) would have combined (in present geographical coordinates) the following.

1. A northern part (Rheic ocean) separated the southern border of the European part of Laurussia and the South Armorican part of the North Gondwanan margin, and subducted toward the south, under the North Gondwanan active margin.

2. A middle part (Galicia ocean), the suture of which is west of Galicia, was the origin of the allochthonous ophiolitic rocks obducted onto the northwestern part of the Iberian autochthon (see references in Martínez Catalán, 1990; Martínez Catalán et al., 1997).

3. A southern part (Pulo de Lobo ocean) subducted northward, as attested by the magmatic polarity within the Ossa Morena zone (Quesada et al., 1994).

Quesada et al. (1994, p. 44) considered that the Pulo de Lobo–Galicia–Rheic ocean, continuous around the Ibero-Armorican Arc, "would have been subducted with a constant sense, toward the inner part of the curved structure." This point remains a problem, because the Galicia ocean is considered to have subducted toward the west (Martínez Catalán et al., 1997).

CONCLUSIONS

The Variscan Belt can be investigated through sedimentological and faunal studies in the regions where fossiliferous sedimentary rocks have been preserved. Those studies allow (1) recognition and delimitation of different domains with distinct faunal, sedimentological, and paleoenvironmental characteristics, and (2) evaluation of their relative positions and mutual relations within the preorogenic paleogeographical configuration. It is possible to distinguish clearly those regions that were part of the North Gondwanan shelf from those that were located in Baltica-Avalonia and later on the southern border of Laurussia,

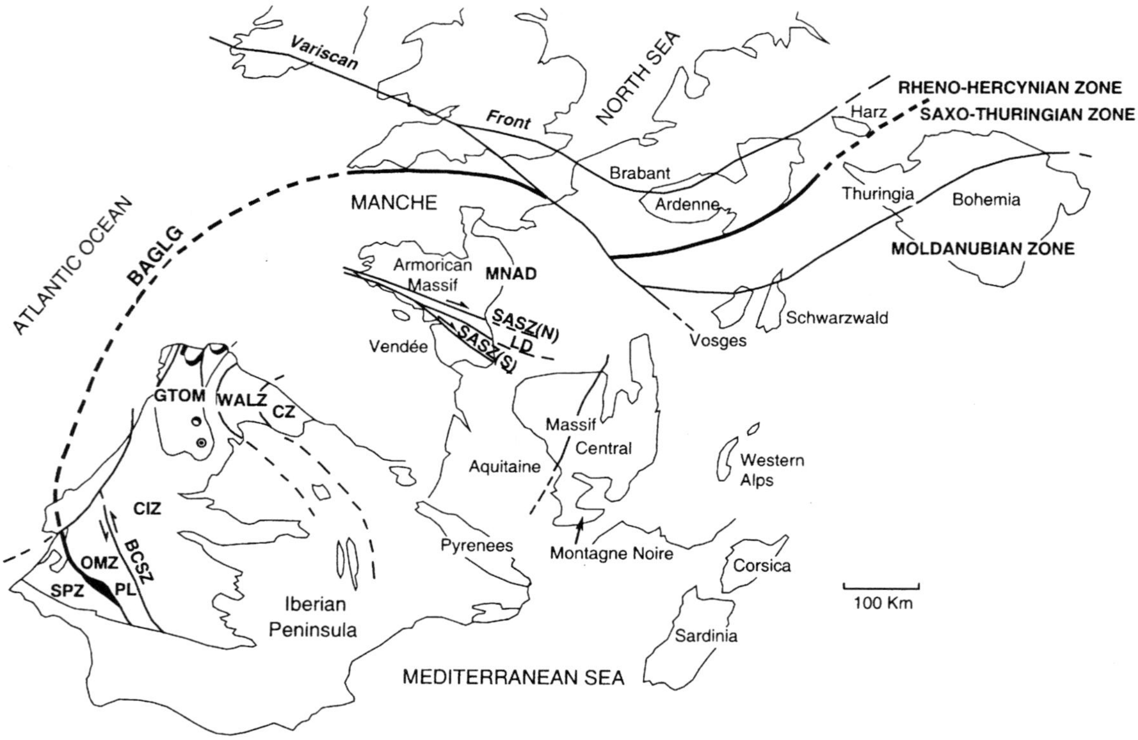

Figure 13. Model with single Variscan ocean connecting Pulo de Lobo, Galicia, and Rheic oceanic areas (after Quesada, 1991; Quesada et al., 1994; Oliveira and Quesada, 1998). Abbreviations as in Figure 1. BAGLG—Beja-Acebuches-Galicia-Lizard-Giesen oceanic suture; black—oceanic rocks of Pulo de Lobo suture and ophiolitic units of allochthonous complexes of northwestern Iberia.

on the other side of the mid-European Rheic ocean. There is no evidence that, between the Early Ordovician and the Devonian, any North Gondwanan region was a microcontinent detached from stable Gondwana and undergoing a distinct proper evolution. It is also possible to characterize, within the North Gondwanan shelf, inner shelf proximal and outer shelf distal areas. In this North Gondwanan pre-Variscan paleogeographical arrangement, the inner shelf Mid-North Armorican–Central Iberian domain appears as a key element.

The results of this paleobiogeographical analysis can be used to test the geodynamical models based on structural, metamorphic, and magmatic data obtained from the active zones of the Variscan Belt.

The pre-Variscan paleogeographical data allow a large number of the geodynamical models that propose distinct continental blocks or oceanic sutures to be dismissed. These are inconsistent with the distribution of benthic faunas and pre-Variscan paleogeographical configuration. This is especially the case in the models that propose two distinct oceanic sutures in the French part of the Variscan Belt and that are strongly influenced and constrained by the present-day geographical configuration.

A model with a single Variscan ocean (mid-European Rheic ocean), as proposed by several authors for the Iberian Peninsula, would much more satisfactorily match the paleobiogeographical data. This model also proves suitable for the French part of the orogen, if it is accepted that, during pre-Variscan time, the Mid-North Armorican regions were not in their present northern position, but formed a direct extension of the Central Iberian regions, a hypothesis based on faunal and sedimentary evidence.

In this model, the Mid-North Armorican domain would have moved (rotation and translation), during the Variscan orogeny, from a pre-Variscan southern position into its present-day northern location. This transfer would most probably have been achieved during early Carboniferous time. It would thus be necessary to view the so-called Ibero-Armorican Arc in a new light, because it is composed of an autochthonous Iberian part and a displaced Mid-North Armorican part.

The pre-Variscan paleogeographical data are not sufficient to enable the geodynamic evolution of the Variscan Belt to be understood completely; however, they impose limitations that should be considered seriously in testing the coherence and the validity of the models based on other geological evidence.

ACKNOWLEDGMENTS

Many of the ideas expressed herein were presented during the Fifteenth International Conference on Basement Tectonics. We thank the organizing committee of the meeting, especially J.R. Martínez Catalán, for their invitation. We also thank Danièle

Bernard (Géosciences-Rennes) for drawing the figures. Euan Clarkson (University of Edinburgh) and an anonymous referee provided helpful and constructive reviews and improved the English text.

REFERENCES CITED

Autran, A., Lefort, J.P., Debeglia, N., Edel, J.B., and Vigneresse, J.L., 1994, Gravity and magnetic expression of terranes in France and their correlation beneath overstep sequences, *in* Keppie, J.D., ed., Pre-Mesozoic geology in France and related areas: Berlin, Springer-Verlag, p. 49–72.

Bachtadse, V., Torsvik, T.H., Tait, J., and Soffel, H.C., 1995, Paleomagnetic constraints on the paleogeographic evolution of Europe during the Paleozoic, *in* Dallmeyer, R.D., Franke, W., and Weber, K., eds., Pre-Permian geology of central and eastern Europe: Berlin, Springer-Verlag, p. 567–578.

Badham, J.P.N., and Halls, C., 1975, Microplate tectonics, oblique collisions, and evolution of the Hercynian orogenic systems: Geology, v. 3, p. 373–376.

Ballèvre, M., Paris, F., and Robardet, M., 1992, Corrélations ibéro-armoricaines au Paléozoïque: Une confrontation des données paléobiogéographiques et tectonométamorphiques: Comptes Rendus de l'Académie des Sciences de Paris, series II, t. 315, p. 1783–1789.

Ballèvre, M., Marchand, J., Godard, G., Goujou, J.C., and Wyns, R., 1994, Eo-Hercynian events in the Armorican Massif, *in* Keppie, J.D., ed., Pre-Mesozoic geology in France and related areas: Berlin, Springer-Verlag, p. 183–194.

Bard, J.-P., 1997, Démembrement anté-mésozoïque de la chaîne varisque d'Europe occidentale et d'Afrique du Nord: Rôle essentiel des grands décrochements transpressifs dextres accompagnant la rotation-translation horaire de l'Afrique durant le Stéphanien: Comptes Rendus de l'Académie des Sciences de Paris, series II, t. 324, p. 693–704.

Bard, J.-P., Burg, J.-P., Matte, Ph., and Ribeiro, A., 1980, La chaîne hercynienne d'Europe occidentale en termes de tectonique des plaques: 26e Congrès Géologique International, Colloque C6 Géologie de l'Europe and Annales de la Société Géologique du Nord, t. 99, p. 233–246.

Bresser, G., and Walter, R., 1999, A new structural model for the SW Irish Variscides: The Variscan front of the NW European Rhenohercynian, *in* Sintubin, M., Vandycke, S., and Camelbeeck, T., eds., Palaeozoic to recent tectonics in the NW European Variscan front zone: Tectonophysics, v. 309, p. 197–209.

Burg, J.-P., 1999, Ductile structures and instabilities: their implication for Variscan tectonics in the Ardennes, *in* Sintubin, M., Vandycke, S., and Camelbeeck, T., eds., Palaeozoic to recent tectonics in the NW European Variscan front zone: Tectonophysics, v. 309, p. 1–25.

Burg, J.-P., Iglesias, M., Laurent, Ph., Matte, Ph., and Ribeiro, A., 1981, Variscan intracontinental deformation: The Coimbra-Cordoba shear zone (SW Iberian Peninsula): Tectonophysics, v. 78, p. 161–177.

Chantraine, J., Lorenz, C., Mégnien, C., Millon, R., and Lienhardt M.L., eds., 1992, Géologie profonde de la France: Forage scientifique de Sancerre-Couy (Cher): Géologie de la France,v. 3-4, p. 1–215.

Cocks, L.R.M., and Fortey, R.A., 1982, Faunal evidence for oceanic separations in the Palaeozoic of Britain: Journal of the Geological Society of London, v. 139, p. 465–478.

Cogné, J., 1977, La Chaîne Hercynienne ouest-européenne correspond-elle à un orogène par collision? Propositions pour une interprétation géodynamique globale: Colloque International CNRS n. 268, Ecologie et Géologie de l'Himalaya, France, Paris, p. 111–129.

Cogné, J., and Lefort, J.P., 1985, The Ligerian Orogeny: A Proto-Variscan event, related to the Siluro-Devonian evolution of the Tethys I ocean, *in* Gee, D.G., and Sturt, B.A., eds., The Caledonide orogen-Scandinavia and related areas: Chichester, Wiley, p. 1185–1193.

Colchen, M., and Poncet, D., 1994, Paleozoic of the Vendean domain, *in* Keppie, J.D., ed., Pre-Mesozoic geology in France and related areas: Berlin, Springer-Verlag, p. 157–161.

Dubreuil, M., 1980, Hypothèse sur la mise en place, au Dinantien, du complexe du Tombeau Leclerc (Bassin d'Ancenis, Sud-Est du Massif armoricain) sous forme d'un olistostrome. Conséquences géodynamiques: Comptes Rendus de l'Académie des Sciences de Paris, series D, t. 290, p. 1455–1458.

Dubreuil, M., 1987, Le bassin en décrochement de Saint-Julien-de-Vouvantes—Angers (Carbonifère inférieur du Sud-Est du Massif armoricain): Bulletin de la Société Géologique de France, series 8, v. 3, p. 215–221.

Faure, M., Leloix, Ch., and Roig, J.Y., 1997, L'évolution polycyclique de la chaîne hercynienne: Bulletin de la Société Géologique de France, v. 168, p. 695–705.

Feist, R., Echtler, H., Galtier, J., and Mouthier, B., 1994, The Massif Central: Biostratigraphy and dynamics of the nonmetamorphic sedimentary record, *in* Keppie, J.D., ed., Pre-Mesozoic geology in France and related areas: Berlin, Springer-Verlag, p. 289–304.

Fortey, R.A., Harper, D.A.T., Ingham, J.K., Owen, A.W., and Rushton, A.W.A., 1995, A revision of Ordovician series and stages from the historical type area: Geological Magazine, v. 132, p. 15–30.

Franke, W., and Engel, W., 1982, Variscan sedimentary basins on the continent, and relations with south-west England: Proceedings of the Ussher Society, v. 5, p. 259–269.

Gutiérrez-Marco, J.C., Robardet, M., and Piçarra, J.M., 1998, Silurian stratigraphy and paleogeography of the Iberian Peninsula (Spain and Portugal): Temas Geológico-Mineros ITGE, v. 23, p. 13–44.

Hammann, W., 1992, The Ordovician trilobites from the Iberian chains in the province of Aragón, NE-Spain, I, The trilobites of the Cystoid Limestone (Ashgill Series): Beringeria, Würzburg, v. 6, p. 3–219.

Havlicek, V., Vanek, J., and Fatka, O., 1994, Perunica microcontinent in the Ordovician (its position within the Mediterranean Province, series division, benthic and pelagic associations): Sbornik geologickych Ved: Geologie (Journal of Geological Sciences: Geology), v. 46, p. 23–56.

Henry, J.L., Nion, J., Paris, F., and Thadeu, D., 1974, Chitinozoaires, ostracodes et trilobites de l'Ordovicien du Portugal (serra de Buçaco) et du Massif Armoricain: essai de comparaison et signification paléogéographique: Comunicações dos Serviços Geológicos de Portugal, t. 57, p. 303–345.

Heuse, T., 2000, Biostratigraphy and biofacies of the Ordovician of Saxo-thuringia: A review, *in* Erdtmann, B.D., and Kraft, P., eds., Pre-Variscan terrane analysis of "Gondwanan Europe": Acta Universitatis Carolinae, Geologica, 1998, v. 42, n. 3/4, p. 423–431.

Jaeger, H., and Robardet, M., 1979, Le Silurien et le Dévonien basal dans le nord de la province de Séville (Espagne): Géobios, v. 12, 5, p. 687–714.

Jegouzo, P., 1980, The South Armorican shear zone: Journal of Structural Geology, v. 2, p. 39–47.

Jones, K.A., 1991, Palaeozoic continental margin tectonics in southern Armorica: Journal of the Geological Society of London, v. 148, p. 55–64.

Julivert, M., and Martinez, F.J., 1983, Estructura de conjunto y vision global de la cordillera hercyniana, *in* Comba, J.A., ed., Geologia de España, Libro jubilar J.M. Rios, Instituto Geologico y Minero de España Madrid, t. I, p. 612–630.

Julivert, M., and Martinez, F.J., 1987, The structure and evolution of the Hercynian Fold Belt in the Iberian Peninsula, *in* Schaer, J.-P., and Rodgers, J., eds., The anatomy of mountain belts, Princeton Series in geology and paleontology: Princeton, New Jersey, Princeton University Press, p. 65–103.

Kent, D.V., and Van der Voo, R., 1990, Palaeozoic palaeogeography from palaeomagnetism of the Atlantic-bordering continents, *in* McKerrow, W.S., and Scotese C.R., eds., Palaeozoic palaeogeography and biogeography: London, England, Geological Society of London Memoir 12, p. 49–56.

Keppie, J.D., ed., 1994, Pre-Mesozoic geology in France and related areas: Berlin, Springer-Verlag, p. 1–514.

Lardeux, H., and Cavet, P., 1994, Paleozoic of the Ligerian domain, *in* Keppie, J.D., ed., Pre-Mesozoic geology in France and related areas: Berlin, Springer-Verlag, p. 152–156.

Lardeux, H., Chauvel, J.J., Henry, J.L., Morzadec, P., Paris, F., Racheboeuf, P., and Robardet, M., 1977, Evolution géologique du Massif Armoricain au cours des temps ordoviciens, siluriens et dévoniens: Colloque International CNRS n. 243, France, Paris, p. 181–192.

Le Corre, C., Auvray, B., Ballèvre, M., and Robardet, M., 1991, Le Massif Armoricain, *in* Piqué A., ed., Les massifs anciens de France: Sciences Géologiques, Bulletin, v. 44, p. 31–103.

Lefort, J.P., 1983, A new geophysical criterion to correlate the Acadian and Hercynian orogenies of western Europe and eastern America, *in* Hatcher, R.D., Williams, H., and Zietz I., eds., Contributions to the tectonics and geophysics of mountain chains: Boulder, Colorado, Geological Society of America Memoir 158, p. 3–18.

Lefort, J.P., 1990, Cinématique de la collision Gondwana-Laurentia entre la Bretagne et la Floride d'après les données du socle submergé: Schweizerische Mineralogische und Petrographische Mitteilungen, v. 70, p. 3–16.

Lefort, J.P., Max, M.D., and Roussel, J., 1988, Geophysical evidence for the location of the NW boundary of Gondwanaland and its relationship with two older satellite sutures, *in* Harris, A.L., and Fettes, D.J., eds., The Caledonian-Appalachian Orogen: London, England, Geological Society of London Special Publication 38, p. 49–60.

Le Menn, J., 1997, Crinoïdes dévoniens d'Afrique du Nord: révision systématique et affinités paléobiogéographiques: Annales de la Société Géologique du Nord, v. 5 (2ème série), p. 129–139.

Martínez Catalán, J.R., 1990, A noncylindrical model for the northwest Iberian allochtonous terranes and their equivalents in the Hercynian belt of Western Europe: Tectonophysics, v. 179, p. 253–272.

Martínez Catalán, J.R., Arenas, R., Díaz García, F., and Abati, J., 1997, Variscan accretionary complex of northwest Iberia: Terrane correlation and succession of tectonothermal events: Geology, v. 25, p. 1103–1106.

Matte, Ph., 1976, Raccord des segments hercyniens d'Europe sud-occidentale (Iles Britanniques, France, Péninsule Ibérique, bloc Corso-Sarde): Les différents marqueurs utilisables: Nova Acta Leopoldina, Abhandlungen der Deutschen Akademie der Naturforscher Leopoldina, n. 224, v. 45, p. 239–262.

Matte, Ph., 1986a, La chaîne varisque parmi les chaînes paléozoïques périatlantiques, modèle d'évolution et position des grands blocs continentaux au Permo-Carbonifère: Bulletin de la Société Géologique de France, series 8, v. 2, p. 9–24.

Matte, Ph., 1986b, Tectonics and plate tectonics model for the variscan belt of Europe: Tectonophysics, v. 126, p. 329–374.

Matte, P., 1991, Accretionary history and crustal evolution of the Variscan belt in Western Europe, in Hatcher, R.D., and Zonenshain, L., eds., Accretionary tectonics and composite continents: Tectonophysics, v. 196, p. 309–337.

Mélou, M., 1982, Brachiopodes articulés, *in* Babin, C., Courtessole, R., Mélou, M., Pillet, J., Vizcaino, D., and Yochelson, E.L., eds., Brachiopodes (articulés) et mollusques (Bivalves, Rostroconches, Monoplacophores, Gastropodes) de l'Ordovicien inférieur (Tremadocien-Arenigien) de la Montagne Noire (France méridionale): Carcassonne, France, Mémoires de la Société d'Etudes Scientifiques de l'Aude, p. 23–35.

Mélou, M., Oulebsir, L., and Paris, F., 1999, Brachiopodes et chitinozoaires ordoviciens dans le NE du Sahara algérien: Implications stratigraphiques et paléogéographiques: Géobios, v. 32, p. 823–839.

Nance, R.D., and Murphy, J.B., 1996, Basement isotopic signatures and Neoproterozoic paleogeography of Avalonian-Cadomian and related terranes in the circum-North Atlantic, *in* Nance, R.D., and Murphy, J.B., eds., Avalonian and related peri-Gondwanan terranes of the circum-North Atlantic: Boulder, Colorado, Geological Society of America Special Paper 304, p. 333–346.

Neugebauer, J., 1989, The Iapetus model: A plate tectonic concept for the Variscan belt of Europe: Tectonophysics, v. 169, p. 229–256.

Oliveira, J.T., and Quesada, C., 1998, A comparison of stratigraphy, structure, and paleogeography of the south Portuguese zone and southwest England, European variscides: Geoscience in Southwest England, v. 9, p. 141–150.

Paris, F., 1993, Evolution paléogéographique de l'Europe au Paléozoïque inférieur: Le test des Chitinozoaires: Comptes Rendus de l'Académie des Sciences de Paris, v. 316, series II, p. 273–280.

Paris, F., 2000, Early Palaeozoic palaeobiogeography of northern Gondwana regions, *in* Erdtmann, B.D., and Kraft, P., eds., Pre-Variscan terrane analysis of "Gondwanan Europe": Acta Universitatis Carolinae, Geologica, 1998, v. 42, n. 3/4, p. 473–483.

Paris, F., and Le Pochat, G., 1994, The Aquitaine Basin, *in* Keppie, J.D., ed., Pre-Mesozoic geology in France and related areas: Berlin, Springer-Verlag, p. 405–415.

Paris, F., and Robardet, M., 1977, Paléogéographie et relations Ibéro-Armoricaines au Paléozoïque anté-Carbonifère: Bulletin de la Société Géologique de France, series 7, v. 19, p. 1121–1126

Paris, F., and Robardet, M., 1990, Early Palaeozoic palaeobiogeography of the Variscan regions: Tectonophysics, v. 177, p. 193–213.

Paris, F., and Robardet, M., 1994, Paleogeographic synthesis, *in* Keppie, J.D., ed., Pre-Mesozoic geology in France and related areas: Berlin, Springer-Verlag, p. 172–176.

Piqué, A., Bogdanoff, S., Quénardel, J.M., Rolet, J., and Santallier, D., 1994, The French Paleozoic Terranes, *in* Keppie, J.D., ed., Pre-Mesozoic geology in France and related areas: Berlin, Springer-Verlag, p. 483–500.

Plusquellec, Y., Boumendjel, K., Morzadec, P., and Paris, F., 1997, Les faunes dévoniennes d'Ougarta dans la paléogéographie des régions maghrébo-européennes: Annales de la Société Géologique du Nord, series 2, v. 5, p. 123–128.

Quesada, C., 1991, Geological constraints on the Paleozoic tectonic evolution of tectonostratigraphic terranes in the Iberian Massif: Tectonophysics, v. 185, p. 225–245.

Quesada, C., Fonseca, P.E., Munha, J., Oliveira, J.T., and Ribeiro, A., 1994, The Beja-Acebuches Ophiolite (Southern Iberia Variscan fold belt): Geological characterization and geodynamic significance: Boletin Geologico y Minero, v. 101, p. 3–49.

Racheboeuf, P., 1990, Paléobiogéographie de la marge nord-gondwanienne au Dévonien inférieur et moyen: Nouvelles données déduites de l'étude des Brachiopodes Chonetacés: Comptes Rendus de l'Académie des Sciences de Paris, series II, v. 310, p. 1481–1486.

Ribeiro, A., Quesada, C., and Dallmeyer, R.D., 1990, Geodynamic evolution of the Iberian Massif, *in* Dallmeyer, R.D., and Martínez García, E., eds., Pre-Mesozoic geology of Iberia: Berlin, Springer Verlag, p. 399–409.

Riding, R., 1974, Model of the Hercynian foldbelt: Earth and Planetary Science Letters, v. 24, p. 125–135.

Robardet, M., 1976, L'originalité du segment hercynien sud-ibérique au Paléozoïque inférieur: Ordovicien, Silurien, et Dévonien dans le nord de la province de Séville (Espagne): Comptes Rendus de l'Académie des Sciences de Paris, series D, v. 283, p. 999–1002.

Robardet, M., and Gutiérrez-Marco, J.C., 1990a, Ossa Morena zone, stratigraphy, passive margin phase (Ordovician-Silurian-Devonian), *in* Dallmeyer, R.D., and Martínez García, E., eds., Pre-Mesozoic geology of Iberia: Berlin, Springer-Verlag, p. 267–272.

Robardet, M., and Gutiérrez-Marco, J.C., 1990b, Sedimentary and faunal domains in the Iberian Peninsula during Lower Paleozoic times, *in* Dallmeyer, R.D., and Martínez García, E., eds., Pre-Mesozoic geology of Iberia: Berlin, Springer Verlag, p. 383–395.

Robardet, M., Paris, F., and Racheboeuf, P.R., 1990, Palaeogeographic evolution of southwestern Europe during early Palaeozoic times, *in* Mc Kerrow, W.S., and Scotese, C.R., eds., Palaeozoic palaeogeography and biogeography: London, England, Geological Society Memoir 12, p. 411–419.

Robardet, M., Blaise, J., Bouyx, E., Gourvennec, R., Lardeux, H., Le Hérissé, A., Le Menn, J., Mélou, M., Paris, F., Plusquellec, Y., Poncet, J., Régnault, S., Rioult, M., and Weyant, M., 1993, Paléogéographie de l'Europe occidentale de l'Ordovicien au Dévonien: Bulletin de la Société Géologique de France, v. 164, p. 683–695.

Robardet, M., Bonjour, J.L., Paris, F., Morzadec, P., Racheboeuf, P.R., 1994a, Ordovician, Silurian, and Devonian of the Medio-North-Armorican domain, *in* Keppie, J.D., ed., Pre-Mesozoic geology in France and related areas: Berlin, Springer-Verlag, p. 131–151.

Robardet, M., Verniers, J., Feist, R., and Paris F., 1994b, Le Paléozoïque anté-varisque de France, contexte paléogéographique et géodynamique: Géologie de la France, 1994, n. 3, p. 3–31.

Robardet, M., Paris, F., and Plusquellec, Y., 2001, Comment: "New Early Devonian paleomagnetic data from NW France: Paleogeography and implications for the Armorican microplate hypothesis," by J. Tait: Journal of Geophysical Research, v. 106 (B7), p. 13,307–13,310.

Sarmiento, G.N., Piçarra, J.M., Rebelo, J.A., Robardet, M., Gutiérrez-Marco, J.C., Storch, P., and Rábano, I., 1999, Le Silurien du synclinal de Moncorvo (NE du Portugal): Biostratigraphie et importance paléogéographique: Géobios, v. 32, p. 749–767.

Scotese, C.R., and Barrett, S.F., 1990, Gondwana's movement over the South Pole during the Palaeozoic: Evidence from lithological indicators of climate, *in* McKerrow, W.S., and Scotese, C.R., eds., Palaeozoic paleogeography and biogeography: London, England, Geological Society Memoir 12: p. 75–85.

Scotese, C.R., and McKerrow, W.S., 1990, Revised world maps and introduction, *in* McKerrow, W.S., and Scotese, C.R., eds., Palaeozoic paleogeography and biogeography: London, England, Geological Society Memoir 12: p. 1–21.

Servais, T., Brocke, R., Erdtmann, B.D., Fatka, O., and Heuse, T., 2000, The importance of acritarchs in Ordovician biostratigraphy and palaeobiogeography in Germany, *in* Erdtmann, B.D., and Kraft, P., eds., Pre-Variscan terrane analysis of "Gondwanan Europe": Acta Universitatis Carolinae, Geologica, 1998, v. 42, n. 3/4, p. 495–500.

Shelley, D., and Bossière, G., 2000, A new model for the Hercynian Orogen of Gondwanan France and Iberia: Journal of Structural Geology, v. 22, p. 757–776.

Spjeldnaes, N., 1961, Ordovician climatic zones: Norsk Geologisk Tidsskrift, v. 41, p. 45–77.

Tait, J., Bachtadse, V., Franke, W., and Soffel, H.C., 1997, Geodynamic evolution of the European Variscan fold belt: palaeomagnetic and geological constraints: Geologische Rundschau, v. 86, p. 585–598.

Vai, J.B., 1991, Palaeozoic strike-slip rift pulses and palaeogeography in the circum-Mediterranean Tethyan realm: Palaeogeography Palaeoclimatology Palaeoecology, v. 87, p. 223–252.

Van der Voo, R., 1993, Paleomagnetism of the Atlantic, Tethys and Iapetus Oceans: Cambridge, Cambridge University Press, 411 p.

Webby, B.D., 1984, Ordovician reefs and climate, *in* Bruton, D.L., ed., Aspects of the Ordovician system: Palaeontological Contributions of the University of Oslo, v. 295, p. 89–100.

Witzke, B.J., 1990, Palaeoclimatic constraints for Palaeozoic palaeolatitudes of Laurentia and Euramerica, *in* McKerrow, W.S., and Scotese, C.R., eds., Palaeozoic palaeogeography and biogeography: London, England, Geological Society Memoir 12, p. 57–73.

Young, T.P., 1989, Eustatically controlled ooidal ironstone deposition: facies relationships of the Ordovician open-shelf ironstones of Western Europe, *in* Young, T.P., and Taylor, W.E.G., eds., Phanerozoic Ironstones: London, England, Geological Society Special Publication 46, p. 51–63.

Young, T.P., 1990, Ordovician sedimentary facies and faunas of Southwest Europe: palaeogeographic and tectonic implications, *in* McKerrow, W.S., and Scotese, C.R., eds., Palaeozoic palaeogeography and biogeography: London, England, Geological Society Memoir 12, p. 421–430.

Manuscript Accepted by the Society February 8, 2002

Printed in the U.S.A.

Geological Society of America
Special Paper 364
2002

Stratigraphic record of preorogenic to synorogenic sedimentation, and tectonic evolution of imbricate units in the Alcañices synform (northwestern Iberian Massif)

Emilio González Clavijo
Instituto Geologico e Mineiro, Estrada da Portela, B. de Zambujal, 2720-461 Alfragide, Portugal
José R. Martínez Catalán
Departamento de Geología, Universidad de Salamanca, 37008 Salamanca, Spain

ABSTRACT

A late Variscan structure, the Alcañices synform, preserves at its core an allochthonous stack of metasediments thrust along a decollement located on Silurian carbonaceous slates. Three internally imbricated units, separated by thrust faults, were placed on top of autochthonous sequences of the northwestern Iberian Massif. The three units consist of Silurian and Devonian clastic sediments, and subordinate carbonate and volcanic rocks. Differences in the lithologic associations and facies, together with the reconstruction of their relative predeformational positions, give information on the evolution of the continental margin of Gondwana prior to and early during its collision with Laurentia. The uppermost unit forms part of a large epizonal thrust sheet that separates the Iberian autochthon from far-traveled allochthonous terranes that include ophiolites and subducted fragments of continental crust.

The Silurian deposits show abundant facies changes, evidence of mass flow and gliding, synsedimentary folds, and small basic sills and volcanic plugs. This points to an extensional regime, and correlates with the prolonged evolution of the northern Gondwanan passive margin, developed during the Ordovician. Extension was succeeded by plate convergence during the Devonian, when the approaching active Laurentian margin supplied terrigenous sequences that filled the synorogenic trough and partly overlapped the passive margin. This was followed by the Variscan thrust and nappe tectonics, characterized by ductile to brittle subhorizontal shearing that placed the allochthonous complexes on top of the former Gondwana margin, while internal deformation induced crustal thickening.

INTRODUCTION

The emplacement of ophiolites and other allochthonous terranes on previously passive continental margins characterizes convergence and, often, the closure of oceanic realms. When a passive margin approaches a destructive plate boundary, it may remain more or less stable until it reaches the subduction zone. Then, partial subduction of the passive margin is followed by obduction of slivers of the oceanic lithosphere, and frequently terranes derived from the other plate become thrust onto the previously stable continental platform when continent-continent or arc-continent collision follows.

In these cases, the evolution of the suture zone involves processes such as the peeling off of the sedimentary cover of the passive margin, and its imbrication beneath the allochthonous stack. Some of these sedimentary sequences represent stages pre-

González Clavijo, E., and Martínez Catalán, J.R., 2002, Stratigraphic record of preorogenic to synorogenic sedimentation, and tectonic evolution of imbricate units in the Alcañices synform (northwestern Iberian Massif), *in* Martínez Catalán, J.R., Hatcher, R.D., Jr., Arenas, R., and Díaz García, F., eds., Variscan-Appalachian dynamics: The building of the late Paleozoic basement: Boulder, Colorado, Geological Society of America Special Paper 364, p. 17–35.

vious to convergence, but others bear the imprint of the onset of subduction and, in some cases, of its subsequent evolution.

The structural analysis of the sedimentary imbricates accreted at the footwall to the suture, coupled with a study of their sedimentary facies and age, permits a palinspastic reconstruction that helps us to understand the mechanisms of the transition of the passive margin to an actively tectonic region. Examples of formerly passive margins that underwent subsequent obduction and accretion, and where a similar methodology has been applied, are the Oman Mountains (Béchennec et al., 1988; Pillevuit et al., 1997) and the Rhenohercynian Massif (Oncken et al., 1999).

The northwestern Iberian Peninsula offers the opportunity to study the evolution of the passive northern margin of Gondwana at the onset of the Variscan orogeny. Ophiolites that originated shortly before their accretion to an orogenic wedge (Díaz García et al., 1999), together with other allochthonous terranes, were thrust onto upper Paleozoic preorogenic and synorogenic clastic sediments. The origin and evolution of the different allochthonous terranes have been discussed (Arenas et al., 1986; Marques et al., 1996; Martínez Catalán et al., 1996, 1997; Ribeiro et al., 1990). Comparatively few studies aiming to establish their meaning in the orogenic process have been carried out in the underlying metasedimentary sequences. This chapter focuses on the Alcañices synform, a structure close to the Portuguese allochthonous complexes, and structurally beneath the suture, where the upper part of the Paleozoic sequence appears to be imbricated and thrust over the Iberian authochthon.

GEOLOGIC SETTING

The Alcañices synform is part of a late Variscan structure that crops out in the northwestern Iberian Massif. The northwest-southeast–striking fold is on the Spain-Portugal border (Figs. 1 and 2) and belongs to the Central Iberian zone of Julivert et al. (1972).

The Central Iberian zone forms part of the Iberian autochthon (Martínez Catalán, 1990; Dallmeyer et al., 1997); its sequences were deposited over the same basement upon which they are now, in spite of the common presence of recumbent folds and thrusts. According to modern views, the Central Iberian zone was situated at the northern margin of Gondwana, and

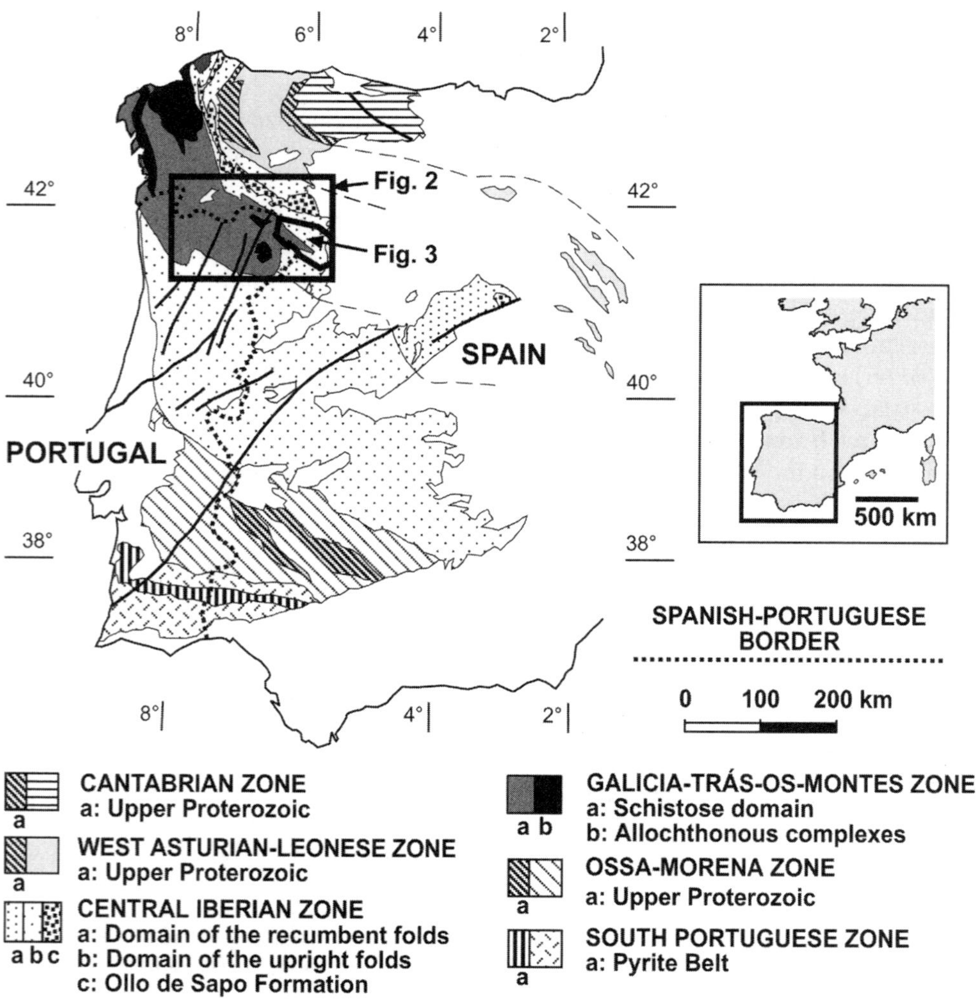

Figure 1. Iberian Massif showing zones established by Julivert et al. (1972) and modified by Farias et al. (1987).

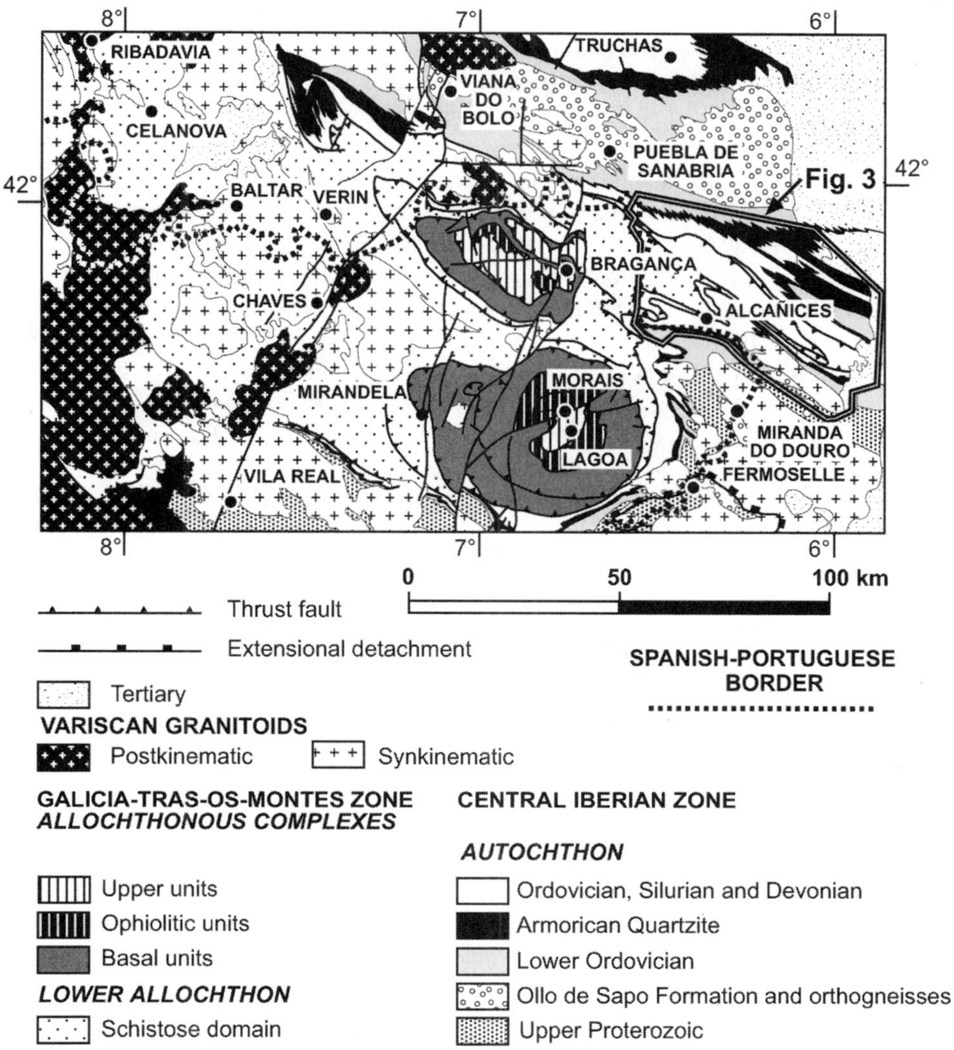

Figure 2. Geological sketch of area around Alcañices synform showing relationships with allochthonous complexes of Bragança and Morais. For location, see Figure 1.

most of its Paleozoic sedimentary sequences were deposited along this passive continental margin (Noblet and Lefort, 1990; Ribeiro et al., 1990; Martínez Catalán et al., 1997).

In Spain, only metasedimentary rocks with subordinate volcanic rocks occur in the Alcañices synform. In Portugal, the fold core is occupied by the Bragança Complex, a stack of allochthonous units including ophiolites and far-traveled terranes, some of which underwent a high-pressure metamorphic event related to subduction (Ribeiro, 1974; Ribeiro et al., 1990; Martínez Catalán et al., 1997). The allochthonous complexes, such as Bragança and Morais (Fig. 2), were thrust over the lower allochthon, a metasedimentary unit also known as the schistose domain (Marquínez García, 1981, 1984; Farias et al., 1987), and the parautochthon (Ribeiro et al., 1990). The ensemble was subsequently placed onto the Paleozoic sedimentary cover of Gondwana.

The allochthonous complexes are included, together with the lower allochthon, in a zone called Galicia-Trás-os-Montes (see Figs. 1 and 2). The lower allochthon occurs beneath the suture marked by the ophiolites, and is viewed as part of the margin of Gondwana, but occupying a marginal position, clearly more distal that the Central Iberian zone (Farias et al., 1987).

To the northwest in Spain, the structure is continued by the Verín synform; its total length exceeds 150 km. In Verín, the core is occupied only by metasedimentary and volcanic rocks (Fig. 2), and there, as in Portugal, parts of these rocks belong to the Galicia-Trás-os-Montes zone. The limit between the Central Iberian zone and Galicia-Trás-os-Montes zone is a thrust that brought both zones into contact during the Variscan compression (Farias Arquer, 1990; Pereira, 1987). This contribution shows that this important boundary can be followed into the Alcañices synform, and that it is one of a complex of imbricate thrusts with varying displacements.

Three main thrust faults separate three units with different stratigraphic sequences. The lower sequence shows Central Iberian zone affinities, whereas the upper crops out as a narrow strip at the core of the synform, and represents the Galicia-Trás-os-Montes lower allochthon. The middle thrust sheet is a transitional unit that includes a thick synorogenic flysch.

STRATIGRAPHY

A metasedimentary sequence ranging in age from late Vendian to Devonian characterizes the Alcañices synform (Figs. 3 and 4). The succession has been described by many (Martínez García, 1972, 1973; Iglesias Ponce de León and Ribeiro, 1981a; Quiroga, 1981, 1982; Vacas and Martínez Catalán, 1987), who gave local names (not always the same) to the mapped formations. The sequence may be divided into three main parts: (1) pre-Ordovician to earliest Ordovician, (2) Ordovician, and (3) Silurian-Devonian (Fig. 4). When different names have been given to the same formation, we retain the subdivision and names used by Vacas and Martínez Catalán (1987).

Pre-Ordovician to lowermost Ordovician sequence

The oldest stratigraphic sequence crops out in an antiformal structure in the southern part of the studied area (Fig. 3). It consists of two nonfossiliferous units, the Villalcampo Schists (Villar Alonso, 1990) and the Villadepera Gneisses (Quiroga, 1981).

Villalcampo Schists. This formation is a monotonous schistose sequence, more than 2000 m thick, with minor graywackes, quartzites, and, in the deeper parts, thin layers of calc-silicate rocks. The formation is part of the Schist-Graywacke Complex, a deposit of late Vendian (to Cambrian?) age, widely represented along the Central Iberian zone (Rodríguez Alonso, 1985; Díez Balda, 1986; Regêncio Macedo, 1988; San José et al., 1990).

Villadepera Gneisses. This formation overlies the older clastic sequence. This is a 250-m-thick unit consisting of fine-grained porphyritic gneisses of volcanic and volcaniclastic origin, strongly foliated and lineated. The gneisses have been correlated to the Ollo de Sapo Formation (Iglesias Ponce de León and Ribeiro, 1981a; Villar Alonso, 1990), a band of porphyritic and augen gneisses that crops out along a large antiformal structure in the northeast part of the Central Iberian zone (Figs. 1 and 2). The gneisses occur stratigraphically below Lower Ordovician schists and quartzites, and U/Pb dating of the Ollo de Sapo Formation outside the studied area gives an Early Ordovician age of 488 Ma (Gebauer, 1993; see also Fernández-Suárez et al., 2000, for less precise but comparable ages).

Ordovician sequence

Four different formations (Figs. 3 and 4) differentiated by Vacas and Martínez Catalán (1987) can be followed all along the Alcañices synform. Quartzites characterize the two lower formations, whereas slates are dominant in the upper part of the sequence.

Santa Eufemia and Peña Gorda Formations. The Santa Eufemia Formation consists of nearly 1000 m of schists and quartzites unconformably overlying the Villadepera Gneisses. The unconformity is seen at map scale, according to Iglesias Ponce de León and Ribeiro (1981a) and Villar Alonso (1990). The conformably overlying Peña Gorda Formation consists of ~400 m of quartzites, and subordinate slates.

Correlation with similar units containing abundant trace fossils gives an Arenig age for the Peña Gorda Formation and suggests a Tremadoc age for the underlying Santa Eufemia Formation. However, because the underlying Ollo de Sapo Formation has yielded a lowermost Arenig age (Gebauer, 1993), according to Haq and Van Eysinga (1998), an Arenig age is also possible for most of the Santa Eufemia Formation.

Villaflor and Campillo Formations. The Villaflor Formation is a 300–600-m-thick sequence of slates with minor quartzites. Some orthids and trilobites found in the upper levels suggest a Llanvirn-Llandeilo age (Gutiérrez Marco et al., 1999), comparable to the Luarca Slates defined by Barrois (1882) in the north of Spain.

The Campillo Formation conformably overlies the previous Ordovician sequence. It is composed of greenish schists and less abundant sandstones. Isolated intrabasinal pebbles of shale and sandstone, normally a few centimeters across but exceptionally to 60 cm, are common (Fig. 5A). The pebbles have been related to glaciogenic activity (Iglesias Ponce de León and Robardet, 1980; Robardet and Doré, 1988; Sarmiento et al., 1999) but many may also be glided blocks (olistoliths). The thickness of this unit ranges from 150 to 450 m, a variability that may be due to tectonism, because the thrust stack commonly overlies this formation. The formation includes brachiopods, bryozoa, trilobites, echinoderms, and conodonts. The latter indicate a Caradoc-Ashgill age (Sarmiento et al., 1997).

Silurian-Devonian sequence

Most of the Silurian and Devonian deposits occur in allochthonous units (Figs. 3, 6, and 7); an important decollement fault and several imbricates developed along carbonaceous slates at the base of the Silurian succession. For that reason, the central area of the Alcañices synform has been divided into three structural units, delimited by thrusts and characterized by different post-Ordovician sedimentary sequences (Figs. 4 and 6). These are named the Bajo Río Esla, the Río Aliste, and the Río Manzanas units. A fourth structural unit, called the Río Duero (described in the preceding), represents the autochthon.

Manzanal del Barco Formation. This was originally defined by Vacas and Martínez Catalán (1987) in the Bajo Río Esla cross section, in the eastern part of the study area. It constitutes the lower part of the Bajo Río Esla and Río Aliste units, where it is overlain by the Almendra and San Vitero Formations, respectively. However, in some places, the Manzanal del Barco Formation conformably overlies the Upper Ordovician Campillo Formation; i.e., it forms part of the autochthonous Río Duero unit (see Figs. 4 and 7, northeast parts of the cross sections).

The formation was initially divided into four members (Vacas and Martínez Catalán, 1987), but this division has proven to be unsuccessful in a regional perspective. The thick-

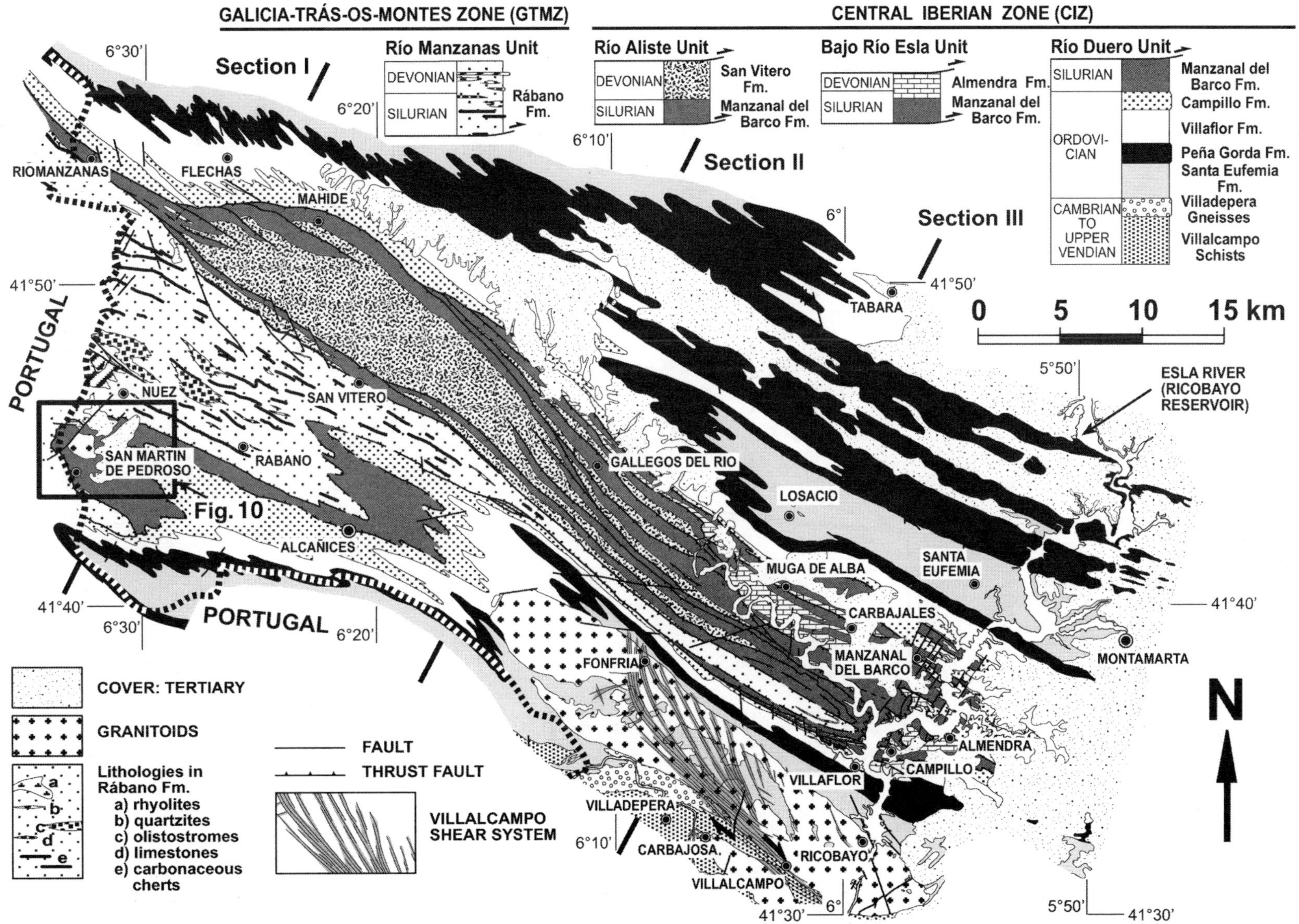

Figure 3. Geological map of Alcañices synform. For location, see Figures 1 and 2. Legend shows stratigraphic formations (Fm.) and structural units within which they appear. For lithologies, see Figure 4.

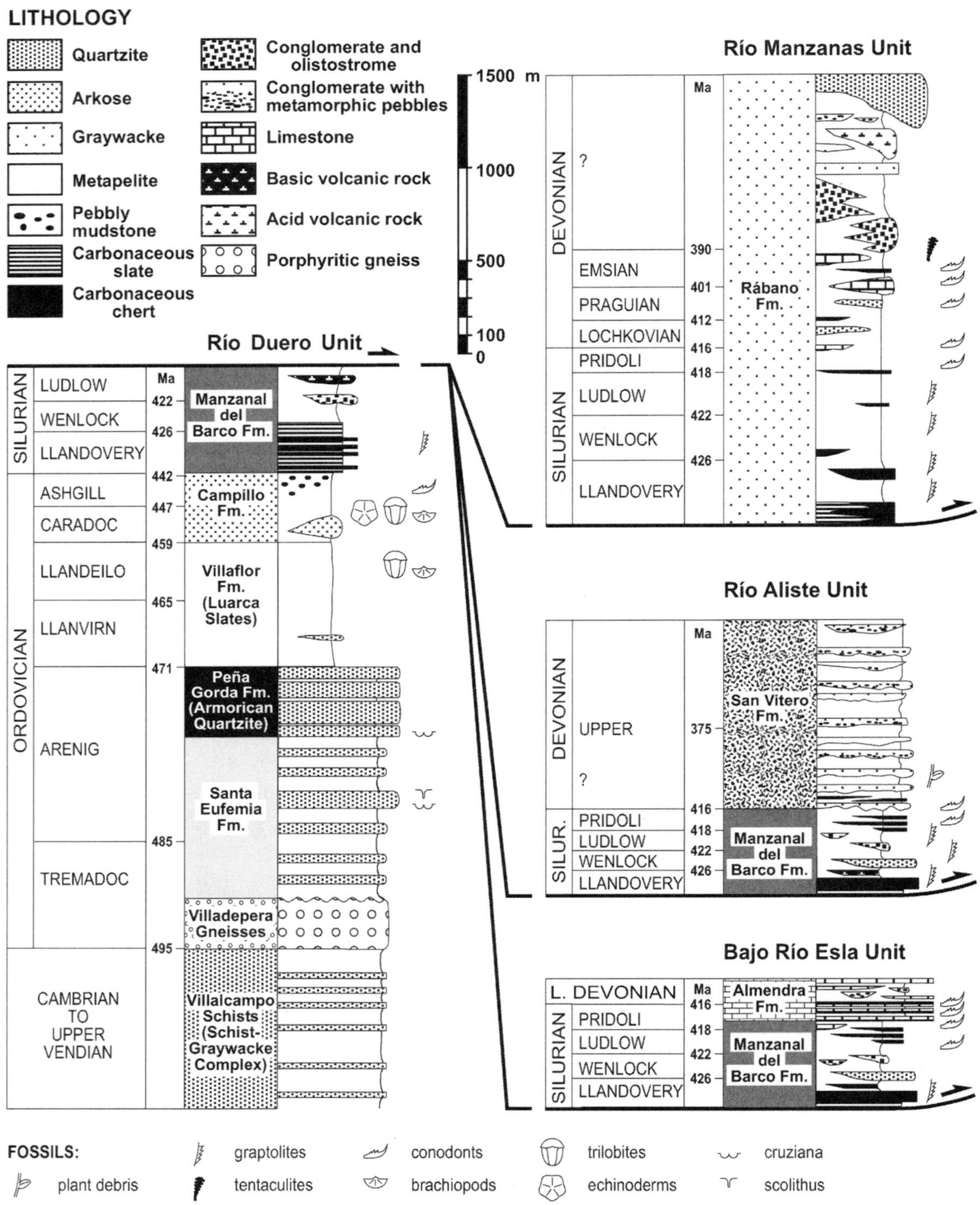

Figure 4. Stratigraphic columns of four structural units. Formations (Fm.) are shown with same patterns and screens as in Figures 3 and 7. Chronology according to Haq and Van Eysinga (1998).

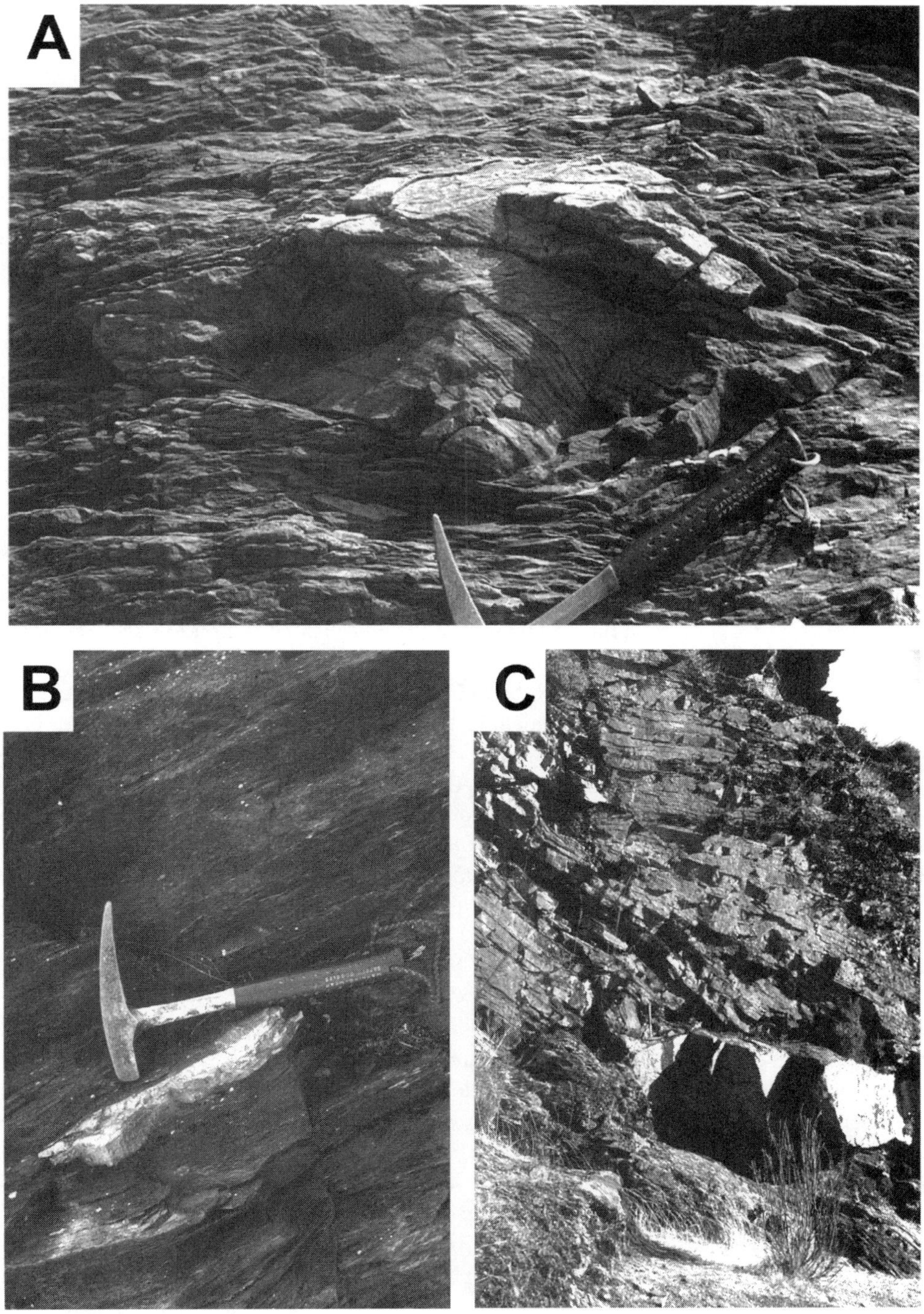

Figure 5. Intrabasinal pebbles in different Paleozoic formations. A: Dropstone or slumped block of greenish layered sandstone in upper schistose part of Campillo Formation. Southern limb of Alcañices synform, east of Fonfría. B: Pebble of light gray chert (presumably from Manzanal del Barco Formation, Silurian) in dark gray slates of Almendra Formation (Upper Silurian–Lower Devonian). Ricobayo Reservoir section, west of Almendra. C: Block of white, nonlayered sandstone in darker layered sandstones of upper part of Rábano Formation near Spanish-Portuguese border west of Nuez. Block is ~3 m long. Hammer length is 33 cm. Localities are shown in Figure 3.

ness of this formation is uniform (~500 m) but its lithology is variable. It consists essentially of slates, commonly gray to black, often rich in carbonaceous matter (ampelites), but they can also be light green and red. Alternations of carbonaceous cherts and sandstones (including quartzites and graywackes) are common. Conglomerates and minor limestones are also present. The stratigraphic succession varies enormously and shows, locally, sills of basic igneous rocks with subophitic texture, rhyolites, and tuffaceous slates. These rocks suggest an unstable basin undergoing tectonism that was probably extensional, due to their bimodal character, and related to the rifting of the passive margin. Graptolite determinations yielded middle Llandovery (Aeronian), Wenlock, and lower Ludlow ages (Quiroga, 1981; González Clavijo et al., 1997), but fossils have not been found in the lowermost and uppermost levels. The fossil record of the lowest levels could have been destroyed by the strong and penetrative deformation related to the basal decollement.

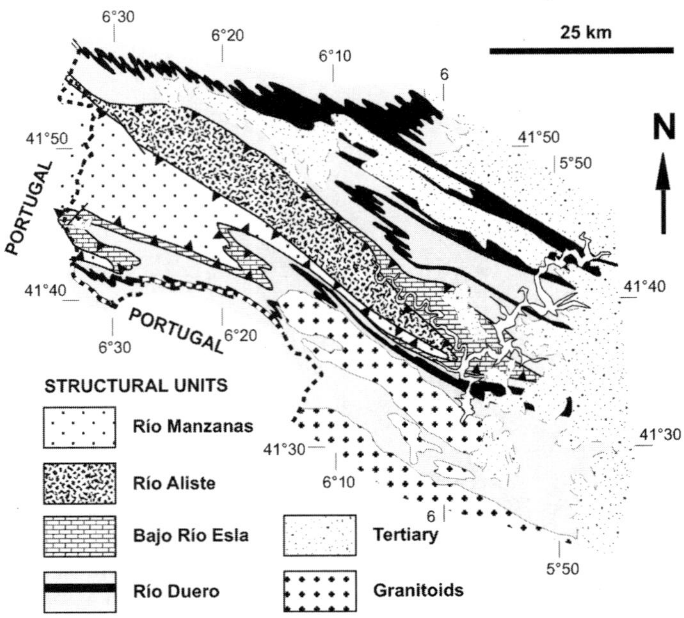

Figure 6. Sketch showing structural units.

Almendra Formation. This formation forms the upper part of the Bajo Río Esla unit (Figs. 3 and 4), is ~200 m thick, and conformably overlies the Manzanal del Barco Formation in a gradual transition. It consists of a rhythmic alternation of limestones and gray slates, with occasional microconglomeratic levels and scattered clasts of Manzanal del Barco lithologies (Fig. 5B). A few large (>1 m^3) blocks of limestone have been found inside the slates. The carbonate beds are discontinuous and include conglomerates, with limestone pebbles up to 10 cm across, and calcarenites. The latter include abundant bioclasts of corals, sciphocrinoides, bivalves, and gastropods. These characteristics suggest a talus deposit, driven by mass sliding. Conodonts yielded a Pridoli-Emsian age (Sarmiento et al., 1997), i.e., latest Silurian and Early Devonian.

San Vitero Formation. Martínez García (1972) defined this formation, which occupies the upper part of the Río Aliste unit, overlies the Manzanal del Barco Formation, and has a flysch character. The unit is made up of at least 1000 m (a thrust fault bounds its upper part) of terrigenous turbidites, including microconglomerates with metamorphic pebbles. Alternations of thin carbonaceous chert are common close to the base.

The formation was interpreted as a Variscan synorogenic flysch related to the central European Culm facies (Antona and Martínez Catalán, 1990). The contact with the underlying formation is a disconformity: the beds are regionally parallel, but in detail, there is an erosive contact at the base of the first turbiditic cycle. The age of the formation is uncertain, because no fossils (other that unclassifiable plant debris) have been found in Spain. However, the equivalent stratigraphic unit in northeastern Portugal (Gimonde Formation) has yielded Frasnian palynomorphs, some of them close to the Givetian-Frasnian boundary (Pereira et al., 1999).

Rábano Formation. The Río Manzanas unit is a very thick sequence mainly composed of gray and greenish slates with minor amounts of limestones, sandstones, graywackes, and olistostromes, and includes some large glided blocks (olistoliths) of quartzites (Fig. 5C) and carbonaceous cherts (>100 m long). According to Garrido and Barberá (1980), these deposits were gravitationally transported as slumps, debris flows, and mud flows. A few bodies of rhyolites were extruded in this succession, mainly in the western part.

It is difficult to distinguish formations inside the Rio Manzanas unit, so that one stratigraphic entity, the newly defined Rábano Formation, encompasses the entire sequence. Its structural thickness attains at least 4000 m (Fig. 7), but tectonic repetitions could be hidden by the monotony of the succession. Graptolites found in carbonaceous slates and cherts cover a time span similar to that of the Manzanal del Barco Formation, i.e., middle Llandovery to early Ludlow (González Clavijo et al., 1997). Conodonts sampled in discontinuous layers of limestone in the middle part of the unit yielded Pridoli to Early Devonian ages (Sarmiento et al., 1997).

The Rábano Formation is therefore considered to be a lateral equivalent to the Manzanal del Barco and Almendra Formations (Bajo Río Esla unit). Its relationship with the San Vitero Formation (Río Aliste unit) is more problematic, and is discussed in the following. The structural position of the Río Manzanas unit, in the uppermost thrust sheet of the Alcañices synform, suggests a more distal position in the passive margin, and its lithology and thickness compare with the Trás-os-Montes domain of Portugal (Ribeiro, 1974), now interpreted as the lower allochthon underlying the far-traveled terranes (Ribeiro et al., 1990).

Presence of metamorphic pebbles. The Almendra, San Vitero, and Rábano Formations share an important peculiarity in that they include discontinuous layers of microconglomerates with metamorphic pebbles. The pebbles have the imprint of one or more tectonic foliations (Fig. 8), and are surrounded by the first regional Variscan cleavage (Aldaya et al., 1976). The microconglomerates occur only in the upper levels of the Almendra and Rábano Formations, and are more common and widespread in the San Vitero Formation (Fig. 4). The metamorphic pebbles have been invoked as proof of an older orogenic cycle (Martínez García, 1972), but Ribeiro and Ribeiro (1974) proposed an early Variscan metamorphic and tectonic origin for these pebbles in a similar deposit in the northeast margin of the Bragança Complex (Fig. 2). Their interpretation is based on the similarity of the low-grade pebbles and the underlying upper Paleozoic (Silurian–Lower Devonian).

That interpretation was supported by Antona and Martínez Catalán (1990) and in this study, because it is compatible with the age of the early Variscan activity in the allochthonous complexes, which began in the Early Devonian according to isotopic data (Peucat et al., 1990; Dallmeyer and Gil Ibarguchi, 1990;

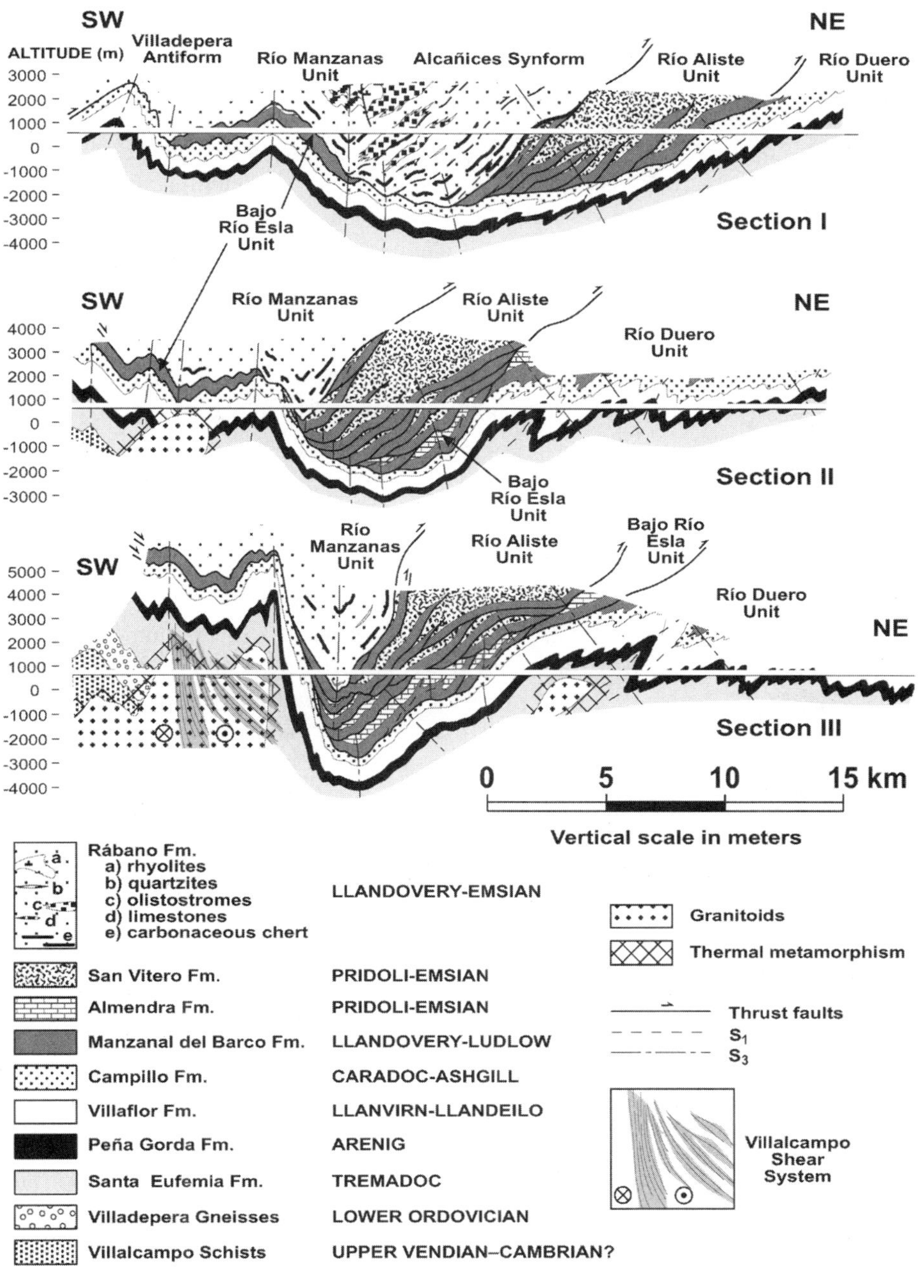

Figure 7. Geological sections across Alcañices synform. For location, see Figure 3.

Dallmeyer et al., 1991; Schäfer et al., 1993; Santos Zalduegui et al., 1996). The migration of the orogenic front toward the east, deduced by Dallmeyer et al. (1997) for the northwestern part of the Iberian Massif, explains the erosion of metamorphic rocks in the internal parts of the orogenic belt, and their deposition in a more external synorogenic trough, subsequently affected by the first and later cleavages.

STRUCTURE

No pre-Variscan structures have been found, other than the cartographic unconformity at the base of the Lower Ordovician (Iglesias Ponce de León and Ribeiro, 1981a; Villar Alonso, 1990). This was interpreted by McDougall et al. (1987) as the expression of pre-Ordovician block tilting. Evidence of related extensional deformation in the Truchas syncline, to the north (Fig. 2), was found by Martínez Catalán et al. (1992), who used a half-graben model to explain the irregular distribution of thickness and sedimentary facies.

The Variscan deformation is unique in the study area in that this is an internal part of the orogen, characterized by internal deformation, metamorphism, and granite intrusion; nevertheless, the core of the Alcañices synform depicts a structure typical of foreland thrust belts.

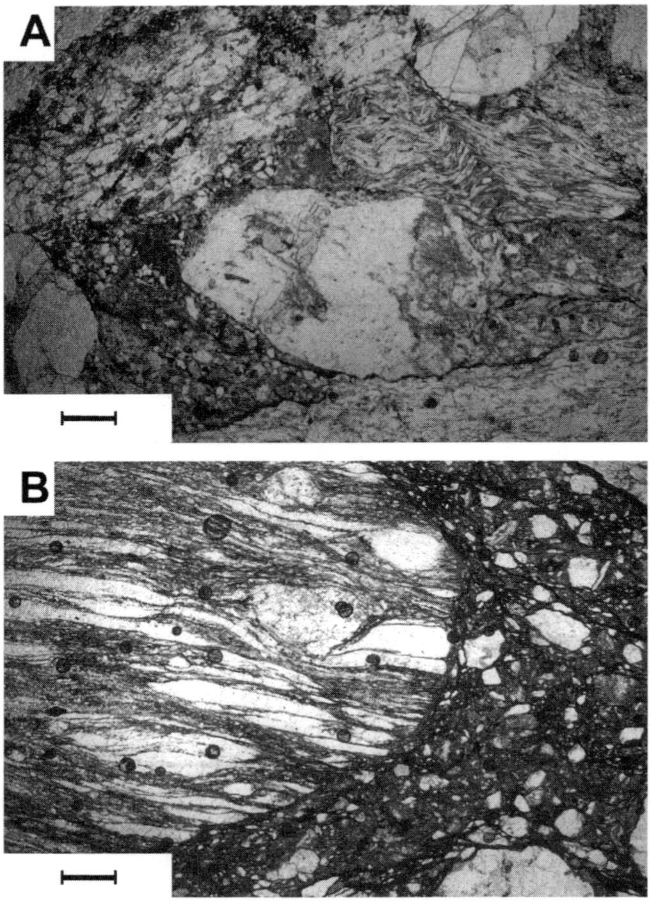

Figure 8. Microphotographs of metamorphic pebbles in conglomerates of San Vitero Formation. A: Tectonic foliation has different disposition in each pebble and displays gentle crenulation in one of them. B: Pebble depicting mylonitic foliation. Both samples were collected south of Gallegos del Río (Fig. 3). Scale bars in lower left corners measures 400 μm.

The area first was affected by recumbent folds and thrust directed northeastward. The associated deformation structures range from ductile to brittle, and can be grouped in events D_1 and D_2, according to a scheme that will facilitate correlation with other areas of the Iberian Massif. Upright folds (D_3) subsequently affected the subhorizontal structures, and subvertical transcurrent ductile shear zones developed.

Fold and thrust tectonics

Recumbent folds of several scales, with northeast vergence, developed during D_1. They have a N120°E attitude and subhorizontal axis. To the north of the synform, they reach map scale (Figs. 3 and 7), whereas in the southwest, only folds of metric scale have been found. The folds have an axial planar cleavage, S_1, widely developed in the area, mainly in the slaty lithologies. The intersection lineation between bedding and S_1 shows a northwest-southeast attitude (Fig. 9A), and is parallel to D_1 fold axes. Total shortening (including buckling and flattening components) ranging from 20% to 60% has been calculated for minor D_1 folds (Vacas and Martínez Catalán, 1987; Antona and Martínez Catalán, 1990).

The S_1 slaty cleavage at Flechas (Fig. 3) was dated by the $^{40}Ar/^{39}Ar$ incremental-release method; a plateau age of 359.3 ± 0.2 Ma (Dallmeyer et al., 1997) was determined around the Devonian-Carboniferous boundary.

Thrust tectonics caused the development of three main allochthonous sheets and numerous imbricates. These structures, ascribed to tectonic event D_2, are common in northwestern Iberia (Bastida et al., 1986; Farias Arquer, 1990; Pérez-Estaún et al., 1991), and their importance in the emplacement of the allochthonous complexes was pointed out by Ries and Shackleton (1971).

The thrusts displace only the Silurian and Devonian sequences, whereas the pre-Ordovician and Ordovician series remained as the autochthon. Thrust faults crosscut D_1 folds and the S_1 foliation, as can be seen in the northwest corner of the map (Fig. 3), and can be followed into Portugal, to the west (Meireles et al., 1995). The main faults separate structural units with different sedimentary sequences (Figs. 6 and 7). All the thrusts merge downward and to the southwest into a floor thrust, which is actually a decollement (Ramsay and Huber, 1987) developed in carbonaceous slates at the base of the Manzanal del Barco and Rábano Formations. The existence of this stratigraphic level, together with the previous existence of a huge D_1 normal limb, several tens of kilometers wide, explains the thin-skinned tectonic characteristics of the core of the Alcañices synform.

Some imbricates are hinterland-dipping duplexes, especially in the Río Aliste and Bajo Río Esla units (Fig. 7). The geometry of the horses, and their cutoff and branch lines, suggests thrusting toward the northeast: this matches the asymmetry of D_1 folds and indicates a tectonic transport toward the external zones (the Cantabrian zone; see Fig. 1) during recumbent folding and thrusting.

Centimeter-thick bands of fault gouge are common at the thrust surfaces, but bands with ductile deformation developed in the footwalls and hanging walls of the thrusts. Their extent and associated strain are usually more important in the hanging wall, close to the thrusts, where a few to a hundred meters of phyllonites developed. These fault rocks have a strong tectonic foliation, S_2, which when it is weakly developed is seen as a crenulation cleavage. The S_2 cleavage was dated by Dallmeyer et al. (1997) in Mahide (northwest of the Alcañices synform; see Fig. 3) as 342.6 ± 0.3 Ma (near the Tournaisian-Visean boundary). This age is interpreted as the final emplacement of the allochthon.

Phyllonites are mainly developed at the basal decollement. They display a stretching lineation, L_2, and often have a microconglomeratic appearance, with elongated pseudoclasts showing aspect ratios of 6:1. L_2 is defined by the mineral alignment of fine-grained quartz and white mica and by the elongation of the pseudoclasts and of pretectonic objects in conglomerates and volcanic rocks. Where quartz ribbons developed, they show quartz rods parallel to that lineation, which trends northwest-

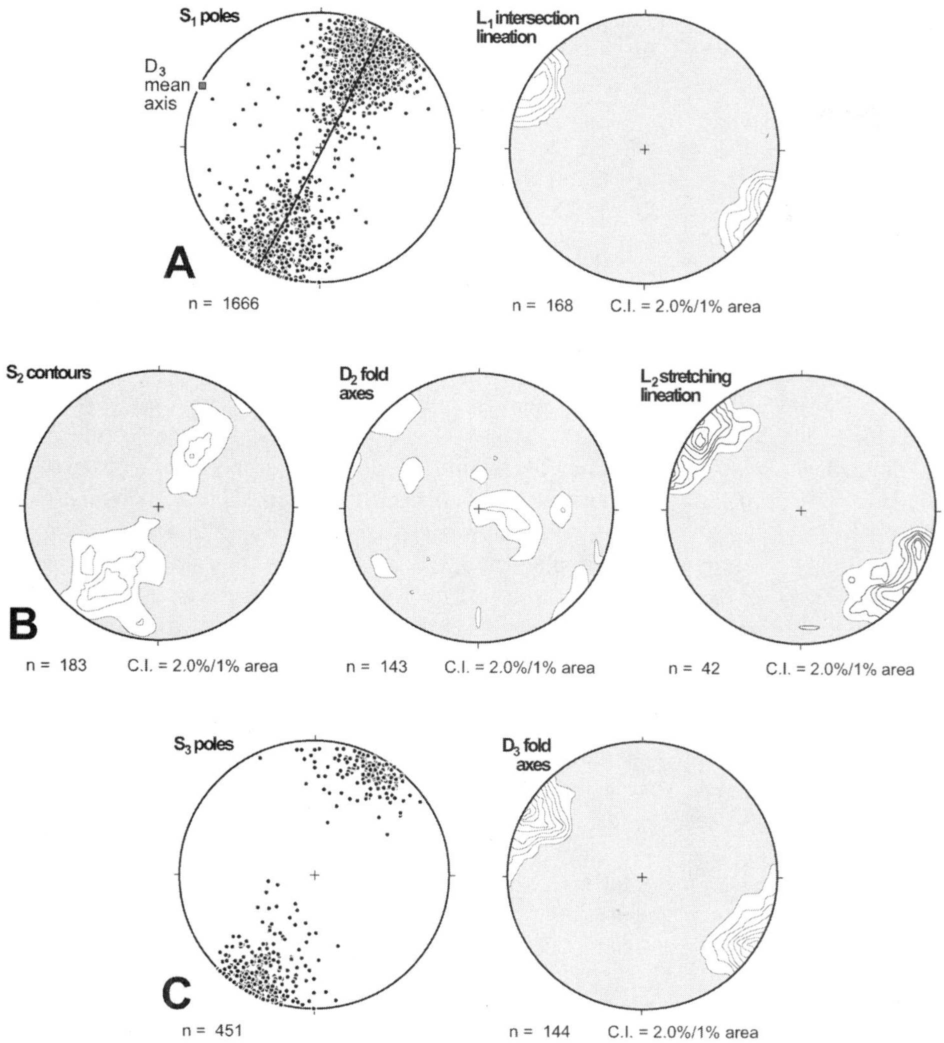

Figure 9. Structural diagrams showing attitude of foliations, fold axes, and lineations. A: Cleavage and intersection lineation associated with D_1 recumbent folds. B: Cleavage, fold axes, and stretching lineation developed during D_2 ductile deformation. C: Cleavage and fold axes associated with D_3 upright folding.

southeast and is subhorizontal (Fig. 9B). Furthermore, and despite the kinematics established from the geometry of the imbricates, bands with type II S-C fabric (Lister and Snoke, 1984) and C′ structures showing displacement of the hanging wall to the southeast, occur close to the floor thrust. In addition, shear bands of type I S-C fabric (Lister and Snoke, 1984) developed in rhyolites, and asymmetric boudinage in cherts and quartz veins, depict the same kinematics. All these criteria point to a southeast-directed mass transport of the entire allochthon during its emplacement.

Other D_2 structures common in the shear zones close to the thrust faults are folds of centimeter to meter scales. They are very flattened, with interlimb angles of 10°–30° (Villar Alonso, 1990), and commonly asymmetric; their flanks are often disrupted. Their hinge lines are sometimes curved, suggesting that they are sheath folds, and depict considerable scattering (Fig. 9B). Hansen's (1971) method was applied in several areas to estimate the kinematics of the shearing event that produced the D_2 folds (González Clavijo, 1997). A top-to-the-northeast sense of movement was established locally. This result agrees with the geometry of the imbricates, but not with the northwest-southeast trend of the stretching lineation L_2, or with the S-C criteria.

The kinematic criteria suggest that two different and near-perpendicular motions took place at the basal decollement. Of them, the southeast-directed mass transport, parallel to D_1 and D_3 fold axes, seems to be the older, because it is associated with ductile deformation: S_2 cleavage and L_2 stretching lineation (Fig. 9B). The northeast-directed motion, normal to fold axes, is mainly deduced from the cutoff and branch lines of the brittle imbricates and, consequently, should be younger.

Stretching lineations parallel to the orogenic trend, which is marked by fold axes, are a common feature in the Central Iberian zone. In a few places not far from the study area, these lineations are linked with top-to-the-southeast kinematic criteria, and associated with extensional detachments, e.g., around Fermoselle, to the south of the Alcañices synform (Fig. 2). There, an extensional detachment places a suite of high-grade metamorphic rocks and granites (the so-called Tormes dome) in contact with epizonal to mesozonal rocks of the Schist-Graywacke Complex (Escuder Viruete et al., 1994). A similar structure occurs ~90 km to the south, where it is called the Salamanca detachment (Díez Balda et al., 1995). Both structures are viewed as driven by gravitational collapse, and it seems that lateral flow of rock masses is a natural response of the thickened crust while orogenic compression is active (Dewey, 1988; Gapais et al., 1992).

Upright folds and transcurrent shear zones

Open folds trending northwest-southeast are ascribed to the D_3 event. These folds, the largest of which is the Alcañices synform, rotated the axial surface and limbs of D_1 folds, to which they are homoaxial, and folded D_2 thrusts (Figs. 3 and 7). Figure 9 shows how the previous cleavages, S_1 and S_2, became folded, whereas the L_1 (intersection) and L_2 (stretching) lineations are homoaxial with D_3 fold axes.

The folds are open, with the axial plane dipping to northeast on the north limb of the Alcañices synform, leading to local overturning of beds and thrusts. On the south limb, D_3 folds are tighter and may have a subvertical axial plane (Fig. 7). An associated cleavage, S_3, axial planar to the folds, can be seen in the minor and major structures, and attains a regional importance in the southern part of the study area. Its character varies according to the material affected. The prevailing type is a crenulation cleavage, more or less developed over previous cleavages. Some rocks, such as cherts, do not display S_3, whereas others, such as quartzites and graywackes, may exhibit a rough S_3 cleavage as the only visible tectonic foliation.

In addition to the folds, steeply dipping ductile shear bands occur to the south of the study area, in what has been called the Villalcampo shear system (González Clavijo et al., 1993). The system crops out for 40 km until covered by the Cenozoic clastic sedimentary rocks of the Duero Basin to the southeast. Its northwest half can be seen in Figure 3, where the system shows a N130°E general trend, and consists of several small shear bands joining together toward the southeast into a single band more than 1 km wide. The splay opens when the Villalcampo shear zone enters the plutonic rocks of Ricobayo, where the system forms a "horse tail," or fan. To the north of the granite, only small areas show a related deformation, and they are close to the intrusive contact (Santa Eufemia Formation). The quartzites of the Peña Gorda Formation (500 m away toward the north) were not deformed (Figs. 3 and 7).

The shear zones are seen when affecting the granite massifs, but can also be identified in the schists and gneisses. Shearing caused a variety of deformation structures. The more common are type I S-C mylonites (Lister and Snoke, 1984), but well-foliated mylonites with quartz ribbons and intrafoliated folds are also present. A subhorizontal stretching lineation developed in the mylonites. The S-C structures provide the most reliable kinematic criteria, but the asymmetry of intrafoliated folds and the steady-state foliation (Means, 1981) seen in thin sections in mylonitic quartz ribbons have also been used to estimate the kinematic framework. A dextral movement with a relative displacement of 3–4 km was calculated by González Clavijo et al. (1993).

Small, conjugate, second-order shear bands are found inside the "horse tail" and to its western part; they trend N70°–110°E and show sinistral displacements. The splay is characterized by the presence of quartz veins and veinlets mineralized with gold, and has been interpreted as an extensional fan at one end of a first-order shear band.

The relative age of the shear system has been established as late D_3, on the basis of shearing of the S_3 foliation in schists and the Ricobayo Granite, which is a syn-D_3 igneous body according to the relationships between the S_3 cleavage and porphyroblast growth related to the induced contact metamorphism.

The Villalcampo shear system forms part of a group of similar ductile strike-slip structures, both dextral and sinistral, described in northwestern Iberia (Iglesias Ponce de León and Choukroune, 1980; Iglesias Ponce de León and Ribeiro, 1981b). Mostly developed in synkinematic Variscan granitoids, they grade into domains with steep crenulation cleavage when entering schistose areas. This suggests that both upright D_3 folds and transcurrent ductile shear zones are genetically related (Iglesias Ponce de León and Choukroune, 1980).

There are several systems of subvertical late faults. The more important are either parallel to the regional northwest-southeast structure or nearly normal to it (northeast-southwest). The latter group is especially well developed in the eastern part of the area, around the Esla River section (Ricobayo Reservoir). These late faults exert important control on the present fluvial system.

METAMORPHISM AND PLUTONISM

During the Variscan orogeny, the area underwent regional metamorphism of the intermediate pressure type (M_1). This was low grade in most of the study area, below the biotite isograd. However, in the Villadepera antiform to the south (Figs. 3 and 7), deeper structural levels crop out, and M_1 reaches a medium grade (Villar Alonso, 1990). Subsequently, the metamorphism evolved toward low-pressure conditions, and an M_2 event gave rise to high-grade metasediments and anatexites (López Plaza, 1982), which are exposed ~30 km south of the area. Following this event, several granite massifs intruded the area, most of them synkinematic to the last deformation event (D_3). They imprinted a thermal metamorphism on the country rocks that is described as M_3, and were themselves overprinted by dynamic metamorphism related to transcurrent shearing.

Regional metamorphism

Chlorite subfacies of the greenschist facies metamorphism prevail in the Alcañices synform and to the north (Vacas and Martínez Catalán, 1987). Antona and Martínez Catalán (1990) measured illite crystallinity to assess the metamorphic conditions, and to compare those of the Manzanal del Barco and San Vitero Formations. They found no differences in metamorphic grade between the preorogenic and synorogenic metasediments of the core of the synform, and calculated a temperature of 350 °C for M_1, which corresponds to the lower epizone, very close to anchizone. The color alteration index (CAI) of conodonts was investigated by Sarmiento et al. (1997). Most of the CAI values they found were in the range 5.5–6.0, implying temperatures higher than 300 °C, and around the anchizone-epizone transition.

In the Villadepera antiform and farther to the south, the biotite, garnet, staurolite, and sillimanite isograds define a metamorphic zoning characteristic of a Barrovian-type gradient. The relative age of M_1 is well defined by S_1 and S_2 tectonic foliations. The diagnostic minerals grew over the S_1 cleavage and were flattened by S_2, supporting an M_1 peak following D_1, but predating D_2.

A low-pressure metamorphic event, M_2, has been identified in the area of Fermoselle, 30 km to the south of the Alcañices synform (Fig. 2). Escuder Viruete et al. (1994) related high-temperature and low-pressure metamorphism there to gravitational collapse of the orogenic crust; according to these authors, the ascent of deep crustal rocks in the footwall of an extensional detachment formed an anatectic dome (the Tormes dome), with partial melting occurring as a consequence of decompression, and with contemporaneous heating of the hanging wall unit.

Traces of M_2 heating are found near San Martín del Pedroso, to the west of the study area (Fig. 10), where a thermal aureole with andalusite prisms several centimeter long crops out and is bounded by a major thrust. The andalusite crystals are surrounded by the S_2 tectonic foliation, which appears strongly flattened around the porphyroblasts, and asymmetric pressure tails developed. In addition, some andalusite crystals show microboudinage. Microstructural relationships indicate a syn-S_2 growth of the large andalusite crystals, unrelated to a late-kinematic granite that intruded to the northwest of the M_2 metamorphic aureole. The granite is undeformed and developed a narrow post-S_3 thermal aureole; small, postkinematic andalusites overprint the S_3 crenulation cleavage.

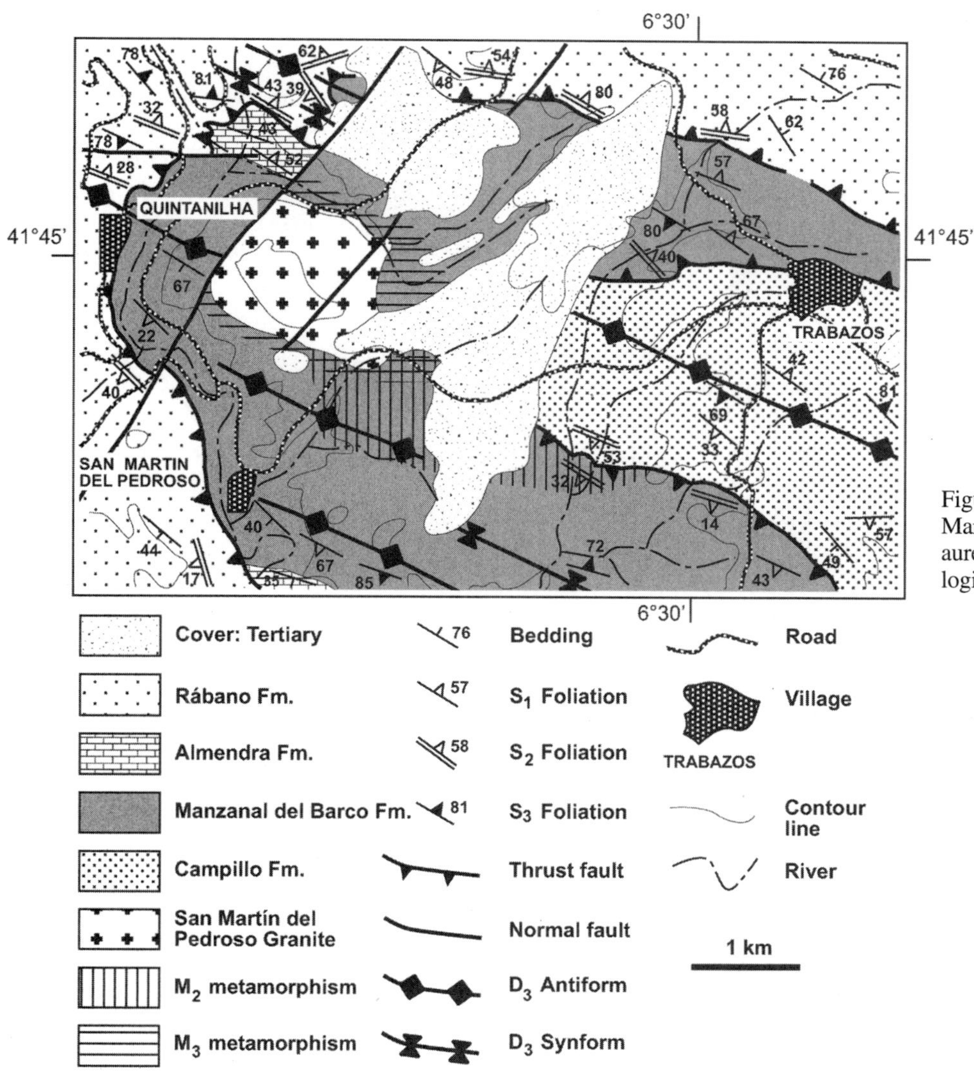

Figure 10. Geological map of area around San Martín del Pedroso, showing two metamorphic aureoles and their relationships to key geological structures. For location, see Figure 3.

The M_2 thermal aureole appears at the base of the allochthonous Manzanal del Barco Formation, but not in the autochthonous Campillo Formation (Fig. 10), implying that the D_2 floor thrust truncated the M_2 thermal aureole and carried a part of it above the authochthon.

Granite intrusion, contact and dynamic metamorphism

Two main intrusive bodies intruded in the southern part of the Alcañices synform, the Ricobayo two-mica granite and the Carbajosa leucogranite (Fig. 3). The Ricobayo Massif is the largest, with 150 km^2 of exposed surface, and includes a wide variety of facies: coarse and fine-grained two-mica granite, biotitic granite with cordierite, leucogranite with garnet and sillimanite, and amphibole-bearing diorite. The massif is elongated following the trend of D_3 folds, but overprints them (Figs. 3 and 7, section II). The Carbajosa leucogranite is smaller than the Ricobayo Massif, and elongated in the same direction. Both massifs have a common aureole of thermal metamorphism (M_3), an elongated area that extends farther to the northwest than the exposed granites, suggesting a continuation at depth that has been proven by drilling (Pierrel et al., 1981; Florido Laraña et al., 2002).

M_3 is contact metamorphism with porphyroblasts of biotite and andalusite overprinting the S_1 and S_2 tectonic foliations. The porphyroblasts relationship with S_3 crenulation cleavage varies from northwest to southeast along the northern side of the Ricobayo pluton. In the southeast the andalusites enclose the S_3 cleavage, whereas in the northwest S_3 is flattened around the porphyroblasts, but the granite appears to cut D_3 folds. Consequently, a synkinematic character in relation to D_3 is suggested for these massifs, different intrusive pulses having taken place during the development of the upright folds.

The Ricobayo and Carbajosa massifs were deformed by the late-D_3 Villalcampo shear system (Figs. 3 and 7), which developed a retrograde metamorphism in the shear bands. Biotites and muscovites were chloritized and the feldspars became sericitized, resulting in an excess of silica, which forms abundant quartz veins inside the deformation bands (González Clavijo et al., 1993).

A late to postkinematic 2-km-diameter intrusive body, the San Martín del Pedroso granite, is exposed to the west of the study area (Figs. 3 and 10). It is a coarse- to medium-grained muscovite granite, clearly not deformed, and crosscuts D_3 folds. It is surrounded by a M_3 thermal aureole where small andalusites that postdate the S_3 crenulation cleavage grew. This M_3 aureole partly overprints a previous one, a remnant of M_2 low-pressure metamorphism (Fig. 10).

EVOLUTION AND TECTONIC SIGNIFICANCE

The study of the Alcañices synform provided data that can be readily integrated with recent work in the allochthonous complexes of northwestern Iberia (Dallmeyer et al., 1991; Arenas et al., 1995; Schäfer et al., 1993; Santos Zalduegui et al., 1996; Martínez Catalán et al., 1996, 1997) and in the Iberian autochthon (Escuder Viruete et al., 1994; Díez Balda et al., 1995; Dallmeyer et al., 1997) to improve the understanding of this piece of the Variscan belt.

The important stratigraphic points are the occurrence of (1) a complete Ordovician sequence, (2) a probably complete Silurian succession bounded by a basal decollement, (3) three different Silurian-Devonian sedimentary successions, each cropping out in a different structural unit, and (4) characteristic synorogenic deposits restricted to the intermediate allochthonous structural unit.

The following structural aspects are important: (1) there is no evidence of pre-Variscan deformation; (2) the start of the Variscan orogeny is indicated by pebbles of metamorphic rocks in Lower Devonian sediments; (3) there is an extensive development of the first Variscan cleavage, while contemporaneous major folds are restricted to the northeast of the Alcañices synform; and (4) thin-skinned tectonics developed in the Silurian-Devonian sequence.

Basin evolution

The oldest sequences that crop out in the Iberian autochthon, the Villalcampo Schists, are essentially Late Proterozoic, and have turbiditic characteristics that have been envisaged as the filling of troughs related to a Pan-African orogenic belt (San José et al., 1990; Murphy and Nance, 1991). Coeval magmatism, mostly intrusive, testifies to the relationship with a magmatic arc (Lancelot et al., 1985; Ochsner, 1993; Fernández-Suárez et al., 1998). The younger porphyritic gneisses of the Ollo de Sapo type, represented by the Villadepera Gneisses, seem to represent postcollisional magmatism (both plutonic and volcanic) generated in the continental crust of Gondwana. The melting leading to the formation of these rocks would be the result of crustal thickening (Gebauer, 1993) or of the subsequent stress relaxation.

The latter took place, and was related to the pulling apart of the Avalon peri-Gondwanan terrane, the opening of the Rheic ocean, and the establishment of a passive margin (Fortey and Cocks, 1988; Soper, 1988). A clastic wedge developed along this passive margin following an Early Ordovician transgression (Fig. 11A), and was linked to crustal extension and thermal subsidence. The Armorican Quartzite (Peña Gorda Formation) records the maximum transgression in the northern margin of Gondwana. According to Gutiérrez Marco et al. (1990), there are different sedimentary facies represented in the Armorican Quartzite of the Central Iberian zone: supratidal sand bar, barrier island, and infratidal high-energy deposit. Partitioning of the basin following fault lines generated distinct sedimentary environments (McDougall et al., 1987). The transgression continued during the Middle Ordovician, when the offshore shales of the Villaflor Formation were deposited. The diversity of facies and lateral thickness variations of the Upper Ordovician (Campillo Formation in Alcañices) in northwestern Iberia have been ex-

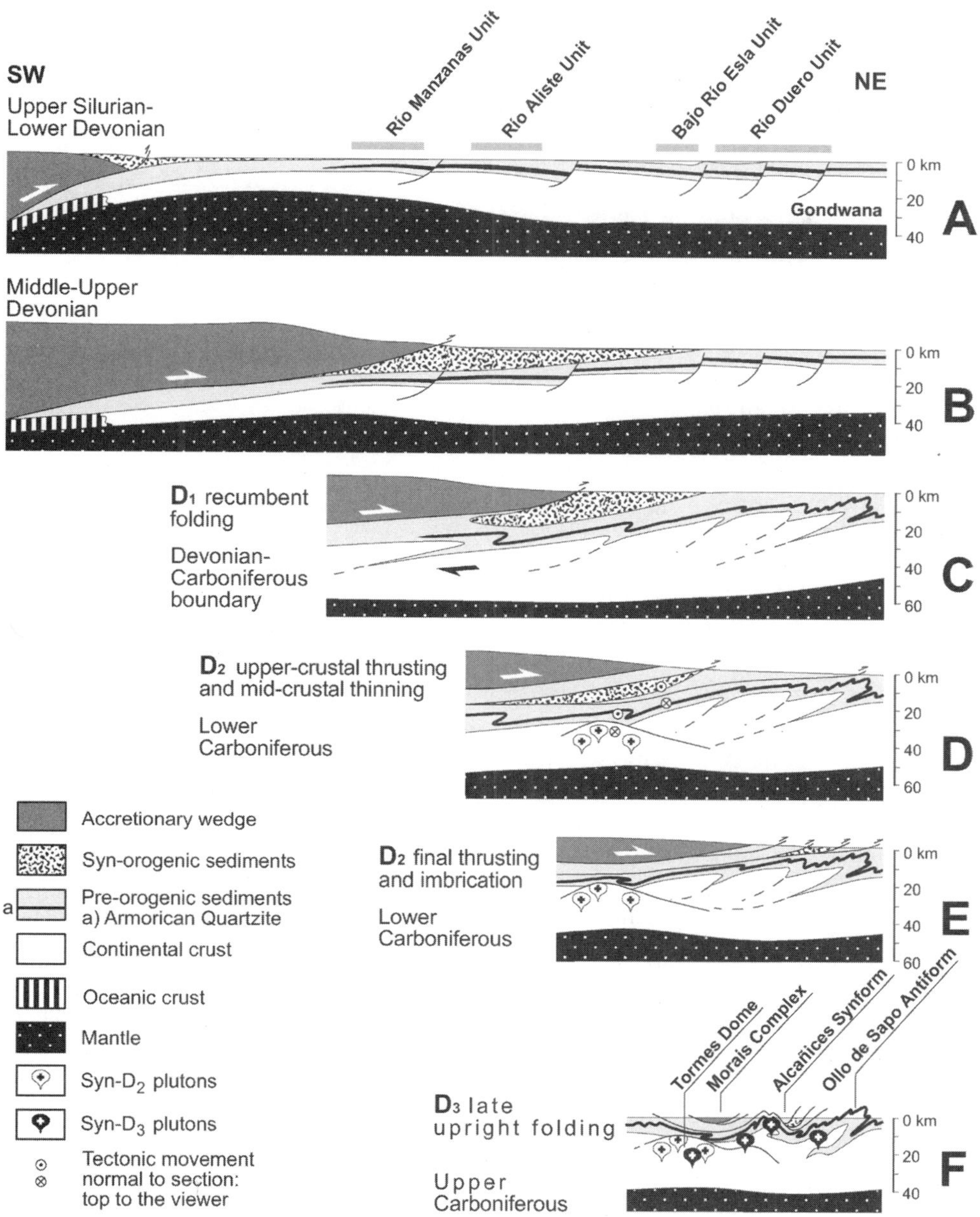

Figure 11. Evolution of area shown in six stages, as deduced from stratigraphic, structural, and metamorphic data. A: Outer margin of Gondwana was being subducted beneath accretionary wedge, which is partly preserved in allochthonous complexes. Note flexural forebulge associated with southwest-directed subduction. Approximate position of four structural units is shown in this reconstruction. B: Synorogenic clastic sedimentary rocks are being deposited in trough in front of advancing accretionary wedge. C: D_1 deformation event. Note crustal thickening and large-scale shearing induced by thrusting of allochthonous complexes. D: Mid-crustal thinning, resulting in exhumation of high-grade metamorphic rocks, heat transfer, and granite intrusion. Mass flow parallel to orogenic trend (top to reader). Contemporaneous development of thrusts and possible along-strike movements on basal decollement. E: Brittle D_2 thrusting and imbrication. F: D_3 folding: main regional structures are indicated.

plained as the result of a fragmentary basin, with changing paleogeography, produced in an area of increasing tectonic instability (Gutiérrez Marco et al., 1990).

The regional unconformity at the base of the Silurian, described in parts of the Iberian Massif (Matte, 1968; Martínez Catalán et al., 1992), has not been identified in the Alcañices synform. The entire Silurian is probably represented in the Manzanal del Barco and Rábano Formations, together with the lower part of the Almendra Formation. These deposits show abundant facies changes, such as evidence of mass flow and slumping, synsedimentary folds, and volcanic sills, plugs, flows, and tuffs. The Almendra Formation was apparently deposited on the lower part of a slope, probably adjacent to one of the normal faults responsible for the thinning of the passive margin. The same environment may explain the sedimentation of the Rábano Formation, with extensional conditions still prevailing, as shown by its volcanic rocks and mass flow episodes (Fig. 4).

Therefore, the instability of the Late Ordovician continued into the Silurian and Early Devonian. However, rather than reflecting continued extension of the passive margin, it seems to witness the transition to plate convergence, because a new and fundamental element, i.e., pebbles of metamorphic rocks, appears in the upper parts of the Almendra and Rábano Formations. They point to an orogenic process in nearby areas, and mark the transition of the sedimentary basin to a synorogenic trough. The synorogenic character is well marked in the San Vitero Formation. A Late Devonian age for similar turbiditic deposits has been established in Portugal (Pereira et al., 1999), and a pre-Carboniferous age is indicated by the fact that the turbidites were affected by the first Variscan cleavage, dated as 359.3 ± 0.2 Ma (Dallmeyer et al., 1997). However, it is clear that the age of the San Vitero Formation could extend to the Middle Devonian, given the fact that the transition from passive margin to synorogenic trough seems to have started earlier.

Isotopic data from the Meguma Group of Nova Scotia (Dallmeyer and Keppie, 1987) indicate that convergence between Laurentia and Gondwana had started by the Late Silurian–Early Devonian. Diachronism in the age of first deformation there, and in the allochthonous complexes of northwestern Iberia, suggests that an accretionary wedge was being constructed along the Laurentian margin. The wedge incorporated peri-Gondwanan terranes, fragments of the Rheic oceanic lithosphere, and, once ocean closure was accomplished, slivers of the subducted Gondwanan margin (Arenas et al., 1995; Martínez Catalán et al., 1997).

The accretionary wedge occupied a position to the west (in present coordinates), and the study area was part of the sedimentary trough between the active margin of Laurentia and the passive margin of Gondwana.

The San Vitero Formation reflects the influence of the colliding wedge (Fig. 11, A and B), the arrival of which is evidenced by the metamorphic pebbles in the upper parts of the Almendra and Rábano Formations. Why the turbidites directly overlie the Manzanal del Barco Formation, instead of some Upper Silurian–Lower Devonian sediments, is a matter of speculation. The San Vitero Formation could include sediments of this age. In this case, the olistostromes and olistoliths of the Rábano Formation could reflect Variscan convergence.

However, coupling the available age data with the known tectonic framework (Martínez Catalán et al., 1996, 1997), it seems more reasonable that the Upper Silurian–Lower Devonian sediments were eroded in a forebulge induced by the lithospheric flexure associated with the subduction toward the west of the passive margin of Gondwana (Fig. 11A). Reactivation of earlier faults, mass flow, slumping, and volcanic activity in the Almendra and Rábano Formations may have been related to this process. The San Vitero turbidites, which disconformably overlie the Manzanal del Barco Formation, would represent the filling of a peripheral depression created subsequently at the place of the earlier bulge (Fig. 11B).

Orogenic evolution

Continental collision proceeded with the development of an orogenic belt, the outer limit of which advanced progressively to the east while the whole belt remained tectonically active. Deformation started with recumbent folding (D_1) affecting the cover (Fig. 11C). The related foliation (S_1) has an age around the Devonian-Carboniferous boundary in the study area (Dallmeyer et al., 1997). Shortening in the cover was accompanied by folding and piling up of Cadomian and lower Paleozoic orthogneisses in the basement (Macaya et al., 1991; Escuder Viruete et al., 1994).

The absence of large D_1 folds in the area underlying the allochthonous complexes, to the south of the Alcañices synform, where the S_1 cleavage is widely developed, is explained as a wide normal limb. This, in turn, may be related to a huge horizontal ductile shear zone: the allochthonous complexes of northwestern Iberia possibly behaved as a rigid body at that time, while the sedimentary pile acted as a soft cushion between this rigid block and the basement (Fig. 11C). The regional metamorphism, characterized by a gradient of the intermediate pressure type, and ranging in grade from epizonal to catazonal, developed synchronously. It was induced by the crustal thickening resulting from deformation of the Gondwana margin plus the emplacement of the allochthonous complexes.

Continuation of the compressional tectonics during the early Carboniferous led to thrusting of the metasediments. A new tectonic foliation (S_2) is associated with D_2 structures. However, stretching lineations and shear criteria point to contemporaneous extension along strike, both in the allochthonous sequences and in their autochthon (Fig. 11D). The previous and continued crustal thickening seems to have induced gravitational collapse, the more plausible cause of longitudinal extension. At mesocrustal levels, heat weakens the crust enough to produce a gravitationally unstable tectonic discontinuity (Lister and Davis, 1989; Malavieille, 1987, 1991). This happened in the Tormes dome, to the south, where a ductile detachment provoked the exhumation of high-grade metamorphic rocks, partial melting, and granite intrusion (Escuder Viruete et al., 1994). A thermal aureole, probably related to this process, was cut by a later thrust, indicating synchronism between compressional tectonics and crustal extension.

The longitudinal stretching lineations in the basal decollement, and several shear criteria there, at odds with the kinematics deduced from cutoff and branch lines in the imbricates, may reflect the same dynamic framework. It may be that D_2 deformation was partitioned between extension along strike (to the southeast) in the shear zone and movement normal to it (to the northeast) along thrust faults. Alternatively, the decollement moved to the southeast at an early stage of its structural history, in response to gravitational or tectonic forces (Fig. 11D).

The more important effect of D_2 convergence was the stacking of thrust sheets derived from different parts of the passive margin (Fig. 11, A–E), permitting its partial reconstruction. A Tournaisian-Visean age seems to date the final emplacement of the allochthonous stack (Dallmeyer et al., 1997).

The final stages of the orogenic shortening gave way to upright folds (D_3) and a related crenulation cleavage (Fig. 11F). Synkinematic granites intruded and became deformed in a transcurrent ductile shear zone, in response to the later shortening events prior to cratonization.

CONCLUSIONS

The study of four different structural units in the Alcañices synform, three of them allochthonous, has improved knowledge of the sedimentary and structural evolution of the northern margin of Gondwana in stages previous to and early during its collision with Laurentia. Three different Silurian-Devonian successions, each cropping out in a different thrust sheet (with internal imbricates), provide useful information when their relative paleogeographical positions are reconstructed from their order of stacking.

The Devonian deposits mark the transition from a passive margin realm to a synorogenic trough. The start of Variscan deformation is represented by rare Lower Devonian layers with pebbles of metamorphic rocks in the upper and lower structural units, where the succession is mostly preorogenic. Conversely, the Upper (and maybe older) Devonian of the middle structural unit is richer in metamorphic clasts and its character is clearly synorogenic. The preservation of these turbiditic deposits in the middle allochthonous unit, and the disconformity separating them from the Silurian deposits, is attributed to its particular paleogeographic position. This was the site of a flexural forebulge induced by the bending of the lithosphere during early subduction to the west, and then the site of a peripheral depression created in more advanced stages of convergence.

Orogenesis prograded eastward during the early Carboniferous, incorporating these deposits into recumbent folds and thrusts, in a deformation that affected the entire crust. The complexes probably behaved as a rigid body at that time, the sedimentary pile acting as a soft cushion between them and the basement and giving rise to a thick shear zone. As orogeny proceeded, orogen-parallel extension accommodated gravitational gradients created by crustal thickening, while compressional deformation became concentrated in discrete shear zones and thrusts, causing stacking and imbrication. The existence of a huge normal limb, where the beds remained low dipping and flat, together with the occurrence of Lower Silurian carbonaceous slates, allowed the decollement of the Silurian-Devonian successions, and the development of thin-skinned tectonics.

ACKNOWLEDGMENTS

This work was funded by project PB93-1149-C03-02 of the Spanish scientific agency Dirección General de Investigación Científica y Técnica (DGICYT). We thank María Antonia Díez Balda, chief of the project, for providing support, and for supervision, advice, and cooperation. We also thank Juan Carlos Gutiérrez Marco and Graciela Sarmiento, who carried out paleontological studies on samples collected in the study area, and Peter Clift and Reinhard O. Greiling for their careful and constructive reviews of the original manuscript, that resulted in a substantial improvement.

REFERENCES CITED

Aldaya, F., Carls, P., Martínez García, E., and Quiroga, J.L., 1976, Nouvelles précisions sur la série de San Vitero (Zamora, Nordouest de l'Espagne): Comptes Rendus de l'Académie des Sciences, Paris, v. 283, p. 881–883.

Antona, J.F., and Martínez Catalán, J.R., 1990, Interpretación de la formación San Vitero en relación con la Orogenia Hercínica: Cuadernos do Laboratorio Xeolóxico de Laxe, v. 15, p. 257–269.

Arenas, R., Gil-Ibarguchi, J.I., González-Lodeiro, F., Klein, E., Martínez-Catalán, J.R., Ortega-Gironés, E. Pablo-Maciá, J.G. de, and Peinado, M., 1986, Tectonostratigraphic units in the complexes with mafic and related rocks of the NW of the Iberian Massif: Hercynica, v. 2, p. 87–110.

Arenas, R., Rubio Pascual, F.J., Díaz García, F., and Martínez Catalán, J.R., 1995, High-pressure micro-inclusions and development of an inverted metamorphic gradient in the Santiago Schist (Ordenes Complex, NW Iberian Massif, Spain): Evidence of subduction and syn-collisional decompression: Journal of Metamorphic Geology, v. 13, p. 141–164.

Barrois, Ch., 1882, Recherches sur les terrains anciens des Asturies et de la Galice: Mémoires de la Société Géologique du Nord, v. 2, p. 1–630.

Bastida, F., Martínez Catalán, J.R., and Pulgar, J.A., 1986, Structural, metamorphic and magmatic history of the Mondoñedo Nappe (Hercynian Belt, NW Spain): Journal of Structural Geology, v. 8, p. 415–430.

Bechennec, F., Le Metour, J., Rabu, D., Villey, M., and Beurrier, M., 1988, The Hawasina Basin: A fragment of a starved passive continental margin, thrust over the Arabian Platform during obduction of the Sumail Nappe: Tectonophysics, v. 151, p. 323–343.

Dallmeyer, R.D., and Gil Ibarguchi, J.I., 1990, Age of amphibolitic metamorphism in the ophiolitic unit of the Morais allochthon (Portugal): Implications for early Hercynian orogenesis in the Iberian Massif: Journal of the Geological Society of London, v. 147, p. 873–878.

Dallmeyer, R.D., and Keppie, J.D., 1987, Polyphase late Paleozoic tectonothermal evolution of the southwestern Meguma Terrane, Nova Scotia: Evidence from $^{40}Ar/^{39}Ar$ mineral ages: Canadian Journal of Earth Sciences, v. 24, p. 1242–1254.

Dallmeyer, R.D., Ribeiro, A., and Marques, F., 1991, Polyphase Variscan emplacement of exotic terranes (Morais and Bragança Massifs) onto Iberian successions: Evidence from $^{40}Ar/^{39}Ar$ mineral ages: Lithos, v. 27, p. 133–144.

Dallmeyer, R.D., Martínez Catalán, J.R., Arenas, R., Gil Ibarguchi, J.I., Gutiérrez Alonso, G., Farias, P., Aller, J., and Bastida, F., 1997, Diachronous Variscan tectonothermal activity in the NW Iberian Massif: Evidence from $^{40}Ar/^{39}Ar$ dating of regional fabrics: Tectonophysics, v. 277, p. 307–337.

Dewey, J.F., 1988, Extensional collapse of orogens: Tectonics, v. 7, p. 1123–1139.

Díaz García, F., Arenas, R., Martínez Catalán, J.R., González del Tánago, J., and Dunning, G., 1999, Tectonic evolution of the Careón ophiolite (Northwest Spain): A remnant of oceanic lithosphere in the Variscan belt: Journal of Geology, v. 107, p. 587–605.

Díez Balda, M.A., 1986, El Complejo Esquisto-Grauváquico, las series paleozoicas y la estructura hercínica al Sur de Salamanca: Acta Salmanticensia, Sección Ciencias, Ediciones Universidad de Salamanca, Spain, v. 52, 162 p.

Díez Balda, M.A., Martínez Catalán, J.R., and Ayarza, P., 1995, Syn-collisional extensional collapse parallel to the orogenic trend in a domain of steep tectonics: the Salamanca Detachment Zone (Central Iberian Zone, Spain): Journal of Structural Geology, v. 17, p. 163–182.

Escuder Viruete, J., Arenas, R., and Martínez Catalán, J.R., 1994, Tectonothermal evolution associated with Variscan crustal extension in the Tormes Gneiss Dome (NW Salamanca, Iberian Massif, Spain): Tectonophysics, v. 238, p. 114–138.

Farias Arquer, P., 1990, La geología de la región del Sinforme de Verín (Cordillera Herciniana, NW de España): A Coruña, Spain, Laboratorio Xeolóxico de Laxe, Nova Terra, v. 2, 201 p.

Farias, P., Gallastegui, G., González Lodeiro, F., Marquínez, J., Martín Parra, L.M., Martínez Catalán, J.R., Pablo Macia, J.G. de, and Rodríguez Fernández, L.R., 1987, Aportaciones al conocimiento de la litoestratigrafía y estructura de Galicia Central: Memórias da Faculdade de Ciências, Universidade do Porto, v. 1, p. 411–431.

Fernández-Suárez, J., Gutiérrez-Alonso, G., Jenner, G.A., and Jackson, S.E., 1998, Geochronology and geochemistry of the Pola de Allande granitoids (northern Spain): Their bearing on the Cadomian-Avalonian evolution of NW Iberia: Canadian Journal of Earth Sciences, v. 35, p. 1439–1453.

Fernández-Suárez, J., Gutiérrez-Alonso, G., Jenner, G.A., and Tubrett, M.N., 2000, New ideas on the Proterozoic-early Paleozoic evolution of NW

Iberia: Insights from U-Pb detrital zircon ages: Precambrian Research, v. 102, p. 185–206.

Florido Laraña, P., González Clavijo, E.J., Ortega Menéndez de Llano, C., Rodríguez Rivas, A.I., and Asensio, B., 2002, Mapa Metalogenético de España, Hoja 28/3–4, Alcañices: Instituto Geológico y Minero de España, 135 p., escala 1:200,000.

Fortey, R.A., and Cocks, L.R.M., 1988, Arenig to Llandovery faunal distributions in the Caledonides, *in* Harris, A.L., and Fettes, D.J., eds., The Caledonian-Appalachian Orogen: Geological Society of London Special Publication 38, p. 233–246.

Gapais, D., Pêcher, A., Gilbert, E., and Ballèvre, M., 1992, Synconvergence spreading of the Higher Himalaya crystalline in Ladakh: Tectonics, v. 11, p. 1045–1056.

Garrido, F., and Barberá, M., 1980, Estudio sedimentológico y metalogenético en el sector de Nuez de Aliste (Zamora), Unpublished memoir: Departamento de Prospección Geológica y Geofísica, Universidad de Barcelona, División de Minería y Geoquímica del Consejo Superior de Investigaciones Científicas, Salamanca, Billiton Española, S.A., Promotora de Recursos Naturales, and Instituto Geológico y Minero de España, Madrid, 111 p.

Gebauer, D., 1993, Intra-grain zircon dating within the Iberian Massif: Ollo de Sapo augengneis, bimodal gneisses from the Massif de Guilleries (Girona), graywacke of the Tentudia Group (Serie Negra, SW Spain) and the HP/HT-rock association at Cabo Ortegal (Galicia), *in* Comunicaçoes da XII Reuniao de Geología do Oeste Penninsular: Evora, Portugal, Universidade de Evora, v. 2, p. 41–46.

González Clavijo, E.J., 1997, La geología del Sinforme de Alcañices, oeste de Zamora [Ph.D. thesis]: Salamanca, Spain, Universidad de Salamanca, 330 p.

González Clavijo, E., Díez Balda, M.A., and Alvarez, F., 1993, Structural study of a semiductile strike-slip system in the Central Iberian Zone (Variscan Fold Belt, Spain): structural control on gold deposists: Geologische Rundschau, v. 82, p. 448–460.

González Clavijo, E., Gutiérrez Marco, J.C., Jiménez Fuentes, E., Moro Benito, M.C., and Storch, P.,1997, Graptolitos silúricos del Sinforme de Alcañices (prov. de Zamora, Zonas Centroibérica y Galaico-trasmontana), *in* Grandal D'Anglade, A., Gutiérrez-Marco, J.C., and Santos Fidalgo, L., eds., Paleozoico Inferior del Noroeste de Gondwana, V Reunión Internacional del Proyecto 351 PICG, Libro de resúmenes y excursiones: Coruña, Spain, Universidade da Coruña, p. 71–74.

Gutiérrez Marco, J.C., Aramburu, C., Arbizu, M., Bernárdez, E., Hacar Rodríguez, M.P., Méndez-Bedia, I., Montesinos López, R., Rábano, I., Truyols, J., and Villas, E., 1999, Revisión bioestratigráfica de las pizarras del Ordovícico Medio en el noroeste de España (zonas Cantábrica, Asturoccidental-leonesa y Centroibérica septentrional): Acta Geologica Hispanica, v. 34, p. 3–87.

Gutiérrez Marco, J.C., San José, M.A., and Pieren, A.P., 1990, Central-Iberian Zone. Post-Cambrian palaeozoic stratigraphy, *in* Dallmeyer, R.D., and Martínez García, E., eds., Pre-Mesozoic geology of Iberia: Berlin, Springer-Verlag, p. 160–171.

Hansen, E., 1971, Strain Facies: Berlin, Springer-Verlag, 207 p.

Haq, B.U., and Van Eysinga, F.W.B., 1998, Geological time table, 5th ed.: Amsterdam, Elsevier.

Iglesias Ponce de León, M., and Choukroune, P., 1980, Shear zones in the Iberian Arc: Journal of Structural Geology, v. 2, p. 63–68.

Iglesias Ponce de León, M., and Ribeiro, A., 1981a, Position stratigraphique de la formation Ollo de Sapo dans la région de Zamora (Espagne)-Miranda do Douro (Portugal): Comunicaçoes dos Serviços Geológicos de Portugal, v. 67, p. 141–146.

Iglesias Ponce de León, M., and Ribeiro, A., 1981b, Zones de cisaillement ductile dans l'arc ibéro-armoricain: Comunicaçoes dos Serviços Geológicos de Portugal, v. 67, p. 85–87.

Iglesias Ponce de León, M., and Robardet, M., 1980, El Silúrico de Galicia Media (Central), su importancia en la paleogeografía Varisca: Cuadernos do Laboratorio Xeoloóxico de Laxe, v. 1, p. 99–115.

Julivert, M., Fonboté, J.M., Ribeiro, A., and Conde, L., 1972, Mapa Tectónico de la Península Ibérica y Baleares: Instituto Geológico y Minero de España, Madrid, 113 p., escala 1:1,000,000.

Lancelot, J.R., Allegret, A., and Iglesias Ponce de León, M., 1985, Outline of Upper Precambrian and lower Paleozoic evolution of the Iberian Peninsula according to U-Pb dating zircons: Earth and Planetary Science Letters, v. 74, p. 325–337.

Lister, G.S., and Davis, G.A., 1989, The origin of metamorphic core complexes and detachment faults formed during Tertiary continental extension in the northern Colorado River region, U.S.A.: Journal of Structural Geology, v. 11, p. 65–94.

Lister, G.S., and Snoke, A.W., 1984, S-C mylonites: Journal of Structural Geology, v. 6, p. 617–638.

López Plaza, M., 1982, Contribución al conocimiento de la dinámica de los cuerpos graníticos en la penillanura Salmantino-Zamorana [Ph.D. thesis]: Salamanca, Spain, Universidad de Salamanca, 332 p.

Macaya, J., González Lodeiro, F., Martínez Catalán, J.R., and Alvarez, F., 1991, Continuous deformation, ductile thrusting and backfolding in the basement of the Hercynian orogen and their relationships with structures in the metasedimentary cover in the Spanish Central System: Tectonophysics, v. 191, p. 291–309.

Malavieille, J., 1987, Kinematics of compressional and extensional ductile shearing deformation in a metamorphic core complex of the northeastern Basin and Range: Journal of Structural Geology, v. 9, p. 541–554.

Malavieille, J., 1991, Kinematic model for postorogenic Basin and Range extension: Geology, v. 19, p. 555–558.

Marques, F.O., Ribeiro, A., and Munhá, J.M., 1996, Geodynamic evolution of the Continental Allochthonous Terrane (CAT) of the Bragança Nappe Complex, NE Portugal: Tectonics, v. 15, p. 747–762.

Marquínez, J.L., 1981, Estudio geológico del área esquistosa de Galicia Central (Zona de Lalín-Forcarei-Beariz): Cuadernos do Laboratorio Xeolóxico de Laxe, v. 1, p. 135–155.

Marquínez García, J.L., 1984, La geología del área esquistosa de Galicia Central (Cordillera Herciniana, NW de España): Memorias del Instituto Geológico y Minero de España, v. 100, 231 p.

Martínez Catalán, J.R., 1990. A noncylindrical model for the northwest Iberian allochthonous terranes and their equivalents in the Hercynian belt of Western Europe: Tectonophysics, v. 179, p. 253–272.

Martínez Catalán, J.R., Arenas, R., Díaz García, F., and Abati, J., 1997, Variscan acretionary complex of northwest Iberia: Terrane correlation and succession of tectonothermal events: Geology, v. 25, p. 1103–1106.

Martínez Catalán, J.R., Arenas, R., Díaz García, F., Rubio Pascual, F.J., Abati, J., and Marquínez, J., 1996, Variscan exhumation of a subducted Paleozoic continental margin: The basal units of the Ordenes Complex, Galicia, NW Spain: Tectonics, v. 15, p. 106–121.

Martínez Catalán, J.R., Hacar Rodríguez, M.P., Villar Alonso, P., Pérez Estaún, A., and González Lodeiro, F., 1992, Lower Paleozoic extensional tectonics in the limits between the Westasturian-Leonese and Central Iberian Zones of the Variscan Fold-Belt in NW Spain: Geologische Rundschau, v. 81, p. 545–560.

Martínez García, E., 1972, El Silúrico de San Vitero (Zamora): Comparación con las series vecinas e importancia orogénica: Acta Geologica Hispanica, v. 4, p. 104–108.

Martínez García, E., 1973, Deformación y metamorfismo en la zona de Sanabria: Stvdia Geologica Salmanticensia, v. 5, p. 7–106.

Matte, Ph., 1968, La structure de la virgation hercynienne de Galice (Espagne): Revue de Géologie Alpine, v. 44, p. 1–128.

McDougall, N., Brenchley, P.J., Rebelo, J.A., and Romano, M., 1987, Fans and fan-delta-precursors to the Armorican Quartzite (Ordovician) in Western Iberia: Geological Magazine, v. 124, p. 347–359.

Means, W.D., 1981, The concept of steady-state foliation: Tectonophysics, v. 78, p. 179–199.

Meireles, C., Ribeiro, A., and Pereira, E., 1995, ContribuçÁo para o conhecimiento de litoestratigrafía e tectónica do Paleozoico a Norte de Bragança: Universidade do Porto-Museu e Laboratório de Minería e Geología, Memórias, v. 4, p. 349–353.

Murphy, J.B., and Nance, R.D., 1991, Supercontinent model for the contrasting character of Late Proterozoic orogenic belts: Geology, v. 19, p. 469–472.

Noblet, Ch., and Lefort, J.P., 1990, Sedimentological evidence for a limited separation between Armorica and Gondwana during the Early Ordovician: Geology, v. 18, p. 303–306.

Ochsner, A., 1993, U-Pb geochronology of the Upper Proterozoic-Lower Paleozoic geodynamic evolution in the Ossa-Morena Zone (SW Iberia): constraints on the timing of the Cadomian orogeny [Ph.D. thesis]: Zurich, Switzerland, Swiss Federal Institute of Technology, 249 p.

Oncken, O., von Winterfeld, C., and Dittmar, U., 1999, Accretion of a rifted passive margin: The late Paleozoic Rhenohercynian fold and thrust belt (Middle European Variscides): Tectonics, v. 18, p. 75–91.

Pereira, E.S., 1987, Estudo Geológico-Estructural da região de Celorico de Basto e sua interpretação geodinâmica, [Ph.D. thesis]: Portugal, Universidade de Lisboa, 274 p.

Pereira, Z., Meireles, C., and Pereira, E., 1999, Upper Devonian palynomorphs of the NE sector of Trás-os-Montes (Central Iberian Zone), *in* XV Reunión de Geología del Oeste Peninsular, International Meeting on Cadomian Orogens: Badajoz, Spain, p. 201–206.

Pérez-Estaún, A., Martínez Catalán, J.R., and Bastida, F., 1991, Crustal thickenning and deformation sequence in the footwall to the suture of the Variscan Belt of NW Spain: Tectonophysics, v. 191, p. 243–253.

Peucat, J.J., Bernard-Griffiths, J., Gil Ibarguchi, J.I., Dallmeyer, R.D., Menot, R.P., Cornichet, J., and Iglesias Ponce de León, M., 1990, Geochemical and geochronological cross section of the deep Variscan crust: The Cabo Ortegal high-pressure nappe (northwestern Spain): Tectonophysics, v. 177, p. 263–292.

Pierrel, L., Gagny, Cl., Hermosa, J.L., and Servajean, G., 1981, La "Mina Dorinda": Un exemple de minéralisation filoniènne en terrain métamorphique liée à un magmatisme leucogranitique (District metallifère de Villadepera, Province de Zamora, Espagne): Cuadernos de Geología Ibérica, v. 7, p. 383–389.

Pillevuit, A., Marcoux, J., Stampfli, G., and Baud, A., 1997, The Oman Exotics: A key to the understanding of the Neotethyan geodynamic evolution: Geodinamica Acta, v. 10, p. 209–238.

Quiroga, J.L., 1981, Estudio geológico del paleozoico del W de Zamora (Alba y Aliste) [Ph.D. thesis]: Oviedo, Spain, Universidad de Oviedo, 210 p.

Quiroga, J.L., 1982, Estudio geológico del Paleozoico del W de Zamora: Trabajos de Geología, Universidad de Oviedo, v. 12, p. 205–226.

Ramsay, J.G., and Huber, M.I., 1987, The techniques of modern structural geology, Volume 2: Folds and factures: London, Academic Press, p. 309–700.

Regêncio Macedo, C.A., 1988, Granitóides, Complexo xisto-grauváquico e Ordovícico na região entre Trancoso e Pinhel (Portugal Central) [Ph.D. thesis]: Coimbra, Portugal, Universidade de Coimbra, 430 p.

Ribeiro, A., 1974. Contribution à l'étude tectonique de Trás-os-Montes Oriental: Serviços Geológicos de Portugal, Memória 24 (Nova Série), 168 p.

Ribeiro, A., Pereira, E., and Dias, R., 1990, Central-Iberian Zone, allochthonous sequences: Structure in the northwest of the Iberian Peninsula, *in* Dallmeyer, R.D., and Martínez García, E., eds., Pre-Mesozoic geology of Iberia: Berlin, Springer-Verlag, p. 220–236.

Ribeiro, M.L., and Ribeiro, A., 1974, Signification paléogéographique et tectonique de la présence de galets de roches métamorphiques dans un flysch d'âge dévonien supérieur du Tras-os-Montes oriental (Nord-Est du Portugal): Comptes Rendus de l'Académie des Sciences, Paris, v. 278, p. 3161–3163.

Ries, A.C., and Shackleton, R.M., 1971, Catazonal complexes of northwest Spain and north Portugal, remnants of a Hercynian thrust plate: Nature Physical Science, v. 234, p. 65–79.

Robardet, M., and Doré, F., 1988, The late Ordovician diamictic formations from southwestern Europe: North-Gondwana glaciomarine deposits: Palaeogeography, Palaeoclimatology, Palaeoecology, v. 66, p. 19–31.

Rodríguez Alonso, M.D., 1985, El Complejo esquisto-grauváquico y el Paleozoico en el Centro-Oeste español: Acta Salmanticensia, Sección Ciencias, Ediciones Universidad de Salamanca, Spain, v. 51, 174 p.

San José, M.A. de, Pieren, A.P., García-Hidalgo, J.F., Vilas, L., Herranz, P., Pelaez, J.R., and Perejón, A., 1990, Central-Iberian Zone, ante-Ordovician stratigraphy, *in* Dallmeyer, R.D., and Martínez García, E., eds., Pre-Mesozoic geology of Iberia: Berlin, Springer-Verlag, p. 147–159.

Santos Zalduegui, J.F., Schärer, U., Gil Ibarguchi, J.I., and Girardeau, J., 1996, Origin and evolution of the Paleozoic Cabo Ortegal ultramafic-mafic complex (NW Spain): U-Pb, Rb-Sr and Pb-Pb isotope data: Chemical Geology, v. 129, p. 281–304.

Sarmiento, G.N., Calvo, A.A., and González Clavijo, E., 1997, Conodontos paleozoicos (Ashgill-Emsiense) del Sinforme de Alcañices (oeste de Zamora, España), *in* Grandal D'Anglade, A., Gutiérrez-Marco, J.C., and Santos Fidalgo, L., eds., Paleozoico Inferior del Noroeste de Gondwana: V Reunión Internacional del Proyecto 351 PICG, Libro de resúmenes y excursiones: Coruña, Spain, Universidade da Coruña, p. 108–111.

Sarmiento, G.N., Gutiérrez Marco, J.C., and Robardet, M., 1999, Conodontos ordovícicos del Noroeste de España: Aplicación al modelo de sedimentación de la región limítrofe entre las Zonas Asturoccidental-leonesa y Centroibérica durante el Ordovícico Superior: Revista de la Sociedad Geológica de España, v. 12, p. 477–500.

Schäfer, H.J., Gebauer, D., Gil Ibarguchi, J.I., and Peucat, J.J., 1993, Ion-microprobe U-Pb zircon dating on the HP/HT Cabo Ortegal Complex (Galicia, NW Spain): Preliminary results: Terra Abstracts, v. 5.4, p. 22.

Soper, N.J., 1988, Timing and geometry of collision, terrane accretion and sinistral strike-slip events in the British Caledonides, *in* Harris, A.L., and Fettes, D.J., eds., The Caledonian-Appalachian Orogen: Geological Society of London Special Publication 38, p. 481–492.

Vacas, J.M., and Martínez Catalán, J.R., 1987, El Sinforme de Alcañices en la transversal de Manzanal del Barco: Stvdia Geologica Salmanticensia, v. 24, p. 151–175.

Villar Alonso, P.M., 1990, Estructura del Antiforme de Villadepera (Zamora) [M.S. dissert.]: Salamanca, Spain, Universidad de Salamanca, 66 p.

Manuscript Accepted by the Society February 8, 2002

Printed in the U.S.A.

Geological Society of America
Special Paper 364
2002

Tectonic setting of the Monte Castelo gabbro (Ordenes Complex, northwestern Iberian Massif): Evidence for an arc-related terrane in the hanging wall to the Variscan suture

P. Andonaegui
J. González del Tánago
R. Arenas
J. Abati
Departamento de Petrología y Geoquímica, Universidad Complutense, 28040 Madrid, Spain
J.R. Martínez Catalán
M. Peinado
Departamento de Geología, Universidad de Salamanca, 37008 Salamanca, Spain
F. Díaz García
Departamento de Geología, Universidad de Oviedo, 33005 Oviedo, Spain

ABSTRACT

The Ordenes Complex is the largest of the allochthonous assemblages containing the Variscan suture in the northwestern Iberian Massif. Its uppermost tectonostratigraphic unit overlies the ophiolitic units, and consists of a thick metasedimentary sequence, the Ordenes Series, intruded by orthogneisses and gabbros. In the lower part of the Ordenes Series, the large Monte Castelo gabbro (~150 km^2) is surrounded by high-grade migmatitic paragneisses. Several shear zones cutting across the gabbro massif depict intermediate-pressure granulite facies, indicating a metamorphic evolution common with the surrounding paragneisses. Recent U-Pb geochronological data prove that the main tectonothermal evolution of the Ordenes Series took place in Late Cambrian–Early Ordovician time. These data suggest that the intrusion of the Monte Castelo gabbro (499 ± 2 Ma; U-Pb in zircons) was immediately followed by a Barrovian metamorphic episode that reached the granulite facies (493–498 Ma; U-Pb in monazites). A later Variscan overprint is indicated by U-Pb rutile ages of 380–390 Ma.

Considering the geochronological evidence for almost coeval magmatism and metamorphism during the Early Ordovician, together with the geochemical characteristics of the Monte Castelo gabbro, an accretionary complex related to a Late Cambrian–Early Ordovician island arc appears as the more probable setting for the uppermost allochthonous terrane in the northwest of the Iberian Massif. This implies the presence of a convergent plate boundary in the oceanic realm between Laurentia and Gondwana, or close to it, during the early Paleozoic. The island arc later became involved in the Variscan convergence and accretion.

Andonaegui, P., González del Tánago, J., Arenas, R., Abati, J., Martínez Catalán, J.R, Peinado, M., and Díaz García, F., 2002, Tectonic setting of the Monte Castelo gabbro (Ordenes Complex, northwestern Iberian Massif): Evidence for an arc-related terrane in the hanging wall to the Variscan suture, *in* Martínez Catalán, J.R., Hatcher, R.D., Jr., Arenas, R., and Díaz García, F., eds., Variscan-Appalachian dynamics: The building of the late Paleozoic basement: Boulder, Colorado, Geological Society of America Special Paper 364, p. 37–56.

INTRODUCTION AND SETTING

The section of the European Variscan Belt exposed in the northwest of the Iberian Massif is almost complete and significant. From east to west (Fig. 1), the orogenic section evolves gradually from a foreland region consisting of a thrust and fold belt, through intermediate zones with Barrovian metamorphism and strong ductile deformation, to a hinterland characterized by high-grade metamorphism and extensive granitic magmatism (Pérez-Estaún et al., 1991; Dallmeyer et al., 1997). An outstanding feature of the internal areas is the presence of the Variscan suture, preserved in several allochthonous complexes with mafic-ultramafic lithologies and high-pressure metamorphism (Arenas et al., 1986): the Cabo Ortegal, Ordenes, and Malpica-Tui Complexes in Galicia (northwestern Spain) and the Bragança and Morais Complexes in the Trás-os-Montes region of northern Portugal.

The allochthonous complexes consist of a stack of thrust sheets, overprinted by extensional detachments, upright folds, and faults. They are underlain and surrounded by autochthonous and parautochthonous sequences, commonly affected by recumbent folds and thrusts (Fig. 1). The autochthon and parautochthon, together with other comparable areas of southern Europe, formed part of the continental margin of Gondwana during the Late Proterozoic and Paleozoic. This margin registered the Cadomian (Pan-African) orogeny, as well as Cambrian-Ordovician continental rifting processes, which resulted in the pulling apart of the Avalon microcontinent (or microcontinents) from the Gondwana mainland and the opening of the Rheic oceanic domain (Fortey and Cocks, 1988; Soper, 1988). For the description of the allochthonous complexes, it is useful to classify their units according to their relative position in the original nappe pile, i.e., basal, intermediate, and upper units (Figs. 2 and 3). Because the intermediate units show clear ophiolitic affinities, they are referred to as the ophiolitic units and are considered to be the suture separating two main paleogeographic realms. Further subdivision can be done based on the pressure-temperature (*P-T*) conditions of the metamorphic evolution of the units.

The allochthonous units (basal units) immediately below the ophiolites are currently considered to be a piece of the deformed continental margin of Paleozoic Gondwana (Martínez Catalán, 1990; Matte, 1991; Marques et al., 1996; Martínez Catalán et al., 1996), but the nature and correlation of the units in the hanging wall to the suture are much more problematic, and their origin and evolution have been discussed. The existence of a pre-Ordovician continental basement in the hanging wall to the suture has been proposed (Marques et al., 1996), although not clearly proved. Sm-Nd pre-Ordovician ages ascribed to magmatism and high-grade metamorphism have been obtained above the ophiolites in the Portuguese complexes (Ribeiro et al., 1993; Santos, 1998), but new and extensive dating, including U-Pb systematics, seems necessary to confirm them.

Conversely, isotopic U-Pb ages around the Cambrian-Ordovician boundary are widely accepted and have been interpreted as dating magmatism (Dallmeyer and Tucker, 1993; Schäfer et al., 1993) and, in some cases, metamorphism (Bernard-Griffiths et al., 1985; Peucat et al., 1990), pointing to the existence of an early Paleozoic tectonic event. In this context, two possibilities have been considered: the magmatism is either related to the rifting of an older continental domain (Marques et al., 1996; Martínez Catalán et al., 1997) or an arc realm (Arenas et al., 1986; Díaz García, 1990). A third possibility, a continental area undergoing extension during the early Paleozoic, coupled with a short-lived Early Ordovician convergent plate boundary, has been proposed (Martínez Catalán et al., 1999; Abati et al., 1999).

The Ordenes Complex is the largest of the allochthonous complexes, and its uppermost unit is an intricate lithological assemblage located in the hanging wall to the suture and overlying ophiolitic units. It consists of a thick metasedimentary sequence, the Ordenes Series, intruded by granitoids and gabbros, and affected by low- to high-grade metamorphism of the intermediate-pressure type. The structural position of this element suggests that it is a terrane that collided in Devonian time to develop the Variscan belt, during the closure of the Rheic ocean (Martínez Catalán et al., 1997). The arc option for the upper allochthonous units has been mainly based on the terrigenous, flyschoid character of the Ordenes Series, and on the existence of important Paleozoic bimodal magmatism, but these criteria fit equally well the outer part of a stable continental margin. New U-Pb data have confirmed a magmatic and metamorphic event close to the Cambrian-Ordovician boundary (493–500 Ma; Abati et al., 1999). The new data provide an age for the mafic and felsic magmatism (large bodies of metagabbros and granitic orthogneisses; Figs. 2 and 3), and the chronology of an early event characterized by penetrative deformation and intermediate-pressure metamorphism that reached the granulite facies in the lower parts of the Ordenes Series. These data support the arc option, because an accretionary setting appears to be the most viable method of integrating the nearly contemporary magmatism, metamorphism, and strong deformation.

There is no recent information concerning the geochemistry of mafic and felsic intrusives in the Ordenes Series, despite its importance for the interpretation of the unit. However, some have described the geochemistry of mafic rocks in the high *P-T* unit of the Cabo Ortegal Complex (Van Calsteren, 1978; Bernard-Griffiths et al., 1985; Peucat et al., 1990; Gil Ibarguchi

→

Figure 1. Geological map and general cross section of northwestern Iberian Massif showing various allochthonous units, main stratigraphic ensembles of Iberian autochthon, and Variscan granitoids. Allochthonous complexes, named after geographical features and localities, are those of Cabo Ortegal, Ordenes, and Malpica-Tui (Spain) and Bragança and Morais (Portugal). Reconstruction of Pangea in area surrounding Iberia is shown in upper part of figure, and is based on Lefort (1989) and Skehan and Rast (1995). *HP, IP, HT,* high pressure, intermediate pressure, high temperature, respectively.

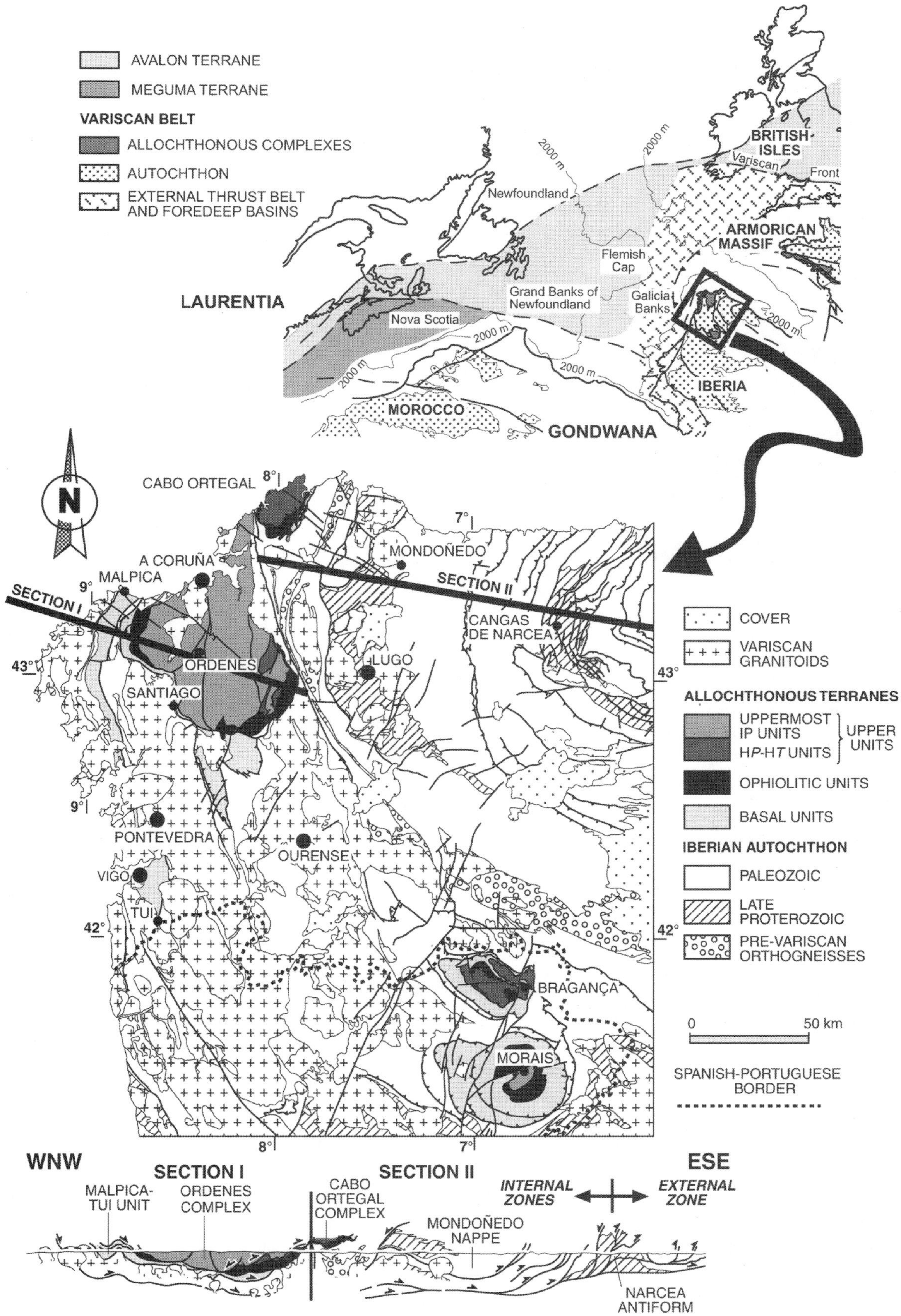

AVALON TERRANE
MEGUMA TERRANE
VARISCAN BELT
ALLOCHTHONOUS COMPLEXES
AUTOCHTHON
EXTERNAL THRUST BELT AND FOREDEEP BASINS
BRITISH ISLES
Variscan Front
Newfoundland
2000 m
ARMORICAN MASSIF
Flemish Cap
LAURENTIA
Grand Banks of Newfoundland
Galicia Banks
Nova Scotia
IBERIA
MOROCCO
GONDWANA
N
CABO ORTEGAL
8°
7°
MONDOÑEDO
A CORUÑA
MALPICA
9°
SECTION I
SECTION II
CANGAS DE NARCEA
ORDENES
LUGO
43°
SANTIAGO
PONTEVEDRA
OURENSE
VIGO
TUI
42°
BRAGANÇA
MORAIS
COVER
VARISCAN GRANITOIDS
ALLOCHTHONOUS TERRANES
UPPERMOST IP UNITS
HP-HT UNITS
UPPER UNITS
OPHIOLITIC UNITS
BASAL UNITS
IBERIAN AUTOCHTHON
PALEOZOIC
LATE PROTEROZOIC
PRE-VARISCAN ORTHOGNEISSES
0 50 km
SPANISH-PORTUGUESE BORDER
WNW
SECTION I
SECTION II
ESE
MALPICA-TUI UNIT
ORDENES COMPLEX
CABO ORTEGAL COMPLEX
INTERNAL ZONES
EXTERNAL ZONE
MONDOÑEDO NAPPE
NARCEA ANTIFORM

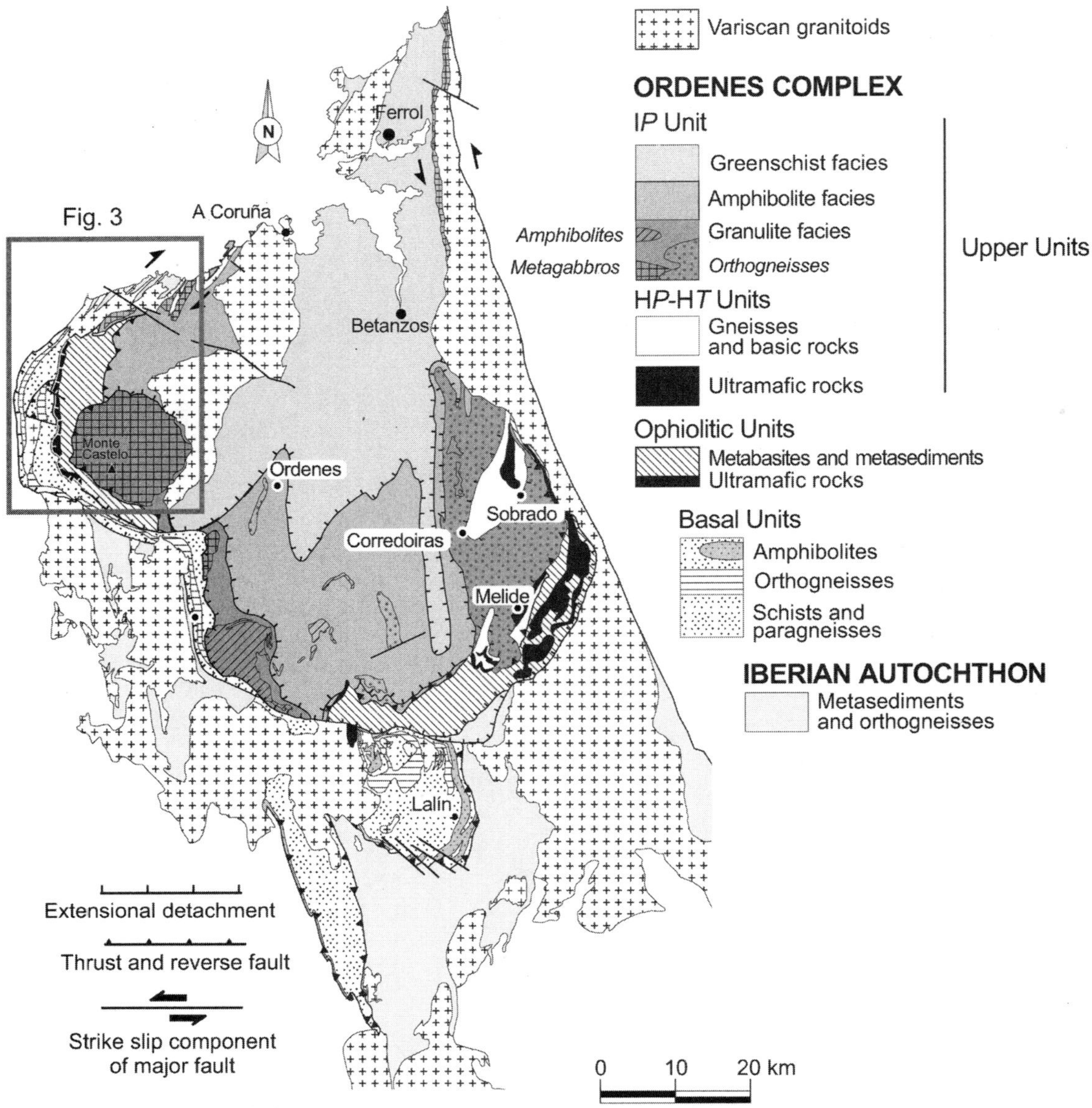

Figure 2. Geological map of Ordenes Complex showing different units of complex and metamorphic zones of uppermost (intermediate-*P*) unit. H*P,* I*P,* H*T,* high pressure, intermediate pressure, high temperature, respectively.

et al., 1990; Gravestock, 1992; Mendia, 1996; Galán and Marcos, 1997). Mafic protoliths of this unit, which directly overlies the ophiolitic unit, exhibit complex geochemical features, but mid-ocean ridge basalt (MORB) and continental tholeiite compositions are the more usual lithological types present. The continental magmatism has been interpreted as related to Early Ordovician extension of a continental realm that progressively led to oceanic spreading and generation of MORB compositions (Galán and Marcos, 1997). A similar high *P-T* unit occurs in the Ordenes Complex, where it is separated from the Ordenes Series by an extensional detachment (Díaz García et al., 1999). The detachment precludes a close correlation between the uppermost intermediate-*P* unit and the high *P-T* unit, occurring at its hanging wall and footwall, respectively (Fig. 2). However, the location of both units above the ophiolites suggests they share at least some part of their geological evolution.

In an attempt to fill in the lack of reliable geochemical data for intermediate-*P* unit, this chapter focuses on the geochemistry and petrologic characteristics of the Monte Castelo gabbro (Warnaars, 1967; Díaz García, 1990), the largest of the gabbro bodies in the Ordenes Series (Fig. 3). The data obtained from this gabbro are used to discuss the tectonic setting of the bimodal magmatism, in order to clarify the origin and Paleozoic evolution of the upper colliding element in the European Variscan Belt.

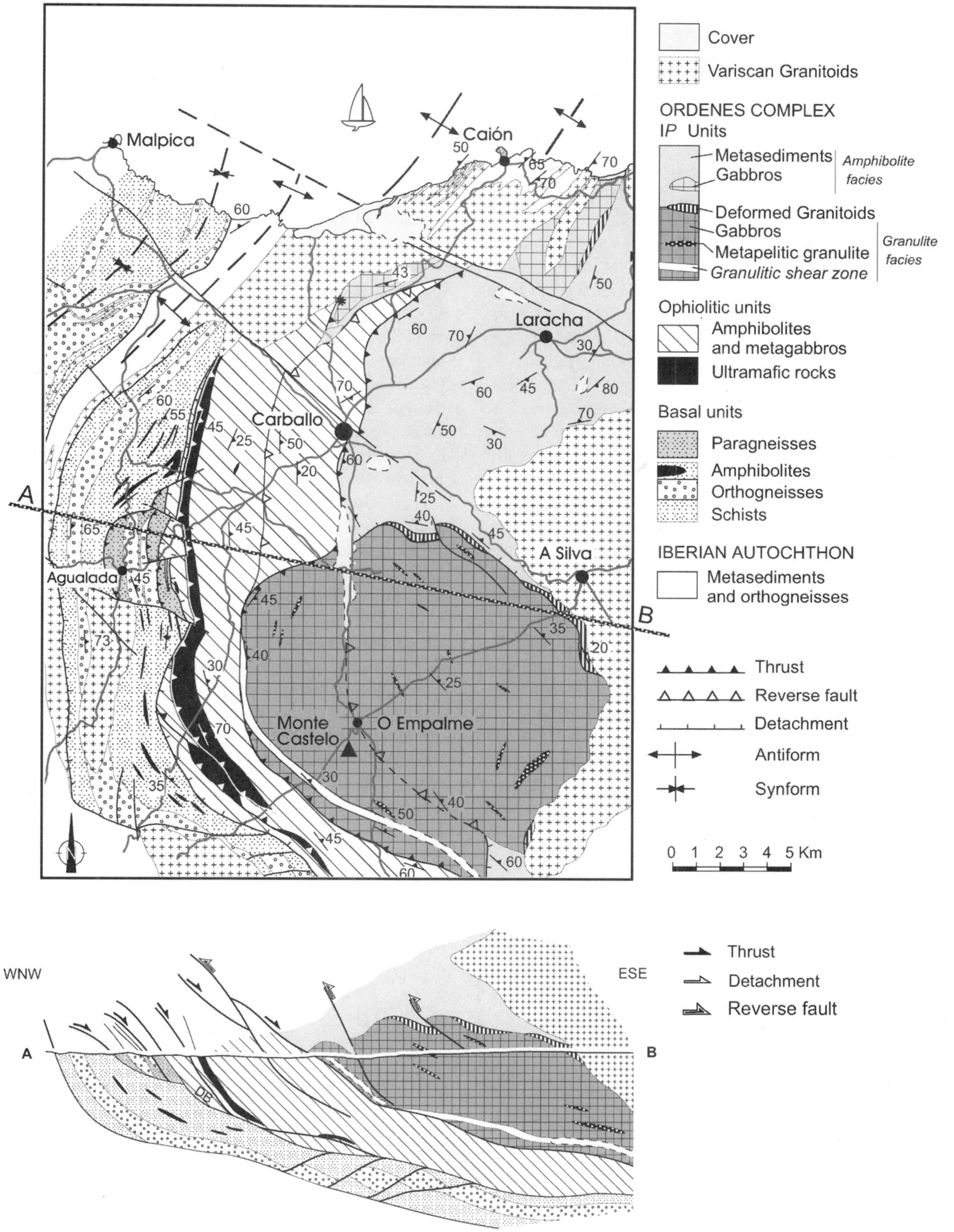

Figure 3. Geological map and cross-section of northwest sector of Ordenes Complex showing structural position of Monte Castelo gabbro.

INTERMEDIATE-*P* UNIT IN THE ORDENES COMPLEX

The intermediate-*P* unit occupies the uppermost structural position in the Ordenes nappe stack. It consists of a very thick sequence of terrigenous metasediments, and large bodies of amphibolites, augengneisses, and gabbros. Low-grade metasediments occupy the upper part of the unit, 2000–3000 m of metapelites and metagraywackes (Matte and Capdevila, 1978). Generalized graded bedding and other abundant sedimentary structures indicate a flyschoid character. The augengneisses show intrusive relationships with the metasediments, and have yielded ages of 500 Ma (U-Pb chronology of the large Corredoiras orthogneiss; Abati et al., 1999). A comparable U-Pb age of 496 Ma was obtained for the similar Lagoa augengneiss, in the Morais Complex of Portugal (Dallmeyer and Tucker, 1993). The gabbros often show subophitic and diabase textures, indicative of a shallow level of emplacement. The largest of these gabbro bodies, the Monte Castelo gabbro, has been dated as 499 ± 2 Ma (U-Pb on zircons; Abati et al., 1999).

Metamorphism ranges from greenschist facies in the structurally upper parts of the unit, to intermediate-*P* granulite facies in the lower parts. The contacts between greenschist, amphibolite, and granulite facies rocks are folded extensional faults (Fig. 2), and another, more important extensional detachment (Martínez Catalán and Arenas, 1992; Díaz García et al., 1999) separates the intermediate-*P* units from the underlying high *P*-*T* units. Several U-Pb ages have been obtained in the Ordenes Series on zircons and monazites in high-grade metasediments and in a granitoid generated during the migmatization of the latter (Abati et al., 1999). These ages range from 493 to 500 Ma and indicate that a metamorphic event affected the upper units during the early Paleozoic. The available ages in the igneous bodies (Corredoiras orthogneiss and Monte Castelo gabbro) and in the metasediments indicate that emplacement (frequently at shallow levels) of igneous bodies in Late Cambrian–Early Ordovician time was immediately followed by burial and regional intermediate-*P* metamorphism that reached granulite facies. The Barrovian character of this metamorphic event points to an orogenic episode involving crustal thickening.

FIELD AND MINERALOGICAL ASPECTS OF THE MONTE CASTELO GABBRO

The large, rounded, Monte Castelo gabbro crops out in the western part of the Ordenes Complex (Figs. 2 and 3), where it intruded the lowermost exposed section of the Ordenes Series. This is a massive metagabbronorite, fine to medium grained, with variable textures from granular to intergranular and ophitic; most of the gabbro is not deformed. In this area igneous mineral assemblages consist of plagioclase, clinopyroxene, orthopyroxene, olivine, amphibole, and biotite; ilmenite, titanite, spinel, and zircon are possible accessory minerals, and quartz rarely appears as primary mineral. Pegmatoid veins consisting of plagioclase, pyroxene, amphibole, minor ilmenite, and occasional quartz are common. The first minerals to crystallize were olivine and plagioclase followed by pyroxenes; cumulate types with plagioclase and olivine as cumulus phases have been occasionally observed. The presence of olivine, pegmatoid veins, and ophitic textures points to a relatively shallow level of emplacement for the gabbro massif.

Three major compositional types have been distinguished in the Monte Castelo gabbro (mineral abbreviations after Kretz, 1983): (1) olivine gabbronorites (Pl + Cpx + Opx + Ol + Hbl); (2) amphibole gabbronorites (Pl + Cpx + Opx + Hbl ± Bt); and (3) biotite gabbronorites (Pl + Cpx + Opx + Bt). A clear areal distribution of these types inside the gabbro body has not been found. Modal mineral contents obtained in 10 representative samples of the different gabbroic types are presented in Table 1. The samples with olivine can contain to 12.4% of this mineral, whereas biotite content does not exceed 4.1%; the proportion of amphibole can be high, but generally is lower than 10%–12%.

TABLE 1. MODAL MINERAL CONTENT OF REPRESENTATIVE GABBRO SAMPLES

Sample	Ol* 93937	Ol* 93985	Ol* 93993	Ol* 93997	Hbl* 91701	Hbl* 91703	Hbl* 93979	Bt* 91709	Bt* 91738	Bt* 93980
Pl	55.5	51.6	55.8	50.5	59.0	39.6	59.1	64.8	63.4	56.0
Ol	4.0	4.7	12.4	7.0	N.A.#	N.A.#	N.A.#	N.A.#	N.A.#	N.A.#
Opx	13.2	2.1	5.1	5.0	13.0	4.7	11.6	10.1	7.8	7.8
Cpx	18.4	28.0	14.3	22.5	20.1	19.7	25.7	22.3	21.1	30.9
Hbl	7.5	6.9	10.8	11.1	5.4	31.2	0.4	N.A.#	N.A.#	N.A.#
Bt	N.A.#	N.A.#	N.A.#	N.A.#	0.7	N.A.#	1.3	0.9	4.1	1.8
Ilm	0.6	0.9	1.2	2.0	1.6	2.0	1.5	1.1	3.1	2.7
Qtz	N.A.#	N.A.#	N.A.#	N.A.#	N.A.#	N.A.#	N.A.#	N.A.#	0.3	0.7
Spl	N.A.#	N.A.#	0.2	N.A.#	N.A.#	N.A.#	N.A.#	N.A.#	N.A.#	N.A.#
Secondary	N.A.#	5.5	N.A.#	1.7	N.A.#	2.1	N.A.#	0.5	N.A.#	N.A.#

Note: Number of points counted in each thin section = 1000.

*Gabbro compositional types: Ol, olivine gabbronorite; Hbl, amphibole gabbronorite; Bt, biotite gabbronorite.

#N.A. = Not applicable.

Cpx values range from 14.3% to 30.9%, whereas those of Opx range between 2.1% and 13.2%. Limited quartz (<1%) was detected in two samples (Table 1).

At the bottom, the Monte Castelo gabbro is crosscut by granulite facies prograde shear zones, which produced a complete recrystallization of igneous minerals and resulted in a new metamorphic assemblage consisting of plagioclase, garnet, orthopyroxene, biotite, quartz, ilmenite, and rutile, arranged in a granonematoblastic fabric (Abati et al., 1999). Other shear zones with similar characteristics are found inside the massif.

Scarce map-scale metapelitic enclaves in the gabbro were also completely recrystallized during the granulite facies metamorphic event, and exhibit a mineral assemblage compatible with that of the host rock, consisting of plagioclase, garnet, orthopyroxene, K-feldspar, biotite, sillimanite, quartz, ilmenite, and rutile. At the outcrop scale or in hand specimen, a high-grade foliation is evident in these granulites, but this fabric is not evident in thin section. The less retrogressed samples contain rounded, coarse garnets (to 1 cm) in a granoblastic polygonal matrix formed of plagioclase, quartz, K-feldspar, subeuhedral elongated orthopyroxene, and reddish biotite. The garnets include numerous ilmenites in their cores and rutile in their rims; the rutile is also present in the matrix, together with ilmenite and hercynite. In the first stages of retrogression, the orthopyroxene breaks down and is replaced by an intergrowth of cordierite, sillimanite, and biotite. Thermobarometric determinations obtained on these granulite facies metapelites suggest temperatures in the range 750–800 °C and intermediate pressures for the high-grade metamorphic event, dated as 493–498 Ma (U-Pb on zircons and monazites; Abati et al., 1999).

A thick pile of monotonous plagioclase-rich paragneisses occurs overlying and surrounding the Monte Castelo gabbro. In some places, the contact between the gabbro and the metasediments of the Ordenes Series contains narrow granitic lenses (50–100 m thick), probably generated by partial melting of the latter. In some outcrops, these granitoids appear deformed and show a planolinear fabric and abundant enclaves (restites) of paragneisses from the surrounding metasediments. In the section of the Ordenes Series exposed to the north of the gabbro massif, the paragneisses show a lower metamorphic grade and contain a mineral assemblage consisting of plagioclase, garnet, biotite, muscovite, sillimanite, quartz, and ilmenite that can be assigned to the first sillimanite zone.

PETROGRAPHY AND MINERAL CHEMISTRY

The compositions of the minerals present in seven representative samples of the different gabbroic types were investigated with electron microprobe. Considering that significant metamorphic recrystallization of the gabbros has only been detected in shear zones, only very fresh and undeformed rocks were considered for chemical analysis. Chemical analysis of minerals was carried out with a JEOL electron microprobe (JXA-8900 M) equipped with four spectrometers at the Universidad Complutense de Madrid. The operating parameters were: 20 s total counting time, 15 kV accelerating voltage, 20 nA beam current, and 1–5 μm beam diameter. Calibration was against standard minerals provided by the Department of Mineral Sciences of the Smithsonian Institution (Washington, D.C.; Jarosewich et al., 1980), and the ZAF correction procedure was used. Detailed analyses were performed for the more characteristic minerals of the gabbros, paying attention to the compositional zoning. Representative chemical analyses of plagioclases, clinopyroxenes, orthopyroxenes, olivines, amphiboles, biotites, and ilmenites are included in Tables 2–8.

Plagioclase

Plagioclase is the dominant mineral in all the gabbro types, the highest modal content being in the biotite gabbronorites (Table 1). Hypidiomorphic tabular crystals with polysynthetic twinning and weak optic zoning are the more common form. Very fine microinclusions of pyroxene have been observed in plagioclases of some samples. Analyzed plagioclases range in composition from An_{51} to An_{74} (Table 2), with very low orthoclase content (<2 mol%). The highest An contents were observed in the olivine gabbronorites (to 74 mol%), and the biotite gabbronorites show the lowest An values (average An content of 52 mol%). Plagioclase crystals generally exhibit a continuous normal zoning, usually not very pronounced (Table 2), although some analyzed crystals show discontinuous zoning characterized by middle areas with the highest An content.

Clinopyroxene

A similar modal content of clinopyroxene characterizes all three gabbronorites types (Table 1). Clinopyroxene crystals exhibit an interstitial character, with frequent exsolutions of orthopyroxene parallel or oblique to the exfoliation. Marginal or symplectitic replacement by brown hornblende or by secondary blue-green amphiboles is a common feature. According to the classification of Morimoto (1988), analyzed clinopyroxenes are diopsides or augites (Fig. 4), with a high $Mg/(Mg + Fe^{2+})$ (MF) ratio (Table 3). The MF ratio in clinopyroxenes is higher than in orthopyroxenes with a positive correlation between both, which can be considered indicative of early crystallization of the monoclinic pyroxenes. The correlation between MF content in clinopyroxenes and An content in plagioclases is positive, which agrees with a predicted fractional crystallization trends in magmas of basaltic composition. Their TiO_2 content is moderate (<1%) and Al_2O_3 content is relatively high (to 4.56%); these two oxides show a positive correlation, also detected in relation to MF (Fig. 5).

Clinopyroxenes of the Monte Castelo gabbro show a moderate content of jadeite molecule (Jd < 10 mol%), which in connection with the Al_2O_3 and MF contents can be considered indicative of an emplacement under moderate pressure (De Bari

TABLE 2. REPRESENTATIVE CHEMICAL ANALYSES OF PLAGIOCLASES

Sample	Ol 93937		Ol 93993		Hbl 91703		Bt 91709		Bt 93979	
Analysis	C#	R#	C#	R#	C#	R#	C#	R#	C#	R#
SiO_2	52.47	52.33	49.46	51.36	55.43	52.63	54.86	55.31	51.42	54.10
Al_2O_3	28.66	28.98	30.76	30.00	28.92	29.28	27.96	27.40	29.91	28.36
FeO*	<0.01	0.15	0.02	0.14	0.02	0.18	0.05	0.06	<0.01	0.09
MnO	0.02	0.04	0.01	<0.01	0.07	0.11	0.01	0.05	0.02	0.09
MgO	<0.01	0.01	<0.01	<0.01	0.02	<0.01	<0.01	<0.01	0.02	<0.01
CaO	11.90	12.06	15.22	13.32	11.71	12.90	11.15	11.15	13.65	11.82
Na_2O	5.15	5.14	3.00	4.06	3.16	4.49	5.62	5.68	4.00	5.01
K_2O	0.07	0.08	0.01	0.02	0.17	0.05	0.20	0.25	0.19	0.15
TiO_2	0.04	<0.01	<0.01	0.02	0.01	<0.01	<0.01	0.04	0.08	0.01
Total	98.31	98.79	98.48	98.92	99.51	99.64	99.85	99.94	99.29	99.63
					Cations per 8 oxygens					
Si	2.42	2.41	2.29	2.36	2.49	2.40	2.48	2.50	2.36	2.46
Al	1.56	1.57	1.68	1.62	1.53	1.57	1.49	1.46	1.62	1.52
Fe^{2+}	<0.01	0.01	<0.01	0.01	<0.01	0.01	<0.01	<0.01	<0.01	<0.01
Mn	<0.01	<0.01	<0.01	<0.01	<0.01	<0.01	<0.01	<0.01	<0.01	<0.01
Mg	<0.01	<0.01	<0.01	<0.01	<0.01	<0.01	<0.01	<0.01	<0.01	<0.01
Ca	0.59	0.59	0.76	0.66	0.56	0.63	0.54	0.54	0.67	0.58
Na	0.46	0.46	0.27	0.36	0.28	0.40	0.49	0.50	0.36	0.44
K	<0.01	<0.01	<0.01	<0.01	0.01	<0.01	0.01	0.01	0.01	0.01
Ab	43.75	43.34	26.28	35.51	32.39	38.55	47.16	47.29	34.24	43.04
An	55.85	56.23	73.67	64.38	66.48	61.18	51.76	51.32	64.69	56.11
Or	0.40	0.43	0.06	0.12	1.13	0.27	1.08	1.39	1.07	0.85

Note: Gabbro compositional types as in Table 1.
*Total iron as FeO.
#C = core analysis; R = rim analysis.

and Coleman, 1989; Aoki and Kushiro, 1968; Elthon, 1987). This conclusion agrees with the common presence of olivine in the gabbros.

Orthopyroxene

Modal contents of orthopyroxene vary, but are significantly lower than those of clinopyroxene. Significant variations in the abundance of Opx are common even in the same compositional group, although the olivine gabbronorites generally show lower Opx contents (Table 1). As in the case of clinopyroxene, orthopyroxene crystals show an interstitial character and frequent exsolutions of clinopyroxene parallel to the exfoliation. Also characteristic in the olivine gabbronorites is the presence of a generation of fine orthopyroxene crystals surrounding olivine grains. Textural relationships in gabbros suggest that orthopyroxene crystallized after clinopyroxene. Replacement by amphiboles, similar to those formed from clinopyroxenes, is also relatively common in orthopyroxenes. All analyzed orthopyroxenes in the Monte Castelo gabbro are enstatites (En = 58–77 mol%; Fig. 4), with $Mg/(Mg + Fe^{2+})$ ratios slightly lower than clinopyroxenes (Table 4). Olivine gabbronorites are characterized by higher MF contents, which show similar values in hornblendic and biotitic gabbronorites.

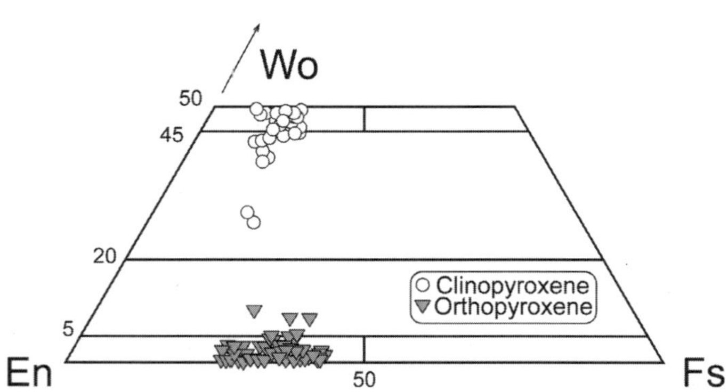

Figure 4. Classification of clinopyroxenes and orthopyroxenes of Monte Castelo massif. Nomenclature of Ca-Mg-Fe pyroxenes according to Morimoto (1988).

Olivine

Olivine gabbronorites in the Monte Castelo gabbro are characterized by a relatively low olivine content (4%–12%; Table 1). Fresh and undeformed gabbros show hypidiomorphic olivine crystals, only replaced by opaque minerals in fracture zones and brown amphiboles in the rim areas. Nevertheless, some slightly deformed gabbros exhibit incipient coronas surrounding olivine crystals. These coronas contain an inner rim of

TABLE 3. REPRESENTATIVE CHEMICAL ANALYSES OF CLINOPYROXENES

Sample	Ol 93937	Ol 93939	Ol 93940	Ol 93993	Hbl 91703	Bt 91709	Bt 93979
SiO_2	50.78	51.80	50.68	51.27	52.59	51.73	53.24
Al_2O_3	4.34	4.12	4.56	4.08	1.85	2.83	1.25
FeO*	7.48	6.29	5.89	7.94	7.22	7.38	7.56
MnO	0.24	0.19	0.13	0.20	0.30	0.19	0.18
MgO	13.13	14.90	13.36	15.81	13.94	14.24	13.89
CaO	21.79	20.28	22.37	19.33	22.93	21.26	22.59
Na_2O	0.60	0.62	0.70	0.44	0.45	0.37	0.25
K_2O	0.02	0.02	<0.01	<0.01	<0.01	0.70	0.02
TiO_2	0.81	0.53	0.89	0.77	0.35	0.61	0.23
Cr_2O_3	0.12	0.23	<0.01	0.19	0.05	0.16	0.11
NiO	0.07	<0.01	<0.01	<0.01	0.03	<0.01	<0.01
Total	99.38	98.98	98.58	100.03	99.71	99.47	99.32
				Cations per 6 oxygens			
Si	1.89	1.92	1.89	1.89	1.95	1.92	1.99
Al^{IV}	0.11	0.08	0.11	0.11	0.05	0.08	0.01
Al^{VI}	0.08	0.10	0.10	0.06	0.04	0.05	0.05
Fe^{3+}	0.02	0.00	0.01	0.03	0.02	0.05	0.00
Fe^{2+}	0.21	0.20	0.17	0.21	0.20	0.18	0.24
Mn	0.01	0.01	<0.01	0.01	0.01	0.01	0.01
Mg	0.73	0.82	0.74	0.87	0.77	0.79	0.77
Ca	0.88	0.81	0.90	0.76	0.91	0.85	0.91
Na	0.04	0.04	0.05	0.03	0.03	0.03	0.02
K	<0.01	<0.01	<0.01	<0.01	<0.01	0.03	<0.01
Ti	0.02	0.01	0.03	0.02	0.01	0.02	0.01
Cr	<0.01	0.01	<0.01	0.01	<0.01	<0.01	<0.01
Ni	<0.01	<0.01	<0.01	<0.01	<0.01	<0.01	<0.01
MF#	0.78	0.80	0.81	0.80	0.79	0.81	0.76
En	40.11	44.81	40.88	47.28	40.92	43.41	40.10
Fs	11.54	10.93	9.39	11.41	10.62	9.89	12.50
Wo	48.35	44.26	49.72	41.31	48.41	46.70	47.40

Note: Gabbro compositional types as in Table 1.
*Total iron as FeO.
#MF = $Mg/(Mg+Fe^{2+})$.

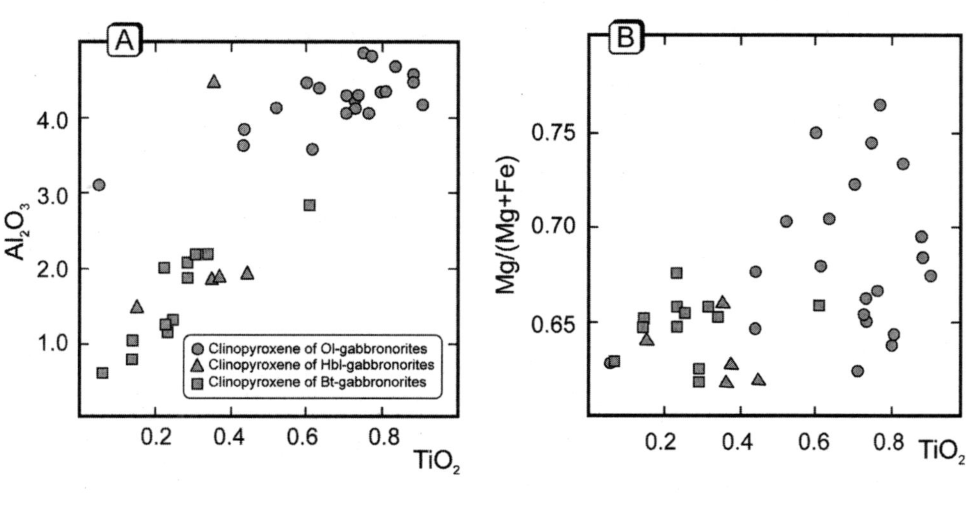

Figure 5. TiO_2-Al_2O_3 (A) and TiO_2-Mg/(Mg + Fe) (B) plots for clinopyroxenes of Monte Castelo massif.

TABLE 4. REPRESENTATIVE CHEMICAL ANALYSES OF ORTHOPYROXENES

Sample	Ol 93937	Ol 93939	Ol 93940	Ol 93993	Hbl 91703	Bt 91709	Bt 93979
SiO_2	52.01	53.87	53.19	53.48	52.06	52.34	53.17
Al_2O_3	2.64	3.01	2.97	2.90	0.96	1.47	0.79
FeO*	21.46	16.83	18.01	14.81	25.10	22.65	23.64
MnO	0.57	0.38	0.38	0.39	0.58	0.55	0.43
MgO	22.25	24.89	24.15	27.26	19.80	21.61	21.24
CaO	0.38	0.47	0.59	0.45	0.58	0.57	0.38
Na_2O	<0.01	0.03	<0.01	<0.01	<0.01	0.03	<0.01
K_2O	0.02	<0.01	<0.01	0.01	0.01	<0.01	0.02
TiO_2	<0.01	0.16	0.08	0.12	0.05	0.10	0.04
Cr_2O_3	<0.01	0.17	<0.01	0.04	<0.01	0.10	<0.01
NiO	0.14	0.03	0.14	<0.01	<0.01	0.04	0.12
Total	99.47	99.84	99.51	99.46	99.14	99.46	99.83
				Cations per 6 oxygens			
Si	1.94	1.96	1.95	1.93	1.99	1.97	2.00
Al^{IV}	0.06	0.04	0.05	0.07	0.01	0.03	0.00
Al^{V}	0.06	0.09	0.08	0.05	0.03	0.03	0.03
Fe^{3+}	0.01	0.00	0.00	0.01	0.00	0.00	0.00
Fe^{2+}	0.66	0.51	0.55	0.44	0.80	0.71	0.74
Mn	0.02	0.01	0.01	0.01	0.02	0.02	0.01
Mg	1.24	1.35	1.32	1.47	1.13	1.21	1.19
Ca	0.02	0.02	0.02	0.02	0.02	0.02	0.02
Na	<0.01	<0.01	<0.01	<0.01	<0.01	<0.01	<0.01
K	<0.01	<0.01	<0.01	<0.01	<0.01	<0.01	<0.01
Ti	<0.01	<0.01	<0.01	<0.01	<0.01	<0.01	<0.01
Cr	<0.01	<0.01	<0.01	<0.01	<0.01	<0.01	<0.01
Ni	<0.01	<0.01	<0.01	<0.01	<0.01	<0.01	<0.01
MF#	0.65	0.73	0.70	0.77	0.59	0.63	0.62
En	64.58	71.81	69.84	76.17	57.95	62.37	61.02
Fs	34.38	27.13	29.10	22.80	41.03	36.60	37.95
Wo	1.04	1.06	1.06	1.03	1.02	1.03	1.03

Note: Gabbro compositional types as in Table 1.
*Total iron as FeO.
#MF = $Mg/(Mg+Fe^{2+})$.

orthopyroxene and external areas of brown amphibole and occasional garnet and spinel. Considering their high grade and the general tectonothermal evolution deduced for the Monte Castelo gabbro and the Ordenes Series, these coronas were probably developed during the Cambrian-Ordovician metamorphic event (Abati et al., 1999). The composition range of analyzed olivines is Fo = 56–73 mol% (Table 5). Their MF values are lower than in clinopyroxenes and show a clear positive correlation with those of the orthopyroxenes.

Amphibole

Two types of amphiboles have been found in the Monte Castelo gabbro, primary large, hypidiomorphic brown amphiboles; and secondary blue-green amphiboles replacing igneous mafic minerals. The modal content of gabbros in brown amphiboles ranges from 31% to >1% in olivine gabbronorites and hornblende gabbronorites (Table 1). In gabbros with higher modal content of brown amphibole, this mineral exhibits an interstitial character in relation to plagioclase crystals. The compositions of analyzed brown amphiboles are characteristic of pargasites with high TiO_2 contents (TiO_2 = 2.5%–4.1%; Table 6), which is considered indicative of a high crystallization temperature (Otten, 1984; Heltz, 1973). The $Mg/(Mg + Fe^{2+})$ ratio ranges between 0.63 and 0.74, with higher values in olivine gabbronorites. The secondary blue-green amphiboles are classified as tremolites and actinolites with low TiO_2 contents, and high FeO, MgO, and Al_2O_3 values.

Biotite

Biotite gabbronorites and hornblende gabbronorites are characterized by accessory biotite contents, which range from 0.7% to 4.1% (Table 1). Biotite crystals exhibit a tabular hypid-

TABLE 5. REPRESENTATIVE CHEMICAL ANALYSES OF OLIVINES

Sample	Ol 93937	Ol 93939	Ol 93940	Ol 93993	
SiO_2	35.63	38.08	37.38	37.66	37.86
Al_2O_3	0.01	<0.01	<0.01	0.05	0.02
FeO*	37.16	27.41	29.92	27.52	24.26
MnO	0.43	0.41	0.44	0.32	0.32
MgO	26.82	33.34	31.48	34.01	37.31
CaO	0.02	0.04	0.03	<0.01	<0.01
Na_2O	<0.01	<0.01	0.05	<0.01	0.01
K_2O	0.03	<0.01	0.02	0.01	<0.01
TiO_2	0.02	0.01	0.04	0.03	<0.01
Cr_2O_3	<0.01	<0.01	<0.01	<0.01	<0.01
NiO	0.14	0.04	0.08	0.09	0.13
Total	100.26	99.33	99.36	99.69	99.91
		Cations per 4 oxygens			
Si	1.00	1.02	1.02	1.01	1.00
Fe^{2+}	0.87	0.61	0.68	0.62	0.53
Mn	0.01	0.01	0.01	0.01	0.01
Mg	1.12	1.33	1.27	1.36	1.46
MF#	0.56	0.69	0.65	0.69	0.73
Fo	56.28	68.56	65.13	68.69	73.77
Fa	43.72	31.44	34.87	31.31	26.63

Note: Gabbro compositional types as in Table 1.
*Total iron as FeO.
#MF = $Mg/(Mg+Fe^{2+})$.

iomorphic shape. Their chemical composition is characterized by a solid solution between the three major end members phlogopite, Fe-eastonite, and Ti-biotite (Table 7). Their high TiO_2 contents (TiO_2 to 5.4%; Ti-biotite to 30 mol%; Table 7) are indicative of elevated temperatures of crystallization, probably similar to those of brown amphiboles.

Ilmenite

Accessory Fe-Ti oxides are anhedral ilmenites with textural relationships indicative of previous to simultaneous crystallization in relation to plagioclase. Analyzed Fe-Ti oxides show high contents in ilmenite end-member (Ilm = 86–94 mol%; Table 8), with minor but significant rutile and pyrophanite molecules. The MgO content decreases from olivine gabbronorites (to MgO = 3.32%), through hornblende gabbronorites (MgO ~0.18%), to the biotite gabbronorites (MgO < 0.1%). An opposite evolution has been observed for MnO content (Table 8).

GEOCHEMISTRY

To study the geochemical characteristics of the Monte Castelo massif, major and trace elements were determined on 15 representative samples of the different gabbronoritic types, including: 7 olivine gabbronorites, 3 amphibole gabbronorites, and 5 biotite gabbronorites (Tables 9 and 10). All selected samples were undeformed gabbros without any evidence of metamorphic recrystallization: 11 samples were analyzed in the laboratories of the Centre National de la Recherche Scientifique and University of Nancy (France), and the remaining four samples (91699, 91706, 93933, and 93980) were determined in ACTLABS laboratories in Ontario (Canada).

Analyzed gabbronorites show a limited compositional range in SiO_2 (46.42%–52.21%), Al_2O_3 (15.81%–19.25%), CaO (8.10%–12.69%), Na_2O (1.73%–3.04%), K_2O (0.01%–0.32%) and TiO_2 (0.44%–1.93%), and more variable contents in MgO (6.60%–12.14%) and FeO (2.87%–8.91%). Magnesium number [M number = $100 \times MgO/(MgO + FeO_t)$] ranges from 31 to 54. In their normative compositions, all olivine gabbronorites and one amphibole gabbronorite are silica saturated and Ol + Hy normative, whereas the rest of the analyzed samples (two amphibole gabbronorites and the five biotite gab-

TABLE 6. REPRESENTATIVE CHEMICAL ANALYSES OF BROWN AMPHIBOLES

Sample	Ol 93937	Ol 93939	Ol 93940	Ol 93993	Hbl 91703
SiO_2	40.31	42.11	42.05	41.32	42.68
Al_2O_3	13.20	14.63	13.78	14.37	12.69
FeO*	11.09	7.81	9.44	9.35	11.76
MnO	0.18	0.13	0.12	0.16	0.10
MgO	11.27	12.67	12.32	13.10	11.67
CaO	11.57	11.69	11.65	11.35	11.78
Na_2O	2.99	3.58	3.18	3.22	2.10
K_2O	0.75	0.39	0.72	0.39	1.01
TiO_2	4.13	4.04	3.94	3.52	2.52
Cr_2O_3	0.41	0.12	0.07	0.02	0.11
NiO	<0.01	0.06	0.03	0.11	0.05
Total	95.90	97.23	97.30	96.91	96.47
		Cations = 13 (except Ca, Na, K)			
Si	6.04	6.22	6.24	6.07	6.38
Al^{IV}	1.96	1.78	1.76	1.93	1.62
Al^{VI}	0.55	0.76	0.64	0.56	0.62
Ti	0.47	0.45	0.44	0.39	0.28
Cr	0.05	0.01	0.01	0.00	0.01
Fe^{3+}	0.00	0.00	0.00	0.03	0.00
Fe^{2+}	1.39	0.96	1.17	1.12	1.47
Mn	0.02	0.02	0.02	0.02	0.01
Mg	2.52	2.79	2.72	2.87	2.60
Ni	0.00	0.01	0.00	0.01	0.01
Ca	1.86	1.85	1.85	1.79	1.89
$Na_{(M4)}$	0.14	0.15	0.15	0.21	0.11
K	0.14	0.07	0.14	0.07	0.19
$Na_{(A)}$	0.73	0.88	0.77	0.70	0.50
ΣCat	15.87	15.95	15.91	15.77	15.69
MF#	0.64	0.74	0.70	0.72	0.63

Note: Gabbro compositional types as in Table 1.
*Total iron as FeO.
#MF = $Mg/(Mg+ Fe^{2+})$.

TABLE 7. REPRESENTATIVE CHEMICAL ANALYSES OF BIOTITES

Sample	Bt 71909		Bt 93979	
SiO_2	37.32	36.93	38.09	37.57
Al_2O_3	15.67	15.06	14.77	14.85
FeO*	16.28	16.66	11.87	13.22
MnO	0.05	0.02	0.03	0.04
MgO	11.96	11.81	15.26	14.68
CaO	0.03	<0.01	0.03	0.06
Na_2O	0.17	0.05	0.12	0.27
K_2O	9.41	9.56	9.57	9.05
TiO_2	4.69	5.43	5.08	4.99
Total	95.58	95.52	94.82	94.73
Cations per 11 oxygens				
Si	2.76	2.73	2.77	2.75
Al^{IV}	1.24	1.27	1.23	1.25
Al^{VI}	0.13	0.05	0.04	0.03
Ti	0.26	0.30	0.28	0.27
Fe^{3+}	0.17	0.30	0.26	0.32
Fe^{2+}	0.84	0.74	0.46	0.49
Mn	<0.01	<0.01	<0.01	<0.01
Mg	1.32	1.30	1.66	<1.60
Ca	<0.01	<0.01	<0.01	<0.01
Na	0.02	0.01	0.02	0.04
K	0.89	0.94	0.89	0.84
MF#	0.61	0.64	0.78	0.77
Talc	8.36	8.99	9.20	11.33
Ti-Biotite	26.14	30.24	27.89	27.44
Fe-Eastonite	17.16	29.62	25.99	32.13
Eastonite	14.65	5.92	5.64	4.13
Phlogopite	33.69	24.22	31.28	24.96

Note: Gabbro compositional types as in Table 1.
*Total iron as FeO.
#MF = Mg/(Mg+Fe^{2+}).

bronorites) are silica oversaturated, with variable contents in Qtz and Hy.

In the total alkalis-silica (TAS) diagram (Le Maitre et al., 1989), olivine gabbronorites and amphibole gabbronorites plot in the field of basalts, whereas biotite gabbronorites contain slightly higher SiO_2 contents, plotting within the basaltic field adjacent to basaltic andesites field (Fig. 6A). On the basis of their very low K_2O contents, analyzed gabbronorites show a tholeiitic affinity in the SiO_2-K_2O diagram of Peccerillo and Taylor (1976) (Fig. 6B). This affinity of gabbroic types of the Monte Castelo massif is substantiated by an AFM diagram (where A = $Na_2O + K_2O$, F = FeO total, and M = MgO), where all samples plot in the tholeiitic field of Irvine and Baragar (1971; Fig. 6C). In this diagram, there is a general increase in FeO content from less-evolved types (olivine gabbronorites) to more differentiated gabbronorites (biotite gabbronorites), the amphibole gabbronorites showing intermediate values.

Harker diagrams for major elements, as well as for some trace elements of significant petrogenetic interest, are shown in Figures 7 and 8, respectively. Al_2O_3, CaO, and MgO contents decrease with differentiation, whereas TiO_2, MnO, and K_2O increase; Na_2O values do not show clear correlation with silica content. The FeO_t variation relative to SiO_2 is complex. If one gabbronorite sample is excluded, based on its unusually high FeO_t content, then the trend shows an initial increase in FeO_t, followed by a decrease within more evolved biotite gabbronorites. The inflection point, i.e., the analyses with higher FeO_t, occurs in gabbronorites with intermediate SiO_2 contents. Variations observed in major elements relative to SiO_2 are indicative of a fractional crystallization process involving plagioclase (decrease of Al_2O_3 and CaO), together with an initial crystallization of olivine and, later, of clinopyroxene-orthopyroxene (continuous decrease of MgO). Cr and Ni behave compatibly, i.e., they show a general decrease with SiO_2 increase, whereas Nb, Zr, Ba, Y, Rb, and, less clearly, Sr behave as incompatible elements (Fig. 8).

Chondrite normalized rare earth element (REE) plots for the three gabbronorite types of the Monte Castelo massif are shown in Figure 9. Olivine gabbronorites contain lower REE total contents (between 13.18 and 21.47); REE abundances range from 5 to 10 times chondrite (Fig. 9A). Moderate REE enrichment, together with relatively unfractionated REE patterns (La/Lu_N = 0.62–1.05), are characteristic of some basaltic types generated in island arcs (Philpotts, 1990). In these diagrams, the generally flat pattern can be interpreted as reflecting with olivine fractionation, whereas the slightly positive Eu anomalies probably indicate plagioclase accumulation. Moreover, the general absence of depletion in heavy REEs suggests that garnet has not been a residual phase during the processes of partial melting (Whalen, 1985; Lucassen and Franz, 1994).

Amphibole gabbronorites show higher REE contents (between 46.26 and 54.84), with abundances ranging around 20 times chondrite (Fig. 9B). Their REE patterns are flat, with very low positive fractionation (La/Lu_N = 1.29–1.64) and with Eu anomalies slightly negative or absent. These patterns can be interpreted as due to fractional crystallization of plagioclase and clinopyroxene, where the ferromagnesian phases would cause low enrichment in light REEs present in the patterns. The biotite gabbronorites show the greatest variation in chondrite-normalized REE patterns (Fig. 9C). Some samples exhibit REE abundances similar to those of olivine gabbronorites (samples 93955, 93933, and 91709); total values range from 15.15 to 36.40, there are positive Eu anomalies (Eu/Eu* = 1.9–2.4), and low fractionation (La/Lu_N = 0.71–1.55). However, other samples are characterized by higher REE contents (93980 and 91699), with total amounts to 60.42, slightly negative Eu anomalies (Eu/Eu* = 0.83–0.88), and low fractionation (La/Lu_N = 1.64–1.79). These REE patterns can also be interpreted as reflecting processes similar to those discussed previously, but with a more important role of olivine, clinopyroxene, and plagioclase.

TABLE 8. REPRESENTATIVE CHEMICAL ANALYSES OF ILMENITES

Sample	Ol 99937	Ol 93939	Ol 93940	Ol 93993	Hbl 91703	Bt 91709	Bt 93979
SiO_2	0.03	0.11	0.05	<0.01	0.07	0.06	<0.01
Al_2O_3	0.06	0.03	<0.01	0.03	<0.01	0.02	0.02
FeO*	44.10	42.47	42.60	43.08	43.47	43.72	45.23
MnO	0.74	0.70	0.65	0.62	1.08	1.72	1.75
MgO	1.69	3.32	2.35	2.29	0.18	0.06	0.04
CaO	<0.01	0.04	0.04	<0.01	0.05	0.12	<0.01
Na_2O	0.04	0.04	0.05	<0.01	0.01	0.03	0.04
K_2O	0.01	<0.01	<0.01	0.03	<0.01	0.01	<0.01
TiO_2	54.15	54.37	53.85	54.20	53.61	53.78	53.37
Total	100.82	101.08	99.59	100.25	98.47	99.52	100.45
				Cations per 3 oxygens			
Si	<0.01	<0.01	<0.01	<0.01	<0.01	<0.01	<0.01
Ti	1.03	1.03	1.01	1.00	1.00	1.01	1.01
Al	<0.01	<0.01	<0.01	<0.01	<0.01	<0.01	<0.01
Fe^{3+}	0.00	0.00	0.00	0.00	0.00	0.00	0.00
Fe^{2+}	0.93	0.93	0.91	0.86	0.89	0.95	0.89
Mn	0.02	0.04	0.02	0.01	0.01	0.04	0.01
Mg	0.06	0.12	0.09	0.08	0.01	<0.01	<0.01
$FeAl_2O_3$	0.00	0.06	0.17	0.09	0.00	0.06	0.09
$FeSiO_3$	0.17	0.15	0.07	0.27	0.12	0.00	0.00
Pyrophanite	2.27	3.59	1.54	1.44	1.36	3.69	1.29
Geikielite	0.66	0.22	6.17	12.05	8.64	0.15	8.37
Ilmenite	89.84	90.01	90.06	86.12	87.74	94.13	88.23
Haematite	0.00	0.00	0.00	0.03	0.00	0.00	0.00
Rutile	7.06	5.97	1.95	0.00	2.14	1.97	2.04

Note: Gabbro compositional types as in Table 1.
*Total iron as FeO.

For the three types of gabbronorites in the Monte Castelo massif, averages for some elements of significant petrogenetic interest have been normalized to the composition of typical N-MORB (Pearce, 1982; see Fig. 10). The diagrams show consistent patterns with low enrichment in large ion lithophile elements (LILE) (Sr, K, Rb, Ba), as well as values close to one or slightly depleted in high field strength elements and REEs. Olivine gabbronorites exhibit an overall depletion in Zr, Ti, and Y, a feature considered characteristic of many tholeiitic magmas generated in island arcs (Wilson, 1989; Lapierre et al., 1992). The gabbronorites compositions have been compared with average values of Pearce (1982) for island-arc tholeiites, volcanic-arc calc-alkaline basalts, tholeiitic within-plate basalts, and within-plate alkaline basalts (Fig. 10). In general, Monte Castelo gabbronorite compositions are very similar to those of island-arc tholeiites (Fig. 10A). Relative to volcanic-arc calc-alkaline basalts, Monte Castelo gabbronorites exhibit a significant depletion in LILEs (Fig. 10B). This depletion extends to the majority of the normalized elements when the gabbronorites are compared with within-plate basalts, both tholeiitic and alkaline.

The Monte Castelo gabbronorites have also been compared (Fig. 11) with other gabbroic rocks, generated either in an island arc (Whalen, 1985) or in more general magmatic-arc settings (Lapierre et al., 1992; De Bari, 1994; Lucassen and Franz, 1994). The compositions of Monte Castelo gabbronorites are very similar to those reported from different magmatic-arc settings, showing multielement diagrams with comparable general patterns and anomalies. Specially, Monte Castelo gabbronorites most closely resemble the magmatic series of Guanajuato (Lapierre et al., 1992), because both rocks series have Nb contents abnormally high for arc tholeiites (in the range of N-MORB). Such high Nb contents have usually been explained by being related to immature stages of arc development.

In order to expand the comparison of Monte Castelo gabbronorites with other basaltic compositions generated in different tectonic settings, all analyzed samples were also plotted using the ($MnO \times 10$) – TiO_2 – ($P_2O_5 \times 10$) diagram of Mullen (1983) (Fig. 12). This plot is effective in discriminating between most of the basaltic types, with the important exception of within-plate basalts, which are very different in composition from the studied gabbronorites. The majority of analyzed samples plot in the island-arc tholeiite field, which confirms the tectonic setting deduced for the Monte Castelo gabbronorites based on multielement diagrams.

TABLE 9. MAJOR AND TRACE ELEMENT ANALYSES OF OLIVINE GABBRONORITES

Sample	Ol 93993	Ol 93939	Ol 93969	Ol 93997	Ol 93940	Ol 93937	Ol 93985
SiO_2	46.42	47.03	47.82	48.09	48.21	48.28	48.59
Al_2O_3	18.48	18.03	17.67	17.46	19.25	17.88	18.67
Fe_2O_3	1.54	1.07	0.93	2.52	1.42	1.74	2.10
FeO	6.27	7.13	6.57	7.05	6.85	7.24	6.58
MnO	0.12	0.13	0.12	0.15	0.12	0.14	0.13
MgO	10.48	12.14	9.41	8.58	9.78	9.00	9.17
CaO	10.57	10.38	12.69	10.85	9.89	10.73	10.50
Na_2O	2.20	2.33	2.27	2.52	2.72	2.58	2.62
K_2O	0.13	0.01	0.05	0.08	0.07	0.08	0.10
TiO_2	0.60	0.64	0.69	1.16	0.60	0.96	0.44
P_2O_5	0.11	0.08	0.10	0.07	0.07	0.07	0.08
LOI	2.17	0.05	0.67	0.45	0.09	0.32	0.14
Total	99.09	99.02	98.99	98.98	99.07	99.02	99.12
Ba	12	4	14	19	7	19	26
Rb	5	3	5	4	3	4	5
Sr	128	169	136	181	191	195	196
Ga	26	20	22	20	23	23	23
Nb	4	3	3	2	3	4	3
Zr	32	17	35	21	16	21	12
Y	16	13	18	16	11	16	12
Cr	469	357	827	356	200	424	277
Ni	175	218	86	118	167	96	208
Co	59	86	69	59	69	60	71
Sc	25.7	25.7	38	33.9	22.2	31.2	26.2
V	154	150	202	210	169	221	147
Cu	45	52	48	39	44	36	37
Zn	55	57	52	67	28	67	64
La	2.08	0.95	1.85	1.74	2.33	1.78	1.42
Ce	5.18	3.59	5.28	4.44	6.71	4.51	3.24
Nd	4.29	2.65	4.54	3.99	3.99	3.76	2.68
Sm	1.94	1.15	2.14	1.94	1.62	1.76	1.22
Eu	0.82	0.65	0.77	0.92	0.89	0.87	0.78
Gd	2.29	1.70	2.54	2.41	1.93	2.35	1.41
Dy	2.12	1.60	2.61	2.31	1.74	2.26	1.43
Yb	1.23	0.89	1.47	1.33	1.00	1.33	0.84
Lu	0.24	0.16	0.27	0.25	0.23	0.24	0.16
Total	20.19	13.34	21.47	19.33	20.44	18.86	13.18

Note: LOI = Loss on ignition.

According to their petrographic characteristics, mineralogical composition, and general geochemical pattern, we conclude that the Monte Castelo gabbronorites show a close similarity to tholeiitic basalts generated in island-arc settings. Major element variations suggest that the gabbronorites originated from an initial magma of tholeiitic basalt composition, through a process of fractional crystallization where olivine, orthopyroxene, clinopyroxene, and plagioclase were the more important removed phases. Olivine gabbronorites are the least evolved lithologies (higher contents in MgO and Cr), whereas amphibole and biotite gabbronorites represent progressively more differentiated compositional types. Variations in major and trace elements and REEs similar to those characteristic of the Monte Castelo gabbronorites have also been reported in other gabbroic rocks assigned to island-arc settings; this is the case for gabbros of the Sierra de Fiambalá in Argentina (De Bari, 1994), for those studied in the Guanajuato region in Mexico (Lapierre et al., 1992), and for gabbroic rocks that crop out in the northern Cordillera of Chile (Lucassen and Franz, 1994), the latter generated in an extensional backarc setting.

TABLE 10. MAJOR AND TRACE ELEMENT ANALYSES OF AMPHIBOLE GABBRONORITES AND BIOTITE GABBRONORITES

Sample	Hbl 91703	Hbl 91706	Hbl 93979	Bt 93933	Bt 93955	Bt 91709	Bt 93980	Bt 91699
SiO_2	48.65	49.66	50.90	50.42	50.42	51.74	51.92	52.21
Al_2O_3	16.63	17.58	16.42	17.16	16.26	17.66	15.81	18.16
Fe_2O_3	2.33	0.94	2.31	1.03	0.92	2.67	9.57	1.10
FeO	7.58	8.46	7.37	8.46	8.91	6.54	2.87	7.25
MnO	0.16	0.16	0.17	0.19	0.20	0.16	0.22	0.15
MgO	8.28	7.74	8.60	6.96	6.91	7.80	6.60	7.02
CaO	10.53	10.49	8.83	10.29	9.46	8.10	8.63	10.27
Na_2O	2.58	1.73	2.45	2.62	2.41	3.04	2.02	1.88
K_2O	0.30	0.31	0.20	0.19	0.30	0.20	0.32	0.24
TiO_2	1.41	1.17	1.12	1.59	0.96	0.86	1.93	1.00
P_2O_5	0.15	0.11	0.11	0.06	0.08	0.08	0.20	0.10
LOI	0.40	0.40	0.50	0.80	2.00	0.22	0.01	0.50
Total	99.00	98.75	98.98	99.77	98.83	99.07	100.10	99.88
Ba	57	57	83	57	125	73	173	121
Rb	7	5	4	4	8	5	5	6
Sr	174	170	151	196	194	200	179	196
Ga	25	17	25	16	25	18	22	18
Nb	4	5	5	2	3	3	7	5
Zr	89	93	56	16	37	31	39	81
Y	29	31	24	13	20	16	40	30
Cr	332	228	215	242	207	307	108	341
Ni	98	86	75	104	85	42	38	110
Co	52	41	56	36	56	38	49	42
Sc	36	36	33	41	40	29	45	36
V	255	162	226	244	291	271	253	142
Cu	39	32	38	23	30	29	23	32
Zn	80	74	86	73	85	86	73	65
La	5.81	6.80	6.19	1.50	4.58	5.69	9.80	9.00
Ce	15.93	19.00	15.56	4.00	10.73	12.89	26.00	22.00
Nd	11.69	12.00	9.43	3.00	7.20	6.37	17.00	13.00
Sm	4.23	3.32	3.27	0.99	2.69	2.41	5.11	3.29
Eu	1.37	1.00	1.05	0.90	1.51	1.55	1.48	1.06
Gd	4.68	4.00	3.47	1.33	2.98	2.69	5.00	3.70
Dy	5.01	5.40	4.17	1.80	3.34	2.49	1.80	4.40
Yb	2.83	2.89	2.66	1.41	2.07	1.93	4.08	3.45
Lu	0.49	0.43	0.46	0.22	0.37	0.38	0.62	0.52
Total	52.04	54.84	46.26	15.15	35.47	36.40	70.89	60.42

Note: LOI = Loss on ignition; Hbl = hornblende; Bt = biotite gabbronorites.

DISCUSSION

Recent U-Pb isotopic ages of the Monte Castelo gabbro, granitoids, and high-grade metasedimentary rocks, clustering between 493 and 500 Ma, indicate that emplacement of igneous bodies in to the uppermost unit of the Ordenes Complex was immediately followed by their burial and by a regional Barrovian metamorphism that reached granulite facies (Abati et al., 1999). A convergent scenario seems to best fit the different data. We suggest that the uppermost intermediate-*P* unit represents an accretionary complex, built with sediments deposited adjacent to a volcanic arc. The gabbros resulted from subduction of oceanic lithosphere, whereas the granitoids developed by partial melting of the thickened wedge, and were, together with the gabbros, buried and metamorphosed shortly after their emplacement. Geochemical characteristics of Monte Castelo gabbronorites

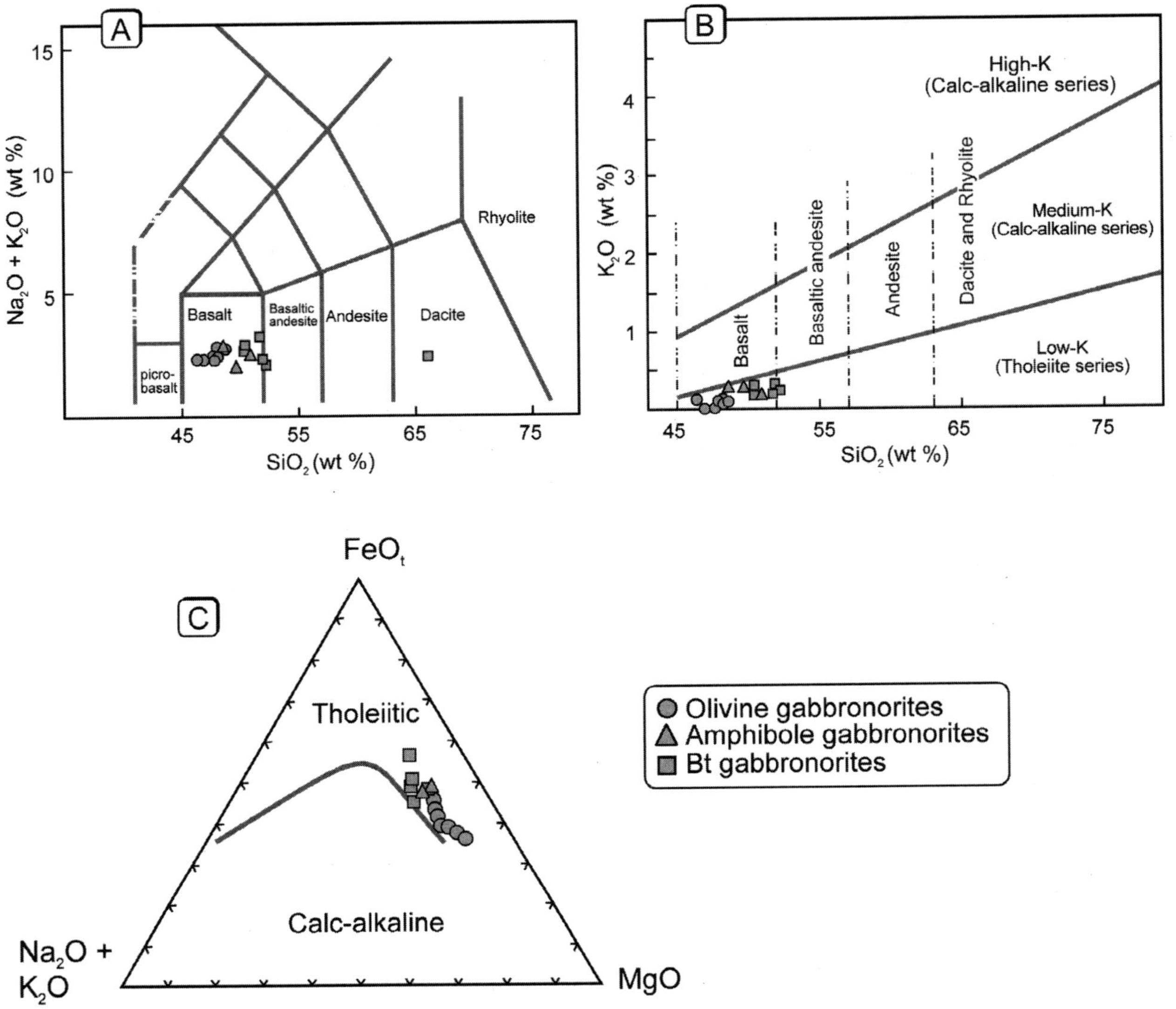

Figure 6. TAS, SiO_2-K_2O, and AFM diagrams with projection of 15 analyzed samples of Monte Castelo gabbroic massif. TAS, total alkalis-silica. AFM, (A = $Na_2O + K_2O$, F = FeO total, M = MgO).

strongly support this interpretation, in particular their close resemblance to island-arc tholeiites.

Recent data obtained in the Variscan basement of central Europe and the Alps were used by Stampfli (1996) to suggest the existence of an island arc in the Rheic domain by Early Ordovician time (ca. 490 Ma). The history of this arc would be at least partially contemporaneous with the development of extension in the margin of Gondwana, which led to the separation of the Avalon microcontinent (Fortey and Cocks, 1988; Soper, 1988). Avalon is the outermost emerged terrane at the margin of the North Atlantic Ocean opposite from the Iberian Peninsula, and was adjacent or very close to western Iberia before the breakup of Pangea (Fig. 1). Characteristic of its Gondwanan provenance and Pan-African deformation and magmatism (Skehan and Rast, 1995), no Late Cambrian–Early Ordovician orogenic events have been reported for Avalon. This fact is additional evidence that a Gondwanan origin does not by itself explain the evolution of the terrane located in the uppermost structural position in the northwest of the Iberian Massif. The data presented herein regarding the petrologic and geochemical characteristics of the gabbronorites of the Monte Castelo Massif, in the Ordenes Complex, together with other metamorphic and geochronological evidence from the uppermost unit of this complex, support an arc-related setting as the most probable origin for this terrane.

ACKNOWLEDGMENTS

Support for this work was provided by grants PB94-1396-CO2 and PB97-0234-CO2 of the Spanish science agency Dirección General de Investigación Científica y Técnica (DGICYT). We

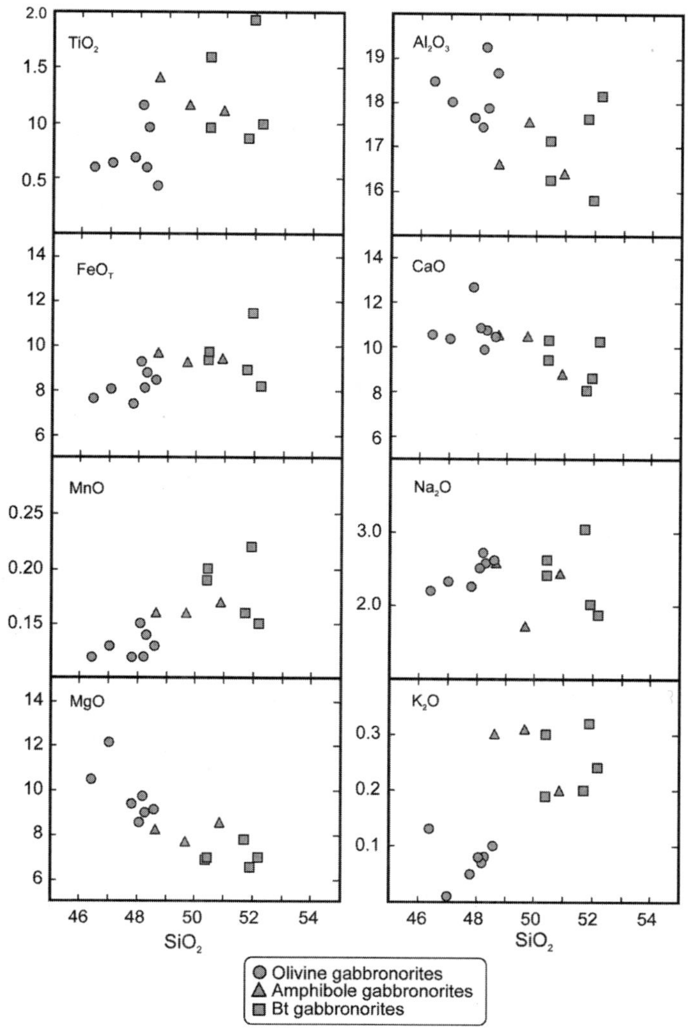

Figure 7. Variation diagrams for major elements vs. SiO_2.

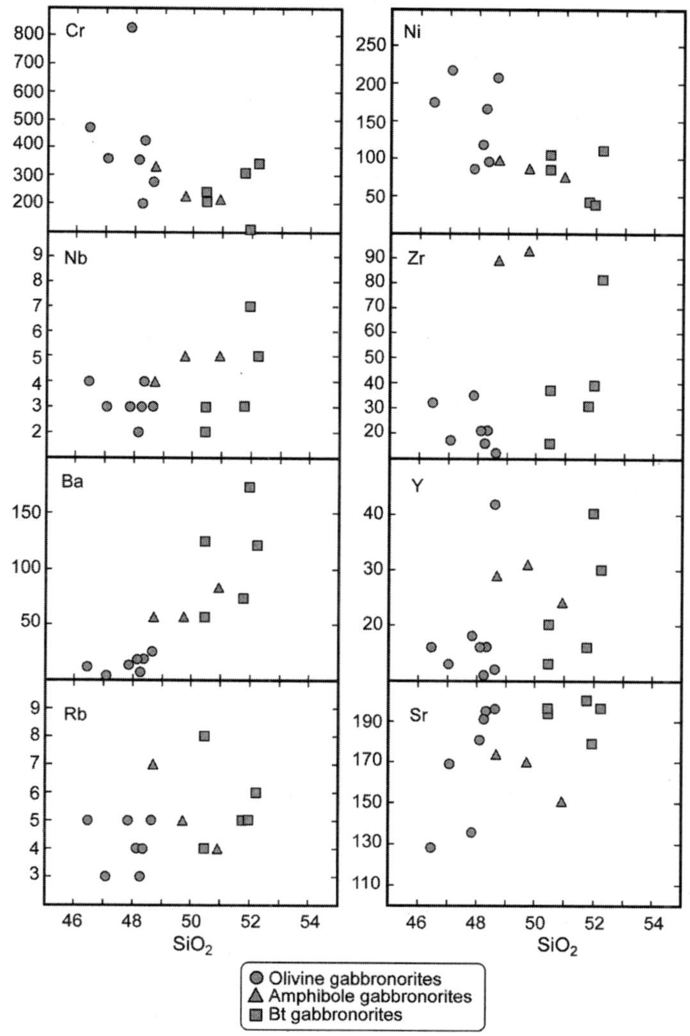

Figure 8. Variation diagrams for some trace elements vs. SiO_2.

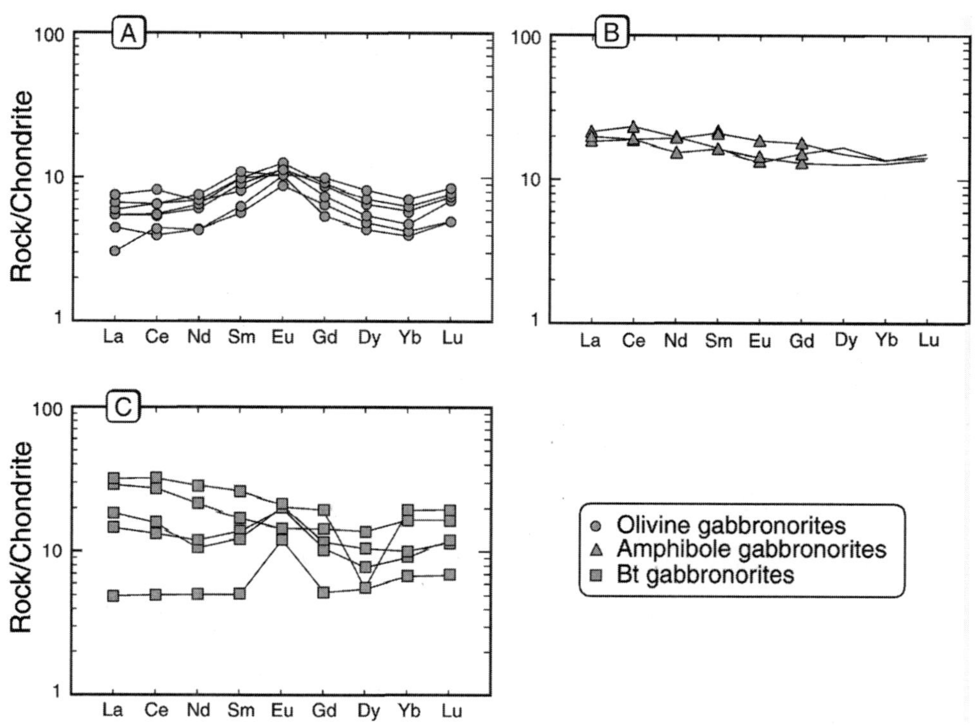

Figure 9. Chondrite normalized rare earth element patterns of three main gabbroic types of Monte Castelo massif.

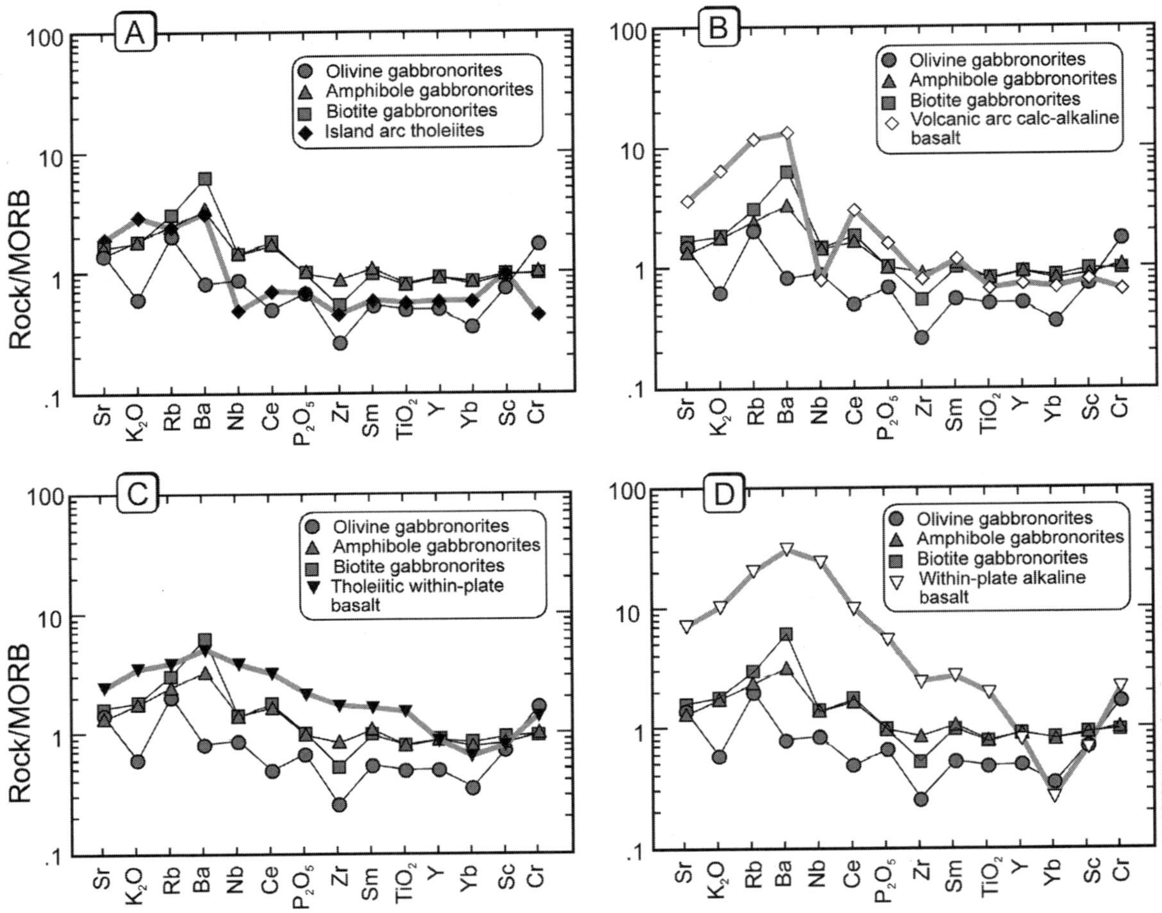

Figure 10. Normal mid-ocean ridge basalt (MORB) normalized multielement diagrams. Composition of gabbroic rocks of Monte Castelo massif is also compared with four basaltic types generated in different tectonic setting: island-arc tholeiites, volcanic-arc calc-alkaline basalts, tholeiitic within-plate basalts, and within-plate alkaline basalts (from Pearce, 1982).

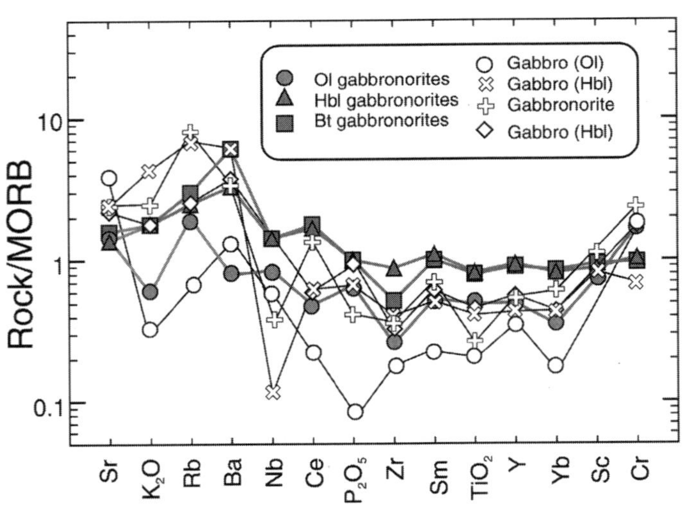

Figure 11. Normal mid-ocean ridge basalt (MORB) multielement diagrams comparing Monte Castelo gabbroic types with other gabbroic rocks generated in magmatic-arc setting. O = CB33 (Lucassen and Franz, 1994); X = 40912 (Whalen, 1985); + = Sf39 (De Bari, 1994); ◊ = MC195 (Lapierre et al., 1992).

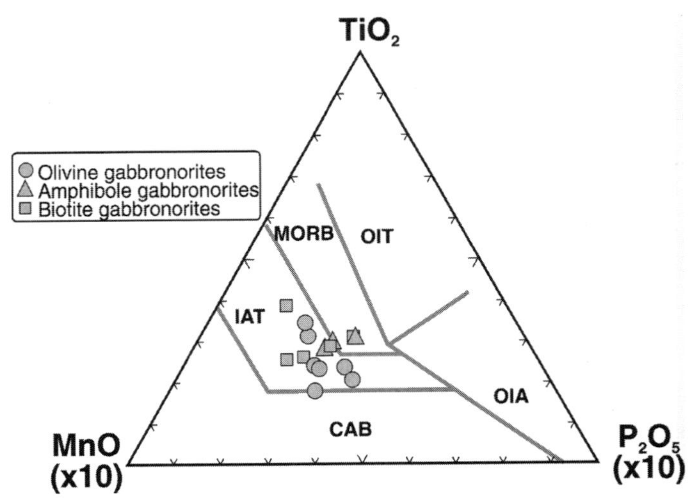

Figure 12. TiO_2-MnO (×10)-P_2O_5 (×10) plot (Mullen, 1983). Diagram shows fields for mid-ocean ridge basalts (MORB), ocean-island tholeiites (OIT), ocean-island alkali basalts (OIA), island-arc tholeiites (IAT), and calc-alkaline basalts (CAB).

thank J.B. Whalen and M.C. Gilbert for reviews that greatly improved the quality and clarity of the manuscript.

REFERENCES CITED

Abati, J., Dunning, G.R., Arenas, R., Díaz García, F., González Cuadra, P., Martínez Catalán, J.R., and Andonaegui, P., 1999, Early Ordovician orogenic event in Galicia (NW Spain): Evidence from U-Pb ages in the uppermost unit of the Ordenes Complex: Earth and Planetary Science Letters, v. 165, p. 213–228.

Aoki, K.I., and Kushiro, Y., 1968, Some clinopyroxenes from ultramafic inclusions in Dreiser Weiher, Eifel: Contributions to Mineralogy and Petrology, v. 18, p. 326–337.

Arenas, R., Gil-Ibarguchi, J.I., González Lodeiro, F., Klein, E., Martínez Catalán, J.R., Ortega Gironés, E., Pablo Maciá, J.G., and Peinado, M., 1986, Tectonostratigraphic units in the complexes with mafic and related rocks of the northwest of the Iberian Massif: Hercynica, v. 2, p. 87–110.

Bernard-Griffiths, J., Peucat, J.J., Cornichet, J., Iglesias Ponce de León, M., and Gil Ibarguchi, J.I., 1985, U-Pb, Nd isotope and REE geochemistry in eclogites from the Cabo Ortegal Complex, Galicia, Spain: An example of REE immobility conserving MORB-like patterns during high-grade metamorphism: Chemical Geology, v. 52, p. 217–225.

Dallmeyer, R.D., and Tucker, R.D., 1993, U-Pb zircon age for the Lagoa augen gneiss, Morais Complex, Portugal: Tectonic implications: Journal of the Geological Society of London, v. 150, p. 405–410.

Dallmeyer, R.D., Martínez Catalán, J.R., Arenas, R., Gil Ibarguchi, J.I., Gutiérrez Alonso, G., Farias, P., Aller, J., and Bastida, F., 1997, Diachronous Variscan tectonothermal activity in the NW Iberian Massif: Evidence from $^{40}Ar/^{40}Ar$ dating of regional fabrics: Tectonophysics, v. 277, p. 307–337.

De Bari, S.M., 1994, Petrogenesis of the Fiambalá gabbroic intrusion, Northwestern Argentina: A deep crustal syntectonic pluton in a continental magmatic arc: Journal of Petrology, v. 35, p. 679–714.

De Bari, S.M., and Coleman, R.G., 1989, Examination of the deep levels of an island arc: Evidence from the Tonsina ultramafic assemblages, Tonsina, Alaska: Journal of Geophysical Research, v. 94 (B4), p. 4373–4391.

Díaz García, F., 1990, La geología del sector occidental del Complejo de Ordenes (Cordillera Hercínica, NW de España): Nova Terra, v. 3, Laxe, Coruña, 230 p.

Díaz García, F., Martínez Catalán, J.R., Arenas, R., and González Cuadra, P., 1999, Structural and kinematic analysis of the Corredoiras detachment: Evidence for early Variscan orogenic extension in the Ordenes Complex, NW Spain: International Journal Earth Sciences, v. 8, p. 337–351.

Elthon, D., 1987, Petrology of gabbroic rocks from the Mid-Cayman Rise spreading center: Journal of Geophysical Research, v. 92 (B1), p. 658–682.

Fortey, R.A., and Cocks, L.R.M., 1988, Arenig to Llandovery faunal distributions in the Caledonides, *in* Harris, A.L., and Fettes, D.J., eds, The Caledonian-Appalachian Orogen: London, Geological Society of London Special Publication 38, p. 233–246.

Galán, G. and Marcos, A., 1997, Geochemical evolution of high-pressure mafic granulites from the Bacariza formation (Cabo Ortegal complex, NW Spain): An example of a heterogeneous lower crust: Geologische Rundschau, v. 86, p. 539–555.

Gil Ibarguchi, J.I., Mendia, M., Girardeau, J., and Peucat, J.J., 1990, Petrology of eclogites and clinopyroxene-garnet metabasites from the Cabo Ortegal Complex (northwestern Spain): Lithos, v. 25, p. 133–162.

Gravestock, P.J., 1992, The chemical causes of uppermost mantle heterogeneities. [Ph.D. thesis]: Milton Keynes, United Kingdom, Open University, 299 p.

Heltz, R.T., 1973, Phase relations of basalts in their melting range at PH_2O=5kb as function of oxygen fugacity. Part I mafic phases: Journal of Petrology, v. 14, p. 249–302.

Irvine, T.N., and Baragar, W.R.A., 1971, A guide to chemical classification of the common volcanic rocks: Canadian Journal of Earth Sciences. v. 8, p. 523–548.

Jarosewich, E.J., Nelen, J.A., and Norberg, J.A., 1980, Reference samples for electron microprobe analysis: Geostandards Newsletters, v. 4, p. 43–47.

Kretz, R., 1983, Symbols for rock-forming minerals: American Mineralogist, v. 68, p. 277-279.

Lapierre, H., Ortiz, L.E., Abouchami, W., Monod, O., Coulon, C., and Zimmermann, J.L., 1992, A crustal section in an intra-oceanic island arc: The Late Jurassic–Early Cretaceous Guanajuato magmatic sequence, central Mexico: Earth and Planetary Science Letters, v. 108, p. 61–77.

Le Maitre, R.W., Bateman, P., Dudek, A., Keller, J., Lameyre Le Bas, M.J., Sabine, P.A., Schmid, R., Sorensen, H., Streckeisen, A., Woolley, A.R., and Zanettin, B., 1989, A classification of igneous rocks and glosssary of terms: Oxford, United Kingdom, Blackwell, 193 p.

Lefort, J.P., 1989, Basement correlation across the North Atlantic: Berlin, Springer-Verlag, 148 p.

Lucassen, F., and Franz, G., 1994, Arc related Jurassic igneous and meta-igneous rocks in the Coastal Cordillera of northern Chile–Region Antofagasta: Lithos, v. 32, p. 273–298.

Marques, F.O., Ribeiro, A., and Munhá, J.M., 1996, Geodynamic evolution of the Continental Allochthonous Terrane (CAT) of the Bragança Nappe Complex, NE Portugal: Tectonics, v. 15, p. 747–762.

Martínez Catalán, J.R., 1990, A non-cylindrical model for the northwest Iberian allochthonous terranes and their equivalents in the Hercynian belt of Western Europe: Tectonophysics, v. 179, p. 253–272.

Martínez Catalán, J.R., and Arenas, R., 1992, Deformación extensional de las unidades alóctonas superiores de la parte oriental del Complejo de Ordenes (Galicia): Geogaceta, v. 11, p. 108–111.

Martínez Catalán, J.R., Arenas, R., Díaz García, F., Rubio Pascual, F.J., Abati, J., and Marquínez, J., 1996, Variscan exhumation of a subducted Paleozoic continental margin: The basal units of the Ordenes Complex, Galicia, NW Spain: Tectonics, v. 15, p. 106–121.

Martínez Catalán, J.R., Arenas, R., Díaz García, F., and Abati, J., 1997, Variscan accretionary complex of northwest Iberia: Terrane correlation and succession of tectonothermal events: Geology, v. 27, p. 1103–1106.

Martínez Catalán, J.R., Arenas, R., Díaz García, F., and Abati, J., 1999, Allochthonous units in the Variscan Belt of NW Iberia, *in* Sinha, A.K., ed., Basement tectonics, v. 13, p. 65–84.

Matte, P., 1991, Accretionary history and crustal evolution of the Variscan belt in Western Europe: Tectonophysics, v. 196, p. 309–337.

Matte, P., and Capdevila, R., 1978, Tectonique en grands plis couchés et plissements superposés d'âge hercynien dans la série de Ordenes-Betanzos (Galice Occidentale): Cuadernos del Seminario de Estudios Cerámicos de Sargadelos, v. 27, p. 193–201.

Mendia, M., 1996, Petrología de la Unidad Eclogítica del Complejo de Cabo Ortegal (NW de España). [Ph.D. thesis]: Leioa, Bilbao, Spain, Universidad del País Vasco, 463 p.

Morimoto, N., 1988, Nomenclature of pyroxenes: Bulletin de Minéralogie, v. 111, p. 535–550.

Mullen, E.D., 1983, $MnO/TiO_2/P_2O_5$: A minor element discriminant for basaltic rocks of oceanic environments and its implications for petrogenesis: Earth and Planetary Science Letters, v. 62, p. 53–62.

Otten, M.T., 1984, The origin of brown hornblende in the Artfjället gabbro and dolerites: Contributions to Mineralogy and Petrology, v. 86, p. 189–199.

Pearce, J.A., 1982, Trace element characteristics of lavas from destructive plate boundaries, *in* Thorpe, R.S., ed., Andesites: New York, John Wiley & Sons, p. 525–548.

Peccerillo, A., and Taylor, S.R., 1976, Geochemistry of Eocene calc-alkaline volcanic rocks from the Kastamonu area, northern Turkey: Contributions to Mineralogy and Petrology, v. 58, p. 63–81.

Pérez-Estaún, A., Martínez Catalán, J.R., and Bastida, F., 1991, Crustal thickening and deformation sequence in the footwall to the suture of the Variscan Belt of northwest Spain: Tectonophysics v. 191, p. 243–253.

Peucat, J.J., Bernard-Griffiths, J., Gil Ibarguchi, J.I., Dallmeyer, R.D., Menot, R.P., Cornichet, J., and Iglesias Ponce de León, M., 1990, Geochemical and geochronological cross section of the deep Variscan crust: The Cabo Ortegal high-pressure nappe (northwestern Spain): Tectonophysics, v. 177, p. 263–292.

Philpotts, A.R., 1990, Principles of igneous and metamorphic petrology: Englewood Cliffs, New Jersey, Prentice-Hall, 497 p.

Ribeiro, M.L., Andriessen, P.A.M., Beetsma, J.J., Ferreira, N., and Oen, I.S., 1993, Precambrian Sm-Nd age of UATC of Morais Massif (NE Portugal): Terra Abstract, v. 5 (1), p. 333.

Santos, J.F., 1998, Geoquímica de litologías básicas e ultrabásicas da Unidade Alóctone Superior do Maciço de Bragança. [Ph. D. thesis]: Aveiro, Portugal, Universidade de Aveiro, 402 p.

Schäfer, H.J., Gebauer, D., Gil Ibarguchi, J.I., and Peucat, J.J., 1993, Ion-microprobe U-Pb zircon dating on the HP/HT Cabo Ortegal Complex (Galicia, NW Spain): preliminary results: Terra Abstracts, v. 5 (4), p. 22.

Skehan, J.V., and Rast, N., 1995, Late Proterozoic to Cambrian evolution of the Boston Avalon Terrane, *in* Hibbard, J.P., Van Staal, C.R., and Cawood, P.A., eds., Current perspectives in the Appalachian-Caledonian Orogen: Toronto, Ontario, Geological Association of Canada Special Paper 41, p. 207–225.

Soper, N.J., 1988, Timing and geometry of collision, terrane accretion and sinistral strike-slip events in the British Caledonides, *in* Harris, A.L., and Fettes, D.J., eds., The Caledonian-Appalachian Orogen: London, Geological Society of London Special Publication 38, p. 481–492.

Stampfli, G.M., 1996, The Intra-Alpine terrain: a Paleotethyan remnant in the Alpine Variscides: Eclogae Geologica Helvetica, v. 89, p. 13–42.

Van Calsteren, P.W.C., 1978, Geochemistry of the polymetamorphic mafic-ultramafic complex at Cabo Ortegal (NW Spain): Lithos, v. 11, p. 61–72.

Warnaars, F.W., 1967, Petrology of a peridotite-amphibolite and gabbro-bearing polyorogenic terrain northwest of Santiago de Compostela (Spain): [Ph.D. thesis] Leiden, Netherlands, University of Leiden, 208 p.

Whalen, J.B., 1985, Geochemistry of an island-arc plutonic suite: The Uasilau–Yau Yau intrusive complex, New Britain, P.N.G: Journal of Petrology, v. 26, p. 603–632.

Wilson, M., 1989, Igneous petrogenesis: London, Unwin Hyman, 466. p.

Manuscript Accepted by the Society February 8, 2002

Printed in the U.S.A.

Geological Society of America
Special Paper 364
2002

Early Devonian suprasubduction-zone ophiolite related to incipient collisional processes in the Western Variscan Belt: The Sierra de Careón unit, Ordenes Complex, Galicia

C. Pin
J.L. Paquette
Département de Géologie, OPGC, Centre National de la Recherche Scientifique–UMR 6524, Université Blaise Pascal, 63038 Clermont-Ferrand, France
J.F. Santos Zalduegui
J.I. Gil Ibarguchi*
Departamento de Mineralogía y Petrología, Universidad del País Vasco, P.O. Box 644, E-48080 Bilbao, Spain

ABSTRACT

The Careón unit consists of imbricate tectonic sheets of serpentinite, metagabbro, and diabase dikes interpreted as a dismembered ophiolitic sequence. Zircons from leucogabbro intruding serpentinites point to an igneous crystallization at 395 ± 3 Ma, in accord with previous results. Combined with a reported 377 ± 1 Ma Ar/Ar cooling age, this date implies that ophiolite generation occurred shortly before its tectonic emplacement. Most samples are enriched in light rare earth elements (REE) relative to heavy REEs and show distinct fractionation of the heavy REEs. Enrichments in Th and La relative to Nb are ubiquitous, while negative anomalies of Zr and Ti occur in most samples. Initial ε_{Nd} values (+6.4 to +9.1), even for samples with high light REE/heavy REE ratios, point to a time-integrated mantle source strongly depleted in Nd relative to Sm, and preclude significant contamination of mafic melts during their ascent through the crust. Combined trace element and Nd-isotope data favor an intraoceanic, suprasubduction-zone setting, where hydrous fluids and silicate melts metasomatized a wedge of depleted mantle and triggered its partial melting. On the basis of magmatic affinities, age constraints, and broad tectonometamorphic context, ophiolite generation is interpreted to reflect oceanic spreading above a subduction zone dipping away from a passive continental margin. Suprasubduction extension might have been associated with subduction hinge retreat caused by instability of old, cold oceanic lithosphere entering the subduction zone. It is suggested that this occurred in the context of early stage arc collision documented at that epoch in the Variscan realm.

*Corresponding author. E-mail: nppgiibi@lg.ehu.es.

Pin, C., Paquette, J.L., Santos Zalduegui, J.F., and Gil Ibarguchi, J.I., 2002, Early Devonian suprasubduction-zone ophiolite related to incipient collisional processes in the Western Variscan Belt: The Sierra de Careón unit, Ordenes Complex, Galicia, *in* Martínez Catalán, J.R., Hatcher, R.D., Jr., Arenas, R., and Díaz García, F., eds., Variscan-Appalachian dynamics: The building of the late Paleozoic basement: Boulder, Colorado, Geological Society of America Special Paper 364, p. 57–71.

INTRODUCTION

Five allochthonous complexes (Cabo Ortegal, Ordenes, Malpica-Tuy, Bragança, and Morais) within the northwestern Iberian Peninsula display well-correlated internal subdivisions: from bottom to top, these are the basal, ophiolitic, and upper units. The upper units have received more attention, principally because of their less common petrological association of eclogitic and high-pressure granulitic, ultramafic, and gneissic rocks (e.g., Gil Ibarguchi et al., 1990; Santos Zalduegui et al., 1996; Abati et al., 1999). Underlying units are not as well studied. Ophiolites represent ancient oceanic rocks, the paleosetting of which (e.g., major ocean basins versus mature backarc basins) has important implications for paleogeographic reconstructions and regional tectonic models. For that reason, there has been a renewed interest in the ophiolitic units of northwest Iberia (e.g., Díaz García et al., 1999a). Previous geochemical data on ophiolitic metabasites from western sectors of the Ordenes Complex pointed to an origin of the protoliths in an arc environment (Díaz García, 1990). Ophiolite and arc complex remnants are scarce within the Variscan fold belt. The eastern sector the Ordenes Complex contains one of the best-preserved Variscan ophiolite in western Europe. Therefore, its geochemical study and age assessment appeared essential.

Among the methods used to identify and constrain the nature of ancient ophiolitic terranes is the comparison of their high field strength element (HFSE, e.g., Zr, Ti, Y, Nb, and Ta), rare earth element (REE), and transition metal (Sc, V, Cr, or Ni) contents with those of present-day oceanic settings. These elements are considered relatively immobile in low-grade or seafloor hydrothermally altered metamorphic rocks, whereas large ion lithophile elements (LILE, e.g., K, Cs, Rb, Ba, and Sr) can be severely disturbed (Pearce, 1975, 1987; Shervais, 1982; Sturchio et al., 1986; Middelburg et al., 1988). In combination with these methods, the use of Nd isotopic data and the precise U-Pb dating of constituent zircons provide key information on the sources and timing of igneous crystallization ages of ophiolitic rocks and, in some cases, their subsequent tectonic evolution by subduction and/or obduction processes.

In order to better determine the geodynamic significance of ophiolite relics within the northwestern Iberian allochthonous complexes, we have examined in detail the major, trace, and REE compositions, including Nd and U-Pb isotope data, of representative mafic lithotypes from the ophiolitic unit of the Sierra de Careón area of the Ordenes Complex, where ophiolitic features are best preserved.

GEOLOGIC SETTING AND PREVIOUS WORK

A prominent feature of the northwestern part of the Iberian massif within the Variscan fold belt of Europe is the occurrence of five allochthonous complexes, Cabo Ortegal, Ordenes, Malpica-Tuy, Bragança and Morais (Fig. 1A); the complexes have distinctive tectono-thermal features as compared to the Variscan granitoids and metamorphic units of their autochthon. The complexes include high-pressure gneisses and metabasites and abundant mafic to ultramafic rocks that have been grouped into three main correlated units, upper, ophiolitic, and basal, separated by thrust and/or ductile extensional detachments (e.g., Arenas et al., 1986; Gil Ibarguchi et al., 1990; Díaz García et al., 1999b). Upper and Lower ophiolitic units are distinguished on the basis of their different petrographic and tectonic features (Fig. 1, A and B).

Upper ophiolitic unit

This corresponds, in the Cabo Ortegal area, to the Purrido-Peña Escrita amphibolites and metagabbros (Gil Ibarguchi et al., 1990; Azcárraga, 2000), to the Careón-Bazar and Carballo–El Pino units of metabasites and ultramafites in the Ordenes Complex (Martínez Catalán et al., 1984; Díaz García, 1990; Díaz García et al., 1999a), and to similar rocks within the Morais and Bragança Complexes, for which various local names have been proposed (e.g., Soeira unit, Izeda amphibolites, Morais gabbros; Anthonioz, 1972; Ribeiro, 1974). The best exposures of this unit are found in the Ordenes and Morais sectors where it includes abundant layered amphibolites, metagabbros, metadiabases, metaplagiogranites, and ultramafic rocks. Considered as a whole, the unit occupies a structural position immediately below the high-temperature, high-pressure allochthon. It is bounded by sharp or gradual ductile thrust contacts at its top (higher pressure rocks above) and bottom (lower grade rocks below). However, the ophiolitic unit does not form a continuous sheet beneath the upper allochthon. Its thickness varies from several hundreds of meters to zero, and it is in contact with various types of gneisses and metabasites of the upper allochthon.

Previous geochemical data suggested low-K tholeiitic affinities for the Carballo–El Pino rocks (Díaz García 1990) and normal to enriched mid-ocean ridge basalt (N- to E-MORB) types in the Careón (Peucat et al., 1990) and Purrido units (Arenas et al., 1986), respectively. In view of their structural position overlying high-pressure (*P*), low- to intermediate-temperature (*T*) metamorphosed units from the former Paleozoic margin of Gondwana (basal units), the rocks of the ophiolitic unit have been interpreted as corresponding to the lower section of an island arc (Díaz García, 1990) or, more likely, an oceanic lithosphere formed during a Silurian to Devonian rifting event, subsequently duplicated through intraoceanic thrusting and obducted onto the continent (Arenas et al., 1986; Ribeiro et al, 1990a, 1990b; Martínez Catalán et al., 1997; Díaz García et al., 1999b). The latter hypothesis would be in agreement with a U-Pb date of 395 ± 2 Ma from zircon from a metagabbro of the Careón unit, interpreted as the age of the mafic protoliths in the Ordenes Complex (Díaz García et al., 1999a).

All these rocks present indications of pervasive synkinematic recrystallization for which medium-*P* amphibolite facies conditions have been generally proposed (Díaz García, 1990; Dallmeyer et al., 1991). However, there is some evidence of rel-

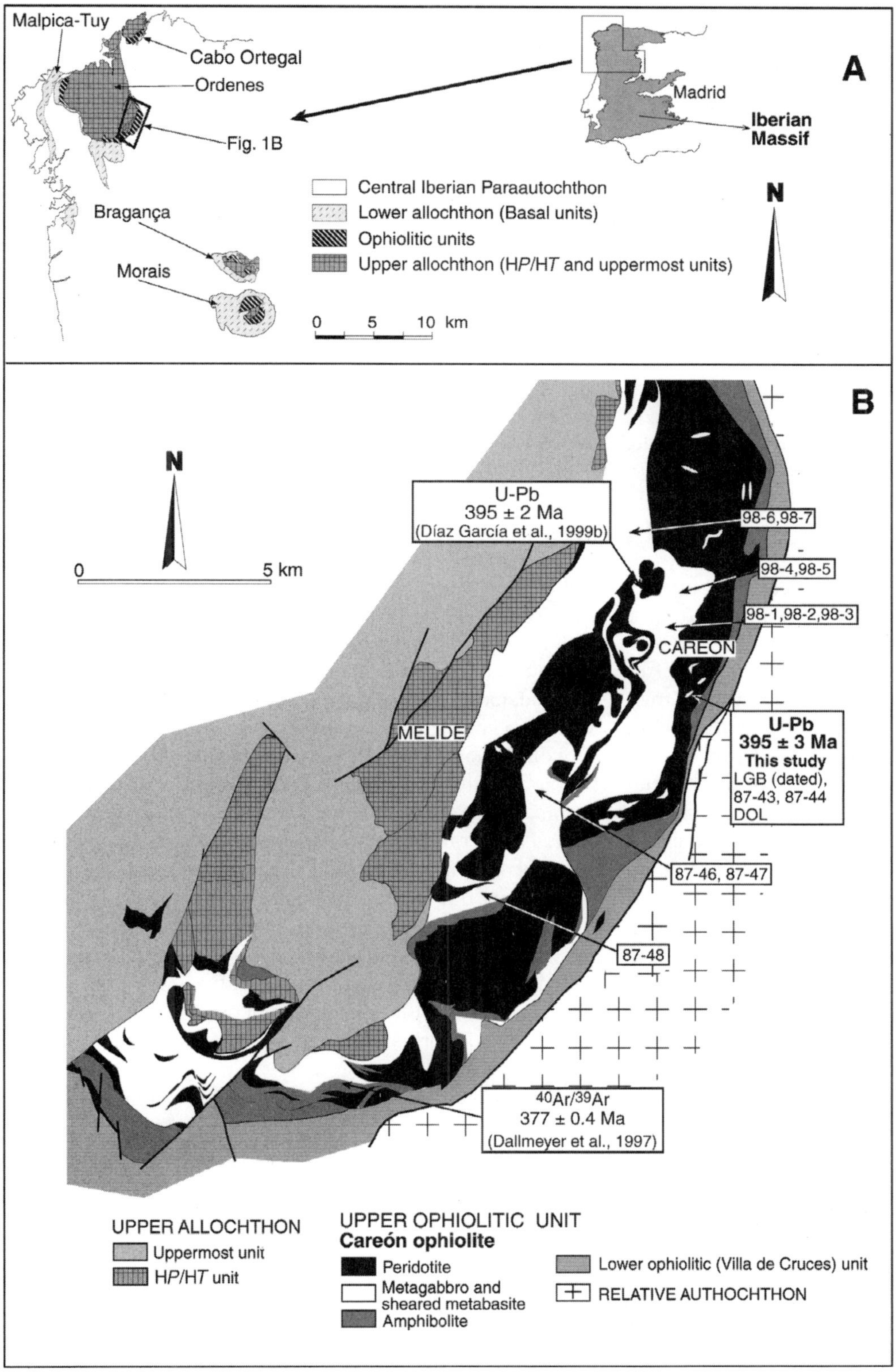

Figure 1. A: Geological map of northwestern Iberian massif showing location of five allochthonous complexes, Cabo Ortegal, Ordenes, Morais, Bragança, and Malpica-Tuy, and their principal lithological units. B: Detail of different units within eastern sector of Ordenes Complex showing extension of main lithotypes within Sierra del Careón ophiolite unit (simplified after Díaz García et al., 1999a). H*P*/H*T* is high pressure, high temperature.

atively high P assemblages destroyed during subsequent amphibolite facies overprinting (Engels, 1972; Arenas, 1991; Azcárraga, 2000). Associated deformation resulted in development of tectonic subunits, which locally show evidence for intraoceanic duplication, and inversion of structural elements separated by thrust faults (Ribeiro et al., 1990a; Díaz García et al., 1999b). The $^{39}Ar/^{40}Ar$ dating of constituent amphibole from various amphibolite units yielded ages of ca. 385–390 Ma for the regional metamorphic event (Dallmeyer and Gil Ibarguchi, 1990; Peucat et al., 1990; Dallmeyer et al., 1991). These ages have been interpreted to date a tectonothermal event reflecting the first stage of Variscan orogenesis in the Iberian Massif. This was most likely related to oceanic closure and associated tectonic emplacement of eclogitic, granulitic, and high-grade ultramafic rocks from overlying units of the upper regional allochthon (Dallmeyer and Gil Ibarguchi, 1990; Dallmeyer et al., 1997). Subsequent episodes of extensional deformation under lower grade conditions have been dated as between 370 and 365 Ma by Ar/Ar in whole rock and minerals of different units (Dallmeyer et al., 1997).

Lower ophiolitic unit

This unit has a lesser extension and its precise limits are not well established. It appears to form part of the strongly deformed units at the base of Cabo Ortegal (Moeche unit) and in the lowermost, eastern sectors of the Ordenes Complex (Villa de Cruces unit, Fig. 1B). This unit would also form the uppermost structural element within the Malpica-Tuy band (Ceán-Razo unit) and, possibly, part of the uppermost units of the Centro-Transmontan series in northern Portugal, where low-grade metabasites of tholeiitic affinity occur immediately below the massive amphibolites and metagabbros of the upper ophiolitic unit (cf. Arenas et al., 1986; Ribeiro et al., 1990a; Díaz García et al., 1999b). The unit bears evidence of strong deformation and dismembering and is composed of low-grade metamorphic rocks, including metapelite, chert, metadiabase, amphibolite, plagiogranite, and serpentinitic rocks with affinities to olivine tholeiites (Arenas et al., 1986). Metamorphic conditions were distinctly lower than those of the upper ophiolitic unit and include evidence (e.g., glaucophane relics and lawsonite pseudomorphs in northern Malpica-Tuy) of relatively high-P–low-T recrystallization predating extensive development of greenschists facies assemblages. A limited number of Ar/Ar mineral and whole-rock ages suggest ca. 385–390 Ma for the prograde regional metamorphism (Dallmeyer et al., 1991) and ca. 340–350 Ma for subsequent emplacement at shallower crustal levels (Dallmeyer et al., 1997; Rodríguez Aller et al., 1997). The unit has been interpreted as corresponding mainly to a dismembered upper portion of a Paleozoic oceanic lithosphere (Arenas et al., 1986; Martínez Catalán et al., 1997). However, in view of its poor preservation state, strong tectonic overprinting, and limited availability of fresh samples, this unit has not been considered in this study.

Petrography and mineral chemistry

Good descriptions of the Sierra de Careón metabasites in the eastern sector of the Ordenes Complex, including details on their petrography, mineralogy, structure, and tectonic imbrication of ophiolite sheets into three main subunits were provided by Díaz García et al. (1999a). Flaser gabbros to metadolerites from the main mass of the Careón metabasites (samples 87-46, 87-47, 87-48, 98-1, 98-2, 98-3, 98-4, 98-5, 98-6, 98-7), together with similar rocks within ultramafics of that same region (samples 87-43, 87-44, LGB, DOL; Fig. 1B) were investigated for their major and trace element and isotope compositions. The studied rocks were selected on the basis of their petrologic characteristics and good preservation. Relic igneous textures are common in these rocks; recrystallization invariably took place under amphibolite facies conditions. Minerals from six polished thin sections of representative samples were analyzed with microprobe. Table 1 presents a selection of mineral analyses with additional details on structural formulas and Fe^{3+} estimates. A brief account of the petrographic varieties studied and their mineralogical features is given in the following.

Gabbroic amphibolites or flaser gabbros ranging to microgabbroic types show generally well preserved igneous texture and, in most cases, amphibolite facies mineralogy. Clinopyroxene, usually replaced by amphibole, appears as stubby prisms or with fibrous habit and is an Al- and Na-poor diopside (Table 1). Amphibole is generally magnesio-hornblende to ferroan pargasitic and, more rarely, tschermakitic hornblende (98-7, Table 1). This amphibole may be variably transformed into secondary actinolitic hornblende, which also occurs forming discrete neoblasts in the matrix (98-2, Table 1). Plagioclase has to An_{46} at some cores, although more often it is albite to oligoclase (Pl, Table 1) either in porphyroclasts or neoblasts. Matrix epidote is unzoned or poorly zoned (98-5, Table 1) and, more rarely, may show a faint rim of clinozoisite (98-2, Table 1). Rare biotite is rich in iron (X_{Mg} 0.19, Table 1). Other minerals present are rutile, either as discrete grains in the matrix or as fine needles within amphibole; ilmenite with generally minor amounts of Mn (Table 1); titanite after ilmenite containing to 1.7% Al_2O_3; apatite, often as short prisms in metagabbro varieties, and zircon (see following).

Díaz García et al. (1999a) described corundum-bearing metabasites forming a thin band beneath ultramafic rocks of one of the ophiolite subunits (Vilouriz). These metabasites exhibit a peculiar mineralogy, including centimeter-size, nearly pure pink corundum porphyroclasts (Cor, Table 1), within a matrix composed essentially of tschermakitic amphibole. The amphiboles have a patchy green tint, the green color being associated with high Cr contents (to 2.78% Cr_2O_3, 95-1/Am, Table 1). Scarce Cr-rich spinel relics (Table 1) are observed in some of these Cr-rich green zones. Other minerals present are margarite (Table 1), which forms fibrous bluish aggregates replacing corundum, and rare clinozoisite within these same aggregates (Czo, Table 1).

METAMORPHIC CONDITIONS

High-T conditions of dynamothermal metamorphism related to tectonic imbrication of ophiolite sheets are evidenced by the occurrence of corundum-bearing assemblages in amphibole-rich metabasites underlying ultramafic rocks of the Vilouriz ophiolite subunit. This would imply conditions of ~750–850 °C for a variable pressure range (Nicollet, 1986). However, the main episode of deformation and recrystallization in the rocks studied took place under lower P-T conditions typical of the high-P and medium- to high-T part of the amphibolite facies field. Temperatures of ~650 ± 55 °C have been deduced for amphibole-plagioclase assemblages in the metabasites using plagioclase-amphibole geothermometer calibrations of Blundy and Holland (1990) and Holland and Blundy (1994). The Al content in hornblende geobarometer (Anderson and Smith, 1995; Ague, 1997) yields pressures of 9.5 ± 1.5 kbar. Garnet does not appear in the rocks studied, although it was described and analyzed from the metamorphic sole of the Careón ophiolite by Díaz García et al. (1999a). Garnet-bearing assemblages studied by these authors provide slightly higher T and P conditions of 640–680 °C and 10.5–12 kbar, which appear to be the maximum P-T conditions recorded by the rocks of the ophiolitic units exposed in the Ordenes Complex. These estimates are consistent with the absence of true high-P assemblages and with the type of metamorphism documented by the compositions of analyzed amphiboles similar to those of medium-P metamorphic areas elsewhere (cf. Laird and Albee, 1981). Following this relatively high P amphibolitic episode, which in most cases is accompanied by the development of pervasive foliation and lineation, the rocks of the ophiolitic unit underwent a decrease in pressure and temperature down to low-grade amphibolite to greenschists facies conditions. This resulted in generalized growth of epidote, replacement of rutile by titanite, local growth of actinolitic amphibole and chlorite after primary amphibole, pyroxene, and garnet in appropriate lithotypes and replacement of corundum by margarite in the corundum-bearing amphibolites. Locally, this event was associated with plastic deformation represented by late, scarce ductile shear zones oblique to the penetrative foliation of amphibolites. Physical conditions for this stage are poorly defined. Temperatures presumably were lower than ~650 °C; in view of the low Al^{VI}/Si ratios and Na^{M4} contents of secondary amphiboles, it can be assumed that the latter structures formed at <5 kbar (cf. Raase, 1974).

Geochemical and isotopic results

In order to determine the igneous crystallization age of the complex, zircons extracted from a dike of plagioclase-rich gabbro, intrusive into serpentinites, have been analyzed for U-Pb isotopes. Moreover, the geochemical characteristics of a set of 14 mafic rocks have been studied, mostly on the basis of selected immobile trace elements (REE, Th, Nb). These samples have also been analyzed for Nd isotopes, in order to further characterize the source reservoirs and petrogenetic evolution of the igneous suite.

U-Pb zircon geochronology

Abundant zircon crystals were extracted from a leucogabbro dike intruding serpentinites (sample LGB). These grains, up to 1 mm long, are light yellow, translucent, and devoid of inclusions. The crystals are either euhedral and short prismatic or anhedral with geometric faces, indicating a late crystallization from a melt. Six air-abraded fractions, four of the euhedral type and two of the anhedral type, have been analyzed (Table 2). The six analytical points are concordant or slightly discordant within analytical uncertainties, defining an upper intercept at 394.6 ± 2.7 Ma, a lower intercept close to the origin (44 ± 160 Ma), and a pooled $^{207}Pb/^{206}Pb$ age of 394.4 ± 1.0 Ma (Fig. 2). The radiogenic Pb contents are very low, between 6 and 8 ppm. Considering the very subordinate degree of Pb loss and the lack of any inheritance, the upper intercept age of 395 ± 3 Ma is interpreted to date the igneous crystallization of the gabbroic dike.

Elemental and Sm-Nd geochemistry

Major, trace, and rare earth element compositions together with significant elemental ratios of analyzed samples from the Sierra de Careón unit of the Ordenes Complex are given in Table 3. Incompatible and trace elements are plotted as chondrite-normalized diagrams in Figure 3. Based on their geochemical patterns, three groups of samples can be defined.

Group 1 comprises some of the fine-grained samples (plagioclase-phyric dolerites; e.g., 98-1, 98-4; 98-7), presumably representing basaltic liquid compositions, that have relatively smooth patterns with an upward convex shape (Fig. 3A). Their REE concentrations are between 10 and 20 times the chondritic abundances, and they display both light REE depletion, as expressed by $(La/Sm)_N \leq 1$ (0.7–1), and depleted heavy REE with $(Sm/Lu)_N > 1$ (1.3-1.5). Despite low concentrations of Th (~5–10 times chondrite) and Th/La ratios between 0.05 and 0.09, a negative anomaly of Nb is apparent on the chondrite-normalized patterns.

Group 2 comprises other doleritic samples (e.g., 87-43, 87-44, 87-46, DOL) that are also characterized by depleted heavy REEs $[(Sm/Lu)_N > 1]$, but show a marked enrichment in light REEs $[(La/Sm)_N > 1.3]$ and in Th (20–50 times chondrite) and a concomitant, strongly negative Nb anomaly (Fig. 3B). Small negative Ti anomalies also occur.

Group 3 includes the rest of samples (metagabbros) that show much more irregular patterns (Fig. 3C) believed to reflect processes of crystal accumulation (specifically, of plagioclase in 87-48, 98-5, and 98-6, as inferred from positive anomalies of Sr, or of ilmenite in 87-47, as shown by a Ti spike in the chondrite-normalized pattern) and removal (e.g., from Fe-Ti

TABLE 1. REPRESENTATIVE CHEMICAL ANALYSES OF MINERALS FROM METABASITES AND RELATED ROCKS OF THE SIERRA DE CAREÓN OPHIOLITE UNIT

	Amphibolite, metagabbro, metadolerite										Corundum-bearing metabasite				
Sample mineral	98-6 Cpx	98-7 Am	98-2 Am	98-7 Pl	DOL Bt	98-5 Ep-core	98-5 Ep-rim	98-2 Ep-core	98-2 Czo-rim	98-6 Ilm	95-1 Amp	95-1 Cor	95-1 Spl	95-1 Mrg	95-1 Czo
SiO_2	51.55	43.12	54.15	68.66	33.61	38.31	38.38	38.18	39.78	bd	46.17	0.01	0.05	31.89	38.90
TiO_2	0.06	0.90	0.25	bd	3.80	0.11	0.15	0.14	bd	53.27	0.19	0.07	0.08	0.12	0.08
Al_2O_3	1.04	12.42	3.35	19.70	15.92	29.56	28.96	29.30	33.44	0.01	14.96	99.81	20.26	49.99	31.88
Cr_2O_3	0.02	0.04	bd	0.02	bd	bd	bd	bd	bd	0.02	2.78	0.17	45.37	0.17	0.19
FeO_t	9.71	15.56	9.04	0.23	29.03	4.91	5.78	5.10	0.04	43.56	4.71	0.17	30.54	0.18	1.71
MgO	12.41	10.83	18.03	0.04	3.38	0.07	0.07	0.11	bd	0.21	15.31	0.04	3.13	0.41	0.04
MnO	0.17	0.39	0.13	bd	0.10	0.16	0.08	0.13	bd	2.21	0.07	bd	0.80	bd	0.06
CaO	24.21	11.14	12.17	0.19	0.09	24.18	24.38	24.14	24.81	0.03	12.14	0.02	0.02	11.88	24.49
Na_2O	0.51	2.17	0.60	11.98	0.13	bd	0.03	bd	0.02	bd	1.55	0.01	bd	1.08	0.03
K_2O	0.03	0.32	0.02	bd	8.53	bd	0.02	bd	bd	bd	0.04	0.03	0.04	bd	0.07
Total	99.71	96.88	97.72	100.81	94.90	97.30	97.85	97.09	98.11	99.30	97.92	100.33	100.29	95.72	97.45
Si	1.935	6.397	7.620	2.981	2.691	2.961	2.956	2.959	3.006	-	6.447	-	0.002	2.084	2.974
Ti	0.002	0.101	0.026	-	0.229	0.007	0.009	0.008	-	1.018	0.020	0.001	0.002	0.006	0.005
Al	0.046	2.171	0.556	1.008	1.502	2.694	2.630	2.677	2.979	-	2.463	1.992	0.789	3.852	2.874
Cr	0.001	0.004	-	0.001	-	-	-	-	-	-	0.307	0.002	1.185	0.009	0.011
Fe^{3+}	0.118	0.510	0.188	-	0.202	0.318	0.372	0.330	0.002	-	0.277	0.002	0.021	0.010	0.109
Fe^{2+}	0.186	1.421	0.875	0.008	1.740	-	-	-	-	0.925	0.273	-	0.823	-	-Mg
0.694	2.394	3.781	0.003	0.403	0.008	0.008	0.012	-	0.008	3.186	0.001	0.154	0.040	0.005	
Mn	0.005	0.049	0.015	-	0.007	0.010	0.005	0.008	-	0.047	0.008	-	0.022	-	0.004
Ca	0.974	1.771	1.835	0.009	0.007	2.002	2.012	2.005	2.009	0.001	1.816	-	0.001	0.832	2.007
Na	0.037	0.624	0.164	1.008	0.021	0.001	0.004	-	0.003	-	0.184	-	-	0.137	0.004
K	0.001	0.061	0.003	-	0.872	-	0.002	-	-	-	0.007	0.001	0.002	-	0.007

Note: 98-6: metadolerite; 98-7: amphibolite; 98-2, DOL; 98-5: metagabbro; 95-1: corundum-bearing metabasite. Cpx: clinopyroxene; Am: amphibole; Pl: plagioclase; Bt: biotite; Ep: epidote; Czo: clinozoisite; Ilm: ilmenite; Cor: corundum; Spl: spinel; Mrg: margarite. Fe^{3+}/Fe^{2+} estimated by regression techniques and charge balance criteria (cf. De Bruiyn et al., 1983; Droop, 1987; Massonne, 1992; Dale et al., 2000) assuming: 4 cations and 6 O (pyroxene), 23 O (amphibole); 8 O (plagioclase); 11 O (biotite, margarite); 8 cations and 12.5 O (epidote, clinozoisite); 2 cations and 3 O (ilmenite, corundum); 3 cations and 4 O (spinel); bd: below detection limits.

TABLE 2. U-PB DATA FROM SAMPLE LEUCOGABBRO (LGB)

		Concentrations		Atomic ratios					Apparent ages (Ma)		
Fraction (μm)	Weight (mg)	U (ppm)	Pb_{rad} (ppm)	$^{206}Pb/^{204}Pb$	$^{208}Pb/^{206}Pb$	$^{206}Pb/^{238}U$	$^{207}Pb/^{235}U$	$^{207}Pb/^{206}Pb$	$^{206}Pb/^{238}U$	$^{207}Pb/^{235}U$	$^{207}Pb/^{206}Pb$
Z1: [5] eu.	0.278	91.1	5.89	2086	0.1335	0.06292 ± 30	0.4733 ± 23	0.05455 ± 6	393	393	394
Z2: [4] eu.	0.194	93.2	5.93	2754	0.1301	0.06257 ± 32	0.4710 ± 25	0.05459 ± 9	391	392	396
Z3: [3] eu.	0.212	104	6.51	3340	0.1177	0.06246 ± 59	0.4697 ± 45	0.05454 ± 12	391	391	393
Z4: [3] eu.	0.208	95.8	6.35	821	0.1577	0.06186 ± 60	0.4655 ± 47	0.05458 ± 15	387	388	395
Z5: [1] an.	0.085	100	6.70	449	0.1921	0.06169 ± 40	0.4644 ± 36	0.05459 ± 22	386	387	396
Z6: [2] an.	0.177	136	8.44	2410	0.1215	0.06117 ± 24	0.4600 ± 20	0.05454 ± 9	383	384	393

Note: The isotopic ratios are corrected for mass discrimination (0.1 ± 0.015 % per amu for Pb and U), isotopic tracer contribution and analytical blanks. Absolute errors corresponding to the last significant digits are given at the 2σ level. Number in brackets is number of grains in fraction. Abbreviations: eu. = euhedral ; an. = anhedral ; rad = radiogenic.

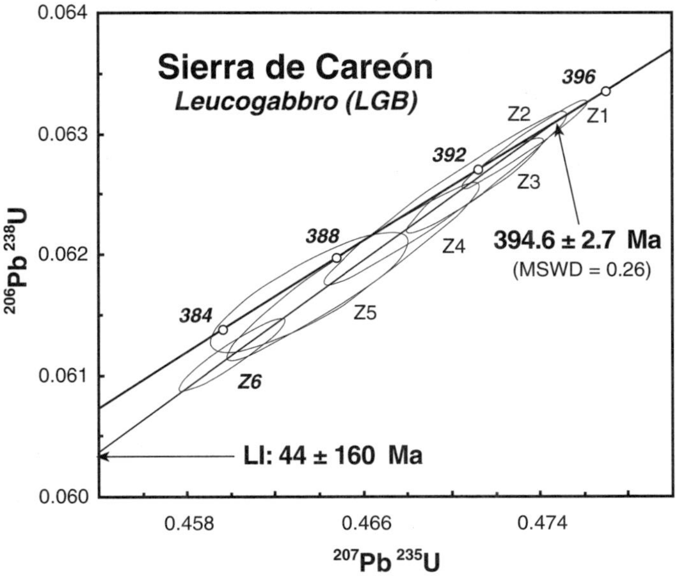

Figure 2. U-Pb concordia diagram for leucogabbro (LGB) of Sierra de Careón ophiolite (ellipse errors are 2σ, cf. analytical techniques). MSWD is mean square of weighted deviates.

oxides and zircon, as deduced from deep negative anomalies of Ti and Zr). Sample LGB has a very high Zr content (1050 ppm) interpreted to reflect the presence of cumulitic zircon in this rock. Negative Nb anomalies also occur in all metagabbros (Fig. 3C).

Sm-Nd analytical data are listed in Table 4, along with $^{143}Nd/^{144}Nd$ initial ratios at 395 Ma expressed using the epsilon notation. All the samples have strongly positive εNd_i values, ranging from +6.4 to +9.1, that reflect derivation from a time-integrated depleted mantle source (i.e., Sm/Nd ratio higher than that of the chondritic, or bulk earth, reservoir during most of its evolution). The highest εNd_i values (+9.1) are observed in the doleritic samples with convex-upward REE patterns and small, albeit distinct, negative Nb anomalies. This Nd isotope composition is slightly less radiogenic (by about one ε unit) than that predicted for the source reservoir of N-MORBs 400 m.y. ago (Zindler, 1982). The new Sm-Nd results do not allow for calculation of a whole-rock isochron for each of the different groups distinguished or for the entire set. A best-fit line for the 14 samples analyzed is close to a meaningless result of 485 ± 110 Ma.

TABLE 3. MAJOR AND TRACE ELEMENT COMPOSITION OF METABASITES FROM THE SIERRA DE CAREÓN OPHIOLITE UNIT

	Group 1			Group 2							Group 3			
Sample	98-1	98-4	98-7	DOL	87-43	87-44	87-46	LGB	87-47	87-48	98-2	98-3	98-5	98-6
SiO_2 (%)	49.6	49.8	48.9	48.97	50.1	50	51.7	51.47	51.2	50.1	50.3	46.5	52.6	52.6
Al_2O_3	15.4	16.3	17.2	14.21	14.5	14.6	18.3	19.85	17.7	16	14.3	14.1	18.3	15.8
Fe_2O_3	10.2	8.31	7.96	11.57	10.9	11.2	5.56	5.19	6.56	6.35	6.9	14.7	7.2	7.88
MgO	7.87	8.2	8.78	6.83	6.81	6.84	6.73	3.57	6.01	10	8.46	7.72	7.72	7.49
CaO	9.7	11	11.4	9.54	9.41	9.22	10.6	9.3	7.88	13.5	10.3	9.19	11.2	10.5
Na_2O	3.13	3.08	3.07	4.57	4.38	4.25	4.15	6.07	5.5	2.03	3.47	3.5	3	3.36
K_2O	0.14	0.1	0.5	0.45	0.24	0.1	0.15	0.12	0.04	0.1	0.1	0.18	0.33	0.28
TiO_2	1.96	1.35	1.21	1.91	1.81	1.82	0.54	1.91	3.53	0.41	1.34	2.91	0.76	0.51
MnO	0.13	0.14	0.12	0.19	0.17	0.17	0.09	0.1	0.08	0.12	0.09	0.25	0.09	0.11
LOI	1.12	0.84	0.81	0.83	1.13	1.14	1.68	1.57	1.35	1.54	4.41	1.44	1.1	1.22
TOTAL	99.25	99.12	99.95	99.07	99.45	99.34	99.5	99.15	99.85	100.15	99.67	100.49	102.3	99.75
Rb (ppm)	6	5	4	6	9	na	5	2	na	na	4	10	5	4
Ba	na	na	na	265	401	na	53	265	na	na	na	na	na	na
Sr	200	178	207	312	236	320	263	671	730	207	313	338	210	212
Zr	104	88	96	133	130	135	92	1050	78	41	100	93	48	27
Nb	2.4	1.8	2.3	4	3.2	3.1	3.3	10	17	0.5	5.5	8	1.3	2
Y	33	28	25	39	36	35	20	63	23	11	40	83	16	16
Th	0.42	0.3	0.25	na	1.1	2.3	0.54	na	16.4	0.2	1.1	1.5	0.16	0.35
La	4.73	3.96	5.42	11.3	10.4	11.4	8.36	26.7	71.2	0.8	7.86	16.8	3.22	4.16
Ce	13.2	11.6	14.4	26.2	25	25.1	18.4	65.4	131	3.32	20.5	47	7.84	8.79
Nd	12.8	10.9	11.2	16.7	17.2	17.8	10.6	39.2	39.5	3.15	16.1	35.5	6.92	6.23
Sm	4.11	3.57	3.17	4.64	5	5.24	2.69	9.4	5.8	1.2	4.68	10.2	2.01	1.74
Eu	1.37	1.27	1.19	1.69	1.68	1.82	0.85	3.15	1.68	0.53	1.33	2.56	0.79	0.71
Gd	4.75	4.16	3.88	6.19	6.14	5.66	2.97	11.4	4.71	1.65	5.46	11.8	2.52	2.2
Dy	5.82	5.14	4.45	6.76	6.76	6.32	3.42	11.1	4.26	1.88	6.83	14.1	2.91	2.83
Er	3.59	3.07	2.48	3.96	4	3.75	2.19	6.56	2.35	1.1	4.09	8.36	1.68	1.7
Yb	3.35	2.93	2.49	4.05	3.85	3.77	2.48	6.46	2.36	1.11	4.05	7.29	1.64	1.67
Lu	0.47	0.43	0.4	0.62	0.6	0.55	0.37	1.02	0.36	0.16	0.59	1	0.25	0.23

Note: na—not analyzed.

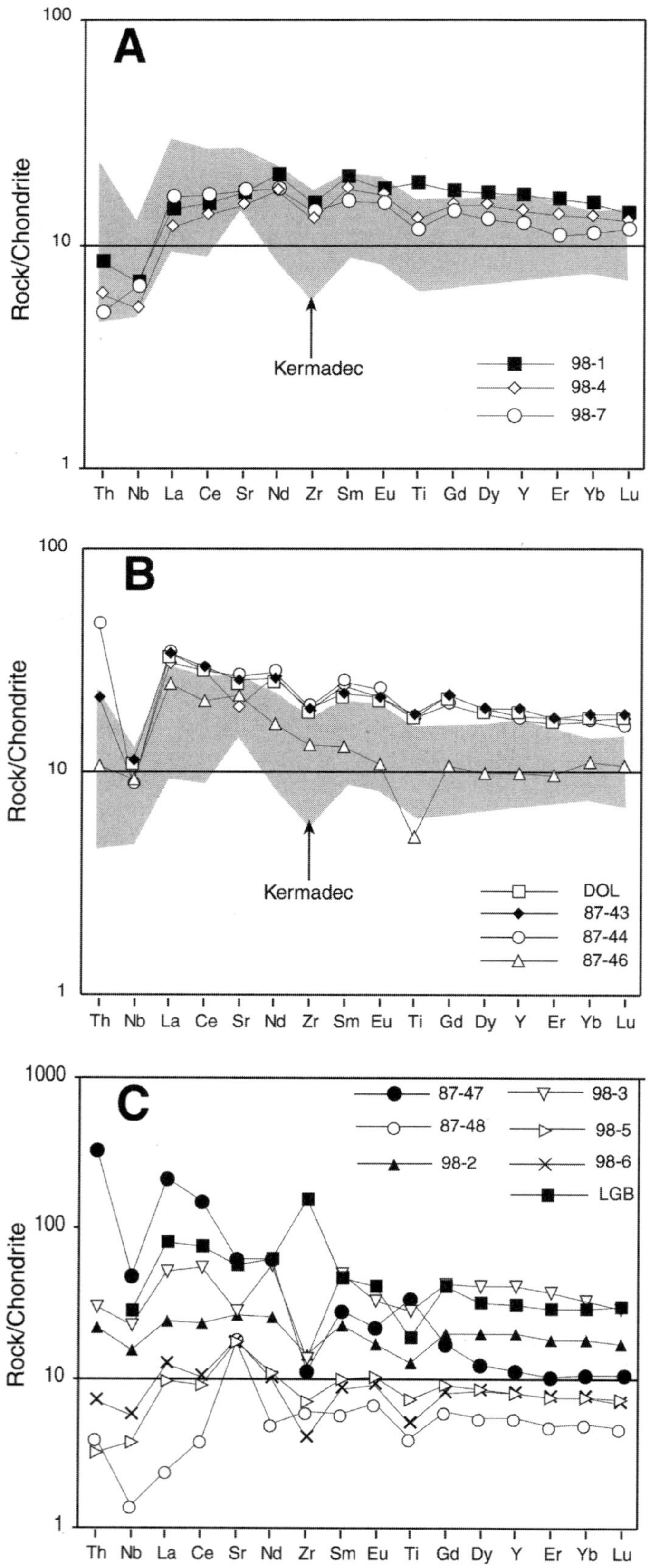

Figure 3. Multielemental chondrite-normalized diagrams (Nakamura, 1974, for REE; Sun, 1980, for other elements) for metabasites of Sierra de Careón ophiolite. A–C: Three groups distinguished (see text). Field for Kermadec rocks is after Gamble and Wright (1995) and Gamble et al. (1996).

DISCUSSION

Age of the ophiolitic unit

The U-Pb age of 395 ± 3 Ma obtained for the leucogabbro within serpentinized peridotite is in perfect agreement with previous U-Pb dating at 395 ± 2 Ma of zircons from metagabbro of the same unit (Díaz García et al., 1999a). This is slightly older than previous ages of amphibole obtained by $^{40}Ar/^{39}Ar$ methods on metabasites from equivalent ophiolitic units within the Morais and Cabo Ortegal Complexes. The amphiboles gave ages of 384 ± 5 and 392 ± 7 Ma in Morais (Dallmeyer and Gil Ibarguchi, 1990) and 389 ± 6 and 391 ± 7 Ma in Ortegal (Peucat et al., 1990). These relatively younger ages are considered representative of the closure age for Ar after the formation of amphiboles during pervasive prograde recrystallization processes that affected the ophiolitic protoliths. The latter ages are in turn consistently older than the $^{40}Ar/^{39}Ar$ 377 ± 0.4 Ma age of hornblende or 365 Ma whole-rock age from associated mylonitic rocks, which represent episodes of deformation and metamorphism related to the tectonic emplacement of the complexes over their autochthon (Dallmeyer et al., 1997). The combined U-Pb and Ar/Ar age data therefore imply that ophiolite generation occurred shortly before its final emplacement, in agreement with limited evidence for subduction of the ophiolitic protoliths.

Significance of geochemical data

The ratios of relatively immobile elements such as La/Nb, in most cases between 2 and 4, are much higher than those of N-MORBs (from 1.0 to 1.4), but similar to those observed in island-arc and backarc basalts (Saunders et al., 1980), or in mafic magmas that underwent large degrees of crustal contamination. Th/Nb ratios (from 0.11 to 0.4) are higher than those observed in all types of mid-ocean ridge or within-plate basalts (average value ~0.07; Sun, 1980). These large ratios point to a strong enrichment of Th and La (LILE) relative to Nb (HFSE), that may be considered as typical of subduction-related magmatism, provided that crustal contamination during magma ascent did not play a significant role (the absence of inherited lead component in analyzed zircons is a strong argument against significant crustal contamination). Moreover, the distinct depletion of the heavy REEs $(Sm/Lu)_N > 1$, observed throughout the metabasaltic samples, suggests that partial melts were in equilibrium with residual garnet, implying that magma generation occurred at a relatively deep level (>90 km; Robinson and Wood, 1998).

The doleritic samples (group 2) characterized by strong Th and light REE enrichment, and concomitant strong Nb anom-

TABLE 4. SM-ND ISOTOPE DATA OF METABASITES FROM THE SIERRA DE CAREÓN OPHIOLITE UNIT

Sample	Sm (ppm)	Nd (ppm)	$^{147}Sm/^{144}Nd$	$^{143}Nd/^{144}Nd$	εNd_t*
Group 1					
98-1	4.04	12.5	0.1959	0.513102 ± 8	9.1
98-4	3.64	11.4	0.1934	0.513100 ± 8	9.2
98-7	3.31	11.3	0.1768	0.513015 ± 8	8.4
Group 2					
DOL	5.02	17.36	0.1746	0.512961 ± 6	7.4
87-43	5.06	17.7	0.1729	0.512970 ± 7	7.7
87-44	5.15	18.1	0.1722	0.512966 ± 8	7.6
87-46	2.72	11.1	0.1481	0.512908 ± 10	7.7
Group 3					
LGB	10.4	42.7	0.1469	0.512860 ± 6	6.9
87-47	6	40.6	0.0892	0.512728 ± 8	7.2
87-48	1.16	3.1	0.2257	0.513126 ± 7	8.1
98-2	4.86	16.3	0.1805	0.512960 ± 8	7.1
98-3	11.3	37.2	0.1842	0.512984 ± 7	7.4
98-5	2.13	7.07	0.1823	0.513029 ± 9	8.4
98-6	2.19	7.5	0.1769	0.512915 ± 8	6.4

Note: εNd(t) at 395 Ma.

alies, display very homogeneous εNd_i, ~+7.6, close to the value expected for intraoceanic island-arc magmas at 400 Ma (+7.3, as calculated using the model of DePaolo, 1981). These highly positive values imply a long-term depletion of Nd relative to Sm in the mantle source of these basaltic melts, contrasting with the observed light REE enrichment. This decoupling between Nd isotope variation and trace element contents suggests that the source reservoir was enriched by the addition of a light REE-rich component shortly before magma generation. The εNd_i values calculated for the gabbroic samples range from +6.4 to +8.3, and encompass the isotope signature of the light REE–enriched dolerites.

The large increase of the degree of Th and light REE enrichment is not accompanied by a concomitant large drop of εNd_i values. This may be highlighted by comparing two samples with identical heavy REE abundances, but having contrasting patterns for the most incompatible trace elements (Fig. 4). Specifically, sample 87-44 (2.3 ppm Th, 11 ppm La, and 18 ppm Nd) has an εNd_i only 1.5 ε units lower (+7.6) than 98-1, with 0.4 ppm Th; 4.7 ppm La, 12 ppm Nd, and εNd = +9.1. This suggests that the additional Th-light REE–rich component, accounting for about one-third of Nd contained in 87-44, was characterized by a radiogenic Nd isotope composition (εNd ~+5). Therefore, the Th and light REE enrichment cannot reflect simple assimilation of old, isotopically evolved crustal materials, for which strongly negative εNd values (<–5) are typical (e.g., DePaolo, 1988).

Bearing in mind the current models for magma generation in subduction-zone settings (e.g., Mcdonald et al., 2000), it is suggested that hydrous, silicic partial melts from the altered igneous crust and, possibly, associated sediments, of a subducted slab were added to the overlying mantle wedge, which was thereby strongly enriched in Th and light REEs. In contrast, the heavy REEs and HFSEs (Zr, Nb) are considered as conservative (Pearce and Peate, 1995) or wedge-controlled (Davidson, 1996) elements, i.e., largely unaffected by the slab contribution. For this reason, their ratios (e.g., Zr/Nb) can be used to evaluate the degree of incompatible element depletion of the mantle wedge prior to the addition of the slab component. The Zr/Nb ratios observed in metadolerites (40–50) tend to be higher than those in N-MORBs (~30–40, Erlank and Kable, 1976), as commonly observed in arc magmas (Davidson, 1996), and suggest derivation

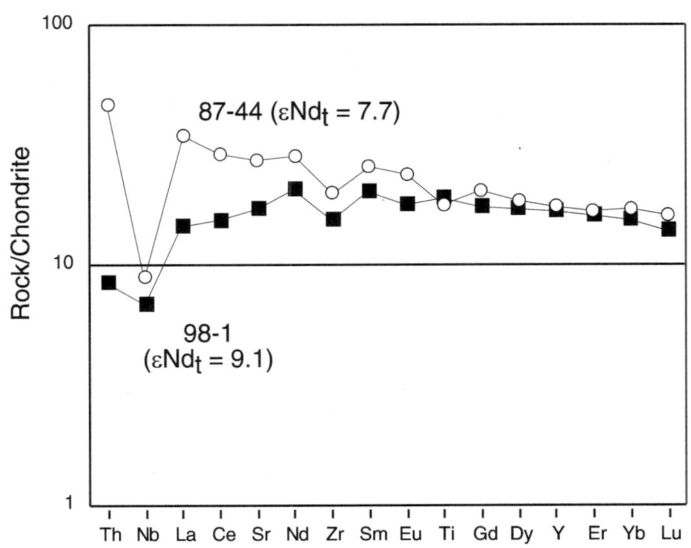

Figure 4. Multielemental chondrite-normalized diagrams (Nakamura, 1974, for REE; Sun, 1980, for other elements) for two samples.

from a mantle wedge depleted by earlier magma extraction. The addition of an H_2O-rich component, acting as a fluxing agent for mantle peridotites (e.g., Stolper and Newman, 1994), and rapid extension of the overlying plate, were probably necessary to trigger partial melting of such a depleted mantle source. The presence of amphibole among primary igneous minerals in metagabbros (Díaz García et al., 1999a) provides circumstantial evidence for H_2O-rich mafic melts.

PALEOTECTONIC SETTING

Any model aiming to infer the geodynamic significance of the Sierra de Careón unit should account for the following observations, pertaining to the timing and tectonic setting of formation of the ophiolitic rocks.

1. As commonly observed in ophiolites, igneous generation and tectonic emplacement, coeval with metamorphism, were separated by a relatively short time span (<20 m.y.), as can be deduced from the 377 Ma $^{39}Ar/^{40}Ar$ cooling age measured on hornblende in the same area (Dallmeyer et al., 1997). This may suggest that the igneous rocks of the Careón unit, albeit formed in an intraoceanic environment, were not generated far away from the passive continental margin on which, eventually, they were tectonically emplaced.

2. On the larger scale of the European Variscan Belt, the generation of the Sierra de Careón ophiolitic rocks provides evidence for a relatively late stage, and probably local, extensional episode, much younger than the major rifting and oceanic spreading event recorded ca. 490–500 Ma throughout the Variscides (e.g., Pin and Marini, 1993). In contrast, igneous generation in the Careón unit was broadly coeval with high-*P*–high-*T* metamorphism documented by eclogites and granulites in the neighboring Cabo Ortegal Complex (Peucat et al., 1990; Santos Zalduegui et al., 1996; Ordóñez et al., 2001) and in many other segments of the Variscides east of the Iberian Peninsula (Brittany, Massif Central, Northern Bohemian Massif; e.g., Pin and Vielzeuf, 1988).

3. The oceanic origin of the Sierra de Careón unit is clearly documented by the strong time-integrated light REE depletion of its mantle source, reflected by high εNd values, and by high Zr/Nb ratios. However, none of the studied samples has trace element characteristics similar to those of MORB. Instead, strong affinities with island-arc and backarc basin basalts are indicated by the most incompatible trace elements, especially Th-Nb-La relationships (Saunders et al., 1980; Saunders and Tarney, 1984), which point to the involvement of a Th- and light REE–rich geochemical component. The highly radiogenic Nd isotope composition rules out direct crustal contamination of basaltic magmas as a suitable mechanism for generating the Th and light REE enrichment. Rather, it is suggested that the enriched component originated from a subducting slab, probably as a silicic partial melt from hydrated oceanic crust, that metasomatized the overlying mantle wedge (e.g., Mcdonald et al., 2000, and references therein). These suprasubduction-zone affinities (Pearce et al., 1984) provide first-order constraints on the tectonic setting in which the Careón mafic-ultramafic complex was generated.

Interpretation of ophiolite emplacement involves a considerable variety of tectonic mechanisms, as discussed by Moores (1982, 1998, see his Fig. 1) and Michard et al. (1985), among others. Among these, models developed to explain the close temporal association of igneous generation and tectonic emplacement of ophiolites onto continental lithosphere basically fall into two categories. The first one involves failed subduction of young, thin, hot, and therefore buoyant oceanic lithosphere during ridge-trench interactions (e.g., Nicolas and Le Pichon, 1980; van den Beukel, 1990, and references therein), whereas the second one favors the collision between the forearc limb of an intraoceanic arc and the passive margin of a continent, during the final evolutionary stage of an intraoceanic subduction dipping away from the continent (e.g., Gealey, 1980; Edelman, 1988).

The first mechanism may operate when a ridge is oblique with respect to a trench, thereby leading to the subduction of progressively younger, lower strength oceanic lithosphere. This may cause the failure of the weak and buoyant subducted slab, and breakup and detachment of a thin oceanic crust sheet, which is transferred to the overriding plate. Upon final closure of the oceanic basin, this oceanic sheet is likely to be emplaced over the opposite continental margin, and eventually incorporated into an orogenic belt, as a result of a two-stage process. In this case, the preserved oceanic crust is created at a normal mid-ocean ridge, and is expected to be mostly composed of depleted (N-MORB) or variably enriched (E-MORB) magmas, similar to those occurring along present-day ridges.

In the second model, in contrast, creation of the ophiolite occurs at a spreading center located above a relatively long lasting subduction zone in the arc or backarc region. This suprasubduction-zone spreading is documented by geochemical signatures similar to those of arc (rather than mid-ocean ridge) magmas, but is not accompanied by the development of an arc if spreading rate is rapid relative to arc generation rate (Edelman, 1988). The generation of suprasubduction-zone ophiolite might occur in an oblique convergence setting, as the result of short spreading segments separated by transform faults above a subduction zone, as observed currently in the Andaman Sea (Moores et al., 1984). Alternatively, suprasubduction spreading might occur as a nonsteady-state tectonic process, during the earliest stage of plate collision. Building on earlier works of Dewey (1980), Le Pichon (1982), Leitch (1984), and Malinverno and Ryan (1986), Edelman (1988) proposed an elegant explanation for suprasubduction spreading as the result of the hinge retreat (or roll-back) of an intraoceanic subduction zone dipping away from a nearby continent (Fig. 5), when the convergence between the oceanic plate and the continental crust-bearing lithospheric plate stops. This situation, where the two plates are locked, may occur when two continental promontories collide (Le Pichon, 1982; Malinverno and Ryan, 1986), or when a salient of an irregular, passive continental margin enters

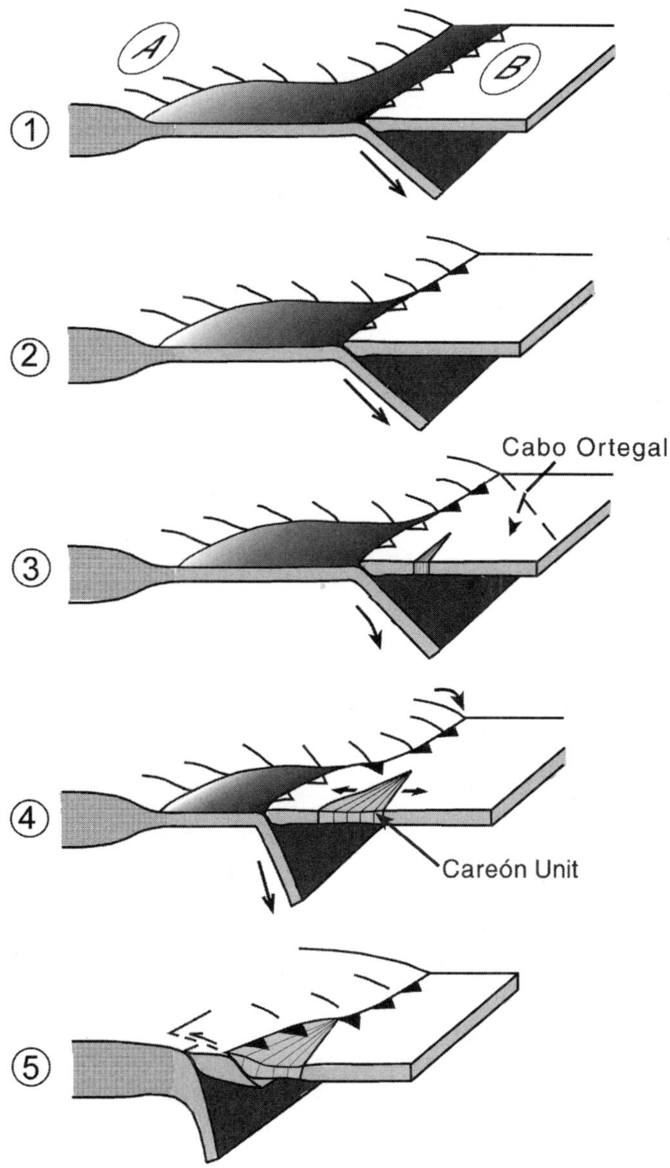

Figure 5. Schematic plate tectonic diagram depicting inferred mode of generation of local, suprasubduction-zone ophiolite related to subduction hinge retreat of old, gravitationally unstable, ocean lithosphere trapped in embayment of passive margin entering subduction zone along strike (adapted from Cawood and Suhr, 1992). 1: Oceanic segment of continent-bearing lithospheric plate A subducts (open barbs) beneath another oceanic plate B. Thereby, irregular passive continental margin approaches intraoceanic subduction zone. 2: Promontory of passive margin has reached subduction zone. This collision results in limited subduction of rifted margin (filled barbs), with concomitant high-pressure metamorphism, as documented in Cabo Ortegal Complex. This also causes strong reduction of convergence rate between plates A and B. 3, 4: continued sinking of negatively buoyant old oceanic lithosphere of plate A (roll-back) is accommodated by extension in overriding plate, reaching spreading stage with generation of ophiolite with suprasubduction-zone geochemical affinities, such as Careón unit. 5: By end of consumption of land-locked remnant oceanic basin of plate A, compressional regime is established in upper plate B; this causes detachment of newly formed suprasubduction-zone ophiolite in weak, high thermal gradient region, and initiates its obduction onto former rifted margin of plate A.

a subduction zone (Cawood and Suhr, 1992). Then, the amount of extension in the overriding plate has to balance the trench retreat, related to the vertical component of subduction. This retreat reflects the negative buoyancy of the subducting oceanic lithosphere (Elsasser, 1971), and is therefore likely to be more rapid when older, colder oceanic lithosphere is involved (Molnar and Atwater, 1978).

The marked suprasubduction-zone affinities displayed by the mafic rocks of the Careón unit, along with their very young igneous age relative to the beginning of collisional tectonics at a broader scale, favor the second, hinge-retreat model of ophiolite generation. More specifically, the Early Devonian history of northwestern Iberia is strongly reminiscent of the situation depicted by Cawood and Suhr (1992) in the northern Appalachians. In their model, the early impingement of a sinuous, passive continental margin with a subduction zone dipping away from the continent leads to collision restricted to salient domains. Although some subduction of the passive margin is likely to occur at this stage (Molnar and Gray, 1979), the convergence rate between the two lithospheric plates should be strongly reduced. By virtue of its gravitational instability, old oceanic lithosphere trapped into recesses of the irregular passive margin would still continue to sink almost vertically in the mantle (roll-back), thereby driving extension in the overlying, oceanic plate (Fig. 5) owing to the greater rate of subduction retreat relative to the plate convergence rate (cf. Dewey, 1980). Therefore, generation of localized, new oceanic crust with suprasubduction-zone affinity is predicted to occur in front of the reentrants of the passive margin (Cawood and Suhr, 1992), shortly after the inception of continental crust subduction along strike.

Applying this model to the Early Devonian geodynamic evolution of Galicia accounts for the close temporal and spatial association between suprasubduction-zone spreading, as documented in the Careón unit, and high-*P*–high-*T* metamorphism, as recorded along strike in the Cabo Ortegal Complex. A ca. 390–400 Ma event is well established at Cabo Ortegal (Peucat et al., 1990; Santos Zalduegui et al., 1996; Ordóñez, 1998). Based on the elevated pressures (15–20 kbar) reached in rocks of both oceanic and attenuated continental crust derivation throughout the Variscides, this event can be interpreted in terms of partial subduction of former rifted continental margins (Pin and Vielzeuf, 1983; Pin and Vielzeuf, 1988). Because continent-continent collision did not occur at that time, as evidenced by the lack of contemporaneous synorogenic sediments, this subduction is interpreted to have occurred beneath oceanic lithosphere (Pin, 1990), in a context of arc-continent collision.

The tectonic emplacement of the young, thin, and still buoyant Careón ophiolitic unit on the adjacent passive continental margin probably occurred when the sinking of the remnant, land-locked oceanic basin was completed, leading to a change from extension to compression in the upper plate (Malinverno and Ryan, 1986; Cawood and Suhr, 1992). The eventual detachment of a thin overthrust sheet of oceanic material is most likely to have occurred in the region of steep geothermal gradient corresponding to the boundary of the newly formed suprasubduction oceanic crust (Armstrong and Dick, 1974). The 377 Ma $^{40}Ar/^{39}Ar$ result published by Dallmeyer et al. (1997) places a younger age limit to this event.

TECTONIC EMPLACEMENT OF A SUPRASUBDUCTION-ZONE OPHIOLITE

While there exists general agreement as to the fact that tangential tectonics dominates the structural style of northwestern Iberia, the recognition and interpretation of observed structures vary among different authors and for each allochthonous complex in this realm. Bastida et al. (1984) proposed that large recumbent folds formed in Cabo Ortegal during a second deformation episode, whereas Azcárraga (2000) related the same structures to the first deformation phase, and recognized discrete, low-angle, amphibolite facies shear zones as D2 thrust structures. In Ordenes, pervasive amphibolite facies ductile deformation in several units, including the Careón ophiolite, has instead been interpreted in relation to extensional deformation during orogenic collapse after lithospheric thickening (Díaz García et al., 1999b). Likewise, stretching lineations interpreted as markers of tectonic transport direction can vary among the complexes, e.g., N160°E trending and south-southeast facing in Bragança and Morais (Anthonioz, 1972; Marques et al., 1992), N100°–120°E trending and east facing in eastern Ordenes (Díaz García et al., 1999a and 1999b), and N0°–30°E trending and north-northeast facing in western Ordenes, Cabo Ortegal, and Malpica-Tuy (Van Zuuren, 1969; Gil Ibarguchi and Ortega Gironés, 1985; Girardeau and Gil Ibarguchi, 1991; Gil Ibarguchi et al., 2000). Consequently, when the ductile strain recorded by both the nappe units and their contacts is considered, a puzzling kinematic picture emerges that has been interpreted by some in the context of large orogen-parallel tectonic displacements due to oblique plate convergence (Vauchez and Nicolas, 1991; Shelley and Bossière, 2000).

Regarding the Careón ophiolite and adjacent units, from east to west the geodynamic configuration proposed here should involve at least: a continent, an old ocean, an intraoceanic subduction zone, and a suprasubduction zone, newly formed basin. This should be reflected in the geology of northwestern Iberia by a superposition of ophiolite units onto subduction-zone units (the rocks from the upper allochthon). Because this is not the case, one must invoke either a second subduction zone to the west or oblique collision processes. The latter possibility is here proposed as a necessary mechanism for the ophiolite emplacement, because while maintaining a unique subduction zone, such a process would allow us to vertically superpose geological units in an order not necessarily coinciding with the geographical order of geotectonic facies belts.

CONCLUSIONS

The results of this geochemical study (combining immobile trace elements and Nd isotopes) support the close relationship between the Sierra de Careón ophiolitic unit and suprasubduction-zone paleosettings, as suggested by Díaz García et al. (1999a, p. 600). Lithospheric extension-related ophiolite generation, likely above a subduction zone, was coeval with high-*P*–high-*T* metamorphism dated as ca. 395 Ma in neighboring domains (Cabo Ortegal). High-*P*–high-*T* metamorphism is thought to be the result of arrested early Paleozoic continental passive margin subduction. This relationship lends support to a tectonic model connecting ophiolite genesis to subduction hinge retreat in an old, land-locked oceanic basin trapped in a continental embayment (Cawood and Suhr, 1992). In this regard, the Sierra de Careón ophiolite would document early (premature?) transient stages of collision between oceanic and continental lithospheres in the western European Variscides during the Early Devonian.

ACKNOWLEDGMENTS

Financial support by Spanish DGES PB97/617 and DGES PB98/143 and French-Spanish cooperation “Picasso” (1997, 1998) grants is acknowledged. This manuscript is a contribution to International Geological Correlation Programme 453. We thank B. Abalos for helpful discussions on the structural and petrological features of the study area and J. Fernández and A. Kerr for thorough and thoughtful reviews.

APPENDIX 1. ANALYTICAL TECHNIQUES

Mineral analyses

Minerals from 6 thin sections of representative samples of metabasite from the Careón unit of Ordenes were analyzed with a SX100 microprobe at the Université Clermont-Ferrand. Operating conditions were 10 s counting time (peak), ~10 nA beam current, and 15 kV accelerating voltage. Calibration was against BRGM (Bureau de Recherches Géologiques et Minières, French Geological Survey) standard minerals, and the ZAF (atomic number, absorption, and fluorescence) correction procedure was used. The average detection limit is 0.1% at 2σ confidence limit.

Whole-rock major and trace element analysis

Major (wt%) and trace element (µg/g) concentrations were measured by simultaneous ICP-AES (inductively coupled plasma–atomic-emission spectrometry; major elements) and/or ICP-MS (mass spectrometry; trace elements). Total iron is given as Fe_2O_3. Analytical uncertainty for trace elements is estimated as 10% or better (see Cantagrel and Pin, 1994; Pin and Joannon, 1997, for additional details).

Isotopic analysis

All isotope analyses were performed at the Centre National de la Recherche Scientifique, Clermont-Ferrand. U-Pb analyses were performed on the least magnetic (2° forward and side tilt at 2.2 A using a Frantz Isodynamic magnetic barrier separator) mechanically abraded and crack-free zircon grains. Zircon dissolution, chemical separation of U and Pb, and isotope analyses were carried out according to methods described by Paquette et al. (1997). Blank analyses were in the region of 5 pg for Pb and 0.5 pg for U. The U and Pb isotopes were analyzed on a Fisons VG Sector 54-30 mass spectrometer in multicollector static mode. Initial common Pb is determined for each fraction using the Stacey and Kramers (1975) two-step model. Data errors (2σ) of the zircon fractions and discordia lines were calculated using the PBDAT 1.24 and ISOPLOT 2.71 programs (Ludwig, 1993, 1994).

We used ~0.1 g of whole rock, spiked with ^{150}Nd–^{149}Sm mixed tracer solution, were used for Sm-Nd analyses. Sample dissolution, chemical separation of Sm and Nd, and isotope analyses were performed following methods described by Pin and Santos Zalduegui (1997). Blank analyses for Sm and Nd were estimated as ~0.1 ng. The isotopic analyses were performed with a fully automated VG ISOMASS 54E mass spectrometer in double collection mode.

REFERENCES CITED

Abati, J., Dunning, G.R., Arenas, R., Díaz García, F., González Cuadra, P., Martínez Catalán, J.R.M., and Andonaegui, P., 1999, Early Ordovician orogenic event in Galicia (NW Spain): Evidence from U-Pb ages in the uppermost unit of the Ordenes complex: Earth and Planetary Science Letters, v. 165, p. 213–228.

Ague, J.J., 1997, Thermodynamic calculation of emplacement pressures for batholithic rocks, California: Implications for the aluminum-in-hornblende barometer: Geology, v. 25, p. 563–566.

Anderson, J.L., and Smith, D.R., 1995, The effects of temperature and fO_2 on the Al-in hornblende barometer: American Mineralogist, v. 80, p. 540–559.

Anthonioz, P.M., 1972, Les complexes polymétamorphiques Precambriens de Morais et Bragança (NE Portugal): Étude pétrographique et structurelle: Memoria no. 20 Serviços Geológicos de Portugal, 192 p.

Arenas, R., 1991, Opposite *P, T, t*, paths of Hercynian metamorphism between the upper units of the Cabo Ortegal complex and their substratum (northwest of the Iberian Massif): Tectonophysics, v. 191, p. 347–364.

Arenas, R., Gil Ibarguchi, J.I., González Lodeiro, F., Klein, E., Martínez Catalán, J.R., Ortega Gironés, E., de Pablo Macia, J.G., and Peinado, M., 1986, Tectonostratigraphic units in the complexes with mafic and related rocks of the NW of the Iberian Massif: Hercynica, v. II (2), p. 87–110.

Armstrong, R.L., and Dick, H.B., 1974, A model for the development of thin overthrust sheets of crystalline rock: Geology, v. 2, p. 35–40.

Azcárraga, J., 2000, Evolución tectónica y metamórfica de los mantos inferiores de grado alto y alta presión del Complejo de Cabo Ortegal: Edicios do Castro, La Coruña, Serie Nova Terra, v. 17, 306 p.

Bastida, F., Marcos, A., Marquínez, J., Martínez Catalán, J.R., Pérez Estaún, A., and Pulgar, J.A., 1984, Mapa y Memoria de la Hoja no. 1 (La Coruña) del Mapa Geológico de España a escala 1:200,000: Instituto Geológico y Minero de España, Madrid.

Blundy, J.D., and Holland, T.J.B., 1990, Calcic amphibole equilibria and a new amphibole-plagioclase geothermometer: Contributions to Mineralogy and Petrology, v. 104, p. 208–224.

Cantagrel, F., and Pin, C., 1994, Major, minor and rare-earth element determinations in 25 rocks standards by ICP-atomic emission spectrometry: Geostandards Newsletter, v. 18, p. 123–138.

Cawood, P.A., and Suhr, G., 1992, Generation and obduction of ophiolites: constraints from the Bay of Islands complex, western Newfoundland: Tectonics, v. 11, p. 884–897.

Dale, J., Holland, T., and Powell, R., 2000, Hornblende-garnet-plagioclase thermobarometry: a natural assemblage calibration of the thermodynamics of hornblende: Contributions to Mineralogy and Petrology, v. 140, p. 353–362.

Dallmeyer R.D., and Gil Ibarguchi, J.I., 1990, Age of amphibolitic metamorphism in the ophiolitic unit of the Morais allochthon (Portugal): Implications for early Hercynian orogenesis in the Iberian Massif: Journal of the Geological Society, London, v. 147, p. 873–878.

Dallmeyer, R.D. Ribeiro, A., and Marques, F., 1991, Polyphase Variscan emplacement of exotic terranes (Morais and Bragança Massifs) onto Iberian successions: Evidence from ^{40}Ar/^{39}Ar mineral ages: Lithos, v. 27, p. 133–144.

Dallmeyer, R.D., Martínez Catalán, J.R., Arenas, R., Gil Ibarguchi, J.I., Alonso, G.G., Farias, P., Bastida, F., and Aller, J., 1997, Diachronous Variscan tectonothermal activity in the NW Iberian Massif: Evidence from ^{40}Ar/^{39}Ar dating of regional fabrics: Tectonophysics, v. 277, p. 307–337.

Davidson, J.P., 1996, Deciphering mantle and crustal signatures in subduction zone magmatism, *in* Subduction: Top to bottom: American Geophysical Monograph 96, p. 251–262.

De Bruiyn, H., Van Der Westhuizen, W.A., and Schoch, A.E., 1983, The estimation of FeO, F and H_2O^+ by regression in microprobe analyses of natural biotite: Journal of Trace Microanalysis Techniques, v. 1, p. 399–413.

DePaolo, D.J., 1981, Neodymium isotopes in the Colorado Front Range and crust-mantle evolution in the Proterozoic: Nature, v. 291, p. 193–196.

DePaolo, D.J., 1988, Neodymium isotope geochemistry: An introduction: Berlin, Springer-Verlag, 187 p.

Dewey, J.F., 1980, Episodicity, sequence, and style at convergent plate boundaries, *in* Strangway, D.W., ed., The Continental Crust and its mineral deposits: Toronto, Ontario, Geological Association of Canada Special Paper 20, p. 553–573.

Díaz García, F., 1990, La geología del sector occidental del Complejo de Ordenes (Cordillera Hercínica, NW de España): Edicios do Castro, La Coruña, Serie Nova Terra, v. 3, 230 p.

Díaz García, F., Arenas, R., Martínez Catalán, J.R., González del Tanago, J., and Dunning, G., 1999a, Tectonic evolution of the Careón Ophiolite (northwest Spain): A remnant of oceanic lithosphere in the Variscan Belt: Journal of Geology, v. 107, p. 587–605.

Díaz García, F., Martínez Catalán, J.R., Arenas, R., and González Cuadra, P.G., 1999b, Structural and kinematic analysis of the Corredoiras detachment: evidence for early Variscan synconvergent extension in the Ordenes complex, NW Spain: International Journal of Earth Sciences, v. 88, p. 337–351.

Droop, G.T.R., 1987, A general equation for estimating Fe^{3+} concentrations in ferromagnesian silicates and oxides from microprobe analyses, using stoichiometric criteria: Mineralogical Magazine, v. 51, p. 431–435.

Edelman, S.H., 1988, Ophiolite generation and emplacement by rapid subduction hinge retreat on a continent-bearing plate: Geology, v. 16, p. 311–313.

Elsasser, W.M., 1971, Sea-floor spreading as thermal convection: Journal of Geophysical Research, v. 76, p. 1101–1112.

Engels, J.P., 1972, The catazonal poly-metamorphic rocks of Cabo Ortegal (NW Spain), a structural and petrographic study: Leidse Geologische Mededelingen, v. 48, p. 83–133.

Erlank, A.J., and Kable, E.J.D., 1976, The significance of incompatible elements in Mid-Atlantic Ridge basalts from 45°N, with particular reference to Zr/Nb: Contributions to Mineralogy and Petrology, v. 54, p. 281–291.

Gamble, J.P., and Wright, I., 1995, The Southern Havre Trough: Geological structure and magma petrogenesis of an active backarc rift complex, *in* Taylor, B., ed., Backarc basins: Tectonics and magmatism: New York, Plenum Press, p. 29–62.

Gamble, J.P., Woodhead, J., Wright, I., and Smith, I., 1996, Basalt and sediment geochemistry and magma petrogenesis in a transect from oceanic island arc to rifted continental margin arc: the Kermadec-Hikurangi margin, SW Pacific: Journal of Petrology, v. 37, p. 1523–1546.

Gealey, W.K., 1980, Ophiolite obduction mechanism, *in* Panayiotou, A., ed., Ophiolites, Proceedings of the Ophiolite Symposium, Cyprus 1979: Cyprus, Geological Survey Department, p. 228–243.

Gil Ibarguchi, J.I., and Ortega Gironés, E., 1985, Petrology, structure and geotectonic implications of glaucophane-bearing eclogites and related rocks from Malpica-Tuy (MT) Unit, Galicia, Northwest Spain: Chemical Geology, v. 50, p. 145–162.

Gil Ibarguchi, J.I., Mendia, M., Girardeau, J., and Peucat, J.J., 1990, Petrology of the eclogites and the clinopyroxene-garnet metabasites from the Cabo Ortegal complex (northwest Spain): Lithos, v. 25, p. 133–162.

Gil Ibarguchi, J.I.,Ábalos, B., Azcárraga, J., Mendia, M., and Puelles, P., 2000, A petrological and structural excursion through the high-grade/high-pressure allochthonous units of the Cabo Ortegal Complex (NW Spain): Basement Tectonics 15 (Galicia 2000, 15th International Conference on Basement Tectonics), Mid-Conference Field Trip Guide, La Coruña, Spain, 59 p., 1 sheet.

Girardeau, J., and Gil Ibarguchi, J.I., 1991, Pyroxenite-rich peridotites of the Cabo Ortegal Complex (Northwestern Spain): Evidence for large-scale upper-mantle heterogeneity: Journal of Petrology, Special Lherzolites Issue, p. 135–154.

Holland, T.J.B., and Blundy, J.D., 1994, Non-ideal interactions in calcic amphiboles and their bearing on amphibole-plagioclase thermometry: Contributions to Mineralogy and Petrology, v. 116, p. 433–447.

Krogh, T.E., 1973, A low-contamination method for hydrothermal decomposition of zircon and extraction of U and Pb for isotopic age determinations: Geochimica et Cosmochimica Acta, v. 37, p. 485–494.

Krogh, T.E., 1982, Improved accuracy of U-Pb zircon ages by the creation of more concordant systems using an air abrasion technique: Geochimica et Cosmochimica Acta, v. 46, p. 637–649.

Laird, J., and Albee, A., 1981, Pressure, temperature, and time indicators in mafic schists: Their application to reconstructing the polymetamorphic history of Vermont: American Journal of Science, v. 281, p. 127–175.

Leitch, E.C., 1984, Island arc elements and arc-related ophiolites: Tectonophysics, v. 106, p. 177–203.

Le Pichon, X., 1982, Land-locked oceanic basins and continental collision: the Eastern Mediterranean as a case example, *in* Hsü, K.J., ed., Mountain building processes: London, Academic Press, p. 201–211.

Ludwig, K.R., 1993, Pbdat: A computer program for processing Pb-U-Th isotope data, version 1.24: U.S. Geological Survey, Open-File Report 88-542, 32 p.

Ludwig, K.R., 1994, Isoplot: A plotting and regression program for radiogenic isotope data, version 2.71: U.S. Geological Survey, Open-File Report 91-445, 44 p.

Malinverno, A., and Ryan, W.B.F., 1986, Extension in the Tyrrhenian Sea and shortening in the Apennines as result of arc migration driven by sinking of the lithosphere: Tectonics, v. 5, p. 227–245.

Marques, F.G., Ribeiro, A., and Pereira, E., 1992, Tectonic evolution of the deep crust: Variscan reactivation by extension and thrusting of Precambrian basement in the Bragança and Morais Massifs (Trás-os-Montes, NE Portuagal): Geodinamica Acta, v. 5, p. 135–151.

Martínez Catalán, J.R., Klein, E., Pablo Maciá, J.G. de, and González Lodeiro, F., 1984, El complejo de Ordenes: Subdivisión, descripción y discusión sobre su origen: Cuadernos do Laboratorio Xeolóxico de Laxe, v. 7, p. 139, 210

Martínez Catalán, J.R., Arenas, R., Díaz Garcia, F., and Abati, J., 1997, Variscan accretionary complex of northwest Iberia: Terrane correlation and succession of tectonothermal events: Geology, v. 25, p. 1103–1106.

Massonne, H.J., 1992, Thermochemical determination of water activities relevant to eclogitic rocks, *in* Kharaka, Y.K., and Maest, A.S., eds., Water-Rock Interaction: Rotterdam, Balkema, p. 1523–1526

Mcdonald, R., Hawkesworth, C.J., and Heath, E., 2000, The Lesser Antilles volcanic chain: a study in arc magmatism: Earth Science Reviews, v. 49, p. 1–76.

Michard, A., Juteau, T., and Whitechurch, H., 1985, L'obduction: Revue des modèles et confrontation au cas de l'Oman: Bulletin de la Société Géologique de France, v. 1, p. 189–198.

Middleburg, J.A., Van der Weijden, C.H., and Woittiez, J.R.W., 1988, Chemical processes affecting the mobility of major, minor, and trace elements during weathering of granitic rocks: Chemical Geology, v. 68, p. 253–273.

Molnar, P., and Atwater, T., 1978, Interarc spreading and Cordilleran tectonics as alternates related to the age of subducted oceanic lithosphere: Earth and Planetary Science Letters, v. 41, p. 330–340.

Molnar, P., and Gray, D., 1979, Subduction of continental lithosphere: Some constraints and uncertainties: Geology, v. 7, p. 58–62.

Moores, E.M., 1982, Origin and emplacement of ophiolites: Review of Geophysics and Space Physics, v. 20, p. 735–760.

Moores, E.M., 1998, Ophiolites, the Sierra Nevada, "Cordilleria," and orogeny along the Pacific and Caribbean margins of North and South America: International Geology Review, v. 40, p. 40–54.

Moores, E.M., Robinson, P.T., Malpas, J., and Xenophontos, C., 1984, Model for the origin of the Troodos massif, Cyprus, and other mideast ophiolites: Geology, v. 12, p. 500–503.

Nakamura, N., 1974, Determination of REE, Ba, Fe, Mg, Na, and K in carbonaceous and ordinary chondrites: Geochimica et Cosmochimica Acta, v. 38, p. 757–775.

Nicolas, A., and Le Pichon, X., 1980, Thrusting of young lithosphere in subduction zones with special reference to structures in ophiolitic peridotites: Earth and Planetary Science Letters, v. 46, p. 397–406.

Nicollet, C., 1986, Saphirine et staurotide riche en magnésium et chrome dans les amphibolites et anorthosites à corindon du Vohibory Sud, Madagascar: Bulletin de Minéralogie, v. 109, p. 599–612.

Ordóñez Casado, B., Gebauer, D., Schäfer, H.J., Gil Ibarguchi, J.I., and Peucat, J.J., 2001, A single Devonian subduction event for the HP/HT metamorphism of the Cabo Ortegal complex within the Iberian Massif: Tectonophysics, v. 332, p. 359–385.

Pearce, J.A., 1975, Basalt geochemistry used to investigate past tectonic environments on Cyprus: Tectonophysics, v. 25, p. 41–67.

Pearce, J.A., 1987, An expert system for the tectonic characterization of ancient volcanic rocks: Journal of Volcanology and Geothermal Research, v. 32, p. 51–65.

Pearce, J.A., Lippard, S.J., and Roberts, S., 1984, Characteristics and tectonic significance of supra-subduction zone ophiolites, *in* Kokelaar, B.P., and Howells, M.F., eds., Marginal basin geology: London, Geological Society of London Special Publication 16, p. 77–94.

Pearce, J.A., and Peate, D.W., 1995, Tectonic implications of the composition of volcanic arc magmas: Annual Review of Earth and Planetary Sciences, v. 23, p. 251–285.

Peucat, J.J., Bernard-Griffiths, J., Gil Ibarguchi, J.I., Dallmeyer, R.D., Menot, R.P., Cornichet, J., and Iglesias Ponce de León, M., 1990, Geochemical and geochronological cross section of the deep Variscan crust: The Cabo Ortegal high-pressure nappe (northwestern Spain): Tectonophysics, v. 177, p. 263–292.

Paquette, J.L., Gleizes, G., Leblanc, D., and Bouchez, J.L., 1997, Le granite de Bassiés (Pyrénées): Un pluton syntectonique d'âge Westphalien. Géochronologie U–Pb sur zircons: Comptes Rendus de l'Académie des Sciences, Paris, v. 324, p. 387–392.

Pin, C., 1990, Variscan oceans: Ages, origins and geodynamic implications inferred from geochemical and radiometric data: Tectonophysics, v. 177, p. 215–227.

Pin, C., and Joannon, S., 1997, Low-level analysis of lanthanides in eleven silicate rock reference materials by ICP-MS after group separation using cation-exchange chromatography: Geostandrads Newsletter, v. 21, p. 43–50.

Pin, C., and Marini, F., 1993, Early Ordovician continental break-up in Variscan Europe: Nd-Sr isotope and trace element evidence from bimodal igneous associations of the Southern Massif Central, France: Lithos, v. 29, p. 177–196.

Pin, C., and Santos Zalduegui, J.F., 1997, Sequential separation of light rare-earth elements, thorium and uranium by miniaturized extraction chromatography: Application to isotopic analyses of silicate rocks: Analytica Chimica Acta, v. 339, p. 79–89.

Pin, C., and Vielzeuf, D., 1988, Granulites de haute-pression d'Europe moyenne témoins d'une subduction éo-hercynienne: Implications sur l'origine des groupes leptyno-amphiboliques: Bulletin de la Société Géologique de France, v. 8, no. IV, p. 13–20.

Raase, P., 1974, Al and Ti contents of hornblende, indicators of pressure and temperature of regional metamorphism: Contributions to Mineralogy and Petrology, v. 45, p. 231–236.

Ribeiro, A., 1974, Contribution à l'étude tectonique de Trás-os-Montes Oriental: Serviços Geológicos de Portugal, v. 24 (Nova Serie), 168 p.

Ribeiro, A., Pereira, E., and Dias, R., 1990a, Structure of Centro-Iberian Allochthon in Northern Portugal, *in* Dallmeyer, R.D., and Martínez, E., eds., Pre-Mesozoic geology of Iberia: Berlin, Springer-Verlag, p. 220–236.

Ribeiro, A., Quesada, C., and Dallmeyer, R.D., 1990b, Geodynamic evolution of the Iberian Massif, *in* Dallmeyer, R.D., and Martínez, E., eds., Pre-Mesozoic geology of Iberia: Berlin, Springer-Verlag, p. 399–409.

Robinson, J.A.C., and Wood, B.J., 1998, The depth of the spinel to garnet transition at the peridotite solidus: Earth and Planetary Science Letters, v. 164, p. 277–284.

Rodríguez Aller, J., Cosca, M., Gil Ibarguchi, J.I., and Dallmeyer, R.D., 1997, Eo-Hercynian HP metamorphism of a subducted continental crust (Malpica-Tui allochthon, NW Spain): New petrological and age constraints: Fifth International Eclogite Conference, Ascona, Terra Nova, v. 9, p. 29.

Santos Zalduegui, J.F., Schärer, U., Gil Ibarguchi, J.I., and Girardeau, J., 1996, Origin and evolution of the Paleozoic Cabo Ortegal ultramafic-mafic complex (NW Spain): U-Pb, Rb-Sr and Pb-Pb data: Chemical Geology, v. 129, p. 281–306.

Saunders, A.D., Tarney, J., Marsh, N.G., and Wood, D.A., 1980, Ophiolites as ocean crust or marginal basin crust: A geochemical approach, *in* Panayiotou, A., ed., Ophiolites, Proceedings of the Ophiolite Symposium, Cyprus 1979: Cyprus, Geological Survey Department, p. 193–204.

Saunders, A.D., and Tarney, J., 1984, Geochemical characteristics of basaltic volcanism within back-arc basins, *in* Kokelaar, B.P., and Howells, M.F., eds., Marginal basin geology: London, Geological Society of London Special Publication 16, p. 59–76.

Shelley, D., and Bossière, G., 2000, A new model for the Hercynian Orogen of Gondwanan France and Iberia: Journal of Structural Geology, v. 22, p. 757-776.

Shervais, J.W., 1982, Ti-V plots and the petrogenesis of modern and ophiolitic lavas: Earth and Planetary Science Letters, v. 59, p. 101–118.

Stacey, J.S., and Kramer, J.D., 1975, Approximation of terrestrial lead isotope evolution by a two-stage model: Earth and Planetary Science Letters, v. 26, p. 207–221.

Steiger, R.H., and Jäger, E., 1977, Subcommission on geochronology: Convention on the use of decay constants in geo- and cosmochronology: Earth and Planetary Science Letters, v. 36, p. 359–362.

Stolper, E., and Newman, S., 1994, The role of water in the petrogenesis of Mariana trough basalts: Earth and Planetary Science Letters, v. 121, p. 293–325.

Sturchio, N.C., Muehlenbachs, K., and Serlz, M.G., 1986, Element redistribution during hydrothermal alteration of rhyolite in active geothermal system: Yellowstone drill cores geothermal system: Yellowstone drill cores Y-7 and Y-8: Geochimica et Cosmochimica Acta, v. 50, p. 1619–1631

Sun, S.S., 1980, Lead isotopic study of young volcanic rocks from mid-ocean ridges, ocean islands and island arcs: Philosophical Transactions of the Royal Society of London, v. A297, p. 409–445.

van den Beukel, J., 1990, Break-up of young oceanic lithosphere in the upper part of a subduction zone: Implications for the emplacement of ophiolites: Tectonics, v. 9, p. 825–844.

Van Zuuren, A., 1969, Structural petrology of an area near Santiago de Compostela (NW Spain): Leidse Geologische Mededelingen, v. 45, p. 1–71.

Vauchez, A., and Nicolas, A., 1991, Mountain building: Strike-parallel motion and mantle anisotropy: Tectonophysics, v. 185, p. 183–201.

Vielzeuf, D., and Pin, C., 1989, Geodynamic implications of granulitic rocks in the Hercynian belt, *in* Daly, J.S., Cliff, R.A., and Yardley, B.W.D., eds., Evolution of metamorphic belts: London, Geological Society of London Special Publication 43, p. 343–348.

Zindler, A., 1982, Nd and Sr isotopic studies of komatiites and related rocks, *in* Arndt, N.T., and Nisbet, E., eds., Komatiites: Boston, Massachusetts, Allen & Unwin, 399 p.

Manuscript Accepted by the Society February 8, 2002

Printed in the U.S.A.

Geological Society of America
Special Paper 364
2002

Prograde development of corona textures in metagabbros of the Sobrado unit (Ordenes Complex, northwestern Iberian Massif)

Ricardo Arenas
Departamento de Petrología y Geoquímica, Universidad Complutense, 28040 Madrid, Spain
José R. Martínez Catalán
Departamento de Geología, Universidad de Salamanca, 37008 Salamanca, Spain

ABSTRACT

A high-pressure and high-temperature unit exposed in the core of the Sobrado tectonic window (Ordenes Complex, northwestern Iberian Massif) contains a lithological ensemble, of continental and/or arc affinity, affected by Early Ordovician metamorphism; the unit reached highest conditions in the high-pressure granulite facies and in the eclogite facies. Alternating with metasediments, map-scale bands of metabasites include a large number of metagabbro bodies ranging from several meters to 2 km in length. The metagabbros were variably affected by the regional metamorphism. Well-preserved gabbros with almost intact igneous textures and mineralogy can be found, as well as metagabbros showing diverse reactional coronitic textures, and highly recrystallized types with no preserved igneous features and entirely transformed into granoblastic rocks. The variety of textures and assemblages in the coronas is interpreted to reflect the prograde reaction history of the metagabbros.

Three groups of coronas have been distinguished: (I) Amp + Opx; (II) (Amp) + (Cpx) + Opx + Grt; (III) (Amp) + Cpx + Opx + Grt (subordinate phases in parentheses). The development and evolution of these coronas follow a complex prograde reaction history, the general characteristics of which are described. Detailed thermobarometry, carried out in groups II and III coronas, together with inferred pressure-temperature conditions for group I coronas, show that the corona textures were generated during a drastic and virtually isothermal increase in pressure. Highest pressure-temperature conditions, registered in group III coronas, reached 660–770 °C and 13–17 kbar. Subduction of a relatively thin slice of continental or arc-related crust below a mantle wedge, either with an intact thermal structure or heated and thermally softened in a relatively inner part of a volcanic arc, is suggested as the probable dynamic setting to explain the pressure-temperature path found in the metagabbros.

INTRODUCTION

The development of corona textures during a complex reaction history has been described in metagabbros exposed in orogenic belts of various ages. Thus, different examples and interpretations have been presented in Precambrian (Indares, 1993), Paleozoic (Griffin and Heier, 1973; Lasnier, 1977), or Alpine (Gómez Pugnaire et al., 1979; Chinner and Dixon, 1973; Meyer, 1983) belts. The mineral assemblages included in the coronas may be complex, although frequently indicative of high-grade conditions. The development and evolution of these assemblages are related to pressure-temperature (*P-T*) changes that destabilize the igneous mineralogy in the gabbro and cause reactions between plagioclase and ferromagnesian phases.

Arenas, R., and Martínez Catalán, J.R., 2002, Prograde development of corona textures in metagabbros of the Sobrado unit (Ordenes Complex, northwestern Iberian Massif), *in* Martínez Catalán, J.R., Hatcher, R.D., Jr., Arenas, R., and Díaz García, F., eds., Variscan-Appalachian dynamics: The building of the late Paleozoic basement: Boulder, Colorado, Geological Society of America Special Paper 364, p. 73–88.

Some preliminary interpretations that related corona development and the cooling history of the gabbro body (Griffin and Heier, 1973) have been revised in order to link corona growth with the regional metamorphic evolution (Mørk, 1986; Meyer, 1983). Furthermore, in many of the described examples, the corona mineral assemblage was used to provide a punctual *P-T* information, generally related to the thermal peak attained in the region (Indares, 1993; Indares and Rivers, 1995). However, corona examples, the reaction histories of which were used to obtain a regional pressure-temperature-time *(P-T-t)* path, have not been described. This probably results from the difficulty of recognizing different *P-T* stages in a single metagabbro, due to the common high-*T* character of the coronas that tends to destroy evidence of the early metamorphic stages (Arenas et al., 1995, 1997). Moreover, it is also difficult to apply thermobarometry to the corona assemblages.

This contribution describes the coronitic types and the associated reaction history of metagabbros of the Sobrado unit (Ordenes Complex, northwestern Iberian Massif). These metagabbros were affected by Early Ordovician metamorphism, which developed a diversity of coronitic types. Textural, chemical, and thermobarometric studies of the mineral assemblages characteristic of the corona types identified suggest different and distinctive *P-T* conditions for the development of the different kinds of coronas. The meaning of these *P-T* differences and their probable connection with the metamorphic evolution and specifically with the prograde *P-T-t* path of the metagabbros will be considered. This contribution also includes a discussion of the dynamic setting of the metamorphism that caused the coronitization of the gabbros.

GEOLOGIC SETTING

The Sobrado unit is a small element of the large Ordenes Complex, one of the allochthonous complexes of northwestern Iberia (Fig. 1). The Ordenes Complex consists of a pile of units stacked and thrust during Variscan deformation (Martínez Catalán et al., 1984; Arenas et al., 1986). The units are divided into three groups according to their relative position in the original nappe pile: basal, intermediate, and upper units (Fig. 1). The intermediate units show oceanic affinities, and are referred to as ophiolitic. They represent a suture separating two main paleogeographic realms. An abridged description of these units and their structure is presented in Martínez Catalán et al. (this volume).

There is a general agreement that the basal units formed the external edge of Gondwana after the opening of the Rheic ocean, and registered an initial high-*P* Variscan metamorphic event (Gil Ibarguchi and Ortega Gironés, 1985; Gil Ibarguchi, 1995; Arenas et al., 1995, 1997), related to westward subduction 380–370 Ma under an accretionary wedge consisting of a pile formed by the ophiolitic and upper units (Martínez Catalán et al., 1996).

The ophiolites depict a wide variety of metamorphic conditions (Arenas et al., 1986; Arenas, 1988; Díaz García et al., 1999a), and a high-*P* metamorphic gradient has been identified in some of them. They appear as allochthonous sheets thrust to the east, and were apparently stacked during the closure of the oceanic domain represented by them. The coeval amphibolite to greenschist facies foliations were formed between 390 and 365 Ma (Dallmeyer et al., 1997), closely following oceanic crust generation as young as 395 Ma (Díaz García et al., 1999a).

The upper units structurally overlie the ophiolites. They can be subdivided into two groups of units: high-*P* and high-*T* units, in the lower position, and uppermost units with intermediate-*P* metamorphism (high-*P*–high-*T* and intermediate-*P*, respectively).

The high-*P*–high-*T* units crop out in the eastern part of the Ordenes Complex, as lenses 1–10 km in length, bounded by thrust and detachment faults. Sobrado is the largest of these units, but a bigger massif occurs to the north, in the Cabo Ortegal Complex (Fig. 1). The high-*P*–high-*T* units include paragneisses and mafic and ultramafic rocks. The metabasites are garnet-clinopyroxene granulites and eclogites, retrograded to the amphibolite facies. Gabbros are common and may appear undeformed with a minor metamorphic imprint, with corona textures (as those studied in this paper), or transformed into high-*P* granulites. Subophitic and diabase textures preserved in the less deformed gabbros suggest an emplacement at relatively shallow crustal levels.

An early high-*P* granulite to eclogite facies metamorphism was followed by decompression and partial melting and then, successively, by a penetrative mylonitization in the amphibolite facies, followed by recumbent folding and thrusting in the greenschist facies (Vogel, 1967; Marcos et al., 1984; Gil Ibarguchi et al., 1990; Arenas, 1991; Girardeau and Gil Ibarguchi, 1991; Martínez Catalán and Arenas, 1992; Mendia, 2000). U-Pb dating of zircon, monazite, titanite, and rutile reveal the presence of an Early Ordovician metamorphic event ca. 500–480 Ma and a later, Early Devonian event ca. 400–380 Ma (Fernández Suárez et al., 2002). The U-Pb ages in conjunction with petrological and structural data indicate that the high-pressure event recorded by these rocks is Early Ordovician in age. This is also the age suggested for corona growth in the metagabbros. A similar age has been proposed for the protoliths of the coronitic metagabbros (Schäfer et al., 1993; Abati et al., 1999; Ordóñez Casado, 2001), which must have been generated just before the high-*P* event, probably in an accretionary setting (Abati et al., 1999).

The overlying intermediate-*P* units include a very thick sequence of terrigenous metasediments, and large bodies of amphibolites, augengneisses, and gabbros, dated as 498–500 Ma (U-Pb on zircon; Abati et al., 1999). As in the previous case, the gabbros show subophitic and diabase textures. Metamorphism ranges from greenschist facies, in the upper parts, to intermediate-*P* granulite facies in the lower parts. Here also, U-Pb data (zircon, monazite, and rutile) in high-grade paragneisses suggest a polyorogenic character for the intermediate-*P* units, with an Early Ordovician (493–498 Ma) Barrovian-style metamorphism up to granulite facies (Abati et al., 1999) and an Early Devonian event localized in areas of intense Variscan tectonothermal reworking (Dallmeyer et al., 1997).

The intermediate-*P* units occupy the hanging wall of a ductile normal fault, the Corredoiras detachment, whereas the So-

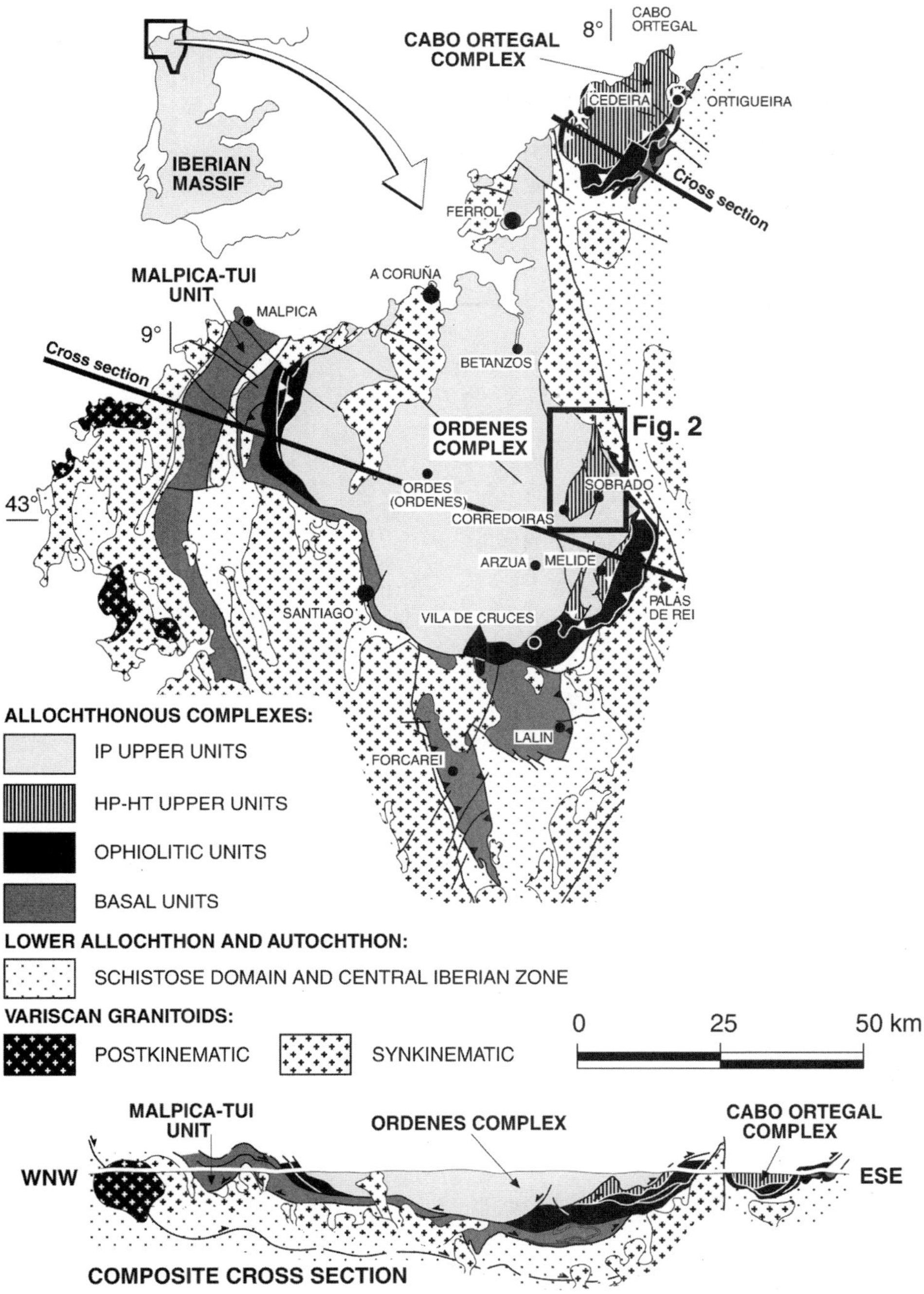

Figure 1. Geological map and cross section of allochthonous complexes of northwestern Spain. *IP, HP, HT,* are intermediate and high pressure, and high temperature.

brado and other high-*P*–high-*T* units occur at its footwall (Fig. 2). The Corredoiras detachment has a top-to-the-north sense of motion (Fig. 3) and was active in early stages of the Variscan deformation (Dallmeyer et al., 1997; Díaz García et al., 1999b).

The origin and significance of the upper units have not been firmly established. Both groups (high-*P*–high-*T* and intermediate-*P* units) share aspects of their tectonothermal evolution and are considered an exotic terrane located in the hanging wall to the Variscan suture. Arenas et al. (1986) and Martínez Catalán (1990) proposed a relationship with an island arc, suggesting that they might represent fragments of either a backarc, the arc, or the forearc. Martínez Catalán et al. (1996, 1997, 1999) considered the arc hypothesis, but recognized the possibility that the units included parts of a rifted continental margin.

According to their lithological and structural characteristics, the upper units were involved in two orogenic episodes and underwent two cycles of burial and exhumation. The high-*P* metamorphism represents an early subduction, which took place in the Early Ordovician in a thick orogenic wedge in which the high-*P*–high-*T* units occupied a deep structural position. Evidence of pre-Variscan crustal thinning and extension comes from the identification of extensional detachments overprinted by the early Variscan folds (Martínez Catalán et al., this volume). These could have resulted from the gravitational collapse

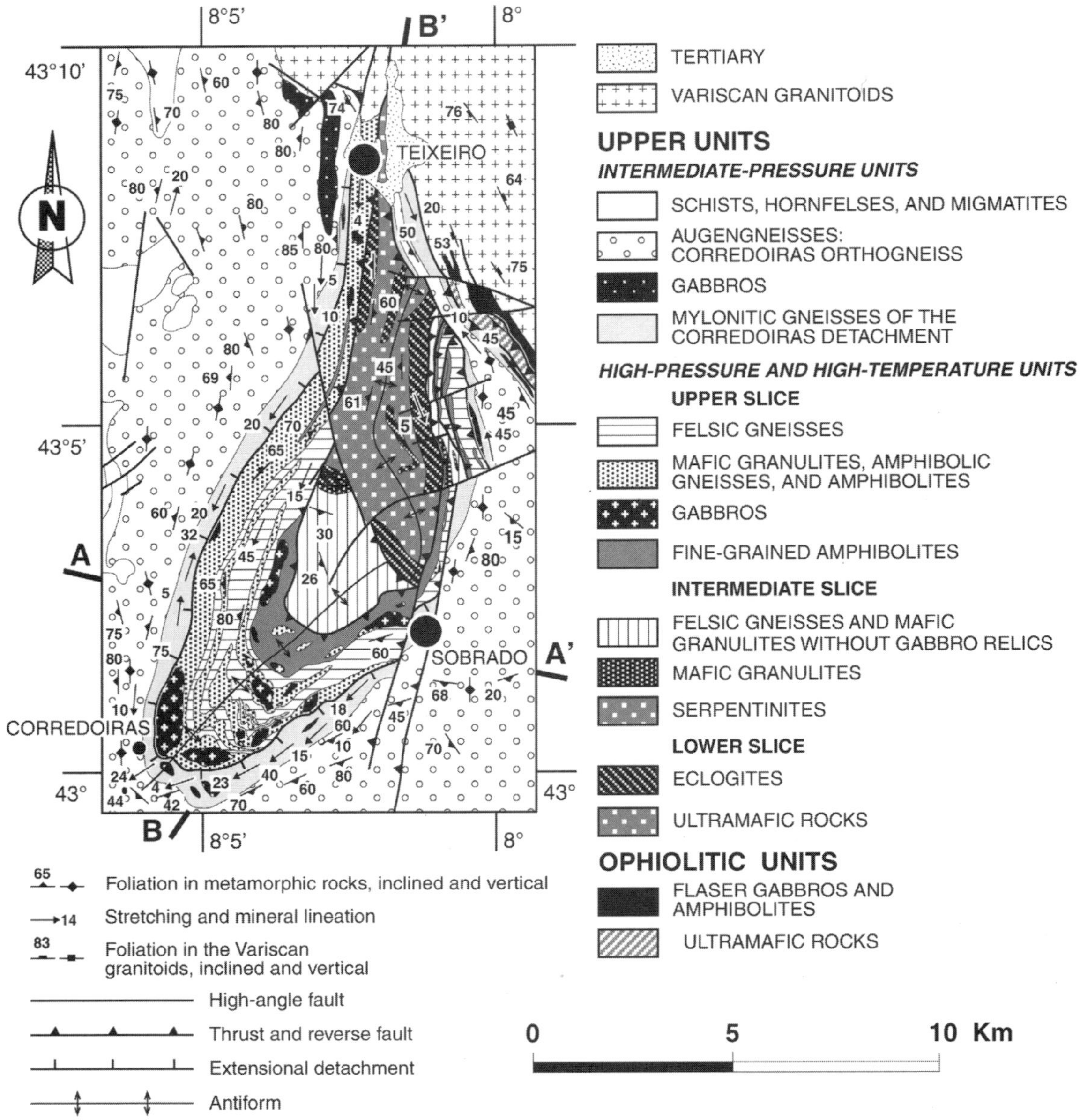

Figure 2. Geological map of Sobrado unit. See location in Figure 1.

of the Ordovician orogenic wedge or from it being involved in an extensional environment.

Subsequently, the upper units underwent Variscan reworking and were included in the Variscan orogenic wedge during early stages of its development. During the Early Devonian episode, developed in intermediate-*P* conditions, the high-*P*–high-*T* units were deformed and thrust upon the ophiolitic units. Coeval amphibolite facies metamorphism, dated as 390–380 Ma, was retrogressive in the high-*P*–high-*T* units and prograde in the underlying ophiolites, where it reflects the westward subduction and understacking of the oceanic lithosphere related to the closure of the Rheic ocean. In turn, this closure was followed by the subduction of the outer edge of Gondwana, represented by the basal units (Martínez Catalán et al., 1996). Finally, the allochthonous units were carried along low-dipping thrust faults over the schistose domain, which constitutes the lower allochthon, and over the autochthonous sequences that form most of the Iberian Massif.

HIGH-*P*– HIGH-*T* SUCCESSION OF THE SOBRADO WINDOW

The Sobrado unit crops out in a tectonic window in the core of a northeast-southwest–striking antiform. It is surrounded by the Corredoiras Orthogneiss; there is a 300–600-m-thick mylonitic band at its base (Martínez Catalán and Arenas, 1992; Díaz García et al., 1999b) that belongs to the overlying intermediate-*P* upper units (Figs. 2 and 3). The

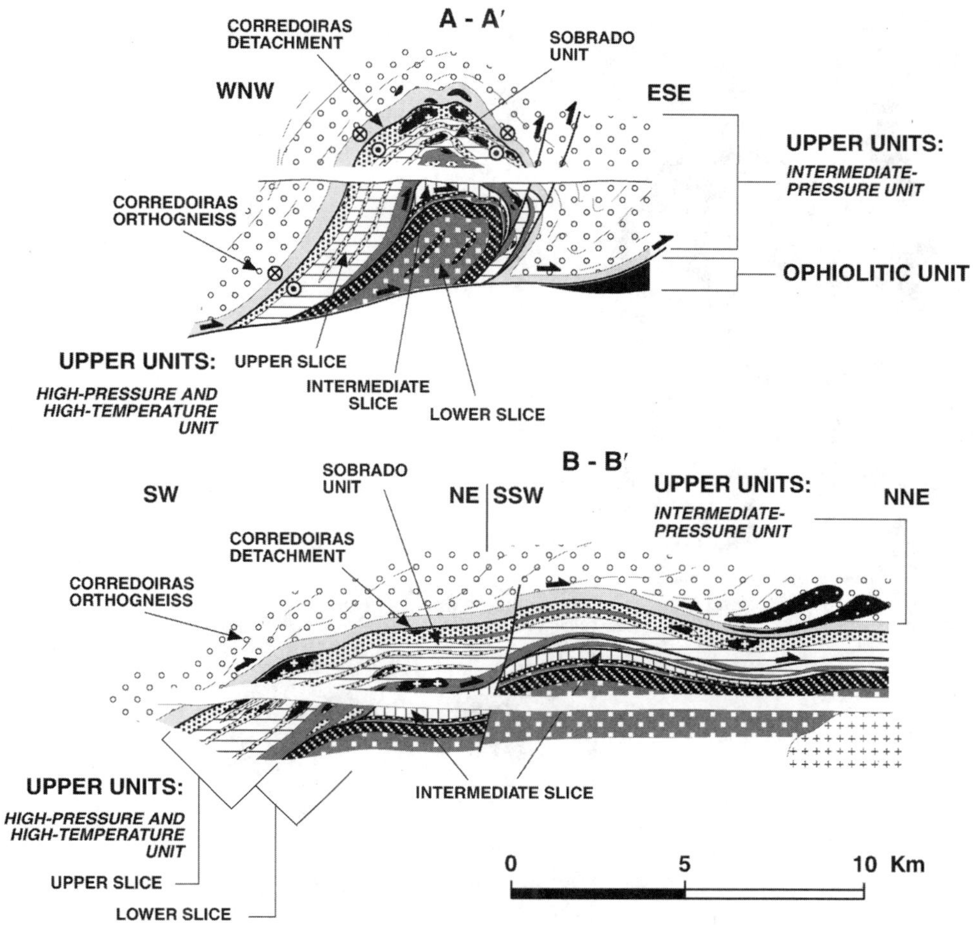

Figure 3. Transverse (A–A′) and longitudinal (B–B′) sections across Sobrado tectonic window. See location in Figure 2. Note top-to-northeast sense of motion of Corredoiras detachment, and subsequent southeast-directed thrusting.

mylonites developed during the emplacement of the intermediate-*P* over the high-*P*–high-*T* upper units along the Corredoiras detachment, of extensional character, and the antiform seems related to thrusting, because the Sobrado unit consists of three tectonic slices forming an antiformal stack (Fig. 3). The three slices are made of high-*P*–high-*T* rock types, but show differences in the lithologic association and the *P-T* conditions reached during metamorphism.

The lowest slice comprises, from bottom to top, highly serpentinized ultramafic rocks with some mafic inclusions and a layer of metabasites with a thickness ~500 m. These mafic rocks include eclogites and related clinopyroxene-garnet rocks without primary plagioclase, as well as other lithological types derived from the retrogression and mylonitization of the early high-*P* types (retrogressed eclogites, amphibolitic gneisses, and amphibolites). The intermediate slice contains a 1000-m-thick package of felsic gneisses (mainly paragneisses with some minor orthogneiss) with frequent inclusions of high-*P* granulite facies mafic rocks. Relicts of the igneous protoliths are not preserved in either the lower or the intermediate slice.

The upper slice (Figs. 2 and 3) includes felsic gneisses and layers of mafic rocks derived from deformed and recrystallized gabbros, that locally grade to undeformed varieties with preserved intact igneous textures (Pablo Maciá and Martínez Catalán, 1984). The metamorphic transformation of the gabbros shows different prograde stages ranging from coronitic textures to total recrystallization of the gabbros to high-*P* granulites. Later, these metabasites were affected by variable rates of partial melting and, finally, developed a regional foliation in the amphibolite facies including amphibolitic gneisses, flaser amphibolites, and fine-grained amphibolites. The partial melting and the later mylonitization also affected the metabasites of the intermediate slice, as well as the felsic gneisses.

The fine-grained amphibolites represent an advanced stage in the mylonitization of the metabasites. They commonly appear as relatively thin layers inside the mafic rocks, but a 300–600-m-thick layer is located at the base of the upper slice (Figs. 2 and 3). Moreover, in the contact area between the lower and the intermediate slices, a level of mylonitic pyroxenite a few meters thick appears, and mylonitic fabrics are commonly found affecting the lithologies of both slices.

The penetrative mylonitic deformation was possibly developed as two stages, the early extensional tectonics and later Variscan thrusting. Mylonitization caused a severe overprinting of the early tectonothermal stages so that very limited information has been obtained about the evolution of the early orogenic wedge. Moreover, the preservation of prograde textures related to early orogenic stages is especially difficult in high-*P*–high-*T*

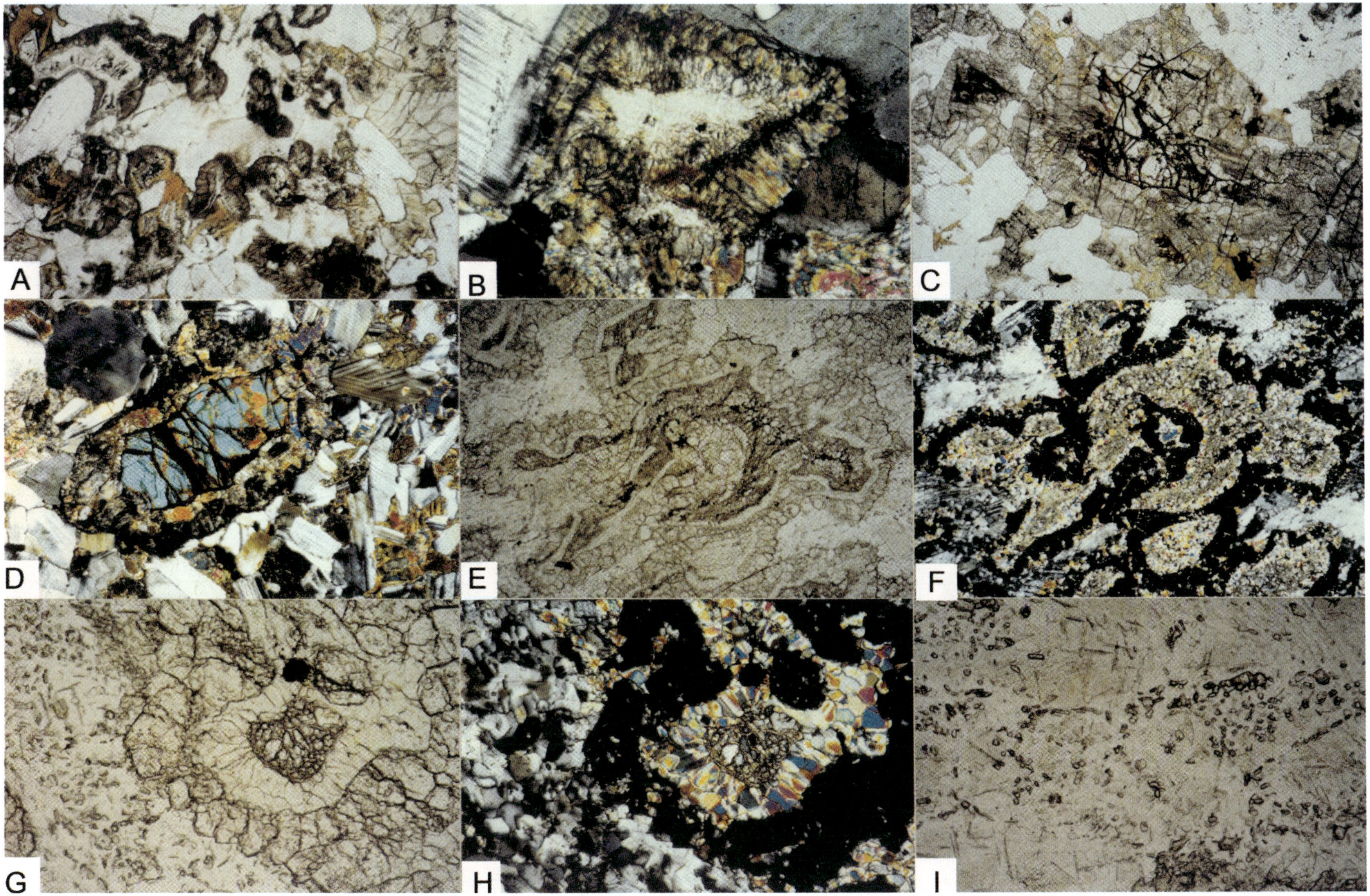

Figure 4. Textural characteristics of coronitic metagabbros: (A, B) group I coronas; (C, D) group II coronas; (E–I) group III coronas. A: General texture of coronitic metagabbro showing Mg-rich anthophyllite-tremolite aggregates pseudomorphic after olivine, surrounded by corona of orthopyroxene and brown pargasite, or by orthopyroxene-pargasite symplectite. B: Olivine pseudomorph with core region replaced by anthophyllite-tremolite aggregate, and external corona of orthopyroxene-pargasite. C, D: Orthopyroxene-clinopyroxene-garnet-pargasite corona between olivine and plagioclase. E–H: Metagabbro showing orthopyroxene aggregates after olivine separated from plagioclase by clinopyroxene (inner) and garnet (outer) coronas. I: Recrystallized plagioclase with abundant inclusions of kyanite and corundum (also visible in G and H). Horizontal length of photomicrographs is 3 mm (A, C, D, E, F), 1.5 mm (G, H), 1 mm (B) and 0.5 mm (I).

units, where the high-*T* recrystallization associated with the thermal peak generally tends to erase previous stages of the *P-T* path. The preservation in the Sobrado unit of metagabbros showing different stages of a metamorphic evolution reflecting the prograde history represents a unique opportunity to obtain data connected with the early tectonothermal evolution of the Early Ordovician wedge.

CORONITIC METAGABBROS: TEXTURES AND MINERAL CHEMISTRY

The metagabbros of the upper slice (Figs. 2 and 3) preserve igneous fabrics including fine- to medium-grained granular, diabasic, and ophitic types. Coarse-grained to very coarse grained pegmatitic gabbros also exist. The metamorphic textures are complex, and include a first recrystallization stage with development of corona textures during a dynamic regime with low to moderate penetrative deformation (Fig. 4). Successive tectonothermal stages developed a regional foliation, which records a pronounced decompression and retrogression of the early mineral assemblages. Coronitic textures are preserved only in low-strain areas inside the gabbros, where coronas initially originated by reaction between plagioclase and the ferromagnesian igneous phases (biotite, orthopyroxene, clinopyroxene, olivine).

In order to study the early metamorphic evolution of the metagabbros, the more significant coronas are those caused by the breakdown of olivine with adjacent plagioclase (Fig. 4). These are the coronas studied in this chapter and they have been divided into three groups based on mineralogy (subordinate phases in brackets; mineral abbreviations are after Kretz, 1983): group I: Amp-Opx.; group II: (Amp)-(Cpx)-Opx-Grt; group III: (Amp)-Cpx-Opx-Grt.

Modal abundance of metamorphic phases and the degree of recrystallization increase from group I to III in agreement with a path that is considered prograde. This interpretation is based upon both the textural evolution and the corona mineral assemblages and reaction history. Bulk compositions of the studied samples exhibit a limited spread (Table 1), and therefore it does not seem to be responsible for differences in mineralogy between samples.

TABLE 1. WHOLE-ROCK ANALYSES OF CHARACTERISTIC METAGABBROS

Sample	I:126	II:130	III:129	III:112
SiO_2	47.87	49.66	49.14	47.29
TiO_2	0.71	0.45	0.10	0.10
Al_2O_3	19.17	18.61	19.56	22.42
FeO*	7.85	7.33	7.88	7.05
MnO	0.12	0.12	0.12	0.10
MgO	9.88	9.41	9.78	10.73
CaO	11.70	11.61	10.37	9.98
Na_2O	2.49	2.59	2.83	2.24
K_2O	0.07	0.14	0.12	0.02
P_2O_5	0.13	0.08	0.10	0.07
LOI	0.41	0.16	0.63	0.88
Total	100.40	100.16	100.63	100.88
X_{Mg}	69.16	69.59	68.86	73.06

*Total iron as FeO.

Group I

Group I textures are characterized by Mg-rich anthophyllite-tremolite (Table 2) aggregates (± Fe-oxide) pseudomorphic after olivine surrounded by a thick corona of orthopyroxene (inner) and brown pargasite (outer) or by orthopyroxene-pargasite symplectite. Orthopyroxene (Table 3) is Mg rich ($X_{Mg} = 65–70$) with Al_2O_3 content varying randomly across the coronas, from 1% to 4%. In the same samples, intercumulus clinopyroxene is Na-poor augite (Table 3) clouded with ilmenite-rutile exsolutions and displays discontinuous rims of brown pargasite. Fe oxides are surrounded by wide pargasite coronas. Plagioclase laths contain sparse corundum inclusions and fall in the compositional range of labradorite with anorthite contents decreasing toward the olivine domains.

The distribution of amphiboles within the olivine domains implies strong chemical gradients, with Al and Na decreasing toward the center of these domains, a result of inefficient diffusion of Al and Na from plagioclase during olivine breakdown.

The development of anthophyllite-tremolite aggregates at the center of the olivine domains can be described by the simplified reaction (disregarding minor amounts of Fe and Al in the amphiboles):

$$\underset{\text{Fo}}{6Mg_2SiO_4} + \underset{\text{An}}{2CaAl_2Si_2O_8} + \underset{\text{Qtz}}{6SiO_2} + V =$$

$$\underset{\text{Ant}}{Mg_7Si_8O_{22}(OH)_2} + \underset{\text{Tre}}{Ca_2Mg_5Si_8O_{22}(OH)_2} + \underset{\text{Crn}}{2Al_2O_3} \quad \text{(reaction 1)}$$

This involves Si and Ca diffusion from plagioclase into the olivine domain, leaving behind excess Al that forms corundum inclusions in plagioclase. However, the reaction cannot be balanced unless extra SiO_2 is available as reactant. The presence of magnetite and the low Fe contents of these amphiboles suggest that reaction 1 was coupled with oxidation of the Fa content of olivine according to a reaction of the type:

$$Fe_2SiO_4 + O_2 = 2\ Fe_3O_4 + 3SiO_2 \quad \text{(reaction 2)}$$

This reaction could release the silica required for the development of the amphibole aggregates (Rivers and Mengel, 1988).

The development of orthopyroxene and pargasite at the rims of the olivine domains can also be explained by more efficient Al and Na diffusion combined with Ca diffusion from the plagioclase domains. Higher Fe contents of these two phases relative to the anthopyllite-tremolite aggregates may be attributed to simultaneous Fe_3O_4 breakdown. These features can be illustrated by the simplified reaction:

$$\underset{\text{Ant}}{11Mg_7Si_8O_{22}(OH)_2} + \underset{\text{Tre}}{2Ca_2Mg_5Si_8O_{22}(OH)_2} + \underset{\text{An}}{4CaAl_2Si_2O_8}$$

$$+ \underset{\text{Ab}}{4NaAlSi_3O_8} + \underset{\text{Mag}}{7.3\ Fe_3O_4} =$$

$$\underset{\text{Opx}}{50Mg_{1.5}Fe_{0.5}Si_2O_6} + \underset{\text{Par}}{4NaCa_2Mg_3FeAl_3Si_6O_{22}(OH)_2}$$

$$+ 9H_2O + 5.65O_2 \qquad \text{(reaction 3)}$$

Reaction 3 does not take into account Al in orthopyroxene (Table 3) that is most likely provided by the excess Al in plagioclase produced by reaction 1, but considers approximately the proportions of Fe and Mg in orthopyroxene and pargasite (Tables 2 and 3).

The formation of orthopyroxene and pargasite in olivine domains is typical of early stages of metamorphic transformation of gabbro in the presence of fluid (Rivers and Mengel, 1988). However, the development of anthopyllite-tremolite aggregates instead of orthopyroxene at the center of the olivine domains is less common. It indicates significant fluid infiltration during breakdown of olivine with extensive Ca diffusion from plagioclase.

Group II

Group II textures are characterized by inner orthopyroxene and outer garnet coronas between olivine and plagioclase. Olivine occurs as a heavily oxidized relict or is pseudomorphosed by orthopyroxene aggregates. The garnet corona is of variable width and locally discontinuous, leaving orthopyroxene in contact with plagioclase. Garnet contains abundant pargasite inclusions. Small clinopyroxene grains occur at the outer rim of the coronas, and as inclusions in garnet or plagioclase. Intercumulus clinopyroxene contains abundant rutile and orthopyroxene exsolutions and is discontinuously rimmed by pargasite.

Orthopyroxene (Table 3; X_{Mg} = 70–76) displays 1.3%–4% Al_2O_3, that tends to increase outward with maximum values at rims adjacent to plagioclase. Garnet coronas (Fig. 5) are moderately zoned, Grs increasing outward from ~15% to 25% compensated by a decrease in Alm (from ~42% to 35%). Plagioclase composition is in the range of labradorite and displays a systematic decrease in An content toward garnet (from ~60–70 to 40–50; see Tables 4 and 5). Both intercumulus and granoblastic clinopyroxene have the composition of sodic augite (Table 3);

TABLE 2. REPRESENTATIVE AMPHIBOLE COMPOSITIONS

Sample analysis	I:211						II:208			III:671
	2	16	4	6	12	42	2-34a*	2-5c*	2-33b*	2-13a*
	Ant		Tre		Par					
SiO_2	57.64	54.19	56.95	57.66	42.05	40.58	41.69	41.74	42.52	43.06
TiO_2	0.08	0.06	0.08	0.00	1.89	0.08	0.00	1.34	3.48	0.80
Al_2O_3	3.88	5.25	0.93	0.39	16.67	18.79	17.22	17.59	16.54	17.36
FeO#	5.48	9.05	6.04	6.10	10.17	10.13	6.13	7.70	6.29	6.33
MnO	0.00	0.00	0.07	0.08	0.11	0.12	0.03	0.00	0.00	0.01
MgO	28.53	25.14	20.74	20.87	12.67	12.68	14.50	13.73	14.35	15.27
CaO	0.00	0.02	12.56	13.22	11.20	11.06	12.38	12.04	12.33	12.56
Na_2O	0.16	0.00	0.23	0.12	3.30	2.81	2.00	2.45	1.89	2.90
K_2O	0.00	0.00	0.00	0.00	0.55	0.63	1.42	1.08	1.32	0.16
Total	95.77	93.71	97.60	98.44	98.61	96.88	95.37	97.67	98.72	98.45
Si	7.816	7.653	7.915	7.956	6.065	5.942	6.119	6.020	6.035	6.089
Ti	0.008	0.006	0.008	0.000	0.205	0.009	0.000	0.145	0.372	0.085
Al^{IV}	0.184	0.347	0.085	0.044	1.935	2.058	1.881	1.980	1.965	1.911
Al^{VI}	0.437	0.527	0.067	0.019	0.898	1.185	1.098	1.010	0.801	0.983
Fe^{3+}	0.000	0.000	0.190	0.078	0.138	0.462	0.054	0.074	0.000	0.117
Fe^{2+}	0.622	1.069	0.512	0.625	1.088	0.779	0.698	0.855	0.747	0.631
Mn	0.000	0.000	0.009	0.010	0.013	0.015	0.004	0.000	0.000	0.001
Mg	5.768	5.293	4.297	4.293	2.723	2.769	3.173	2.952	3.036	3.219
Ca	0.000	0.003	1.870	1.954	1.730	1.735	1.947	1.861	1.875	1.903
Na	0.041	0.000	0.063	0.031	0.923	0.799	0.569	0.685	0.520	0.795
K	0.000	0.000	0.000	0.000	0.102	0.117	0.266	0.199	0.239	0.029
Total	14.876	14.898	15.016	15.010	15.820	15.870	15.809	15.781	15.590	15.763

*a, Inclusion in Grt; b, inclusion in Pl; c, rim adjacent to both Grt and Pl.
#Total iron as FeO.

TABLE 3. REPRESENTATIVE PYROXENE COMPOSITIONS

Sample	I:211	II:208	III:129	I:211	II:208		III:129		III:671	
analysis	28	1a*	13	24b*	36c*	23b*	2-12a*	2-13a*	4-16	4-41
		Opx					Cpx			
SiO_2	53.36	55.28	55.94	52.38	53.20	52.88	54.65	54.29	54.33	53.86
TiO_2	0.02	0.00	0.00	0.80	0.33	0.47	0.00	0.06	0.00	0.03
Al_2O_3	2.87	1.76	0.77	4.34	5.23	3.50	2.62	3.46	5.48	5.18
$FeO^{\#}$	20.38	15.99	14.86	5.92	4.12	5.57	5.07	5.27	3.14	3.27
MnO	0.40	0.25	0.15	0.06	0.00	0.09	0.03	0.12	0.00	0.01
MgO	23.48	27.76	28.87	13.94	14.43	14.73	15.02	14.47	14.43	14.16
Cr_2O_3	N.D.	N.D.	N.D.	0.16	0.12	0.20	0.15	0.17	0.25	0.15
NiO	N.D.	N.D.	N.D.	0.00	0.05	0.00	0.00	0.04	0.00	0.00
CaO	0.61	0.29	0.27	22.99	22.44	22.58	22.00	22.15	22.00	22.29
Na_2O	N.D.	N.D.	N.D.	1.01	1.31	1.04	1.57	1.47	1.83	1.62
Total	101.12	101.33	100.86	101.60	101.23	101.06	101.11	101.50	101.46	100.57
Si	1.940	1.956	1.984	1.894	1.912	1.916	1.967	1.951	1.936	1.941
Ti	0.001	0.000	0.000	0.022	0.009	0.013	0.000	0.002	0.000	0.001
Al^{IV}	0.050	0.044	0.016	0.079	0.074	0.051	0.001	0.023	0.063	0.059
Al^{VI}	0.063	0.029	0.016	0.106	0.147	0.098	0.110	0.123	0.166	0.161
Fe^{3+}	0.000	0.028	0.000	0.050	0.024	0.061	0.060	0.043	0.017	0.004
Fe^{2+}	0.620	0.445	0.441	0.129	0.100	0.108	0.092	0.112	0.077	0.094
Mn	0.012	0.007	0.005	0.002	0.000	0.003	0.001	0.004	0.000	0.000
Mg	1.273	1.464	1.527	0.751	0.773	0.795	0.806	0.775	0.767	0.761
Cr	N.D.	N.D.	N.D.	0.005	0.003	0.006	0.004	0.005	0.007	0.004
Ni	N.D.	N.D.	N.D.	0.000	0.001	0.000	0.000	0.001	0.000	0.000
Ca	0.024	0.011	0.010	0.891	0.864	0.877	0.848	0.853	0.840	0.861
Na	N.D.	N.D.	N.D.	0.079	0.091	0.073	0.109	0.103	0.127	0.113
Total	3.983	3.984	3.999	4.008	3.998	4.001	3.998	3.998	4.000	3.999
X_{Mg}	0.6725	0.7669	0.7759	0.8534	0.8855	0.8804	0.8976	0.8737	0.9088	0.8901
X_{Jd}	N.A.	N.A.	N.A.	0.0710	0.0910	0.0730	0.1090	0.1040	0.1090	0.1080
X_{Cats}	N.A.	N.A.	N.A.	0.0350	0.0560	0.0250	0.0010	0.0200	0.0640	0.0570

*a, Rim adjacent to Grt; b, rim adjacent to Pl; c, rim adjacent to both Grt and Pl.

#Total iron as FeO.

higher Jd and Cats contents are measured in granoblastic grains adjacent or included in plagioclase.

The local presence of relic olivine and the absence of amphibole aggregates at the center of the olivine domains suggest that group II textures did not evolve via the anthophyllite-tremolite stage. Chemical gradients across the olivine domains are expressed by increase in Al, Ca, and Na toward plagioclase with orthopyroxene (Ca-Na free and containing little Al) forming at the central parts of these domains and garnet (Al-Ca bearing) with pargasite inclusions (Ca-Na-Al bearing) occupying the rims. Orthopyroxene aggregates and coronas may have been formed together with pargasite according to a reaction such as:

$$\underset{Ol}{9(MgFe)_2SiO_4} + \underset{An}{6CaAl_2Si_2O_8} + \underset{Ab}{3NaAlSi_3O_8} + V =$$

$$\underset{Opx}{6(MgFe)_2Si_2O_6} + \underset{Par}{3NaCa_2(MgFe)_4Al_3Si_6O_{22}(OH)_2} + \underset{Crn}{3Al_2O_3}$$

(reaction 3a)

Development of garnet, with pargasite inclusions and minor clinopyroxene at the outer rim, may be attributed to pargasite + orthopyroxene breakdown due to further Ca and Al diffusion from the plagioclase sites, according to a reaction of the type:

$$\underset{Opx}{3(MgFe)_2Si_2O_6} + \underset{Par}{NaCa_2(MgFe)_4Al_3Si_6O_{22}(OH)_2}$$

$$+ \underset{An}{2CaAl_2Si_2O_8} + \underset{Crn}{Al_2O_3} + \underset{Qtz}{0.9SiO_2} =$$

$$\underset{Prp,\ Alm}{3(MgFe)_3Al_2Si_3O_{12}} + \underset{Grs}{Ca_3Al_2Si_3O_{12}} + \underset{Di}{CaMgSi_2O_6}$$

$$+ \underset{Jd}{0.1NaAlSi_2O_6} + \underset{Ab}{0.9NaAlSi_3O_8}$$

(reaction 4)

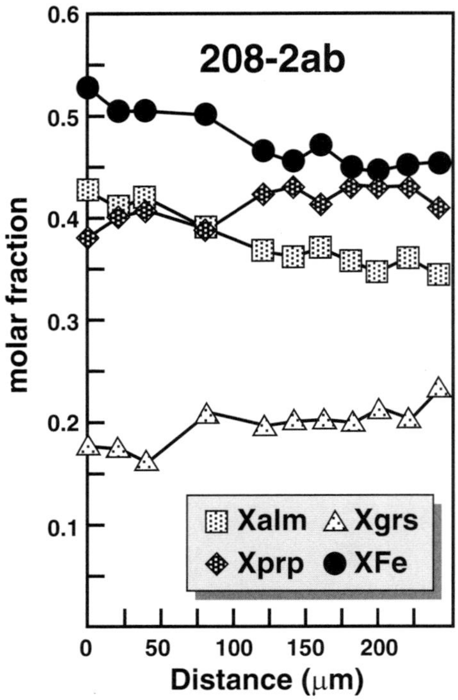

Figure 5. Chemical profiles across garnet coronas of group II.

This reaction is balanced by using quartz released by the oxidation of olivine (reaction 2), and corundum formed by reaction 3 in the plagioclase domains. It accounts for the proportion of grossular in garnet (Table 4), and for the jadeite content, as great as 10%, of clinopyroxene (Table 3). Development of clinopyroxene at the outer rims of the coronas and as inclusions in plagioclase may be explained by excess of Fe and Mg diffused in the plagioclase site combined by depletion of Al available for reaction; this is consistent with the absence of corundum in plagioclase in these samples.

Group III

Group III textures are the most recrystallized. They display orthopyroxene aggregates, locally surrounding olivine relics, separated from plagioclase by clinopyroxene (inner) and garnet (outer) coronas. Garnet coronas contain inclusions of pargasite, clinopyroxene, and, at the outer rim, corundum. In the same samples intercumulus clinopyroxene is replaced by aggregates of fine-grained granoblastic clinopyroxene surrounded by a discontinuous garnet corona. Locally, pargasite occurs between garnet and clinopyroxene, and/or at the garnet-plagioclase interfaces. Plagioclase laths are partly recrystallized and contain abundant inclusions of kyanite and corundum.

TABLE 4. REPRESENTATIVE GARNET COMPOSITIONS

Sample	I:208				III:129			III:671		
analysis	2-13c*	2-41d*	2-56d*	2-12c*	2-14a*	2-8a*	2-20b*	4-13a*	4-38a*	4-27b*
SiO_2	40.30	39.90	40.17	40.07	40.49	40.34	40.54	40.84	41.01	40.52
TiO_2	0.00	0.10	0.05	0.03	0.08	0.00	0.02	0.00	0.03	0.00
Al_2O_3	22.43	22.52	22.64	22.62	23.09	22.45	22.99	22.58	22.82	22.73
FeO#	19.86	20.30	19.76	20.12	17.24	18.80	15.17	15.70	15.67	14.27
MnO	0.43	0.76	0.44	0.43	0.28	0.30	0.22	0.26	0.53	0.18
MgO	10.70	9.97	10.41	10.48	11.49	11.79	9.80	12.57	12.74	9.42
CaO	7.33	7.95	7.51	7.93	8.94	7.02	12.38	8.34	8.47	13.50
Total	101.05	101.50	100.98	101.68	101.61	100.70	101.12	100.29	101.27	100.62
Si	3.003	2.978	2.996	2.977	2.976	2.999	2.993	3.013	3.000	3.004
Ti	0.000	0.006	0.003	0.002	0.005	0.000	0.001	0.000	0.002	0.000
Al^{IV}	0.000	0.022	0.004	0.023	0.024	0.001	0.007	0.000	0.000	0.000
Al^{VI}	1.970	1.956	1.985	1.958	1.975	1.965	1.995	1.964	1.968	1.987
Fe^{3+}	0.030	0.035	0.012	0.040	0.020	0.035	0.004	0.036	0.030	0.013
Fe^{2+}	1.208	1.233	1.221	1.211	1.039	1.133	0.932	0.933	0.929	0.872
Mn	0.027	0.048	0.028	0.027	0.018	0.019	0.014	0.016	0.033	0.011
Mg	1.188	1.109	1.158	1.161	1.259	1.306	1.079	1.383	1.389	1.041
Ca	0.585	0.636	0.600	0.632	0.704	0.559	0.980	0.659	0.664	1.073
Total	8.011	8.023	8.007	8.031	8.020	8.017	8.005	8.004	8.015	8.001
X_{Alm}	0.4015	0.4074	0.4060	0.3995	0.3442	0.3756	0.3103	0.3118	0.3081	0.2908
X_{Prp}	0.3950	0.3666	0.3851	0.3832	0.4169	0.4329	0.3591	0.4623	0.4608	0.3474
X_{Grs}	0.1796	0.1901	0.1924	0.1878	0.2208	0.1680	0.3235	0.2024	0.2043	0.3514
X_{Adr}	0.0150	0.0200	0.0073	0.0207	0.0123	0.0174	0.0026	0.0181	0.0159	0.0066
X_{Sps}	0.0089	0.0159	0.0092	0.0089	0.0059	0.0062	0.0046	0.0054	0.0109	0.0038

*a, Rim adjacent to Cpx; b, rim adjacent to Pl; c, rim adjacent to both Opx and Pl; d, rim adjacent to both Cpx and Pl.
#Total iron as FeO.

TABLE 5. REPRESENTATIVE PLAGIOCLASE COMPOSITIONS

Sample	II:208	III:129		III:671	
analysis	2-29r*	2-5c*	2-15r*	4-19c*	4-15r*
SiO_2	57.46	53.16	58.06	54.46	58.51
Al_2O_3	27.06	29.55	26.21	28.23	26.15
CaO	8.95	12.33	8.14	11.50	8.10
Na_2O	6.83	4.92	7.12	5.05	7.54
K_2O	0.00	0.01	0.00	0.05	0.00
Total	100.30	99.97	99.53	99.29	100.30
Si	2.540	2.409	2.603	2.458	2.607
Al	1.423	1.578	1.385	1.534	1.373
Ca	0.428	0.598	0.391	0.556	0.387
Na	0.591	0.432	0.619	0.442	0.651
K	0.000	0.001	0.000	0.003	0.000
Total	4.982	5.018	4.998	4.993	5.018
X_{An}	0.420	0.580	0.387	0.557	0.373

*c, Core; r, rim adjacent to Grt.

Orthopyroxene (Table 3; X_{Mg} = 75–80) is less aluminous than in the other groups (Al_2O_3 ~0.5%–1.5%). Clinopyroxene coronas and aggregates are composed by Na-augite (Jd_{7-14}; Table 3), with maximum Jd contents increasing outward. Ca-Ts contents are variable, and range between 1% and 10%, with maximum values in clinopyroxene aggregates. Garnet coronas display moderate to strong zoning, with Grs increasing from ~15%–20% to ~33%–40% toward plagioclase, compensated by antithetic Prp and to minor extent Alm trends, so that XFe increases slightly outward (Fig. 6). Plagioclase displays a decrease in An toward garnet, from 40%–50% to 30%–38%.

Orthopyroxene pseudomorphs after olivine, inclusions of pargasite in garnet, and the presence of clinopyroxene in the coronas imply that group III textures in olivine domains likely developed via the same type of reactions as group II textures. However, high grossular contents at the outer garnet rims (Fig. 6), together with the presence of kyanite inclusions in plagioclase, imply that an additional reaction also took place, i.e.,

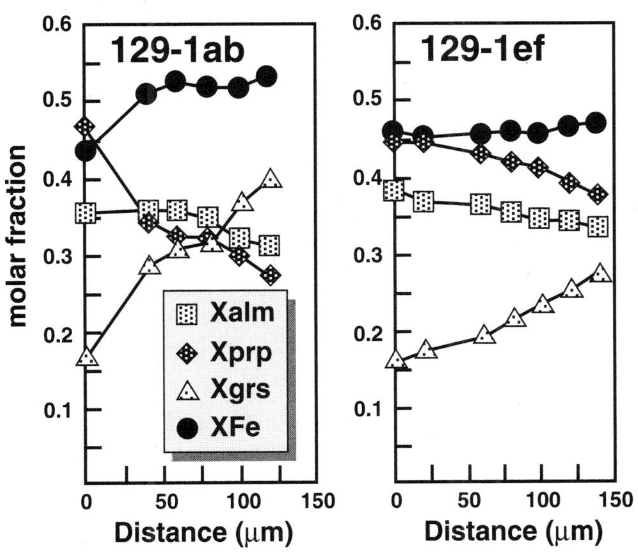

Figure 6. Chemical profiles across garnet coronas of group III.

$$3CaAl_2Si_2O_8 + Al_2O_3 = Ca_3Al_2Si_3O_{12} + Al_2SiO_5$$

An Crn Grs Ky

(reaction 5)

This reaction requires the presence of corundum in the plagioclase sites that is still abundant in group III samples, in contrast to those of group II, where all corundum seems to have been consumed by reaction 4. Therefore, an alternative reaction is proposed for the breakdown of pargasite in type III coronas:

$$(MgFe)_2Si_2O_6 + 2NaCa_2(MgFe)_4Al_3Si_6O_{22}(OH)_2 +$$

Opx Par

$$2CaAl_2Si_2O_8 =$$

An

$$2(MgFe)_3Al_2Si_3O_{12} + Ca_3Al_2Si_3O_{12} + 2\ CaMgSi_2O_6$$

Prp, Alm Grs Di

$$+ 2NaAlSi_3O_8 + 2\ Al_2O_3$$

Ab Crn

(reaction 4a)

Reaction 4a produces corundum instead of consuming it, and is balanced in a system with no excess silica (Rivers and Mengel, 1988). In addition, reaction 4a produces Grs-richer garnet and more diopside than reaction 4, in accordance with high Grs contents in garnet (Table 4) and higher amounts of diopside in these samples. Excess Al_2O_3 may also account for the development of discontinuous garnet coronas around granoblastic clinopyroxene aggregates that can be described by the reaction:

$$3\ CaMgSi_2O_6 + 2\ Al_2O_3 = Ca_3Al_2Si_3O_{12} + Mg_3Al_2Si_3O_{12}$$

Di Crn Grs Prp

(reaction 6)

However, additional kyanite may have been produced together with the jadeite component of clinopyroxene, according to the reaction:

$$NaAlSi_3O_8 + Al_2O_3 = NaAlSi_2O_6 + Al_2SiO_5$$

Ab Crn Jd Ky

(reaction 7)

Interpretation

Cpx-Grt coronas around ferromagnesian domains and kyanite inclusions in plagioclase are typical of incomplete transformation of Mg-Al gabbro under high-*P* metamorphic conditions, due to limited diffusion of elements between the plagioclase and the ferromagnesian textural domains (Indares, 1993; Indares and Rivers, 1995). In the context of the upper slice of the Sobrado unit, the presence of high-*P* granoblastic granulites and the low Na contents of clinopyroxene in the coronites indicate that the latter represent arrested stages during transformation to granulite. This evolution, the prograde transformation of the gabbros into high-*P* granulites, can be recognized even at outcrop scale. The preservation of the early metamorphic stages could result from kinetic factors in low-strain domains.

Concerning differences in metamorphic assemblages between the three groups of textures, higher proportions of garnet and metamorphic clinopyroxene in group III samples relative to those of group II may be a result of low silica activity (due to reduced oxidation of fayalite) during the breakdown of pargasite in the former (cf. reactions 4 and 4a). Furthermore, it is possible that reactions in rocks of group III were arrested at later stages of the metamorphic evolution relative to group II rocks. In addition, the absence of garnet and metamorphic clinopyroxene in group I samples does not have an obvious explanation, because theoretically these two minerals could have been produced by reaction 4a, unless reactions in group I rocks were arrested at lower *P-T* conditions than in groups II and III. It is therefore probable that different stages in the prograde metamorphic evolution are preserved in the three different groups of coronas, which recorded intensity in *P-T* conditions increasing from group I to group III. Given similarities in mineralogy between group III coronas and the granoblastic granulites, it is assumed that this group reflects conditions close to the metamorphic peak.

The absence of textures related to subsequent cooling and/or decompression suggests that the coronites have not recorded any of these stages of the metamorphic evolution. In that respect, note that high-grade coronitic metagabbros commonly display secondary plagioclase collars between clinopyroxene and garnet coronas interpreted to have formed during decompression at the initial stages of unloading (Indares, 1993). Such textures may develop during isothermal decompression or during thermal relaxation if peak temperatures postdated peak pressures (Indares and Dunning, 1997). The absence of plagioclase collars in group III samples that display the most advanced stages of metamorphic transformation is, therefore, compatible with simultaneous achievement of maximum *P-T* conditions. Hence, the prograde metamorphic evolution inferable from group I to group III coronas is compatible with a simple *P-T* path resulting from the burial of the unit, including the metagabbros.

P AND *T* ESTIMATES

This chapter deals with the evaluation of the metamorphic conditions recorded by the coronites. Although preservation of igneous fabrics and corona development attest for disequilibrium at the sample scale, local equilibrium may be assumed between touching rims of adjacent corona minerals, or rims of corona minerals and adjacent phases pseudomorphic after igneous phases. Therefore, thermobarometry may be applied to specific microtextural settings, providing that *P-T*–dependent mineral equilibria can be written (Indares and Rivers, 1995) and provided that mineral compositions were not reset during subsequent retrogression. In the coronites of the Sobrado unit, *P-T* calculations can be performed on samples of groups II and III only.

Temperature determinations may be obtained by application of Grt-Cpx thermometry and Grt-Opx thermometry (group II only) to adjacent garnet and pyroxene rims. No appropriate equilibria can be written for group II samples for estimating pressure. However, upper pressure limits may be calculated by using equilibria with a missing phase. Given the coexistence of Opx-Grt-Pl and Cpx-Grt-Pl in the outer rims of the ferromagnesian domains, the following reactions can be used: (1) 3An + 3Di = 2Grs + Prp + 3Qtz (Cpx-Grt-Pl); and (2) 3An + 3En = Grs + 2Prp + 3Qtz (Opx-Grt-Pl), with quartz as the missing phase. In group III samples coexistence between garnet (outer rims of the ferromagnesian domains) and plagioclase with kyanite and corundum inclusions allows calculation of pressures with the equilibrium: 3An + Crn = Grs + 3Ky (Grt-Pl-Ky-Crn) (Indares and Rivers, 1995).

Ideally, analyses used for temperature calculations should be derived from the same microtextural domain as those used for pressure calculations. This is the case for group II samples, because some Grt-Cpx and Grt-Opx pairs used for thermometry coexist with plagioclase and are also used for Pl-Opx/Cpx-Grt barometry. However, in group III samples, temperature estimates can only be obtained at the inner rims of garnet coronas in contact with clinopyroxene, whereas pressure estimates can only be obtained at the outer rims of those coronas, in contact with plagioclase (+ Ky + Crn). In both microtextural settings of this group, chemical equilibrium was most likely achieved close to the maximum temperature conditions. Consequently, *P-T* estimates can be combined, provided that mineral compositions were not affected by subsequent resetting. Chemical resetting due to Fe-Mg exchange during cooling would result in decrease of the Mg/Fe ratio in garnet and increase of the Mg/Fe ratio in clinopyroxene toward their mutual contacts. In the absence of these features, the compositions of adjacent Grt-Cpx rims are interpreted to reflect maximum temperature equilibrium. Resetting of adjacent Grt-Pl pairs would be sensitive to decompression, leading to decrease of grossular in garnet and/or increase of anorthite plagioclase toward their mutual contacts. However, these features were not observed. Therefore, it is concluded that adjacent rims of plagioclase and garnet maintained the compositions attained during maximum pressure conditions and that peak pressure conditions were coincident with peak temperature conditions.

Method

Thermobarometry was applied to three representative samples (208, group II, and 129 and 671, group III) by using Berman's database TWQ202 (updated from Berman, 1991). Activity models used were those incorporated in TWQ202, i.e., Fuhrman and Lindsley's (1988) for plagioclase and that of Berman et al. (1995) for garnet, orthopyroxene, and clinopyroxene. Activity models for pyroxenes take into account Fe^{2+}-Al and Mg-Al interactions and constitute an important improvement over the ideal two-site mixing model used in previous calibrations. Temperatures were also calculated with the Krogh (1988) calibration of the Grt-Cpx thermometer that is widely used in high-pressure rocks, in order to compare the results with those of the TWQ202.

In each thin section 20–30 temperature determinations and 15–25 pressure determinations were made. *P-T* ranges calculated with TWQ202 are shown in Figure 7. For all samples, temperatures calculated with TWQ202 (Grt-Cpx equilibrium) range between 660–770 °C. The ~100 °C spread of the *T* values is attributed to variations in the estimated Fe^{3+}/Fe^{tot} ratio of the clinopyroxene analyses. Because the analyzed clinopyroxenes are low in Fe (Table 3), and Fe^{3+} is calculated by charge balance (see Appendix), small analytical errors that are most likely responsible for the calculated variations in Fe^{3+} can have a significant effect on the calculated temperatures. In contrast, temperatures calculated by the Opx-Grt equilibria (sample 208) cover a narrower range, 670–745 °C. Orthopyroxene has more Fe than clinopyroxene; therefore variations in Fe^{3+}/Fe^{tot} due to analytical error have less effect on calculated temperatures.

Temperatures calculated with Krogh's (1988) calibration range between 600 and 770 °C, with more estimates on the high-*T* side of the range. The larger spread of *T* values obtained with this calibration relative to TWQ202 may be due to the fact that the former empirically attributes all effects of nonideality to the Grs content of garnet and does not account for interactions in clinopyroxene.

For the estimated temperature range, pressures calculated with Grs-Pl-Ky-Crn equilibrium range between 12.5–16 kbar (sample 129) and 13–17 kbar (sample 671). Maximum pressures calculated in sample 208 range between 12–15 kbar (Grt-Pl-Cpx-Qtz barometry), and 9.5–11 kbar (Grt-Pl-Opx-Qtz barometry). Within individual samples *P-T* isopleths calculated by the barometers (with TWQ202) show limited dispersion with maximum spread of 1 kbar. Therefore, spread in the calculated pressures is due to the dependence of the barometers on temperature.

An important problem for the deduction of the *P-T* path followed by the metagabbros arises because precise *P-T* determinations cannot be made in the garnet-free mineral assemblages of group I coronas. According to the reaction history proposed for group I coronas, the mineral assemblages with Ant + Tre + Crn were probably developed in a relatively early metamorphic stage and so predate the mineral assemblages with Opx + Par, generated during the higher grade metamorphic stage preserved in group I coronas. Although there is limited experimental information relative to the stability of pyroxenes in metabasites, what is available shows that development of Opx in mafic compositions is only possible in high-*T* conditions, ever inside of the granulite facies (Spear, 1981, 1993). Hence, this is the probable *T* range for group I coronas.

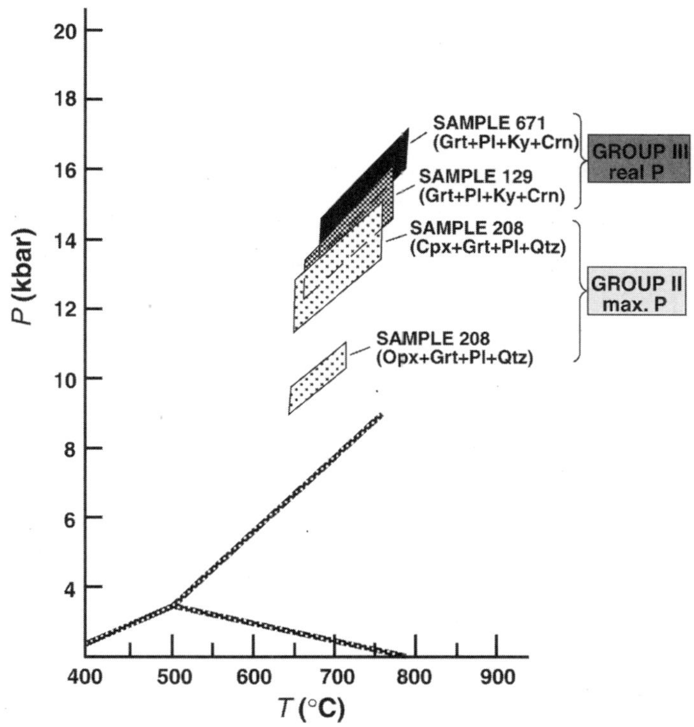

Figure 7. Pressure-temperature (*P-T*) diagram showing thermobarometric data for mineral assemblages of coronas of groups II and III. Stability of Al_2SiO_5 polymorphs according to Holdaway (1971).

The pressure range of the group I coronas, although highly imprecise, would be lower than the baric range calculated for group II, which varies between 12–15 kbar (Grt-Pl-Cpx-Qtz barometry) and 9.5–11 kbar (Grt-Pl-Opx-Qtz barometry). This lower *P* range (consistent with the absence of garnet in group I coronas and with their lower degree of recrystallization in relation to group II coronas) is also compatible with the available experimental data. These data clearly show how the development of garnet-free mineral assemblages with Opx + Cpx in metabasites, although very dependent of protolith composition, is possible in general below 8 kbar (Ringwood and Green, 1966; Green and Ringwood, 1967).

P-T PATH AND TECTONIC INTERPRETATION

Although mineral assemblages of group I coronas do not allow precise *P-T* determinations, and despite the maximum character of *P* determinations obtained in group II coronas, data presented herein imply that the three corona groups define a prograde *P-T* path for the metagabbros of the upper slice. This path explains the transition from group I to group III coronas. This transition represents a drastic increase in pressures, almost isothermal, that operated at high temperatures, in the low-*T* range of the granulite facies (Fig. 8). This prograde *P-T* path accounts for the ensemble of textural, mineralogical, and chemical characteristics of the coronas that developed during the early metamorphic evolution of the high-*P*–high-*T* Sobrado unit. The

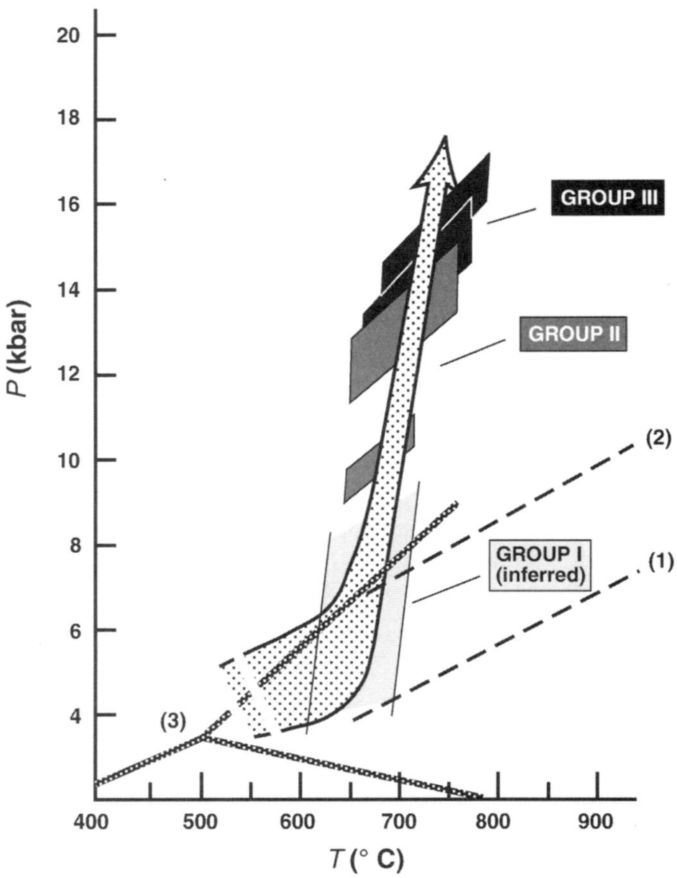

Figure 8. Pressure-temperature (*P-T*) path suggested for coronitic metagabbros of Sobrado unit. *P-T* conditions for development of groups I, II, and III coronas are also shown, together with lower (1) and upper (2) conditions for beginning of garnet stability in mafic granulites (after Ringwood and Green, 1966; Green and Ringwood, 1967). Curve 3 shows Al_2SiO_5 polymorphs stability (after Holdaway, 1971).

prograde evolution advanced in the lower parts of the high-*P*–high-*T* unit, through the upper baric range of the granulite facies, up to eclogite facies (Martínez Catalán and Arenas, 1992).

Metamorphic *P-T* paths characterized by a drastic increase in pressure under high-*T* conditions, such as the one deduced for the coronitic metagabbros of the Sobrado unit, are typical of subduction settings. Numerical simulations of the *P-T-t* paths characteristic of subduction zones developed in normal oceanic lithospheres (Peacock, 1990, 1991), suggest that prograde paths similar to that deduced in the Sobrado coronitic metagabbros are possible in two specific contexts: (1) subduction of young (i.e., relatively hot) oceanic crust; and (2) subduction zones very recently initiated, where subduction operates below a mantle wedge not yet refrigerated by the continuous underplating.

The lithological ensemble of the Sobrado unit does not seem to represent a normal oceanic crust. Either a continental margin, a forearc, or a backarc setting is suggested by the large amount of metasediments occurring in the intermediate and upper parts of the unit. Nevertheless, it can be expected that the simulations of Peacock (1990, 1991) are also valid for the thermal evolution of subducted thin continental or arc-related crustal slices, providing that the mantle wedge below which the subduction takes place is hot enough to account for the initial heating of the slab. Accordingly, the second possibility appears to explain the prograde, burial-related path affecting the metagabbros. A subduction below a mantle wedge with an intact thermal structure (i.e., a mantle wedge not refrigerated by previous underplating or possibly heated and thermally softened in a relatively inner part of a volcanic arc) is considered a reasonable explanation of the peculiar *P-T* path obtained.

The research on the coronitic metagabbros of the Sobrado unit clearly shows that it is still possible to use the complex reaction history of the corona textures to reconstruct the early metamorphic evolution of a high-*P*–high-*T* unit. In this way our research suggests that the study of the complex metamorphic reactions characteristic of the coronitic metagabbros can be used to deduce the *P-T* conditions for their development, and therefore represents an important tool for the elucidation of the early (and generally lost) parts of the orogenic activity.

APPENDIX: ANALYTICAL TECHNIQUES

Mineral analyses were performed on a CAMECA SX50 electron probe in ED mode at Memorial University with 20 nA specimen current (except for plagioclase, for which 10 nA current was used), 15 kV accelerating potential, and 75 s counting time. Data were reduced by a ZAF correction program. In each type of microtextural domain, several traverses were performed with analysis spacing ranging from 25 μm to 50 μm. Additional analyses of adjacent garnet-clinopyroxene, garnet-orthopyroxene, and garnet-plagioclase rims were performed in samples used for thermobarometry. Fe^{3+} in clinopyroxene was determined by the charge balance scheme of Droop (1987) and the end members were calculated according to the method of Cawthorn and Collerson (1974).

ACKNOWLEDGMENTS

This study was financed by the projects PB94-1396-CO2 and PB97-0234-CO2 of the Spanish agency Dirección General de Investigación Científica y Técnica (DGICYT) and by the NATO Collaborative Research Grant 950190. Some of the ideas presented in this paper were developed during long and stimulating discussion with Aphrodite Indares (Memorial University of Newfoundland, Canada). The Sobrado tectonic window was carefully mapped by José de Pablo Maciá in the mid-1980s. By using his map we pay tribute to his memory and intend to acknowledge his friendship and cooperation. Reviews by J. Alcock and M. Mendia resulted in considerable improvement of the manuscript, and are very much appreciated.

REFERENCES CITED

Abati, J., Dunning, G.R., Arenas, R., Díaz García, F., González Cuadra, P., Martínez Catalán, J.R., and Andonaegui, P., 1999, Early Ordovician orogenic event in Galicia (NW Spain): Evidence from U-*P*b ages in the uppermost unit of the Ordenes Complex: Earth and Planetary Science Letters, v. 165, p. 213–228.

Arenas, R., 1988, Evolución petrológica y geoquímica de la unidad alóctona inferior del complejo metamórfico básico-ultrabásico de Cabo Ortegal (Unidad de Moeche) y del Silúrico paraautóctono, Cadena Hercínica Ibérica (NW de España): Corpus Geologicum Gallaeciae, v. 4, 543 p.

Arenas, R., 1991, Opposite *P, T, t* paths of Hercynian metamorphism between the upper units of the Cabo Ortegal Complex and their substratum (northwest of the Iberian Massif): Tectonophysiscs, v. 191, p. 347–364.

Arenas, R., Gil Ibarguchi, J.I., González Lodeiro, F., Klein, E., Martínez Catalán, J.R., Ortega Gironés, E., Pablo Maciá, J.G. de, and Peinado, M., 1986, Tectonostratigraphic units in the complexes with mafic and related rocks of the NW of the Iberian Massif: Hercynica, v. 2, p. 87–110.

Arenas, R., Rubio Pascual, F.J., Díaz García, F., and Martínez Catalán, J.R., 1995, High-pressure micro-inclusions and development of an inverted metamorphic gradient in the Santiago Schists Ordenes Complex, NW Iberian Massif, Spain): Evidence of subduction and syncollisional decompression: Journal of Metamorphic Geology, v. 13, p. 141–164.

Arenas, R., Abati, J., Martínez Catalán, J.R., Díaz García, F., and Rubio Pascual, F.J., 1997, *P-T* evolution of eclogites from the Agualada unit (Ordenes Complex, northwest Iberian Massif, Spain): Implications for crustal subduction: Lithos, v. 40, p. 221–242.

Berman, R.G., 1991, Thermobarometry using multi-equilibrium calculations: A new technique, with petrological applications: Canadian Mineralogist, v. 29, p. 833–855.

Berman, R.G., Aranovich, L.Y., and Patisson, D.R.M., 1995, Reassessment of the garnet-clinopyroxene Fe-Mg thermometer: II Thermodynamic analysis: Contributions to Mineralogy and Petrology, v. 119, p. 30–42.

Cawthorn, A.F., and Collerson, K.D., 1974, The recalculation of pyroxene endmember parameters and estimation of ferrous and ferric iron content from microprobe analysis: American Mineralogist, v. 59, p. 1203–1208.

Chinner, G.A., and Dixon, J.E., 1973, Some high-pressure parageneses of the Allalin gabbro, Valais, Switzerland: Journal of Petrology, v. 14, p. 185–202.

Dallmeyer, R.D., Martínez Catalán, J.R., Arenas, R., Gil Ibarguchi, J.I., Gutiérrez-Alonso, G., Farias, P., Aller, J., and Bastida, F., 1997, Diachronous Variscan tectonothermal activity in the NW Iberian Massif: Evidence from $^{40}Ar/^{39}Ar$ dating of regional fabrics: Tectonophysics, v. 277, p. 307–337.

Díaz García, F., Arenas, R., Martínez Catalán, J.R., González del Tánago, J., and Dunning, G.R., 1999a, Tectonic evolution of the Careón ophiolite (northwest Spain): A remnant of the oceanic lithosphere in the Variscan Belt: Journal of Geology, v. 107, p. 587–605.

Díaz García, F., Martínez Catalán, J.R., Arenas, R., and González Cuadra, P., 1999b, Structural and kinematic analysis of the Corredoiras detachment: Evidence for early Variscan synconvergent extension in the Órdenes Complex, NW Spain: International Journal of Earth Sciences, v. 88, p. 337–351.

Droop, G.T. R., 1987, A general equation for estimating Fe^{3+} concentrations in ferromagnesian silicates and oxides from microprobe analysis, using stoichiometric criteria: Mineralogical Magazine, v. 51, p. 431–435.

Fernández-Suárez, J., Corfu, F., Arenas, R., Marcos, A., Martínez Catalán, J.R., Díaz García, F., Abati, J., and Fernández, F.J., 2002, U-Pb evidence for a polyorogenic evolution of the high-*P*–high-*T* units of the NW Iberian Massif: Contributions to Mineralogy and Petrology, v. 143, p. 236-253.

Fuhrman, M.L., and Lindsley, D.H., 1988, Ternary feldspar modelling and thermobarometry: American Mineralogist, v. 73, p. 201–215.

Gil Ibarguchi, J.I., 1995, Petrology of jadeite-metagranite and associated orthogneiss from the Malpica-*T*uy allochthon (northwest Spain): European Journal of Mineralogy, v. 7, p. 403–415.

Gil Ibarguchi, J.I., and Ortega Gironés, E., 1985, Petrology, structure and geotectonic implications of glaucophane-bearing eclogites and related rocks from the Malpica-Tuy (MT) unit, Galicia, Northwest Spain: Chemical Geology, v. 50, p. 145–162.

Gil Ibarguchi, J.I., Mendia, M., Girardeau, J., and Peucat, J.J., 1990, Petrology of eclogites and clinopyroxene-garnet metabasites from the Cabo Ortegal Complex (northwestern Spain): Lithos, v. 25, p. 133–162.

Girardeau, J., and Gil Ibarguchi, J.I., 1991, Pyroxene-rich peridotites of the Cabo Ortegal Complex (northwestern Spain): Evidence for large-scale upper mantle heterogeneity, *in* Menzies, M.A., Dupuy, C., and Nicolas, A., eds., Orogenic lherzolites and mantle processes: Journal of Petrology Special Volume, p. 135–154.

Gómez Pugnaire, M.T., Mottana, A., Bocchio, R., Liborio, G., and Abraham, K., 1979, Coronitic eclogites in the Sierra de Baza (Betic Cordilleres, Spain): Neues Jahrbuch für Mineralogie Abhandlungen, v. 136, p. 42–62.

Green, D.H., and Ringwood, A.E., 1967, An experimental investigation of the gabbro-to-eclogite transformation and its petrological applications: Geochimica et Cosmochimica Acta, v. 31, p. 767–833.

Griffin, W.L., and Heier, K.S., 1973, Petrological implications of some corona structures: Lithos, v. 6, p. 315–335.

Holdaway, M.J., 1971, Stability of andalusite and the aluminum silicate phase diagram: American Journal of Science, v. 271, p. 97–131.

Indares, A., 1993, Eclogitized gabbros from the eastern Grenville Province: Textures, metamorphic context, and implications: Canadian Journal of Earth Science, v. 30, p. 159–173.

Indares, A., and Dunning, G., 1997, Coronitic metagabbro and eclogite from the Grenville Province of western Quebec: Interpretation of U-Pb geochronology and metamorphism: Canadian Journal of Earth Science, v. 34, p. 891–901.

Indares, A., and Rivers, T., 1995, Textures, metamorphic reactions and thermobarometry of eclogitized metagabbros: A Proterozoic example: European Journal of Mineralogy, v. 7, p. 43–56.

Kretz, R., 1983, Symbols for rock-forming minerals: American Mineralogist, v. 68, p. 277–279.

Krogh, E.J., 1988, The garnet-clinopyroxene Fe-Mg geothermometer—A reinterpretation of experimental data: Contributions to Mineralogy and Petrology, v. 99, p. 44–48.

Lasnier, B., 1977, Persistance d'une série granulitique au coeur du Massif Central Francais (Haut-Allier). Les termes basiques, ultrabasiques et carbonatés [Ph.D. thesis]: Nantes, France, Université de Nantes, 351 p.

Marcos, A., Marquínez, J., Pérez-Estaún, A., Pulgar, J., and Bastida, F., 1984, Nuevas aportaciones al conocimiento de la evolución tectonometamórfica del Complejo de Cabo Ortegal (NW de España): Cuadernos del Laboratorio Geológico de Lage, v. 7, p. 125–137.

Martínez Catalán, J.R., 1990, A noncylindrical model for the northwest Iberian allochthonous terranes and their equivalents in the Hercynian belt of Western Europe: Tectonophysics, v. 179, p. 253–272.

Martínez Catalán, J.R., and Arenas, R., 1992, Deformación extensional de las unidades alóctonas superiores de la parte oriental del Complejo de Órdenes (Galicia): Geogaceta, v. 11, p. 108–111.

Martínez Catalán, J.R., Klein, E., Pablo Maciá, J.G. de, and González Lodeiro, F., 1984, El Complejo de Órdenes: Subdivisión, descripción y discusión sobre su origen: Cuadernos del Laboratorio Geológico de Lage, v. 7, p. 139–210.

Martínez Catalán, J.R., Arenas, R., Díaz García, F., Rubio Pascual, F.J., Abati, J., and Marquínez, J., 1996, Variscan exhumation of a subducted Paleozoic continental margin: The basal units of the Órdenes Complex, Galicia, NW Spain: Tectonics, v. 15, p. 106–121.

Martínez Catalán, J.R., Arenas, R., Díaz García, F., and Abati, J., 1997, Variscan accretionary complex of northwest Iberia: Terrane correlation and succession of tectonothermal events: Geology, v. 27, p. 1103–1107.

Martínez Catalán, J.R., Arenas, R., Díaz García, F., and Abati, J., 1999, Allochthonous units in the Variscan belt of NW Iberia: Terranes and accretionary history, *in* Sinha, A.K., ed., Basement Tectonics 13: Dordrecht, Netherlands, Kluwer, p. 65–84.

Mendia, M., 2000, Petrología de la unidad eclogítica del Complejo de Cabo Ortegal (NW de España): Nova Terra, v. 16, 424 p.

Meyer, J., 1983, The development of the high-pressure metamorphism in the Allalin metagabbro (Switzerland): Terra Cognita, v. 3, p. 187.

Mørk, M.B., 1986, Coronite and eclogite formation in olivine gabbro (western Norway): Reaction paths and garnet zoning: Mineralogical Magazine, v. 50, p. 417–426.

Ordóñez Casado, B., Gebauer, D., Schäfer, H.J., Gil Ibarguchi, J.I., and Peucat, J.J., 2001, A single subduction event for the high-*P*/high-*T* metamorphism of the Cabo Ortegal complex within the Iberian Massif: Tectonophysics, v. 332, p. 359–385.

Pablo Maciá, J.G. de, and Martínez Catalán, J.R., 1984, Estructura, Petrología y evolución de la región de Sobrado de los Monjes (La Coruña): Cuadernos del Laboratorio Geológico de Lage, v. 7, p. 103–124.

Peacock, S.M., 1990, Numerical simulation of metamorphic pressure-temperature-time paths and fluid production in subducting slabs: Tectonics, v. 9, p. 1197–1211.

Peacock, S.M., 1991, Numerical simulation of subduction zone pressure-temperature-time paths: Constraints on fluid production and arc magmatism: Philosophical Transactions of the Royal Society of London, v. A-335, p. 341–353.

Ringwood, A.E., and Green, D.H., 1966, An experimental investigation of the gabbro-eclogite transformation and some geophysical implications: Tectonophysics, v. 3, p. 383–427.

Rivers, T., and Mengel, F.C., 1988, Contrasting assemblages and petrogenetic evolution of corona and noncorona gabbros in the Grenville Province of western Labrador: Canadian Journal of Earth Science, v. 25, p. 1629–1648.

Schäfer, H.J., Gebauer, D., Gil Ibarguchi, J.I., and Peucat, J.J., 1993, Ion-microprobe U-Pb zircon dating on the high-*P*/high-*T* Cabo Ortegal Complex (Galicia, NW Spain): Preliminary results: Terra Abstracts, v. 5–4, p. 22.

Spear, F.S., 1981, An experimental study of hornblende stability and compositional variability in amphibolite: American Journal of Science, v. 281, p. 697–734.

Spear, F.S., 1993, Metamorphic phase equilibria and pressure-temperature-time paths: Mineralogical Society of America, Monograph, 799 p.

Vogel, D.E., 1967, Petrology of an eclogite- and pyrigarnite-bearing polymetamorphic rock complex at Cabo Ortegal, NW Spain: Leidse Geologische Mededelingen, v. 40, p. 121–213.

Manuscript Accepted by the Society February 8, 2002

Printed in the U.S.A.

Geological Society of America
Special Paper 364
2002

Variscan pressure-temperature evolution of garnet pyroxenites and amphibolites in the Baie d'Audierne metamorphic series, Brittany (France)

Harald Lucks
Institut für Geologie und Mineralogie, Schloßgarten 5, D-91054 Erlangen, Germany
Bernhard Schulz*
Institut für Mineralogie, Technische Universität Bergakademie, Brennhausgasse 14, D-09596 Freiberg/Sachsen, Germany
Claude Audren
Géosciences Rennes, Laboratoire de Tectonique, Unité Mixte de Recherche, Centre National de la Recherche Scientifique 6118, Avenue du Général Leclerc, F-35042 Rennes Cedex, France
Claude Triboulet
Laboratoire de Pétrologie, Unité de Recherche Associée, Centre National de la Recherche Scientifique, 736, Université Pierre et Marie Curie, 4 place Jussieu, F-75252 Paris Cedex 05, France

ABSTRACT

In the Baie d'Audierne of the South Armorican domain, an intermediate metabasic-ultrabasic complex with isolated garnet pyroxenite lenses is exposed between upper and lower metasedimentary units. Metabasites in all units show geochemical characteristics similar to mid-ocean ridge magmatites. Garnets in pyroxenites are zoned with increasing *X*Mg from cores to rims, and coexist with clinopyroxene (Jd_{10} to Jd_{33}) and plagioclase (An_{17} to An_{35}) of a mineral assemblage M1. Maximal pressure-temperature (*P-T*) conditions of the M1 metamorphism range between 800 and 950 °C and 14 and 19 kbar in different garnet pyroxenite lenses. The high-*P*–high-*T* rocks were overprinted by an assemblage M2 with Ca-amphiboles. Brown-green tschermakitic Ti-hornblende is in some cores and there are core to rim zonations from actinolite to tschermakite. Increasing Al^{IV} and Al^{VI} toward the rims of green amphiboles signal increases of *P* and *T* from 3 to 9 kbar and <500 to 680 °C. In the metabasites of the lower metasedimentary unit, the Ca-amphiboles display similar zonation trends from actinolite cores to tschermakite rims, and a similar prograde *P-T* path up to 630 °C/6 kbar. Amphibolites in the upper metasedimentary unit show brown-green amphibole, interpreted as magmatic relict. Zonations from actinolite to magnesio-hornblende surround these cores and recorded prograde *P-T* paths to 5 kbar and 550 °C. The high-*P*–high-*T* rocks were welded with the metasedimentary units at the end of the first decompression path M1-M2. Subsequently the units underwent new burial and heating along an amphibolite facies *P-T* cycle M2. The present metamorphic configuration could be the consequence of thrusting at a late stage of M2. This two-stage metamorphic history predates late Variscan granite intrusions and differs from *P-T* paths of other Armorican high-*P* units.

*Corresponding author. E-mail: bschulz@geol.uni-erlangen.de.

Lucks, H., Schulz, B., Audren, C., and Triboulet, C., 2002, Variscan pressure-temperature evolution of garnet pyroxenites and amphibolites in the Baie d'Audierne metamorphic series, Brittany (France), *in* Martínez Catalán, J.R., Hatcher, R.D., Jr., Arenas, R., and Díaz García, F., eds., Variscan-Appalachian dynamics: The building of the late Paleozoic basement: Boulder, Colorado, Geological Society of America Special Paper 364, p. 89–103.

INTRODUCTION

The Variscides in Europe are composed of numerous lithostratigraphic units that had complex metamorphic histories. Lenses of eclogite facies metabasites in this orogenic belt provide considerable insight into the pressure-temperature-time (*P-T-t*) evolution during subduction, continental collision, and uplift and/or exhumation processes. They are considered important for the reconstruction of ancient plate boundaries, especially when an ophiolitic character can be recognized. The Baie d'Audierne region in Brittany is part of the South Armorican domain of the Armorican Massif. In the models of Matte (1986) and Dias and Ribeiro (1995) this domain is part of a Devonian inner suture zone of the Variscan Belt. In Baie d'Audierne, garnet pyroxenites occur with serpentinites, metagabbros, amphibolites, and greenschists of an ophiolite complex between upper and lower metasedimentary units. This chapter deals with the petrography, mineral chemistry, and Variscan *P-T* evolution of the metabasites in the stacked units. Thermobarometry based on zonations in Ca-amphiboles provided arguments for a complex uplift history. It will be shown that after a first decompression, the high-*P*–high-*T* rocks underwent a second burial together with the metasediments along a greenschist to amphibolite facies metamorphic cycle.

Regional geological setting

The Armorican Massif in western France is subdivided into the four segments of the North, Central, and South Armorican domains, and the Léon (Fig. 1A). The South Armorican domain, limited by the South Armorican shear zone to the north, is part of the internal Variscan Belt. It is composed of the following upper Proterozoic to lower Paleozoic lithostratigraphic units (Cogné, 1960, 1988; Ballèvre et al., 1994; Brown and Dallmeyer, 1996). (1) A high-grade Lower Group (Golfe de Morbihan Group or Southern Brittany Migmatite Belt) with predominant migmatized gneisses is interbedded with numerous amphibolites and some kinzigite layers (Jones and Brown, 1990; Audren and Triboulet, 1993). (2) An amphibolite facies Middle Group (Rhuys–La Vilaine Group) comprises micaschists and amphibolite layers (Triboulet and Audren, 1985). (3) Greenschist facies volcanosedimentary rocks are found in the Upper Group (Belle-Ile-en-Mer Group, Audren, 1987). (4) A high-*P*–low-*T* rock association with glaucophane-eclogites, blueschists, greenschists, and metapelites is exposed only in the Ile de Groix and at Bois de Cené (Triboulet, 1991; Audren et al., 1993; Schulz et al., 2001) and represents the uppermost unit of the pile.

Eclogite facies rocks and their relicts occur in the Léon region (Cabanis and Godard, 1987; Paquette et al., 1987), in the Essarts Complex (Godard, 1988), in the Champtoceaux Complex between the northern and southern branches of the South Amorican shear zone (Ballèvre et al., 1987, 1989), and in the Baie d'Audierne (Peucat, 1973; Velde, 1973; Cogné, 1988). It is not known if all these Armorican eclogites belong to the Lower Group, because some of the contacts are obscured by late Variscan faults. Among other criteria, the Champtoceaux eclogites with their low-*T*–high-*P* trajectory during retrogression are considered equivalents of Malpica-Tuy in Galice (Ballèvre et al., 1992). The Léon eclogites may rather belong to a Cabo Ortegal–Vendée–Bas-Limousin–Essarts group of high-*P* rocks (Godard, 1988; Godard and Mabit, 1998).

The high-grade Lower Group in the South Armorican domain is intruded by preorogenic granites (orthogneiss at 463 Ma; Peucat et al., 1978), then by synorogenic anatectic granites (376 Ma; Vidal, 1973). Late orogenic leucogranites (330–300 Ma; Vidal, 1973) occur in the Middle and Upper Groups. These intrusions allow the distinction of eo-Variscan metamorphic events between 463 and 376 Ma from Variscan events between 330 and 300 Ma (Peucat et al., 1982; Peucat, 1983; Paquette et al., 1985; Audren and Triboulet, 1993; Ballevre et al., 1994; Brown and Dallmeyer, 1996). A multistage eo-Variscan metamorphism is observed in the migmatitic Lower Group (Jones and Brown, 1990; Audren and Triboulet, 1993; Schulz et al., 1996). A complex polymetamorphic evolution with a high-*P* stage followed by amphibolite and greenschist facies metamorphism has been described in the uppermost high-*P*–low-*T* unit of Ile de Groix (Audren et al., 1993; Schulz et al., 2001). This contrasts with the monometamorphic Middle and Upper Groups with single clockwise *P-T* paths (Triboulet and Audren, 1985, 1988). Within this framework, the Baie d'Audierne region and its high-*P*–high-*T* rocks provide important information for the reconstruction of the Variscan tectonometamorphic evolution in the South Armorican domain and in larger scale models (e.g., Shelley and Bossière, 2000).

Geological setting in the Baie d'Audierne region

The Baie d'Audierne region (Fig. 1B) is bounded by the dextral South Armorican shear zone to the north and by a conjugate sinistral shear zone limiting the Pont l'Abbé granite to the south. An extension offshore is known (Lefort and Peucat, 1974; Poulpiquet, 1988). Because the Baie d'Audierne metamorphic sequences are entirely surrounded by late Variscan granites, it is difficult to relate them to the other principal units of the South Armorican domain mentioned here. According to Peucat (1973) and Peucat and Cogné (1974), several lithostratigraphic units in two main metamorphic series can be recognized. A lower metasedimentary unit consists of micaschists (Trunvel Formation) with albite, garnet, chloritoid, and muscovite, and contains some amphibolite layers and lenses. An intermediate metamagmatic unit or metabasic-ultrabasic complex includes: (1) greenschists of volcanic origin (Tréogat Formation) with Ca-amphibole, epidote, plagioclase and chlorite; (2) amphibolites (Peumerit Formation) with Ca-amphiboles, plagioclase, epidote, chlorite, and sometimes garnet; (3) serpentinites (Ty-Lan Formation) with rare olivine, clinopyroxene and locally

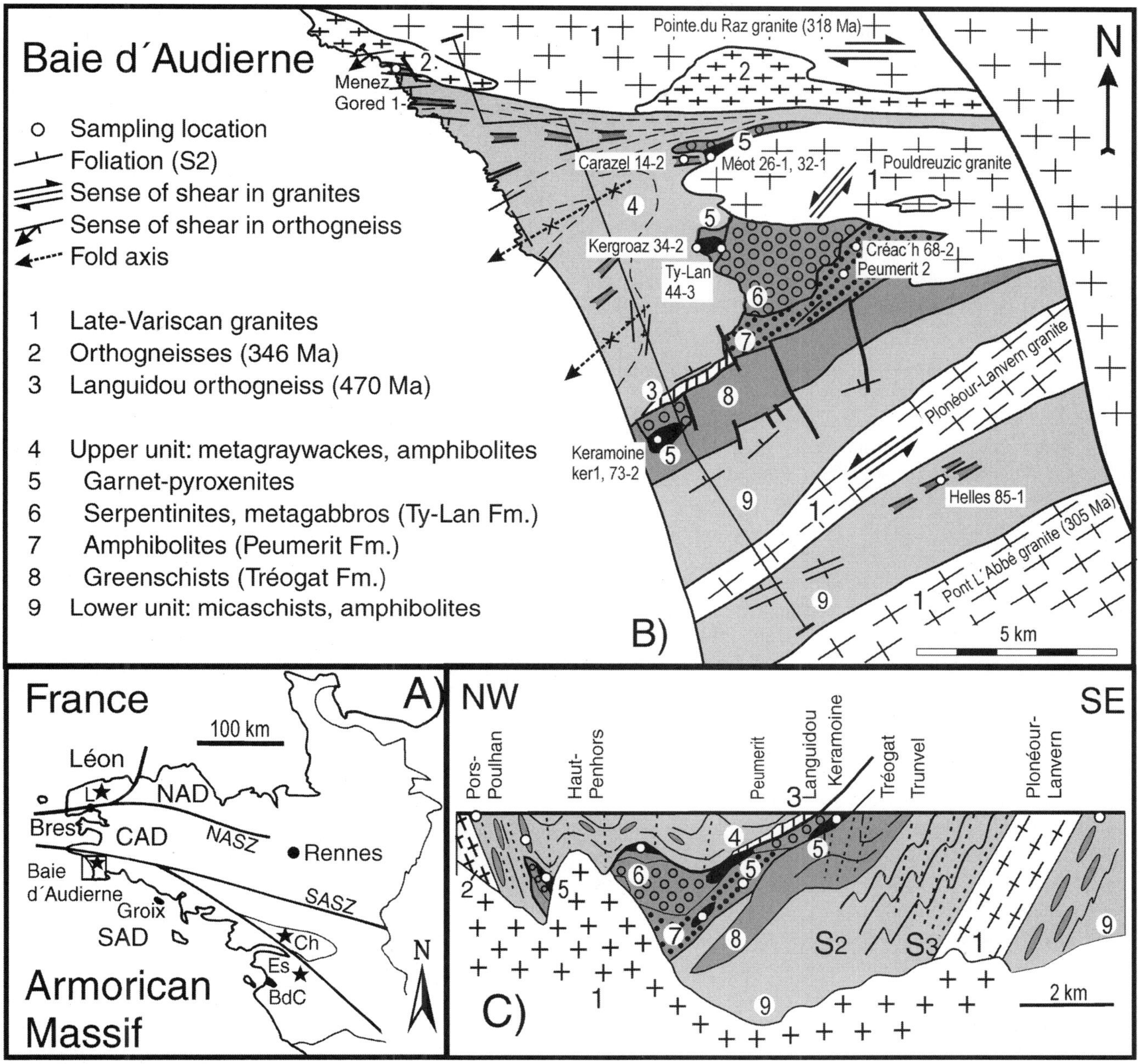

Figure 1. A: Regional geology in Armorican Massif. Tectonic subdivison and occurrences of high-pressure rocks (stars) in addition to those from Baie d'Audierne. BdC—Bois de Cené; Ch—Champtoceaux; Es—Essart; Groix—Ile de Groix; L— Lesneven; CAD—Central Armorican domain; NAD—North Armorican domain; SAD—South Armorican domain; NASZ—North Armorican shear zone; SASZ—South Armorican shear zone. B: Geological map of Baie d'Audierne region, modified from Peucat (1973) and Plaine and Peucat (1978) and with our observations. C: Geological cross section, modified from Peucat (1973). Sampling locations (partly projected) for pressure-temperature path reconstructions are marked. Sampling locations (Gauβ-Krüger coordinates to the west of Greenwich): Méot: 1399640/5312995; Kergroaz: 1399325/5310495; Keramoine: 1398248/5306303; Creac'h: 1403235/5310780; Ty-Lan: 1399945/5310573; Peumerit–Pont Illis quarry: 1402770/5310048; Helles: 1406210/5305665; Menez Gored: 1391970/5315495; Carazel: 1399105/5313218.

chromite; (4) metagabbros rarely exhibiting coronitic structures (Lasnier, 1970); and (5) garnet pyroxenites with garnet, clinopyroxene, plagioclase, rutile, and quartz.

The upper metasedimentary units are metagraywackes and micaschists with some amphibolite layers. At the base of the upper unit is the mylonitic orthogneiss of Languidou. Rare and thin layers of garnet- and kyanite-bearing micaschists and a boudin of kyanite-bearing eclogite (Marchand, 1982) mark the discontinuity between the upper unit and the metabasic-ultrabasic complex (Fig. 1, B and C). From comparison of structures and metamorphic mineral assemblages, the lower metasedimentary unit and the metabasic-ultrabasic complex were assigned to a lower metamorphic series; the upper metasedimentary unit represents an upper metamorphic series. The allochthonous character of the upper unit with a thrust displacement upon the metabasic-ultrabasic complex has been suspected (Cogné, 1988). The boundary between the Peumerit amphibolites and the Tréogat greenschists has been considered as the basal tectonic discontinuity of the metabasites and ultrabasites within the lower metamorphic series (Peucat and Cogné, 1974). By modeling a depth profile from magnetic and seismic data, Poulpiquet (1988) concluded an allochthonous position of the metabasic-ultrabasic complex, and pointed to a similar lithostratigraphic situation in the allochthonous Morais Complex of the Iberian Meseta.

Protolith ages of the Baie d'Audierne serpentinites were interpreted from an 840 Ma U-Pb zircon upper intercept (Paquette et al., 1985). U-Pb zircon data from the garnet pyroxenites give a lower intercept age of 384 ± 6 Ma (Peucat, 1983), interpreted as the age of a high-*P*–high-*T* event. The orthogneiss of Languidou yielded a discordant U-Pb zircon age of 470 ± 7 Ma, indicating the protolith age (Peucat, 1983). According to geochronological data (Peucat et al., 1982), the metabasic-ultrabasic complex was affected by a metamorphic event ca. 385–370 Ma. A younger metamorphic event then affected both units and was recorded in the Pors-Poulhan orthogneiss, emplaced 346 Ma. Late Variscan granites intruded both sequences and were deformed in shear zones by late transcurrent movements.

In the metasediments of the lower unit in the Baie d'Audierne, the dominant structural element is a main S2 foliation, which is axial planar to isoclinal F2 folds (S1 and P1 in Peucat, 1973) and parallel to the lithological banding. Planar quartz layers are considered as relict S1 and are deformed in isoclinal intrafolial F2 folds. Internal foliation (Si) in garnet and plagioclase porphyroblasts (Peucat, 1973) indicates that the main planar anisotropy developed during progressive deformation. The main foliation is deformed in open to tight F3 folds (P2 in Peucat, 1973) with S3 subvertical axial planes. The presence of chlorite in metapelites, and chlorite, hornblende, and epidote in metabasites, as well as the planar anisotropy in flasergabbro, underline the S3 axial planes. F3 fold axes plunge 10°–25° to the west. The F3 fold axes are parallel to a lineation (L3) defined by mineral aggregates or preferentially oriented long prismatic minerals. Peucat (1973) found evidence that L3 lineation could be coaxial to an older stretching lineation (L2), associated with the F2 folding. The small-scale F3 folds are minor structures of kilometer-scale synforms and antiforms.

In the upper unit, the main S2 foliation (S1 in Peucat, 1973) varies in dip toward the north and the south. This is due to F3 folding. Small-scale F3 folds are minor structures of a kilometer-scale synform with a steeply dipping northern limb at Pors Poulhan and a flatly dipping southern limb at Languidou (Fig. 1C). Axial planes of F3 are in general subvertical, but horizontal axial planes occur as well, explained by progressive shortening of F3 folds (Peucat, 1973). A lineation (L3) marked by a preferential orientation of amphibole in metabasite layers and of quartz-feldspar aggregates in metagraywackes plunges 20°–40° to the west and is parallel to the F3 fold axes. The similar F3 fold geometry and orientation provide an argument for a common deformation of the upper and lower units. Foliation and lineation parallel to F3 axial planes and L3 in the 335 ± 20 Ma Pors-Poulhan (Peucat, 1973) and the 280 Ma Plonéour-Lanvern, Plozévet, and Pouldreuzic (Vidal, 1973) granites indicate a post-granitic age for this deformation.

SC-structures in the granites and shear bands that cut across the main foliation in micaschists give evidence of postgranitic and late Carboniferous tectonism. To the north in the Pointe du Raz granite, dextral shear along west-east–striking planes is observed. In contrast, sinistral shear along northeast-southwest–striking planes is evident in the Pouldreuzic granite. Sinistral shear also occurs in the Plonéour-Lanvern granite to the south (Fig. 1B). Micaschists of the upper unit show a top-to-the-west shearing parallel to the lineation. In the lower metasedimentary unit, a top-to-the-east sense of shear along shear bands can be stated. Shear senses and the differently oriented SC-planes and shear bands can be understood as synthetic and antithetic Riedel shears and conjugated low-angle normal faults generated by progressive deformation in a general west-east–striking dextral shear regime. This system probably has been coupled to the South Armorican shear zone and resulted from a northwest-southeast–directed compression.

Geochemistry of metabasites

For this study 40 whole-rock samples of metabasites from the three lithological units were analysed by X-ray fluorescence. The data match results from earlier studies (Fig. 2A). In the TiO_2-Al_2O_3 screen (Pearce, 1983), the garnet pyroxenites have signatures of clinopyroxene gabbro. Amphibolites mostly plot in the basalt field. Previous studies (Peucat, 1973; Hanmer, 1977; Bernard-Griffiths and Cornichet, 1985) revealed a tholeiitic trend and trace element contents indicative of an ocean-floor tectonic environment for the metabasitic-ultrabasic intermediate unit, contrasting a calc-alkaline trend in the upper unit amphibolites. The upper unit amphibolites show elevated $K_2O + Na_2O$, which results in a position inside the calc-alkaline field of the FeO-MgO-($Na_2O + K_2O$) diagram. However, K and Na are supposed to be mobile during metamorphism, especially when the host rocks of the basites have metapelitic compositions. In conse-

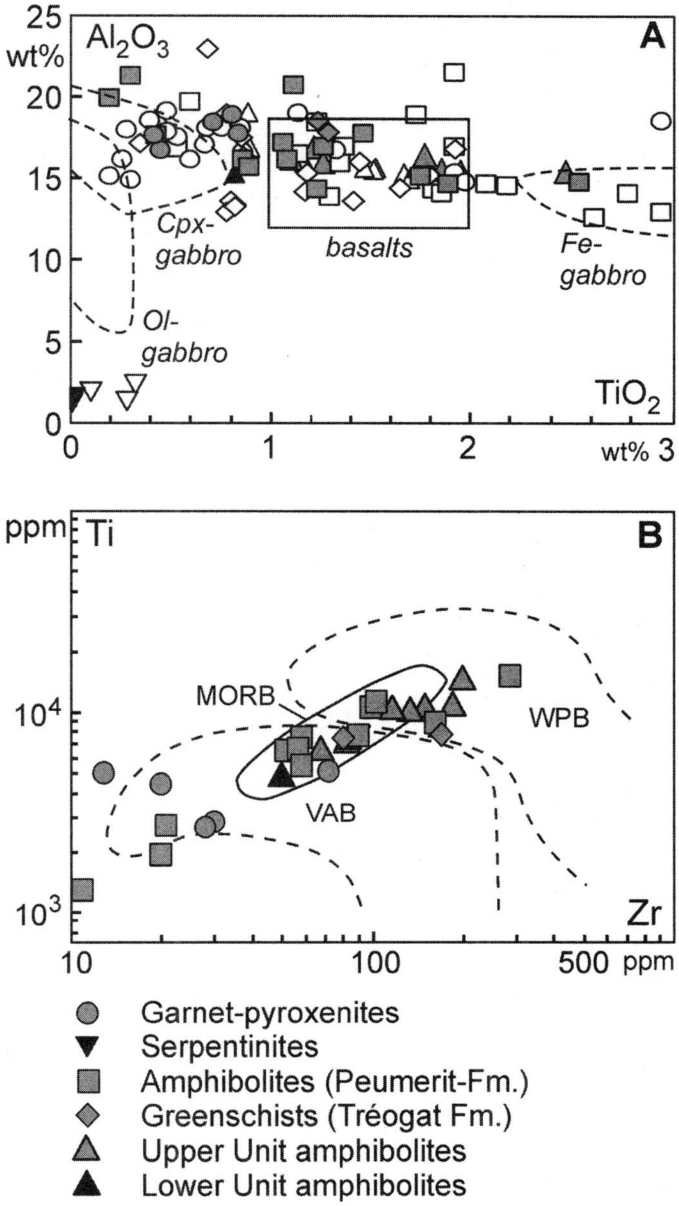

Figure 2. Whole-rock X-ray fluorescence data from Baie d'Audierne metabasites in TiO_2-Al_2O_3 screen of Pearce (1983) and Zr-Ti discrimination diagram of Pearce (1982). Marks with no filling refer to published data of others (see text). MORB—mid-ocean ridge basalt; WPB—within plate basalt; VAB—volcanic arc basalt.

quence, when discrimination diagrams involving less mobile elements such as Al, Y, Ti, and Zr are considered, a subalkaline and tholeiitic character of the upper unit metabasites is obvious and matches the signatures of the basic-ultrabasic suite. Similar correlations of elements with a fractionation index Zr are observed from both groups. The range of MgO is 3–11 wt% in the metabasic-ultrabasic unit and 6%–8% in the amphibolites of the metasedimentary units. Ni and MgO show negative, and Y, Ti, Sr, Ba, and K_2O show positive linear correlations with increasing Zr. Similar and overlapping correlation trends as well as absolute abundances are observed from the units. This is also true when the rocks are compared in mid-ocean ridge basalt (MORB) and chondrite-normalized variation diagrams. In the Zr-Ti discrimination diagram the data define a single fractionation trend, matching the element abundances of MORB (Fig. 2B). Following the major and the limited trace element data, amphibolites from the metasedimentary units and rocks from the metabasic-ultrabasic complex apparently belong to a single magmatic suite with the signatures of mid-ocean ridge magmatites.

Mineral chemistry and petrography

Microprobe analyses (700 points, CAMECA SX 50 and SX 51, 15 kV, 20 nA) were performed in 20 selected thin sections from the metabasites. Cations in amphiboles were normalized to 23 oxygen and sum(T1 + T2 + M1 + M2 + M3) = sumMg = 13.000 or 13 (exCa, Na, K) (described in detail by Triboulet, 1992). $Fe3^+$ has been estimated according to Papike (1974). The zonations in the amphiboles are best described in terms of trends in Al^{IV}-Al^{VI} and Ti-Al^{IV}, but Si, Mg, and Na display the corresponding variations. Garnet pyroxenites were sampled at Méot (samples 26-1, 32-1), Kergroaz (34-2), and Keramoine (ker1, 73-2) (see Fig. 1 for geographic coordinates). They appear as lenses or layers in the uppermost part of the metabasic-ultrabasic complex immediately beneath the upper metasedimentary unit. Strongly retrogressed garnet pyroxenites occur away from this contact at location Creac'h within the Peumerit Formation. The petrography of the garnet pyroxenites was detailed in Lasnier (1970), Peucat (1973), and Peucat et al. (1982). Garnets to 1 cm in diameter occur in these rocks. They enclose clinopyroxene, oligoclase, quartz, and rutile. The matrix is composed of clinopyroxene, plagioclase, zoisite, quartz, rutile, and opaques (Fig. 3A). Thus, the mineral assemblage of the high-*T* metamorphism M1 is Grt + Cpx + Pl + Zoi + Qtz + Rt. The garnets show slight growth zonation with decreasing spessartine and grossular and increasing pyrope contents from cores to rims. The grossular content range is ~20–30 mol% and pyrope contents range from 15% in the cores to 45% in the rims (Fig. 4, A–D). When the garnet compositions in several samples are compared in grossular-pyropespessartine coordinates, coherent zonation trends are found. Garnets in samples Keramoine ker1, 73-2, and Kergroaz 34-2 display segments of the overall zonation trend in sample Méot 26-1 (Fig. 4E; Table 1). An exception is sample Creac'h, where garnet displays an opposite (retrograde) zonation trend with decrease of Mg toward the rim. Primary clinopyroxene in the matrix or enclosed in garnet and coexisting with garnet is poorly zoned. The compositions in samples Méot 26-1, ker1, 73-2, and Creac'h 68-2 range from 10 to 20 mol% jadeite and plot outside the omphacite compositional field in Ca-Fe-Mg-pyroxene–jadeite–aegirine coordinates (Fig. 4F). Omphacite with as much as 33 mol% jadeite occurs in sample Kergroaz 34-2. The plagioclases are calcic with An_{17} to An_{20} in Kergroaz 34-2, and An_{25} to An_{36} in the other garnet pyroxenites. In amphibolitic garnet pyroxenites, relict pyroxene and kelyphitic garnet are metastable

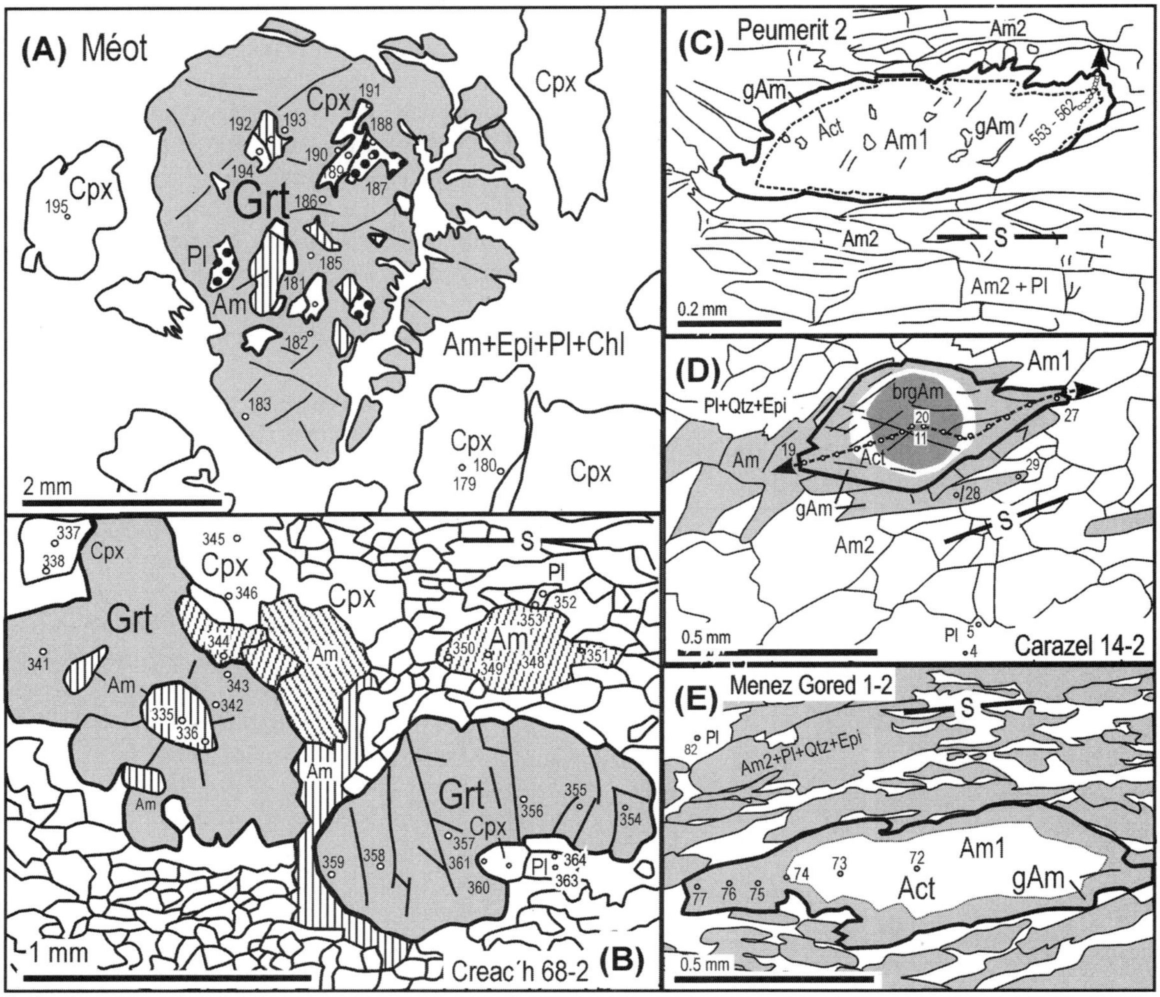

Figure 3. Microstructures in garnet pyroxenites and amphibolites (drawn from thin sections under polarized light). Numbers and arrows are microprobe analyses and profiles. A: Garnet pyroxenite with garnet-clinopyroxene-plagioclase assemblage. B: Garnet pyroxenite with brown-green amphibole. C: Amphibolite of Peumerit Formation with large zoned amphibole 1 and small zoned amphibole 2 of M2 assemblage. Broken line marks zone with actinolite. See Figure 5D for amphibole zonation. D: Coarse-grained amphibolite from upper metasedimentary unit. In amphibole 1, brown-green cores are surrounded by actinolite and green amphibole rims. See Figure 6 (C and F) for zonation profiles. E: Large zoned amphibole with actinolite core and green amphibole rim in fine-grained metabasite from upper unit. See Figure 6 (B and E) for zonation profiles. Act—actinolite, Am—amphibole, brgAm—brown-green amphibole, Cpx—clinopyroxene, Cc—calcite, Epi—epidote, gAm—green amphibole, Pl—plagioclase.

and preserved within an inequigranular polygonal fabric of brown-green amphibole (tschermakitic Ti-hornblende), plagioclase, epidote, chlorite, calcite, titanite, and quartz. Some brown-green amphibole occurs inside garnet, possibly due to intergrowth with the matrix (Fig. 3, A and B). It is not clear whether the brown-green amphiboles belong to a pre-M1 or a post-M1 retrograde assemblage.

Invariably, the M1 assemblage was partially to almost completely replaced by a greenschist to amphibolite facies assemblage M2; garnets are surrounded by coronas of radially arranged acicular, tschermakitic hornblende or magnesiohornblende and plagioclase. Clinopyroxene has been partly to completely replaced by fine- and coarse-grained symplectites with plagioclase and secondary clinopyroxene or amphibole. Late nematoblastic green amphibole grew along garnet cracks, in symplectitic coronas, and as matrix porphyroblasts parallel to a planar anisotropy. Ca-amphiboles of the M2 assemblage can be defined in the Si-Mg ($Ti < 0.5$, $[Na + K]_A < 0.5$) coordinates according to Leake et al. (1997). They display zonations in samples from Méot and Ty-Lan 44-3. There are tschermakitic Ti-hornblendes with large brown-green cores with $Al^{IV} = 1.5$, $Al^{VI} = 0.6$, and $Ti = 0.15$ pfu. Rims of these amphiboles show actinolite composition (Fig. 5, A and G). Amphiboles of M2 with preferentially oriented long axes define a mineral lineation.

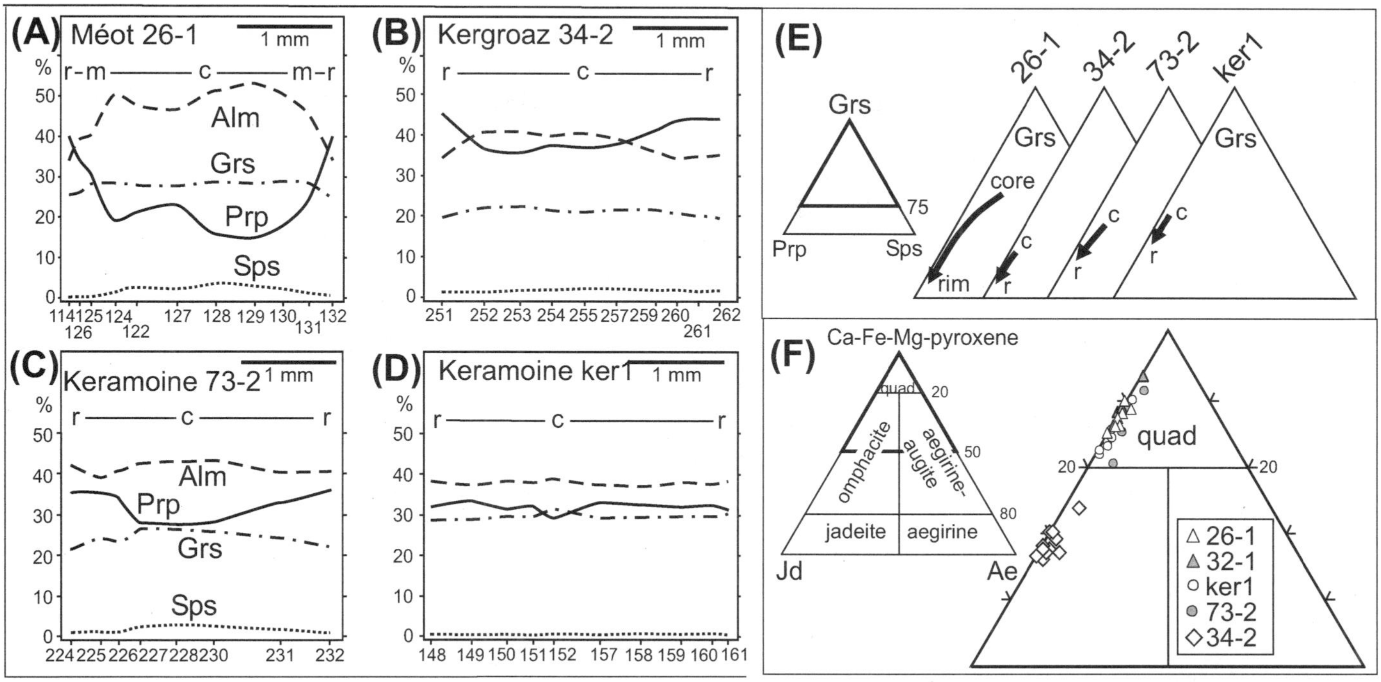

Figure 4. Mineral compositions in garnet pyroxenites and eclogites. A–D: Garnet zonation profiles from core-mid-rim (c, m, r) in almandine (Alm), grossular (Grs), pyrope (Prp), and spessartine (Sps) end members. E: Garnet zonation trends in grossular-pyrope-spessartine coordinates. F: Compositions of clinopyroxenes in Ca-Fe-Mg-pyroxenes and jadeite-aegirine coordinates. See Table 1.

TABLE 1. GARNET PYROXENITES

	Méot 26-2*		Kergroaz 34-2**		26-2	34-2
Sample	Grt 129	Grt 125	Grt 288	Grt 283	Cpx 142	Cpx 293
SiO_2	38.26	39.10	39.88	40.72	51.46	53.45
TiO_2	0.12	0.11	0.02	0.04	0.54	0.38
Al_2O_3	21.56	22.63	21.81	22.32	7.82	11.05
FeO	24.75	19.83	20.73	17.51	4.34	3.76
MnO	1.38	0.27	0.77	0.40	0.13	0.02
MgO	3.84	7.97	9.25	11.68	12.35	9.73
CaO	10.18	10.33	8.05	8.40	20.61	16.85
Na_2O	0.06	0.00	0.01	0.01	1.79	4.55
Cr_2O_3	0.11	0.04	0.00	0.00	0.10	0.00
Total	100.25	100.28	100.53	101.08	98.48	99.43
Si	2.991	2.966	3.014	3.010	1.891	1.920
Ti	0.007	0.007	0.001	0.002	0.015	0.010
Al	1.986	2.023	1.943	1.944	0.338	0.468
Fe^{3+}	0.020	0.029	0.029	0.034	0.000	0.000
Fe^{2+}	1.588	1.215	1.267	1.030	0.133	0.113
Mn	0.092	0.017	0.049	0.025	0.004	0.001
Mg	0.447	0.901	1.042	1.287	0.676	0.521
Ca	0.853	0.840	0.652	0.665	0.812	0.649
Na	0.000	0.000	0.000	0.000	0.127	0.317
Cr	0.006	0.002	0.000	0.000	0.003	0.000

Note: Microprobe analyses of garnet (normalized to 12 oxygens), clinopyroxene (normalized to 6 oxygens and 4 cations; Fe^{3+} is estimated after Droop, 1987), and plagioclase (normalized to 8 oxygen).
* Plagioclase 120: Ca 0.241; Na 0.740; K 0.002.
** Plagioclase 290: Ca 0.182; Na 0.883; K 0.006.

Their cores have actinolite compositions and toward the rims a zonation with increasing Al^{IV} and Al^{VI} but constant low Ti is developed. The green tschermakite rims of these amphiboles have higher Al^{VI} (1.0–1.5 pfu) and lower Ti (0.025) when compared to the brown-green amphiboles (Table 2; Fig. 5, B and G). In sample Ty-Lan 44-3, the complete zonation trend is developed in single porphyroblasts with brown-green amphibole cores, pale green actinolitic inner rims, and dark green amphibole rims (Fig. 5, C and H).

Amphibolites of the Peumerit Formation (sample Peumerit 2) in the metabasic-ultrabasic complex come from the Pont Illis quarry. No relicts of garnet and clinopyroxene or decomposition textures of these phases were found. A M2 assemblage with amphibole, epidote, chlorite, plagioclase (albite and oligoclase), calcite, quartz, and titanite is observed. The coarse-grained rocks contain Ca-amphiboles. Long axes of strictly preferentially oriented amphibole 2 grains with small actinolite cores define a mineral lineation, associated with a more or less weak planar foliation. They surround large amphibole 1 grains. Amphibole 1 is strongly zoned. In the green amphibole cores, Ti (0.05 pfu) is low, with Al^{IV} = 0.8 and Al^{VI} = 0.25 pfu giving a magnesio-hornblende composition. The elliptical cores are surrounded along a sharp contact by a narrow pale green zone. In this zone, Ti, Al^{IV}, and Al^{VI} decrease significantly toward compositions of actinolitic hornblende and actinolite. Then Al^{IV} and Al^{VI} increase stepwise in the zonation profiles toward the dark green rim of the amphiboles, whereas Ti contents remain at low

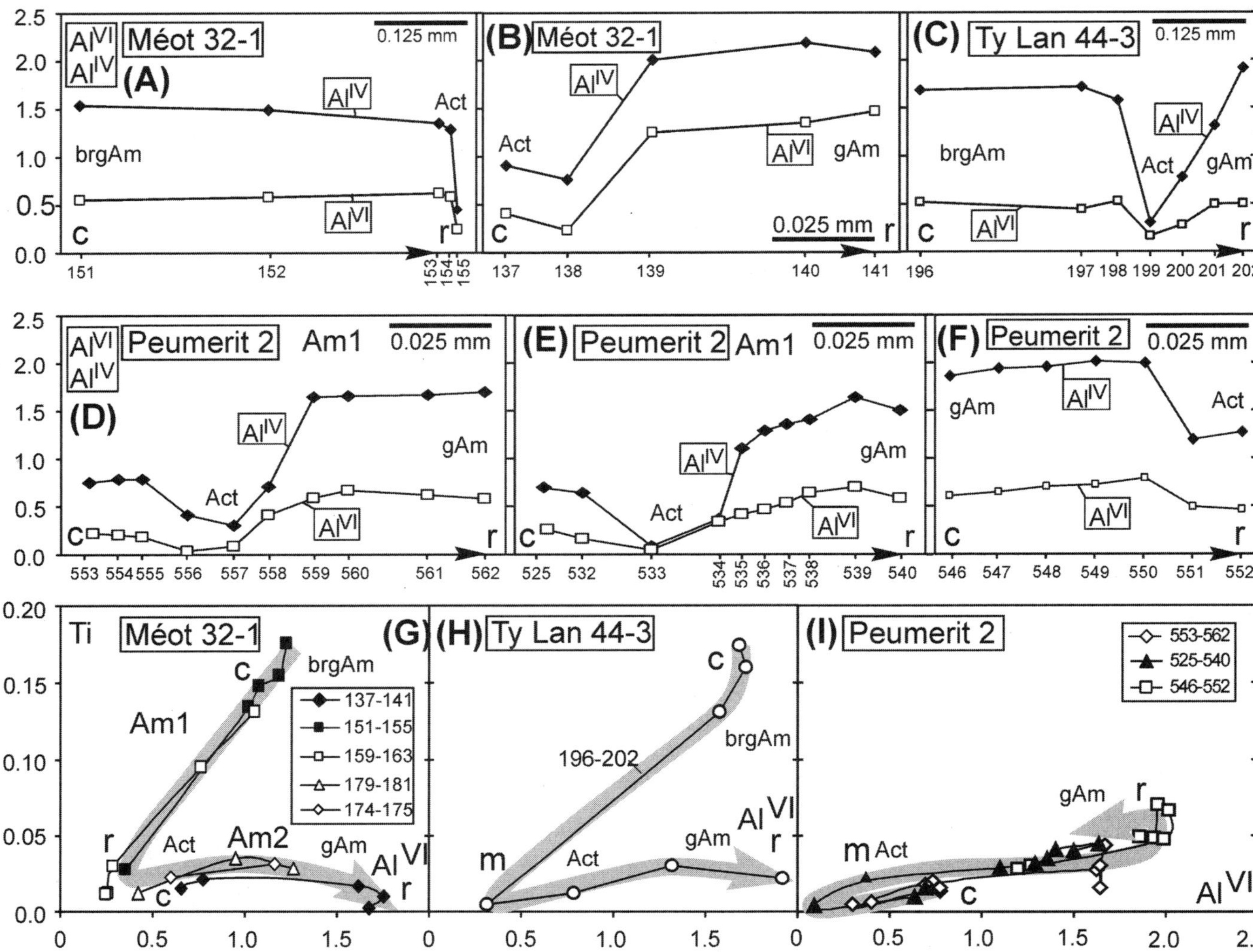

Figure 5. Amphibole zonations in garnet pyroxenites and Peumerit amphibolites. A–F: Amphibole zonation profiles in Al^{IV} and Al^{VI} from cores (c) to rims (r). G–I: Amphibole zonation trends in Al^{IV}-Ti coordinates; arrows mark core to mid to rim zonations (c, m, r). Brown-green amphibole (brgAm), green amphibole (gAm), actinolite (Act). in amphibole porphyroblast generations (Am1, Am2). See Table 2.

levels. In these dark green amphibole rims, compositions of green magnesio-hornblende to tschermakitic hornblende are observed. Some amphibole porphyroblasts display a final slight decrease of Al^{IV} and Al^{VI} in the outermost part of the rim (Table 2; Figs. 3C and 5 [D–F, I]). Plagioclase porphyroblasts (both albite and oligoclase) contain straight and curved foliation S_i, which mostly does not line up with external S. The fabrics are interpreted as the result of successive foliation development during a deformation with progressive character. Late extensional shear bands are underlined by chlorite.

Amphibolites from the lower metasedimentary unit were sampled (Helles 85-1) from isolated lenses to the southeast of Plonéour-Lanvern. An assemblage M2 with amphibole, epidote, chlorite, plagioclase (albite and oligoclase), calcite, quartz, and titanite and no relicts of garnet or clinopyroxene are observed. Amphibole 1 is surrounded by preferentially oriented long prismatic amphibole 2, defining a mineral lineation. Cores of amphibole 1 consist of pale green or colorless actinolite with Al^{IV} and Al^{VI} lower than 0.3 pfu. Toward the dark green amphibole rims, a continuous zonation to compositions of magnesio-hornblende (Al^{IV} = 1.5, Al^{VI} = 0.6, Ti = 0.05 pfu) is developed (Table 2; Fig. 6, A and D). These zonation trends and actual Al^{IV}, Al^{VI}, and Ti contents are similar to those observed in amphibolites from the Peumerit Formation.

Amphibolites from the upper unit come from various outcrops in the northern and southern part near Pouldreuzic and Plouvan. Assemblages M2 with Ca-amphibole, epidote, chlorite, plagioclase (albite and oligoclase), calcite, quartz, and titanite are abundant. Amphibole 1 is surrounded by preferentially oriented long prismatic amphibole 2 defining a mineral lineation (Fig. 3, D and E). In fine-grained rock types as in Menez Gored 1-2, cores of amphibole 1 consist of pale green or colorless actinolite with Al^{IV} and Al^{VI} lower than 0.3 pfu. Toward the green amphibole rims, a continuous zonation toward compositions of

TABLE 2. AMPHIBOLITES

	Méot 32-1			Peumerit 2			Helles 85-1		Menez Gored 1-2				Carazel 14-2	
Sample	152	137	140	558	559	560	137	139	72	74	75	77	21	26
SiO_2	44.79	48.56	39.19	50.19	43.07	42.80	48.38	41.91	53.30	46.21	46.17	46.59	43.37	44.78
TiO_2	1.41	0.19	0.08	0.16	0.24	0.14	0.36	0.45	0.46	0.41	0.47	0.17	1.31	0.34
Al_2O_3	12.09	8.56	20.62	6.38	12.64	13.12	7.35	11.88	2.79	9.62	10.09	7.63	13.50	11.41
FeO	10.19	13.68	15.79	13.73	17.31	16.72	15.92	19.53	13.17	15.28	16.56	15.97	11.74	14.60
MnO	0.14	0.18	0.21	0.44	0.35	0.44	0.24	0.30	0.30	0.32	0.32	0.46	0.36	0.35
MgO	13.93	13.13	7.43	13.28	10.12	10.04	12.45	11.09	14.67	10.92	10.72	11.36	10.91	10.19
CaO	12.08	11.45	11.30	11.54	11.11	11.50	11.55	11.09	12.15	11.83	12.05	11.76	11.83	12.13
Na_2O	1.48	1.03	2.03	1.24	2.01	2.02	1.25	1.90	0.39	1.38	1.40	1.12	1.74	1.10
K_2O	0.65	0.18	0.20	0.20	0.31	0.26	0.13	0.37	0.02	0.12	0.23	0.12	0.24	0.28
Total	96.76	97.01	96.90	97.17	97.21	97.10	97.72	96.58	96.86	96.18	98.14	95.20	95.12	95.26
Si	6.514	7.028	5.798	7.305	6.373	6.355	7.046	6.322	7.731	6.898	6.788	7.025	6.509	6.763
Al^{IV}	1.486	0.972	2.202	0.695	1.627	1.645	0.954	1.678	0.269	1.102	1.212	0.975	1.491	1.237
Al^{VI}	0.586	0.489	1.395	0.398	0.579	0.652	0.309	0.435	0.209	0.591	0.536	0.382	0.898	0.795
Ti	0.154	0.020	0.009	0.018	0.027	0.015	0.039	0.051	0.005	0.046	0.052	0.019	0.148	0.038
Fe^{2+}	0.933	1.065	1.347	1.373	1.283	1.384	1.339	1.515	1.414	1.682	1.690	1.585	1.453	1.766
Fe^{3+}	0.289	0.568	0.580	0.276	0.830	0.664	0.574	0.915	0.162	0.200	0.318	0.401	0.002	0.054
Mn	0.017	0.022	0.027	0.054	0.044	0.056	0.031	0.039	0.038	0.040	0.039	0.059	0.046	0.045
Mg	3.020	2.833	1.638	2.881	2.234	2.221	2.702	2.033	3.172	2.430	2.349	2.552	2.440	2.293
Ca	1.882	1.776	1.792	1.799	1.762	1.830	1.803	1.793	1.888	1.892	1.897	1.900	1.902	1.963
Na_{M4}	0.118	0.224	0.208	0.201	0.238	0.170	0.197	0.207	0.112	0.108	0.103	0.100	0.098	0.037
Na_A	0.299	0.064	0.375	0.150	0.340	0.411	0.158	0.349	0.001	0.292	0.297	0.229	0.409	0.286
K	0.121	0.033	0.038	0.036	0.058	0.049	0.025	0.072	0.003	0.023	0.044	0.022	0.045	0.054
Y	0.579	0.903	0.588	0.813	0.602	0.540	0.817	0.580	0.996	0.685	0.659	0.749	0.545	0.661
*X*Mg	0.764	0.727	0.549	0.677	0.635	0.616	0.669	0.573	0.692	0.591	0.582	0.617	0.615	0.565
Class	Ti-MgH	MgH	Ts	ActH	TsH	TsH	MgH	TsH	Act	MgH	MgH	MgH	Ti-MgH	MgH

Note: Microprobe analyses of amphiboles in metabasites (normalized to 23 oxygens and cations 13CNK; Fe^{3+} is estimated after Papike, 1974). Classification after Leake et al. (1997). For positions of analyses, see Figures 3, 5, and 6.

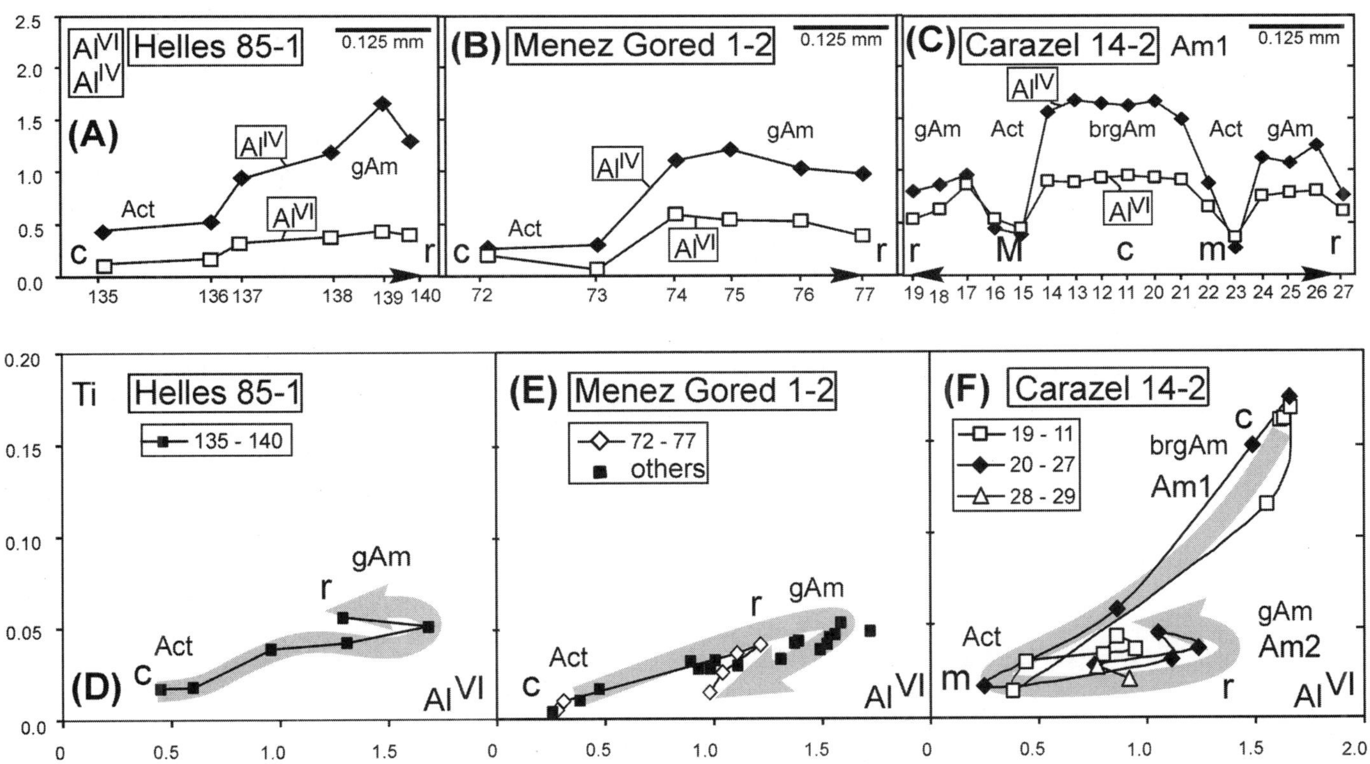

Figure 6. Amphibole zonations in amphibolites of upper and lower metasedimentary units. A–C: Amphibole zonation profiles in Al^{IV} and Al^{VI} from cores (c) to rims (r). D–F: Amphibole zonation trends in Al^{IV}-Ti coordinates; arrows mark core to mid to rim zonations (c, m, r). Brown-green amphibole (brgAm), green amphibole (gAm), actinolite (Act) in amphibole porphyroblast generations (Am1, Am2). See Table 2.

magnesio-hornblende (Al^{IV} = 1.2, Al^{VI} = 0.6) is developed. The outer margins of the porphyroblasts possibly display slight decrease of Al^{IV} (1.0) and Al^{VI} (0.6) (Figs. 3E and 6B). Small amphiboles 2 are poorly zoned with compositions matching the magnesio-hornblende rims of the amphibole porphyroblasts (Figs. 3E and 6E). In coarse-grained amphibolites, as from Carazel 14-2, cores of amphibole 1 are sometimes brown-green amphibole. They are rich in Ti (0.17) and Al^{IV} (1.6) with Al^{VI} = 1.0 pfu. The brown-green tschermakitic Ti-hornblendes have euhedral crystal faces and are surrounded by narrow zones of pale green or colorless actinolite. A sharp optical contact is developed between the brown cores and the pale green zones. Toward the rims of these amphiboles, there are more or less continuous zonations with increase of Al^{IV} and Al^{VI} toward the green magnesio-hornblende rims (Figs. 3D and 6, C and F).

Geothermobarometry and **P-T** *evolution*

Two episodes of metamorphic evolution can be distinguished from the mineral assemblages and microstructural relationships in the metabasites. In garnet pyroxenites of the metabasic-ultrabasic complex, the garnet-clinopyroxene-plagioclase assemblages (enclosing quartz and Ti-bearing phases) are indicators of M1 metamorphism. Amphibole-bearing assemblages overprinting the assemblage M1 in garnet pyroxenites, prevailing in the amphibolites of the Peumerit Formation and abundant in amphibolite lenses in the lower and upper metasedimentary units, are related to M2 metamorphism. It is tempting to consider the M1 and M2 assemblages as steps along a single clockwise *P-T* evolution, with a high-*P*–high-*T* stage and a subsequent decompression passing through amphibolite facies conditions. However, *P-T* paths of tectonic units could deviate from the general clockwise type. Additional and precise information could be expected by an examination of the *P-T* evolution within the metamorphic episodes M1 and M2 in each lithostratigraphic unit.

P-T conditions of the metamorphism M1 in garnet- and clinopyroxene-bearing rocks have been estimated from several samples and outcrops (Fig. 1A) using the calibrations of the garnet-clinopyroxene Mg-Fe cation-exchange thermobarometer (Ellis and Green, 1979; Krogh, 1988), the related geobarometers based on Jd_{SS} + Qtz = Ab_{SS} (Holland, 1980, 1983), and equilibria involving garnet, clinopyroxene, and plagioclase (Newton and Perkins, 1982). Analyses from Mg-rich rims of zoned garnet were combined with analyses from enclosed clinopyroxene and plagioclase. When no inclusions occurred in the garnet rims, analyses from cores or rims of adjacent unzoned clinopyroxene and plagioclase were used. These combinations yielded the maximal temperatures and corresponding pressures for the M1 event. A systematic decrease of Mn from cores to rims in most of the analyzed garnets indicates a growth zonation, whether or not they are slightly modified by high-temperature diffusion. The increase of Mg from garnet cores to rims in many samples can be interpreted in a semiquantitative way to indicate garnet growth during increasing temperatures. It is not clear whether the cores of slightly zoned clinopyroxene crystallized with garnet cores or have been homogenized during plastic deformation at high temperatures. When these arguments are accepted, the M1 assemblages in the garnet pyroxenites recorded *P-T* segments of increasing temperature, to 800 °C and 14 kbar in Méot (32-1), and to 950 °C and 19 kbar in Kergroaz (34-2), as outlined in Figure 7A. Reported values for the maximum temperatures represent the mean of the applied geothermometers and barometers for each mineral pair. The equivalent garnet zonation trends in samples Kergroaz, Méot, and Keramoine signal similar prograde *dP-dT* trends, but passing different *P-T*max (Fig. 7A). In contrast, the strongly retrogressed garnet pyroxenite from location Creac'h with retrograde garnet zonation trends shows a retrograde segment at 700 °C and 10 kbar at the end of the high-*P*–high-*T* evolution (Fig. 7A).

Slightly zoned brown-green amphiboles are abundant in some of the garnet pyroxenites. Furthermore, they occur with similar compositions in cores of green amphiboles in the upper unit. The Ti contents are considerably higher, but Al^{IV} and Al^{VI} are within the range as in the green amphiboles. An empirically calibrated Ti-Al^{IV} geothermobarometer (Colombi, 1989) with T (°C) = Ti * 980 + 600; for Ti > 0.08, P is given by a diagram; applied to these amphiboles it gives results of ~700–750 °C and 5–7 kbar. Chivas (1981), Czamanske et al. (1981), and Pe-Piper (1988) studied amphiboles with similar mineral chemistry from plutonic complexes. They argued that a subsolidus transformation of magmatic clinopyroxene during cooling could be a formation mechanism of such Ti-rich amphiboles. Zenk (2001) described comparable Ti-rich amphiboles enclosed by Ca-amphiboles from chlorite-zone metabasites. Because the upper unit amphibolites and their metapelitic host rocks show no further and independent indications of a metamorphism at high temperatures, the brown-green amphiboles could be considered as altered magmatic relicts that recrystallized prior to the M2 metamorphism. In the garnet pyroxenites one cannot discriminate whether the brown-green amphiboles result from transformation of magmatic minerals, or crystallized at high temperatures as a part of a pre-M1 or a retrograde M1-M2 metamorphic assemblage.

Ca-amphiboles of the M2 assemblages in garnet pyroxenites, or amphibolitized types of these rocks, in the amphibolites of the Peumerit Formation and in the metasedimentary units display systematic zonation trends. Amphiboles are solid solutions with a wide range of combined end members in different structural crystal sites (Spear, 1993; Leake et al., 1997). Because they can crystallize in different solid-solution series and over a wide range of pressure, temperature, and bulk-rock compositions, they are suitable for studying the progressive metamorphism of basic rocks. The distribution of Si, Al, and Na in several amphibole structural sites depends on the physical conditions of crystallization: Al^{IV} and Na_A increase with temperature and Al^{VI} and Na_{M4} increase with pressure (e.g., Raase, 1974; Holland and Richardson, 1979; Laird and Albee, 1981). The following observations provide arguments that the green Ca-amphibole

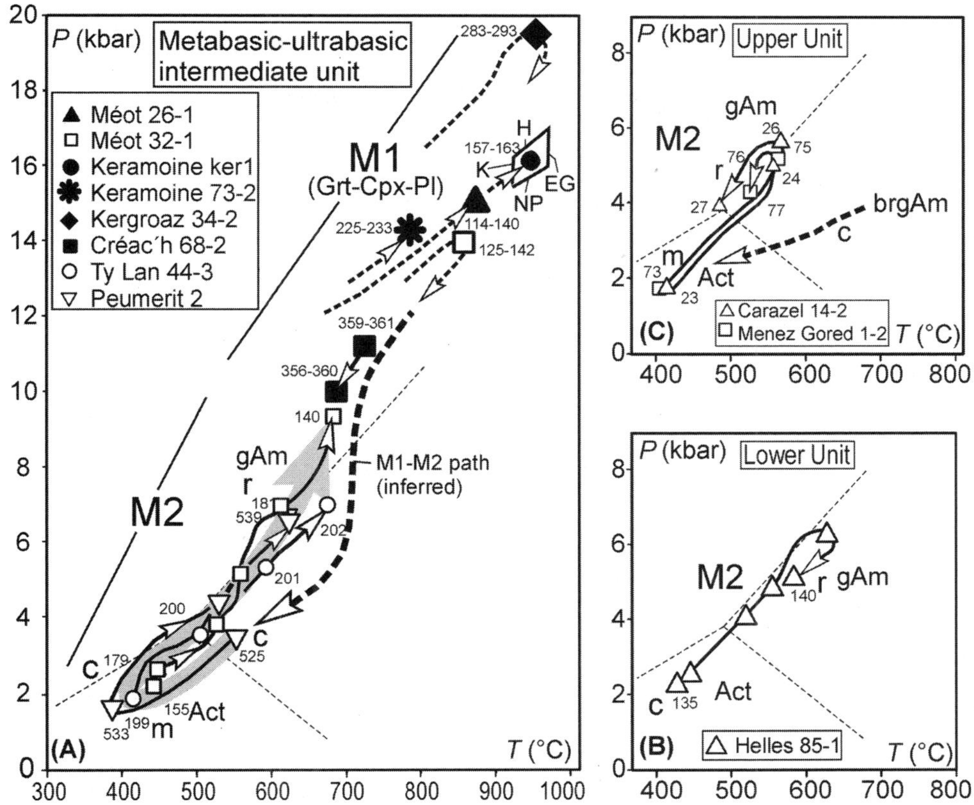

Figure 7. Pressure-temperature (*P-T*) data and *P-T* paths from Baie d'Audierne metabasites. A: *P-T* data from garnet-clinopyroxene-plagioclase equilibria (M1 mineral assemblage) from metabasic-ultrabasic intermediate unit. Marks represent mean from several calibrations (EG: Ellis and Green, 1979, H: Holland, 1980; K: Krogh, 1988; NP: Newton and Perkins, 1982). Thin broken lines indicate semiquantitative *P-T* evolution derived from Mg zonation in garnet (see text). Numbers are garnet-clinopyroxene equilibrium pairs. Thick broken line marks inferred first decompression path M1-M2 of garnet pyroxenites. *P-T* paths from mineral assemblage M2 with actinolites (Act) and green amphiboles (gAm) according to amphibole thermobarometer of Gerya et al. (1997). Zonations from core (c) to mid (m) and rim (r) in green amphiboles indicate second metamorphic evolution prograde in *T* and *P.* B: Lower metasedimentary unit: prograde *P-T* paths from zoned actinolites (Act) and green amphiboles (gAm) of mineral assemblage M2. C: Upper metasedimentary unit: *P-T* paths from mineral assemblage M2 with zoned actinolite (Act) and green amphibole (gAm). Thick broken line marks supposed pre-M2 subsolidus magmatic evolution (see text). Thermobarometer of Gerya et al. (1997). Core (c), mid (m), and rim (r) of zoned amphiboles.

mineral zonations reflect compositional variations during crystallization or growth and were not caused by retrograde replacement or later diffusion:

1. The rim-core-rim zonations are symmetrical (Fig. 6C).
2. Actinolite cores are not irregular or patchy but have elliptical shapes, matching the long prismatic shape of the amphiboles (Fig. 3, C and E).
3. Brown-green amphibole cores display no patchy alteration to colorless amphibole, but are surrounded by discrete zones of actinolite (Fig. 3D).
4. Similar zonation trends or segments of them occur in various porphyroblasts within a thin section.
5. Zonation profiles at the amphibole margins are not affected by contacts of other phases.

One could argue that increasing Al and Na in zoned Ca-amphiboles of M2 have been controlled by successive breakdown of metamorphic clinopyroxene and garnet during decreasing pressure. However, the similar type of Ca-amphibole with systematic increase of Al and Na is observed in the Peumerit and metasedimentary units, metabasites, which contain no clinopyroxene and garnet. This provides an argument that the amphibole zonations reflect cation exchange and net transfer of continuous reactions controlled by changing metamorphic conditions.

In the M2 assemblages, epidote compositions slightly vary with the microstructural site and zonations are abundant. As a re-equilibration is not necessarily evident in most cases, a change in the modal ratios between minerals involved in mass-transfer reactions during metamorphism would reduce the compositional variations of the other minerals during the exchange. In a closed chemical system this could enable a zoning of amphiboles stronger than in coexisting phases. From the evolution of amphibole compositions in core-rim zonation profiles, semiquantitative and relative P-T trends during amphibole crystallization in equilibrium with plagioclase, chlorite, and epidote of M2 can be inferred. Because plagioclase is involved in M2 assemblages, for the detailed thermobarometric interpretation of the amphibole zonation trends, the amphibole end-member equilibria requiring albite in the assemblage and chosen by Triboulet (1992) and Triboulet et al. (1992) cannot be applied. For the evaluation of the P-T variations recorded by the continuously zoned green Ti-poor Ca-amphiboles of M2, an experimentally calibrated geothermobarometer based on equilibria in the assemblage Ca-amphibole–plagioclase (An > 10)–quartz–zoisite–calcite has been used. Expressions for the P-T calculations, derived from experimental data, are according to Gerya et al. (1997):

$$T\,(\mathrm{K}) = (6119 - 28.4P + 114XMg_{Hbl})/8.181 - R\ln(8.489 - Si_{Hbl});$$

and

$$P\,(\mathrm{kbar}) = [2543 - 4.744T + 175XMg_{Hbl} + RT\ln(Al_{Hbl} + 1.433)]/148.1,$$

with XMg_{Hbl} = Mg/(Fe + Mg) of bulk Mg mole fraction in amphibole; Si_{Hbl} and Al_{Hbl} are Si and Al in amphibole recalculated assuming sumMg = 13.000 cations. The thermobarometric estimates involve a minimum uncertainty of ±37 °C and 1.2 kbar, whatever calibration and thermodynamic data set is used. An empirically calibrated version of this geothermobarometer exists and considers Fe^{3+} in amphibole (Zenk, 2001). It gives equivalent temperatures and slightly higher pressures. Minor uncertainty arises from microprobe analyses. Pressure estimates from actinolite are possibly afflicted by larger uncertainty, as their compositions are at the limits of the calibration, but are still restricted to the range of low pressures.

Rims of brown-green amphibole in garnet pyroxenite samples are actinolite, which reflect low-*T* and low-*P* conditions of <500 °C and 3 kbar. When combined with *P-T* data from the M1 garnet-clinopyroxene-plagioclase assemblages, this defines a retrograde M1-M2 evolution. An evaluation of the *P-T* evolution recorded by the M2 green amphiboles with zonations from actinolite cores to tschermakite rims in the garnet pyroxenites by the Gerya et al. (1997) calibration, yielded, in contrast, prograde *P-T* paths from 450 °C and 2 kbar to 680 °C and 9 kbar (Fig. 7A).

Amphibolites of the Peumerit Formation display slightly decreasing *P* and *T* within low-*T*–low-*P* conditions recorded by the green amphibole cores (Figs. 3C, 5, D and E, and 7A; analyses 525–533 and 553–556). Amphibole zonations with significant increase of Al^{IV} and Al^{VI} toward the rims signal a subsequent prograde *P-T* path to 620 °C and 7 kbar (Figs. 3C, 5, D and E, and 7A; analyses 533–539 and 556–562). From the zoned amphiboles of the lower metasedimentary unit, a similar *P-T* evolution with a prograde path to 620 °C and 6 kbar has been obtained (Fig. 7B).

In the upper unit amphibolites, continuous zonation from actinolite core and actinolite inner rim mantling brown-green amphibole, toward magnesio-hornblende rim is due to a prograde *P-T* evolution from low-*P*–low-*T* conditions to 550 °C and 5 kbar. Some of the amphibole porphyroblasts display slightly decreasing Al^{IV} and Al^{VI} in the outermost parts of the rims, which points to a subsequent retrograde evolution (Fig. 7C).

From the zonation profiles and Ti-Al^{IV} diagrams, it is obvious that the chemical evolution from brown-green amphibole cores to actinolite rims or inner rims is rather abrupt when compared to the subsequent continuous chemical record from actinolite toward the magnesio-hornblende or tschermakite rim compositions (Figs. 5, A, C, G, and H, and 6, C and F). Such a steep or discontinuous chemical gradient could signal a miscibility gap of amphibole solid solutions or slow and inhibited amphibole crystallization during the retrograde evolution, which followed the brown-green amphibole crystallization. In contrast, the subsequent prograde zonations involve intermediate compositions between actinolite and hornblende, and give no hint about a miscibility gap.

CONCLUSIONS

Even when precise, *P-T* estimates are afflicted by considerable uncertainties resulting from imprecise knowledge of thermodynamic properties and activity-composition relations, and limited bases of empirical calibrations. This problem is pertinent when *P-T* estimates result from different mineral assemblages in one sample, as in the garnet pyroxenites, or from different thermobarometric calibrations applied to a single mineral assemblage, as in the amphibolites. However, the results from the amphibole thermobarometer of Gerya et al. (1997) are in line with *P-T* estimates from alternative calibrations (Colombi, 1989; Holland and Blundy, 1994), even when actual results do not always overlap. Therefore, it is convenient to discuss the *P-T* conditions in a semiquantitative way and to consider relative *P-T* variations or significantly different *P-T* conditions resulting from evidently differing mineral compositions only. *P-T* data from several single samples within a lithotectonic unit could provide some control, because samples and assemblages record *P-T* segments out of the actual *P-T* evolution.

The high-*P*–high-*T* metamorphism at 384 ± 6 Ma (Peucat, 1983) has been recorded only by the garnet pyroxenites. Other rock types show no relicts of this event. Preserved prograde garnet zonation and *P-T* records indicate that these rocks underwent considerable burial and heating after their magmatic emplacement. This is typical of ophiolites involved in subduction and/or collision processes and contrasts with the direct tectonic accretion of isolated high-*P* mafic bodies from the upper mantle, as described by Gardien et al. (1990) from the Massif Central. There are different *P*max and *T*max recorded by the studied rocks, even when a considerable uncertainty in *P-T* estimates is considered. This could indicate different tectonic slices or lenses of magmatic bodies that underwent a burial history down to different levels of depth. A first uplift of the Baie d'Audierne high-*P*–high-*T* rocks is characterized by significant decompression and minor cooling, with formation of symplectites from clinopyroxene. A part of the retrograde *P-T* path was recorded in the garnet pyroxenite sample Creac'h 68-2. The final step of the first decompression of the high-*P*–high-*T* rocks is marked by crystallization of actinolite. An overprinting of the M1 assemblage by the amphibole-bearing M2 assemblage requires H_2O. It appears likely that a tectonic emplacement of the high-*P*–high-*T* rocks between metasedimentary units could have provided such fluid. Zoned green amphibole cores with decreasing Al^{IV} and Al^{VI} in the Peumerit amphibolites indicate that these rocks at least partly participated in the late stage of this first decompression (Fig. 7A).

The M1 metamorphism of the Baie d'Audierne high-*P*–high-*T* rocks shows marked differences when compared with Armorican eclogites. Maximum temperatures are considerably higher than in the Léon and Champtoceaux eclogites. This indicates burial along a comparably high geothermal gradient and a specific geotectonic position of the Baie d'Audierne rocks previous to and during subduction. In the course of decompression, the Champtoceaux eclogites passed low-*T*–high-*P* conditions with crystallization of glaucophane along a low geothermal gradient (Godard et al., 1981; Ballèvre et al., 2000), and in the Léon eclogites the formation of sapphirine indicates high-*T*–low-*P* conditions along a high geothermal gradient after the high-*P* event (Godard and Mabit, 1998). The first decompression path, M1-M2, of the Baie d'Audierne garnet pyroxenites is poorly defined by thermobarometric data, but it evolved apparently intermediate between these extremes (Fig. 7A). It could be compared to general decompression paths described from the allochthonous complexes in the Iberian Meseta (Martinez Catalán et al., 1999) or the Bohemian Massif (O'Brien and Carswell, 1993).

Growth zonations with increasing Al^{IV} and Al^{VI} toward the rims of green amphiboles from strongly retrograded garnet pyroxenites and from the Peumerit amphibolites point to increasing *T* and *P* during a moderate new burial and heating of the metabasic-ultrabasic complex. According to the thermobarometry, this second metamorphism started at greenschist facies conditions and is characterized by a geotherm along the sillimanite-kyanite transition (Fig. 7A). Thus, the second metamorphism does not differ much from *P-T* conditions passed during the preceding first decompression. Actinolitic hornblendes in the outer rims of green amphiboles provide the record of a (second) decompression during M2.

From zonations and related thermobarometry of green amphiboles in amphibolites of the upper and lower metasedimentary units, it is evident that these units participated in the second prograde greenschist to amphibolite facies metamorphism. Peak temperatures and pressures of the prograde M2 metamorphism are lower (550 °C, 5 kbar) in the upper unit, intermediate (620 °C, 7 kbar) in the Peumerit Formation and the lower metasediments (630 °C, 6 kbar), and highest (680 °C, 9 kbar) in the garnet pyroxenites. In combination with the similar structural evolution, these differences indicate a metamorphic zonation during the M2 event. Therefore, even when all three units were involved together in the second metamorphism, their different *P*max-*T*max during M2 indicate different levels of the crust.

The metamorphic zonation during M2 appears to be inverted within the metabasic-ultrabasic complex. This suggests a tectonic contact between the base of the metabasites-ultrabasites and the lower metasediments, as outlined in Peucat and Cogné (1974). Discontinuity within inverted metamorphic zonations cannot be explained by low-angle normal detachment. The base of, and contacts within the metabasic-ultrabasic unit, should be thrusts, operative after the M2 thermal peak.

Between the metabasic-ultrabasic complex and the upper unit, a normal metamorphic zonation for M2 can be recognized. Due to the lack of M1 metamorphism in the upper unit, the contact to the metabasic-ultrabasic complex could be interpreted as a thrust. Furthermore, the contact between the upper unit and the metabasites-ultrabasites is a metamorphic discontinuity, separating units of different crustal levels within an M2 evolution. When the different peak *P-T* conditions of M2 in the upper unit and the metabasites-ultrabasites are considered in combination with their small vertical distance, the contact represents a considerable section of omitted crust. Such a reduced metamorphic profile could be explained by an extensional low-angle normal fault that was active during the decompression after the thermal peak of the M2 metamorphism.

Is it possible to relate the prograde M2 *P-T* evolution to contact metamorphism along the widespread Variscan granites? This can be rejected by several arguments. (1) According to the thermobarometry, the prograde M2 metamorphism involved increase of pressure, whereas a contact metamorphism should be characterized by temperature increase without marked pressure variations. This is true even when pressures from actinolite are underestimated. (2) Differences in peak temperatures during M2 are related to the lithotectonic unit and not to the distance to the granites. (3) One of the late granites (Pouldreuzic) cuts across the tectonic and metamorphic discontinuity in the upper unit and the metabasic-ultrabasic complex (Fig. 1, B and C). This indicates a pregranitic age of the tectonic contact.

Therefore, a pre-late Carboniferous age of at least the prograde M2 evolution should be taken into account. In contrast, the final retrograde evolution of M2 is associated to a postgranitic formation of foliation S3 at greenschist facies conditions. Amphibole zonations and related thermobarometry indicate a two-stage pre-late Variscan metamorphism and a complex decompression history of the Baie d'Audierne metamorphic units. Geodynamic models involving two distinct pre-late Carboniferous principal stages of crustal contraction, i.e., subduction followed by collisional stacking, during the Variscan orogeny in the Armorican Massif are supported by these data.

ACKNOWLEDGMENTS

Microprobe analyses were performed with H.-P. Meyer at Mineralogisches Institut der Universität Heidelberg, with H. Remy and M. Fialin at Laboratoire Pétrologie Minéralogique, Université Pierre et Marie Curie in Paris, and with M. Bohn at IFREMER in Brest. S. Nicholson and A. Baier helped with whole-rock X-ray fluorescence analyses at Erlangen. The constructive reviews by G. Godard and P. O'Brien are gratefully acknowledged. Field and analytical work in Brittany have been funded by grants from the DAAD of Germany, the University of Erlangen-Nürnberg, the Deutsche Forschungsgemeinschaft (Schu 676/6) and supported by further funds of the CNRS and BRGM of France.

REFERENCES CITED

Audren, C., 1987, Évolution structurale de la Bretagne Méridionale au Palaeozoïque: Mémoires Société Géologique Minéralogique de la Bretagne, v. 31, p. 1–365.

Audren, C., and Triboulet, C., 1993, *P-T-t* deformation paths recorded by kinzigites during diapirism in the western Variscan belt (Golfe du Morbihan, southern Brittany, France): Journal of Metamorphic Geology, v. 11, p. 337–356.

Audren, C., Triboulet, C., Chauris, L., Lefort, J.-P., Vigneresse, J.L., Audrain, J., Thiéblemont, D., Goyallon, J., Jégouzo, P., Guennoc, P., Augris, C., and Carn, A., 1993, Note explicative de la feuille Ile de Groix (Carte Géologique de la France 1:25,000): Orléans, Éditions du B.R.G.M., 101 p.

Ballèvre, M., Paris, F., and Robardet, M., 1992, Corrélations Ibéro-Armoricains: Une confrontation des données paléobiogéographiques et tectonométamorphiques: Comptes Rendus de l'Académie Sciences Paris, Série II, v. 315, p. 1783–1789.

Ballèvre, M., Marchand, J., Godard, G., Goujou, J.C., and Wyns, R., 1994, Eo-Hercynian events in the Armorican Massif, *in* Keppie, J.D., ed., Pre-Mesozoic geology in France and related areas: Berlin, Springer-Verlag, p. 183–194.

Ballèvre, M., Kiénast, J.R., and Paquette, J.-L., 1987, Le métamorphisme éclogitique dans la nappe hercynienne de Champtoceaux (Massif Armoricain): Comptes Rendus de l'Académie Sciences Paris, Série II, v. 305, p. 127–131.

Ballèvre, M., Kiénast, J.R., Pinardon, J.L., and Vuichard, J.P., 1989, Reversal of Fe-Mg partitioning between garnet and staurolite in eclogite facies metapelites from the Champtoceaux nappe (Brittany, France): Journal of Petrology, v. 30, p. 1321–1349.

Ballèvre, M., Bosse, V., Capdevila, R., de Jong, K., Féraud, G., Guerrot, C., Peucat, J.J., Ruffet, G., 2000, Structure and thermochronology of the Champtoceaux complex (France): A key for Ibero-Armorican correlations: Variscan-Appalachian dynamics: The building of the Upper Paleozoic basement. Basement Tectonics 15, A Coruna, Spain, Program and Abstracts, p. 52–55.

Bernard-Griffiths, J., and Cornichet, J., 1985, Origin of eclogites from south Brittany, France: A Sm-Nd isotopic and REE study: Chemical Geology (Isotope Geosciences Section), v. 52, p. 185–201.

Brown, M., and Dallmeyer, R.D., 1996, Rapid Variscan exhumation and the role of magma in core complex formation: Southern Brittany metamorphic belt, France: Journal of Metamorphic Geology, v. 14, p. 361–379.

Cabanis, B., and Godard, G., 1987, Les éclogites du pays de Léon (Nord-Ouest du Massif armoricain): Étude pétrologique et géochimique; implications géodynamiques: Bulletin Société Géologique de France, v. 1987 série 8 III no. 6, p. 1133–1142.

Chivas, A.R., 1981, Geochemical evidence for magmatic fluids in porphyry copper mineralization, Part 1: Mafic silicates from the Koloula Igneous Complex: Contributions to Mineralogy and Petrology, v. 78, p. 389–403.

Cogné, J., 1960, Schistes cristallins et granites en Bretagne méridionale. Le domaine de l'anticlinal de Cornouaille: Mémoire Explicative Carte Géologique de France, p. 1–382.

Cogné, J., 1988, The Precambrian in the metamorphic terrains of the Ligerian domain and the problem of Precambrian geodynamic evolution in the mobile Variscan belt of Western France, *in* Zoubek, V., ed., Precambrian in younger fold belts: European Variscides, the Carpathians and Balkans: Chichester, John Wiley & Sons, p. 529–552.

Colombi, A., 1989, Métamorphisme et géochimie des roches mafiques des Alpes ouest-centrales (géoprofil Viège–Domodossola–Locarno): Mémoires de Géologie (Lausanne), v. 4, p. 1–216.

Czamanske, G.K., Ishihara, S., and Atkin, S.A., 1981, Chemistry of rockforming minerals of the Cretaceous-Paleogene batholith in southwestern Japan and implications for magma genesis: Journal of Geophysical Research, v. 86, p. 10,431–10,469.

Dias, R., and Ribeiro, A., 1995, The Ibero-Armorican Arc: A collision effect against an irregular continent: Tectonophysics, v. 246, p. 113–128.

Droop, G.T.R., 1987, A general equation for estimating Fe^{3+} concentrations in ferromagnesian silicates and oxides from microprobe analyses, using stoichiometric criteria: Mineralogical Magazine, v. 51, p. 431–435.

Ellis, D.J., and Green, D.H., 1979, An experimental study of the effect of Ca upon garnet-clinopyroxene Fe-Mg exchange equilibria: Contributions to Mineralogy and Petrology, v. 71, p. 13–22.

Gardien, V., Tegyey, M., Lardeaux, J.M., Misseri, M., and Dufour, E., 1990, Crust-mantle relationships in the French Variscan chain: The example of the Southern Monts du Lyonnais unit (eastern French Massif Central): Journal of Metamorphic Geology, v. 8, p. 477–492.

Gerya, T.V., Perchuk, L.L., Triboulet, C., Audren, C., and Sez'ko, A.I., 1997, Petrology of the Tumanshet Zonal Metamorphic Complex, Eastern Sayan: Petrology, v. 5, p. 503–533.

Godard, G., 1988, Petrology of some eclogites in the Hercynides: The eclogites from the Southern Armorican massif, *in* Smith, D.C., ed., Eclogites and eclogite facies rocks: Amsterdam, Elsevier, p. 451–519.

Godard, G., and Mabit, J.-L., 1998, Peraluminous sapphirine formed during regression of a kyanite-bearing eclogite from Pays de Léon, Armorican Massif, France: Lithos, v. 43, p. 15–29.

Godard, G., Kienast, J.-R., and Lasnier, B., 1981, Retrogressive development of glaucophane in some eclogites from "Massif Armoricain" (east of Nantes, France): Contributions to Mineralogy and Petrology, v. 78, p. 126–135.

Hanmer, S.K., 1977, Age and tectonic implications of the Baie d'Audierne basic-ultrabasic complex: Nature, v. 270, p. 336–338.

Holland, T.J.B., 1980, The reaction albite = jadeite + quartz determined experimentally in the range 600–1200 °C: American Mineralogist, v. 65, p. 129–134.

Holland, T.J.B., 1983, The experimental determination of activities in disordered and short-range ordered jadeitic pyroxene: Contributions to Mineralogy and Petrology, v. 82, p. 214–220.

Holland, T.J.B., and Blundy, J., 1994, Non-ideal interactions in calcic amphiboles and their bearing on amphibole-plagioclase thermometry: Contributions to Mineralogy and Petrology, v. 116, p. 433–447.

Holland, T.J.B., and Richardson, S.W., 1979, Amphibole zonation in metabasites as a guide to the evolution of metamorphic conditions: Contributions to Mineralogy and Petrology, v. 70, p. 143–148.

Jones, K.A., and Brown, M., 1990, High-temperature "clockwise" *P-T* paths and melting in the development of regional migmatites: An example from Southern Brittany, France: Journal of Metamorphic Geology, v. 8, p. 551–578.

Krogh, E.J., 1988, The garnet-clinopyroxene Fe-Mg geothermometer—A reinterpretation of existing experimental data: Contributions to Mineralogy and Petrology, v. 99, p. 44–48.

Lasnier, B., 1970, Le métamorphisme régional des gabbros d'après la littérature internationale: Etude préliminaire des gabbros coronitiques du Massif Armoricain et du Massif des Maures (France) [Thèse Doctorat 3ème cycle]: Nantes, Université de Nantes, 295 p.

Laird, J., and Albee, A.L., 1981, Pressure, temperature, and time indicators in mafic schists: Their application to reconstructing the polymetamorphic history of Vermont: American Journal of Science, v. 281, p. 127–175.

Leake, B.E., Woolley, A.R., Arps, C.E.S., Birch, W.D., Gilbert, M.C., Grice, J.D., Hawthorne, F.C., Kato, A., Kisch, H.J., Krichovichev, V.G., Linthout,

K., Laird, J., Mandarino, J.A., Maresch, W.V., Nickel, E.H., Rock, N.M.S., Schumacher, J.C., Smith, D.C., Stephenson, N.C.N., Ungaretti, L., Whittaker, E.J.W., and Youzhi, G., 1997, Nomenclature of amphiboles: Report of the subcommittee on amphiboles of the International Mineralogical Association, Commission on New Minerals and Mineral Names: American Mineralogist, v. 82, p. 1019–1037.

Lefort, J.P., and Peucat, J.J., 1974, Le socle antémésozoïque submergé à l'Ouest de la baie d'Audierne (Finistère): Comptes Rendus de l'Académie Sciences Paris, Série D, v. 279, p. 635–637.

Marchand, J., 1982, Une véritable eclogite en Bretagne occidentale (Baie d'Audierne–France): Terra Cognita, v. 2/3, p. 312.

Martinez Catalán, J.R., Arenas, R., Diaz Garcia, F., and Abati, J., 1999, Allochthonous units in the Variscan belt of NW Iberia: Terranes and accretionary history, *in* Sinha, A.K., ed., Basement Tectonics 13: Amsterdam, Kluwer Academic Publishers, p. 65–84.

Matte, P., 1986, Tectonics and plate tectonics model for the Variscan belt of Europe: Tectonophysics, v. 126, p. 329–374.

Newton, R.C., and Perkins, D., 1982, Thermodynamic calibration of geobarometers based on the assemblage garnet-plagioclase-orthopyroxene-clinopyroxene-quartz: American Mineralogist, v. 67, p. 203–222.

O'Brien, P.J., and Carswell, D.A., 1993, Tectonometamorphic evolution of the Bohemian Massif: Evidence from high pressure metamorphic rocks: Geologische Rundschau, v. 82, p. 531–555.

Paquette, J.-L., Peucat, J.J., Bernard-Griffiths, J., and Marchand, J., 1985, Evidence for old Precambrian relics shown by U-Pb zircon dating of eclogites and associated rocks in the Hercynian belt of South Brittany, France: Chemical Geology (Isotope Geosciences Section), v. 52, p. 203–216.

Paquette, J.-L., Balé, P., Ballèvre, M., and Georget, Y., 1987, Géochronologie et géochimie des éclogites du Léon: nouvelles contraintes sur l'évolution géodynamique du Nord-Ouest du Massif Armoricain: Bulletin de Minéralogie, v. 110, p. 683–696.

Papike, J.J., 1974, On the chemistry of clinoamphiboles: Eos (Transactions, American Geophysical Union), v. 55, p. 469.

Pearce, J.A., 1982, Trace element characteristics of lavas from destructive plate boundaries, *in* Thorpe, R.S., ed., Andesites: Chichester, John Wiley & Sons, p. 525–548.

Pearce, J.A., 1983, Role of the sub-continental lithosphere in magma genesis at active continental margins, *in* Hawkesworth, C.J., and Norry, M.J., eds., Continental basalts and mantle xenoliths: Nantwich, Shiva, p. 230–249.

Pe-Piper, G., 1988, Calcic amphiboles of mafic rocks of the Jeffers Brook plutonic complex, Nova Scotia, Canada: American Mineralogist, v. 73, p. 993–1006.

Peucat, J.J., 1973, Les schistes cristallins de la Baie d'Audierne: Étude pétrographique et structurale [Thèse 3eme cycle]: Rennes, Université de Rennes, 108 p.

Peucat, J.J., 1983, Géochronologie des roches métamorphiques (Rb-Sr et U-Pb). Exemples choisis au Groenland, en Laponie, dans le massif Armoricain et en Grande Kabylie: Mémoires Société Géologique et minéralogique de la Bretagne, v. 28, p. 1–158.

Peucat, J.J., and Cogné, J., 1974, Les schistes cristallins de la Baie d'Audierne (Sud-Finistère): Un jalon intermédiaire dans le socle antécambrien entre la Meseta ibérique et les régions sud-armoricaines: Comptes Rendus de l'Académie Sciences Paris, Série D, v. 278, p. 1809–1812.

Peucat, J.J., Le Metour, J., and Audren, C., 1978, Arguments géochronologiques en faveur de l'existence d'une double ceinture métamorphique d'âge siluro-devonien en Bretagne méridionale: Bulletin Société Géologique de la France, v. 1978-20, no. 7, p. 163–167.

Peucat, J.J., Vidal, P., Godard, G., and Postaire, B., 1982, Precambrian U-Pb zircon ages in eclogites and garnet pyroxenites from South Brittany (France): An old oceanic crust in the West European Hercynian Belt?: Earth and Planetary Sciences Letters, v. 60, p. 70–78.

Plaine, J, and Peucat, J.J., 1978, Note explicative de la feuille Pont-Croix (Carte Géologique de la France 1:50,000): Orléans, Éditions du B.R.G.M., 56 p.

Poulpiquet, J. de, 1988, Étude magnétique et sismique du massif basique et ultrabasique de la Baie d'Audierne (Massif Armoricain): Géologie de la France, v. 4/1988, p. 11–22.

Raase, P., 1974, Al and Ti contents of hornblende, indicators of pressure and temperature of regional metamorphism: Contributions to Mineralogy and Petrology, v. 45, p. 231–236.

Schulz, B., Audren, C., Triboulet, C., Thiéblemont, D., and Moisan, J., 1996, Prograde *P-T* path of garnet-orthoamphibole-gneiss from the Rhuys peninsula (Southern Brittany, France): Evidence for a pre-Carboniferous collisional stage in the Variscan orogeny: Comptes Rendus de l'Académie Sciences Paris, Série II, v. 322, p. 943–950.

Schulz, B., Triboulet, C., Audren, C., Pfeifer, H.-R., and Gilg, A., 2001, Two-stage prograde and retrograde Variscan metamorphism of glaucophane-eclogites, blueschists and greenschists from Ile de Groix (Brittany, France): International Journal of Earth Sciences, v. 90, p. 871–889.

Shelley, D., and Bossière, G., 2000, A new model for the Hercynian orogen of Gondwanan France and Iberia: Journal of Structural Geology, v. 22, p. 757–776.

Spear, F.S., 1993, Metamorphic phase equilibria and pressure-temperature-time paths: Washington, Mineralogical Society of America Monograph Series, 799 p.

Triboulet, C., 1991, Étude géothermo-barométrique comparée des schistes bleus paléozoïques de l'Ouest de la France (Ile de Groix, Bretagne méridionale et Bois-de-Céné, Vendée): Comptes Rendus de l'Académie Sciences Paris, Série II, v. 312, p. 1163–1168.

Triboulet, C., 1992, The (Na-Ca)amphibole-albite-chlorite-epidote-quartz geothermobarometer in the system S-A-F-M-C-N-H_2O, 1: An empirical calibration: Journal of Metamorphic Geology, v. 10, p. 545–556.

Triboulet, C., and Audren, C., 1985, Continuous reactions between biotite, garnet, staurolite, kyanite-sillimanite-andalusite, and *P-T*-time-deformation path in micaschists from the estuary of the river Vilaine, South Brittany, France: Journal of Metamorphic Geology, v. 3, p. 91–105.

Triboulet, C., and Audren, C., 1988, Controls on *P-T-t* deformation path from amphibole zonation during progressive metamorphism of basic rocks (estuary of the River Vilaine, South Brittany, France): Journal of Metamorphic Geology, v. 6, p. 117–133.

Triboulet, C., Thiéblemont, D., and Audren, C., 1992, The (Na-Ca)amphibole-albite-chlorite-epidote-quartz geothermobarometer in the system S-A-F-M-C-N-H_2O, 2: Applications to metabasic rocks in different metamorphic settings: Journal of Metamorphic Geology, v. 10, p. 557–566.

Velde, B., 1973, The origin of some granulite facies rocks from the Baie d'Audierne (Finistère). Bulletin Société Géologique de la France, v. 1972 IV, no. 2, p. 91–95.

Vidal, P., 1973, Premières données géochronologiques sur les granites hercyniens du sud du Massif Armoricain: Bulletin Société Géologique de la France, v. 1973 XV, no. 7, p. 239–245.

Zenk, M., 2001, Mikrostrukturen, Mineralchemie und Geothermobarometrie der Metabasite in den Mineralzonen nach Barrow im schottischen Dalradian: Erlanger geologische Abhandlungen, v. 133, 199 p.

Manuscript Accepted by the Society February 8, 2002

Printed in the U.S.A.

Geological Society of America
Special Paper 364
2002

Contrasting high-pressure metabasites from the Santiago unit (Ordenes Complex, northwestern Iberian Massif, Spain)

F.J. Rubio Pascual
R. Arenas
Departamento de Petrología y Geoquímica, Facultad de Geología, Universidad Complutense, 28040 Madrid, Spain
F. Díaz García
Departamento de Geología, Universidad de Oviedo, 33005 Oviedo, Spain
J.R. Martínez Catalán
Departamento de Geología, Universidad de Salamanca, 37008 Salamanca, Spain
J. Abati
Departamento de Petrología y Geoquímica, Facultad de Geología, Universidad Complutense, 28040 Madrid, Spain

ABSTRACT

The Santiago unit forms part of the Basal unit of the Ordenes Complex. It is part of the Gondwana continental paleomargin, partially subducted under Laurentia at the beginning of the collisional episode that gave rise to the European Variscan chain. It is composed mainly of pelitic-semipelitic schists, orthogneisses, and intercalations of metabasites. The metabasites intercalated in orthogneisses preserve evidence of their transformation into low-temperature (T) eclogites during an initial syn-D_1 high-pressure (P) episode (P_{min} = 14.8 ± 0.3 kbar; T = 502 ± 10 °C). The metabasites intercalated in the schists conserve an internal schistosity included in albite porphyroblasts, corresponding to garnet amphibolites developed during the same high-P episode (P = 15.8 ± 0.3 kbar; T = 466 ± 14 °C). Petrographic evidence, the chemical zoning of the syn-D_1 garnets, and the thermometry indicate that the basic rocks included in the pelitic schists underwent metamorphic recrystallization prior to the peak pressure, whereas those included in the orthogneisses went directly from slightly hydrated igneous conditions to eclogitization.

The subsequent extensional relaxing of the accretional prism produced the superposition of a hot mantle material layer over the Santiago unit. The development of an inverted metamorphic gradient is reflected in the metabasites included in the schists by the appearance of three metamorphic syn-D_2 zones: (1) tschermakite + oligoclase + chlorite; (2) tschermakite + andesine + biotite; and (3) ferrotschermakite + labradorite + garnet. The peak conditions reached by the top of the unit during D_2 are P = 9.3 ± 0.8 kbar and T = 528 ± 34 °C.

INTRODUCTION

The development of contrasting mineral assemblages in high-pressure rocks has traditionally been interpreted as a response to differences in bulk-rock chemistry (Ridley, 1984), in aH_2O or that of other components of the fluid phase (Green and Ringwood, 1967; Newton and Fyfe, 1976; Black, 1977), or to the influence of the deformation as catalyst of the metamorphic reactions (Rubie, 1986, 1990; Koons et al., 1987), but always taking into account its direct effect on the reactions that form

Rubio Pascual, F.J., Arenas, R., Díaz García, F., Martínez Catalán, J.R., and Abati, J., 2002, Contrasting high-pressure metabasites from the Santiago unit (Ordenes Complex, northwestern Iberian Massif, Spain), *in* Martínez Catalán, J.R., Hatcher, R.D., Jr., Arenas, R., and Díaz García, F., eds., Variscan-Appalachian dynamics: The building of the late Paleozoic basement: Boulder, Colorado, Geological Society of America Special Paper 364, p. 105–124.

eclogites. This study shows that rocks with the same bulk chemistry, following approximately the same pressure-temperature-time (*P-T-t*) path, but under different aH_2O conditions and different intensities of the deformation, have followed different reactional pathways during their syn-D_1 pressurization, leading to the development of eclogites and garnet amphibolites. The role of the deformation in the homogenization of the rocks and the establishing of chemical equilibrium conditions among the constituents of the mineral assemblage is also reflected in the perfect syn-D_2 zoning of the inverted metamorphic gradient registered by the metabasites included in the schists, where it is possible to use index associations that include minerals of such complex chemistry as amphiboles or plagioclase. However, the metabasites included in the orthogneisses, less deformed and recrystallized according to the pronounced core-rim gradients of the boudins, show extreme variability in mineral composition due to the persistence of microdomainal equilibria, which invalidate them as precise markers of metamorphic conditions.

GEOLOGICAL SETTING

The Ordenes Complex (Fig. 1) is one of the allochthonous complexes with mafic and related rocks (Arenas et al., 1986) that are found emplaced over the Galicia–Trás-os-Montes Schistose domain (Farias et al., 1987; Arenas et al., 1988), in the northwestern Iberian Massif. Most of these structures (Cabo Ortegal, Ordenes, Bragança, and Morais) are formed by the stacking of units with very different tectonic-metamorphic histories; these units can be grouped in three main sets. The Ophiolitic units correspond to the suture between two continental masses, Gondwana and Laurentia, represented by the Basal units, together with the relative autochthon and by the Upper units, respectively (Martínez Catalán et al., 1997, 1999).

An important sector of the Basal unit of the Ordenes Complex is formed by what is known as the Santiago unit (Martínez Catalán et al., 1984; Díaz García, 1990), a narrow strip of pelitic and semipelitic schists (Fig. 2) with a stratiform body of orthogneisses located near its base (Santiago Gneiss, 470.8 ± 5.8 Ma, Rb-Sr; García Garzón et al., 1981). The set presents abundant intercalations of mafic rocks of centimeters to meters in size. Its chemical composition is of tholeiitic basalts and no significant geochemical differences can be observed between the amphibolites intercalated in the orthogneisses and those found in the schists; hence its protoliths have been interpreted as belonging to the same intrusive episode (Díaz García, 1990). The metabasites intercalated in the orthogneisses can show quartz veins with large rutile crystals enclosing the boudins. Neither calcite nor aragonite has been found in these veins. The

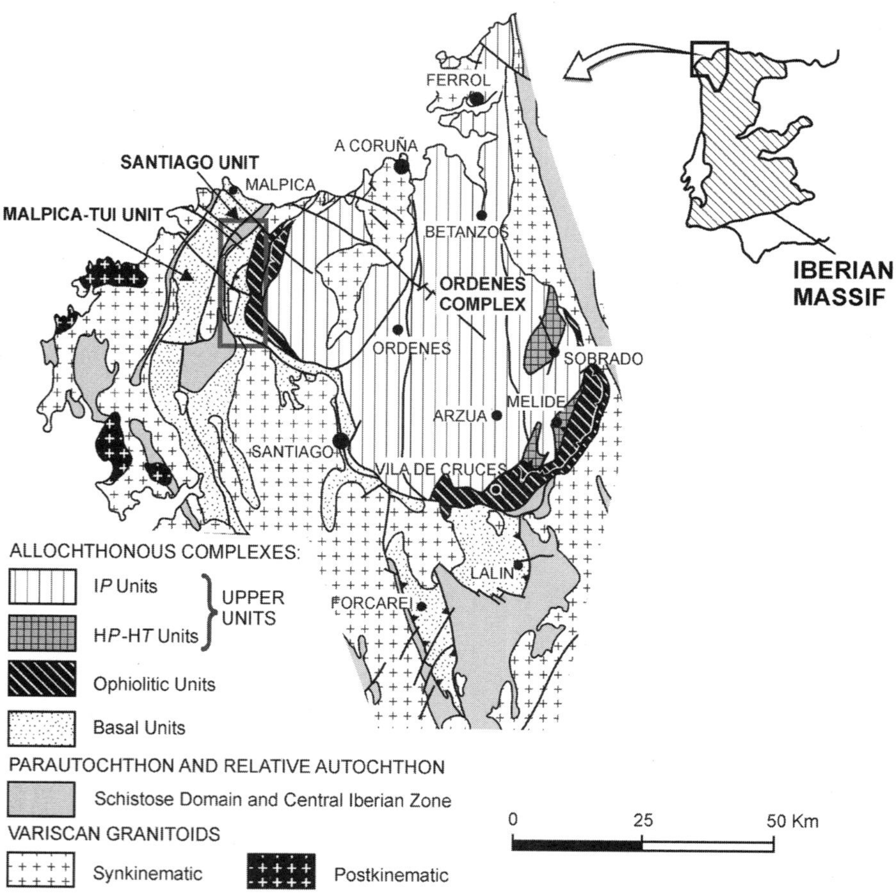

Figure 1. Geological sketch of Ordenes Complex with location of studied area in Santiago unit. Based on Martínez Catalán et al. (1996). *IP, HP, HT*—intermediate pressure, high pressure, high temperature, respectively.

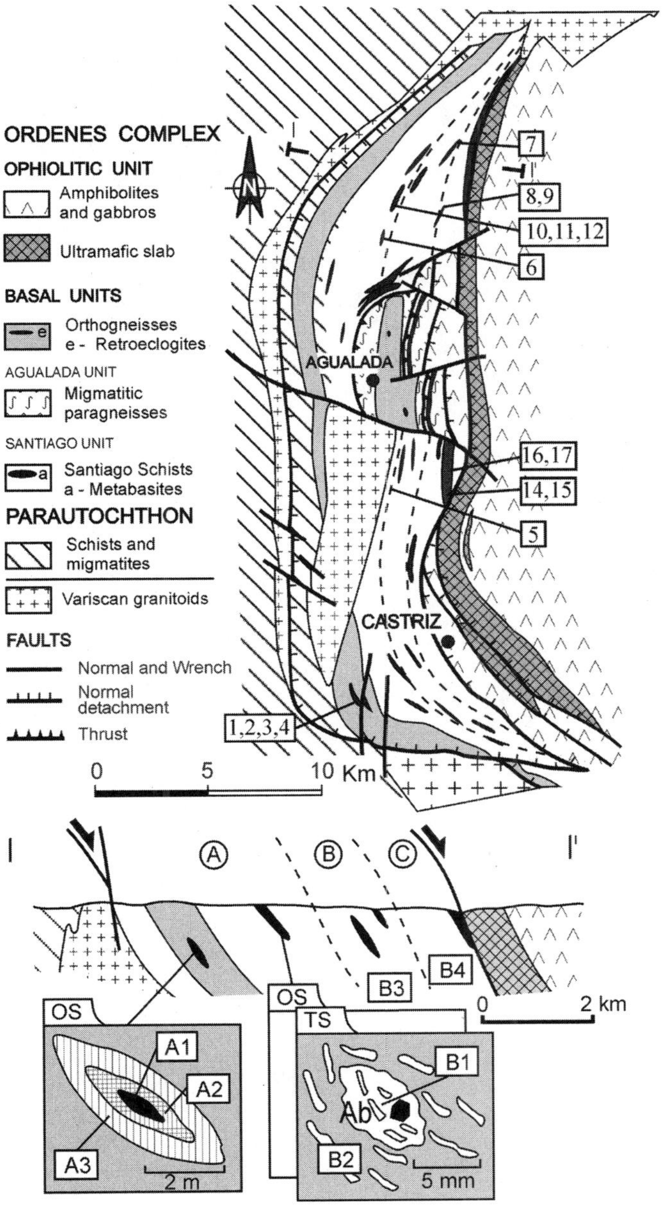

Figure 2. Simplified geological map of Santiago unit; situation of metabasites studied (1–17), and outline of spatial relations between their different assemblage (A1–A3, B1–B4). OS, outcrop scale; TS, thin-section scale; A, B, C, metamorphic zones. For description of assemblage and metamorphic zones, see text.

pelitic and semipelitic schists have abundant quartz segregation veins, but they do not appear to be linked to the mafic boudins.

The Santiago unit has been correlated (see Fig. 1) to the Malpica-Tui unit (Arenas et al., 1986; Martínez Catalán et al., 1996), as well as to the materials that crop out in the core of the Lalín and Forcarei synclines (Díaz García, 1993; Martínez Catalán et al., 1996), and to the Agualada unit (Martínez Catalán et al., 1996; Arenas et al., 1997), which forms other segments of the Basal unit of the complex. The Basal units as a whole have been interpreted as a fragment of the Gondwana continental margin subducted in eo-Hercynian time (Arenas et al., 1995; Martínez Catalán et al., 1996), according to the Rb-Sr age of 374 Ma obtained in immediately posteclogitic phengites in Malpica-Tui (Van Calsteren et al., 1979). The polarity of decreasing pressures from Malpica-Tui to Santiago and from the latter to Lalín and Forcarei, registered by the high-pressure metamorphic episode, marks a general westward direction of the subduction of the lithospheric layer (Martínez Catalán et al., 1996).

The tectonothermal evolution of the Santiago unit, calculated from the metapelitic lithologies, shows a first episode of pressurization (D_1) to eclogitic conditions (14.7 ± 0.7 kbar and 493 ± 9 °C, Arenas et al., 1995), followed by a rapid syndeformational decompression (D_2) in low-temperature amphibolite facies (Arenas et al., 1995). This study analyses this tectonic-metamorphic evolution by means of the record preserved by the metabasic lithologies of the unit, which range from different types of amphibolites included in the pelitic and semipelitic schists to variably retrograded eclogites included in the orthogneisses.

PETROGRAPHY

The eclogitic and retroeclogitic types only appear as inclusions in the orthogneisses; the intercalations in the schists are garnet amphibolites extensively recrystallized in the epidote amphibolite facies or in the low-T and high-P part of the amphibolite facies, showing no evidence of an initial eclogitic transformation. This may be due to a differential eclogitization rather than to a differential retroeclogitic overprinting, given the major textural and mineral and chemical differences between the two types of rocks. This differential eclogitization is considered to be linked to the different preeclogitic peak mineral assemblages. In the following descriptions, the numbering of phases means different textural (or chemical-textural) types along both eclogites and amphibolites. Correlative numbers have a sequential sense only inside each rock type.

The eclogites with preserved omphacitic clinopyroxene (Fig. 3) show a coarsely oriented porphyroblastic texture, defined by coarse colorless amphibole crystals (amphibole 1) that replace the original clinopyroxene and include small allotriomorphic relicts of colorless to pale green omphacite; the inclusions of ilmenite and idiomorphic garnet (garnet EC) tend to be marginal and scarce. These porphyroblasts of colorless amphibole usually have bluish-green rims (amphibole 4), more intense in the contact with garnet and possibly caused by late mass diffusion. Amphiboles of this type (amphibole 4) also grow as monomineral rims of keliphytoid aspect around the garnet (EC) of the matrix.

The mesostasia is mainly formed by small garnets of the same type (0.20–0.25 mm), but subidiomorphic and having a tendency to form small anastomosed chains (honeycomb), and by a dark microaggregate of colorless amphibole (amphibole 2) and plagioclase (1) that form a microcrystalline to cryptocrystalline symplectite aggregate. In the matrix some coarse id-

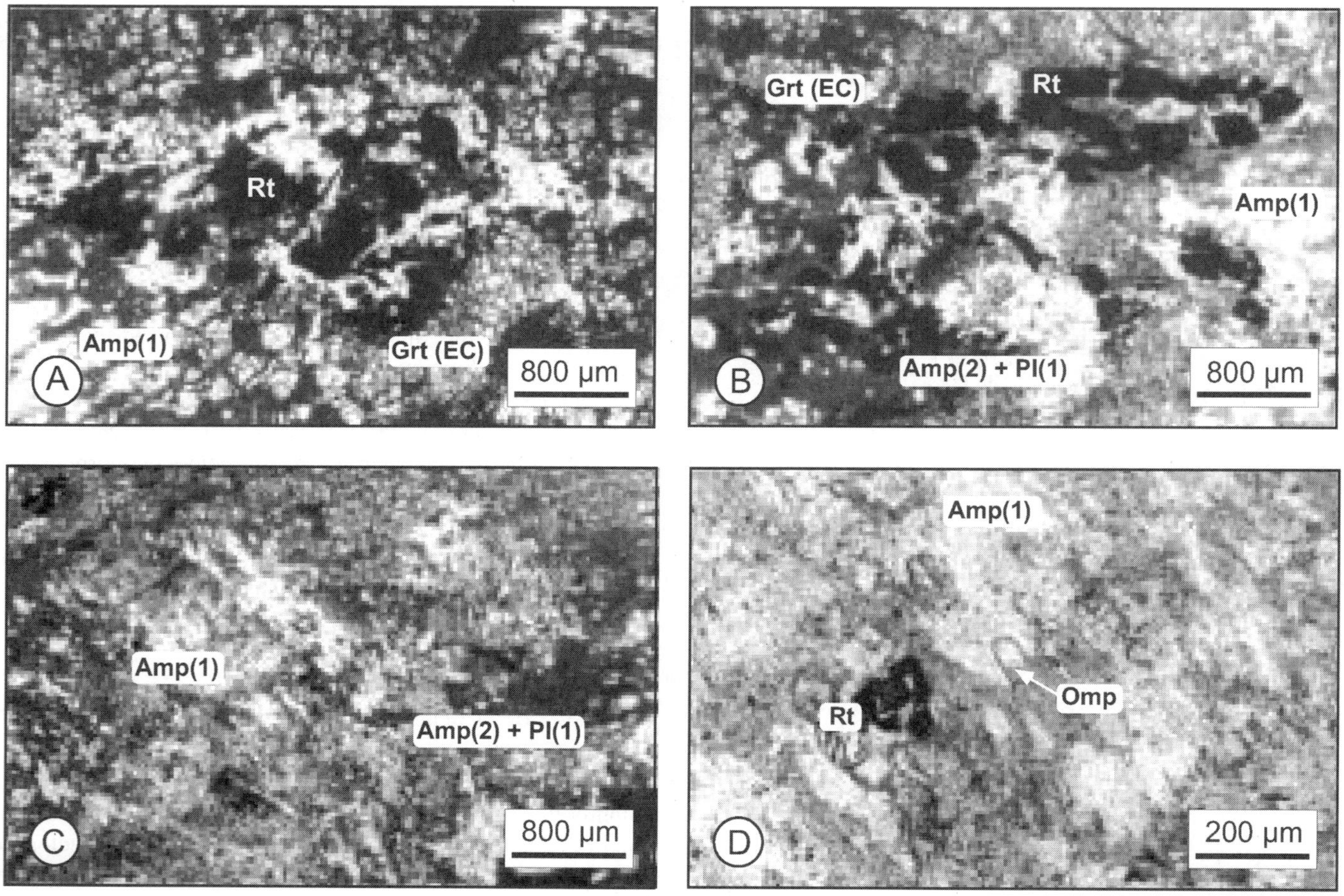

Figure 3. Microphotographs showing some petrographic aspects of eclogites. Coarse colorless amphibole crystals (Amp 1) present tiny omphacite relicts, maximum size of which is visible in D. Dark microaggregate of amphibole and plagioclase (Amp 2 + Pl 1) and garnet (EC) masses surround Amp (1) crystals. Rutile grains usually show ilmenite and sphene rims.

iomorphic rutile crystals encased in ilmenite rims are outstanding, as well as short zoisite prismatic crystals passing to more elongated clinozoisites; there are also white micas (paragonite and phengite I) with clinozoisite rims, and quartz.

The more deformed outer zones of the metabasite boudins included in the orthogneisses correspond to retroeclogites of smaller grain size. They show a nematoblastic to porphyronematoblastic texture with pale green amphibole (amphibole 3) with darker bluish-green edges (sometimes with chemical composition similar to type 4), and clinozoisite prismatic crystals defining the schistosity of the rock (S_2). The garnet (EC) is scarcer, allotriomorphic, and often surrounded by plagioclase; the plagioclase rims around the garnet (EC) form a more advanced keliphytoid trait than the amphibole rims (4) and keep the garnet from reacting with the amphibole and being totally consumed. The plagioclase is more abundant than in the less retrograde eclogites and forms poikiloblasts with zoning in irregular patches (plagioclases 2 and 3). Ilmenite, sphene, and quartz complete the mineral assemblage.

The metabasites included in the pelitic and semipelitic schists (Fig. 4) show no evidence of eclogitization. Crystallization-deformation differences between retro-eclogites and amphibolites are shown by comparing Figures 5 and 6. The rocks located toward the lower part of the unit, that of lower metamorphic grade, present a porphyronematoblastic texture in which the second phase schistosity (S_2), linked to the decompressive emplacement of the unit (Arenas et al., 1995; Martínez Catalán et al., 1996), appears marked by the orientation of green to bluish-green amphibole crystals (amphibole 6) and epidote, ilmenite transforming to sphene, some chlorite (chlorite 2, which, closer to the top of the unit, passes to biotite), and quartz. The oldest plagioclase (plagioclase 4) is found forming the porphyroblasts of the lesser grade zone. These albitic porphyroblasts can be thick (4–8 mm), with internal schistosity from straight to sigmoidal formed by amphibole (amphibole 5), epidote, ilmenite, and sphene, and they sometimes preserve rutile relicts, some chlorite (chlorite 1), white mica (phengite 2), and small-sized idiomorphic garnet (garnet I, 0.15–0.30 mm). They are similar to those described by Arps (1981) in the Malpica-Tui unit, where relicts of a high-pressure assemblage are also preserved (Gil Ibarguchi and Ortega Gironés, 1985). A second type of plagioclase (plagioclase 5) is formed by recrystallization of the former in zones of the blasts that are rich in epidote inclusions, or else appears as small grains in the matrix.

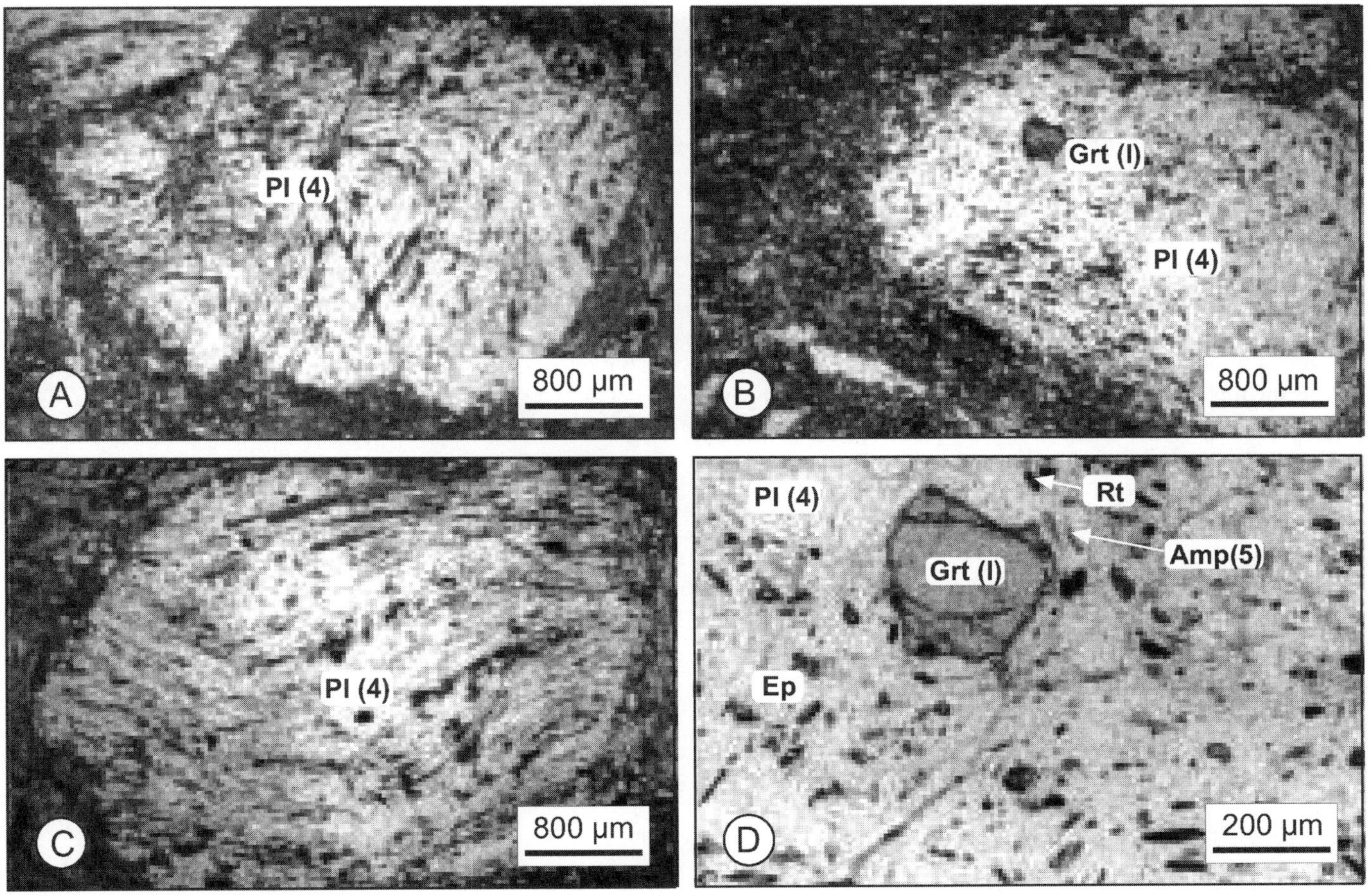

Figure 4. Microphotographs showing some petrographic aspects of amphibolites. Albite porphyroblasts (Pl 4) exhibit internal schistosity mainly formed by epidote, amphibole (type 5), and ilmenite. Garnet, rutile, and white mica are often present.

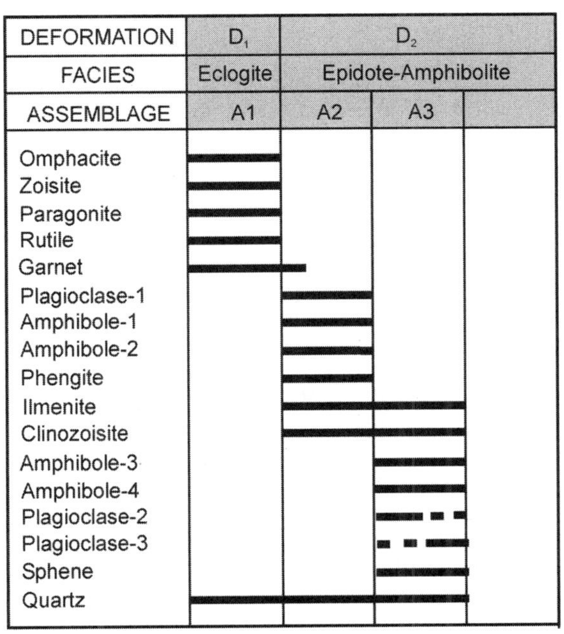

Figure 5. Crystallization-deformation diagram for eclogites and retro-eclogites included in orthogneisses of Santiago unit.

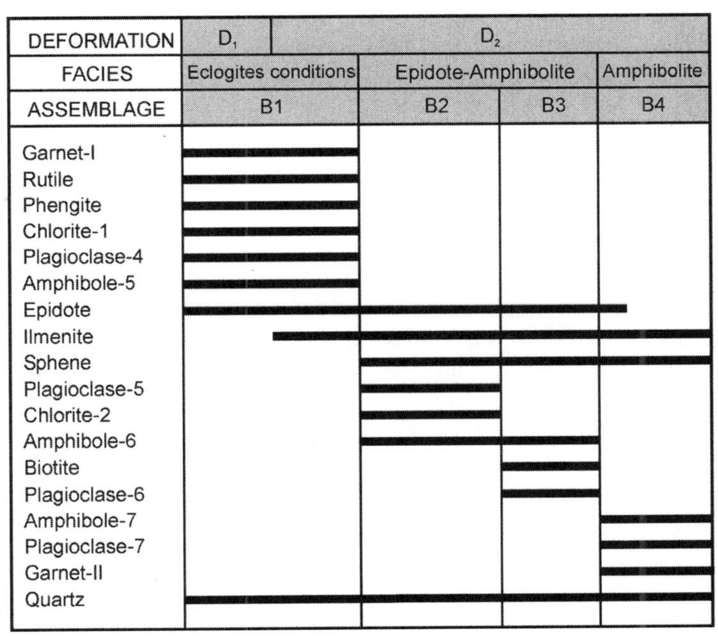

Figure 6. Crystallization-deformation diagram for amphibolites included in metapelites of Santiago unit.

The schistosity included in the plagioclase porphyroblasts (4–5) is in general oblique to the external schistosity, but at times it appears concordant with the S_e, hence it could correspond to a S_1, as in the case of the albite porphyroblasts characteristic of the semipelitic schists encasing these amphibolites (Arenas et al., 1995), or to a very early S_2. As in the Malpica-Tui unit, part of the mineralogy contained in the porphyroblasts indicates conditions characteristic of high pressure, regionally attributed to the first deformation phase. The plagioclase porphyroblasts in the medium and upper structural zones of the unit (type 6 and 7 plagioclases, respectively) are scarcer and lose this internal schistosity.

At the top of the unit the amphibolites pass from being homogeneously porphyronematoblastic to frequently showing a banded texture parallel to the foliation of the second deformation phase. They show an alternation of melanocratic nematoblastic levels, with predominant dark green amphibole (amphibole 7) together with type II garnets (0.5–1 mm) and granoblastic leucocratic levels with a predominance of quartz and a non-porphyroblastic plagioclase chemically similar to plagioclase 5.

MINERAL CHEMISTRY

Seven samples representing eclogites and retroeclogites and nine of the amphibolites situated among the pelitic schists (location of the samples in Fig. 2) were selected for study by electron microprobe. The analyses were carried out at the University of Oviedo (Spain) with a three-spectrometer CAMEBAX SX50 microprobe. The working conditions were: integration time = 10 s, acceleration potential = 15 kV, and beam current = 15 nA; the French Bureau de Recherches Géologiques et Minières patterns were used and the ZAF correction method was followed. The mean analytic error for major elements is considered to be <2%. The first number of the signature of each analysis corresponds to the number of the sample, the second number corresponds to the crystal number of that mineral analyzed in thin section, and the third number corresponds to the point of analysis within each specimen, plus core and rim, where necessary.

Clinopyroxene

The structural formula of the pyroxenes was standardized to four cations using the Cebriá Gómez (1990) program, and the Fe^{3+} values were obtained by charge balance. The clinopyroxene relicts included in colorless amphibole (amphibole 1) are omphacites according to the classification of Essene and Fyfe (1967). They have a fairly homogeneous composition (Table 1), the maximum jadeite content (X_{jd}) range being 0.38–0.43 according to the rule of Cawthorn and Collerson (1974). The mean molecular formula of the crystals analyzed is ($Acm_{1.90}Jd_{40.13}Cfs_{0.49}Wo_{28.07}En_{23.93}Fs_{5.48}$).

The Malpica-Tui unit eclogites clinopyroxenes systematically show higher values in jadeite content (X_{jd} = 0.43–0.54), in accordance with Gil Ibarguchi and Ortega Gironés (1985), and may even reach practically pure jadeite (X_{jdmax} > 0.92) in some acid metagranite lithologies poor in Fe and Mg (Gil Ibarguchi, 1995). Comparison with the Santiago omphacites implies a greater P_{max} in Malpica-Tui, according to the westward polarity of the eo-Hercynian subduction. In the Agualada unit, which must have been farther west than the Santiago unit in the subducting slab (Martínez Catalán et al., 1996), the jadeite contents of a first generation of clinopyroxenes included in atoll garnets of the eclogites (Jd_{max} = 40 mol%) are very similar to those of Santiago (see Figs. 5 and 6). This first generation of clinopyroxenes represents the eclogitic peak in the Agualada unit, whereas the clinopyroxenes in the matrix of the eclogites (Jd_{max} = 36 mol%) already represent the decompressive process (Arenas et al., 1997).

Garnet

The structural formula of the garnets was calculated on a basis of 12 oxygens; calculation of the end members was made following the rule by Knowles (1987).

Type EC garnet. The syn-D_1 garnets of the eclogites included in the orthogneisses are rich in almandine, grossular, and pyrope (Table 2). The mean core composition is ($Alm_{62.62}$

TABLE 1. REPRESENTATIVE ANALYSES OF OMPHACITES

	1.1.1 Omp	1.2.1 Omp	1.3.1 Omp
SiO_2	55.42	56.51	56.86
TiO_2	0.02	0.08	0.08
Al_2O_3	10.05	8.73	9.80
Cr_2O_3	0.00	0.21	0.15
FeO(t)*	4.01	4.12	4.15
MnO	0.00	0.00	0.00
MgO	8.93	8.86	8.22
CaO	13.86	14.81	13.69
Na_2O	6.57	6.07	6.62
Total	98.86	99.39	99.57
Si	1.99	2.03	2.03
Al^{IV}	0.01	0.00	0.00
Al^{VI}	0.41	0.37	0.41
Ti	0.00	0.00	0.00
Cr	0.00	0.01	0.00
Fe^{3+}	0.06	0.00	0.00
Fe^{2+}	0.06	0.12	0.12
Mn	0.00	0.00	0.00
Mg	0.48	0.47	0.44
Ca	0.53	0.57	0.52
Na	0.46	0.42	0.46
Jd	0.42	0.38	0.43
Acm	0.03	0.01	0.00
Cfs	0.01	0.00	0.00
Wo	0.26	0.30	0.27
En	0.24	0.25	0.23
Fs	0.03	0.06	0.06

Note: Cations on a basis of 6 oxygens.
*(t) = total.

TABLE 2. REPRESENTATIVE ANALYSES OF SYN-D_1 GARNETS OF ECLOGITES

	1.1.1c* EC	1.1.5r* EC	1.2.1c* EC	1.2.5r* EC	2.1.1c* EC	2.1.5r* EC	3.1.1c* EC	3.1.5r* EC
SiO_2	37.53	37.70	38.47	38.24	37.91	38.11	37.67	38.15
Al_2O_3	20.67	21.35	19.92	19.88	20.74	21.46	20.79	21.19
FeO(t)†	29.53	26.33	28.60	28.10	27.30	27.57	28.37	26.29
MnO	0.10	0.10	0.77	0.05	1.99	0.00	1.89	0.00
MgO	2.41	4.78	2.89	2.95	0.86	3.24	1.41	2.75
CaO	10.16	9.45	9.40	9.77	11.58	10.11	10.40	11.72
Total	100.40	99.71	100.05	98.99	100.38	100.49	100.53	100.10
Si	2.97	2.95	3.05	3.06	3.02	2.98	2.99	3.00
Al^{IV}	0.03	0.05	0.00	0.00	0.00	0.02	0.01	0.00
Al^{VI}	1.89	1.92	1.91	1.87	1.94	1.97	1.94	1.96
Fe^{3+}	0.14	0.13	0.04	0.01	0.02	0.05	0.07	0.04
Fe^{2+}	1.82	1.59	1.86	1.86	1.79	1.76	1.81	1.69
Mn	0.01	0.01	0.05	0.00	0.13	0.00	0.13	0.00
Mg	0.28	0.56	0.34	0.35	0.10	0.38	0.17	0.32
Ca	0.86	0.79	0.80	0.84	0.99	0.85	0.88	0.99
Alm	61.20	54.02	60.93	61.02	59.44	58.90	60.59	56.34
Prp	9.57	18.90	11.20	11.50	3.38	12.67	5.58	10.74
Grs	22.09	20.28	24.31	26.72	31.55	26.03	25.99	30.98
Adr	6.91	6.58	1.87	0.66	1.18	2.40	3.59	1.94
Sps	0.23	0.22	1.70	0.11	4.45	0.00	4.25	0.00

Note: Cations on a basis of 12 oxygens.
*c = core; r = rim.
†(t) = total.

$Grs_{27.05}Prp_{7.60}Adr_{2.42}Sps_{0.29}$), and the mean rim composition is ($Alm_{57.57}Grs_{26.00}Prp_{13.45}Adr_{2.90}Sps_{0.08}$). The garnets are poorer in pyrope than the garnets in the Malpica-Tui unit eclogites described by Van der Wegen (1978) and Gil Ibarguchi and Ortega Gironés (1985) and the non-atoll garnets in the Agualada unit (Arenas et al., 1997); inversely, they are richer in almandine than the latter, which could be interpreted as a greater pressure and/or temperature of these other units during D_1. Their zonings follow two models. The first and most characteristic (illustrated in Fig. 7A) shows a continuous decrease of MnO, from initial values close to 2% until total exhaustion, and an initial increase in FeO and a drastic decrease toward the rims, at the same time as the $Mg/Fe_{(t)}$ ratio rises from <0.1 to 0.2. The CaO shows an initial decreasing evolution; its increase near the rim (not noted in all the garnets and in others culminating in a new collapse) may be linked to the participation of a calcic phase in the garnet growth reaction (Råheim, 1975), probably the destruction of a calcisodic plagioclase to give omphacite and an excess of calcium as grossular. The second zoning model (Fig. 7B) is also normal for the MnO and the $Mg/Fe_{(t)}$ ratio (prograde growth zoning; Tracy, 1982), although the composition of the cores is systematically poorer in MnO (<1%) and has a higher $Mg/Fe_{(t)}$ ratio (>0.1). The increase in the $Mg/Fe_{(t)}$ ratio is more uniform in these garnets, which can show significant increases in MnO% in the last microns of garnet, probably owing to a retrograde effect by mass diffusion.

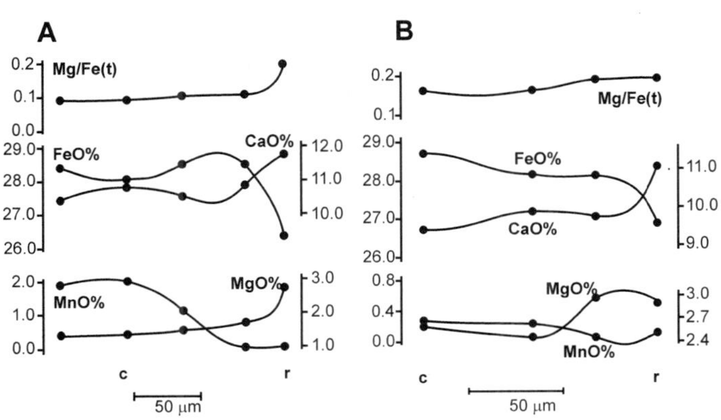

Figure 7. Chemical zoning profiles across garnets (EC). A is model a; B is model b; c, center; r, rim.

The projections of type EC garnets in the Mn-Fe-Mg triangle (Fig. 8) describe two branches, one marked by the preferential fractioning of Fe and Mn until what can be interpreted as the virtual exhaustion of the reservoir in the latter element, and another by the preferential fractioning of Mg. The first of these branches is short and may even be missing. This can be interpreted as indicative of the scarcity or absence of an Mn-rich phase in the garnet-forming reacting assemblage, which is consistent with the absence of chlorite in the assemblage. Hence we are inclined to think that the garnets (EC) crystallize from a basically igneous mineralogy. Similar behavior is shown by garnets in high-grade blue schists and eclogites of the Franciscan belt, where Oh et al. (1991) established a limit near 450 °C (the ap-

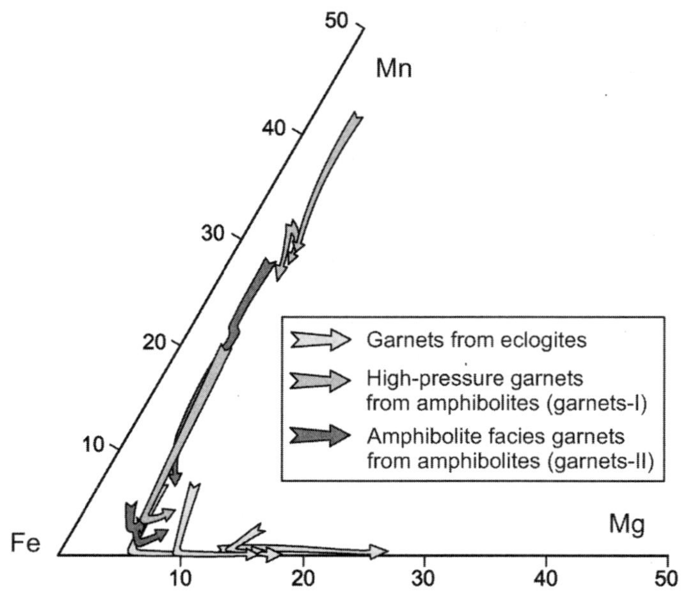

Figure 8. Projection Mn-Fe-Mg showing zoning of different generations of garnets in Santiago unit. Arrows indicate compositional variation of garnets from center to rim (arrowheads).

proximate temperature of the almost total consumption of Mn to form garnet in the metabasites) between garnets with Mn preferential partitioning and those with Mg preferential partitioning.

Type I garnet. The high-*P* syn-D_1 garnets of the metabasites intercalated in the schists (Table 3) are rich in almandine, grossular, and spessartine. The mean molecular composition of their cores is ($Alm_{45.82}Grs_{29.12}Sps_{21.25}Prp_{2.53}Adr_{1.28}$) and that of the rims is ($Alm_{50.99}Grs_{28.22}Sps_{15.31}Prp_{3.75}Adr_{1.73}$). The somewhat asymmetrical chemical profile of a garnet I (Fig. 9A) shows normal zoning for the MnO, the same as in the EC garnets, but with much higher values and a much higher initial chemical gradient. The Mg/$Fe_{(t)}$ ratio evolves somewhat differently than in the garnets of the eclogites, growing at first to become stabilized near the rim. The CaO can show initial growth, which collapses in the last microns of garnet, although in other crystals analyzed it only underwent a decrease in the chemical gradient. However, in both cases it indicates exhaustion in the reservoir of the reacting calcic phase, probably the An component of the plagioclase, to give grossular, epidote, and a plagioclase rich in Ab (type 4 plagioclase).

The reaction history of the type I garnet (Fig. 8) is defined by a marked decrease in manganese and, in some garnets, a strong final increase in the Mg/$Fe_{(t)}$ ratio. These patterns suggest a prograde increase based on a continuous reaction that preferentially fractionates Mn and Fe, probably from chlorite, giving as a result a new chlorite enriched in Mg (type 1 chlorite) and that could begin to be consumed subsequently. That is, the type I garnet is probably growing from a pre-P_{max} assemblage that is already metamorphic, with chlorite and albite-rich plagioclase. Furthermore, the high MnO content of the type I garnets, which possibly begin to grow at <450 °C (Oh et al., 1991), evidence the beginning of metamorphic recrystallization of the amphibolites prior to that in the eclogites.

Type II garnet. The syn-D_2 garnets that appear in the amphibolites situated at the top of the Santiago unit are almandines

TABLE 3. REPRESENTATIVE CHEMICAL ANALYSES OF GARNETS OF AMPHIBOLITES

	8.1.1c*	8.1.4r*	13.1.1c*	13.1.3r*	9.1.1c*	9.1.7r*	16.1.1c*	16.1.7r*
	I	I	I	I	II	II	II	II
SiO_2	38.28	38.09	37.52	38.09	37.95	38.59	37.38	37.57
Al_2O_3	21.74	21.71	20.19	20.08	21.48	21.62	19.92	20.18
FeO(t)†	23.84	29.10	18.35	21.24	30.36	28.70	24.02	29.41
MnO	5.86	1.29	12.82	9.15	1.49	0.71	8.72	2.24
MgO	0.65	1.26	0.62	0.88	0.70	1.33	0.61	1.02
CaO	11.09	10.47	10.35	10.48	9.72	10.47	9.81	10.89
Total	101.46	101.92	99.85	99.92	101.70	101.42	100.46	101.31
Si	3.01	2.98	3.01	3.05	2.99	3.03	3.00	2.97
Al^{IV}	0.00	0.02	0.00	0.00	0.00	0.00	0.00	0.02
Al^{VI}	2.02	1.98	1.91	1.90	1.99	2.00	1.88	1.86
Fe^{3+}	0.00	0.03	0.06	0.00	0.01	0.00	0.12	0.17
Fe^{2+}	1.57	1.87	1.18	1.42	1.99	1.88	1.49	1.78
Mn	0.39	0.09	0.87	0.62	0.10	0.05	0.59	0.15
Mg	0.08	0.15	0.07	0.10	0.08	0.16	0.07	0.12
Ca	0.93	0.88	0.89	0.90	0.82	0.88	0.84	0.92
Alm	52.81	62.75	39.04	46.68	66.48	63.49	49.69	59.85
Prp	2.57	4.93	2.46	3.45	2.75	5.24	2.43	4.05
Grs	31.48	27.74	26.76	29.51	26.81	29.68	22.04	22.62
Adr	0.00	1.72	2.80	0.00	0.64	0.00	6.08	8.43
Sps	13.15	2.87	28.94	20.37	3.33	1.59	19.76	5.05

Note: Cations on a basis of 12 oxygens.
*c = core; r = rim.
†(t) = total.

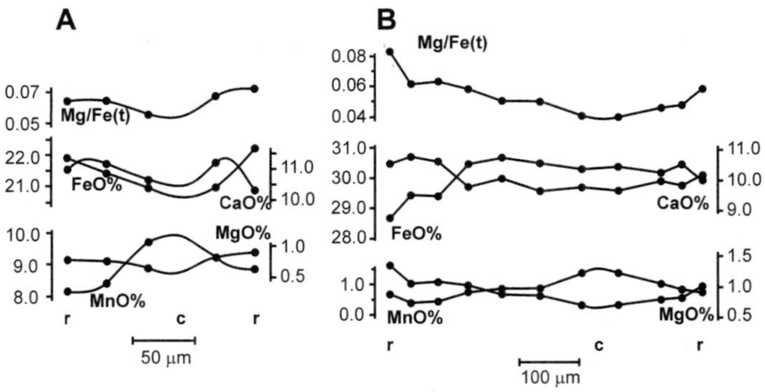

Figure 9. Chemical zoning profiles across garnets of amphibolites. A is type I garnet; B is type II garnet; c, center; r, rim.

rich in grossular (Table 3). The mean composition of their cores is ($Alm_{60.08}Grs_{26.26}Sps_{8.72}Prp_{2.92}Adr_{2.03}$) and that of their rims is ($Alm_{61.75}Grs_{27.96}Prp_{5.05}Adr_{2.81}Sps_{2.43}$). Their chemical profiles (Fig. 9B) show normal zonings for MnO and MgO, and, as in the type EC garnets, FeO decreases and CaO increases near the rims. The increase in MgO coupled with the impoverishment in FeO and MnO indicates the consumption of a phase with a smaller Fe/Mg ratio than the garnet, probably chlorite and/or biotite. The ternary diagram in Figure 8 shows that there are type II garnets with long trajectories of decrease in Mn and increase in Fe^{2+}, which grew according to a continuous reaction, probably at the expense of the ferric end member of the chlorite, in the same way as the type I garnets. The crystallization of these Mn-rich garnets did not necessarily occur under T_{max} of 450 °C, because the MnO concentrated in the chlorite (chlorite 2) with the destabilization of the type I garnet during the beginning of D_2. Other type II garnets indicate more complex histories; their branches of decrease in Mn and increase in Fe^{2+} are very short, and toward the rim they are replaced by trajectories of increase in Mn and decrease in Fe^{2+}. These trajectories indicate that the garnet grows at the expense of a phase with high Mg content, probably a chlorite enriched in its magnesian end member or a biotite, although they could also be due to an Fe/Mg exchange reaction between the garnet and the amphibole, which is increasingly ferric.

Amphiboles

The structural formula of the amphiboles was calculated on a basis of 23 oxygens and standardized for a mean Fe^{3+} content, using the RECAMP program (Spear and Kimball, 1984). We identified seven types of calcic amphiboles according to their textural and mineral-chemical characteristics. The first four correspond to amphiboles present in the variably retrograded eclogites of the unit (Table 4), whereas the other three types are found in the amphibolites (Table 5).

TABLE 4. REPRESENTATIVE CHEMICAL ANALYSES OF AMPHIBOLES OF THE ECLOGITES

	1.1.1c* Type 1	1.1.2r* Type 4	1.2.1 Type 1	1.3.1 Type 4	2.1.1c* Type 1	2.1.2r* Type 4	2.2.1 Type 2	3.1.1 Type 2	3.2.1 Type 2	4.1.1 Type 3	7.1.1c* Type 3	7.1.2r* Type 3
SiO_2	54.05	36.75	53.90	44.69	54.65	38.45	49.28	54.48	47.23	48.08	49.25	43.66
TiO_2	0.10	0.24	0.07	0.03	0.16	0.11	0.02	0.03	0.20	0.18	0.08	0.26
Al_2O_3	5.64	20.79	5.83	9.51	4.82	19.86	8.47	3.79	10.58	10.45	9.64	13.67
Cr_2O_3	0.04	0.00	0.05	0.06	0.01	0.09	0.09	0.02	0.12	0.00	0.03	0.08
FeO(t)†	9.48	21.44	8.42	21.84	9.05	19.73	14.05	11.03	15.27	16.65	16.18	17.70
MnO	0.00	0.00	0.00	0.00	0.00	0.00	0.00	0.00	0.00	0.00	0.02	0.17
MgO	16.55	3.97	16.56	7.36	17.44	5.34	12.31	16.95	11.99	10.68	11.54	8.33
CaO	10.03	11.22	9.89	11.62	10.52	9.96	12.70	10.90	10.32	9.78	8.80	11.17
Na_2O	1.36	1.98	1.61	0.67	0.93	1.93	0.59	0.59	1.19	0.73	0.87	1.19
K_2O	0.09	0.17	0.17	0.46	0.11	0.67	0.17	0.10	0.35	0.26	0.18	0.34
Total	97.30	96.56	96.47	96.18	97.68	96.05	97.59	97.87	97.13	96.81	96.56	96.49
Si	7.61	5.63	7.62	6.84	7.66	5.85	7.17	7.70	6.88	7.07	7.21	6.54
Al^{IV}	0.39	2.37	0.38	1.15	0.34	2.15	0.83	0.30	1.12	0.93	0.79	1.46
Al^{VI}	0.55	1.39	0.60	0.56	0.46	1.41	0.62	0.33	0.70	0.88	0.88	0.95
Ti	0.01	0.03	0.01	0.00	0.02	0.01	0.00	0.00	0.02	0.02	0.01	0.03
Cr	0.01	0.00	0.01	0.01	0.00	0.01	0.01	0.00	0.01	0.00	0.00	0.01
Fe^{3+}	0.10	0.47	0.10	0.40	0.05	0.49	0.02	0.06	0.39	0.10	0.06	0.29
Mg	3.47	0.91	3.49	1.68	3.64	1.21	2.67	3.57	2.60	2.34	2.52	1.86
Fe^{2+}	1.01	2.28	0.89	2.40	1.01	2.01	1.68	1.24	1.47	1.94	1.92	1.93
Mn	0.00	0.00	0.00	0.00	0.00	0.00	0.00	0.00	0.00	0.00	0.00	0.02
Ca	1.51	1.84	1.50	1.91	1.58	1.62	1.98	1.65	1.61	1.54	1.38	1.79
Na^{M4}	0.34	0.08	0.41	0.05	0.24	0.24	0.01	0.14	0.21	0.17	0.23	0.13
Na^{A}	0.03	0.50	0.03	0.15	0.01	0.33	0.15	0.02	0.13	0.03	0.02	0.22
K	0.02	0.03	0.03	0.09	0.02	0.13	0.03	0.02	0.06	0.05	0.03	0.06

Note: Cations on a basis of 23 oxygens.
*c = core; r = rim.
†(t) = total.

TABLE 5. REPRESENTATIVE CHEMICAL ANALYSES OF AMPHIBOLES OF THE AMPHIBOLITES

	8.1.1 Type 5	13.1.1 Type 5	14.1.1 Type 5	17.1.1 Type 5	8.2.1 Type 6	13.2.1 Type 6	14.2.1 Type 6	15.1.1 Type 6	9.1.1 Type 7	9.2.1 Type 7	11.1.1 Type 7	16.1.1 Type 7
SiO_2	43.17	46.98	46.44	46.81	43.34	47.68	43.36	43.88	41.07	42.90	43.36	43.89
TiO_2	0.32	0.40	0.25	0.28	0.35	0.21	0.35	0.31	0.39	0.41	0.41	0.39
Al_2O_3	12.67	9.94	10.37	9.52	13.23	8.25	13.43	12.89	15.46	14.35	13.90	10.97
Cr_2O_3	0.02	0.01	0.00	0.03	0.00	0.04	0.04	0.10	0.00	0.05	0.00	0.01
FeO(t)*	20.48	13.72	19.25	17.21	19.75	18.40	18.59	17.60	21.36	17.09	16.06	20.41
MnO	0.34	0.68	0.40	0.20	0.41	0.53	0.39	0.23	0.29	0.42	0.25	0.33
MgO	8.67	11.80	10.60	9.99	8.82	12.20	9.84	10.39	6.57	9.73	10.65	8.78
CaO	10.81	11.78	9.78	12.43	10.37	8.88	10.21	11.35	10.77	11.01	11.42	10.96
Na_2O	1.32	1.26	1.11	1.23	1.76	1.06	1.38	1.71	1.33	1.38	1.96	1.81
K_2O	0.24	0.18	0.21	0.31	0.22	0.15	0.49	0.34	0.34	0.37	0.42	0.15
Total	98.02	96.74	98.41	97.98	98.25	97.36	98.04	98.70	97.58	97.66	98.43	97.69
Si	6.39	6.91	6.79	6.94	6.40	7.00	6.36	6.41	6.16	6.31	6.34	6.57
Al^{IV}	1.61	1.09	1.21	1.06	1.60	1.00	1.64	1.59	1.84	1.69	1.66	1.43
Al^{VI}	0.60	0.64	0.57	0.61	0.70	0.43	0.68	0.63	0.89	0.79	0.74	0.51
Ti	0.04	0.04	0.03	0.03	0.04	0.02	0.04	0.03	0.04	0.04	0.04	0.04
Cr	0.00	0.00	0.00	0.00	0.00	0.01	0.01	0.01	0.00	0.01	0.00	0.00
Fe^{3+}	0.81	0.16	0.55	0.02	0.67	0.49	0.79	0.58	0.69	0.63	0.42	0.53
Mg	1.91	2.59	2.31	2.21	1.94	2.67	2.15	2.26	1.47	2.13	2.32	1.96
Fe^{2+}	1.72	1.53	1.81	2.12	1.77	1.77	1.49	1.56	1.99	1.47	1.54	2.02
Mn	0.04	0.08	0.05	0.02	0.05	0.07	0.05	0.03	0.04	0.05	0.03	0.04
Ca	1.71	1.86	1.53	1.97	1.64	1.40	1.60	1.78	1.73	1.73	1.79	1.76
Na^{M4}	0.15	0.10	0.16	0.02	0.19	0.15	0.20	0.12	0.14	0.14	0.11	0.13
Na^{A}	0.23	0.26	0.16	0.33	0.31	0.15	0.20	0.36	0.24	0.25	0.44	0.40
K	0.04	0.03	0.04	0.06	0.04	0.03	0.09	0.06	0.06	0.07	0.08	0.03

Note: Cations on a basis of 23 oxygens.
*(t) = total.

Type 1 amphibole. These coarse colorless porphyroblasts compose the first amphibole formed during the recrystallization of the eclogites. The omphacite relicts preserved inside it suggest that the amphibole grew by the destabilization of the clinopyroxene in a hydrated medium in apparent textural equilibrium with the EC garnet. According to the nomenclature by Leake et al. (1997) given in the diagram in Figure 10, these are actinolites and edenites. They generally have weak prograde zoning with an increase in the Al_2O_3 and TiO_2 contents and in the $Fe^{2+}/(Fe^{2+} + Mg)$, $Na/(Ca + Na)$, and $Al/(Si + Al)$ ratios toward the rim of the crystal. Their $Na/(Ca + Na)$:$Al/(Si + Al)$ and Al^{VI}/Al^{IV} ratios, expressed in Figures 11 and 12, are always >1, in accordance with crystallization in high-*P* conditions (Laird and Albee, 1981). Likewise, the diagram Ti versus Al^{VI}/Al^{IV} in Figure 12 shows the decompressive and prograde evolution of the Santiago unit amphiboles, of which only some of the amphiboles of the retroeclogites originally had high-*P* compositions. These types of high pressure coincide with the minimum Ti values, which in a broad sense are equivalent to lower crystallization temperatures (Raase, 1974; Spear, 1981; Ernst and Liou, 1998). In the classification diagram of Leake et al. (1997), the actinolite-hornblende compositional gap can be seen around a value of 7.45 (Si p.f.u.); an apparent second gap of miscibility around 6.70 (Si p.f.u.) could be a random effect. High-*P* calcic amphiboles, texturally similar to type 1 in the Santiago unit and with compositions that range from tremolites to edenites or Mg-hornblendes, have been described in other eclogites of the Ibero-Armorican arc (Godard, 1988), Bohemia Massif (O'Brien, 1993), and the Alps (Messiga et al. 1992), and closer, such as in the Malpica-Tui unit (Van der Wegen, 1978).

Type 2 amphibole. The amphiboles of the simplectoid microaggregate show a certain compositional variety (Fig. 10), from high-*P* actinolitic types similar to those of type 1 to individuals of later recrystallization, chemically similar to the type 3 amphiboles classifiable as Mg-hornblendes. Because texturally they define a single crystallization episode, these variations are interpreted as subsequent reequilibria controlled by the microtextural position, and it is conjectured that the original composition must have been closest to that of the type 1 amphiboles.

Type 3 amphibole. The more recrystallized the retroeclogite, the more abundant are the green nematoblastic amphiboles. Their compositions (Fig. 10) range from actinolites (Figs. 11 and 12) corresponding to slightly recrystallized type 2, to magnesiohornblendes and ferrohornblendes; some of these amphiboles corresponding to strongly colored grains and grain rims can be classified as alumino-ferrotschermakites or ferropargasites, chemically more similar to type 4 amphiboles.

Type 4 amphiboles. The amphiboles linked to rims around garnets or to irregular rims of type 1 amphiboles are of alumino-ferrotschermakitic or ferropargasitic composition. This type of

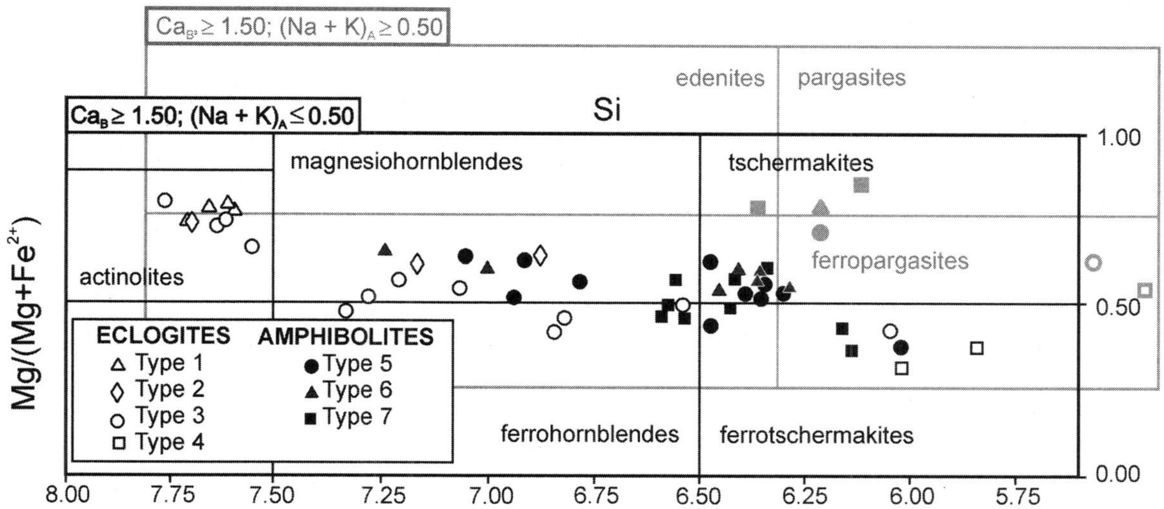

Figure 10. Classification diagram (IMA 97) for calcic amphiboles. Note actinolite-hornblende compositional gap around Si = 7.45 c.p.f.u.; second gap appears around Si = 6.70 c.p.f.u.; to right of this second gap amphiboles seem to undergo greater edenitic substitution.

amphibole is often aluminous, which seems to be related to the destabilization of the garnet, the microtextural position of the amphibole, and the effectiveness of the deformation; similar relations were described in Janák et al. (1996).

Type 5 amphibole. The amphiboles included in the internal schistosity of the plagioclase porphyroblasts show great compositional variability, no differences being found between the different metamorphic zones. According to IMA 97 classification, they are mainly magnesiohornblendes and tschermakites, with Na/(Ca + Na):Al/(Si + Al) and Al^{VI}/Al^{IV} ratios (Figs. 11 and 12, respectively) that, according to Laird and Albee (1981), would be proper to crystallization amphiboles in medium-pressure conditions, contrary to what can be deduced for the texturally older amphiboles of the amphibolites. Laird

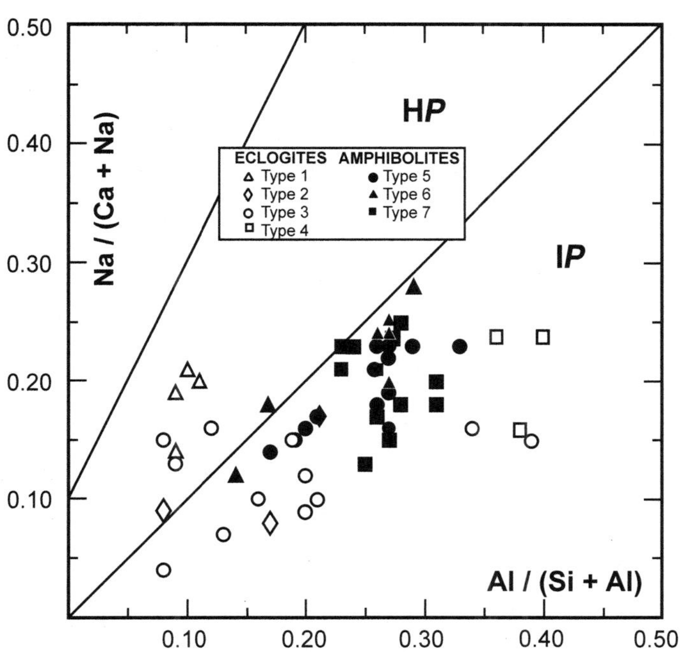

Figure 11. Na/(Ca + Na)–Al/(Si + Al) diagram for amphiboles of metabasites in Santiago unit. High-pressure–intermediate-pressure (H*P*, I*P*) patterns from Laird and Albee (1981).

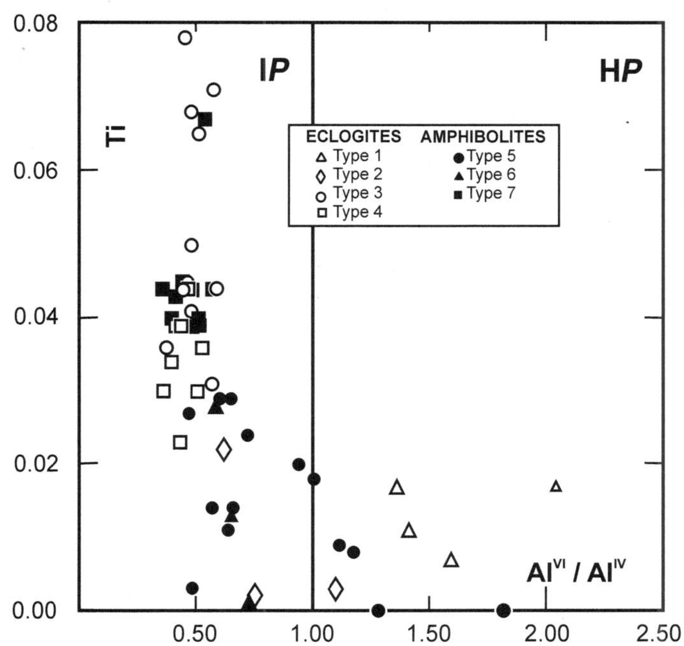

Figure 12. Ti-Al^{VI}/Al^{IV} diagram for amphiboles of metabasites in Santiago unit. High pressure–intermediate-pressure (H*P*, I*P*) patterns are from Laird and Albee (1981).

and Albee's diagrams are probably not applicable to this type of amphibole, but a certain reequilibrium is also possible at the beginning of the decompressive trajectory of the unit (D_2).

Type 6 amphibole. The amphiboles of the second phase schistosity in the structurally lower zones of the Santiago unit are mainly tschermakites (Fig. 10), although there are also magnesiohornblendes and pargasites. The chemical relations of Figures 11 and 12 show that these are medium-pressure amphiboles with moderate contents in titanium.

Type 7 amphibole. In the structurally higher zone of the unit the amphiboles of the D_2 schistosity have compositions of ferrotschermakites, ferrohornblendes, and, to a lesser degree, tschermakites, magnesiohornblendes, ferropargasites, and edenites. These are medium-pressure amphiboles (Figs. 11 and 12) with medium to high Ti contents, higher than those of the syn-D_2 amphiboles of the structurally lower zones. The $Mg/(Mg + Fe^{2+})$ ratios tend to be lower, and the $Al_{(tot)}$ contents higher, than in the type 6 amphiboles; this can also be interpreted as indicative of a prograde effect (Ernst and Liou, 1998) toward the top of the unit.

Plagioclases

We identified seven chemical-textural types of plagioclases in the metabasites of the Santiago unit (Table 6). The first three types developed during the recrystallization of the eclogites included in the basal orthogneisses, whereas the other four appear in the amphibolite boudins between the pelitic schists.

Type 1 plagioclase. This is the first plagioclase to appear during the recrystallization of the eclogites included in the orthogneisses. This cryptocrystalline to microcrystalline plagioclase that forms simplectoid aggregates together with the type 2 amphibole has compositions of intermediate oligoclase (An_{19-21}). Compositions of intermediate to acid oligoclase are consistent with high-P conditions, because it is the plagioclase with the least molar volume and is therefore the typical composition of the last stable preeclogitic plagioclase and the first stable posteclogitic plagioclase (Godard, 1988). It is only preserved in the least deformed and recrystallized central zones of the boudins.

Type 2 and 3 plagioclases. Type 2 plagioclases respond to almost pure albite compositions (An_{02}) and comprise the poikiloblasts of the innermost zones of the boudins, whereas the type 3 plagioclases are basic oligoclases (An_{24-30}) of the outermost and recrystallized zones of the boudins.

Messiga et al. (1992) described a symplectitic plagioclase of composition An_{10-22} and subsequent plagioclases (in rims around the garnets) with two modes, An_{02} and An_{12-34}, in the Variscan eclogite-amphibolites in the Savona crystalline massif (western Ligurian Alps, Italy).

Type 4 plagioclase. The plagioclase of albitic composition ($An_{<3}$), formed by the porphyroblasts of the lowest structural and least metamorphic grade zone, is texturally the oldest plagioclase in the amphibolites and possibly preserves the composition of the syn-D_1 baric peak and may even be prior to P_{max}.

TABLE 6. REPRESENTATIVE ANALYSES OF PLAGIOCLASES

	1.1.1 Type 1	1.2.1 Type 1	6.1.1 Type 2	7.1.1 Type 2	2.1.1 Type 3	2.2.1 Type 3	8.1.1 Type 4	13.1.1 Type 4	8.1.2 Type 5	13.2.1 Type 5	10.1.1 Type 6	11.1.1 Type 7
SiO_2	63.18	65.35	68.77	67.93	62.25	63.06	68.68	68.82	63.99	64.84	57.16	55.66
Al_2O_3	22.27	19.02	19.80	20.21	23.90	23.49	19.66	19.43	22.04	21.55	27.16	28.45
FeO(t)*	1.14	2.02	0.02	0.06	0.09	0.41	0.05	0.13	0.05	0.15	0.03	0.00
MgO	0.65	1.56	0.01	0.00	0.00	0.26	0.00	0.01	0.00	0.01	0.00	0.00
CaO	4.18	3.43	0.17	0.40	5.65	4.78	0.12	0.35	3.41	3.18	9.44	10.77
Na_2O	9.08	8.45	11.29	11.50	8.28	8.71	11.08	11.22	9.38	9.56	5.95	5.21
K_2O	0.07	0.08	0.11	0.01	0.11	0.08	0.00	0.07	0.16	0.06	0.04	0.03
Total	100.57	99.91	100.17	100.11	100.28	100.79	99.59	100.03	99.03	99.35	99.78	100.12
Si	2.79	2.90	2.99	2.97	2.75	2.77	3.00	3.00	2.85	2.87	2.56	2.50
Al	1.16	1.00	1.02	1.04	1.24	1.22	1.01	1.00	1.16	1.12	1.44	1.50
Fe^{2+}	0.04	0.07	0.00	0.00	0.00	0.01	0.00	0.00	0.00	0.01	0.00	0.00
Mg	0.04	0.10	0.00	0.00	0.00	0.02	0.00	0.00	0.00	0.00	0.00	0.00
Ca	0.20	0.16	0.01	0.02	0.27	0.22	0.01	0.02	0.16	0.15	0.45	0.52
Na	0.78	0.73	0.95	0.97	0.71	0.74	0.94	0.95	0.81	0.82	0.52	0.45
K	0.00	0.00	0.01	0.00	0.01	0.00	0.00	0.00	0.01	0.00	0.00	0.00
Ab	0.79	0.81	0.98	0.98	0.72	0.76	0.99	0.98	0.82	0.84	0.53	0.47
An	0.20	0.18	0.01	0.02	0.27	0.23	0.01	0.02	0.17	0.15	0.47	0.53
Or	0.00	0.01	0.01	0.00	0.01	0.00	0.00	0.00	0.01	0.00	0.00	0.00

Note: Cations on a basis of 8 oxygens.
*(t) = total.

Type 5 plagioclase. The small granoblastic plagioclases formed in the amphibolites of the lower structural zone and some recrystallized zones in the albite porphyroblasts, typically around the epidote crystals of the internal schistosity, have an oligoclase (An_{11-24}) composition.

Type 6 plagioclase. The scarce analyses of plagioclases coexisting with biotite in the structurally intermediate zone show compositions of andesine around An_{47}. Some porphyroblasts in the upper zone also preserve a composition of andesines.

Type 7 plagioclase. The plagioclases in the porphyroblasts of the upper zone show a predominant composition of labradorites (to An_{59}). Nevertheless, in this zone there are also plagioclases with other compositions; thus, the presence of granoblastic plagioclases with oligoclase or andesine composition seems to suggest that not all the plagioclases underwent recrystallization with the increase of the metamorphic grade, or that it may be due to the consumption of the anorthitic end member by the crystallization of the type II garnet (Spear et al., 1991).

The compositions of the plagioclases of the metabasites intercalated in the pelitic schists increased their basicity toward the top of the unit, compatible with the inversion of the metamorphic gradient registered in the Santiago unit during its exposure (Arenas et al., 1995).

White Micas

In addition to the scant paragonite crystals, probably of primary nature (Table 7), the eclogites in the Santiago unit contain discrete amounts of potassic white mica (type 1 phengite; Table 7), both frequently encased in clinozoisite rims. These are phengites with a moderate molecular celadonite content (22.4–24.6 mol%), with Si values of between 3.23 and 3.27 c.p.f.u. Unlike the phengites in the amphibolites, they have noticeable CaO contents.

In the amphibolites, the white micas included in the albite-oligoclase-porphyroblasts (type 2 phengite; Table 7) show a greater dispersal in the celadonite contents (which oscillate from 12.9 to 33.6 mol%) and the Si content (3.14–3.36 c.p.f.u.). They show a greater content in Ti than the type 1 phengites. In both cases the less phengitic crystals must correspond to the recrystallization of the types richer in celadonite, following a decompressive and/or prograde pattern. Although K-feldspar and biotite are not present, the Si content of the phengites indicates

TABLE 7. REPRESENTATIVE ANALYSES OF WHITE MICAS

	2.1.1 Pg	3.1.1 Phe (1)	7.1.1c* Phe (1)	7.1.2r* Phe (1)	15.1.1 Phe (2)	15.2.1 Phe (2)	16.1.1 Phe (2)	16.2.1 Phe (2)
SiO_2	48.78	50.63	50.62	50.42	49.32	50.73	47.41	46.49
TiO_2	0.07	0.01	0.01	0.03	0.25	0.21	0.24	0.54
Al_2O_3	39.42	34.68	32.32	32.22	29.96	28.98	31.86	32.18
Cr_2O_3	0.00	0.00	0.05	0.10	0.00	0.00	0.00	0.00
FeO(t)†	0.46	0.89	1.84	2.18	2.39	2.41	2.98	2.92
MnO	0.00	0.00	0.12	0.23	0.05	0.01	0.01	0.00
MgO	0.17	1.64	2.43	2.31	2.31	2.66	1.37	1.41
CaO	0.26	0.07	0.20	0.20	0.00	0.00	0.00	0.00
Na_2O	5.18	0.06	0.44	0.27	0.55	0.42	0.75	0.77
K_2O	1.57	10.45	9.38	9.64	10.96	10.77	10.81	10.47
Total	95.91	98.43	97.41	97.60	95.79	96.19	95.43	94.78
Si	3.08	3.23	3.27	3.26	3.29	3.36	3.19	3.14
Ti	0.00	0.00	0.00	0.00	0.01	0.01	0.01	0.03
Al^{IV}	0.92	0.77	0.73	0.74	0.71	0.64	0.81	0.86
Al^{VI}	2.01	1.83	1.73	1.72	1.64	1.62	1.71	1.71
Cr	0.00	0.00	0.00	0.00	0.00	0.00	0.00	0.00
Fe^{2+}	0.02	0.05	0.10	0.12	0.13	0.13	0.17	0.16
Mn	0.00	0.00	0.01	0.01	0.00	0.00	0.00	0.00
Mg	0.02	0.16	0.23	0.22	0.23	0.26	0.14	0.14
Ca	0.02	0.00	0.01	0.01	0.00	0.00	0.00	0.00
Na	0.63	0.01	0.05	0.03	0.07	0.05	0.10	0.10
K	0.13	0.85	0.77	0.79	0.93	0.91	0.93	0.90
Mrg	2.26	0.55	1.64	1.64	0.00	0.00	0.00	0.00
Pg	81.49	0.86	6.55	4.02	7.09	5.60	9.54	10.05
Cel	1.27	22.37	24.66	24.64	26.75	33.59	16.83	12.93
Ms	14.98	76.22	67.15	69.70	66.16	60.82	73.63	77.02

Note: Cations on a basis of 11 oxygens.
*c = core; r = rim.
†(t) = total.

high minimum pressures for both assemblages (Massone and Schreyer, 1987).

Chlorites

The chlorite is only present in the amphibolites included in the schists of the lower zone of the unit, that of lesser metamorphic grade. Both the type 1 chlorites in the internal schistosity of the plagioclase porphyroblasts (syn-D_1 or early syn-D_2) and the syn-D_2 chlorites in the outer schistosity (type 2 chlorite) are moderately magnesian (Table 8), with Fe/(Fe + Mg) ratios that oscillate between 0.35 and 0.50, although the type 1 chlorites tend to be richer in Si (pycnochlorites) than the type 2 chlorites (ripidolites).

PARAGENETIC STAGES

Metabasites included in orthogneisses

Stages A1–A3 mark the evolution of the rocks from eclogite conditions to epidote amphibolites of progressively higher temperature. The recrystallization of the small bodies of metabasites is zonal and progresses from the outside to the inside of the boudins, such that in the core of the larger ones, even the oldest assemblage are preserved (Fig. 2).

Stage A1 corresponds to the episode of higher pressure, which, according to the absence of plagioclase in the oldest assemblage preserved, took place in eclogite facies conditions. This episode has been related to the subduction of the unit under Laurentia (Martínez Catalán et al., 1996) during a first deformation episode (D_1); the eclogitic metabasites preserve a coarse orientation of the garnet chains as the only microstructural evidence of this. The A1 assemblage (Fig. 5) is formed by: omphacite + EC garnet + zoisite + paragonite + rutile + quartz. The quartz and rutile veins that surround the boudins and the presence of hydrated phases, although not very important volumetrically, indicate the existence of a hydrous fluid phase during its eclogitization (Essene and Fyfe, 1967; Ernst, 1973; Brown and Bradshaw, 1979; Holland, 1979a).

Stage A2 (Fig. 5) is a decompressive stage (syn-D_2) marked by the destabilization of the omphacitic clinopyroxene, which gave rise to coarse porphyroblasts of amphibole (type 1) or else a symplectic microaggregate of plagioclase and amphibole (type 2), and by the transformations of rutile to ilmenite and zoisite to clinozoisite. During this stage, the development of rims with inverse zoning in the garnets by mass diffusion could already have begun, but they are not be included in the assemblage. The stable assemblage comprises amphibole (types 1 and 2) + plagioclase (type 1) + ilmenite + clinozoisite + phengite (type 1) + quartz.

The relationship between amphibolitization and decomposition of the omphacite during the evolution of many retroeclogites is not clear. Heinrich (1982) related the development of Cpx-Pl symplectites to areas of low aH_2O, whereas in areas of greater aH_2O the transformation would give rise to Amp-Pl symplectites. Van der Wegen (1978) maintained that, in the Malpica-Tui eclogitic types similar to those of Santiago, the de-

TABLE 8. REPRESENTATIVE ANALYSES OF CHLORITES

	15.1.1 Chl (1)	15.2.1 Chl (2)	16.1.1 Chl (2)	13.1.1 Chl (1)	13.2.1 Chl (2)	13.3.1 Chl (2)	14.1.1 Chl (1)	14.2.1 Chl (2)
SiO_2	26.46	26.01	26.04	30.36	26.20	26.05	27.17	26.85
TiO_2	0.05	0.07	0.06	0.18	0.08	0.24	0.04	0.06
Al_2O_3	20.56	22.00	21.52	22.87	21.18	21.27	20.33	20.21
FeO(t)*	23.45	23.87	24.10	21.66	21.71	22.30	18.76	20.71
MnO	0.17	0.11	0.15	0.15	0.11	0.09	0.04	0.13
MgO	17.76	16.99	16.98	12.36	17.78	16.50	19.67	18.36
Na_2O	0.00	0.00	0.14	0.68	0.00	0.00	0.00	0.02
K_2O	0.00	0.00	0.02	0.08	0.00	0.00	0.00	0.03
Total	88.45	89.05	89.01	88.34	87.06	86.45	86.01	86.37
Si	2.73	2.67	2.68	3.06	2.72	2.73	2.81	2.80
Ti	0.00	0.00	0.00	0.01	0.01	0.02	0.00	0.00
Al^{IV}	1.27	1.33	1.32	0.94	1.28	1.27	1.19	1.20
Al^{VI}	1.23	1.33	1.29	1.78	1.32	1.37	1.29	1.28
Fe^{2+}	2.02	2.05	2.07	1.83	1.89	1.96	1.62	1.80
Mn	0.01	0.01	0.01	0.01	0.01	0.01	0.00	0.01
Mg	2.73	2.60	2.61	1.86	2.75	2.58	3.03	2.85
Na	0.00	0.00	0.03	0.13	0.00	0.00	0.00	0.00
K	0.00	0.00	0.00	0.01	0.00	0.00	0.00	0.00

Note: Cations on a basis of 14 oxygens.
*(t) = total.

velopment of Amp-Pl symplectites seems, on the basis of microtexture, to be subsequent to the forming of the colorless amphibole porphyroblasts that replace the omphacite. In the Santiago eclogites, although the textural relationships are not clear, the centers of the colorless amphibole porphyroblasts (type 1) and many of the amphiboles of the symplectite (type 2) have the same chemical composition, suggesting that the development of these amphiboles corresponds to one single process.

Stage A3 (Fig. 5) corresponds to the greater degree of syn-D_2 recrystallization of the retroeclogites, which took place in epidote amphibolite facies conditions, being formed by: amphibole (type 3) + amphibole (type 4) + plagioclase (type 2 or 3) + clinozoisite + ilmenite and/or sphene + quartz. Amphibole (type 3) and amphibole (type 4) coexist, but in different microtextural positions (the most aluminous and ferric type 4 amphiboles preferentially next to garnets), because the not very intense deformation of the lithologies included in the orthogneisses reduces diffusion of the elements and homogenization by mass transfer in the rocks. Plagioclase (type 3) is formed in the outermost areas of the boudins at the expense of less calcic plagioclase (type 2).

Metabasites included in schists

The amphibolites included in the pelitic series show more intense and uniform recrystallization; only in the structurally lower positions do they preserve evidence of a first high-*P* tectonometamorphic event, the assemblage associated with stage B1. The rest of the reactional history of the rocks (B2–B4) corresponds to a decompressive and slightly prograde evolution, developed under an inverted temperature gradient that defines three metamorphic zones at cartographic scale (Fig. 2).

Stage B1 is defined by the mineral phases included as internal schistosity in the plagioclase porphyroblasts of the amphibolites and by the host albite (Fig. 6): amphibole (type 5) + plagioclase (type 4) + garnet (type I) + rutile and/or ilmenite + phengite (type 2) + epidote + chlorite (type 1) + quartz.

The assemblage of B1 is the oldest one showed by these rocks: the mineralogy resembles high-*P* conditions and could correspond approximately to the same tectonic-metamorphic event as the eclogitic assemblage of the metabasites included in the orthogneisses (A1), and probably to part of its retroeclogitic history still at high pressure (A2), as is indicated by the presence of inclusions of ilmenite from rutile. However, evidence such as the textural equilibrium between syn-D_1 garnet (type I garnet) and plagioclase, the existence of epidote instead of zoisite or clinozoisite (more oxidant conditions), the presence of chlorite, and the mineral-chemical differences between garnets (types I and EC) and between plagioclases (types 1 and 4) indicate that the metabasites intercalated in the schists probably underwent metamorphic recrystallization prior to the high-*P* peak of the unit, did not reach eclogitization, and maintained plagioclase-bearing assemblages during the high-*P* event that produced the eclogitization of the metabasites included in the orthogneisses.

Stage B2 (Fig. 6) includes the mineral assemblage developed by the metabasites intercalated in the schists of the structurally lower zone of the unit during the second deformation phase: amphibole (type 6) + plagioclase (type 5) + epidote + ilmenite and/or sphene + chlorite (type 2) + quartz, and would therefore be approximately correlatable to the A3 stage of the retroeclogites, the outcrops of which remain in the same structural zone. As in the garnets of the eclogites, the last microns next to the rim show inverse zoning by mass diffusion probably developed during the decompression of the rocks, which would not be included in the assemblage of B2 and B3.

Stage B3 (Fig. 6) includes the mineral assemblage developed by the metabasites intercalated in the schists of the structurally intermediate zone of the unit during the second deformation phase: amphibole (type 6) + plagioclase (type 6) + epidote + ilmenite and/or sphene + biotite + quartz.

Chlorite and biotite are scarce and almost mutually exclusive in these metabasites, chlorite and plagioclase (type 5) appearing in the lower zone, and biotite in the company of a more calcic plagioclase (type 6 plagioclase) in the intermediate zone. Stage B4 (Fig. 6) corresponds to the higher grade recrystallization (thermal climax) reached during D_2 by metabasites of the unit. It is represented by the assemblage present in the amphibolites intercalated in the pelitic schists of the upper part of the unit. B4 comprises amphibole (type 7) + plagioclase (type 7) + garnet (type II) + ilmenite and/or sphene + quartz ± epidote. The destabilization of the epidote in some of the rocks and the appearance of almandine-rich garnet indicate that this stage took place under amphibolite facies conditions.

METAMORPHIC ZONES

During the second deformation phase zoning was developed in the Santiago unit according to an inverted metamorphic gradient (Fig. 2). The assemblage used for defining the metamorphic zones were exclusively those of the amphibolites, because the retroeclogites (given that they are included in less ductile and pervasive rocks) underwent the syn-D_2 deformation and recrystallization processes much less intensely, such that the chemical composition of some mineral phases (particularly the amphiboles) was strongly controlled by their microtextural situation (mosaic equilibrium) and cannot be directly correlated to the equivalent phases in the amphibolites. The composition of three mineral phases (amphibole, plagioclase, and another Fe-Mg phase in equilibrium) was used to define three syn-D_2 metamorphic zones; from bottom to top, these are (1) the tschermakite-oligoclase-chlorite zone (Ts-Olg-Chl); (2) the tschermakite-andesine-biotite zone (Ts-Ads-Bt); and (3) the ferrotschermakite-labradorite-garnet zone (Fts-Lab-Grt).

These three mineral zones show a metamorphic field gradient with temperature increasing upward. If we take into account that this increase in metamorphic grade takes place in <3 km, the inverted geothermal gradient under which the zoning developed must have been very pronounced.

The origin of this inversion of the metamorphic gradient during D_2 has been related to the displacement on the unit of a portion of hot mantle wedge by means of an extensional detachment. The development of this detachment took place through the extension of the orogenic wedge, under the mantle of which part of the Gondwana margin subducted (Martínez Catalán et al., 1996). The S_2 schistosity of the Santiago unit is related to its emplacement in the crustal level that it now occupies. The decompressive and locally prograde syn-D_2 evolution of the Santiago unit is compatible with the progress of the subduction of the relative autochthon under the basal units, accreted to the base of the orogenic wedge, at the same time as the upper level of the accretional prism (including a portion of hot mantle) collapsed to maintain the gravitational equilibrium of the whole, in accordance with the model of Platt (1986, 1987).

THERMOBAROMETRY

Of the conventional methods of thermobarometry used, only those based on the cation exchange of garnet with other phases proved to be accurate and reliable. Therefore, the stages quantified were limited to the episodes of higher pressure, both in the eclogites (A1) and in the amphibolites (B1), as well as to the stage of higher temperature registered for the latter (B4). Only garnet rim compositions were used and the *P-T* values expressed correspond to the minimum and maximum results of the application of each method.

The *P-T* conditions of the eclogitic assemblage (stage A1) were estimated by Grt-Cpx thermometry (Krogh, 1988) and Ab-Jd-Qtz barometry (Holland 1980, 1983), using only Jd_{max} values. The absence of plagioclase in the A1 assemblage implies that the pressures obtained are minimum values: the values obtained were 11.15–11.25 kbar at 492–512 °C. The Pg = Omp + Ky + H_2O curve was taken as representative of maximum pressure values (Fig. 13). These results are very similar to those obtained from the assemblage included in albite porphyroblasts of the semipelitic schists in the lower zone of the unit (14.7 ± 0.7 kbar and 493 ± 9 °C; episode M1 in Arenas et al., 1995), but the range of pressures is too broad.

The high-*P* episode preserved as an internal schistosity in the albite porphyroblasts of the amphibolites less recrystallized during D_2 (B1) reflects conditions of higher *P* but lower *T* than those obtained in the eclogites (Fig. 13). Using thermometry of Fe-Mg exchange between garnet and hornblende (calibration of Graham and Powell, 1984) and Grt-Rt-Ilm-Pl-Qtz barometry (GRIPS) of Bohlen and Liotta (1986), conditions of 13.2–15.8 kbar at 452 °C and 13.8–16.5 kbar at 480 °C were obtained (~15.8 ± 1.4 kbar at 466 ± 14 °C). The application of Amp-Grt-Pl-Qtz barometers (calibrations of Kohn and Spear, 1990) resulted in pressures between 19 and 20 kbar for the same range of temperatures. These pressures are considered to be overestimated due to the high Mn content of the type I garnets and to the low Ca content of plagioclase (type 4), outside the compositional limits recommended by the previous authors, but they seem to confirm the high-*P* nature of the B1 assemblage, although type 5 amphibole compositions could be readjusted.

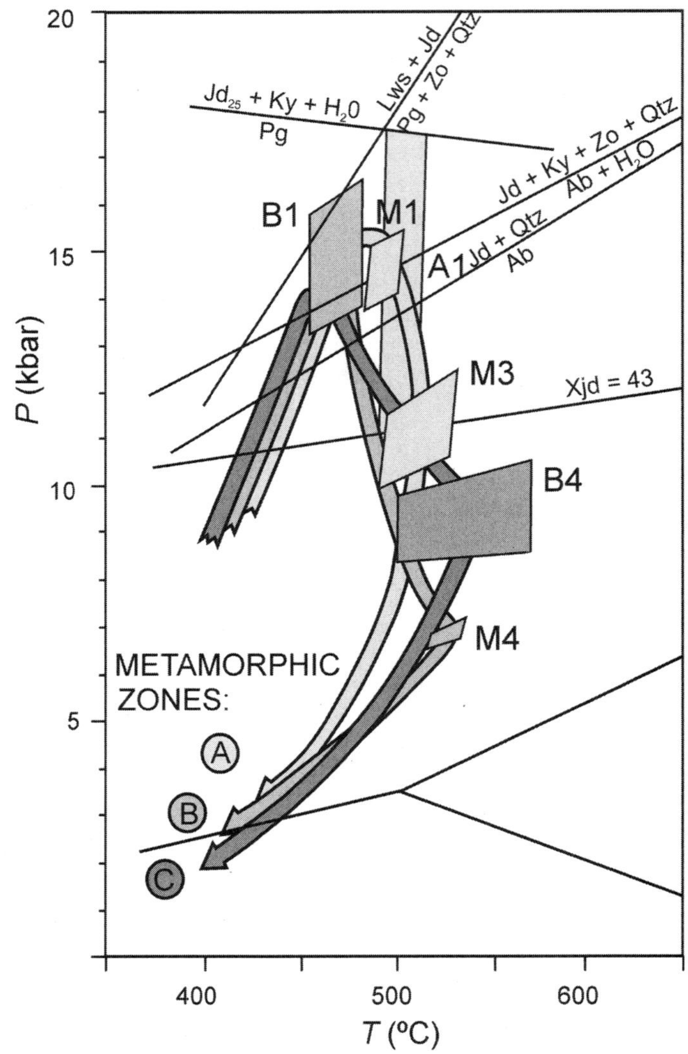

Figure 13. Calculated pressure-temperature (*P-T*) conditions for some of different metamorphic stages in eclogites (A1) and amphibolites (B1, B4). Metamorphic stages in metapelites (M1, M3, M4) are from Arenas et al. (1995). Pg = Jd_{25} + Ky + H_2O reaction curve is directly extrapolated after Holland (1979b); Ab = Jd + Qtz and Jd_{43} curves are from Holland (1980, 1983); Lws + Jd = Pg + Zo + Qtz is from Heinrich and Althaus (1980); Pl + H_2O = Jd + Zo + Ky + Qtz curve is from Wayte et al. (1989) and aluminosilicate phase diagram is from Holdaway (1971). *P-T*-time paths: A, possible evolution for uppermost part of unit; B, upper-central part; C, lower part. See text for discussion.

The conditions of the syn-D_2 (B4) peak episode were calculated using the same thermometer as in the previous stage and Amp-Grt-Pl-Qtz barometry (P_{Mg} and P_{Fe} calibrations of Kohn and Spear, 1990). The activity models used for the components of the garnet were those of Hodges and Spear (1982) and the activity model used for the anorthite was that of Hodges and Royden (1984). For the pressures, the zone of maximum overlapping was used, which is above the minimum value of application of P_{Mg} calibration and below the maximum value of the application of P_{Fe} calibration. The resulting values were 8.3–9.8 kbar at 499 °C and 8.6–10.6 kbar at 568 °C (~9.3 ± 0.8 kbar at 528 ± 34 °C) and are consistent with the results of the application of the barometer of Bohlen and Liotta (1986) in the B4 assemblage (7.0–10.5 kbar at 499 °C and 7.4–10.9 kbar at 568 °C), from which P_{max} (absence of rutile) values were obtained; these values are not in Figure 13 because of their high uncertainty (±1.75 kbar). In comparison with the values obtained for the equivalent stage in pelitic rocks (M4 in Arenas et al., 1995), the amphibolites give somewhat higher pressures and a wider range of temperatures for the thermal peak of the unit.

DISCUSSION AND CONCLUSIONS

The Santiago unit is a fragment of the continental margin of Gondwana that subducted under Laurentia ca. 374 Ma. The polarity of this subduction was westward, the unit reaching minimum pressures that could oscillate around 14.7 ± 0.7 kbar at 493 ± 9 °C, thermobarometrically calculated in the semipelitic schists (Arenas et al., 1995), the 14.8 ± 0.3 kbar at 502 ± 10 °C calculated in the eclogites, and the 15.8 ± 1.4 kbar at 466 ± 14 °C obtained in the thermobarometry of the garnet amphibolites. During this tectonic process (D_1), the basic igneous rocks emplaced in the continental margin underwent a high-pressure metamorphism that gave rise to an eclogitic assemblage, with omphacitic pyroxene and without plagioclase, in the metabasites included in orthogneisses. The metabasites included in the pelitic schists, however, developed a high-pressure assemblage without omphacite and with albitic plagioclase. The importance of this is in the fact that rocks with assemblages of albite, idiomorphic garnet, and rutile and lacking ompacite may have reached eclogite facies conditions without development of the diagnostic assemblage.

In Figure 13, the maximum pressures reached by the *P-T-t* paths are on the lower pressure side of the Pg = Omp + Ky + H_2O reaction and on the higher temperature side of the Lws + Jd = Pg + Zo + Qtz reaction, consistent with the presence of paragonite and the absence of kyanite and lawsonite observed in the A1 assemblage. They are also on or slightly above the Pl = Jd + Zo + Ky + Qtz reaction in hydrated conditions, such that, whether there was kyanite (subsequently replaced by phengite), the *P-T* conditions obtained could be considered valid.

The absence of omphacite in the high-pressure assemblage of the amphibolites included in schists (stage B1) cannot be attributed to differences in bulk composition, because both types of metabasites show similar compositions with regard to whole-rock major elements (Díaz García, 1990). The presence of hydrous minerals shows the existence in the eclogites of a fluid rich in H_2O during its eclogitization, the same as the quartz veins with rutile (Ernst, 1973); therefore, it does not seem to be a matter of absence of H_2O in the protolith material, in the sense of Green and Ringwood (1967) or Newton and Fyfe (1976), although a clear difference of aH_2O between eclogites and garnet amphibolites is patent. Because the orthogneisses and eclogites are found at the base of the unit, the two different high-pressure assemblages of the metabasites could reflect the thermal inversion associated with the Benioff plane during its subduction (Ernst, 1973), the garnet amphibolites remaining on the higher temperature side of the Ab = Jd + Qtz reaction. However, the separation between one type of rock and another is only a few hundred meters and, in any case, the thermometry tends to indicate slightly higher temperatures for the eclogites. Neither the positive effect of the deformation nor the pressure of fluids on the kinetics of the metamorphic reactions, including those forming eclogites (Rubie, 1986, 1990; Koons et al., 1987), is a direct explanation for the noneclogitization of the metabasites included in the schists. On the contrary, the stabilization of eclogitic assemblages at low temperatures seems to require low water pressures (Yardley, 1989).

The analysis of the zonings of the syn-D_1 garnets of both types of rocks gives considerable information. The type I garnets in the garnet amphibolites, with their strong fractionation of Mn and Fe^{2+}, grew from an already metamorphosed protolith, rich in chlorite (syn-D_1 chlorite is still preserved as inclusions in the albite porphyroblasts) and are probably earlier than the garnets of the eclogites, whereas the basaltic rocks included in the orthogneisses must have been relatively fresh when they recrystallized as eclogites. The tendency of the EC garnets to form honeycomb structures and the absence of inclusions in the core of the omphacite and/or amphibole (1) crystals seem to indicate that the original igneous texture was still preserved.

The greater water content of the pelitic encasing rock, its pervasiveness to the circulation of fluid, and its greater ductility in the face of deformation were probably the factors that operated in favor of the pre–high-*P* peak metamorphism of the metabasites intercalated in the schists and led to different reaction pathways. Heinrich (1982) showed that, during progressive metamorphism, the dehydration of the pelites causes hydration in the enclosed mafic bodies. This early metamorphism of the eclogites-amphibolites transformed the igneous basaltic mineralogy into a rock with increasingly albitic plagioclase and the greater part of the Ca-forming epidote, X_{Grs} in garnet and amphibole, in a process similar to that described in Koons et al. (1987). When eclogitic conditions were reached, the igneous plagioclase of the basic rocks included in the orthogneisses reacted to form Omp (min Jd_{38-43}) + Qtz + Grt + Zo + Pg, whereas in the amphibolites included in schists, the almost pure albite remained stable below Jd_{100} conditions. Thus, the peak pressure conditions reached by the Santiago unit should be between the Jd_{43} and Jd_{100} stability fields. However, the X_{Jd} isopleths calculated by Holland (1980, 1983) show lower pressures than those predicted by the thermobarometry in the eclogites and in the amphibolites as well as in the schists (Fig. 13); thus, it seems necessary to admit a metastable equilibrium of the pure albite above Jd_{100}.

A comparable case is that of the high-*P* metagranites in the Western Alps. The undeformed metagranites of the Sesia zone developed jadeite plus other minerals, including hydrated phases such as zoisite and phengite (Hy, 1984; Rubie, 1986), whereas the deformed metagranites of the Gran Paradiso nappe contain albite (primary) instead of jadeite. Le Goff and Ballèvre (1990) supposed that if conditions above Ab = Jd + Qtz were reached, the growth of Jd could have been prevented by kinetic factors. In the Bergen arc nappe complex (western Norway) some pegmatite veins preserve pure albite without growth of Cpx at pressures at least close to those of Jd_{100} stability (Austrheim and Griffin, 1985; Andersen et al., 1991), if not higher (Jamtveit et al., 1990).

In our opinion, it is possible that a primary albitic composition of the plagioclase or an early albitization of igneous Ca-Na plagioclases delayed the kinetics of the Jd-forming reaction, in particular inhibiting the nucleation of omphacite, thus hindering the development of the jadeite substitution Ca Mg, Fe $\leftrightarrow$ Na Al, Fe^{3+} when higher pressures are reached. The direct nucleation of Jd_{100} (the only composition possible from almost pure albite) probably requires higher pressures than the minimum pressures of jadeite stability.

Subsequent evolution of the unit (syn-D_2) is decompressive and slightly prograde (Fig. 13). Whereas the subduction spread backward by means of the underthrusting of the innermost zones of the continental margin under the Santiago unit, the upper levels of the prism collapsed laterally and eastward, decompressing and exposing the unit, according to the model of Platt (1986, 1987). One of these extensional structures transported over Santiago a portion of the mantle wedge previously accreted (Arenas et al., 1995; Martínez Catalán et al., 1996). The hot rocks of the mantle induced the inversion of the thermal profile in the unit, reaching peak temperature conditions at its top. The recognizable zoning in the basic rocks is, from bottom to top, Ts-Olg-Chl, Ts-Ads-Bt, and Fts-Lab-Grt. Toward the top of the unit, the field metamorphic gradient shows a progressive increase of X_{An} in the plagioclase, a greater effectiveness of the edenitic, pargasitic, and tschermakitic substitutions in the amphiboles, and the destabilization of the epidote in the presence of plagioclase; all this is consistent with the inversion of the metamorphic gradient. The peak conditions calculated on the amphibolites directly under the detachment are 9.3 ± 0.8 kbar at 528 ± 34 °C. The situation of the thermal maximum reached by the top of the unit toward the middle of its decompressive history reflects the passage of the hot mantle rocks above the Santiago unit to finally be emplaced more to the southeast. Outcrops of these mantle rocks (the so-called Campo-Marzo slice) are currently ~60 km southeast of the zone (Martínez Catalán et al., 1996).

ACKNOWLEDGMENTS

This study was financed by projects PB88-0145-CO2 and PB91-0192-CO2 of the Dirección General de Investigación Científica y Técnica (DGICYT) (Spain). The microprobe analyses were performed at the University of Oviedo. B. Knowles is kindly acknowledged for help in the English version.

REFERENCES CITED

Andersen, T., Austrheim, H., and Burke, E.A.J., 1991, Mineral-fluid-melt interactions in high pressure shear zones in the Bergen Arcs nappe complex, Caledonides of W. Norway: Implications for the fluid regime in Caledonian eclogite-facies metamorphism: Lithos, v. 27, p. 187–204.

Arenas, R., Gil Ibarguchi, J.I., González Lodeiro, F., Klein, E., Martínez Catalán, J.R., Ortega Gironés, E., Pablo Macía, J.G. de, and Peinado, M., 1986, Tectonostratigraphic units in the complexes with mafic and related rocks of the Iberian Massif: Hercynica, v. II, p. 87–110.

Arenas, R., Farias, P., Gallastegui, G., Gil Ibarguchi, J.I., González Lodeiro, F., Klein, E., Marquínez, J., Martín Parra, L.M., Martínez Catalán, J.R., Ortega Gironés, E., Pablo Macía, J.G. de, Peinado, M., and Rodríguez Fernández, L.R., 1988, Características geológicas y significado de los dominios que componen la Zona de Galicia Tras-os-Montes, *in* Proceedings, Congreso Geológico de España, II, Granada: Sociedad Geológica de España-Universidad de Granada, p. 75–84.

Arenas, R., Rubio Pascual, F.J., Díaz García, F., and Martínez Catalán, J.R., 1995, High-pressure micro-inclusions and development of an inverted metamorphic gradient in the Santiago Schists (Ordenes Complex, NW Iberian Massif, Spain): Evidence of subduction and syncollisional decompression: Journal of Metamorphic Geology, v. 13, p. 141–164.

Arenas, R., Abati, J., Martínez Catalán, J.R., Díaz García, F., and Rubio Pascual, F.J., 1997, P-T evolution of eclogites from the Agualada unit (Ordenes Complex, NW Iberian Massif, Spain): Implications for crustal subduction: Lithos, v. 40, p. 221–242.

Arps, C.E. S., 1981, Amphibolites and other metamorphic mafic rocks of the blastomylonitic graben in western Galicia, NW Spain: Field relations and petrography: Leidse Geologische Mededelingen, v. 52, p. 57–71.

Austrheim, H., and Griffin, W.L., 1985, Shear deformation and eclogite formation within granulite-facies anorthosites of the Bergen Arcs, Western Norway: Chemical Geology, v. 50, p. 267–281.

Black, P.M., 1977, Regional high-pressure metamorphism in New Caledonia: Phase equilibria in the Ouegoa district: Tectonophysics, v. 43, p. 89–107.

Bohlen, S.R., and Liotta, J.J., 1986, A barometer for garnet amphibolites and garnet granulites: Journal of Petrology, v. 27, p. 1025–1034.

Brown, E.H., and Bradshaw, J.Y., 1979, Phase relations of pyroxene and amphibole in greenstone, blueschist, and eclogite of the Franciscan complex, California: Contributions to Mineralogy and Petrology, v. 71, p. 67–83.

Cawthorn, R.G., and Collerson, K.D., 1974, The recalculation of pyroxene end-member parameters and the estimation of ferrous and ferric iron content from electron microprobe analyses: American Mineralogist, v. 59, p. 1203–1208.

Cebriá Gómez, J.M., 1990, PX: A program for pyroxene classification and calculation of end-members: American Mineralogist, v. 75, p. 1426–1427.

Díaz García, F., 1990, La Geología del sector Occidental del Complejo de Ordenes (Cordillera Hercínica, NW de España): Nova Terra, v. 3, 230 p.

Díaz García, F., 1993, Análisis comparativo de la foliación regional y estructuras asociadas en el Dominio de Santiago y su autóctono relativo: Revista de la Sociedad Geológica de España, v. 6, p. 105–114.

Ernst, W.G., 1973, Blueschist metamorphism and *P-T* regimes in active subduction zones: Tectonophysics, v. 17, p. 255–272.

Ernst, W.G., and Liou, J., 1998, Experimental phase-equilibrium study of Al- and Ti-contents of calcic amphibole in MORB—A semiquantitative thermobarometer: American Mineralogist, v. 83, p. 952–969.

Essene, E.J., and Fyfe, W.S., 1967, Omphacite in Californian metamorphic rocks: Contributions to Mineralogy and Petrology, v. 15, p. 1–23.

Farias, P., Gallastegui, G., González Lodeiro, F., Marquínez, J., Martín Parra,

L.M., Martínez Catalán, J.R., Pablo Macía, J.G. de, and Rodríguez Fernández, L.R., 1987, Aportaciones al conocimiento de la litoestratigrafía y estructura de Galicia Central: Memorias do Museu e Laboratório mineralógico e geológico da Faculdade de Ciências do Porto, v. 1, p. 411–431.

García Garzón, L., Pablo Macía, J.G. de, and Llamas, J. de, 1981, Edades absolutas obtenidas mediante el método Rb/Sr en dos cuerpos de ortoneises en Galicia Occidental: Boletín Geológico y Minero, v. 92–94, p. 463–466.

Gil Ibarguchi, J.I., 1995, Petrology of jadeite metagranite and associated orthogneiss from the Malpica-Tuy allochthon (Northwest Spain): European Journal of Mineralogy, v. 7, p. 403–415.

Gil Ibarguchi, J.I., and Ortega Gironés, E., 1985, Petrology, structure and geotectonic implications of glaucophane-bearing eclogites and related rocks from the Malpica-Tuy (MT) unit, Galicia, Northwest Spain: Chemical Geology, v. 50, p. 145–162.

Godard, G., 1988, Petrology of some eclogites in the Hercynides: The eclogites from the Southern Armorican Massif, France, *in* Smith, D.C., ed., Eclogites and eclogite-facies rocks: Developments in petrology, Volume 12: Amsterdam, Elsevier, p. 451–519.

Graham, C.M., and Powell, R., 1984, A garnet-hornblende geothermometer: calibration, testing and application to the Pelona Schist, Southern California: Journal of Metamorphic Geology, v. 2, p. 13–31.

Green, D.H., and Ringwood, A.E., 1967, An experimental investigation of the gabbro to eclogite transformation and its petrological applications: Geochimica et Cosmochimica Acta, v. 31, p. 767–833.

Heinrich, C.A., 1982, Kyanite-eclogite to amphibolite facies evolution of hydrous mafic and pelitic rocks, Adula Nappe, central Alps: Contributions to Mineralogy and Petrology, v. 81, p. 30–38.

Heinrich, W., and Althaus, E., 1980, Die obere Stabilitatsgrenze von Lawsonit plus Albit bzw. Jadeit: Fortschrifte für Mineralogie, v. 58, p. 49–50.

Hodges, K.V., and Royden, L., 1984, Geologic thermobarometry of retrograde metamorphic rocks: An indication of the uplift trajectory of a portion of the northern Scandinavian Caledonides: Journal of Geophysical Research, v. 89, p. 7077–7090.

Hodges, K.V., and Spear, F.S., 1982, Geothermometry, geobarometry and the Al_2SiO_5 triple point at Mt. Moosilauke, New Hampshire: American Mineralogist, v. 67, p. 1118–1134.

Holdaway, M.J., 1971, Stability of andalusite and the aluminum silicate phase diagram: American Journal of Science, v. 271, p. 97–131.

Holland, T.J. B., 1979a, High-water activities in the generation of high-pressure kyanite eclogites of the Tauern Window, Austria: Journal of Geology, v. 87, p. 1–27.

Holland, T.J. B., 1979b, Experimental determination of the reaction paragonite = jadeite + kyanite + water, and internally consistent thermodinamic data for part of the system Na_2O-Al_2O_3-SiO_2-H_2O, with application to eclogites and blueschists: Contributions to Mineralogy and Petrology, v. 68, p. 293–301.

Holland, T.J. B., 1980, The reaction albite = jadeite + quartz determined experimentally in the range 600–1200 °C: American Mineralogist, v. 65, p. 129–134.

Holland, T.J. B., 1983, The experimental determination of activities in disordered and short-range ordered jadeitic pyroxenes: Contributions to Mineralogy and Petrology, v. 82, p. 214–220.

Hy, C., 1984, Métamorphisme polyphasé et évolution tectonique dans la croûte continentale éclogitisée: les séries granitiques et pélitiques du Monte Mucrone (zone Sesia-Lanzo, Alpes italiennes) [Ph.D. thesis]: Paris, Université de Paris VI, 198 p.

Jamtveit, B., Bucher-Nurminen, K., and Austrheim, H., 1990, Fluid controlled eclogitization of granulites in deep crustal shear zones, Bergen Arcs, western Norway: Contributions to Mineralogy and Petrology, v. 104, p. 184–193.

Janák, M., O'Brien, P.J., Hurai, V., and Reutel, C., 1996, Metamorphic evolution and fluid composition of garnet-clinopyroxene amphibolites from the Tatra Mountains, Western Carpathians: Lithos, v. 39, p. 57–79.

Knowles, C.R., 1987, A basic program to recast garnet end members: Computers & Geosciences, v. 13, p. 655–658.

Kohn, M.J., and Spear, F.S., 1990, Two new geobarometers for garnet amphibolites, with applications to southeastern Vermont: American Mineralogist, v. 75, p. 89–96.

Koons, P.O., Rubie, D.C., and Frueh-Green, G., 1987, The effects of disequilibrium and deformation on the mineralogical evolution of quartz diorite during metamorphism in the eclogite facies: Journal of Petrology, v. 28, p. 679–700.

Krogh, E.J., 1988, The garnet-clinopyroxene Fe-Mg geothermometer—a reinterpretation of existing data: Contributions to Mineralogy and Petrology, v. 99, p. 44–48.

Laird, J., and Albee, A.L., 1981, Pressure, temperature, and time indicators in mafic schist: Their application to reconstructing the polymetamorphic history of Vermont: American Journal of Science, v. 281, p. 127–175.

Le Goff, E., and Ballèvre, M., 1990, Geothermobarometry in albite-garnet orthogneisses: A case study from the Gran Paradiso nappe (Western Alps): Lithos, v. 25, p. 261–280.

Leake, B.E., Woolley, A.R., Arps, C.E.S., Birch, W.D., Gilbert, M.C., Grice, J.D., Hawthorne, F.C., Kato, A., Kisch, H.J., Krivovichev, V.G., Linthout, K., Laird, J., Mandarino, J.A., Maresch, W.V., Nickel, E.H., Rock, N.M.S., Schumacher, J.C., Smith, D.C., Stephenson, N.C.N., Ungaretti, L., Whittaker, E.J.W., and Guo, Y., 1997, Nomenclature of amphiboles; report of the Subcomittee on Amphiboles of the International Mineralogical Association, Commission on New Minerals and Mineral Names: The Canadian Mineralogist, v. 35, p. 219–246.

Martínez Catalán, J.R., Klein, E., Pablo Macía, J.G. de, and González Lodeiro, F., 1984, El Complejo de Ordenes: Subdivisión, descripción y discusión sobre su origen: Cuadernos del Laboratorio Xeolóxico de Laxe, v. 7, p. 139–210.

Martínez Catalán, J.R., Arenas, R., Díaz García, F., Rubio Pascual, F.J., Abati, J., and Marquínez, J., 1996, Variscan exhumation of a subducted Paleozoic continental margin: The basal units of the Ordenes Complex, Galicia, NW Spain: Tectonics, v. 15, p. 106–121.

Martínez Catalán, J.R., Arenas, R., Díaz García, F., and Abati, J., 1997, The Variscan accretionary complex of northwest Iberia: Terrane correlation and succesion of tectonothermal events: Geology, v. 27, p. 1103–1106.

Martínez Catalán, J.R., Arenas, R., Díaz García, F., and Abati, J., 1999, Allochthonous units in the Variscan belt of NW Iberia: Basement Tectonics, v. 13, p. 65–84, Kluwer.

Massone, H.J., and Schreyer, W., 1987, Phengite geobarometry based on the limiting assemblage with K-feldspar, phlogopite, and quartz: Contributions to Mineralogy and Petrology, v. 96, p. 212–224.

Messiga, B., Tribuzio, R., and Caucia, F., 1992, Amphibole evolution in Variscan eclogite-amphibolites from the Savona crystalline massif (Western Ligurian Alps, Italy): Controls on the decompressional *P-T-t* path: Lithos, v. 27, p. 215–230.

Newton, R.C., and Fyfe, W.S., 1976, High-pressure metamorphism, *in* Bailey, D.K., and Macdonald, R., eds., The evolution of the crystalline rocks: London, Academic Press, p. 101–186.

O'Brien, P.J., 1993, Partially retrograded eclogites of the Münchberg Massif, Germany: Records of a multistage Variscan uplift history in the Bohemian Massif: Journal of Metamorphic Geology, v. 11, p. 241–260.

Oh, C.W., Liou, J.G., and Maruyama, S., 1991, Low-temperature eclogites and eclogite schists in Mn-rich metabasites in Ward Creek, California; Mn and Fe effects on the transition between blueschist and eclogite: Journal of Petrology, v. 32, p. 275–301.

Platt, J.P., 1986, Dynamics of orogenic wedges and the uplift of high-pressure metamorphic rocks: Geological Society of America Bulletin, v. 97, p. 1037–1053.

Platt, J.P., 1987, The uplift of high-pressure–low-temperature metamorphic rocks: Philosophical Transactions of the Royal Society of London, v. A321, p. 87–103.

Raase, P., 1974, Al and Ti contents of hornblende, indicators of pressure and temperature of regional metamorphism: Contributions to Mineralogy and Petrology, v. 45, p. 231–236.

Råheim, A., 1975, Mineral zoning as a record of *P, T* history of Precambrian eclogites and schists in western Tasmania: Lithos, v. 8, p. 221–236.

Ridley, J., 1984, Evidence of a temperature-dependent "blueschist" to "eclogite" transformation in high-pressure metamorphism of metabasic rocks: Journal of Petrology, v. 25, p. 852–870.

Rubie, D.C., 1986, The catalysis of mineral reactions by water and restrictions on the presence of aqueous fluid during metamorphism: Mineralogical Magazine, v. 50, p. 399–415.

Rubie, D.C., 1990, Role of kinetics in the formation and preservation of eclogites, *in* Carswell, D.A., ed., Eclogite facies rocks: Glasgow, Blackie, p. 111–140.

Spear, F.S., 1981, An experimental study of hornblende stability and compositional variability in amphibolite: American Journal of Science, v. 281, p. 697–734.

Spear, F.S., and Kimball, K.L., 1984, RECAMP-a FORTRAN IV program for estimating Fe^{3+} contents in amphiboles: Computers & Geosciences, v. 10, p. 317–325.

Spear, F.S., Kohn, M.J., Florence, F.P., and Menard, T., 1991, A model for garnet and plagioclase growth in pelitic schists: Implications for thermobarometry and *P-T* path determinations: Journal of Metamorphic Geology, v. 8, p. 683–696.

Tracy, R.J., 1982, Compositional zoning and inclusions in metamorphic minerals, *in* Ferry, J.M., ed., Characterization of metamorphism through mineral equilibria: Washington, D.C., Mineralogical Society of America, p. 355–397.

Van Calsteren, P.W.C., Boelrijk, N.A.I.M., Hebeda, E.H., Priem, H.N.A., Den Tex, E., Verdurmen, E.A.T.H., and Verschure, R.H., 1979, Isotopic dating of older elements including the Cabo Ortegal mafic-ultramafic Complex) in the Hercynian Orogen of NW Spain; manifestations of a presumed Early Paleozoic Mantle-Plume: Chemical Geology, v. 24, p. 35–36.

Van der Wegen, G., 1978, Garnet-bearing metabasites from the Blastomylonitic Graben, western Galicia, Spain: Scripta Geologica, v. 45, p. 1–95.

Wayte, G.J., Worden, R.H., Rubie, D.C., and Droop, G.T.R., 1989, A TEM study of disequilibrium plagioclase breakdown at high pressure: The role of infiltrating fluid: Contributions to Mineralogy and Petrology, v. 101, p. 426–437.

Yardley, B.W. D., 1989, An introduction to metamorphic petrology: Harlow, Longman Scientific & Technical, 248 p.

Manuscript Accepted by the Society February 8, 2002

Printed in the U.S.A.

Geological Society of America
Special Paper 364
2002

Structural record during exhumation and emplacement of high-pressure–low- to intermediate-temperature rocks in the Malpica-Tui unit (Variscan Belt of Iberia)

S. Llana-Fúnez*
A. Marcos
Departamento de Geología, Universidad de Oviedo, Arias de Velasco s/n, 33005 Oviedo, Spain

ABSTRACT

Upper crustal rocks of the Malpica-Tui unit were buried to high-pressure (*P*) and low- to intermediate-temperature (*T*) conditions during the Variscan orogeny in northwestern Iberia and subsequently exhumed in the same tectonic cycle. The structural record of the exhumation is presented in this chapter as a sequence of structures ordered according to superposing relations and decreasing grade of ductility. Synthetically, two different stages during the deformation process are identified corresponding to two major types of structures: a first general but heterogeneous ductile episode, seen in the rocks by the development of a regional tectonic fabric, and a second and subsequent ductile episode recorded in more discrete structures that include the bounding and discrete basal shear zone of the Malpica-Tui unit. The first event accounts for the exhumation of the high-*P* rocks to the middle crust and the second for their final emplacement, prior to the later Variscan intracontinental tectonics. From geological observations, consistent with the tectonic setting of indentation tectonics in the development of the Ibero-Armorican arc of southwestern Europe, we suggest that in both stages general shortening and associated tectonic transport are initially highly transverse to the orogenic trend. The data set presented here also emphasizes the relevance of internal deformation during ascent of high-*P* nappes.

INTRODUCTION

Metamorphism at high-pressure (*P*) and low- to intermediate-temperature (*T*) conditions occurs in geodynamic settings associated with convergent margins, either in narrow belts close to plate junctions (Cloos, 1982, 1986; Ernst, 1988) or in wider areas that cover great parts of the continental crust involved in collision (Platt, 1986; Escher and Beaumont, 1997). In the Variscan Belt of southwestern Europe, high-*P*–low- to intermediate-*T* metamorphism was first reported in rocks from the allochthonous complexes of the Armorican Massif (Carpenter and Civetta, 1976; Audren and Triboulet, 1986, 1989; Guiraud et al., 1987; Ballèvre et al., 1994), the Iberian Massif (van der Wegen, 1978; Gil Ibarguchi and Ortega, 1985; Azor et al., 1994; Gil Ibarguchi, 1995), and more recently in the suture zone in southwestern Iberia, placed between the Ossa-Morena and South-Portuguese zones, also within the Iberian Massif (Fonseca, 1995; Pedro and Munhá, 1997a, 1997b; Fonseca et al., 1999) (Fig. 1).

This chapter presents the tectonic evolution during the exhumation of one of these high-*P* tectonic units, the Malpica-Tui unit, belonging to the allochthonous complexes in northwestern

*Present address:Department of Earth Sciences, University of Manchester, Manchester M13 9PL, UK. E-mail: sfunez@fs1.ge.man.ac.uk.

Llana-Fúnez, S., and Marcos, A., 2002, Structural record during exhumation and emplacement of high-pressure–low- to intermediate-temperature rocks in the Malpica-Tui unit (Variscan Belt of Iberia), *in* Martínez Catalán, J.R., Hatcher, R.D., Jr., Arenas, R., and Díaz García, F., eds., Variscan-Appalachian dynamics: The building of the late Paleozoic basement: Boulder, Colorado, Geological Society of America Special Paper 364, p. 125–142.

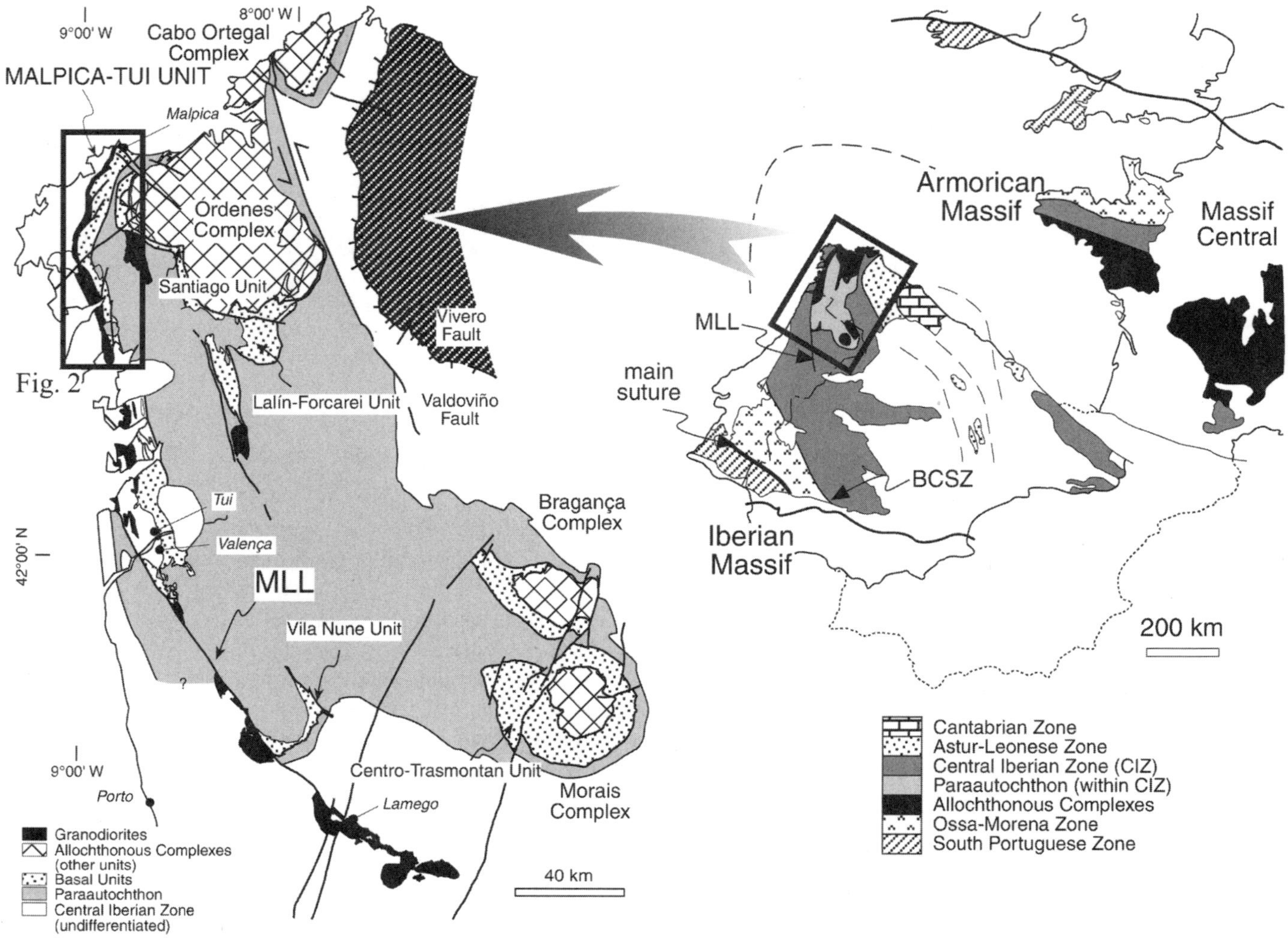

Figure 1. Location of Iberian and Armorican Massifs in Ibero-Armorican arc structure in Variscan Belt of southwestern Europe. Tectonic zones of Variscan Belt are based on Julivert et al. (1972) for southern branch and on Martínez Catalán (1990) for its correlation in France. Reconstruction of position of Iberian Peninsula in Barremian–early Aptian time (ca. 125 Ma) is from Olivet (1996). High-pressure nappes are also found in Badajoz-Córdoba Shear Zone (BCSZ) and in boundary between Ossa-Morena and Central Iberian zones (see text for basic references). MLL is Malpica-Lamego Line.

Iberia (Figs. 1 and 2). Mineral assemblages equilibrated at high-*P*–low- to intermediate-*T* conditions were found by several authors in different lithological units in the northern half of the Malpica-Tui unit, from eclogites to orthogneisses and metasediments (van der Wegen, 1978; Gil Ibarguchi and Ortega, 1985; Gil Ibarguchi, 1995). Peak pressure conditions of 2.5 GPa and 640 °C (Rodríguez Aller et al., 1997b) were followed by an isothermal decompression during their ascent (Matte, 1998; Llana-Fúnez, 2001) in a similar way as other equivalent tectonic units in Brittany (Ballèvre et al., 1994). All the allochthonous high-*P* units are bounded by greenschist facies shear zones and show internal deformation in the form of a generalized tectonic fabric, that in the Malpica-Tui unit is associated with a strong stretching lineation oriented parallel to the structural trend of the belt, i.e., that defined by the boundaries of the different tectonic elements, including the suture zone.

Several significant structures that affect the regional foliation in the rocks of the Malpica-Tui unit in its northern half are described here: a recumbent fold of hectometric scale, a west-dipping and discrete basal shear zone below the Malpica-Tui unit, shear bands associated with low-angle normal faults, and open to close folds with subvertical axial planes (often associated with later subvertical strike-slip shear zones). In agreement with the geometric relations and the *P-T* conditions during their development, the mentioned structures are associated with two distinct tectonic processes that are inferred to have occurred at different depths: exhumation (foliation, lineation, and recumbent folding) and emplacement (basal shear zone, shear bands, and low-angle normal faults).

While the kinematics during the last tectonic stage, i.e., emplacement, indicate movement of units roughly toward the foreland (east), the previous exhumation stage has some particulari-

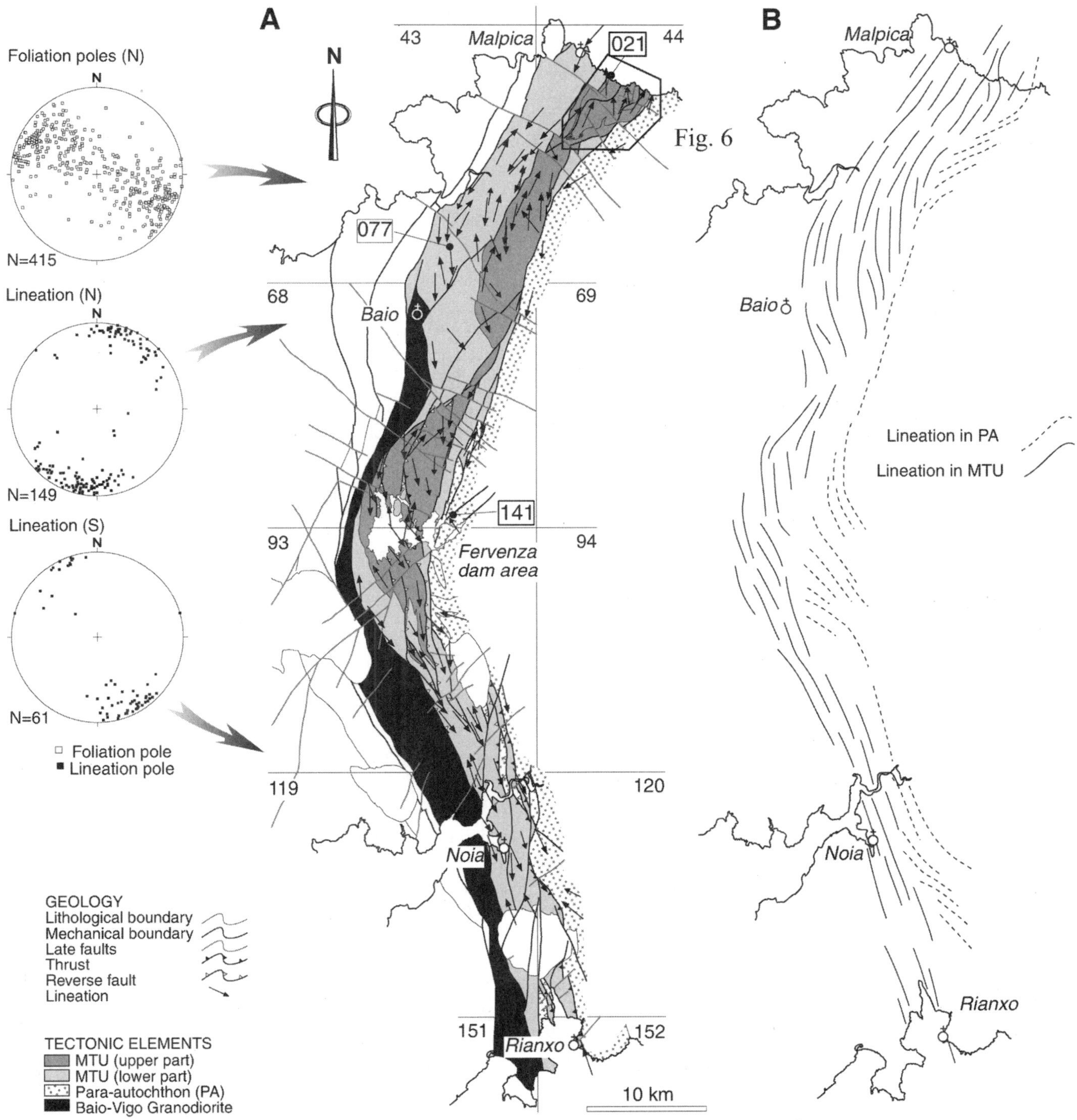

Figure 2. Lineation pattern in relation with main fabric in northern half of Malpica-Tui unit (MTU) and in underlying rocks of paraautochthon. A: Structural map simplified from Llana-Fúnez (2001). B: Sketch showing flow lines according to lineations in A. In latter, late oblique strike-slip shear zones produce reorientation of lineation in shear zones and antithetic rotation in domains bounded by them. Clear oblique relations with lineation in underlying rocks are seen in some domains, sometimes related to late faulting. Numbered grid corresponds to Spanish Geographical Institute 1:50,000 scale maps. Numbers in boxes refer to sample locations.

ties as the stretching lineation forms approximately parallel to the trace of major tectonic element boundaries. We present in the discussion an alternative interpretation for the development of extensive fabrics distinct to simple shear, suggesting that internal deformation of tectonic slices during the exhumation process departs strongly from the simple shear regime and approaches more general noncoaxial type of flow, with the stretching (maximum finite deformation) oblique or perpendicular to tectonic transport, the latter being related to the noncoaxial component of deformation. This has some significant implications in the interpretation of generalized deformation in other parts of the hinterlands of the Variscan Belt and in equivalent geodynamic settings (i.e., indentation tectonics) in other collisional orogens.

TECTONIC FRAMEWORK: THE VARISCAN BELT IN NORTHWESTERN IBERIA

The most prominent tectonic feature of the Variscan Belt in southwestern Europe is the Ibero-Armorican arc, a bend of more than 180° in the Variscan orogen (Ries et al., 1980; Iglesias and Ribeiro, 1981; Matte and Burg, 1981; Brun and Burg, 1982; Burg et al., 1987; Dias and Ribeiro, 1995). This megaarc structure is associated with the Variscan collision, formed during the indentation of the Iberian plate against Laurentia, and it has been compared to similar structures (i.e., syntaxis) in the Himalayas (Matte, 1986, 1991). It is formed by two branches, the Armorican Massif to the north and the Iberian Massif to the south, that are discussed in more detail here (Fig. 1).

In the Iberian Peninsula the Variscan Belt has been divided into zones (Julivert et al., 1972): from the east, where the foreland fold and thrust belt is situated (the Cantabrian zone), and to the west, where the inner parts of the orogen are exposed (Fig. 1). A general increase in deformation, metamorphism, and magmatism occurs in the same direction.

According to the preservation of ophiolitic rock sequences, two major Variscan suture zones have been identified in this massif: (1) the boundary between the Ossa-Morena and South Portuguese zones in southwestern Iberia (Ribeiro et al., 1990; Crespo-Blanc and Orozco, 1991; Fonseca and Ribeiro, 1993) and (2) the allochthonous complexes of northwestern Iberia (Ries and Shackleton, 1971; Ribeiro et al., 1990; Pérez-Estaún et al., 1991; Díaz García et al., 1999; Marcos and Farias, 1999). Both have been related with the closure of an ocean, the Rheic ocean, formerly to the west in present-day coordinates (Martínez Catalán et al., 1999). The ordering of nappes and the asymmetry and vergence of the structures seen in the footwall of the suture in northwestern Iberia (Pérez-Estaún et al., 1991) and the likely location of the root zone of the allochthonous complexes to the west (Ries and Shackleton, 1971) seem to indicate that the subduction zone during the Variscan event dipped to the west, although a more complex subduction zone with an east-dipping Wadati-Benioff plane has been suggested by others (Ribeiro et al., 1990; Dias and Ribeiro, 1995). The tectonic evolution of the Variscan orogen in the Iberian section is far more complicated than this, in part due to the strong strike-slip reworking during ongoing indentation that obscures lithological correlations between different zones, and led some authors to propose as a previous suture zone another crustal-scale shear zone in the boundary between the Central Iberian and Ossa-Morena zones, the Badajoz-Córdoba shear zone, which also contains allochthonous units with high-*P*–low- to intermediate-*T* metamorphic records (Ábalos et al., 1991; Azor et al., 1994).

Allochthonous complexes

The allochthonous complexes of northwestern Iberia (Cabo Ortegal, Órdenes, Morais, Bragança, and Malpica-Tui) are crystalline nappes with high-*P* and related rocks above the Central Iberian zone (Fig. 1). They are composed of a number of units, or tectonic slices, that can be simplified into three groups according to their tectonometamorphic affinities. From bottom to top of the nappe pile, these are: (1) the paraautochthon (Ribeiro et al., 1990) and the basal units, which belong to the Gondwana realm; (2) the ophiolithic units, relics of a former oceanic crust that now delineate the suture zone; and (3) the exotic terranes, of western provenance, in the highest position (a thinned upper mantle–lower crustal section at the bottom and a thinned medium to upper crustal section at the top). The present distribution of the allochthonous complexes seen in Figure 1 evidences a minimum overlap of 200 km during the Variscan convergence, regarding the limit between the Ossa-Morena zone and South Portuguese zone as the main suture zone and the root zone for the nappes.

Malpica-Tui unit in the basal units

The term "basal units" (after Martínez Catalán et al., 1996) includes a number of tectonic slices within the allochthonous complexes with a metamorphic record of high *P* and low- to intermediate *T* (a high *P*/*T* metamorphic ratio) that is unequally distributed and preserved. In northwestern Iberia the only eclogites found in such units are in the Malpica-Tui unit, but other high-*P* parageneses have been described in the Lalín-Forcarei unit (Arenas et al., 1995) and the Centro-Trasmontan unit (Munhá et al., 1984; Gil Ibarguchi and Dallmeyer, 1991). The basal units are regarded as allochthonous over their relative autochthon situated underneath, the paraautochthon, without a metamorphic record of peak high-*P*–low- to intermediate-*T* conditions.

The evidences of high-*P* metamorphism in these rocks are preserved as relics and are mostly reequilibrated during amphibolite and greenschist facies regional deformation (presumably related to the exhumation process) or during later lower *P*–high-*T* metamorphism associated with pervasive granite intrusions and related deformation. Peak high-*P*–low- to intermediate-*T* metamorphic conditions are not recorded in all the rocks forming part of the Malpica-Tui unit and the basal units as a whole (e.g., as documented by Floor [1966] in the southern half of the Malpica-Tui unit). As shown in Figure 3, the *P*-*T* paths in different parts of

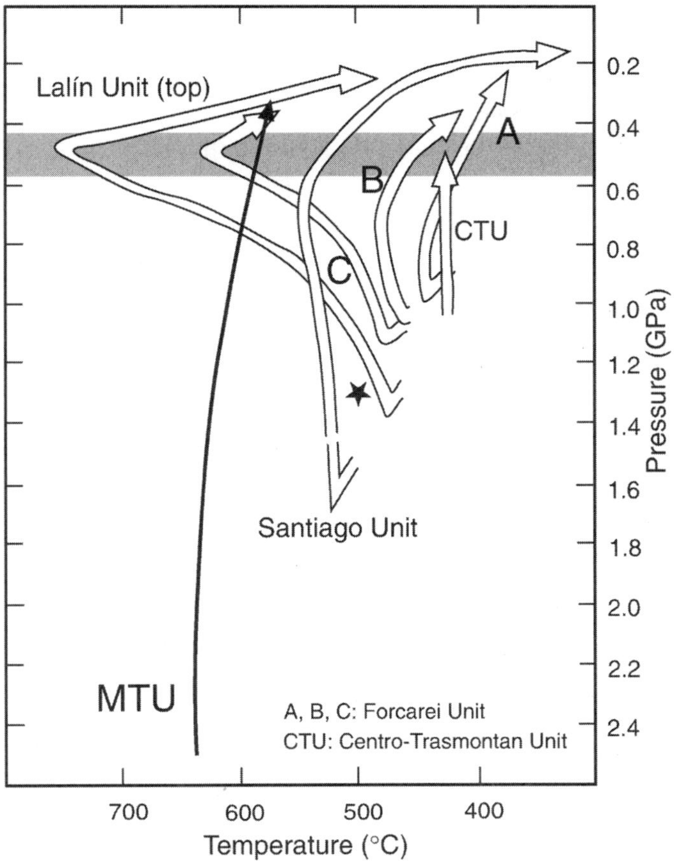

Figure 3. Metamorphic pressure-temperature (*P-T*) paths estimated from available published geothermobarometric data for tectonic evolution in Variscan high-*P*–low- to intermediate-*T* units in northwestern Iberia. Paths from Santiago unit, Lalín unit, and Forcarei unit (A, B, C) are from Martínez Catalán et al. (1996); data from Gil Ibarguchi and Dallmeyer (1991) are used to infer path in Centro-Trasmontan unit (CTU); and data from van der Wegen (1978), Gil Ibarguchi and Ortega (1985), Gil Ibarguchi (1995), and Rodríguez Aller et al. (1997b) are used to build a bulk path for Malpica-Tui unit (MTU) (note that there is one discordant estimate of *P* and *T* from Rodríguez Aller et al. [1997b] in schists from Malpica-Tui unit, shown in graph with black star).

the basal units of northwestern Iberia indicate that not all rocks were subjected to the same pressure (to a certain extent representing depth), and therefore the metamorphic and structural records (or finite strain history) during their exhumation might not be the same (also inferred from the distinct *P-T* trajectories of Malpica-Tui unit and similar units in the Órdenes Complex, with more complex tectonic history; see Martínez Catalán et al., 1996).

The rock sequences that formed the Malpica-Tui unit and the rest of basal units within northwestern Iberia are composed of metasediments alternating schists and feldspathic sandstones (metagraywackes; Floor 1966) and include a number of metavolcanic and intrusive bodies, from basic to acid. Martínez Catalán et al. (1997) associated the basal units with the passive margin of Gondwana.

Phases of deformation to the footwall of the suture

Traditionally, three deformation phases are used in the description of structures related to the Variscan Belt of northwestern Iberia in the transition from the external to the internal parts of the orogen (Matte, 1968; Marcos, 1973; Bastida et al., 1984; Pérez-Estaún et al., 1991): D1 refers to extensive foliation development (at regional scale) and recumbent folding, D2 refers to thrusting of nappes, and D3 refers to late subvertical folding with local crenulation. Although it might be too simple, we can generally apply this scheme to the hinterlands of the orogen, in particular to the footwall of the suture, introducing the appropriate qualifications. According to this, D1 refers to internal deformation of nappes and tectonic slices accompanied by extensive regional metamorphism from high pressure to medium pressure and includes as major structures a regional tectonic fabric (with variably developed lineation) and recumbent folding. D2 includes discrete shear zones as thrusts. D3 refers to subvertical shear zones (with significant strike-slip history) as well as subvertical folding (usually in domains bounded by the shear zones). Although with some exceptions, for example whether recumbent folds are associated to thrusts or whether normal faulting occurs before, during, or after thrusting, it applies reasonably well at the orogen scale. The major difference with the tectonic framework in the external zones is the role of metamorphism and magmatism in the geometry of structures and the diachronism of deformation phases across the orogen: D1 occurs earlier in the hinterlands than in the external zones (see Dallmeyer et al., 1997).

In general, D1 is contemporaneous with extensive metamorphism, which has a relatively high *P/T* ratio, and with some exceptions D2 is often too discrete to include in a discussion about regional metamorphism; however, deformation usually occurs in greenschist facies, as does deformation in D3 shear zones. Magmatism during D2-D3, first with the localized intrusion of calc-alkaline batholiths in shear zones and second with the extensive intrusion of two-mica granites, modifies locally this scheme for the latest deformation phase, and some D3 structures are strongly ductile.

Time limitations on deformation events

The available geochronological data on certain Variscan events help to establish broad limits to the different deformation episodes. The age of the high-*P*–low- to intermediate-*T* event from Malpica-Tui unit rocks has been calculated by several authors and yielded ages in the range 365–360 Ma, although higher age values (370 Ma) have been estimated (van Calsteren et al., 1979; Santos Zalduegui et al. 1995; Rodríguez Aller et al., 1997a, 1997b). Except for Santos Zalduegui et al. (1995), who used U-Pb zircon dating, these ages were obtained using white mica in eclogites and eclogitic rocks (Rb-Sr, K-Ar, and Ar-Ar methods). The isothermal decompression during exhumation of the Malpica-Tui unit (Fig. 3), above the closure temperature for the different methods in white mica (McDougall and Harrison, 1999), and the strong retrogression seen in the rocks (biotite rims around white mica) lead us to treat these data with caution, because they may indicate minimum ages for the high-*P* metamorphic event (seen in the results of Rodríguez Aller et al., 1997a).

The final emplacement of high-*P*–low- to intermediate-*T* rocks by means of a greenschist facies shear zone (i.e., basal shear zone) must be older than the overprinting strike-slip deformation and the intrusion of I-type granodiorites related to some of the crustal-scale shear zones that nucleated strike-slip tectonics (Llana-Fúnez and Marcos, 2001). Variscan granodiorite batholiths intruded ca. 350 Ma (Serrano Pinto et al., 1987; Bellido et al., 1992; Gallastegui, 1993), and therefore represent upper time boundaries for the final emplacement of the Malpica-Tui unit.

In summary, the available geochronological data, with minimum ages (?) of 370–365 Ma for the high-*P*–low- to intermediate-*T* event and 350 Ma for the beginning of postnappe tectonics, allow us to broadly define a minimum period of 15–20 m.y. for the process of exhumation of Malpica-Tui unit from high-*P* conditions to the middle crust.

MALPICA-TUI UNIT: GENERAL STRUCTURE, LITHOLOGY, AND METAMORPHISM

The Malpica-Tui unit extends 150 km approximately north-south, parallel to the trend of structures in the Variscan orogen in northwestern Iberia with an average width of 10 km (Fig. 1). It was previously defined as the Complejo Antiguo (Parga Pondal, 1960) and Blastomylonitic graben (den Tex and Floor, 1967) due to the strong development of a stretching lineation in orthogneisses and the fact that it was not migmatized in most of its parts in contrast with the surrounding regions. Strain and metamorphic changes associated with the major regional deformation stage (D1) are better preserved to the north of the unit, and thermal reequilibration related to (late) Variscan migmatization and two-mica granite intrusions increase their effect on previous structures to the south (syn-D3). Because the sequence dips slightly to the north, this means that effects of D1 deformation increase toward the top of the pile and D3 thermal reequilibration increases toward the lower parts.

The cartographic pattern seen in Figure 1 is determined by the development of the Malpica-Lamego line, which is currently its western boundary (Llana-Fúnez and Marcos, 2001). This subvertical crustal-scale shear zone formed after the emplacement of the crystalline nappes in relation with intracontinental deformation within the Variscan Belt and it is regarded as a typical D3 structure in the previously mentioned tectonic scheme.

Lithologically, the Malpica-Tui unit is composed of a sequence of predominantly schistose metasedimentary rocks wherein centimetric paragneiss and micaschist bands alternate. The relative abundance of plagioclase blasts determines this compositional banding and might reflect original sedimentary layering; however, the development of isoclinal folds (Fig. 4A) and the strong recrystallization observed in most of the sections cannot preclude a tectonic origin for the banding. Several amphibolite lenses and different types of acidic rocks of igneous origin are included in the sequence and are parallel to the general foliation. The strong ductile deformation affecting most of these rocks obscured their original relationships with the metasediments. At least three types of quartz-feldspathic gneisses have been differentiated in the literature according to their geochemical features: fine-grained gneisses of calc-alkaline nature (Pin et al., 1992); A-type alkaline and peralkaline orthogneisses (Ribeiro and Floor, 1987); and coarse-grained calc-alkaline orthogneisses with primary biotite (Pin et al., 1992). Both calc-alkaline and peralkaline orthogneisses show a strong mylonitic fabric, variously recrystallized, that is parallel to the foliation in the metasediments; they appear as layered bodies within them (Floor, 1966). In contrast, the coarse-grained calc-alkaline gneisses with primary biotite appear as elongated thick granitic bodies that preserve the inner parts nearly undeformed and the edges strongly foliated (see the latter in Fig. 4B). The nature of the metasediments and the different magmatic pulses that originated the intrusions were related to a rifting episode during the Ordovician by Ribeiro and Floor (1987).

*High-*P *metamorphism*

High-*P*–low- to intermediate-*T* mineral assemblages have been found partially preserved in eclogite lenses within the fine-grained orthogneisses (van der Wegen, 1978; Gil Ibarguchi and Ortega, 1985) (Fig. 4C), in the undeformed parts of coarse-grained calc-alkaline orthogneisses (Gil Ibarguchi, 1995), in the micaschists (Rodríguez Aller et al., 1997b), and as inclusions within albite blasts (in the Lalín-Forcarei unit, an equivalent unit of Malpica-Tui unit in the Órdenes Complex; Arenas et al., 1995). However, most of the rocks forming the Malpica-Tui unit are reequilibrated in amphibolite facies or in greenschist facies, and high-*P* assemblages are not easy to identify in the field. The heterogeneity of the distribution of high-*P* metamorphism and its subsequent deformation during the exhumation allowed in the Malpica-Tui unit the preservation of primary sedimentary structures (Llana-Fúnez, 2001), acritarchs (Fombella Blanco, 1984), and a wide range of intermediate-*P*–low-*T* metamorphic assemblages extended regionally. Within the Malpica-Tui unit, the boundaries among what can be regarded as subunits, with different peak pressures, are not easily identified in the field, due to poor exposure conditions and to the strong strike-slip reworking; however by analogy with the discrete lower boundary of the Malpica-Tui unit, the basal shear zone, they are thought to be also developed in greenschist facies during the final emplacement of nappes at mid-crustal depths, and therefore after the exhumation process.

STRUCTURES IN MALPICA-TUI UNIT

In this section we describe the structures found in the rocks of the Malpica-Tui unit according to their likely order of appearance and the grade of ductility, thought to be developed dur-

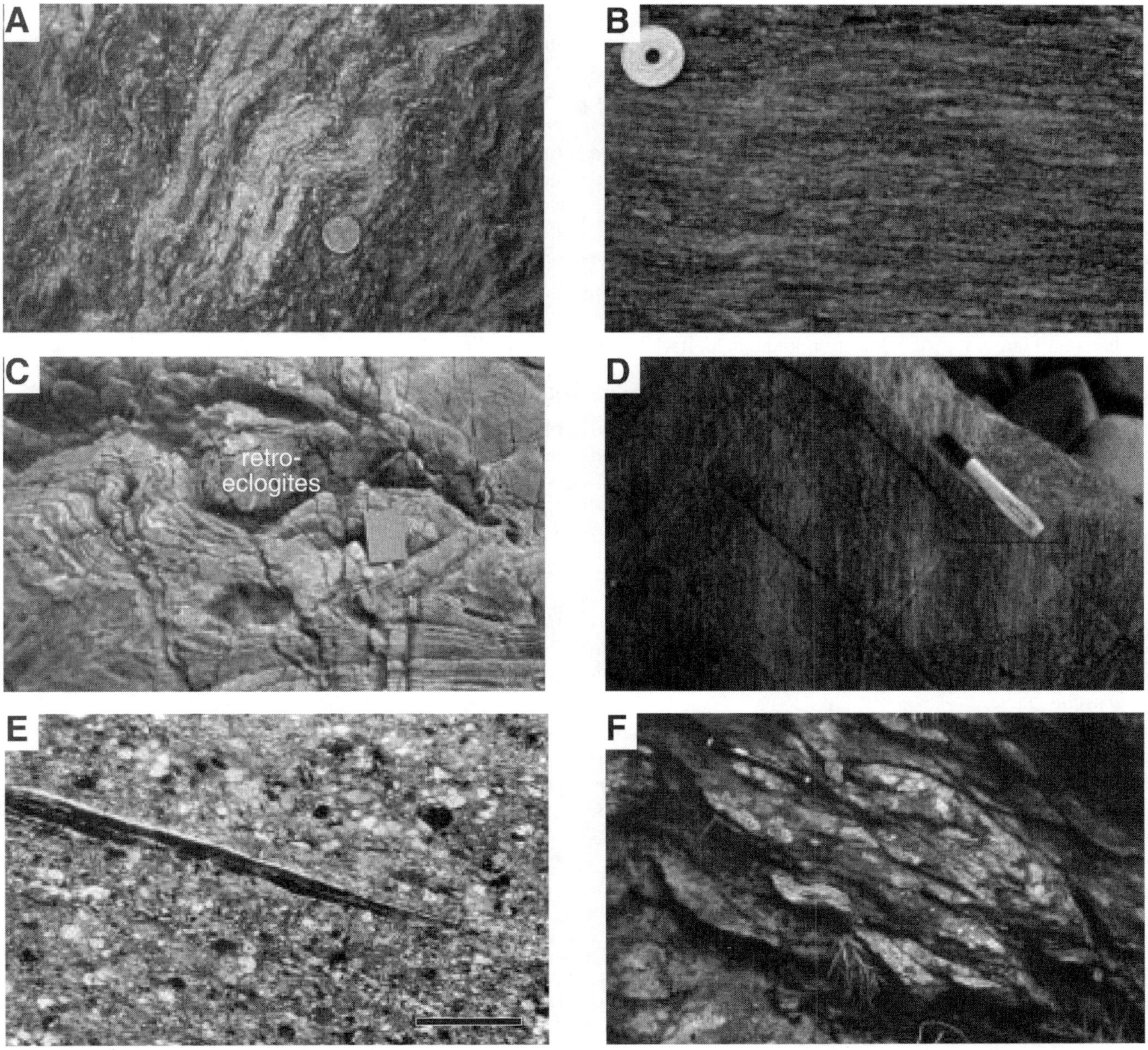

Figure 4. A: Isoclinal folding in metasediments at centimeter scale. B: Field appearance of main fabric (mylonitic) in coarse-grained felsic orthogneisses. C: Minor folds in fine-grained felsic gneisses nucleating in retroeclogite inclusions. D: Stretching lineation in fine-grained felsic gneisses. E: Deformation of metasediments at relatively low-temperature conditions in basal shear zone (scale bar = 200 μm). F: C′ shear bands developed in schists in hanging-wall rocks of low-angle normal faults (top to west) (see Figs. 6 and 8).

ing their ascent from high-*P* conditions. This order is not always straightforward and is treated with more detail in the discussion in connection with the geodynamic setting.

Internal deformation of the Malpica-Tui unit: The regional tectonic fabric

The rocks of the Malpica-Tui unit and those underlying it, the upper part of the paraautochthon, show a pervasive and extensive tectonic foliation that is parallel to the compositional banding and lithological boundaries. The foliation has an associated lineation of different nature depending on the type of rock: stretching of polycrystalline aggregates of quartz, plagioclase, and K-feldspar in orthogneisses (Fig. 4, B and D), mineral orientation of blue-green amphibole in amphibolites, and stretching (quartz) or intersection (white mica and late mimetic biotite) in schists and paragneisses. Tectonic fabrics are varied: linear to planolinear in orthogneisses (Fig. 4D), planar to planolinear in schists and paragneisses (respectively), and planolinear in amphibolites. All these different lineation types (stretching, mineral orientation, and intersection) in extensive foliated domains are consistently oriented parallel to the trend of the major structures in the orogen (Fig. 2).

A range of mineral associations can be described in relation or in equilibrium with the main fabric in the different lithological types. High-*P*–low- to intermediate-*T* assemblages have been described in foliated eclogites (omphacite, garnet, rutile, quartz, white mica, and clinozoisite; van der Wegen, 1978; Gil Ibarguchi and Ortega, 1985) and schists (white mica, garnet, rutile, cloritoid; Rodríguez Aller et al., 1997b). In both cases, fabrics are planar and present a weak lineation in the field, if visible at all. However, more commonly mylonitic fabrics have developed under amphibolite facies conditions in the orthogneisses (LS and SL) or reequilibrated in greenschist facies in schists and paragneisses (S and SL). In the former, quartz, K-feldspar, and plagioclase show simultaneously evidence of dynamic recrystallization, indicating a range of temperatures during deformation of ~500–550 °C (Srivastava and Mitra, 1996), consistent with the temperature during the exhumation path of the Malpica-Tui unit shown in Figure 3. In the latter, massive growth of plagioclase, development of biotite coronas around metamorphic white micas, and appearance of secondary white mica and chlorite and ilmenite after rutile indicate retrogression in greenschist facies.

In general, the evidence of high-*P*–low- to intermediate-*T* metamorphism in these rocks appears as relic mineral assemblages in domains surrounded by the main fabric, which shows deformation in amphibolite facies conditions during its main development and further retrogression in greenschist facies. The changing equilibrium *P-T* conditions during the formation of the regional foliation and the regional consistency of its orientation pattern along the Malpica-Tui unit led us to relate this structure with the exhumation of this unit from high-*P* conditions.

An internal foliation, seen in porphyroblasts of plagioclase (in schists and paragneisses) and garnet (in layered amphibolites), often shows continuity with the main fabric outside the blasts. This preserved foliation probably represents an earlier stage in the main fabric development, protected from changes associated to deformation by crystal growth of plagioclase and garnet, as shown by Arenas et al. (1995) in plagioclase blasts from the Lalín-Forcarei unit, farther to the east in the Órdenes Complex.

The kinematics during regional deformation, with regard to sense of shear, is difficult to establish in a setting such as this, where recumbent folds occur (though not always is recognized locally due to poor exposure conditions). Because this microstructural study is treated in more detail in another paper (Llana-Fúnez, 2002), we will only give some key aspects of it here. Some observed structures and strain features in the rocks such as the west-dipping foliation, the plastic deformation in quartz (with development of crystallographic preferred orientation, CPO), and some C′ shear bands in orthogneisses do not provide enough useful information about the regional kinematics. The mapping of quartz *c*-axis textures in relation with the regional foliation shows patterns with the presence of two perpendicular asymmetries represented synthetically in Figure 5. This apparent composite nature of textures might be interpreted as due to the accumulation of two consecutive and distinct noncoaxial flows during deformation of quartz (the first predominant with the noncoaxial component of deformation perpendicular to the lineation and the second with the noncoaxial component parallel to the lineation) or as a result of (triclinic) deformation under three-dimensional general flow (Llana-Fúnez, 2002). Asymmetry in quartz *c*-axis textures parallel to the lineation in the long limb of a recumbent fold indicates that top moves to the north, in agreement with the sense indicated by C′ shear bands in orthogneisses in the same structural orientation. This probably represents a late stage in the exhumation process and not the major part of it (further discussion following).

Internal deformation of the Malpica-Tui unit: Recumbent folding

The presence of a recumbent fold of hectometric scale is inferred from the geological map of the northernmost coastal section of the Malpica-Tui unit and from a cross section along this coast (the fold axis plunges an average 7° to N23E and therefore almost perpendicular to the coast, more or less east-west; Fig. 6, A and C; Llana-Fúnez, 1997). The sequence of rocks in the eastern long limb, formed by fine-grained calc-alkaline orthogneisses at the base and interlayered amphibolites and schists at the top, appears inverted in the western short limb over ~1 km (Fig. 6, B

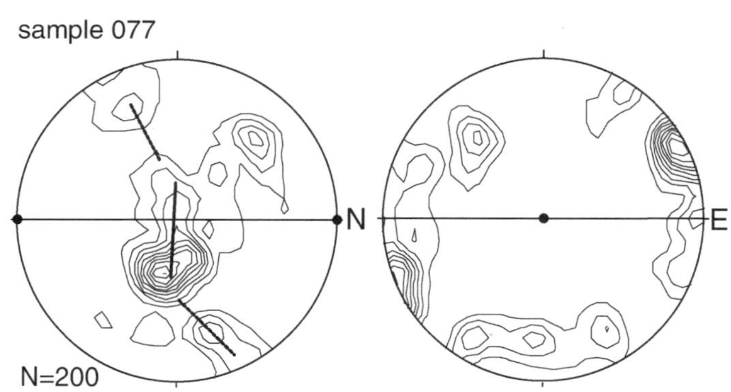

Figure 5. Example of two distinct asymmetries in same pattern of *c*-axis textures in relation to quartz deformation during development of main foliation. In section plane parallel to lineation (left) stronger development of one of arms of type I crossed girdle seem to indicate presence of noncoaxial component with sense of shear to south. This asymmetry is not in same plane as strong maximum defined by *c*-axes in prismatic position (close to foliation plane), which is better envisaged in sections perpendicular to lineation (right). Quartz *c*-axes, measured with U-stage, have been plotted in lower hemisphere equal-area projections and contoured in 1% intervals (uniform distribution) in all figures. Horizontal line represents foliation and black dots represent lineation in rock.

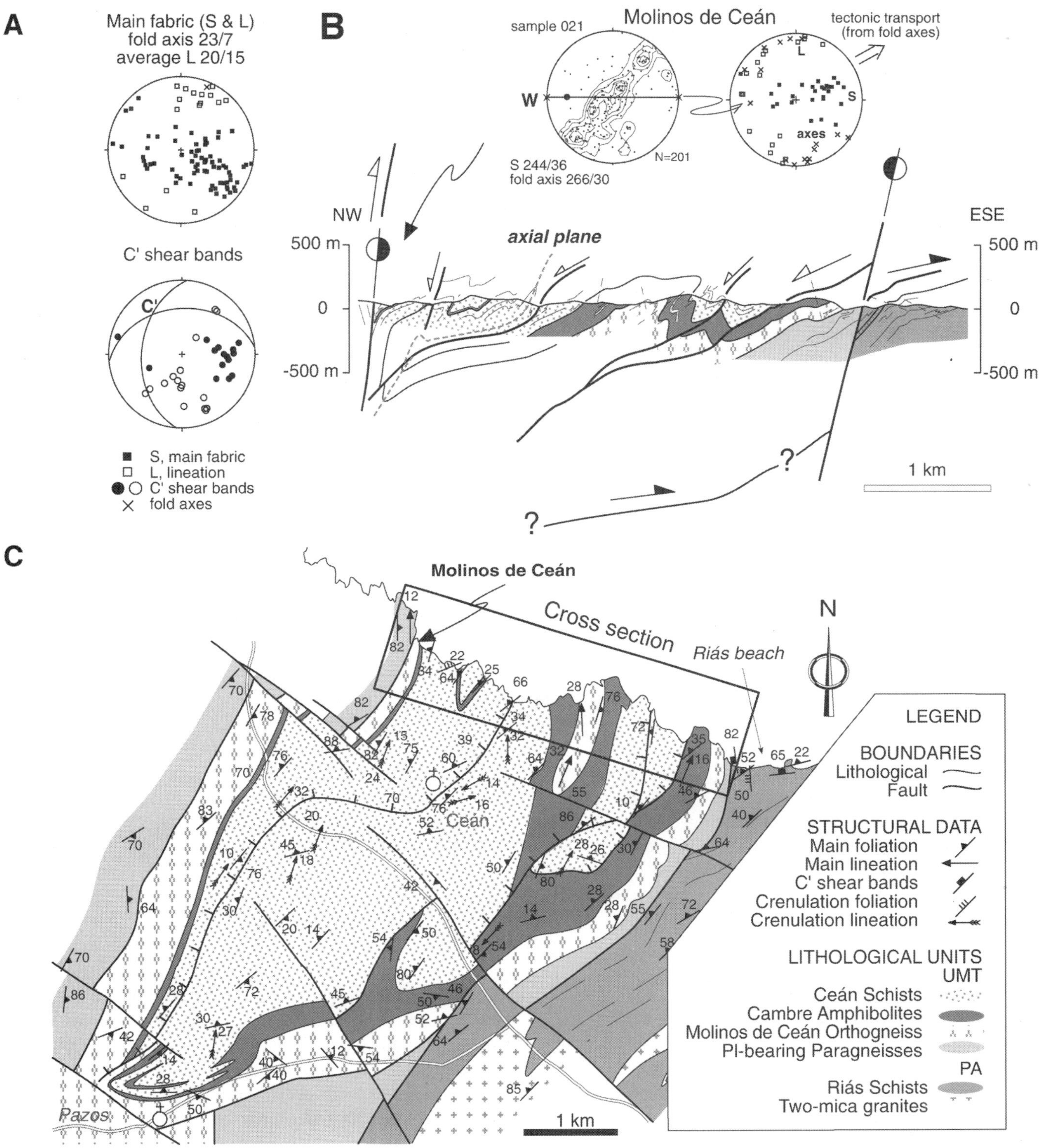

Figure 6. Hectometric recumbent fold formed in upper part of Malpica-Tui unit affected by later homoclinal open folds and by oblique subvertical shear zone (see also Fig. 2). Stereoplots in A (lower hemisphere equal area) show relation between foliation (black squares), inferred fold axis (cross), and lineation in rocks (white squares). General situation of map (C) is in Figure 2. Stereoplots in B show quartz *c*-axis to the left and lineation, foliation pole, and fold axes plots to the right.

and C). The fold axis plunges ~7°–10° to the north-northeast, subparallel to the trend of other structures (e.g., lineations), allowing the observation of its periclinal closure to the south (Fig. 6C). Because there is only one hinge and no way-up criteria have been found in these rocks, no vergence could be inferred from the shape of the fold. In contrast to other large-scale recumbent folds described in the high-*P*–low- to intermediate-*T* units in the Órdenes Complex (Martínez Catalán et al., 1996), no transect cleavage has been observed in relation to this recumbent fold.

Some decimetric to metric folds occurred in the finer grained calc-alkaline orthogneisses and nucleated around the mafic inclusions (Fig. 4C), where high-*P*–low- to intermediate-*T* mineral assemblages are partially preserved (Gil Ibarguchi and Ortega, 1985). These folds show opposite asymmetries in the limbs of the recumbent fold and are therefore associated here with it as minor folds. In contrast to the major fold axis, the minor fold axes, plunging north-northwest to south-southeast, are curved in one locality (with a dispersion to 40°) (Molinos de Ceán; see Fig. 6, B and C) and fold slightly the previous stretching lineation in the orthogneisses (also indicated by Alonso and González, 1982).

By analogy with the recumbent folds that affect a similar sequence of rocks farther to the east in the Lalín-Forcarei unit (see Figs. 4 and 6 in Martínez-Catalán et al., 1996), the general tectonic transport is thought to be perpendicular to the fold axis, i.e., to the east in present coordinates. The tectonic transport direction inferred from the dispersion of minor fold axes in Molinos de Ceán is west-southwest to east-northeast (Fig. 6B). The validity of this direction as tectonic transport is corroborated (with a slight difference in orientation) by the orientation of the noncoaxial component of plastic deformation in quartz in a late vein, parallel to the previous mylonitic fabric, that have been folded during this episode. The texture pattern of the measured quartz *c*-axes shows a clear asymmetric crossed girdle oblique to the fold axis, which is oriented east-west, indicating a strong noncoaxial component during deformation in that particular orientation (Fig. 6B). However, the sense of shear that can be inferred from the texture and quartz shape fabric is not considered as reliable because it is probably related to the folding mechanism that affected the vein (not necessarily related to tectonic transport; e.g., see a case in Stünitz, 1991).

Discrete structures: The basal shear zone of the Malpica-Tui unit (D2)

The Malpica-Tui unit is separated from the underlying rocks, the paraautochthon, by a shear zone named the basal shear zone. It is strongly reworked by late subvertical strike-slip shear zones and intruded by two-mica granites and pegmatite veins, but it preserves in certain key locations a previous structural geometry. In such cases, a deformation zone of several tens of meters in width, dipping 40°–50° to the west, is seen with foliations in both hanging wall and footwall parallel to the tectonic boundary. This deformation affects the metasediments and is characterized by retrograde features: white mica grains are recrystallized and reoriented with the foliation; undeformed pseudoclasts of quartz float and coexist with a plastically deformed matrix of fine-grained quartz (Fig. 4E); and plagioclase porphyroblasts show pressure solution in surfaces parallel to the foliation.

The tectonic foliation in the shear zone has an associated stretching lineation defined by quartz. This lineation in most cases is oriented subparallel to the trace of the structure, following the trace of the shear zone (parallel to strike of foliation). However, some oblique orientations have been found, especially in the rocks belonging to the footwall, that may indicate a previous oblique orientation of the stretching lineation (Fig. 2). Effects of plastic deformation in quartz were studied measuring *c*-axes in a section parallel to this stretching lineation. The *c*-axis texture patterns in Figure 7 show a bulk texture dominated by an asymmetric small circle girdle around the Z axis that indicates a top-to-the-southeast sense of shear. There is a strong domainal deformation of quartz, with even different kinematics (as the orientation of the crossed girdle in domain O seems to suggest, perpendicular to the structural reference frame, i.e., foliation and lineation) that indicates deformation in greenschist facies conditions.

The observed kinematic criteria, the dip of the shear zone, and the metamorphic differences between hanging and footwall rocks indicate that the basal shear zone is a thrust directed toward the foreland, to the east (Ortega and Gil Ibarguchi, 1983; González-Lodeiro et al., 1984; Llana-Fúnez, 2001). Similar obliquity in the orientation of the lineation and in conditions of deformation has been observed in similar shear zones bounding tectonic units in other allochthonous complexes (Matte, 1986; Burg et al., 1987; Martínez Catalán et al., 1996; Marcos and Farias, 1999).

Discrete structures: C′ shear bands and low-angle normal faults (D2–D3)

C′ shear bands with millimetric and centimetric spacing appear in micaschists in the northernmost section of Malpica-Tui unit rocks, in a 100-m-thick deformation band (Fig. 4E), associated with relatively low-angle normal faults (in the sense of McClay and Ellis, 1987) with similar kinematics (top to the west; Figs. 6 and 8) (Llana-Fúnez, 1997). These structures were only found in the long limb of the recumbent fold, so no clear relative time relations with this fold and the basal shear zone can be established with certainty. Faults have basically developed in schists and the tectonites according to the classification of Sibson (1977), characterized by crush breccias in the brittle-ductile transition.

The cross section in Figure 6B shows open folds with very gently dipping axial planes and a strong vergence to the east, affected by later upright ones. Our interpretation is that these folds formed initially due to the presence of flats and ramps in the low-angle normal faults (Fig. 8) (Llana-Fúnez, 2001).

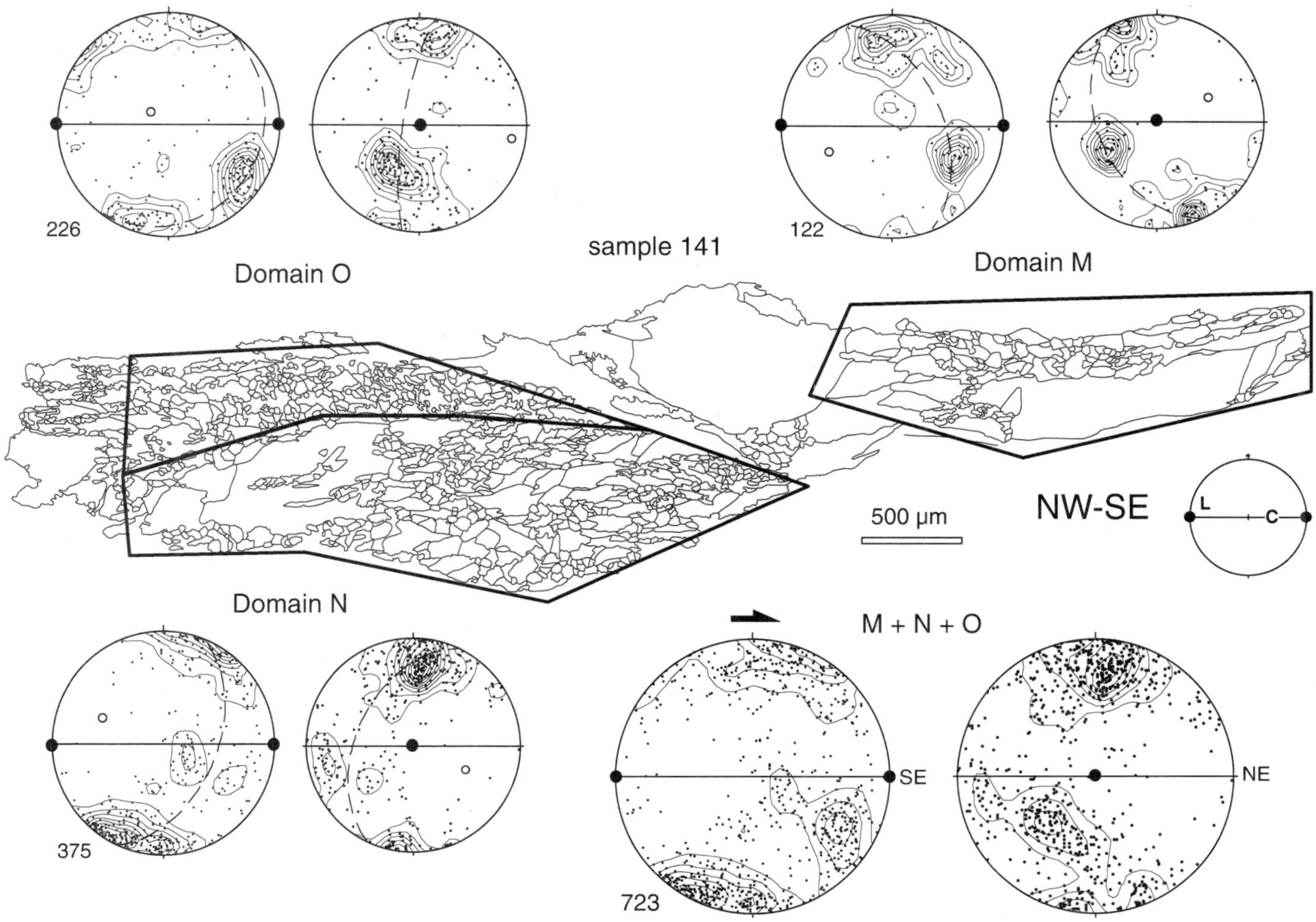

Figure 7. Microstructure of polycrystalline aggregate of quartz in mylonitized granite from basal shear zone of Malpica-Tui unit. Stretching lineation is oblique to general trend of belt (sample locality 141 in Fig. 2). Bulk quartz *c*-axis texture shows predominance of *c*-axis in basal positions and weak crossed girdle indicating asymmetry parallel to lineation and tectonic transport of top-to-southeast. Domainal deformation of quartz aggregate (domains M, N, and O) accounts for heterogeneity of deformation in shear zone; crossed girdles in separate domains are oblique to reference frame, foliation, and lineation in rock.

Open folds with subvertical axial planes (D3)

Late open and upright folds affect the structures described until D3. These folds are widespread in northwestern Iberia and are considered as a late deformation phase (D3) in the Variscan Belt (Matte, 1968; Bastida et al., 1984). The axes of these folds are parallel or subparallel to the previous linear structures in this part of the chain, indicating homoaxiality during polyphase deformation in these rocks (e.g., see Bastida et al., 1993). In the study area, these folds appear in domains bounded by subvertical strike-slip shear zones and have been associated with the same tectonic event (Llana-Fúnez and Marcos, 2001). As expected, they produce at different scales type-3 fold interferences (Ramsay, 1967) with the previous recumbent folds (Figs. 4A and 6B). According to this late compressive event, a little inversion effect in the previous low-angle normal faults cannot be discarded.

DISCUSSION

Tectonic features of the exhumation and emplacement processes

The number of structures in Malpica-Tui unit rocks described herein during the evolution of the Variscan orogen were separated into two classes: a first type of structures that reflect internal deformation of thrust sheets (D1) and a second one, where strain is concentrated or partitioned in discrete deformation zones (and related structures) that bound the thrust sheets or crystalline nappes (D2). Both stages are thought to be separate but consecutive in time, and they are envisaged to occur in the same tectonic process as a consequence of progressive deformation in a somehow changing environment during the ascent through the crust of Malpica-Tui unit rocks, from high-*P*–low- to intermediate-*T* conditions to middle crustal depths.

Figure 8. Development of asymmetric folds with gently dipping axial plane in relation with low-angle normal faults. These folds are affected by later upright folding. Traces of axial plane of two types of folds, with gently and upright dipping axial planes, have been drawn for clarity. Cross section is mirror image of one in Figure 6 in order to facilitate comparison with hinge zone in photograph.

The mechanism that triggers the change in style from extensive and pervasive deformation within the nappes or tectonic slices to a more discrete type in their bounding shear zones is not known with certainty. At the present state of knowledge, the isothermal decompression recorded in Malpica-Tui unit rocks during their exhumation, as illustrated in Figure 3, excludes a strong change in temperature as the main cause of deformation partitioning and leaves the decreasing pressure and particularly the influx of fluids and high rates of deformation as the most probable causes of this change of behavior, occurring likely in the brittle-ductile transition in the middle crust. This transition can be associated with an increase of heterogeneity in deformation that ends with concentration of deformation in discrete shear zones. Heterogeneity is also a major feature of regional deformation in the Malpica-Tui unit and is probably favored by the speed of these processes, from burial to exhumation, that in some orogens reach values of the same order of magnitude as plate motions (Duchene et al., 1997; Gebauer et al., 1997).

The process of the exhumation of Malpica-Tui unit high-*P*–low- to intermediate-*T* rocks gives rise to some other interesting points that we try to address. Some of these are as follows.

1. The tectonic fabric observed in Malpica-Tui unit eclogites evidences that deformation was occurring at a deep part of the Variscan orogenic wedge.

2. The distribution of strain features during burial and exhumation is strongly heterogeneous, allowing the preservation not only of high-*P*–low- to intermediate-*T* mineral assemblages but also of previous primary features (as igneous textures and sedimentary structures).

3. During the stage of extensive ductile deformation, rocks have to ascend from deep levels in the wedge (at pressures to 2.5 GPa) to middle crustal depths while traveling to the east (the root zone is situated to the west) in a relatively short period of time. This is also true in the tectonic models that suggest oblique collision for the Variscan Belt (Badham and Halls, 1975; Badham, 1982; Martínez Catalán, 1990).

4. Although some shortening occurs perpendicular to the rock sequence (thus implying thinning of the pile of rocks, not necessarily extension at horizontal orientation), maximum finite stretching within the Malpica-Tui unit is consistently subhorizontal and parallel to the structural grain of the orogen and therefore, to a certain extent, parallel to the boundaries of the colliding continents.

5. Stretching lineations during exhumation and emplacement are oriented at high angles (almost perpendicular) to each other, but there is no evidence to support such a substantial change in direction of compression and tectonic transport for this period (if developed in simple shear).

The rest of the structural and kinematic history involves the development of discrete structures bounding different thrust sheet–like bodies. To this stage is ascribed the formation of the basal shear zone and its likely associated minor structures on the hanging wall (including low-angle normal back-faulting), prior to the later strong intracontinental tectonics.

Kinematics of deformation during the exhumation and emplacement processes

In simple shear flow, the maximum stretching axis after strain accumulation (lineation) is oriented or tends to rotate within the XZ plane toward parallelism with the direction of tectonic transport (Passchier, 1998). This type of flow, however, is a particular case and is "strongly sensitive to small deviations from ideal conditions," according to recent theoretical simulations (Passchier, 1998): the orientation of the stretching lineation toward the tectonic transport direction depends on the vorticity of the flow, the extrusion factor, and the mode of accumulation of finite strain in the rocks (Passchier, 1997). The homoaxiality of the deformation phases on Malpica-Tui unit rocks (also in northwestern Iberia as a whole) makes acceptable the assumption that the regional flow parameters are constant (orientation of compression and shortening during D1, D2, and D3) and that the accumulation of finite strain (finite strain ellipsoid) would only be dependent on the value of these parameters and on the time of accumulation (following Passchier, 1998).

If flow parameters are steady during progressive deformation, then the accumulation of finite strain leads to two stable types of fabrics in rocks: Xi and Yi shear-zone segments (Passchier, 1998). In the first type, maximum finite strain will be perpendicular to the direction of tectonic transport (b lineation), while in the second it will be parallel (a lineation). The latter would be closer to the simple shear model applied by default in deformed rocks (although developed for shear zones by Ramsay and Graham, 1970). It is not yet possible to quantify the vorticity and the extrusion number of nondiscrete deformation in Malpica-Tui unit rocks, but some qualifications are feasible that lead us to suggest that the flow during extensive deformation is closer to Xi (b lineation) than to Yi. First, there is the strong stretching parallel to the orogen, normal or to a certain degree oblique to tectonic transport, inferred from the allochthonous character of the complexes and the location of their root zone. Second, there is the development of an asymmetry in quartz *c*-axis textures oriented east-west (perpendicular to the stretching lineation) during extensive deformation at the beginning of the exhumation (Llana-Fúnez, 2002).

If we assume that the internal deformation in the Malpica-Tui unit occurs under simple shear flow regime, the directions of tectonic transport change dramatically from the exhumation to the emplacement, as lineations are oriented at high angles to each other: structures related with an earlier exhumation stage would show maximum stretching and transport to be parallel to the orogenic wedge (therefore to colliding continent boundaries), while subsequent emplacement would show a strong transverse east-west tectonic component of movement. The first episode does not give any evidence of a component of tectonic transport tranverse to the orogen and it would not be consistent with the location of the allochthonous complexes at a minimum distance of 200 km to the east from their root zone in the suture situated to the west, regardless that a major transverse tectonic transport perhaps attained during the emplacement stage. Because strike-slip tectonics would not be a geodynamic setting were such high-*P*–low- to intermediate-*T* metamorphism developed and preserved (Thompson et al., 1997), we think that some of the initial assumptions are not valid. We suggest instead that extensive ductile deformation during the exhumation process develops under a general noncoaxial flow regime that departs from plane strain (close to b lineation) and therefore from the simple shear model, which is an extreme case in noncoaxial deformation (see Passchier, 1997, 1998). As a consequence, the regional lineation in Malpica-Tui unit rocks that would indicate flow direction, as suggested by others (Burg et al., 1987; Dias and Ribeiro, 1994, 1995), cannot be used as indicator of tectonic transport during exhumation of high-*P* nappes.

The assumptions given by Ramsay and Graham (1970) to the application of the simple shear model to natural SL-tectonites, i.e., discrete plane-strain deformation in planar parallel-sided shear zones without wall rocks being involved, are therefore not fulfilled in the regional fabrics observed in the Malpica-Tui unit: we are examining the internal deformation of a thrust sheet–like body bounded by discrete shear zones.

With regard to the kinematic indicators showing tectonic transport to the north (some asymmetries in quartz *c*-axis textures and C′ shear bands in orthogneisses), a more careful detailed and widespread analysis is needed. With the scarce available data we can only suggest that it is related, at a final stage of the exhumation process, to lateral spreading or displacement parallel to the orogenic trend, still under a general compressive regime, simultaneously or immediately prior to the final emplacement of the allochthonous nappes. This is a common late feature in general compressive regimes in other orogenic belts such as the Himalayas or the Alps (Tapponier et al., 1986; Ratsbacher et al., 1991), but it is usually accompanied by major transverse structures (e.g., Simplon line; Mancktelow, 1985; Merle et al., 1989). These have not yet been described in northwestern Iberia at this stage of the orogenic history, although they are present as later features in relation to the late Variscan high-*T* and low-*P* metamorphism (Burg et al., 1994; Escuder Viruete et al., 1994; Díez Balda et al., 1995).

Normal faulting, prior to the intracontinental event, is registered in the C′ shear bands and low-angle normal faults described here (Llana-Fúnez, 1997), indicating normal sense with the hanging wall down to the west, therefore back-directed movement (Figs. 6 and 8) (see also Ribeiro et al., 1990). Such back-directed kinematics in relatively low angle detachments in upper parts of high-*P* nappes in junction with thrusts in the lower parts of these nappes have been related to the extrusion of high-*P* rocks in other collisional settings (Burchfield et al., 1992; Michard et al., 1993; Chemenda et al., 1995, 1997; Grujic et al., 1996; Escher and Beaumont, 1997). However, in our case, the style of deformation in the faults, which is typical of upper crustal levels, and the magnitude of the displacements, would not support such interpretations.

Tectonic setting

Our hypothesis for the kinematics during the exhumation of high-*P* rocks in the Malpica-Tui unit becomes more meaningful when we consider the tectonic setting in which the unit is placed, that provides a clear evidence of an irregular continental collision; i.e., the hinge zone of the Ibero-Armorican arc (Fig. 1) (Matte and Burg, 1981; Brun and Burg, 1982; Matte, 1986; Burg et al., 1987; Dias and Ribeiro, 1994, 1995). We give some evidence of the structural record at an early stage of the evolution of this arc, consistent with the model proposed by Dias and Ribeiro (1995).

In recent mountain belts the indentation of stiff irregular continental plates in collisional orogens is a frequent phenomenon that leads to extrusion of material to the sides of the indentor (Tapponier and Molnar, 1976; Ratsbacher et al., 1991), because indentation produces lateral strain and pressure gradients that cause these movements (see Mancktelow, 1993). In such cases, deformation is not restricted to plane strain (and in consequence is different from simple shear). We suggest that the structure indicating finite stretching related with widespread deformation at depth, i.e., the regional lineation, would delineate the flow lines of the extrusion of material from the hinge but not necessarily the transport direction of nappes, which would be indicated at this stage by the noncoaxial component of internal deformation within the nappes (related to the asymmetry of structures) (Fig. 9).

One consequence of ongoing indentation is that it produces rotation in the emplacement direction of nappes along discrete shear zones as they become inactive. In the Ibero-Armorican arc such phenomena have been reported in the external zone, in the Cantabrian zone (Julivert and Arboleya, 1984; Pérez-Estaún et al., 1988), and in the hinterlands, in the suture zone in Cabo Ortegal, where a progressive rotation of the emplacement direction was noted from a present east-west orientation at higher temperature deformation (highly transverse to orogen) to an oblique northwest-southeast orientation at lower temperature deformation (Marcos and Farias, 1999). This general rotation in the southern branch of the Ibero-Armorican arc is not necessarily related to a general oblique collisional orogen, but it could be simply due to the progressive indentation of the Iberian plate against Laurentia, as envisaged very synthetically in Figure 9 and in Dias and Ribeiro (1995).

Features of internal deformation in other areas related to the footwall of the suture zone

To give more consistency to the hypothesis that internal deformation in the Malpica-Tui unit did not occur under a simple shear regime but in a more general noncoaxial flow, we briefly review how the kinematics of the internal deformation (D1) of Variscan crystalline nappes in northwestern Iberia is described in the footwall of the suture zone in the same orogenic wedge and discuss the data.

In the Central Iberian zone (Fig. 1), in the hinterland of the Variscan orogen, two structures are related with the first phase of Variscan deformation: the regional tectonic foliation and the isoclinal and recumbent folds of kilometric scale (Matte, 1968). Given the dimensions of this zone of the Variscan basement, the tectonic fabric presents different aspects due to different metamorphic imprints, but has structural features in common all along its extension: the fabric is parallel to lithological boundaries in the limbs of the kilometric recumbent folds and it shows stretching, defined by quartz and deformed preorogenic objects, parallel to the structural trend of the orogen defined by major boundaries (Matte, 1968; Díez Balda, 1986; Dias and Ribeiro, 1994, 1995; Bastida et al., 1993). The axes of the kilometric and minor recumbent folds are also parallel to the orogen and therefore to the D1 stretching lineation. Whether stretching is previous to folding or folding has stretching parallel to fold axes has not been studied in this context of indentation tectonics; however, folds with stretching parallel to fold axes have been reproduced experimentally with analogue models (Grujic and Mancktelow, 1995) and are consistent with the tectonic framework seen before. In this context tectonic transport, approximately east-west, is deduced from asymmetry of recumbent folds (Matte, 1968) and orientation of stretching lineation in thrust planes (D2).

Moving to the outer part of the orogen, in the West Asturian Leonese zone (Fig. 1), deformation occurs at progressively higher crustal levels, although it is still recognized as ductile in extensive domains of this slate belt. Here, fold axes of kilometric recumbent folds are still oriented parallel to the orogen, but stretching lineation does not have the same orientation and is found perpendicular to fold axes (Marcos, 1973; Marcos et al., 1980; Martínez Catalán, 1985), indicating that rheological contrasts of layers being folded have already changed. The orientation of shortening and tectonic transport is again in this case highly transverse to the belt and it is inferred from asymmetry of fold profiles.

In summary, very similar structural features have been described in relation to the first deformation phase in the Variscan Belt of northwestern Iberia. They are geometrically comparable to what has been described until now in relation to internal deformation of the Malpica-Tui unit: stretching lineations and axes of kilometric recumbent folds parallel to the arc. Tectonic trans-

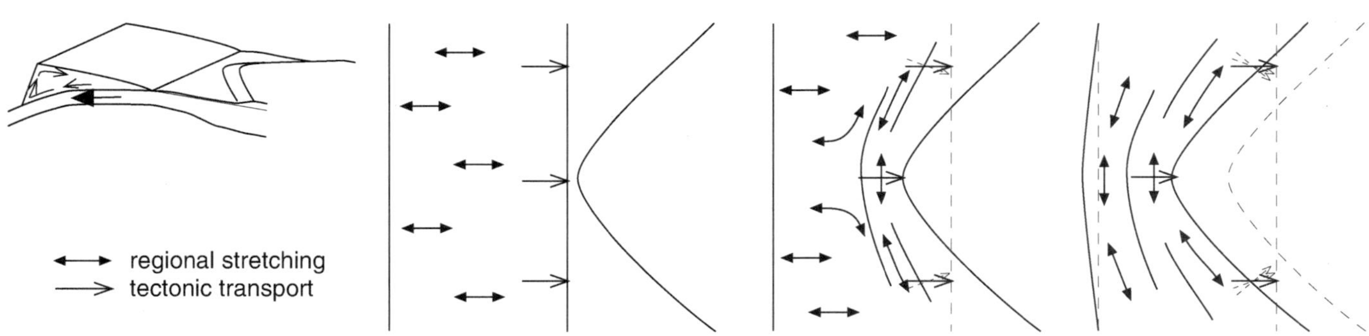

Figure 9. Sketch illustrating ideal development and accumulation of finite deformation in regional lineations and likely orientation of emplacement directions in discrete shear zones (thrusts) in geodynamic settings of indentation during collision. This requires indentor (Iberia) and buttress (Laurentia) to be stiff and unconstrained margins to sides (drawings based on Ratsbacher et al., 1991). In such tectonic setting, gradient of pressure and strain is developed from hinge of arc to outer parts; finite stretching in widespread deformation would follow flow lines of this gradient.

port transverse to the orogenic trend during D1 is therefore consistent at the scale of the section of the Variscan Belt studied.

CONCLUSION

In order to define the type of deformation that affected high-*P*–low- to intermediate-*T* rocks and its evolution in time during their exhumation and emplacement, a general sequence of structures recorded in Malpica-Tui unit rocks, divided in two stages, was described herein. Despite this artificial subdivision, deformation evolves progressively from one to the other and the structures show a decreasing grade of ductility with time as the rocks ascend through the crust. Perhaps the most relevant aspect of the ascent of these initially high pressure rocks is the style of deformation during the first stage. The flow regime during this first deformation stage is inferred to be not restricted to plane strain; there is a strong finite stretching orientation parallel to the orogenic belt that is not parallel to the direction of exhumation inferred from the regional setting and is supported by some of the features of the microstructural record. This flow regime would not only characterize the internal deformation of Malpica-Tui unit but of most nappes below the suture of the Variscan orogen in at least its northwestern section. The main tectonic implication of such a type of flow in rocks is that regional lineations (including pure stretching ones) cannot be used as kinematic indicators of tectonic transport in this particular tectonic setting. This is not a new concept in structural geology (b lineations of Sander, 1950), or in the Variscan Belt (Burg et al., 1987; Ribeiro et al., 1990; Dias and Ribeiro, 1994, 1995), and it has been described to occur in other deep sections of collisional orogens, e.g., the Caledonides (Gilotti and Hull, 1993).

The subsequent deformation phases show homoaxiality with the earlier tectonic evolution (an east-west oriented compression) and indicate a general emplacement of nappes and thrust sheets toward the foreland, roughly to the east, as described in other allochthonous units in northwestern Iberia (Burg et al., 1987; Martínez Catalán et al., 1996; Díaz García et al., 1999; Marcos and Farias, 1999).

The data presented here and the sequence of structures deduced from their geometrical relationships put some constraints on the processes of ascent of deeply buried upper crustal rocks in relation with collisional orogens and may be of great help when reproducing these processes in experiments that usually do not take into account any of the features of the internal deformation during the exhumation of high-*P* nappes.

ACKNOWLEDGMENTS

This study was supported by research projects DGE92-PB1022 and DGE95-PB1052 from the Spanish Dirección General de Investigación Científica y Técnica (DGICYT) and by a PhD grant to Llana-Fúnez from the Spanish Ministry of Education. We thank D. Brown and P. Floor for their helpful reviews of a previous version of this paper.

REFERENCES CITED

Ábalos, B., Gil Ibarguchi, J.I., and Eguíluz, L., 1991, Cadomian subduction/collision and Variscan transpression in the Badajoz-Córdoba shear belt, southwest Spain: Tectonophysics, v. 199, p. 51–72.

Alonso, J.L., and González, J.C., 1982, Geological Map of Spain (Sisargas-Carballo, sheet 20–44): Plan MAGNA Instituto Geo-Minero de España (IGME), scale 1:50,000, 1 sheet and report, p. 1–61.

Arenas, R., Rubio Pascual, F.J., Díaz García, F., and Martínez Catalán, J.R., 1995, High-pressure micro-inclusions and development of an inverted metamorphic gradient in the Santiago Schists (Ordenes Complex, NW Iberian Massif, Spain): Evidence of subduction and syncollisional decompression: Journal of Metamorphic Geology, v. 13, p. 141–164.

Audren, C., and Triboulet, C., 1986, Métamorphisme et déformation dans la ceinture de haute pression de L'Ile de Groix, Bretagne Méridionale: Bulletin Société Géologique du Minéralogie Bretagne, v. (C) 16, 1, p. 61–70.

Audren, C., and Triboulet, C., 1989, Pressure-temperature-time-deformation paths in metamorphic rocks and tectonic processes, as exemplified by the Variscan orogeny in South Brittany, France, *in* Daly, J.S., Cliff, R.A., and Yardley, B.W.D., eds., Evolution of metamorphic belts: London, Geological Society of London Special Publication 43, p. 441–446.

Azor, A., González-Lodeiro, F., and Simancas, F., 1994, Tectonic evolution of the boundary between the Central Iberian and Ossa-Morena zones (Variscan Belt, southwest Spain): Tectonics, v. 13, p. 45–61.

Badham, J.P.N., 1982, Strike-slip orogens—An explanation for the Hercynides: Journal of the Geological Society of London, v. 139, p. 493–504.

Badham, J.P.N., and Halls, C., 1975, Microplate tectonics, oblique collisions, and evolution of the Hercynian orogenic systems: Geology, v. 3, p. 373–376.

Ballèvre, M., Marchand, J., Godard, G., Goujou, J.-C., and Wyns, R., 1994, Eo-Hercynian events in the Armorican Massif, *in* Keppie, J.D., ed., Pre-Mesozoic geology in France and related areas: Berlin, Springer-Verlag, p. 183–194.

Bastida, F., Marcos, A., Marquínez, J., Martínez Catalán, J.R., Pérez-Estaún, A., and Pulgar, J.A., 1984, Geological Map of Spain (La Coruña, sheet 1): Instituto Geo-Minero de España (IGME), scale 1:200,000, 1 sheet and report, p. 1–155.

Bastida, F., Aller, J., and Fernández Viejo, G., 1993, The structure of the Ollo de Sapo antiform in the Cantabrian coast (NW Spain): Revista de la Sociedad Geológica de España, v. 6, p. 93–103.

Bellido Mulas, F., Brandle, J.L., Lasala, M., and Reyes, J., 1992, Consideraciones petrológicas y cronológicas sobre las rocas graníticas hercínicas de Galicia: Cuadernos del Laboratorio Xeolóxico de Laxe, v. 17, p. 241–261.

Brun, J.-P., and Burg, J.-P., 1982, Combined thrusting and wrenching in the Ibero-Armorican arc: A corner effect during continental collision: Earth and Planetary Science Letters, v. 61, p. 319–332.

Burchfield, B.C., Zhiliang, C., Hodges, K.V., Yuping, L., Royden, L.H., Changrong, D., and Jiene, X., 1992, The South Tibetan Detachment System, Himalayan Orogen: Extension contemporaneous with and parallel to shortening in a collisional mountain belt: Geological Society of America Special Paper 269, p. 1–41.

Burg, J.-P., Bale, P., Brun, J.-P., and Girardeau, J., 1987, Stretching lineation and transport direction in the Ibero-Armorican arc during the siluro-devonian collision: Geodinamica Acta, v. 1, p. 71–87.

Burg, J.-P., van den Driessche, J., and Brun, J.-P., 1994, Syn- to post-thickening extension in the Variscan Belt of Western Europe: Modes and structural consequences: Géologie de la France, v. 3, p. 33–51.

Carpenter, M.S.N., and Civetta, L., 1976, Hercynian high pressure/low metamorphism temperature in the Ile de Groix Blueschists: Nature, v. 262, p. 276–277.

Chemenda, A.I., Mattauer, M., Malavielle, J., and Bokun, A.N., 1995, A mechanism for syn-collisional rock exhumation and associated normal faulting: Results from physical modelling: Earth and Planetary Science Letters, v. 132, p. 225–232.

Chemenda, A.I., Matte, P., and Sokolov, V., 1997, A model of Palaeozoic obduction and exhumation of high-pressure/low-temperature rocks in the southern Urals: Tectonophysics, v. 276, p. 217–227.

Cloos, M., 1982, Flow melanges: Numerical modeling and geologic constraints on their origin in the Franciscan subduction complex, California: Geological Society of America Bulletin, v. 93, p. 330–345.

Cloos, M., 1986, Blueschists in the Franciscan Complex of California: Petrotectonic constraints on uplift mechanisms, *in* Evans, B.W., and Brown, E.H., eds., Blueschists and Eclogites: Boulder, Colorado, Geological Society of America Memoir 164, p. 77–93.

Crespo-Blanc, A., and Orozco, M., 1991, The boundary between the Ossa-Morena and Southportuguese zones (Southern Iberian Massif): A major suture in the European Hercynian Chain: Geologische Rundschau, v. 80, p. 691–702.

Dallmeyer, R.D., Martínez Catalán, J.R., Arenas, R., Gil Ibarguchi, J.I., Gutiérrez Alonso, G., Farias, P., Bastida, F., and Aller, J., 1997, Diachronous Variscan tectonothermal activity in the NW Iberian Massif: Evidence from $^{40}Ar/^{39}Ar$ dating of regional fabrics: Tectonophysics, v. 277, p. 307–337.

den Tex, E., and Floor, P., 1967, A blastomylonitic and polymetamorphic "graben" in Western Galicia (NW Spain): Institut de Géologie de L'Université de Neuchâtel, Etages Tectoniques, p. 169–178

Dias, R., and Ribeiro, A., 1994, Constriction in a transpressive regime: An example in the Iberian branch of the Ibero-Armorican arc: Journal of Structural Geology, v. 16, p. 1543–1554.

Dias, R., and Ribeiro, A., 1995, The Ibero-Armorican arc: A collision effect against an irregular continent?: Tectonophysics, v. 246, p. 113–128.

Díaz García, F., Arenas, R., Martínez Catalán, J.R., González del Tánago, J., and Dunning, G.R., 1999, Tectonic evolution of the Careón Ophiolite (Northwest Spain): A remnant of oceanic lithosphere in the Variscan Belt: Journal of Geology, v. 107, p. 587–605.

Díez Balda, M.A., 1986, El Complejo Esquisto-Grauváquico, las series paleozoicas y la estructura hercínica al Sur de Salamanca: Ediciones Univ. de Salamanca, Acta Salmanticensia, v. 52, p. 1–162.

Díez Balda, M.A., Martínez Catalán, J.R., and Ayarza Arribas, M.P., 1995, Syn-collisional extensional collapse pararell to the orogenic trend in a domain of steep tectonics: The South of Salamanca detachment zone (Central Iberian zone, Spain): Journal of Structural Geology, v. 17, 2, p. 163–182.

Duchene, S., Lardeaux, J.-M., and Albarede, F., 1997, Exhumation of eclogites: Insights from depth-time path analysis: Tectonophysics, v. 280, p. 125–140.

Ernst, W.G., 1988, Tectonic history of subduction zones inferred from retrograde blueschist *P-T* paths: Geology, v. 16, p. 1081–1084.

Escher, A., and Beaumont, C., 1997, Formation, burial, and exhumation of basement nappes at crustal scale: A geometric model based on the Western Swiss-Italian Alps: Journal of Structural Geology, v. 19, p. 955–974.

Escuder Viruete, J., Arenas, R., and Martínez Catalán, J.R., 1994, Tectonothermal evolution associated with Variscan crustal extension in the Tormes Gneiss Dome (NW Salamanca, Iberian Massif, Spain): Tectonophysics, v. 238, p. 117–138.

Floor, P., 1966, Petrology of an aegirine-riebeckite gneiss-bearing part of the Hesperian Massif: The Galiñeiro and surrounding areas, Vigo, Spain: Leidse Geologische Mededelingen, v. 36, p. 1–204.

Fombella Blanco, M.A., 1984, Age palynologique du Blastomilonitic Graben, zone Occidentale de la Galice: Revue Micropaléontologie, v. 27, no. 2, p. 113–117.

Fonseca, P., and Ribeiro, A., 1993, Tectonics of the Beja-Acebuches ophiolite: A major suture in the Iberian Variscan Foldbelt: Geologische Rundschau, v. 82, p. 440–447.

Fonseca, P., Munhá, J., Pedro, J., Rosas, F., Moita, P., Araújo, A., and Leal, N., 1999, Variscan ophiolites and high-pressure metamorphism in southern Iberia: Ofioliti, v. 24, no. 2, p. 259–268.

Fonseca, P.E., 1995, Estudo da Sutura Varisca no SW Ibérico nas Regiões de Serpa-Beja-Torrão e Alvito-Viana do Alentejo [Ph.D. Thesis]: Lisbon, Portugal, GeoFCUL, University of Lisboa, 325 p.

Gallastegui, G., 1993, Petrología del macizo granodiorítico de Baio-Vigo (Pontevedra, España) [Ph.D. Thesis]: Oviedo, Spain, Universidad de Oviedo, 356 p.

Gebauer, D., Schertl, H.-P., Brix, M., and Schreyer, W., 1997, 35 Ma old ultrahigh-pressure metamorphism and evidence for very rapid exhumation in the Dora Maira Massif, Western Alps: Lithos, v. 41, p. 5–24.

Gil Ibarguchi, J.I., 1995, Petrology of jadeite metagranite and associated orthogneiss from the Malpica-Tuy allochthon (Northwest Spain): European Journal of Mineralogy, v. 7, p. 403–415.

Gil Ibarguchi, J.I., and Dallmeyer, R.D., 1991, Hercynian blueschist metamorphism in North Portugal: Tectonothermal implications: Journal of Metamorphic Geology, v. 9, p. 539–549.

Gil Ibarguchi, J.I., and Ortega, E., 1985, Petrology, structure and geotectonic implications of glaucophan-bearing eclogites and related rocks from the Malpica-Tui unit, Galicia, Northwest Spain: Chemical Geology, v. 50, p. 145–162.

Gilotti, J.A., and Hull, J.M., 1993, Kinematic stratification in the hinterland of the central Scandinavian Caledonides: Journal of Structural Geology, v. 15, p. 629–646.

González-Lodeiro, F., Hernández Urroz, J., Martínez Catalán, J.R., Naval Balbín, A., Ortega Gironés, E., and de Pablo Maciá, J.G., 1984, Geological map of Spain (Santiago de Compostela, sheet 7): Instituto Geo-Minero de España (IGME), scale 1:200,000, 1 sheet and report, p. 1–99.

Grujic, D., and Mancktelow, N.S., 1995, Folds with axes parallel to the extension direction: An experimental study: Journal of Structural Geology, v. 17, p. 279–291.

Grujic, D., Casey, M., Davidson, C., Hollister, L.S., Kündig, R., Pavlis, T., and Schmid, S., 1996, Ductile extrusion of the Higher Himalayan crystalline in Bhutan: Evidence from quartz microfabrics: Tectonophysics, v. 260, p. 21–43.

Guiraud, M., Burg, J.-P., and Powell, R., 1987, Evidence for a Variscan suture zone in the Vendée, France: A petrological study of blueschist facies rocks from Bois de Cené: Journal of Metamorphic Geology, v. 5, p. 225–237.

Iglesias Ponce de León, M., and Ribeiro, A., 1981, Zones de cisaillement ductile dans l'arc ibéro-armoricain: Comunicações Serviços Geologicos Portugal, v. t 67, 1, p. 85–87.

Julivert, M., and Arboleya, M.L., 1984, A geometrical and kinematical approach to the nappe structure in an arcuate fold belt: The Cantabrian nappes (Hercynian chain, NW Spain): Journal of Structural Geology, v. 6, p. 499–519.

Julivert, M., Fontboté, J.M., Ribeiro, A., and Conde, L., 1972, Tectonic map of Iberian Peninsula and Balear Islands, Instituto Geo-Minero de España (IGME), scale 1:1,000,000, 1 sheet.

Llana-Fúnez, S., 1997, Estructura del Borde Oriental de la Unidad de Malpica-Tuy en la transversal Molinos de Ceán-Razo (Galicia, NO España) [abs.], *in* Pires, C.C., Gomes, M.E.P., and Coke C., eds., Vila Real, Portugal: Comunicações XIV Reunião de Geología do Oeste Peninsular, p. 283–288.

Llana-Fúnez, S., 2001, La estructura de la unidad de Malpica-Tui (Cordillera Varisca en Iberia): Madrid, Spain, Instituto Geológico y Minero de España, Serie Tesis Doctorales, v. 1/2001, p. 1–295, plus color map.

Llana-Fúnez, S., 2002, Quartz *c*-axis texture mapping of a Variscan regional foliation (Malpica-Tui Unit, NW Spain): Journal of Structural Geology, v. 24, p. 1299–1312.

Llana-Fúnez, S., and Marcos, A., 2001, The Malpica-Lamego line: A major crustal-scale shear zone in the Variscan Belt of Iberia: Journal of Structural Geology, v. 23, p. 1015–1030.

Mancktelow, N.S., 1985, The Simplon Line: A major displacement zone in the western Lepontine Alps: Eclogae Geologicae Helvetiae, v. 78, p. 73–96.

Mancktelow, N.S., 1993, Tectonic overpressure in competent mafic layers and the development of isolated eclogites: Journal of Metamorphic Geology, v. 11, p. 801–812.

Marcos, A., 1973, Las Series del Paleozoico Inferior y la estructura herciniana del occidente de Asturias (NW de España): Universidad de Oviedo, Trabajos de Geología, v. 6, p. 1–113.

Marcos, A., and Farias, P., 1999, La estructura de las láminas inferiores del Complejo de Cabo Ortegal y su autóctono relativo (Galicia, NW España): Universidad de Oviedo, Trabajos de Geología, v. 21, p. 201–220.

Marcos, A., Pérez-Estaún, A., and Martínez, F.J., 1980, Geological map of Spain (Vegadeo, sheet 25): Plan MAGNA Instituto Geo-Minero de España (IGME), scale 1:50,000, 1 sheet and report, p. 1–25.

Martínez Catalán, J.R., 1985, Estratigrafía y estructura del Domo de Lugo: Corpus Geologicum Gallaeciae, v. 2, p. 1–291.

Martínez Catalán, J.R., 1990, A non-cylindrical model for the northwest Iberian allochthonous terranes and their equivalents in the Hercynian belt of Western Europe: Tectonophysics, v. 179, p. 253–272.

Martínez Catalán, J.R., Arenas, R., Díaz García, F., and Abati, J., 1997, Variscan accretionary complex of northwest Iberia: Terrane correlation and succession of tectonothermal events: Geology, v. 25, p. 1103–1106.

Martínez Catalán, J.R., Arenas, R., Díaz García, F., and Abati, J., 1999, Allochthonous units in the Variscan Belt of NW Iberia (Terranes and Accretionary History), *in* Sinha, A.K., ed., Basement tectonics 13: Dordrecht, Kluwer Academic Publishers, p. 65–84.

Martínez Catalán, J.R., Arenas, R., Díaz García, F., Rubio Pascual, F.J., Abati, J., and Marquínez, J., 1996, Variscan exhumation of a subducted Paleozoic continental margin: The basal units of the Ordenes Complex, Galicia, NW Spain: Tectonics, v. 15, p. 106–121.

Matte, P., and Burg, J.-P., 1981, Sutures, thrusts and nappes in the Variscan arc of western Europe: Plate tectonic implications, *in* McClay, K.R., and Price, N.J., eds., Thrust and nappe tectonics: London, Geological Society of London Special Publication 9, p. 353–358.

Matte, Ph., 1968, La structure de la virgation hercynienne de Galice (Espagne): Geologie Alpine, v. 44, p. 155–280.

Matte, Ph., 1986, Tectonics and plate tectonics model for the Variscan Belt of Europe: Tectonophysics, v. 126, p. 329–374.

Matte, Ph., 1991, Accretionary history and crustal evolution of the Variscan Belt in Western Europe: Tectonophysics, v. 196, p. 309–337.

Matte, Ph., 1998, Continental subduction and exhumation of high-pressure rocks in Paleozoic orogenic belts: Uralides and Variscides: GFF Journal of the Geological Society of Sweden, v. 120, p. 209–222.

McClay, K.R., and Ellis, P.G., 1987, Geometries of extensional fault systems developed in model experiments: Geology, v. 15, p. 341–344.

McDougall, I., and Harrison, T.M., 1999, Geochronology and Thermochronology by the $^{40}Ar/^{39}Ar$ Method: Oxford, Oxford University Press, 269 p.

Merle, O., Cobbold, P.R., and Schmid, S., 1989, Tertiary kinematics in the Lepontine dome, *in* Coward, M.P., Dietrich, D., and Park, R.G., eds., Alpine tectonics: London, Geological Society of London Special Publication 45, p. 113–134.

Michard, A., Chopin, C., and Henry, C., 1993, Compression versus extension in the exhumation of the Dora-Maira coesite-bearing unit, Western Alps, Italy: Tectonophysics, v. 221, p. 173–193.

Munhá, J., Ribeiro, A., and Ribeiro, M.L., 1984, Blueschists in the Iberian Variscan Chain (Tras-os-Montes: NE Portugal): Comunicaçoes Serviços Geológicos Portugal, v. 70, p. 31–53.

Olivet, J.L., 1996, Kinematics of the Iberian Plate: Bulletin des Centres de Recherches Exploration-Production ELF Aquitaine, v. 20, no. 1, p. 131–195.

Ortega, E., and Gil Ibarguchi, J.I., 1983, La unidad de Malpica-Tui ("Complejo Antiguo"–"Fosa Blastomilonítica"), *in* Libro Jubilar J.M. Ríos, Geología de España: Madrid, Instituto Geológico Minero de España, v. 1, p. 430–444.

Parga Pondal, I., 1960, Observación, interpretación, y problemas geológicos de Galicia: Notas y Comunicaciones del Instituto Geo-Minero de España (IGME), v. 59, p. 333–358.

Passchier, C.W., 1997, The fabric attractor: Journal of Structural Geology, v. 19, p. 113–127.

Passchier, C.W., 1998, Monoclinic model shear zones: Journal of Structural Geology, v. 20, p. 1121–1137.

Pedro, J., and Munhá, J., 1997a, Caracterizaçao do metamorfismo de alta pressão na área de Safira (Montemor-o-Novo, Zona de Ossa Morena) durante a Orogenia Varisca, [abs.], *in* Pires, C.C., Gomes, M.E.P., and Coke, C., eds., Vila Real, Portugal: Comunicações XIV Reunião de Geología do Oeste peninsular, p. 173.

Pedro, J., and Munhá, J., 1997b, Metamofismo de alta pressão durante a Orogenica Varisca em gnaisses félsicos ortoderivados na área de Safira (Montemor-o-Novo), Zona de Ossa Morena, [abs.], *in* Pires, C.C., Gomes, M.E.P., and Coke, C., eds., Vila Real, Portugal: Comunicações XIV Reunião de Geología do Oeste peninsular, p. 305.

Pérez-Estaún, A., Bastida, F., Alonso, J.L., Marquínez, J., Aller, J., Alvarez-Marrón, J., Marcos, A., and Pulgar, J.A., 1988, A thin-skinned tectonics model for an arcuate fold and thrust belt: The Cantabrian zone: Tectonics, v. 7, p. 517–538.

Pérez-Estaún, A., Martínez Catalán, J.R., and Bastida, F., 1991, Crustal thickening and deformation sequence in the footwall to the suture of the Variscan Belt of northwest Spain: Tectonophysics, v. 191, p. 243–253.

Pin, C., Ortega Cuesta, L.A., and Gil Ibarguchi, J.I., 1992, Mantle-derived, early Paleozoic A-type metagranitoids from the NW Iberian massif: Nd isotope and trace-element constraints: Bulletin de la Société Géologique de France, v. 163, p. 483–494.

Platt, J.P., 1986, Dynamics of orogenic wedges and the uplift of high-pressure metamorphic rocks: Geological Society of America Bulletin, v. 97, p. 1037–1053.

Ramsay, J.G., 1967, Folding and fracturing of rocks: New York, McGraw-Hill, 568 p.

Ramsay, J.G., and Graham, R.H., 1970, Strain variation in shear belts: Canadian Journal of Earth Sciences, v. 7, p. 786–813.

Ratsbacher, L., Merle, O., Davy, P., and Cobbold, P., 1991, Lateral extrusion in the Eastern Alps, Part 1: Boundary conditions and experiments scaled for gravity: Tectonics, v. 10, p. 245–256.

Ribeiro, A., Pereira, E., and Dias, R., 1990, Structure in the Northwest of the Iberian Peninsula (within the Allochthonous Sequences in the Centro-Iberian zone), *in* Dallmeyer, R.D., and Martínez García, E., eds., Pre-Mesozoic geology of Iberia: Berlin, Springer-Verlag, p. 220–236.

Ribeiro, M.L., and Floor, P., 1987, Magmatismo peralcalino no Maciço Hesperico: Sua distribuçao e significado geodinamico, *in* Bea, F., Carnicero, A., Gonzalo, J.C., López Plaza, M., and Rodríguez Alonso, M., eds., Geología de los granitoides y rocas asociadas del Macizo Hespérico: Madrid, Editorial Rueda, p. 211–221.

Ries, A.C., and Shackleton, R.M., 1971, Catazonal complexes of northwest Spain and north Portugal: Remnants of a Hercynian thrust plate: Nature, v. 234, no. 47, p. 65–68.

Ries, A.C., Richardson, A., and Shackleton, R.M., 1980, Rotation of the Iberian arc: Palaeomagnetic results from North Spain: Earth and Planetary Science Letters, v. 50, p. 301–310.

Rodríguez Aller, J., Cosca, M., and Gil Ibarguchi, J.I., 1997a, $^{40}Ar/^{39}Ar$ dating of Hercynian metamorphism in the HP Malpica-Tuy Allochthon and LP Parautochthon, Iberian Massif, NW Spain: Terra Nova, v. 9, no. 1, p. 497.

Rodríguez Aller, J., Cosca, M., Gil Ibarguchi, J.I., and Dallmeyer, R.D., 1997b, Eo-hercynian HP metamorphism of a subducted continental crust (Malpica-Tui allochthon, NW Spain): New petrological and age constraints [abs.]: Eclogite Conference Abstracts Volume, Ascona Italy.

Sander, B., 1950, Einführung in die Gefügekunde der geologischen Körper: Wien, Springer.

Santos Zalduegui, J.F., Schärer, U., and Gil Ibarguchi, J.I., 1995, Isotope constraints on the age and origin of magmatism and metamorphism in the Malpica-Tuy allochthon, Galicia, NW Spain: Chemical Geology, v. 121, p. 91–103.

Serrano Pinto, M., Casquet, C., Ibarrola, E., Corretgé, L.G., and Portugal Ferreira, M., 1987, Síntesis geocronológica dos granitoides do Maciço Hespérico, *in* Bea, F., Carnicero, A., Gonzalo, J.C., López Plaza, M., and Rodríguez Alonso, M., eds., Geología de los granitoides y rocas asociadas del Macizo Hespérico: Madrid, Editorial Rueda, p. 69–86.

Sibson, R.H., 1977, Fault rocks and fault mechanisms: Journal of the Geological Society of London, v. 133, p. 191–213.

Srivastava, P., and Mitra, G., 1996, Deformation mechanisms and inverted thermal profile in the North Almora Thrust mylonite zone, Kumaon Lesser Himalaya, India: Journal of Structural Geology, v. 18, p. 27–39.

Stünitz, H., 1991, Folding and shear deformation in quarzites, inferred from crystallographic preferred orientation and shape fabrics: Journal of Structural Geology, v. 13, p. 71–86.

Tapponier, P., and Molnar, P., 1976, Slip-line field theory and large-scale continental tectonics: Nature, v. 264, p. 319–324.

Tapponier, P., Peltzer, G., and Armijo, R., 1986, On the mechanics of the collision between India and Asia, *in* Coward, M.P., and Ries, A.C., eds., Collision tectonics: Boulder, Colorado, Geological Society Special Publication 19, p. 115–157.

Thompson, A.B., Schulmann, K., and Jezek, J., 1997, Thermal evolution and exhumation in obliquely convergent (transpressive) orogens: Tectonophysics, v. 280, p. 171–184.

van Calsteren, P.W.C., Boelrijk, N.A.I.M., Hebeda, E.H., Priem, H.N.A., den Tex, E., Verdurmen, E.A.T.H., and Verschure, R.H., 1979, Isotopic dating of older elements (including the Cabo Ortegal mafic-ultramafic complex) in the hercynian orogen of NW Spain: Manifestations of a presumed Early Paleozoic Mantle-plume: Chemical Geology, v. 24, p. 35–56.

van der Wegen, G., 1978, Garnet-bearing metabasites from the Blastomylonitic Graben, western Galicia, Spain: Scripta Geologica, v. 45, 95 p.

Manuscript Accepted by the Society February 8, 2002

Printed in the U.S.A.

Geological Society of America
Special Paper 364
2002

Tectonic framework of the Cabo Ortegal Complex: A slab of lower crust exhumed in the Variscan orogen (northwestern Iberian Peninsula)

A. Marcos*
P. Farias
Departamento de Geología, Universidad de Oviedo, Arias de Velasco s/n, 33005 Oviedo, Spain
G. Galán
Departamento de Geología, Universidad Autónoma de Barcelona,
Edificio C (Sur), 08193 Bellaterra, Barcelona, Spain
F.J. Fernández
S. Llana-Fúnez*
Departamento de Geología, Universidad de Oviedo, Arias de Velasco s/n, 33005 Oviedo, Spain

ABSTRACT

The Cabo Ortegal Complex is a composite allochthonous terrane that was thrust onto the western edge of Gondwana during the Variscan orogeny. It is formed of two main tectonic units: the Upper Tectonic unit, comprising rocks affected by high-pressure (*P*)–high-temperature (*T*) metamorphism, and the Lower Tectonic unit, which represents the resulting suture of the Variscan collision. The suture preserves remnants of strongly deformed and metamorphosed ophiolitic rocks overriding the parautochthon, and the lower Paleozoic sequence of the Ollo de Sapo antiform, regarded as the autochthonous sequence of the Iberian plate. The Upper Tectonic unit is formed by layered ultramafic, mafic, and quartzo-felspathic rocks that were buried at levels in excess of 50 km (~1.56 GPa) before the Variscan collision in a convergent plate boundary within the Rheic ocean domain ca. 490–480 Ma (Early Ordovician). They have been interpreted either (1) as an earlier thinned continental crust, underlain by a lithospheric mantle and oceanic spreading, or (2) as independent terranes, formed in different geodynamic settings (island arc, oceanic). Most structures observed in these rocks are ductile and are associated to their exhumation process. It started with the development of a persistent horizontal foliation in granulite facies conditions, which equilibrated in amphibolite facies conditions ca. 385 Ma, and ended in higher crustal levels with the progressive development of noncoaxial structures, such as east-verging asymmetrical isoclinal folds and thrusts, leading to the emplacement of the Upper Tectonic unit over the Lower Tectonic unit ca. 365 Ma (Late Devonian).

*E-mail, Marcos: marcos@geol.uniovi.es. Current address, Llana-Fúnez: Rock Deformation Laboratory, Department of Earth Sciences, University of Manchester, Oxford Road, Manchester M13 9PL, UK.

Marcos, A., Farias, P., Galán, G., Fernández F.J., and Llana-Fúnez, S., 2002, Tectonic framework of the Cabo Ortegal Complex: A slab of lower crust exhumed in the Variscan orogen (northwestern Iberian Peninsula), *in* Martínez Catalán, J.R., Hatcher, R.D., Jr., Arenas, R., and Díaz García, F., eds., Variscan-Appalachian dynamics: The building of the late Paleozoic basement: Boulder, Colorado, Geological Society of America Special Paper 364, p. 143–162.

INTRODUCTION

The hinterland of the Variscan orogen in the Iberian Peninsula preserves several allochthonous complexes as klippen, including exotic ultrabasic and basic rocks, orthogneisses, and various metasediments interpreted by some as indicative of a polymetamorphic and polyorogenic evolution (den Tex, 1981; Arenas et al., 1986; Ribeiro et al., 1990; Fernández Suárez et al., 2000) (Fig. 1; Plate 1). The Cabo Ortegal Complex is the easternmost of these structures and defines the suture zone of the orogen. It comprises rock units with entirely different geodynamic settings, ages, and metamorphic and structural evolutions, providing valuable information on the tectonic evolution of the collisional processes.

In the hanging wall of the suture zone, a sequence of crystalline rocks records a metamorphic event of high-pressure (*P*) (~1.5 GPa minimum), high-temperature (*T*) conditions. This sequence is considered the relic of an earlier thinned lower crust (probably with a portion of lithospheric mantle) (Galán and Marcos, 1997), although some authors (Peucat et al., 1990) considered some of the mafic rocks related to a volcanic arc close to a continental margin. Deformation started in shear zones under eclogite facies conditions, but was reequilibrated during their exhumation under retrograde granulite and amphibolite facies conditions. The presence of such lower crustal slabs in relation with thickening processes in collisional orogens has been widely reported in other mountain belts (Windley, 1983; Windley and Tarney, 1986). Examples occur in the Kohistan island arc of the Himalayas (Coward et al., 1982; Coward and Windley, 1986; Bard, 1983), the Ivrea zone of the Italian Alps (Rutter et al., 1993), and the Variscan sutures of western-central Europe (Matte, 1983, 1998).

The origin of the complexes of northwestern Iberia has been a matter of debate in recent years. Some authors suggested an autochthonous "mantle plume" model for their origin (Keasberry et al., 1976; Calsteren, 1977, 1981; Calsteren and den Tex, 1978; Kuijper and Arps, 1983; Kuijper et al., 1985), whereas others consider them as allochthonous terranes emplaced on the Gondwana margin during the Variscan orogeny (e.g., Ribeiro et al., 1964; Anthonioz, 1970; Ries and Shackleton, 1971; Bayer and Matte, 1979; Iglesias et al., 1983; Bastida et al., 1984; Arenas et al., 1986; Matte, 1986; Pérez Estáun et al., 1991; Dallmeyer et al., 1997; Martínez Catalán et al., 1997, 1999). Although at present the allochthonous hypothesis is generally accepted, there is still controversy regarding the origin and metamorphic evolution of the different lithological and structural units.

There are two major groups of hypotheses about the metamorphic evolution of the high-*P*–high-*T* rocks of the Cabo Ortegal Complex: polyorogenic and single-cycle models. Polyorogenic metamorphic models suggest that the eclogite and granulite facies metamorphism occurred during Precambrian time (Vogel, 1967; Anthonioz, 1970; Engels, 1972; Marques et al., 1996), and a strong retrogression in amphibolite facies is related to pre-Variscan and/or Variscan events. Single-cycle metamorphic models propose the subduction or underthrusting of both oceanic (ophiolitic rocks) and continental crust and later obduction of them during the Variscan orogeny (Gil Ibarguchi et al., 1990; Peucat et al., 1990; Arenas, 1991; Schäfer et al., 1993; Marcos and Galán, 1994; Àbalos et al., 1996; Santos Zalduegui et al., 1996; Matte, 1998; Galán and Marcos, 1999, 2000). The eclogite and quasisimultaneous granulite facies metamorphism is related to the subduction or underthrusting process either ca. 480 Ma (Peucat et al., 1990) or ca. 395 Ma (Schäfer et al., 1993; Santos Zalduegui et al., 1996; Ordoñez Casado et al., 1996). Pre-Variscan and Variscan exhumation caused the retrograde metamorphism in amphibolite and greenschist facies. More recent U-Pb dating (Fernández Suárez et al., 2002) indicates that these rocks record a polyorogenic evolution with a high-*P*–high-*T* episode ca. 490–480 Ma and a younger event dated as ca. 390–385 Ma. According to this model, the lower crustal rocks of the Cabo Ortegal Complex would have been exhumed to the surface from >50 km depth (1.5 GPa) in two consecutive tectonic events, separated in time by ~100 m.y. The exhumation of deep crustal rocks (granulites formed in a thickened crust) during two unrelated cycles of tectonism is consistent with the model proposed by Ellis (1987).

Based on this model, the aims of this chapter are (1) to study the ductile deformation in the high-*P*–high-*T* rocks, where pervasive ductile deformation is retrogressive and related to the process of their exhumation; and (2) to assess the effects of the emplacement of the lower crustal section onto the Iberian plate during the Variscan orogeny. A progressive change in the emplacement direction is discussed with geodynamic models of the Variscan orogeny in the Iberian Belt.

GEOLOGIC SETTING

The Galicia-Trás-os-Montes zone (Farias et al., 1987) is the most internal zone of the Variscan Belt in the northwest of the Iberian Peninsula (Fig. 1). It was thrust over Paleozoic metasediments and porphyroids of the neighboring Central Iberian zone, which is represented in northwest Spain by the Ollo de Sapo anticlinorium, a narrow structure built by the interference between large recumbent and late Variscan open subvertical (upright) folds. Two domains have been distinguished within the Galicia-Tras os Montes zone: (1) the Schistose domain (also named parautochthon; Ribeiro et al., 1990), mainly composed of a metasedimentary sequence more than 4000 m thick and probably of early Paleozoic age; and (2) the Allochthonous complexes, tectonically emplaced over the parautochthon and composed of mafic-ultramafic and quartzo-feldspathic rocks with different origins (e.g., oceanic, ophiolitic, and continental). The Schistose domain has been interpreted as the most external sediments of the continental margin of Gondwana (Ribeiro et al., 1990; Martínez Catalán et al., 1997, 1999; Marcos and Farias, 1999).

The rocks forming the Allochthonous complexes are grouped in three major tectonostratigraphic units characterized by their different rock assemblages and metamorphic evolution;

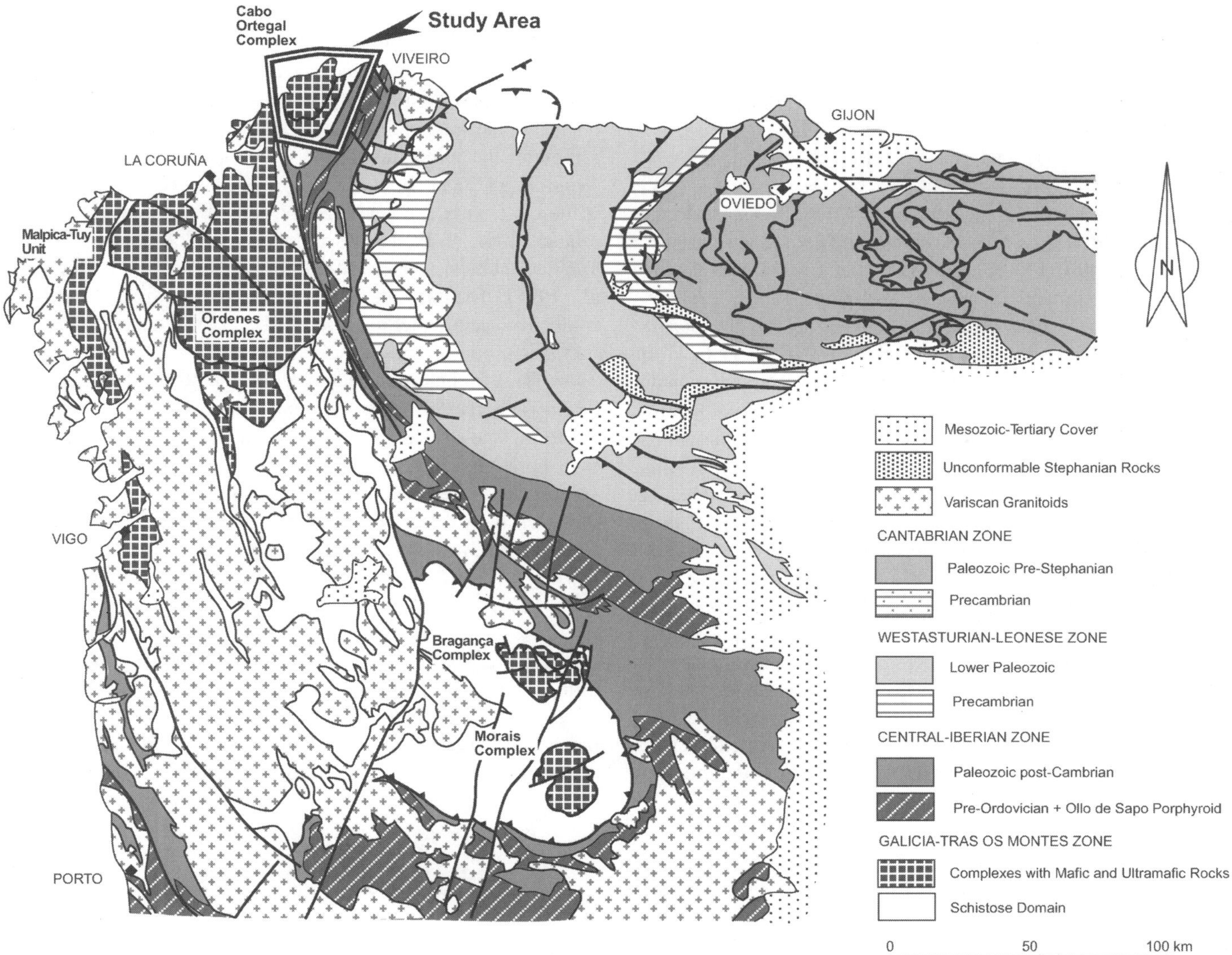

Figure 1. Geological map of northwestern Iberia, simplified after Julivert et al. (1974), Parga Pondal et al. (1982), and Farias et al. (1987). Various allochthonous units and main paleogeographic and structural ensembles of Iberian autochthonous units are shown.

those are, from bottom to top, the Basal, Ophiolitic, and Upper units (Arenas et al., 1986). These units have been correlated with the paleogeographic realms involved in the Variscan collision (Martínez Catalán et al., 1997, 1999). The Basal units are formed of orthogneisses, paragneisses, amphibolites, and schists. These rocks, related to the external continental margin of Gondwana, intruded during an Ordovician rifting event (Ribeiro and Floor, 1987; Martínez Catalán et al., 1996; Marcos and Farias, 1999), and later were involved in a westward subduction episode at the beginning of the Variscan orogeny. They record metamorphism under high-*P* and low- to intermediate-*T* conditions (Arenas et al., 1995; Martínez Catalán et al., 1996). The Ophiolitic units include partially mylonitized amphibolites of basaltic origin and a tectonic melange. They represent the suture zone formed by the closure of the Paleozoic Rheic ocean (Martínez Catalán et al., 1997, 1999). The Upper units are placed in the hanging wall to the suture as an exotic terrane and include high-*P*–high-*T* rocks and mainly intermediate-*P* metasediments, as a whole seen as the remnants of an earlier thinned lower crustal section.

All these units are found in the Cabo Ortegal Complex (Vogel, 1967); the closest crystalline klippe to the foreland belt (Fig. 1), but we have grouped them slightly differently for simplification. Thus, we distinguish the Upper Tectonic unit (containing the high-*P*–high-*T* rocks that formed the thinned lower crust) and the Lower Tectonic unit, which includes both the ophiolitic rocks and the basal units and represents the suture of the Variscan collisional belt in northwest Iberia. The thin Rio

Baio thrust sheet (regarded as parautochthonous) separates the Lower Tectonic unit from the underlying lower Paleozoic metasediments of the Ollo de Sapo antiform (Plate 1).

PARAAUTOCHTHON AND AUTOCHTHON OF THE CABO ORTEGAL COMPLEX

The Rio Baio thrust sheet forms part of the Schistose domain of the Galicia-Tras os Montes zone (Farias et al., 1987; Marcos and Farias, 1999). It is composed of a thick succession of siliciclastic metasediments (quartzites, feldspathic sandstones, graywackes, shales) with some interbedded layers of volcanic-sedimentary and volcanic rocks (rhyolites and dacites), all included within the Loiba and Queiroga Series by Marcos and Farias (1999) (Plate 1). This lithostratigraphy, of unknown age, has several differences with respect to the underlying autochthon.

The rocks of the Rio Baio thrust sheet form a monoclinal sequence with scarce folds, but are affected by a slaty cleavage developed under greenschist facies conditions. The basal thrust is parallel to the bedding in the hanging-wall rocks, and is a 50–120-m-thick zone with intense phyllonitization. The phyllonites at the base of the Rio Baio thrust sheet show a weak stretching lineation with a constant northeast orientation and the development of C′ shear bands, which indicate movement in the same direction, top to the northeast. However, the last tectonic pulses produced discontinuous structures in the phyllonites, i.e., centimeter-scale tectonic wedges, with opposite kinematics (top to the southeast). This late sense of movement is in agreement with the cutoff lines of Silurian rocks placed in the autochthon (Plate 1) (Marcos and Farias, 1999).

The Rio Baio thrust sheet is tectonically placed on the lower Paleozoic (Lower Ordovician to Silurian) metasedimentary sequence in the western limb of the Ollo de Sapo antiform, which represents the autochthon of the Cabo Ortegal Complex (Matte, 1968; Pérez Estaún et al., 1991) (Plate 1). Structures developed in these rocks under greenschist and amphibolite facies have been related to two distinct Variscan phases of deformation (Matte, 1968; Capdevila, 1968; Bastida et al., 1984, 1993; Réche et al., 1998). The first phase is represented by tight recumbent folds, from centimeter to kilometer scale, with horizontal north-northeast–south-southwest axes, vergent to the east (Matte, 1968). This folding is associated with axial planar slaty cleavage, which is the regional tectonic foliation in the western limb of the antiform. The structures are overprinted by straight, open folds, with horizontal or slightly north-dipping fold axes, which develop a crenulation cleavage parallel to the axial planes. The basal thrust of the Rio Baio thrust sheet cuts the first folds but is deformed by the later open folds (Marcos and Farias, 1999) (Fig. 2).

LOWER TECTONIC UNIT: THE SUTURE ZONE

The rocks of the Lower Tectonic unit that were initially interpreted as a sequence with ophiolitic affinities (Martínez García et al., 1975; Arce Duarte et al., 1977; Bayer and Matte, 1979; den Tex, 1981; Iglesias et al., 1983; Arenas and Peinado, 1984; Arenas et al., 1986; Arenas, 1988; Ribeiro et al., 1990) are var-

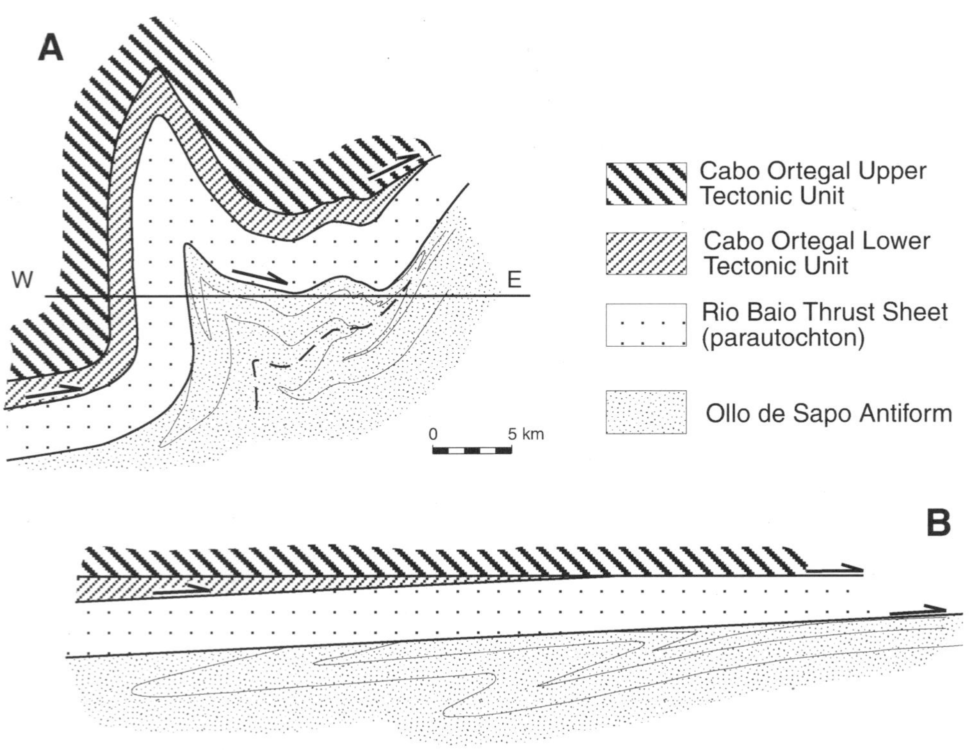

Figure 2. A: Schematic geological cross section of Cabo Ortegal Complex showing relation between allochthonous thrust sheets and their autochthonous sequence. B: Partial restoration of cross-section A (after Marcos and Farias, 1999).

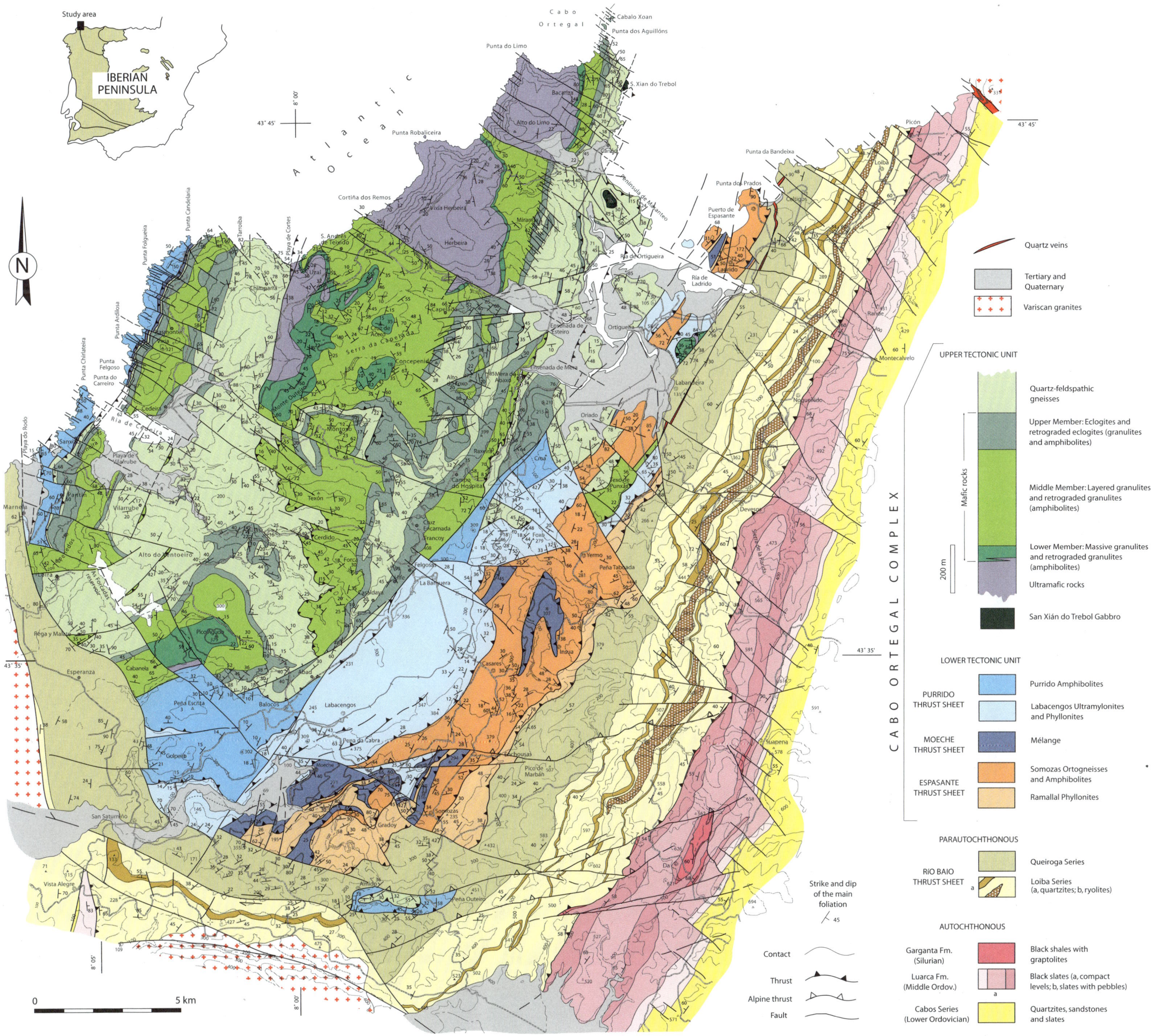

Plate 1. Geological map of Cabo Ortegal Complex (based on Marcos and Farias, 1999; Marcos et al., 2000).

ied lithological elements arranged heterogeneously (see geological maps in Fernández Pompa and Monteserín López, 1976; Arce Duarte et al., 1977; Arenas, 1988). However, detailed geological mapping by Marcos and Farias (1997, 1999) show that these rocks can be grouped in three superposed and separate thrust sheets: the Purrido, Moeche, and Espasante units (Plate 1; Fig. 3). They represent the suture zone of the Variscan orogen in the Cabo Ortegal Complex and separate two distinct terranes: the rocks from the Iberian plate below, and the high-*P* rocks of Upper Tectonic unit above.

Purrido thrust sheet

The ~1100-m-thick Purrido thrust sheet is composed of two formations, the Purrido amphibolites (Engels, 1972; Vogel, 1967; Arenas, 1988, 1991), and the Labacengos ultramylonites

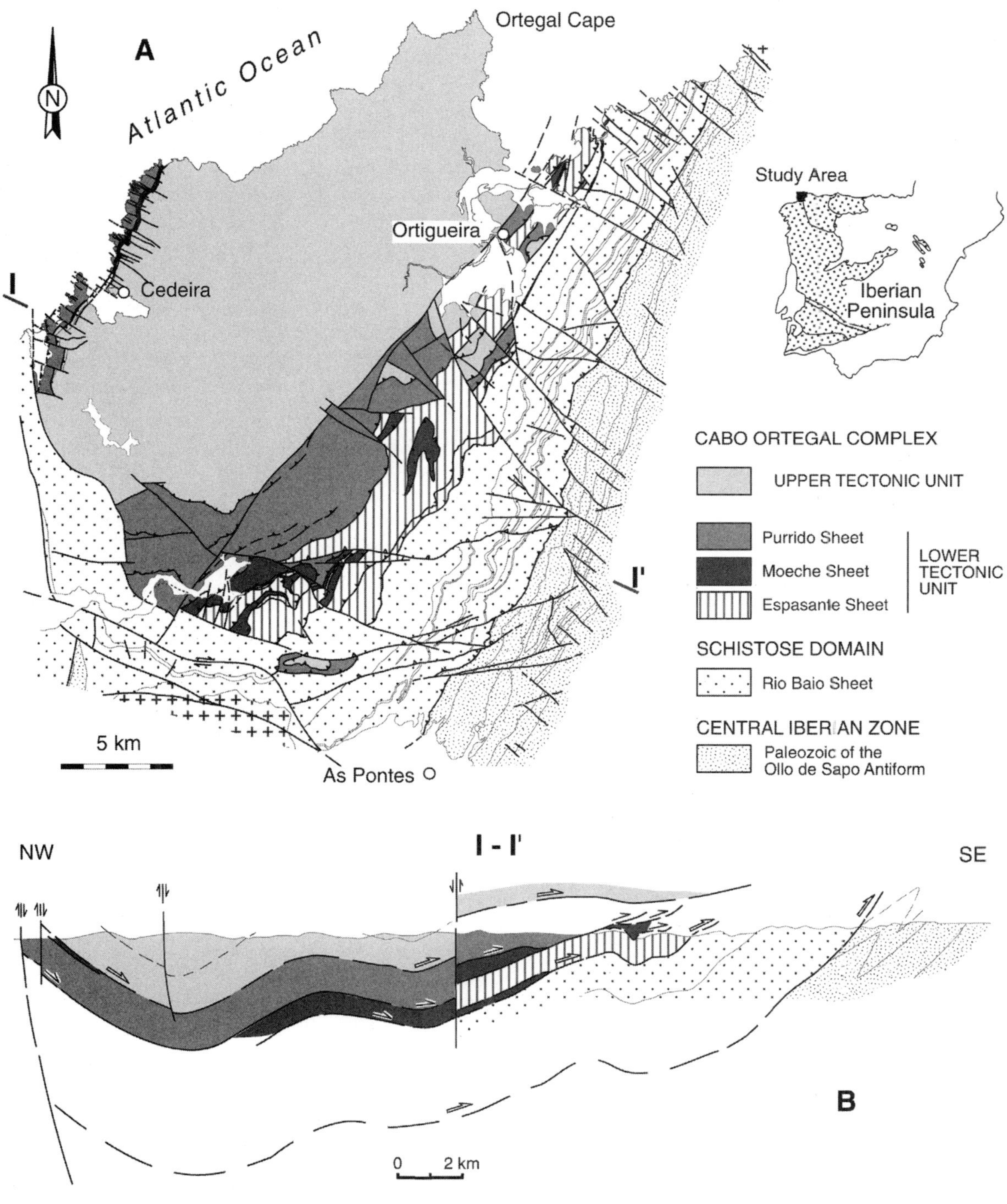

Figure 3. Structure and geological cross section through Lower Tectonic unit (after Marcos and Farias, 1999).

and phyllonites at the base (Marcos and Farias, 1997). The Purrido amphibolites (~300 m thick) form a homogeneous succession of green-grayish amphibolites that have a fine and discontinuous compositional banding defined by plagioclase. The amphibolites are composed of Hbl, Pl, and Ep-Czo, and Grt, Rt, and Ttn are accessory minerals (mineral symbols as in Kretz, 1983); they are probably the metamorphic equivalent of gabbroic rocks (Vogel, 1967). The rocks present a well-developed tectonic fabric defined by the shape-preferred orientation of hornblende, giving the rock fissility (Fig. 4A). The amphibolite facies metamorphism (500–600 °C) was dated by Peucat et al. (1990) as ca. 390 Ma, using $^{40}Ar/^{39}Ar$ in Hbl. The development of the tectonic fabric is probably of the same age.

The Labacengos ultramylonites and phyllonites are made of rocks with the field appearance of phyllites that are affected by a strong and penetrative tectonic foliation (Fig. 4, B and C). The rocks are composed of Qtz, Chl, Act, Ab, Ms, and Bt and derived tectonites from the extreme deformation of the Purrido amphibolites (Arenas, 1983, 1988). This deformation was dated by Dallmeyer et al. (1997) as ca. 364 Ma ($^{40}Ar/^{39}Ar$ whole-rock age).

The Purrido amphibolites have the chemical composition of olivine tholeiitic basalts, whereas the Labacengos ultramylonites and phyllonites have a quartz tholeiitic composition (Arenas, 1988), which points to some chemical changes during mylonitization. Available trace element compositions of these rocks do not discriminate between normal mid-ocean ridge basalt (N-MORB) and within-plate basalt; Arenas (1988) suggested an average enriched-MORB composition for them, although it is uncertain whether it is a primary or a secondary character.

Moeche thrust sheet

This tectonic unit forms a thin, <500-m-thick sheet made of scattered metric to decametric fragments of exotic rocks (e.g., marble, quartzite, basic rocks), but also of native rocks (serpentinized ultramafic rocks), all of them in a matrix of serpentinite, talc-schists, and chlorite-schists. Meer Mohr (1975) found coral fragments, crinoidal debris, and foraminifers not older than Middle Ordovician in the marbles.

All the rocks that form the Moeche thrust sheet are intensely deformed. The serpentinites, talc-schists, and chlorite-schists have ultramylonitic and phyllonitic textures, indicating that deformation took place under greenschist facies conditions. This deformation obliterated previous structures and metamorphism. The structural features of the rocks suggest that they are a tectonic melange (in the sense of, e.g., Hsü, 1968; Raymond, 1984; Cowan, 1985; Rast and Wright Horton, 1989).

Espasante thrust sheet

This 900-m-thick sheet is composed of acid and basic igneous rocks that have been grouped in one lithological unit, named the Somozas orthogneisses and amphibolites (Marcos and Farias, 1997). The lower part of this formation, located close to the basal thrust, is extremely deformed and represented by ultramylonites and phyllonites named the Ramallal phyllonites by Marcos and Farias (1997). The Somozas orthogneisses and amphibolites are ~400 m thick. Quartzo-feldspathic gneisses with ±Ep ± Am predominate in the lower part of the formation, and epidote amphibolites are important in the upper part. This lithological unit includes also dikes of basic rocks (Arenas and Peinado, 1984; Arenas, 1988). According to Fernández Pompa and Monteserín López (1976) and Arenas (1988), the Somozas orthogneisses and amphibolites derive from high-K calc-alkaline and shoshonitic volcanic rocks that are related to a continental margin setting with incipient subduction.

The rocks in the Espasante thrust sheet show a continuous planar tectonic fabric without a clear stretching lineation. The textural features are those of mylonites (sensu lato) developed under a progressive deformation regime, allowing the separation of different deformation stages in the rocks seen as protomylonites, mylonites, ultramylonites, and phyllonites (Fig. 4, D–F). The last two types are geometrically related to the planes of the main basal thrust. Thus, the Ramallal phyllonites represent the last deformational stage in the reactivation of the mylonites and ultramylonites.

In the orthogneisses, Qtz, Pl, Ms, ± Ep-Czo, ± Hbl, ± Grt, ± Stp, ± Aln are stable during mylonitic deformation. In the amphibolites, the most common mineral association defining the main fabric contains Hbl, Pl, Qtz, Ep-Czo, Ms, ± Grt. In ultramylonites and phyllonites only Chl is synkinematic.

UPPER TECTONIC UNIT: THE SLAB OF LOWER CRUSTAL ROCKS

The Upper Tectonic unit is located at the uppermost part of the Cabo Ortegal Complex. It comprises ultramafic, mafic, and quartzo-feldspathic rocks that record high-*P*–high-*T* metamorphic events and polyorogenic deformation. It crops out in a young tectonic graben bounded by rocks of the Lower Tectonic unit and/or the Rio Baio thrust sheet (Plate 1). Within the Upper unit, the reconstruction of structures is partially masked by a system of high-angle normal faults striking N130E dipping to the north-northeast. These faults are several kilometers long, straight, and have vertical separations of hundreds of meters. Due to this late tectonic event, lithological boundaries exhibit frequently discontinuous and complicated cartographic patterns, changing from the footwall to the hanging wall of these normal faults, which were probably formed during the Alpine deformation events in northwestern Iberia (Ferrús, 1998).

The ductile structures of the Upper unit can be related to deformation under prograde granulite and eclogite facies conditions, although the more conspicuous structures result from retrogressive deformation associated with the exhumation processes. The fabric used as reference for the structural reconstruction is a widespread mylonitic fabric, which is observed throughout the Upper unit rocks. This mylonitic fabric is affected by isoclinal folds with axes striking north-northeast–south-

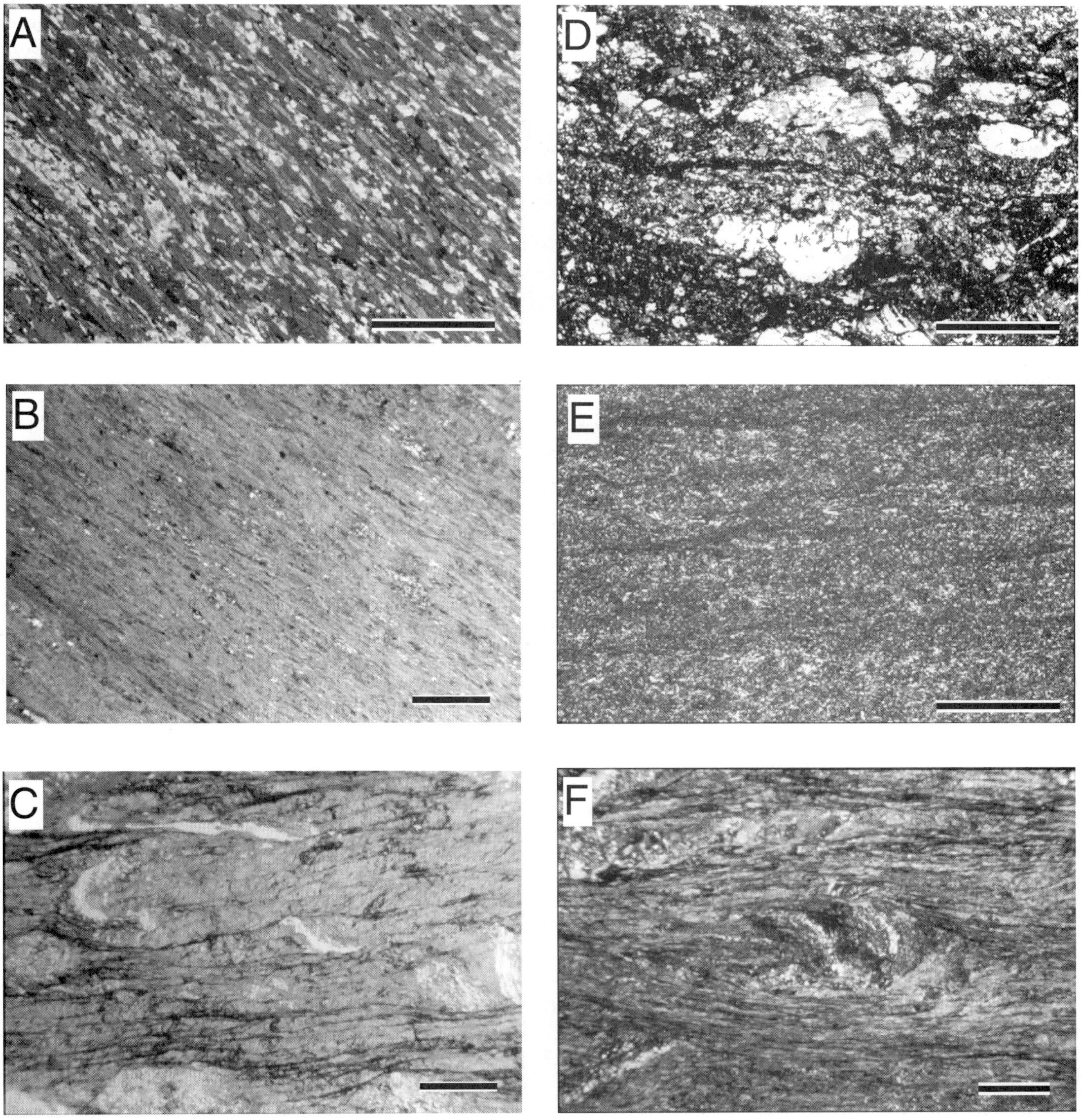

Figure 4. Microstructures in rocks from Lower Tectonic unit in Purrido thrust sheet. A: Tectonic fabric in Purrido amphibolites defined by shape-preferred orientation of Hbl and Pl. B and C: Ultramylonites and phyllonites from Labacengos Formation. Note very fine grain size and its uniform distribution in ultramylonites. Evolution of tectonic fabric in quartzo-feldspathic rocks in Espasante thrust sheet is observed from mylonites (D) to ultramylonites (E) and phyllonites (F). In all photomicrographs, scale bar is 1 mm (figure taken from Marcos and Farias, 1999).

southwest and verging to the east-southeast. The major folds are isoclinal and recumbent and have reverse limbs reaching 6 km; the folds are crosscut by thrusts. Two of them, the named Basal and Upper thrusts, reach several kilometers of displacement. The Upper thrust separates two tectonic sheets: the Cedeira sheet below, and the Capelada sheet above. Both sheets have the same tectonostratigraphic sequence, but are differentiated essentially by their relative degree of deformation and retrograde characteristics, which are stronger in the lower unit, where granulite and eclogite rocks are only found as relics.

The late Variscan refolding phase caused the characteristic elliptical outcrop pattern of the Cabo Ortegal Complex. Open folds with subvertical axial planes and fold axes oriented north-northeast–south-southwest form at all scales, giving Ramsay type 3 intersection patterns with the previous isoclinal folds at outcrop and cartographic scale.

Tectonostratigraphic sequence

Regardless of their detailed metamorphic and deformational history, the rocks that form the Upper Tectonic unit can be divided into three major lithological units: (I) the ultramafic rocks, (II) the mafic rocks, and (III) the quartzo-feldspathic gneisses (see legend in Plate 1).

The ultramafic rocks (I) represent the base of the sequence and are formed of more than 1000 m of alternating serpentinized peridotites (spinel-amphibole peridotites, harzburgites, and dunites) and pyroxenites (websterites and clinopyroxenites) (Vogel, 1967; Girardeau et al., 1989; Gravestock, 1992; Moreno, 1999). They build the core of a huge recumbent fold. The restoration of the structure allows the recognition of a sequence that is formed, from bottom to top, of ~ 300 m layered piroxenites, ~ 300 m dunites (alternating with centimeter to decameter Grt ± Spl pyroxenitic bands) and harzburgites. The origin of the ultramafic rocks is controversial. Girardeau and Gil Ibarguchi (1991) suggested that the peridotites have strong affinities with residual oceanic peridotites and the pyroxenites are intruding magma segregates. Moreno (1999) interpreted this ultramafic-layered complex as the root of an oceanic arc similar to the Himalayan Jijal Complex. The contact with the overlying mafic rock unit (II) is gradational (Galán and Marcos, 1997).

Three members can be distinguished within the mafic rocks (II). The lower member is formed by massive or weakly foliated ultramafic-mafic granulites (50–100 m thick), including the "hornblende pyrigarnites" (Grt, Cpx, ± Hbl, ± Pl < 5%, ± Qtz, Rt, ± Aln, ± Czo, ± Ap) and the "hornblende plagiopyrigarnites" (Grt, Cpx, ± Hbl, Pl, ± Qtz, Rt, ± Aln, ±Czo, ± Ap) of Vogel (1967). Some of these rocks lack Pl in equilibrium with Grt-Cpx and others show relict Ky and/or Zo, suggesting an earlier eclogite facies metamorphism (Galán and Marcos, 2000). Their composition is essentially ultrabasic-basic and equivalent to Fe-Ti rich (picro) basalts-(mela) gabbros (Galán and Marcos, 1997). In the middle member (400 m thick), these rocks alternate with garnet amphibolites (Grt, Hbl, Pl, Qtz, ± Bi, Ilm, Ttn, Czo, Ep, ± Aln) retrogressed from granulites, the composition of which is basic to intermediate. Both types of rocks define a centimetric to metric layering in the less deformed outcrops. Layers of garnet hornblende schist, garnet tonalitic gneisses, and scapolite-rich and carbonate-rich mafic granulites are also present in much subordinated proportion. These two members of mafic rocks are interpreted in two different ways. Peucat et al. (1990) favored either a continental volcanic arc margin or a continental rifting (attenuated crust), and Galán and Marcos (1997) favored the latter hypothesis. Galán and Marcos (1997) interpreted these two members as derived from a layered gabbro-type protolith related to continental within-plate magmatism in an extensional setting. In the latter model, both the ultramafic rocks (II) and the mafic rocks (II) would be part of a thinned lower crust. However, the westernmost outcrop band of this mafic middle member (Plate 1) has some peculiar lithological features: there are observed layered ± garnet amphibolites throughout, retrogressed after granulites (Vogel, 1967), and metagabbros and amphibolites after metagabbros, which were previously affected by granulite-amphibolite facies metamorphism in less severe conditions (Vogel, 1967; Gil Ibarguchi et al., 1990). The geochemistry of these rocks was correlated to N- and transitional (T)-MORB by Peucat et al. (1990).

A sharp contact separates the middle member from the upper member made of massive mafic rocks (100–200 m). The upper member is basically formed of eclogites, but fine-grained dark garnet amphibolites, retrogressed after eclogites-granulites (Vogel, 1967), are also present, especially in the northwestern outcrops of the Cabo Ortegal Complex (Plate 1). The most common type of eclogite contains Omp, Grt, Zo, Qtz, Amph, and Rt (Concepenido-type eclogites of Vogel, 1967), but kyanite-bearing eclogites (Vogel, 1967; Gil Ibarguchi et al., 1990; Mendia, 1996) and Fe-Ti–rich eclogites are also observed (Mendia, 1996). The Ky-bearing eclogites are mainly exposed as a thin band, <25 m thick, along the contacts. All the eclogites are more or less retrogressed in granulite-amphibolite facies conditions. Van Calsteren (1978) proposed a continental quartz-normative tholeiite parentage for the eclogites on the basis of major and trace element geochemistry. However, most authors agree on a N-MORB protolith for the common eclogites (Bernard-Griffiths et al., 1985; Peucat et al., 1990; Gravestock, 1992; Mendia, 1996).

The last unit (III) is composed of >600 m of quartzo-feldspathic gneisses in sharp contact with the eclogites. Near this contact the augen gneisses are fine or medium grained, composed of Qtz, Pl, Grt, Bt, Ms, and Ky (Vogel, 1967), and have a characteristic compositional banding resulting from changes in the amount of mafic minerals. They belong to the Banded Gneiss Formation of Vogel (1967) and include blocks or pods, lenses, and bands of eclogites, more or less retrogressed, and subordinate mafic and ultramafic rocks, giving the appearance of a block-in-matrix formation, in the sense of Raymond et al. (1989). Metamorphic mineral associations in both mafic rocks and host gneisses indicate a high-*P*–high-*T* metamorphism (Vogel, 1967; Gil Ibarguchi et al., 1990), causing partial melting of the gneisses. The thickness of these lower gneisses is ~150 m.

The rest of the unit (III) is formed of two-mica layered quartzo-feldspathic gneisses (often with Grt, St, and Ky), chemically and mineralogically similar to the Banded Gneiss Formation, but with different appearance in the field. The Cariño Gneisses and most of the Chímparra Gneisses of Vogel (1967) are included. They are interpreted as metamorphic graywackes (Vogel, 1967; Fernández and Marcos, 1997) with calc-silicate and mafic inclusions. The mafic inclusions have mineral assemblages formed in amphibolite facies in the Cariño Gneisses and in granulite facies in the Chimparra Gneisses, which were also affected by migmatization, as was the Banded Gneiss Formation (Vogel, 1967). Available geochemical data for some of the mafic inclusions indicate T-MORB compositions with important contributions of continental material (Peucat et al., 1990), which suggests that they intruded in an extensional continental setting.

Metamorphism and age

In general, the metamorphic evolution of both the ultramafic (I) and mafic (II) rocks is similar. The ultramafic rocks record a first episode of high-*P*–high-*T* metamorphism at ~800 °C, 1.6 Gpa, followed by a pervasive amphibolitization at >750 °C (Girardeau and Gil Ibarguchi, 1991). The metamorphic evolution of mafic rocks suggests a first high-*P*–high-*T* metamorphic event, followed by the development of different structures related to exhumation processes. They consecutively equilibrated in high-*P*–high-*T* eclogite-granulite, amphibolite, and greenschist facies, defining a decompression-type *P-T* path (Mendia, 1996; Galán and Marcos, 2000). The estimated metamorphic peak conditions, based on thermobarometric methods, range from 838–876 °C, ≥1.2–1.4 GPa (in the Bacariza-type eclogites within lower and middle mafic member; Galán and Marcos, 1999, 2000) to 770–800 °C, ≥1.4–1.7 GPa (common Concepenido-type eclogites within the upper mafic member; Gil Ibarguchi et al., 1990; Mendia, 1996).

Although the metamorphic evolution of the quartzo-feldspathic gneisses (III) is less studied, available thermobarometric data also indicate a high-*P*–high-*T* event for some of them. The metamorphic evolution of the quartzo-feldspathic gneisses in close contact with the upper mafic rocks is similar to them, as inferred from their eclogitic mafic inclusions. Some of these mafic inclusions, with plagioclase already coexisting with Grt-Cpx, record 880–620 °C, 1.35 ± 0.13 Gpa (Gil Ibarguchi et al., 1990), comparable to values estimated for the host gneisses (800–700 °C, 1.60–1.50 Gpa) (Basterra et al., 1989; Fernández Rodríguez, 1997). Other types of gneisses from the same unit (III) (e.g., the Cariño Gneisses) only present evidence of metamorphism under amphibolite facies conditions, subsequently retrogressed in greenschist facies (Vogel, 1967).

The absolute age of the protoliths and of the high-*P*–high-*T* metamorphism are a matter of debate (see reviews in Ábalos et al., 1996; Martínez Catalán et al., 1999). Peucat et al. (1990) consider an Early Ordovician age, ca. 490–480 Ma, for both the protholiths of the lower and middle mafic member and for the high-*P*–high-*T* event in eclogite and granulite facies. Ordoñez Casado (1998) agreed on this age for the protoliths and obtained a comparable age (507–473 Ma) for the protoliths of the Concepenido eclogites (i.e., for the upper mafic member). However, Ordoñez Casado and others (Schäffter et al., 1993; Santos Zalduegui et al., 1996) estimated the age of the high-*P*–high-*T* event in eclogite and granulite facies as Early Devonian (ca. 405–380 Ma). Schäffter et al. (1993) and Ordoñez Casado (1998) also provided similar ages for the high-*P*–high-*T* episode in the quartzo-feldspathic gneisses (397–391 Ma). However, more recent data, from U-Pb dating on zircon, monazite, titanite, and rutile (Fernández Suárez et al., 2002) indicate that the rocks in the Upper Tectonic unit record a polyorogenic evolution with a high-*P*–high-*T* episode ca. 490–480 Ma (Early Ordovician) and a younger Early Devonian event dated ca. 390–385 Ma. The age of the late deformation in greenschists facies conditions would occur ca. 365 Ma (Late Devonian) (Dallmeyer et al., 1997).

*Structures associated with high-*P*–high-*T* metamorphism*

Most of the structures found in the high-*P*–high-*T* rocks of the Upper Tectonic unit are developed during exhumation under retrogressive metamorphic conditions. However, preferred orientation of omphacite in eclogites and granulites (Engels, 1972; Godard and van Roermund, 1995; Ábalos, 1997) and the development of ductile shear zones (Marcos and Galán, 1994) under high-*P* conditions provide evidence of intracrystalline plasticity of clinopyroxene caused by deformation during the earlier high-*P*–high-*T* metamorphism. The crystallographic preferred orientation of omphacite from the Concepenido-type eclogites (Upper Member of the mafic rocks unit) show the distribution of c[001] axes in girdles and b[100] in maximum, indicating the development of planar fabrics (S-type according to Helmsteadt et al., 1972), which suggests that coaxial flow is predominant during high-*P* deformation (Engels, 1972; Godard and van Roermund, 1995; Ábalos, 1997). Ábalos (1997), however, reported local noncoaxial components during the deformation of omphacite, relating the crystallographic preferred orientation fabrics with the structural reference framework in the rock, i.e., the foliation and the lineation. However, at the regional scale the lineation in omphacite scatters in the foliation plane (Mendia, 1996) and does not provide a consistent orientation for noncoaxial deformation, suggesting that deformation is predominantly coaxial (Llana-Fúnez et al., 2001). This is not incompatible with local noncoaxial components at thin-section scale, because deformation is probably more heterogeneous and might be influenced by the presence of rigid garnets. The development of these microstructures could be related to the crustal thickening produced by pre-Variscan continental collision.

Deformation during exhumation and emplacement of the Upper Tectonic unit

The first structure that appears is a planar generalized tectonic fabric (RD1), which at the map scale is parallel to the lithological boundaries. It is a mylonitic foliation, sometimes showing a gneissic banding, which started to form in granulite facies conditions and finally equilibrated in amphibolite facies conditions (Fernández, 1993; Marcos and Galán, 1994; Fernández and Marcos, 1996) (Fig. 5). At the outcrop scale, the foliation shows an anastomosing aspect nucleated around pods in which the rocks preserved a previous migmatic fabric, or a clear and completely formed planar fabric. In places a disrupted or truncated foliation is observed, which suggests the development of more than one consecutive set of mylonitic fabric during the same progressive deformation event (Fig. 6A). A heterogeneous component of noncoaxial strain is indicated by some nonpervasive, asymmetrical elements, such as folds and sigma and delta microstructures (Fig. 6B). A detailed analysis of the macrostructures and microstructures associated with the mylonitic foliation in quartzo-feldspathic gneisses leads us to propose a heterogeneous coaxial flow regime during the extensive ductile deformation. This model is supported by the lack of a well-developed stretching lineation, the orthorhombic symmetry of quartz *c*-axis textures, and the absence of a consistent sense of shear inferred from kinematic criteria (Fernández, 1993; Fernández and Marcos, 1996) (Fig. 7). Persistent foliations not related to any particular type of tectonic structure (e.g., a detachment) are common in granulites. Mechanisms restricted to the lower crust, such as regional coaxial flow, repeated transposition, and advective or convective flow, have been proposed by several authors for the development of these foliations (see Dirks et al., 1997).

Locally, in mafic rocks of La Capelada sheet, the main mylonitic fabric (RD1) is affected by centimeter- to decameter-size discrete shear bands where blastomylonitic biotite gneisses (Galán and Marcos, 1997) are formed (Fig. 8). These tectonites, mapped by Vogel (1967) as "doubtful paragneisses," locally enclose centimeter blocks of ultramafic-mafic granulites and garnet amphibolites in a chaotic arrangement. These blastomylonitic gneisses contain Grt, Am, Pl, Scp, and Czo porphyroclasts set in a very fine grained mylonitic matrix made of Bi, Pl, Qtz, and Ttn.

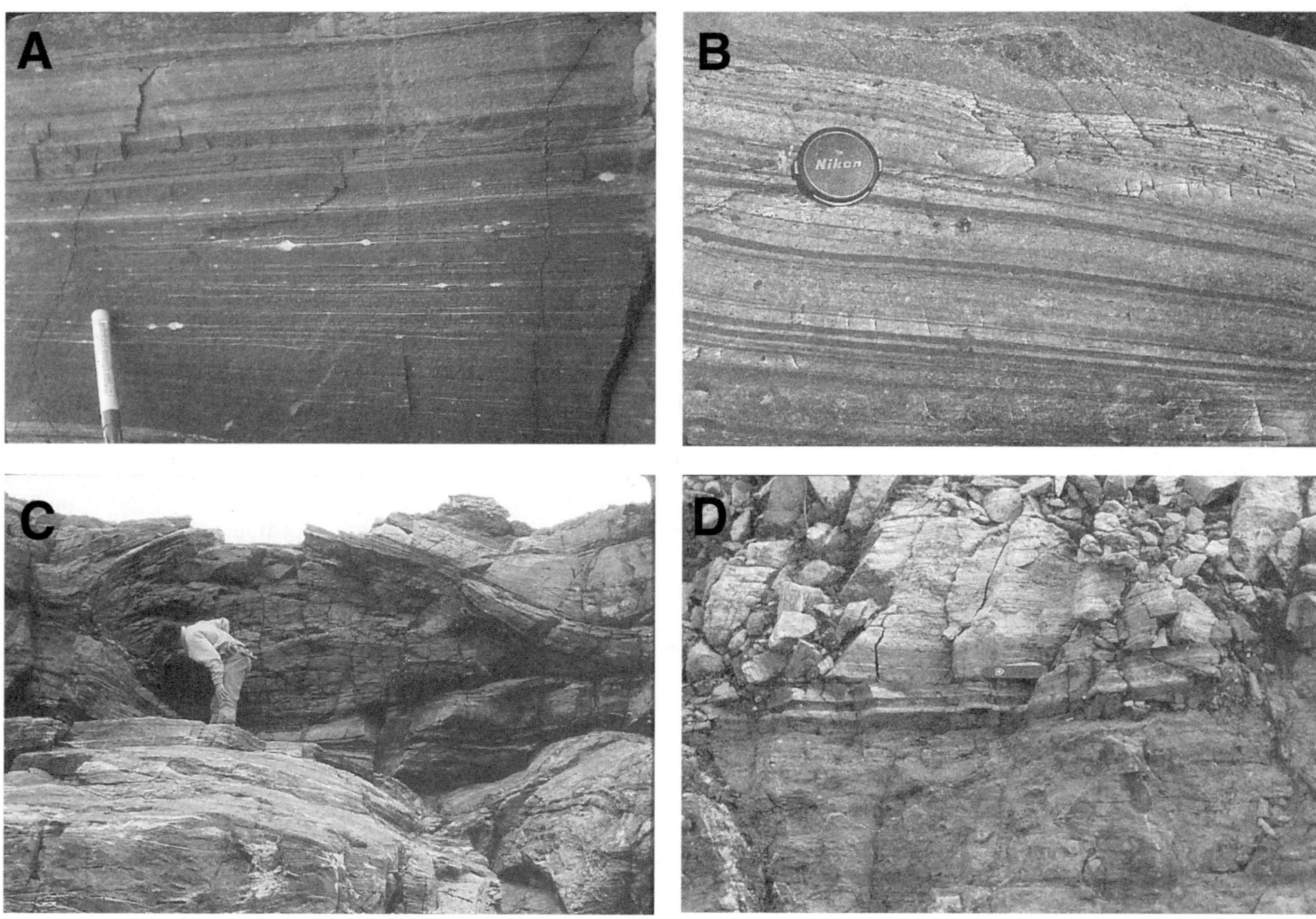

Figure 5. Field aspects of mylonitic foliation RD1. Planar fabric is shown in quartzo-feldspathic gneisses (Ría de Cedeira, immediately east of Punta Chirlateira) (A) and in mafic rocks (B) from Upper Member (garnet amphibolites, Tarroiba boulder beach). White part of the marker in A is 10 cm and diameter of lens cap in B is about 5 cm. C: In mafic granulites of Middle Member, main foliation surrounds pods of pyriagarnites. West of San Andrés de Teixido. D: Contact zone between foliated mafic (below in picture) and ultramafic rocks in reverse limb of recumbent fold of La Capelada is shown. Knife is 6 cm in length. West of Miranda hill.

Figure 6. Truncated foliations (A) and intrafolial folds (B) in relation to main mylonitic fabric RD1 (mafic granulites). In A, hammer is about 32 cm in length. In B, magnifier is 2 cm wide. Outeiro hill.

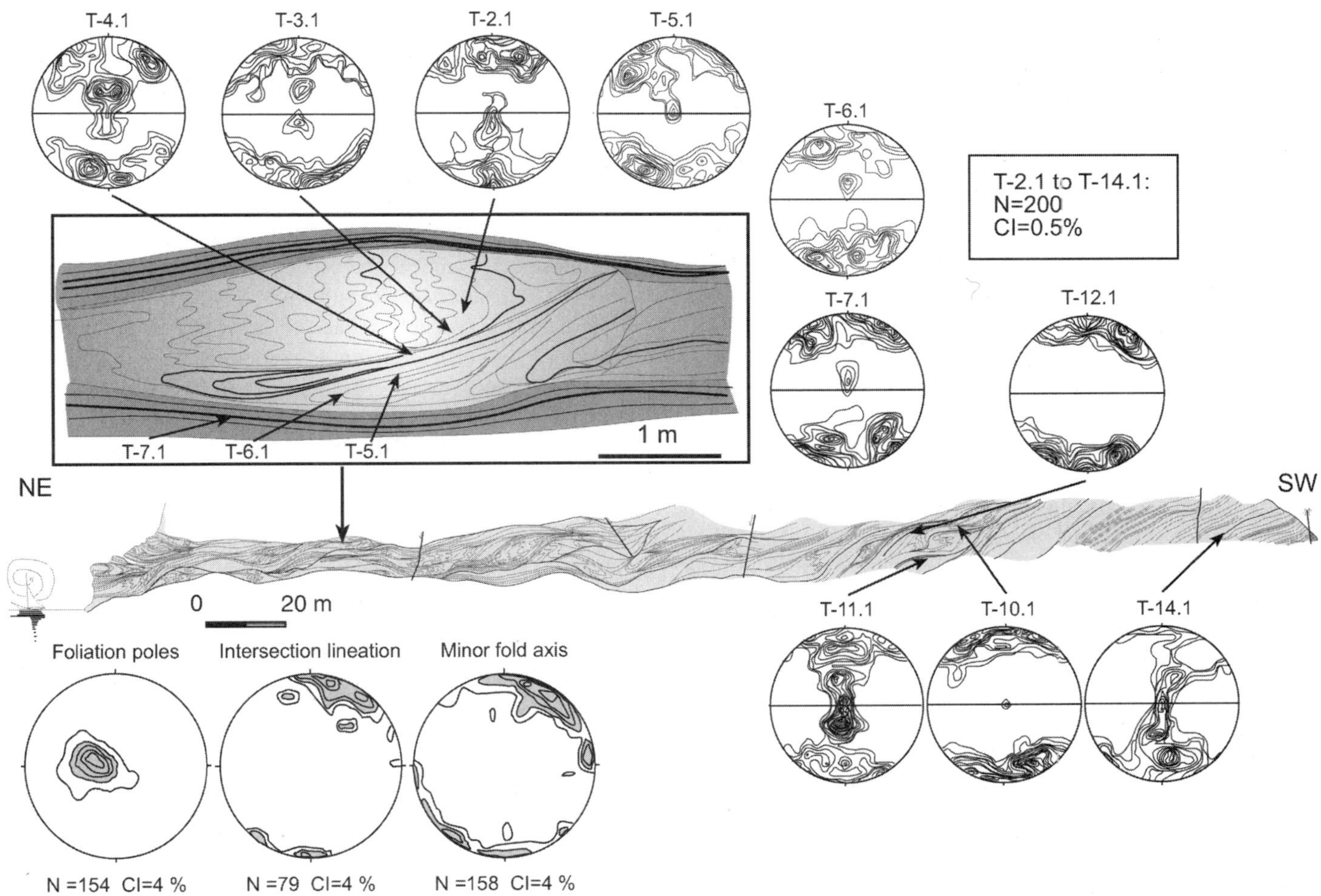

Figure 7. Structural transition in quartzo-feldspathic gneisses from homogeneous planar mylonitic fabrics to inhomogeneous anatomizing pattern is observed in Tarroiba cliff. Intersection between foliation and compositional banding in gneisses (Chímparra Gneisses) is predominant lineation observed in field, having same orientation as fold axes of minor folds. Quartz *c*-axis fabrics in these rocks have orthorhombic symmetry, indicating that heterogeneous coaxial flow is predominant during quartz deformation. Only sample T-14.1, collected from southwest sector, has asymmetric girdle and deviates from this common pattern. Half-opening angles of small-circle girdles vary between 20° and 40°. All plots are Schmidt lower hemisphere projection (CI is contour interval).

Figure 8. Biotite blastomylonitic gneisses in meter-size discrete shear zone (lower part of picture). Hammer handle is 60 cm in length. West of San Andrés de Teixido.

The general foliation is affected by asymmetric similar folds (RD2 folds), trending to isoclinal folds, with low interlimb angles. Their axes are oriented north-northeast–south-southwest and their vergence is to the east-southeast. The major folds are isoclinal and recumbent and their reverse limbs reach kilometric scale (>6 km) (Plate 1; Fig. 9). Related minor folds are of decametric size and generally also asymmetric (Fig. 10). Refolding of previous intrafoliar folds by these RD2 folds resulted in type 3 interference patterns of Ramsay (1967) and sometimes type 2 patterns. In the field, no axial plane cleavage is observed, although a slight orientation of hornblende (in amphibolites) or phyllosilicates (in gneisses) can be seen in some RD2 fold hinges. The lack of foliation associated with RD2 folding makes it difficult to establish the exact *P-T* conditions at which deformation occurred. However, both the similar geometry of folds and their strong flattening imply a generally ductile behavior of the rocks, which suggests deformation under amphibolite facies conditions. In consequence, there would be a relevant reworking of the RD1 foliation in the limbs of the RD2 folds. The inferred reworking of the main previous foliation during this folding stage is confirmed by the development of a stretching lineation parallel to the RD2 fold axes. The colinearity of lineation and fold axes points to folding having occurred in a constrictional environment.

The major RD2 folds are crosscut by thrusts (RD3), two of them reaching kilometer size (Plate 1; Fig. 9). One of them, the Basal thrust, places the Upper Tectonic unit over the Lower Tectonic unit. It is parallel to the reverse limb of a major recumbent fold and gives rise to the thrusting of high-grade gneisses over the Purrido amphibolites. In its westernmost section, it is highlighted by strongly sheared and serpentinized ultramafic rocks, probably belonging to the Moeche thrust sheet (Marcos and Farias, 1999). The related shear zone, with associated minor structures such as thrusts, folds, and crenulation cleavage, is in some locations >50 m thick and is developed mainly in the footwall rocks. The minor folds are asymmetric, reach meter scale, and have curved hinges that indicate the direction of the tectonic transport, approximately east-southeast. The Upper thrust imbricates the rocks of the Upper Tectonic unit and produces the stacking of the La Capelada tectonic sheet over the Cedeira sheet. The geometrical relationships between the Basal and Upper thrusts and the existence of klippen of the La Capelada sheet over the underlying Lower Tectonic unit (Plate 1) point to a relative younger age for the Upper thrust, which would be consistent with an inward sequence of emplacement of these thrust sheets (Marcos and Farias, 1999).

The following RD4 deformation episode is related to the general postnappe refolding that affected both the allochthonous and autochthonous rocks at all scales and gave the Cabo Ortegal Complex its typical elliptical cartographic shape. The RD4 folds are open with subvertical axial planes oriented west-northwest–east-southeast. In places they show axial crenulation cleavage and a mineral lineation parallel to fold axes defined by chlorite. The distribution of RD4 structures is heterogeneous. The interference between RD2 and RD4 folds produces the type 3 patterns of Ramsay (1967), both at cartographic (Fig. 11) and outcrop scale. Contemporaneously or subsequent to the folding, some discrete north-south–oriented shear zones, equilibrated in greenschist facies, appear and usually reactivate previous structures. These shear zones are subvertical, with left-lateral shear sense. They are probably related to other major structures in northwest Iberia characterizing intracontinental postcollisional deformation, such as the Valdoviño fault (Iglesias Ponce de León and Choukroune, 1980) (see Fig. 1).

Kinematic interpretation of the main tectonic fabric. Most stretching and/or intersection lineations associated with the main tectonic RD1 foliation in the Upper Tectonic unit rocks show a constant south-southwest–north-northeast orientation. Kinematic criteria, such as delta and sigma porphyroblasts systems and intrafoliar folds, do not indicate a consistent sense of shear at outcrop scale. This seems to suggest that the mylonitic foliation is developed in a general noncoaxial flow (Fernández Rodríguez, 1997; Fernández and Marcos, 1996).

The general kinematic interpretation is difficult because of the superimposition of RD1 and RD2 structures. Thus, it is not easy to evaluate the effects on the main foliation caused by the development of (RD2) isoclinal folds in ductile conditions. In addition, the best developed lineation in the rocks is defined by the hornblende and plagioclase, which could grow during RD1 or RD2. Interpretation of kinematic criteria must be made with caution. For example, one assumption might be that RD1 lineation is intimately related to the noncoaxial component of deformation and indicates the direction of tectonic transport (Ábalos et al., 1996). However, it must be kept in mind that deformation of RD1 lineation by homoaxial isoclinal folds results in opposite sense of shear in normal and reverse limbs.

KINEMATICS OF THE SUTURE-ZONE THRUST SHEETS

RD2 and RD3 structures can be considered the result of progressive deformation that led to the emplacement of the Upper unit. Both the vergence of the RD2 folds and the sense of shear in RD3 thrusts indicate a general tectonic transport of the nappe toward the east-southeast (Marcos and Farias, 1999). This is consistent with the geometry of the structures developed in the underlying sheets (Lower Tectonic unit, Rio Baio thrust sheet) and in the Paleozoic autochthonous sequence (Fig. 11).

The lower tectonic sheets of the Cabo Ortegal Complex that form the suture zone are, from top to bottom, the Purrido thrust sheet, the Moeche sheet, and the Espasante thrust sheet (Plate 1; Fig. 3). The internal structure of the Moeche thrust sheet is that of a tectonic melange. A mylonitic foliation is developed parallel to the boundaries of the Purrido and Espasante sheets, and no folds have been observed in relation to the fabric. In both sheets, ultramylonites and phyllonites occur toward the bottom as the result of the extreme deformation associated with their emplacement (Fig. 4). The main textural change in the rocks is the grain-size reduction. The transition between mylonites and ultramylonites is gradational and corresponds to a band of ductile shear. This observed reworking of the mylonites and ultramylonites near the basal thrusts (i.e., basal induced shear) implies the development of a new foliation simultaneously with the general rock retrogression. The new fabric is either a slaty cleavage or a closely spaced crenulation cleavage, with synkinematic Chl and opaque minerals in the cleavage plane. Previous mylonites and ultramylonites are preserved in some lenticular bodies or pods surrounded by the new foliation, suggesting a relatively late formation of the basal thrust. Some of the structural features already mentioned, e.g., the development of a tectonic fabric parallel to the sheet boundaries, the changes of textures approaching the lower parts, and the lack of folds associated with the main fabric, provide evidence that these thrusts represent major tectonic boundaries and support the existence of a basal resistance to forward flow (Rathbone et al., 1983; Ramsay and Lisle, 2000).

Available data on the kinematics of the different thrust sheets are scarce and most of them are from the Purrido thrust sheet. In this sheet, a well-developed amphibole lineation is consistently orientated at outcrop scale, but its orientation differs from place to place, from N30E in the western coast to N40E–N60E in the southernmost inland exposures. However, there are several kinematic indicators, such as rotated porphyroblasts and C′ shear bands, that point to displacement of the hanging wall toward the northeast. In the Labacengos ultramylonites and phyllonites, also within the Purrido sheet, both the mineral lineation and the curved hinges of meter-scale folds indicate tectonic transport to N70E. The last structures (phyllonites, folds with curved hinges) formed in relation to the thrust movement under greenschist facies conditions, indicating tectonic transport to the southeast.

In summary, during the ascent and emplacement of the thrust sheets under retrograde metamorphic conditions from amphibolite to greenschist facies, a progressive clockwise rotation of transport direction is recorded in the tectonites (Marcos and Farias, 1999), in a similar way to the changes or rotations of overthrust shear that were reported in the Helvetic nappes of the Alps by Dietrich and Durney (1986).

CONCLUSIONS

The Cabo Ortegal Complex is a composite allochthonous terrane that was thrusted onto the western edge of Gondwana during the Variscan orogeny. It is formed of two main tectonic units: the Upper Tectonic unit and the Lower Tectonic unit, each of which includes several sheets. Another thin thrust sheet (the Rio Baio sheet or parautochthonous sheet) separates the Lower Tectonic unit from the underlying metasediments of the lower Paleozoic sequence in the Ollo de Sapo antiform, which is considered the autochthonous sequence of the Iberian plate.

The Upper Tectonic unit includes ultramafic rocks, mafic rocks classified into three members, and quartzo-feldspathic gneisses, all of them affected by high-*P*–high-*T* metamorphism in eclogite and granulite facies. Most structures observed in both mafic rocks and quartzo-feldspathic gneisses are related to their exhumation and concomitant retrogression. An exception would be the preferred orientation of omphacite in eclogites and granulites, retrogressed after eclogites. This orientation indicates predominately coaxial flow during the earlier stages of the high-*P*–high-*T* event, probably caused by crustal thickening during pre-Variscan continental collision. During the exhumation, these rocks developed a persistent horizontal foliation (RD1) in a constrictional environment, that started to form in granulite facies and was finally equilibrated in amphibolite facies conditions (Table 1). Later, also in amphibolite facies conditions, the rocks recorded the progressive development of noncoaxial structures, such as east-verging asymmetrical isoclinal folds and thrusts. These thrusts led to the emplacement of the Upper Tectonic unit over the Lower Tectonic unit and of both of them over the parautochthon and autochthon (Fig. 11).

The Lower Tectonic unit is believed to represent the resulting suture of the Variscan collision and the closure of the Rheic ocean. Available geochemical data indicate that the rocks in each sheet are derived from different geodynamic settings: E-MORB for the Purrido rocks, tectonic melange for the Moeche sheet, and high-K calc-alkaline volcanics for the Espasante sheet. All of them were equilibrated in amphibolite or greenschist facies conditions. Deformation is concentrated at the bottom of each sheet, indicating that their boundaries constitute major tectonic limits. Kinematic criteria provide a sense of tectonic transport for these thrusts toward the east, but show a clockwise rotation from northeast to southeast in progressively shallower

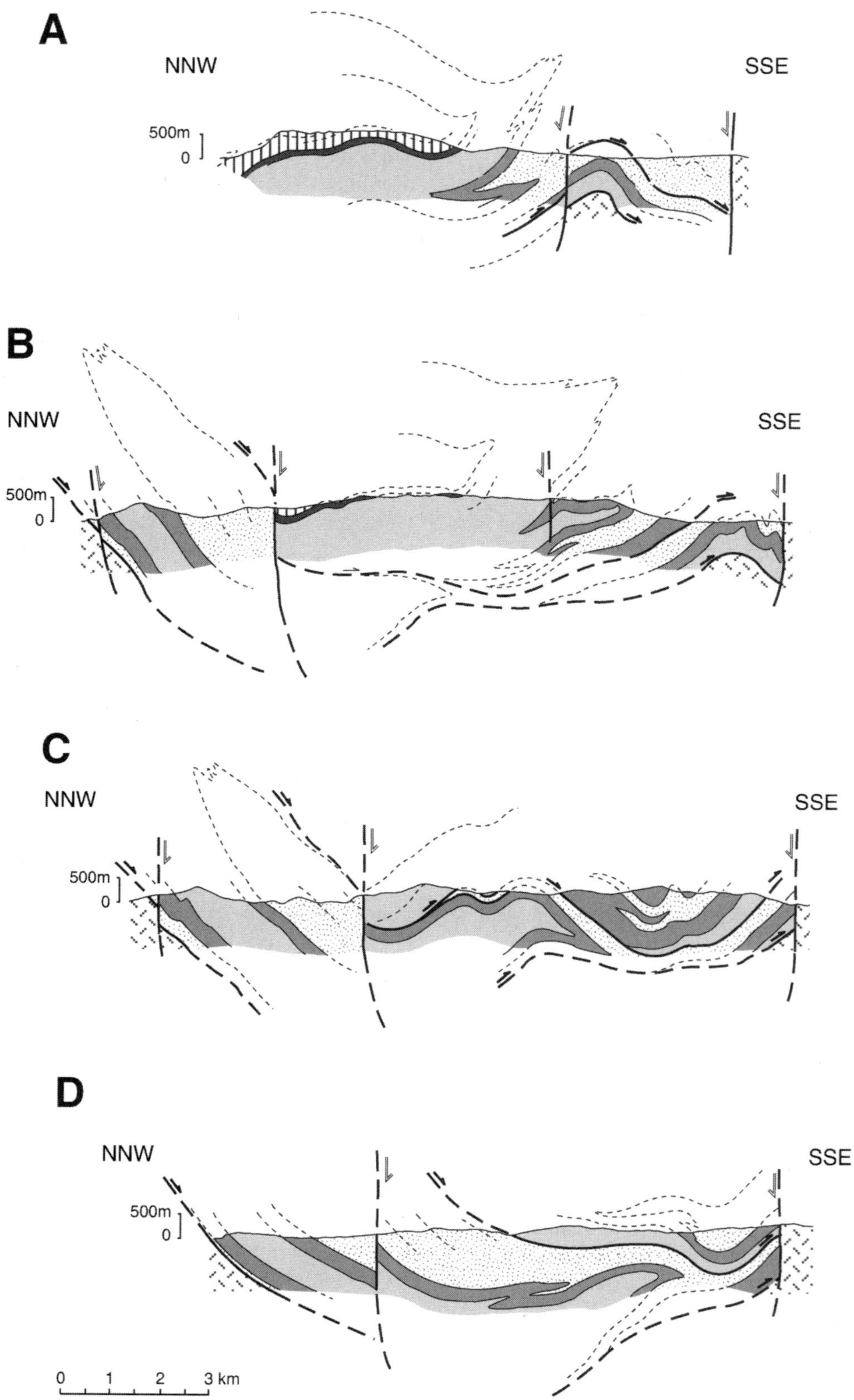
A
NNW
SSE
500m
0
B
NNW
SSE
500m
0
C
NNW
SSE
500m
0
D
NNW
SSE
500m
0
0 1 2 3 km

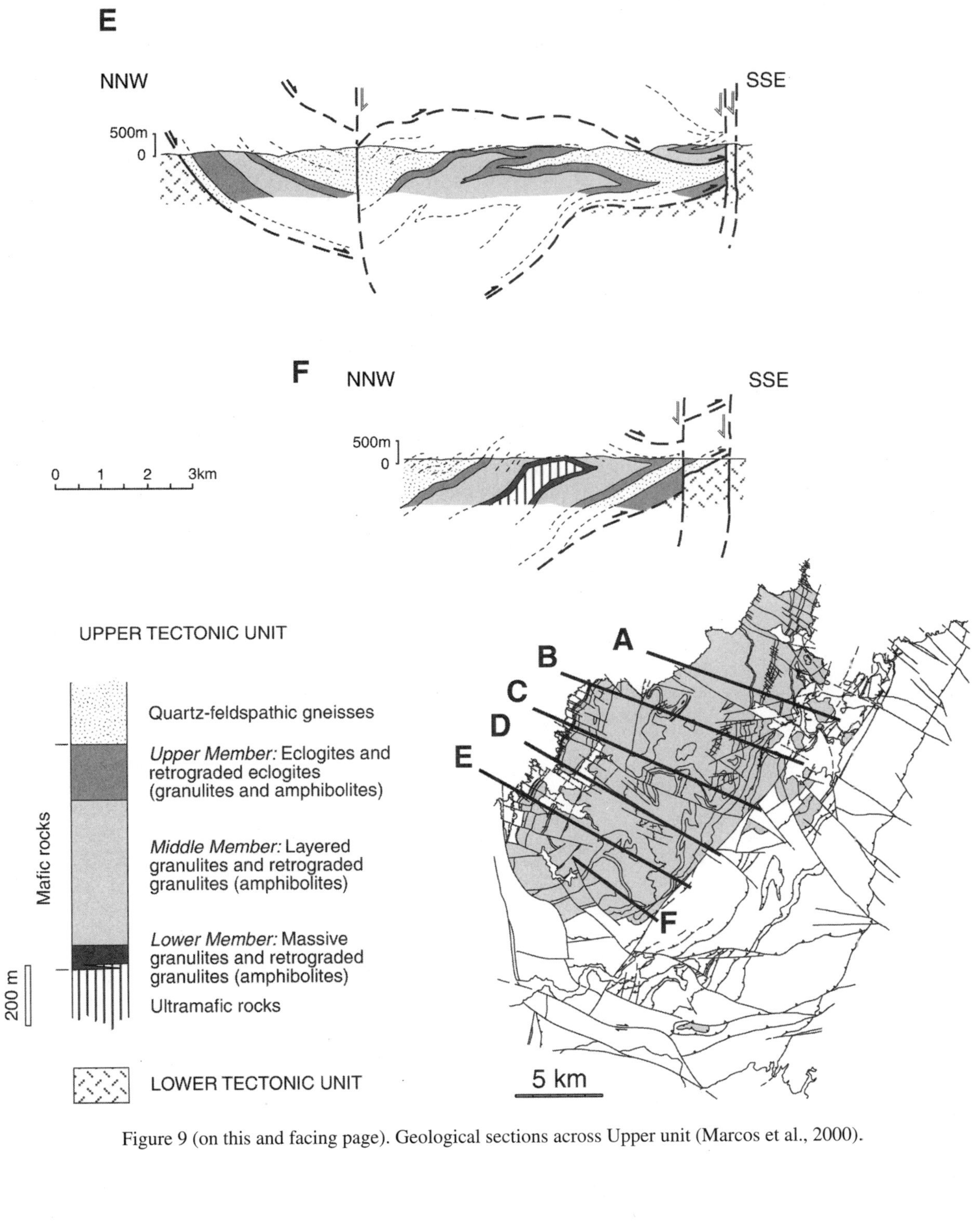

Figure 9 (on this and facing page). Geological sections across Upper unit (Marcos et al., 2000).

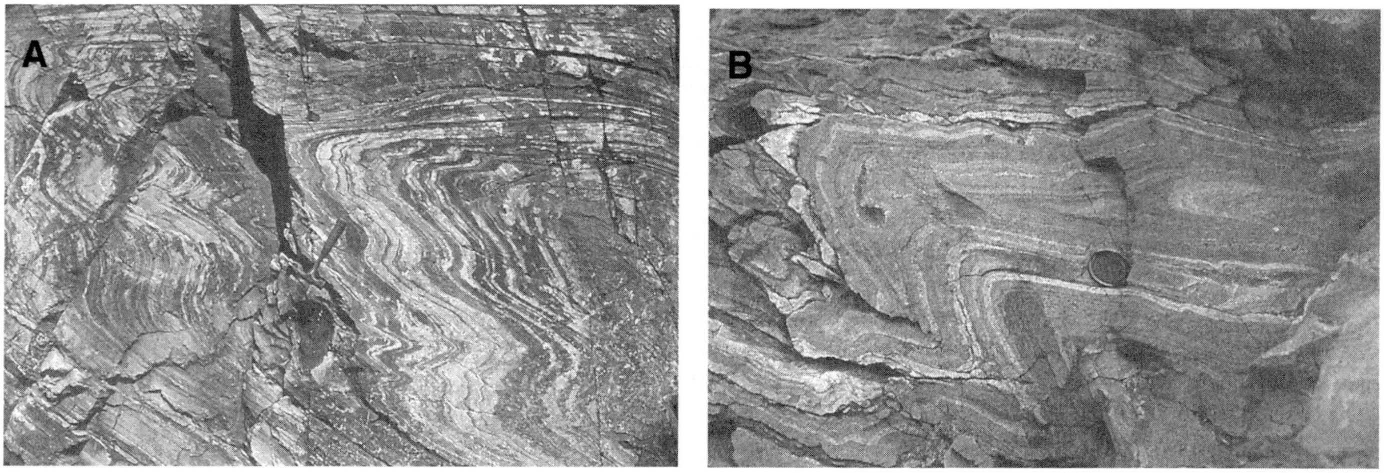

Figure 10. A: RD2 folds in quartzo-feldspathic gneisses affect previous mylonitic fabric. Hammer is 28 cm in length. Punta do Carreiro. B: Interference pattern between intrafolial folds, related to development of main tectonic fabric, and RD2 folds is shown. Diameter of the lens cap as a scale is about 5 cm. Masanteo Peninsula.

deformation conditions. Kinematic criteria in the Rio Baio sheet also indicate an earlier tectonic transport (top to the northeast), changing later toward the southeast.

An attempt to integrate the various tectonothermal events observed in the Cabo Ortegal Complex and underlying units is presented in Table 1. However, this type of modeling is poorly defined, among other reasons, by the fact that available radiometric ages show a high degree of scatter and contradictions. This must be regarded as normal due to the limitations of the dating methods and to the complex mineral reactions during the different tectonothermal events. Thus, the age of the high-*P*–high-*T* metamorphic event remains as a major question (Table 1). One set of data points to Early Ordovician, while the other set of data point to Early Devonian. Both ages are U-Pb dates, the former

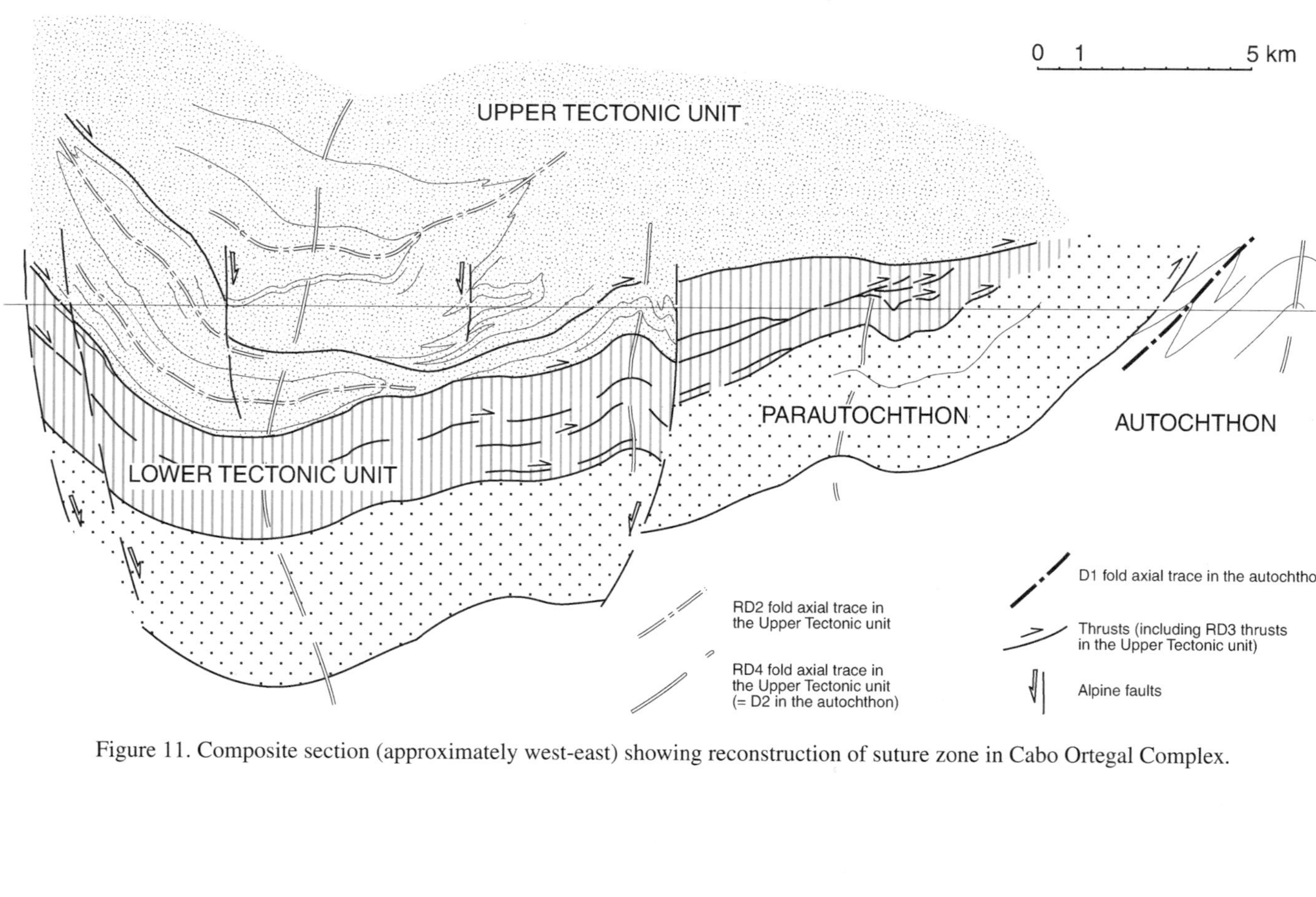

Figure 11. Composite section (approximately west-east) showing reconstruction of suture zone in Cabo Ortegal Complex.

TABLE 1. GENERALIZED TECTONIC HISTORY OF THE CABO ORTEGAL COMPLEX

Ages (in Ma)	Events
490–480 (Early Ordovician) or 390 (Early Devonian) ?	Subduction or underthrusting of a thinned mafic crust and metasediments in a convergent setting within the Rheic Ocean, clearly before or during the Eo- Variscan collision. H*P*-H*T* metamorphism in eclogite and granulite facies, reaching temperatures of 770–876 °C and minimum pressure of 1.4 Gpa. Partial melting of some of the mafic rocks and quartzo-feldspathic gneisses. Orientantion of omphacite could be related to this episode. It indicates coaxial flow, which is likely related to crustal thickening.
<480 ?	Partial exhumation and reequilibration, in granulite facies, of lower crustal rocks up to the base of a normal continental crust (35–40 km; ca. 1.2 GPa). Development of a conspicuous horizontal foliation (RD1) in a constrictional environment. This RD1 foliation was initially formed in HP granulite facies, but later reequilibrated in amphibolite facies conditions (720–600 °C, P: 1.25–0.50 Gpa).
390–385 (Early-Middle Devonian)	Laurentia-Gondwana Collision (Variscan orogeny). Formation of non-coaxial structures in the Upper Tectonic Unit rocks (RD2 recumbent folds) in a constrictional environment. Reworking of previous RD1 foliation in the fold limbs. Strain partitioning: RD3 thrust onset from about 15 km (ca. 0.5 GPa).
ca. 365 Ma (Late Devonian–Early Carboniferous)	Formation of ultramylonites and filonites related to RD3 thrust in the rocks of the Lower Tectonic Unit. First Variscan deformation event in the autochthon (Ollo de Sapo Antiform).
ca. 315 Ma (Late Carboniferous)	General refolding yielding folds with vertical axial planes (RD4) and subvertical shear zones with predominant strike-slip displacement.
Late Oligocene–Early Miocene	Alpine faulting.

on multigrain or single-grain fractions and the latter on single crystals, using sensitive high-resolution ion microprobe.

This contradiction is a crucial point because an Early Ordovician high-*P*–high-*T* event implies the exhumation of these rocks in two orogenic cycles, while a single orogenesis would account for the second age. Structures clearly related to such an important tectonic event during the Early Ordovician have never been reported in the neighboring areas. Ellis (1987) discussed that high-*P* granulites, formed as a result of doubly thickened crust and therefore similar to rocks from the Upper Tectonic unit, could have been exhumed in two orogenic events. The pressure-temperature-time (*P-T-t*) path of such granulites would present an important isobaric cooling (~400 °C), after the metamorphic peak, at pressures equivalent to a normal lower crust (~1.2 Gpa). This does not seem to fit the available *P-T-t* paths for the eclogites and granulites of the Upper Tectonic unit. In both rocks, the *P-T-t* paths show both decreasing *P* and *T* or clear isothermal decompression. However, if these rocks were uplifted to the base of a normal crust and stayed there for a period of ~100 m.y., they would have been reequilibrated to conditions of intermediate-*P* granulites. Such mineral associations have never been reported in the rocks of the Upper Tectonic unit. Eclogites were retrograded either to eclogites or to high-*P* granulites and later to amphibolite facies.

The likely age of the onset of retrogression in granulite and amphibolite facies conditions is 390–385 Ma. This seems to be supported by $^{40}Ar/^{39}Ar$ data on amphibole, which yield a similar age (385 Ma) (Peucat et al., 1990; Dallmeyer et al., 1997). However, temperatures for the first stages of the retrograde path are only slightly lower with respect to the peak metamorphic conditions, and if the closure temperature of zircons for the U-Pb system is within this narrow range or slightly lower, it would be difficult to distinguish the peak age from the age of the earlier retrograde stage. Therefore, 390–385 Ma would be a minimum age for the earlier high-*P*–high-*T* metamorphic event. In summary, a further detailed study is needed in order to understand the meaning of all the radiometric ages in relation to other geological data from the Cabo Ortegal Complex.

From Early to Middle Devonian time onward, exhumation of the Upper Tectonic unit is correlated to the Variscan orogeny (sensu stricto) and continental collision. This led to final emplacement of the Upper Tectonic unit in upper crustal levels over the Lower Tectonic unit. The rocks recorded the progressive development of noncoaxial structures, east-verging asymmetrical isoclinal folds, and thrusts. These thrusts caused the final emplacement of the Cabo Ortegal Complex over the parautochthon and autochthon ca. 365 Ma.

ACKNOWLEDGMENTS

This study was supported by CICYT research project PB98-1545, and the writing of it was planned within the projects PB95-1052 and PB97-0198-C02-01. We thank F. Nebauer and an anonymous referee for their detailed and constructive review. We are grateful to H. Stoll for reviewing the English version.

REFERENCES CITED

Ábalos, B., 1997, Omphacite fabric variation in the Cabo Ortegal Eclogite (NW Spain); relationships with strain symmetry during high-pressure deformation: Journal of Structural Geology, v. 19, no. 5, p. 621–637.

Ábalos, B., Azcárraga, J., Gil Ibarguchi, J.I., Mendia, M.S., and Santos Zalduegui, J.F., 1996, Flow stress, strain rate and effective viscosity evaluation in a high-pressure metamorphic nappe (Cabo Ortegal, Spain): Journal of Metamorphic Geology, v. 14, no. 2, p. 227–248.

Anthonioz, P.M., 1970, Etude des Complexes polymétamorphiques Précambriens de Morais et Bragança: Sciences de la Terre, v. 15, p. 147–166.

Arce Duarte, J.M., Fernández Tomás, J., and Monteserín López, V., 1977, Mapa Geológico de España E. 1:50,000, Hoja no. 7 (7–2) (Cillero) y Memoria explicativa: Instituto Geológico Minero España, Madrid, 47 p.

Arenas, R., 1983, Milonitización y retrometamorfismo en los cizallamientos del Macizo básico/ultrabásico alóctono de Cabo Ortegal (Cadena Herciniana Ibérica, NW de España): Estudios Geológicos, v. 39, p. 31–39.

Arenas, R., 1988, Evolución petrológica y geoquímica de la Unidad alóctona inferior del Complejo metamórfico básico-ultrabásico de Cabo Ortegal (Unidad de Moeche) y del Silúrico paraautóctono, Cadena Herciniana Ibérica (NW de España): Corpus Geologicum Gallaeciae, v. 4, p. 1–545.

Arenas, R., 1991, Opposite *P, T, t* paths of Hercinian metamorphism between the upper units of the Cabo Ortegal Complex and their substratum (northwest of the Iberian Massif): Tectonophysics, v. 191, p. 347–364.

Arenas, R., and Peinado, M., 1984, Petrología de alta presión y de las eclogitas de tipo B del macizo de Cabo Ortegal, Galicia, NW de España: Congreso Español de Geología, v. 2, p. 27–48.

Arenas, R., Gil Ibarguchi, J.I., Gonzalez Lodeiro, F., Klein, E., Martínez Catalán, J.R., Ortega Girones, E., de Pablo Macia, J.G., and Peinado, M., 1986, Tectonostratigraphic units in the complexes with mafic and related rocks of the NW of the Iberian Massif: Hercynica, v. 2, no. 2, p. 87–110.

Arenas, R., Rubio Pascual, F.J., Díaz-García, F., and Martínez-Catalán, J.R., 1995, High-pressure micro-inclusions and development of an inverted metamorphic gradient in the Santiago Schists (Ordenes Complex, NW Iberian Massif, Spain): Evidence of subduction and syn-collisional decompression: Journal of Metamorphic Geology, v. 13, p. 141–164.

Bard, J.P., 1983, Metamorphism of an obducted island arc: Example of the Kohistan sequence (Pakistan) in the Himalayan collided range: Earth and Planetary Science Letters, v. 65, p. 133–144.

Basterra, R., Cassi, J.M., Pérez San Roman, L., Tascón, A., and Gil Ibarguchi, J.I., 1989, Evolución metamórfica de las rocas pelíticas y semipelíticas de las formaciones "Banded Gneisses" y "Gneises de Cariño" (Cabo Ortegal, NW España): Studia Geologica Salmanticensia, v. 4, p. 131–144.

Bastida, F., Marcos, A., Marquínez, J., Martínez Catalán, J.R., Pérez-Estaún, A., and Pulgar, J.A., 1984, Geological map of Spain, La Coruña and report: Instituto Geológico Minero España, Madrid, p. 1–155, scale 1:200,000, sheet 1.

Bastida, F., Aller, J., and Fernández Viejo, G., 1993, The structure of the Ollo de Sapo antiform in the Cantabrian coast (NW Spain): Revista de la Sociedad Geológica de España, v. 6, p. 93–103.

Bayer, A., and Matte, Ph., 1979, Is the mafic/ultramafic massif of Cabo-Ortegal (Northwest Spain) a nappe emplaced during Variscan obduction?: A new gravity interpretation: Tectonophysics, v. 57, p. T9-T18.

Bernard-Griffiths, J., Peucat, J.J., Cornichet, J., Iglesias Ponce de León, M., and Gil Ibarguchi, J.I., 1985, U-Pb, Nd isotope, and REE geochemistry in eclogites from the cabo Ortegal Complex, Galicia, Spain: An exasmple of REE immobility conserving MORB-like patterns during high-grade metamorphism: Chemical Geology, v. 52, p. 217–225.

Calsteren, P.W.C. Van, 1977, A mantle-plume–model interpretation for the Paleozoic geology of Galicia with emphasis on the Cabo Ortegal area (NW Spain): Proceedings Koninklijke Nederlandse Akademie van Wetenschappen, Series B, Physical Sciences, v. 80, no. 3, p. 156–178.

Calsteren, P.W.C. Van, 1978, Geochemistry of the polymetamorphic mafic-ultramafic complex at Cabo Ortegal (NW Spain): Lithos, v. 11, p. 61–72.

Calsteren, P.W.C. Van, 1981, The mantle-plume model, its feasibility and consequences: Leidse Geologische Mededelingen, v. 52, no. 1, p. 73–76.

Calsteren, P.W.C. Van, and den Tex, E., 1978, An early Paleozoic continental rift system in Galicia (NW Spain), *in* Ramberg, I.B., and Neumann, E.R., eds., Tectonics and geophysics of continental rifts: Dordrecht, Holland, Reidel Pub. Co., p. 125–132.

Calsteren, P.W.C. Van, Boelrijk, N.A.I.M., Hebeda, E.H., Priem, H.N.A., den Tex, E., Verdurmen, E.A.T.H., and Verschure, R.H., 1979, Isotopic dating of older elements (including the Cabo Ortegal mafic-ultramafic complex) in the hercynian orogen of NW Spain: Manifestations of a presumed early Paleozoic Mantle-plume: Chemical Geology, v. 24, p. 35–56.

Capdevila, R., 1968, Les tipes de métamorphisme intermédiaires de basse pression dans le segment hercynien de Galice nord-orientale (Espagne): Compte Rendu Académie Sciences Paris, v. 266, p. 1924–1927.

Cowan, S., 1985, Structural styles in Mesozoic and Cenozoic melanges in the Western Cordillera of North America: Geological Society of America Bulletin, v. 96, no. 4, p. 451–462.

Coward, M.P., Jan, M.Q., Rex, D., Tarney, J., Thirlwall, M., and Windley, B.F., 1982, Geotectonic framework of the Himalaya of N. Pakistan: Journal of the Geological Society of London, v. 139, p. 299–308.

Coward, M.P., and Windley, B.F., 1986, Collision tectonics in the NW Himalaya, *in* Coward, M.P., and Ries, A.C., eds., Collision tectonics: Oxford, Blackwell Scientific, Geological Society of London Special Publication 19, p. 203–219.

Dallmeyer, R.D., Martínez Catalán, J.R., Arenas, R., Gil Ibarguchi, J.I., Gutiérrez Alonso, G., Farias, P., Bastida, F., and Aller, J., 1997, Diachronous Variscan tectonothermal activity in the NW Iberian Massif; evidence from $^{40}Ar/^{39}Ar$ dating of regional fabrics: Tectonophysics, v. 277, no. 4, p. 307–337.

den Tex, E., 1981, A geological section across the Hesperian massif in western and central Galicia: Geologie Mijnbouwn, v. 60, p. 33–40.

Díaz García, F., Arenas, R., Martínez Catalán, J.R., González del Tánago, J., and Dunning, G.R., 1999, Tectonic evolution of the Careón ophiolite (northwest Spain): A remnant of oceanic lithosphere in the Variscan Belt: Journal of Geology, v. 107, p. 587–605.

Dietrich, D., and Durney, D.W., 1986, Change in direction of overthrust shear in the Helvetic nappes: Journal of Structural Geology, v. 8, p. 389–398.

Dirks, P.H.G.M., Zhang, J.S., and Passchier, C.W., 1997, Exhumation of high-pressure granites and the role of lower crustal advection in the North China Craton near Datong: Journal of Structural Geology, v. 19, p. 1343–1358.

Ellis, D.J., 1987, Origin and evolution of granulites in normal and thickened crusts: Geology, v. 15, p. 167–170.

Engels, J.P., 1972, The catazonal polymetamorphic rocks of Cabo Ortegal (NW Spain): A structural and petrofabric study: Leidse Geologische Mededelingen, v. 48, no. 1, p. 83–133.

Farias, P., Gallastegui, G., González Lodeiro, F., Marquínez, J., Martín Parra, L.M., Martínez Catalán, J.R., Pablo Macia, J.G. De, and Rodríguez Fernández, L.R., 1987, Aportaciones al conocimiento de la litoestratigrafía y estructura de la Galicia Central: Memorias Facultad Ciencias, Universidad Porto, v. 1, p. 411–431.

Fernández, F., 1993, La deformación de los gneises de Chimparra en Punta Tarroiba (Cabo Ortegal, NW de España): Revista Sociedad Geológica España, v. 6, no. 3–4, p. 77–91.

Fernández, F.J., and Marcos, A., 1996, Mylonitic foliation development by heterogeneous pure shear under high-grade conditions in quartzofeldspathic rocks (Chímparra Gneiss Formation, Cabo Ortegal Complex, NW

Spain), *in* Oncken, O., and Janssen, C., eds.: Basement tectonics 11: Dordrecht, Kluwer Academic Publishers, p. 17–34.

Fernández, F.J., and Marcos, A, 1997, Datos geoquímicos de los gneises de Chímparra y consideraciones sobre el origen de los gneises cuarzofeldespáticos del Complejo de Cabo Ortegal (NW de España), *in* Pires, C.C., Gomes, M.E. P., and Coke, C., eds., Evoluçao geologica do Maciço Ibérico e seu enquadramento continental: Comunicações do XIV Reunião de geologia do Oeste Peninsular, Reunião anual do PICG-376, Vila Real, Portugal, p. 43–48.

Fernández Pompa, F., and Monteserín López, V., 1976, Mapa geológico de España: Hoja no. 7 (6–3) (Cillero) y Memoria explicativa: Instituto Geológico Minero España, Madrid, 73 p., scale 1:50,000.

Fernández Rodríguez, F.J., 1997, Estructuras desarrolladas en gneises bajo condiciones de alta *P* y *T* (genises de Chímparra, Cabo Ortegal, A Coruña, Galicia, España): Serie Nova Terra, O Castro, v. 13, 250 p.

Fernández Suárez, J., Corfu, F., Arenas, R., Marcos, A. Martínez Catalán, J.R., Díaz García, F., Abati, J., and Fernández, F.J., 2000, U-Pb evidence for a polyorogenic evolution of the HP-HT units of the NW Iberian Massif: Variscan-Appalachian dynamics: The building of the Upper Paleozoic basement: Contributions in Mineralogy and Petrology, v. 143, p. 236–253.

Ferrús, B., 1998, Análisis de cuenca y relaciones tectónica-sedimentación en la Cuenca de As Pontes (Galicia). [Ph.D. thesis]: Barcelona, Spain, Universidad de Barcelona, 365 p.

Galán, G., and Marcos, A., 1997, Geochemical evolution of high-pressure mafic granulites from the Bacariza formation (Cabo Ortegal complex, NW Spain): An example of a heterogeneous lower crust: Geologische Rundschau, v. 86, no. 3, p. 539–555.

Galán, G., and Marcos, A., 1999, The high-pressure granulites of the Bacariza Formation: An earlier stage in the exhumation of other eclogites in the Cabo Ortegal complex (Hercynian belt, NW Spain): Trabajos de Geología, Universidad de Oviedo, v. 21, p. 141–157.

Galán, G., and Marcos, A., 2000, The metamorphic evolution of the high-pressure mafic granulites of the Bacariza Formation (Cabo Ortegal Complex, Hercynian belt, NW Spain): Lithos, v. 54, no. 3-4, p. 139–171.

Gil Ibarguchi J.I., Mendia M., Girardeau, J., and Peucat, J.J., 1990, Petrology of the Cabo Ortegal Complex (Northwestern Spain): Lithos, v. 25, p. 133–162.

Girardeau, J., and Gil Ibarguchi, J.I., 1991, Pyroxenite-rich peridotites of the Cabo Ortegal Complex (northwestern Spain): Evidence for large-scale upper-mantle heterogeneity: Journal of Petrology, Special Lherzolites Issue, p. 135–154.

Girardeau, J., Gil Ibarguchi, J.I., and Ben Jamaa, N., 1989, Evidence for a heterogeneous upper mantle in the Cabo Ortegal Complex, Spain: Science, v. 245, no. 4923, p. 1231–1233.

Godard, G., and van Roermund, H.L.M., 1995, Deformation-induced clinopyroxene fabrics from eclogites: Journal of Structural Geology, v. 17, p. 1425–1443.

Gravestock, P.J., 1992, The chemical causes of uppermost mantle heterogeneities [Ph.D. Thesis]: Open University, Milton Keynes, UK, 269 p.

Helmsteadt, H., Anderson, O.L., and Gavasci, A.T., 1972, Petrofabric studies of eclogite, spinel-websterite, and spinel-lherzolite xenoliths from kimberlite-bearing breccia pipes in southeastern Utah and northeastern Arizona: Journal of Geophysical Research, v. 77, p. 4350–4365.

Hsü, K.J., 1968, The principles of mélanges and their bearing on the Franciscan-Knoxville paradox: Geological Society of America Bulletin, v. 79, p. 1063–1064.

Iglesias, M., Ribeiro, M.L., and Ribeiro, A., 1983, La interpretación aloctonista de la estructura del Noroeste peninsular: Instituto Geológico Minero España, Libro Jubilar J.M. Rios, Geología de España, Madrid, v. 1, p. 459–467.

Iglesias Ponce de León, M., and Choukroune, P., 1980, Shear zones in the Iberian Arc: Journal of Structural Geology, v. 2, p. 63–68.

Julivert, M., Fontboté, J.M., Ribeiro, A., and Conde, L., 1974, Mapa tectónico de la Península Ibérica y Baleares: Instituto Geológico Minero España, 113 p.

Keasberry, E.J., Van Calsteren, P.W.C., and Kuijper, R.P., 1976, Early Paleozoic mantle diapirism in Galicia: Tectonophysics, v. 31, p. 61–65.

Kretz, R., 1983, Symbols for rock-forming minerals: American Mineralogists, v. 68, p. 277–279.

Kuijper, R.P., and Arps, C.E.S., 1983, Un modelo de evolución basado en el desarrollo de una mantle plume: Instituto Geológico Minero España, Libro Jubilar J.M. Rios, Geología de España, v. 1, p. 455–458.

Kuijper, R.P., Vogel, D.E., and den Tex, E., 1985, Eclogite-plagiopyrigarnite relations in the catazonal complexes of Northwest Spain: Chemical Geology, v. 50, p. 163–171.

Llana-Fúnez, S., Marcos, A., Kunze, K., and Burg, J.-P., 2001, Deformation features in Concepenido eclogites (Cabo Ortegal Complex, Variscan belt of Iberia) [abs.]: Deformation Mechanisms, Rheology and Tectonics, Noordwijkerhout (The Netherlands), April 2–4, 2001, Abstract volume, p. 105.

Marcos, A., and Farias, P., 1997, La estructura de la sutura varisca en la transversal de Cabo Ortegal (NW de España), *in* Pires, C.C., Gomes, M.E. P., and Coke, C., eds., Evoluçao geologica do Maciço Ibérico e seu enquadramento continental: Comunicações do XIV Reunião de geologia do Oeste Peninsular, Reunião anual do PICG-376, Vila Real, Portugal, p. 109–114.

Marcos, A., and Farias, P., 1999, La estructura de las láminas inferiores del Complejo de Cabo Ortegal y su autóctono relativo (Galicia, NW España): Trabajos de Geología, Universidad de Oviedo, v. 21, p. 201–220.

Marcos, A., and Galán, G., 1994, Structural and metamorphic evolution of high-pressure bassic granulites during basement exhumation: An example from the Bacariza formation, Cabo Ortegal complex (NW Sapin): Proceedings, 11th International Conference on Basement Tectonics, International Basement Tectonics Association, Postdam, p. 89–90.

Marcos, A., Fernández Rodríguez, F.J., and Llana Fúnez, S., 2000, Structure of the Cabo Ortegal nappe: Basement tectonics 15, Mid-Conference Field Trip, A Coruña, Spain, p. 1–15.

Marques, F.O., Ribeiro, A., and Munhá, J.M., 1996, Geodynamics evolution of the Continental Allochthonous Terrane (CAT) of the Bragança Nappe Complex, NE Portugal: Tectonics, v. 15, p. 747–762.

Martínez Catalán, J.R., 1990, A non-cylindrical model for the northwest Iberian allochthonous terranes and their equivalents in the Hercynian belt of Western Europe: Tectonophysics, v. 179, p. 253–272.

Martínez Catalán, J.R., Arenas, R., Díaz García, F., Rubio Pascal, F.J., Abati, J., and Marquínez, J., 1996, Variscan exhumation of a subducted Paleozoic continental margin; the basal units of the Ordenes Complex, Galicia, NW Spain: Tectonics, v. 15, no. 1, p. 106–121.

Martínez Catalán, J.R., Arenas, R., Díaz García, F., and Abati, J., 1997, Variscan accretionary complex of northwest Iberia: Terrane correlation and succession of tectonothermal events: Geology, v. 25, p. 1103–1106.

Martínez Catalán, J.R., Arenas, R., Díaz García, F., and Abati, J., 1999, Allochthonous units in the Variscan Belt of NW Iberia (Terranes and Accretionary History), *in* Sinha, A.K., ed., Basement tectonics 13: Dordrecht, Kluwer Academic Publishers, p. 65–84.

Martínez García, E., Fernández Pompa, F., Arce, M., Fernández Martínez, F., Fernández Tomás, J., and Monteserín, V., 1975, Nuevos datos para la interpretación del complejo básico de Cabo Ortegal (Galicia, noroeste de España): Tecniterrae, v. 7, p. 8–12.

Matte, Ph., 1968, La structure de la virgation hercynienne de Galice (Espagne): Geologie Alpine, v. 44, p. 155–280.

Matte, Ph., 1983, Two geotraverses across the Ibero-Armorican Variscan arc of western Europe, *in* Rast W., and Delany, M., eds., Profiles of orogenic belts: Washington, D.C., American Geophysical Union, Geodynamic series, v. 10, p. 53–81.

Matte, Ph., 1986, Tectonic and plate tectonic model for Variscan belt of Europe: Tectonophysics, v. 126, p. 329–374.

Matte, Ph., 1998, Continental subduction and exhumation of HP rocks in Paleozoic orogenic belts: Uralides and Variscides: G.F.F., v. 120, p. 209–222.

Meer Mohr, C.G. Van Der, 1975, The Palaeozoic strata near Moeche in Galicia, NW Spain: Leidse Geologische Mededelingen, v. 49, p. 32–37.

Mendia, M., 1996, Petrología de la unidad eclogítica del Complejo de Cabo Ortegal (NW de España) [Ph.D. thesis]: Universidad del País Vasco, Leioa, Pais Vasco, Spain, 463 p.

Moreno, T., 1999, Platinum-group elements and chromite mineralisation in ultramafic rock: a case study from the Cabo Ortegal Complex, Northwest Spain [Ph.D. thesis]: Wales, Cardiff University, 224 p.

Ordoñez Casado, B., 1998, Geochronological studies of the Pre-Mesozoic basement of the Iberian Massif: The Ossa-Morena zone and the allochthonous Complexes within the Central Iberian zone [Dissertation]: ETH, Zürich, no. 12.940, 235 p.

Ordoñez Casado, B., Gebauer, D., Shäfer, H.J., Gil Ibarguchi, J.I., and Peucat, J.J., 1996, A single subduction event ca. 392 Ma for the ultramafic-mafic HP-HT-rocks of the Cabo Ortegal Complex: Geogaceta, v. 20, p. 489–490.

Parga Pondal, I., Parga Peinador, J.R., Vegas, R., and Marcos, A., 1982, Mapa geológico del Macizo Hespérico: Publicacios Area Xeoloxía Minería, Seminario Estudos Galegos, Edicios do Castro, La Coruña, scale 1:500,000.

Pérez-Estaún, A., Martínez Catalán, J.R., and Bastida, F., 1991, Crustal thickening and deformation sequence in the footwall to the suture of the Variscan belt of northwest Spain: Tectonophysics, v. 191, p. 243–253.

Peucat, J.J., Bernard-Griffiths, J., Gil Ibarguchi, J.I., Dallmeyer, R.D., Menot, R.P., Cornichet, J., and Iglesias Ponce de Leon, M., 1990, Geochemical and geochronological cross section of the deep Variscan crust; the Cabo Ortegal high-pressure nappe (northwestern Spain): Tectonophysics, v. 177, p. 263–292.

Ramsay, J.G., 1967, Folding and fracturing of rocks: New York, McGraw-Hill, 568 p.

Ramsay, J.G., and Lisle, R.J., 2000, The techniques of modern structural geology. Volume 3: Applications of continuum mechanics in structural geology: London, Academic Press, p. 701–1061.

Rast, N., and Wright Horton, J. Jr., 1989, Mélanges and olistostromes in the Appalachians of the United States and mainland Canada, *in* Wright Horton, J. Jr., and Rast, N., eds., Mélanges and Olistostromes of the U.S. Appalachians: Boulder, Colorado, Geological Society of America Special Paper 228, p. 1–15.

Rathbone, P.A., Coward, M.P., and Harris, A.L., 1983, Cover and basement: A contrast in style and fabrics, *in* Hatcher, R.D., Jr., Williams, H., and Zietz, I., eds., Contributions to the tectonics and geophysics of mountain chains: Boulder, Colorado, Geological Society of America Memoir 158, p. 213–223.

Raymond, A., 1984, Classification of mélanges, *in* Raymond, L.A., ed., Mélanges: Their nature, origin and significance: Boulder, Colorado, Geological Society of America Special Paper 198, p. 7–20.

Raymond, L.A., Yurkouich, S.P., and McKinney, M., 1989, Block-in matrix structures in the North Caroline Blue Ridge Belt and their significance for the tectonic history of the northern Applachian orogen: Geological Society of America Bulletin, v. 228, p. 195–215.

Réche, J., Martínez, F.J., Arboleya, M.L., Dietsch, C., and Briggs, W.D., 1998, Evolution of a kyanite-bearing belt within a HT-LP orogen: The case of NW Variscan Iberia: Journal of Metamorphic Geology, v. 16, no. 3, p. 379–394.

Ribeiro, A., Cramez, C., Rebelo, J.A., 1964, Sur la structure de Tras-os-Montes (Nord-Est du Portugal): Paris, Compte Rendu Académie Sciences, v. 53, p. 233–244.

Ribeiro, M.L., and Floor, P., 1987, Magmatismo peralcalino no Maciço Hespérico: Sua distribução e significado geodinámico, *in* Bea, F., Carnicero, A., Gonzalo, J.C., López-Plaza, M., and Rodríguez Alonso, M.D., eds., Geología de los granitoides y rocas asociadas del Macizo Hespérico: Madrid, Spain, Rueda, p. 211–221.

Ribeiro, A., Pereira, E., and Dias, R., 1990, Structure in the northwest of the Iberian Peninsula (within the Allochthonous Sequences in the Centro-Iberian zone), *in* Dallmeyer, R.D., and Martínez García, E., eds., Pre-Mesozoic geology of Iberia: Berlin, Springer-Verlag, p. 220–236.

Ries, A.C., and Shackleton, R.M., 1971, Catazonal complexes of northwest Spain and North Portugal: Remnants of a Hercynian thrust plate: Nature, v. 234, no. 47, p. 65–68.

Rutter, E.H., Brodie, K.H., and Evans, P.J., 1993, Structural geometry, lower crustal magmatic underplating and lithospheric stretching in the Ivrea-Verbano zone, northern Italy: Journal of Structural Geology, v. 15, p. 647–662.

Santos Zalduegui, J.F., Schaerer, U., Gil Ibarguchi, J.I., and Girardeau, J., 1996, Origin and evolution of the Paleozoic Cabo Ortegal ultramafic-mafic complex (NW Spain); U-Pb, Rb-Sr and Pb-Pb isotope data: Chemical Geology, v. 129, no. 3–4, p. 281–304.

Schäfer, H.J., Gebauer, D., Gil Ibarguchi, J.I., and Peucat, J.J., 1993, Ion-microprobe U-Pb zircon dating on the HP/HT Cabo Ortegal Complex (Galicia, NW Spain), preliminary results: Terra Abstracts, v. 22, p. 54.

Vogel, D.E., 1967, Petrology of an eclogite- and pyrigarnite-bearing polymetamorphic rock complex at Cabo Ortegal, NW Spain: Leidse Geologische Mededelingen, v. 40, p. 121–213.

Windley, B.F., 1983, Metamorphism and tectonics of the Himalaya: Journal of the Geological Society of London, v. 140, p. 711–725.

Windley, B.F., and Tarney, J., 1986, The structural evolution of the lower crust of orogenic belts, past and present, *in* Dawson, J.B., Carswell, D.A., Hall, J., and Wedepohl, K.H., eds., The nature of lower continental crust: Geological Society Special Publication, no. 24, p. 221–230.

Manuscript Accepted by the Society February 8, 2002

Printed in the U.S.A.

Geological Society of America
Special Paper 364
2002

Thrust and detachment systems in the Ordenes Complex (northwestern Spain): Implications for the Variscan-Appalachian geodynamics

José R. Martínez Catalán
Departamento de Geología, Universidad de Salamanca, 37008 Salamanca, Spain
Florentino Díaz García
Departamento de Geología, Universidad de Oviedo, 33005 Oviedo, Spain
Ricardo Arenas, Jacobo Abati, Pedro Castiñeiras
Departamento de Petrología y Geoquímica, Universidad Complutense, 28040 Madrid, Spain
Pablo González Cuadra, Juan Gómez Barreiro
Departamento de Geología, Universidad de Salamanca, 37008 Salamanca, Spain
Francisco J. Rubio Pascual
Instituto Geológico y Minero de España, Ríos Rosas 23, 28003 Madrid, Spain

ABSTRACT

The allochthonous complexes of northwestern Iberia consist of a pile of units of Gondwanan and peri-Gondwanan provenance, and include oceanic lithosphere. The units are classified into upper, intermediate (ophiolitic), and basal. We present a detailed geological map and sections across the Ordenes Complex, together with a brief description of its units and a discussion of its structures.

In the upper units, two complete cycles of burial and exhumation have been identified. The first cycle, of Early Ordovician age, records a convergent plate margin, possibly in a peri-Gondwanan domain. The second is Variscan, and the structural evolution of the three groups of allochthonous units reflects progressive accretion to an active orogenic wedge. Continuous understacking of continental and oceanic fragments toward the west began with the upper units and ended with the basal units. The latter represent the outermost margin of Gondwana, and their subduction marked the closure of the intervening ocean, and the change from subduction to a collisional regime. Terrane accretion took place in the Devonian and, during the Late Devonian and the Carboniferous, the deformation progressed inboard of the Gondwana margin.

Variscan emplacement of the allochthonous units occurred in two successive thrusting episodes. The first placed the basal units over the sedimentary cover of the Gondwana margin in what seems to follow a normal sequence of thrusting. The second carried the upper and ophiolitic units on top of the previous nappe pile and has an out-of-sequence character.

A possible correlation of the Early Ordovician convergence, early Variscan accretion, subsequent oceanic closure, continent-continent collision, and renewed thrust activity during the late Carboniferous in northwestern Iberia is established with the Taconian, Acadian, and Alleghanian orogenies in the Appalachians.

Martínez Catalán, J.R., Díaz García, F., Arenas, R., Abati, J., Castiñeiras, P., González Cuadra, P., Gómez Barreiro, J., and Rubio Pascual, F.J., 2002, Thrust and detachment systems in the Ordenes Complex (northwestern Spain): Implications for the Variscan-Appalachian geodynamics, *in* Martínez Catalán, J.R., Hatcher, R.D., Jr., Arenas, R., and Díaz García, F., eds., Variscan-Appalachian dynamics: The building of the late Paleozoic basement: Boulder, Colorado, Geological Society of America Special Paper 364, p. 163–182.

INTRODUCTION

The Iberian Massif is part of a tight orocline known as the Ibero-Armorican arc (Bard et al., 1971; Ribeiro et al., 1995), which represents the western part of the European Variscan Belt. Variscan structures were once continuous between northern Iberia and the Armorican Massif, and, apparently, the belt extended southward into northern Africa. A late Paleozoic reconstruction of the continental masses (i.e., at the end of the Variscan cycle) places Iberia adjacent to the Grand Banks of Newfoundland (Fig. 1) and close to the junction of three important Paleozoic belts: the Appalachians, the Caledonides, and the Variscides (Lefort, 1989).

Although the position of Iberia at the end of the Proterozoic is not precisely located, there is general agreement that it occupied a marginal area in northern Gondwana. During the Late Proterozoic, this area was an active continental margin (Murphy and Nance, 1991; Ochsner, 1993) and underwent the Cadomian orogeny. Cadomian activity was followed by Cambrian-Ordovician continental rifting, which resulted in the pulling

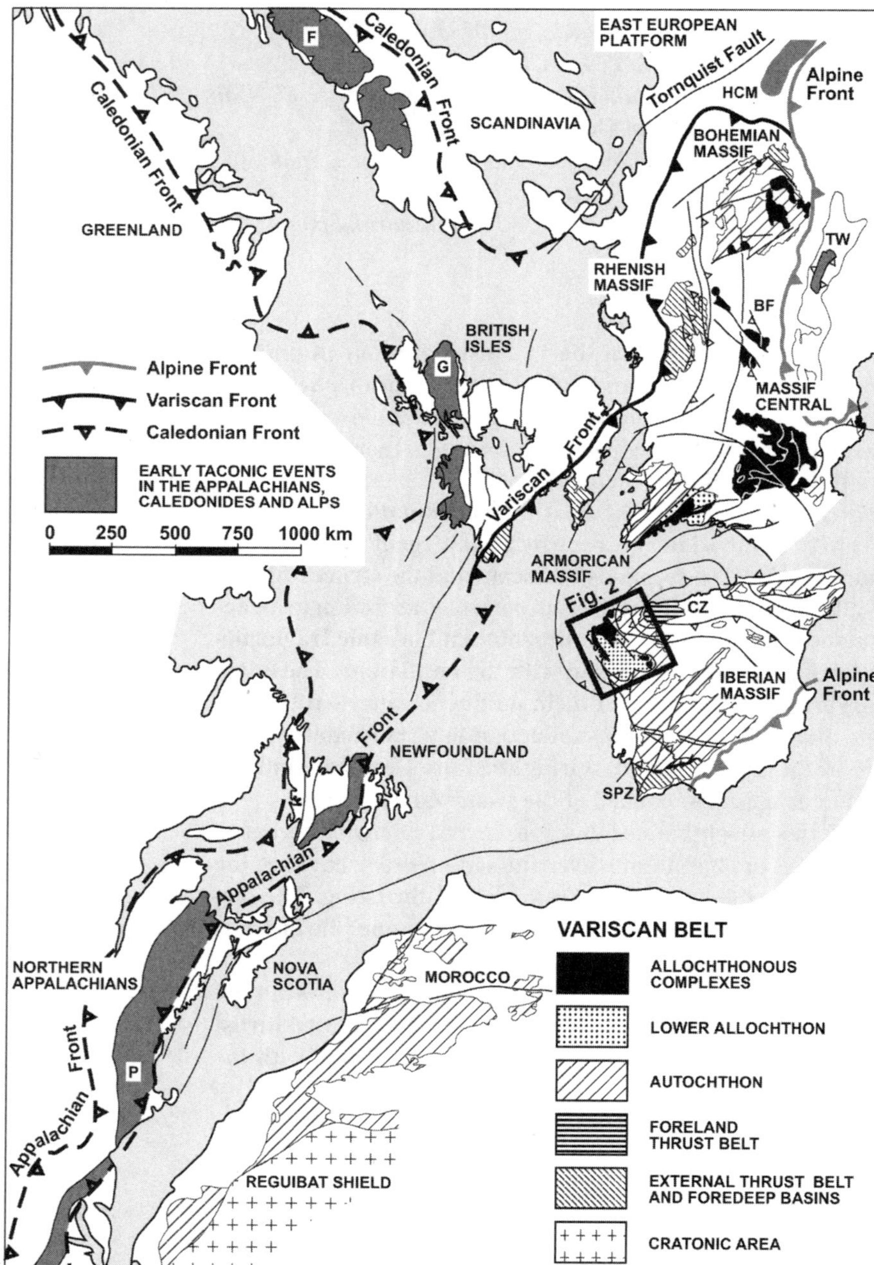

Figure 1. Sketch showing position of Iberia in relation with Appalachian, Caledonian, and Variscan Belts. Abbreviations: BF, Black Forest; CZ, Cantabrian zone; F, Finnmarkian; G, Grampian; HCM, Holy Cross Mountains; P, Penobscottian; SPZ, South Portuguese zone; TW, Tauern Window. Modified from Neuman and Max (1989), with data from Lefort (1989), Franke (1989), and Martínez Catalán et al. (1999).

apart of the Avalon microcontinent and other peri-Gondwanan terranes from the Gondwana mainland and the opening of the Rheic oceanic domain (Fortey and Cocks, 1988; Soper, 1988).

The Rheic oceanic realm became subsequently closed during convergence between Laurentia and Gondwana. Convergence started in the Early Ordovician in the Appalachians (Keppie, 1989; Neuman and Max, 1989; Cawood et al., 1995; Hibbard and Samson, 1995) and proceeded during the Paleozoic, finally resulting in the collision of both continents. Convergence and collision gave rise to the Variscan and Appalachian orogenic belts.

Relicts of the peri-Gondwanan terranes and of the Rheic lithosphere have been preserved in five allochthonous complexes in northwestern Iberia (Fig. 2). This chapter describes and compares the structural history of the individual tectonic units of one of them, the Ordenes Complex, and focuses on their stacking and emplacement on top of the autochthonous northern Gondwana margin. Ductile to brittle structures are described and interpreted, and the role of extensional structures is evaluated. Particular attention is paid to the large-displacement thrust faults responsible for the final emplacement of most of the complex.

The tectonometamorphic history of the different units sheds light on the geodynamic history of the Gondwana-Laurentia convergence. The implications in the orogenic evolution on both sides of the present North Atlantic are discussed.

GEOLOGIC SETTING

Most of the Iberian Massif has Neoproterozoic–early Paleozoic basement derived from northern Gondwana, as shown by the identification of Cadomian (Pan-African) events, and by the similarity in the oldest ages obtained from upper intercepts and inherited zircons from Iberian orthogneisses and northern Gondwana basement. Ages range between 2.7 and 1.8 Ga (Kuijper, 1980; Lancelot et al., 1985; Peucat et al., 1990; Gebauer, 1993) and are similar to those of the western African craton (Bessoles, 1977; Caby, 1989).

The Iberian Massif is essentially composed of rocks with Gondwanan affinity: the Cadomian basement and the Paleozoic sedimentary sequences, intrusives, and volcanics formed at its northern continental margin. This margin seems to have been predominantly passive from the Cambrian to the Silurian-Devonian boundary. This is suggested by the shallow-water platform character of most stratigraphic units, and by the continuity in the sedimentation, which in places continued to the Early Devonian.

The Gondwanan provenance seems clear for the foreland thrust belt of northern Iberia (the Cantabrian zone; see Figs. 1 and 2), and for most of the internal zones, characterized by penetrative ductile deformation, Barrovian-type metamorphism, and large volumes of synkinematic to postkinematic granitoids. Conversely, a Gondwanan provenance seems doubtful for an amphibolitic belt in southwestern Iberia (the Beja-Acebuches ophiolites), for the basement of the terrane occurring south of the ophiolite (the South Portuguese zone), having age and facies similar to the Rhenohercynian zone of central Europe (Franke, 1989), and for some of the units that crop out in the allochthonous complexes in the northwest, the object of this contribution.

The allochthonous complexes crop out in five synforms or structural basins as megaklippen in Galicia (northwest Spain) and northern Portugal (Fig. 2). They consist of a pile of units that were stacked, thinned, and dismembered during the Variscan deformation (Arenas et al., 1986; Martínez Catalán et al., 1997; Dallmeyer et al., 1997).

Ordenes (135 by 75 km) is the largest of the klippen and contains representatives of all different types of allochthonous units. Detailed descriptions and interpretations of many of its units have been published elsewhere (Van Zuuren, 1969; Marquínez García, 1984; Díaz García, 1990; Arenas et al., 1995, 1997; Martínez Catalán et al., 1996; Abati et al., 1999; Díaz García et al., 1999a, 1999b). Plate 2 shows a detailed geological map of the complex, and Plate 3 depicts an idealized composite of nine geological sections through the different allochthonous units, and gives an idea of the structural style, showing the main crosscutting relationships among structures.

The allochthonous complexes were thrust over the sedimentary cover of the passive continental margin of Gondwana. This cover was in turn detached, giving rise to the lower allochthon (the paraautochthon of Ribeiro et al., 1990), which is commonly referred to as the schistose domain (Marquínez García, 1984; Farias et al., 1987). The ensemble formed by the lower allochthon and the complexes was carried inboard onto the autochthonous Paleozoic sequence of Gondwana (see González Clavijo and Martínez Catalán, this volume). The autochthon is commonly repeated by recumbent folds and thrusts. However, we maintain the term autochthon because the sequences in it were deposited over the same basement upon which they now rest. The lower allochthon was part of the same continental margin, but occupied an external position relative to the autochthon (Farias et al., 1987).

ALLOCHTHONOUS UNITS AND THEIR INTERNAL STRUCTURE

The complexes consist of several allochthonous units, which are characterized by a lithologic association and a particular tectonometamorphic evolution. The units are tectonic in character: they are separated from each other by faults, either thrusts or extensional detachments. The units are grouped according to their lithologic affinities and their relative position in the original nappe pile. Further subdivision can be done based on the lithologies and the pressure-temperature (*P-T*) conditions of the metamorphic evolution of the units.

The Ordenes Complex shares with other Iberian complexes the types of allochthonous units and their order of stacking, which although generally consistent, may be locally modified by tectonic mixing in thrust zones. Three groups, basal, intermediate, and upper, are described from bottom to top in ascending

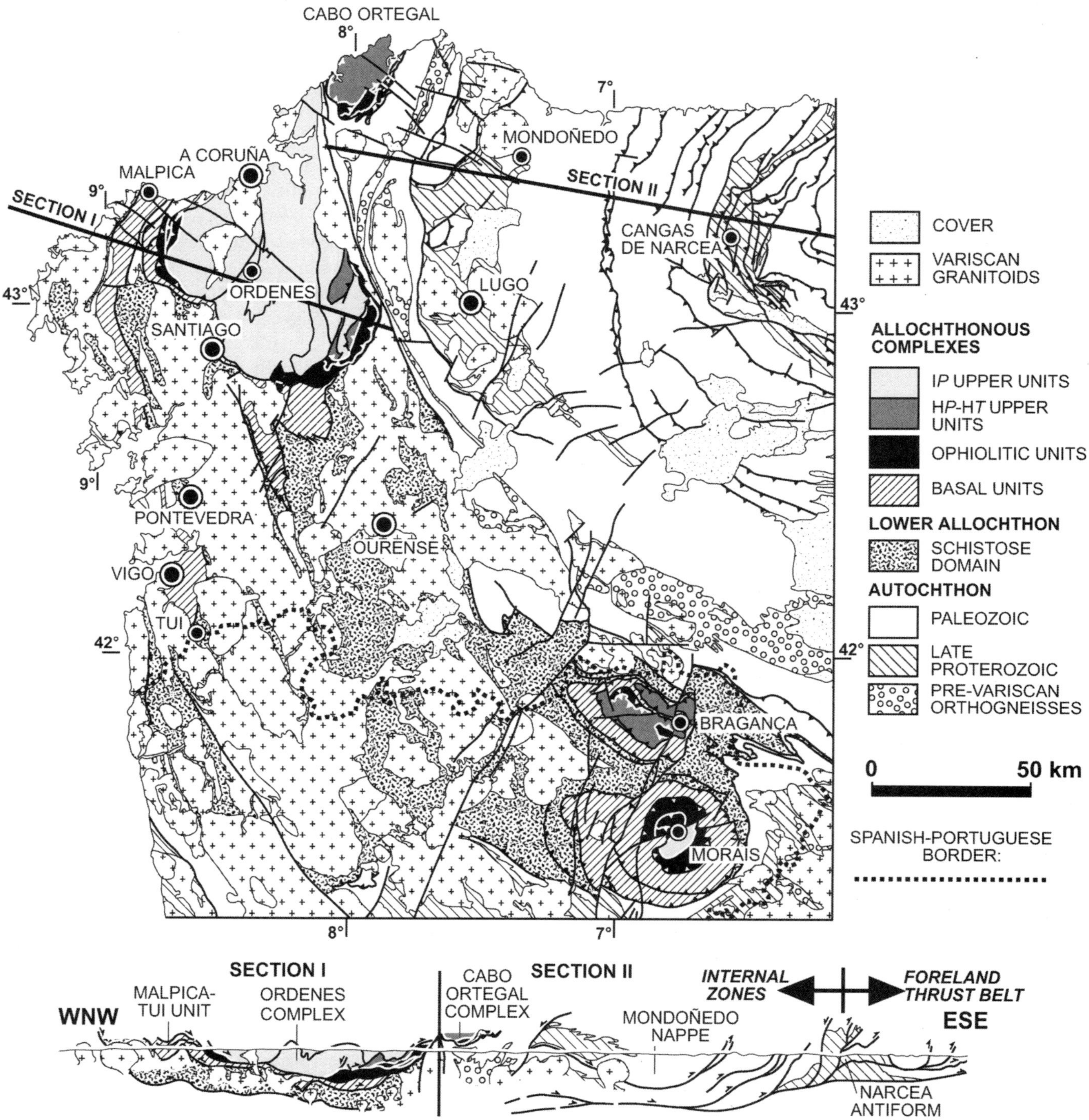

Figure 2. Geological map and composite cross section of northwestern Iberia, showing allochthonous complexes and their groups of units. For location, see Figure 1. I*P,* intermediate pressure; H*P*-H*T,* high-pressure–high temperature.

structural order, together with the individual units constituting them. The intermediate units show ophiolitic affinities and are referred to as the ophiolitic units. They are interpreted as the suture separating two main paleogeographic realms.

Basal units

The lowermost units crop south and west of the Ordenes Complex, and continue to the west in the Malpica-Tui unit (Fig. 2), the structure and evolution of which are described in Llana-Fúnez and Marcos (this volume).

The basal units consist of schists and paragneisses alternating with felsic and mafic igneous rocks. The metasediments are derived from graywackes and pelites, which, together with their monotonous character, suggest a relatively distal depositional environment. Granitic and peralkaline orthogneisses have yielded ages of 480–460 Ma (Rb-Sr whole rock, Van Calsteren et al., 1979; García Garzón et al., 1981; U-Pb, zircon, Santos Zalduegui et al., 1995). The magmatism reflects an Ordovician episode of rifting (Ribeiro and Floor, 1987; Pin et al., 1992) that probably resulted in the pulling apart of a peri-Gondwanan terrane. However, the basal units are thought to have remained attached to Gondwana, because no ophiolites separate them from the underlying sequences with Gondwanan affinities.

The basal units are considered allochthonous because there is a dramatic change in the metamorphic grade and evolution between them and their substratum: the basal units record a high-*P* regional metamorphism not found in the lower allochthon or in the autochthon. Representing the edge of the margin formed after the opening of an ocean, these units underwent westward subduction (in present coordinates) between 380 and 365 Ma, during the closure of the oceanic domain. The subduction of the outermost edge of Gondwana took place under an accretionary wedge, built just before and consisting of a pile formed by the ophiolitic and upper units (Arenas et. al., 1995; Martínez Catalán et al., 1996).

Forcarei, Lalín, and Santiago units. In the south of the Ordenes Complex, the basal units form the core of two late synforms in Lalín and Forcarei, where their lower limit is the Lalín-Forcarei thrust, and the upper one, the Pico Sacro detachment. In the west, they surround the complex from Santiago to the coast. Here, the Pico Sacro detachment is the lower limit, whereas a thrust fault constitutes its upper boundary. This map pattern permits the identification of three main basal units: Forcarei, Lalín, and Santiago, separated by granitoids and the Pico Sacro detachment. The Santiago unit was overthrust by a small tectonic slice, called the Agualada unit, and the Lalín unit is locally structurally overlain by an ultramafic sheet, the Campo Marzo unit (Plates 2 and 3; Fig. 3).

The basal units are folded by a large recumbent anticline (Plate 3, section 5), the lower limb of which is affected by the Lalín-Forcarei thrust, a ductile fault with eastward movement. The fold hinge occurs north of Lalín (Plate 2), and its axis plunges 10°–20° to the northwest. A band of amphibolites can be mapped to the south and southwest of Lalín, and to the west of Forcarei (Plate 2), delineating the overturned limb of the anticline, which attains a length of nearly 30 km. Its normal limb is also defined by the amphibolites and felsic orthogneisses (Plates 2 and 3, section 5; Fig. 4) and can be traced westward to the Santiago unit, in the west border of the complex.

In the hinge zone, a medium-grained foliation (S_2) is parallel to the bedding and is folded with it. This foliation is characterized by the synkinematic growth of albite porphyroblasts. An earlier, fine-grained cleavage (S_1) is preserved as oriented microinclusions in the albite porphyroblasts. *P-T* estimates for these microinclusions (500 °C, 15–17 kbar) indicate that the early fine-grained foliation developed in high-*P* metamorphic conditions (Fig. 5), and is interpreted as the fabric contemporaneous with the subduction of the basal units (Arenas et al., 1995). Similar *P-T* conditions have been calculated in eclogites of the Santiago unit (Rubio Pascual et al., this volume). The high-*P* event ended ca. 374–365 Ma (Van Calsteren et al., 1979; Santos Zalduegui et al., 1995; Rb-Sr age of posteclogitic white micas in the Malpica-Tui unit). The different *P-T* conditions estimated from the two limbs of the recumbent anticline were interpreted by Martínez Catalán et al. (1996) to reflect a westward polarity for the early Variscan subduction.

Other foliations overprint S_2, the fabric folded by the recumbent anticline. They were mostly developed at its reverse and normal limbs, and an axial planar foliation can be identified at its hinge zone. The younger foliations are collectively termed S_3, because they are roughly contemporaneous, even though their origin varies.

At the hinge zone, S_3 is probably related to the recumbent folding. In the overturned limb of the anticline S_3 transects S_2, dipping 10°–20° steeper to the west (when the effects of late folding are removed). In the basal part of Lalín, this fabric is fine grained and is rotated into parallelism with the Lalín-Forcarei thrust (Plate 3, section 5), indicating a relationship between the S_3 and thrusting. The thrust is a 40–80-m-thick shear zone with mylonites and ultramylonites, and separates the basal units in the Lalín and Forcarei synforms from the lower allochthon. Rocks of both units are mylonitized. Thin layers of mylonitic quartzite occur in the shear zone and were considered to mark the thrust surface by Hilgen (1971). They have an elongated texture and show a stretching lineation, which is variably oriented due to later folding. Quartz c-axis fabrics are of the single girdle type or show the axes at high angles to the mylonitic foliation, with maxima normal to it (Díaz García, 1993). The single girdle types, when oblique to the foliation, provide kinematic criteria for the motion of the thrust sheet (Bouchez et al., 1983). The east-west attitude of the lineation southwest of Lalín is considered little modified by younger folding, because it occurs at the hinge zone of the Lalín synform, which is a rather open fold. There, the quartz c-axis fabrics indicate a top-to-the-east sense of motion, which is consistent with S-C structures and asymmetric sigmoidal boudinage in the shear zone.

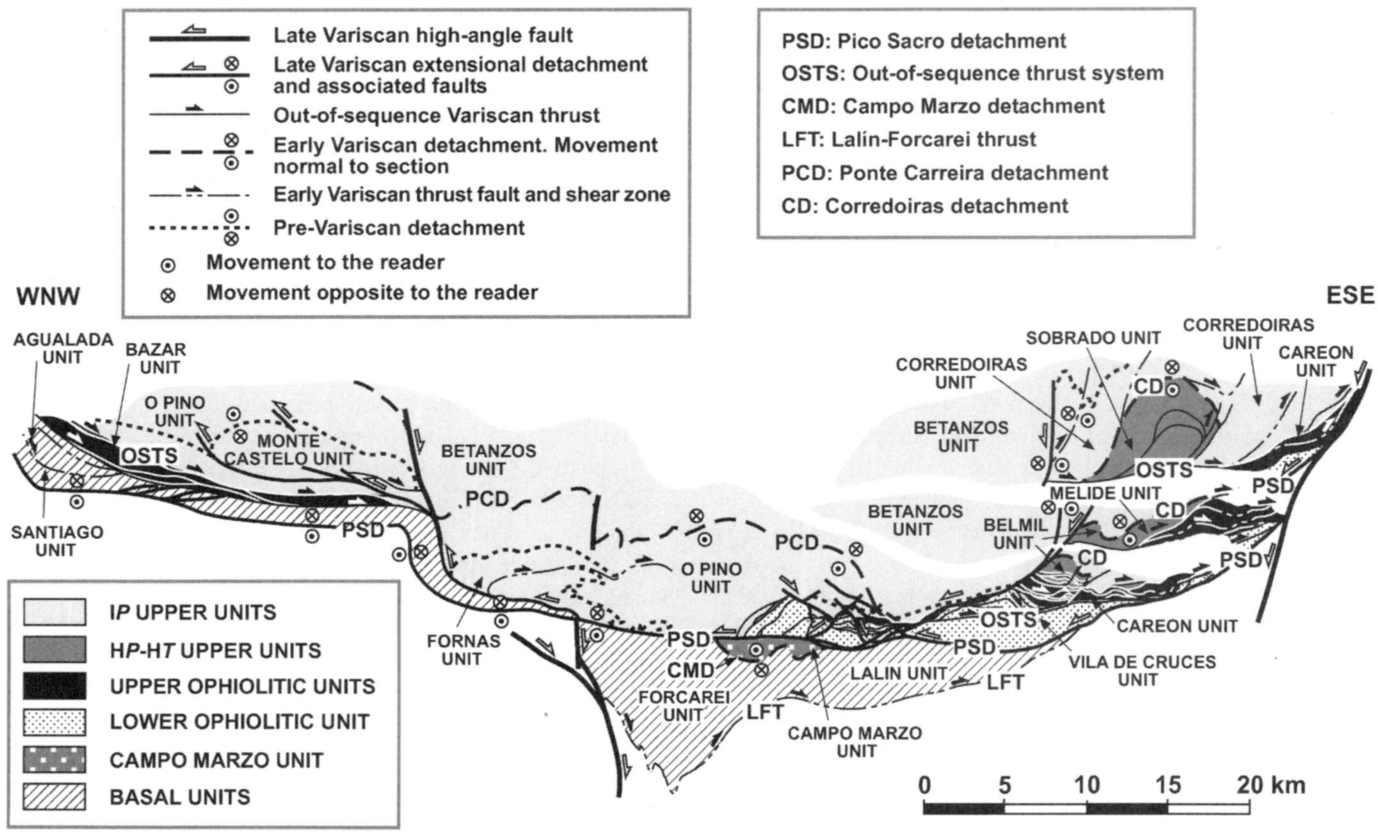

Figure 3. Structural sketch of Ordenes Complex based on Plate 3, showing groups of allochthonous units and different types of faults separating units. I*P*, intermediate pressure; H*P*-H*T*, high-pressure–high-temperature.

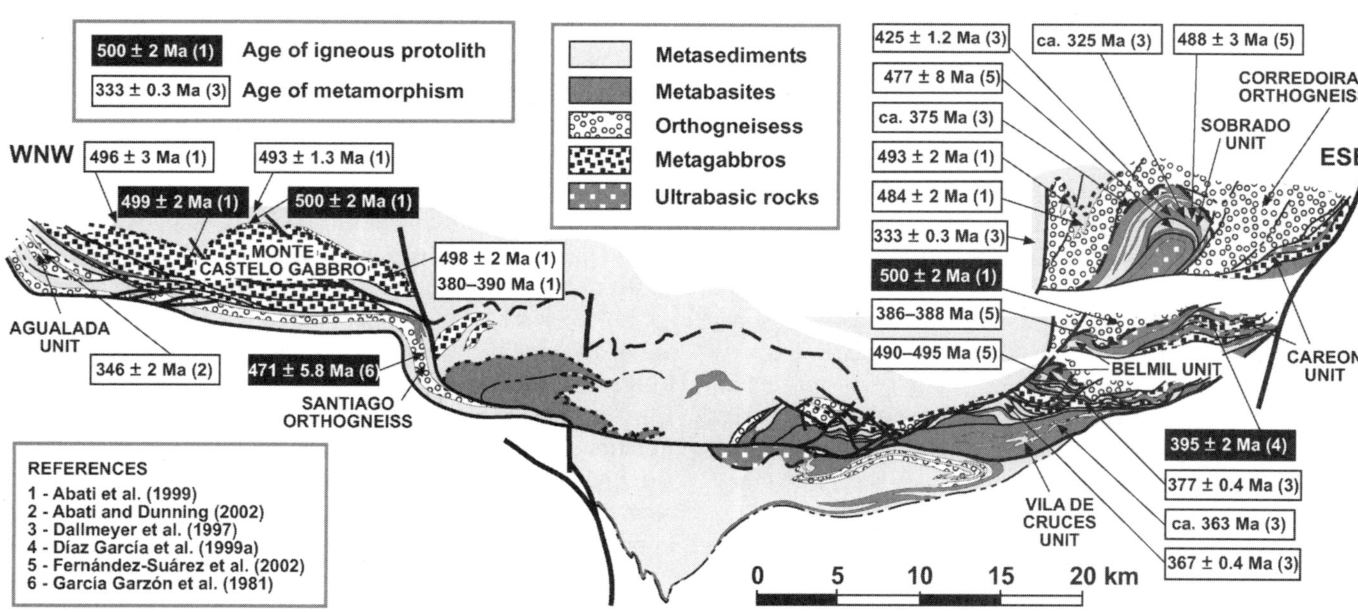

Figure 4. Distribution of main lithologic types in Ordenes Complex, and published isotopic ages of igneous protoliths (black) and regional fabrics (white). Fault types as in Figure 3.

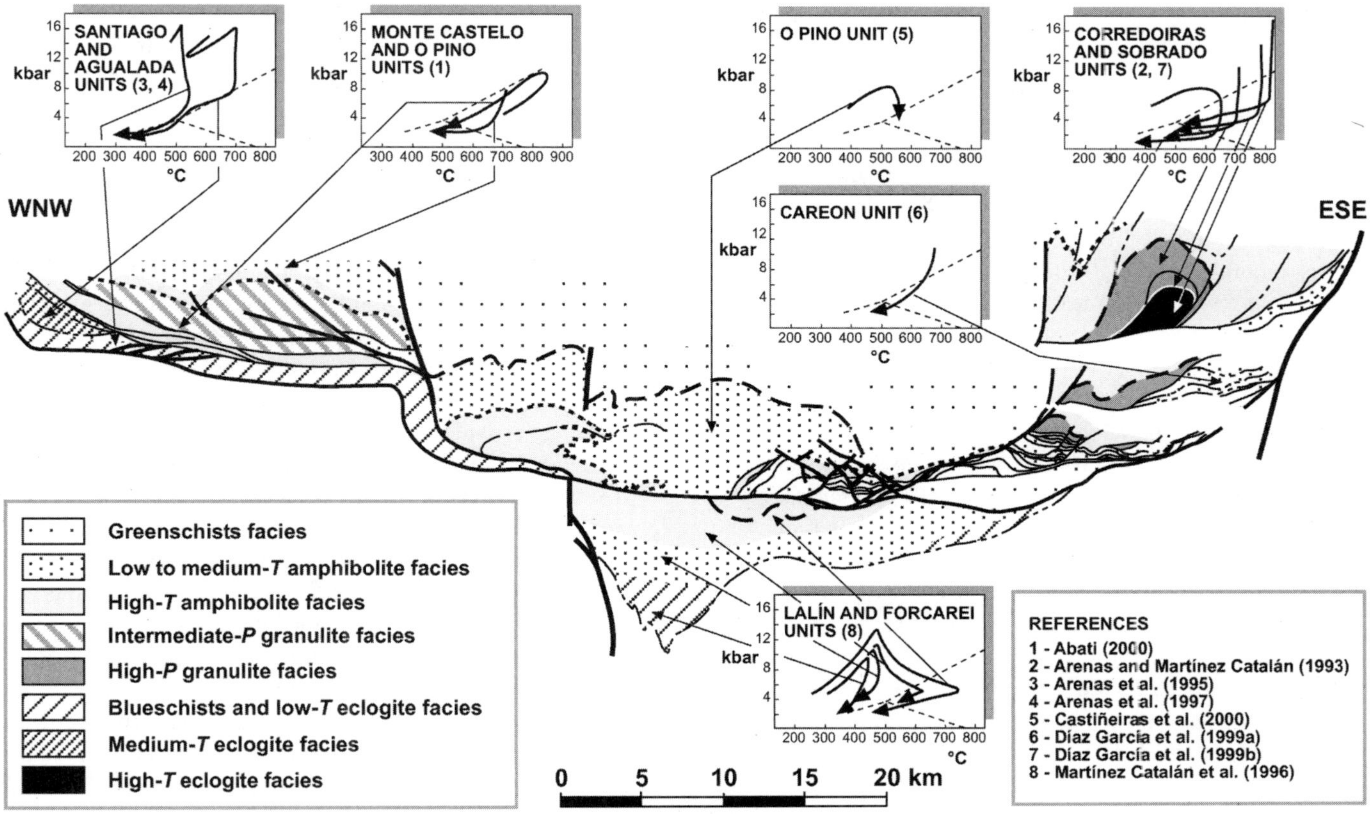

Figure 5. Distribution of main types of metamorphic facies in Ordenes Complex, and published pressure-temperature (*P-T*) paths for several units. Dashed lines in *P-T* diagrams represent aluminum silicate phase diagram after Holdaway (1971). Fault types as in Figure 3.

In the normal limb of the recumbent anticline, the foliation is a low- to high-*T* fabric that shows an upward increase in metamorphic grade (Arenas et al., 1995). The lower grade S_2 is folded by the anticline, but at structurally higher levels a synkinematic recrystallization took place after the fold had been formed, giving rise to a medium- to coarse-grained foliation (S_3). This foliation is parallel to the lower grade fabric, and seems induced by the emplacement of a hot overlying slice, the Campo Marzo unit. The metamorphic gradient is inverted (Fig. 5) in the northern parts of the Lalín and Forcarei units, and the isograds crosscut the recumbent folds and the Lalín-Forcarei thrust (Martínez Catalán et al., 1996). The *P-T* paths close to the Campo Marzo unit show the strong heating undergone by the upper part of the Lalín and Forcarei units during decompression (Fig. 5).

Agualada and Campo Marzo units. In the northwest, around Agualada, a migmatitic basal unit was transported on top of another basal unit of lower metamorphic grade by a thrust fault. *P-T* estimates on eclogitic lenses yield temperatures of 720 °C and a minimum pressure of 15 kbar (Arenas et al., 1997; see Fig. 5). Discontinuous lenses of ultramafics at the bounding fault and in a narrow shear zone inside the high-grade unit point to the imbricate character of thrusting and indicate that the mantle was involved with it.

A larger outcrop of ultramafic rocks occurs in Campo Marzo, northwest of Lalín (Plates 2 and 3, section 5). These rocks were probably derived from the hot mantle wedge above the subducted basal units (Arenas et al., 1995, 1997), and the contact with the underlying basal units is a folded low-dipping fault, the Campo Marzo detachment. Martínez Catalán et al. (1996) interpreted it as an extensional fault, based on its direction of motion, which is roughly north-south, parallel to the orogenic trend (Fig. 3). Its emplacement on top of the relatively cold basal units provoked a heat transfer that gave rise to an inverted metamorphic gradient, as seen in the Santiago unit, and in the northern parts of the Lalín and Forcarei units. The mantle rocks induced the widespread development of migmatites (Plates 2 and 3, sections 5–7). U-Pb dating of monazites from the Agualada migmatitic paragneisses has yielded a concordant age of 346 ± 2 Ma (Abati and Dunning, 2002).

To provoke such a thermal imprint, the ultramafic slice must have been extensive. However, this slice was dismembered by subsequent thrusting, well characterized in the overlying units (see following). In addition, a small part could be hidden by the Pico Sacro detachment (Plate 3; Fig. 3), a late extensional fault.

Ophiolitic units

The units occupying a structurally intermediate position crop out in the periphery of the Ordenes Complex, in three different and tectonically imbricated slices known as the Vila de Cruces (south), Careón (southeast), and Bazar (west) units (Plates 2 and 3; Fig. 3). The latter two have similar rock types and, because Careón overlies the Vila de Cruces unit, they are often called the upper ophiolitic units. Vila de Cruces is then referred to as the lower ophiolitic unit (Fig. 3).

Vila de Cruces unit. The lowermost ophiolitic unit consists of greenschist facies mafic rocks (greenstones) and metapelites, with rare felsic orthogneisses, serpentinites, and cherts. Mesozonal relics in the mafic rocks indicate that an earlier amphibolite facies regional metamorphic event affected parts of the unit. In addition, a low- to intermediate-T and high-P metamorphic gradient has been identified in the metapelites.

Two phases of recumbent folding, the first with isoclinal structures of kilometer scale, were followed by the development of an east-southeast–directed thrust, which resulted in structural repetition of lithologies. Several duplexes developed during thrusting (Plate 3; Figs. 3 and 6). The greenschist facies foliation related to the second-phase recumbent folds has been dated as 363–367 Ma (Dallmeyer et al., 1997; $^{40}Ar/^{39}Ar$, whole-rock and muscovite concentrates in phyllites; see Fig. 4).

The deformation is so penetrative that it is difficult to establish the protoliths of the greenstones. However, for those interbedded with the metasediments, a volcanic origin seems reasonable. On these grounds, their association with seemingly distal metasediments and ultramafics suggests that they may belong to an ophiolitic suite, even though this unit may not represent a typical ophiolite. The presence of orthogneisses and the metamorphic evolution of the metapelites, which compares to that of the basal units, suggest that the Vila de Cruces unit represents the transition between the outermost continental crust of Gondwana and the oceanic lithosphere of the overlying ophiolitic units.

Careón unit. This unit represents the basal parts of an ophiolitic sequence, characterized by the abundance of pegmatitic gabbros and diabase dikes, and by the lack of layered gabbros. The mafic rocks, their igneous textures either preserved or transformed into flasergabbros and amphibolites, are underlain by ultramafics, and this sequence was repeated by east-directed thrust faults (Plate 3, sections 1 and 2; Fig. 3) in three main thrust sheets with abundant internal imbrication (Díaz García et al., 1999a).

Although the ultramafic section is characterized by a high degree of serpentinization, a harzburgitic composition can be inferred where alteration is minor. The transition between harzburgite and gabbro is abrupt, although locally the mafic-ultramafic boundary is intruded by numerous gabbroic and diabase dikes. It is considered the "petrological" paleo-Moho and occurs along a surface without significant deformation and mostly parallel to the foliation in the mafic and ultramafic rocks.

The gabbroic section consists of a complex network formed by multiple intrusions of gabbroic rocks, wehrlite, and wehrlite-gabbro transitional compositions, and diabase to pegmatoid dikes. Gabbroic rocks outside ductile shear zones remain mostly undeformed and retain their igneous texture. The most common type is a medium- to coarse-grained isotropic metagabbro, commonly intruded by pegmatoid and diabase dikes. A weakly foliated plagioclase-rich gabbro gave a concordant age of 395 ± 2 Ma (Díaz García et al., 1999a; U-Pb geochronology in zircons), which is considered a protolith age (Fig. 4). Because plagioclase-rich gabbros are common components of the ophiolite, the age provides evidence for oceanic crust generation in this peri-Gondwanan domain by Early Devonian time.

Complete transformation of gabbroic rocks into well-foliated amphibolites occurs at the boundary between the main thrust sheets, where garnetiferous amphibolites appear discontinuously in the hanging wall. In the footwall, garnetiferous amphibolites occur attached to the allochthonous ultramafic rocks showing granonematoblastic textures and a penetrative foliation. Garnet growth is synkinematic with the foliation. Inside each of the sheets, the garnetiferous amphibolites give way with increasing depth to epidote-amphibolite facies metabasic rocks. Corundum-bearing metabasic rocks occur immediately beneath the ultramafic rocks of the uppermost thrust sheet (Díaz García et al., 1999a). The preservation of a high-T mineral such as corundum in amphibole-rich metabasic rocks attests to an intense dynamothermal metamorphism (sanidinite facies of Yardley, 1989) close to the overlying ultramafic sheet. The vertical distribution of both garnet and corundum points to an inverted metamorphic gradient in the upper parts of the sheets, similar to that described in thrust-related metamorphic soles (Jamieson, 1986).

The P-T conditions ranging between 625 and 680 °C and 11.1 and 11.9 kbar have been obtained on the peak mineral assemblage of the garnet amphibolites by Díaz García et al. (1999a). A well-foliated garnet amphibolite was dated by the $^{40}Ar/^{39}Ar$ method (Dallmeyer et al., 1997), obtaining a plateau isotopic correlation age of 376.8 ± 0.4 Ma, which represents a cooling age following the amphibolite facies metamorphism (Fig. 4). This age is similar to that obtained in Cabo Ortegal and the Portuguese Complexes by the same method (390–380 Ma; Dallmeyer and Gil Ibarguchi, 1990; Dallmeyer et al., 1991).

Subsequently, retrogressive greenschist facies metamorphism developed throughout the Careón unit (Fig. 5) and is especially concentrated close to the thrust surfaces. It developed serpentinite mylonites in the hanging walls, and well-foliated metabasites with fine-grained nematoblastic textures and pervasive shear bands in the footwalls.

Thrust-related shear zones are characterized by intensely foliated rocks, meter-scale tight isoclinal folds, and two systems of shear bands. An east-west stretching lineation is defined by recrystallized olivine in the ultramafic rocks, the foliation of which developed at the highest temperatures, in the amphibolites and in the serpentinite mylonites. Fold asymmetry, σ-type porphyroclast systems, and shear bands give a consistent top-to-the-east sense of displacement for the two main thrusts repeating the mafic-ultramafic sections. The existence of mantle-rooted thrusts and related metamorphic soles points to a relationship with a subduction environment, the kinematics of which agree with the westward polarity inferred in the basal units from the metamorphic conditions. In contrast, the later thrusts separating the Careón unit from the adjacent ones, and several related imbricate faults, show top-to-the-southeast motion.

Bazar unit. The Bazar unit crops out in the west part of the Ordenes Complex (Plates 2 and 3; Fig. 3); this unit shares many characteristics with those of Careón. It represents the lower part of an ophiolitic sequence composed of serpentinized dunites and isotropic gabbros with minor intrusions of pegmatoid gabbros. The isotropic gabbros exhibit a tholeiitic character, with flat chondrite-normalized rare earth element patterns, and show similarities to volcanic arc basalts, based on positive anomalies in light lithophile elements and marked negative Nb anomaly (Díaz García, 1990). The deformation is heterogeneous, with ductile shear zones separating weakly deformed gabbros and repeating the mafic and ultramafic association. Metamorphic conditions reached the granulite facies and changed along the deformation to the amphibolite and greenschist facies.

Upper units

The units that structurally overlie the ophiolites are collectively termed the upper units, but can be subdivided into two subgroups of unequal area extent. High-*P*–high-*T* units occur as small massifs and kilometer-scale lenses in the southeast of the Ordenes Complex (Plate 2). These units appear discontinuously overlying the ophiolites of the Careón unit, and occupy a structurally low position among the upper units considered as an ensemble (Fig. 3). The more important are the Sobrado, Belmil, and Melide units (Plates 2 and 3, sections 1–3). The high-*P*–high-*T* units are well represented in the Cabo Ortegal Complex, and their structure is described in Marcos et al. (this volume).

Structurally above the high-*P*–high-*T* units is a group of intermediate-*P* units. They are by far the more extensive, occupying the central part of the Ordenes Complex and representing two-thirds of its outcropping surface.

The boundary between intermediate-*P* and high-*P*–high-*T* upper units is a low-dipping fault known as the Corredoiras detachment, that appears folded and limits the tectonic windows of Sobrado and Belmil (Plate 3; Fig. 3). The fault includes a 300–600-m-thick mylonitic shear zone, with kinematic criteria indicating top-to-the-north or northeast motion (Díaz García et al., 1999b). The Corredoiras detachment has an extensional character, as shown by the difference in the peak pressure between its hanging wall (the intermediate-*P* Corredoiras unit) and its footwall (the high-*P*–high-*T* Sobrado, Melide, and Belmil units; see Fig. 5). A hornblende $^{40}Ar/^{39}Ar$ analysis carried out in an amphibolite of the mylonitic zone yielded an age of 375 Ma (Dallmeyer et al., 1997).

The ensemble of the upper units was thrust over the ophiolitic units. However, earlier detachments, folds, and thrusts have been identified, making the structural evolution of the upper units one of the most complicated found so far in the Iberian Massif. This complexity derives from the fact that the upper units have undergone an older, Early Ordovician tectonothermal event, overprinted by a long Variscan deformational history.

High-*P*–high-*T* upper units: Sobrado, Melide, and Belmil. These units consist of paragneisses and basic and ultrabasic metaigneous rocks. The characteristic rocks are metabasites, commonly garnet-clinopyroxene granulites and eclogites, retrograded to the amphibolite facies (Vogel, 1967; Hubregtse, 1973). Gabbros occur in several stages of transformation, from practically undeformed and scarcely affected by the metamorphism, to coronitic metagabbros and high-*P* granulites. In the less deformed gabbros, subophitic and diabase textures have been preserved, indicating an emplacement at relatively shallow crustal levels (Arenas and Martínez Catalán, this volume).

The tectonothermal evolution of these units include an older high-*P* granulite to eclogite facies metamorphism (475–495 Ma, Fernández-Suárez et al., 2002; see Figs. 4 and 5), followed by decompression and partial melting and then, successively, by a penetrative mylonitization in the amphibolite facies, recumbent folding, and thrusting in the greenschist facies (Vogel, 1967; Marcos et al., 1984; Gil Ibarguchi et al., 1990; Arenas, 1991; Girardeau and Gil Ibarguchi, 1991; Mendia Aranguren, 1996).

Two thrust faults have been identified within the Sobrado unit, separating three tectonic slices (Plate 3, section 1; Fig. 3) with different peak metamorphic conditions (Martínez Catalán and Arenas, 1992; Arenas and Martínez Catalán, this volume). The upper slice underwent the lower metamorphic conditions, in the high-*P* part of the granulite facies, whereas the lower slice attained high-*T* eclogitic conditions. A low-grade ultramylonitic gneiss probably marking a thrust fault within the upper slice yielded a muscovite $^{40}Ar/^{39}Ar$ age of ca. 325 Ma (Dallmeyer et al., 1997).

In the Cabo Ortegal Complex, early Paleozoic ages (490–480 Ma) were obtained by Peucat et al. (1990; U-Pb on zircons) in metabasic rocks, and interpreted as dating the high-*P*–high-*T* metamorphism. This interpretation has been confirmed by U-Pb dating of monazites in paragneisses and zircons in a vein formed during the partial melting that followed the high-*P*–high-*T* event (Fernández-Suárez et al., 2002). Furthermore, U-Pb ages between 405 and 390 Ma have been obtained on zircons, monazites, titanites, and rutile (Schäfer et al., 1993; Santos Zalduegui et al., 1996; Ordóñez Casado et al., 2001; Fernández-Suárez et al., 2002), pointing to an Early Devonian metamorphic event that was followed by a subsequent retrograde amphibolite facies metamorphism at 390–380 Ma (Dallmeyer et al., 1991, 1997; Valverde Vaquero and Fernández, 1996).

Intermediate-*P* upper units: Fornás, Monte Castelo, Corredoiras, O Pino, and Betanzos. The intermediate-*P* upper units occupy the higher structural positions in the nappe pile. They include a thick sequence of terrigenous metasediments, and bodies of amphibolites, gabbros, and orthogneisses. Metamorphism ranges from the granulite facies in the lower parts to the greenschist facies above. Changes in metamorphic grade are abrupt and occur in detachments, tentatively viewed as extensional due to their subtractive character (Díaz García, 1990, 2000). High-grade rocks occupy the lowermost position among these units and occur as massifs of metaigneous rocks: the Fornás unit (amphibolites), the Monte Castelo unit (gabbro), and the Corredoiras unit (orthogneiss). These are separated from the overlying O Pino unit, which is essentially mesozonal, by subtractive detachments, and a major low-dipping fault, the Ponte Carreira detachment, separates the mesozonal schists and paragneisses from the uppermost slates and graywackes of the Betanzos unit (Plates 2 and 3; Fig. 3).

A magmatic and metamorphic event affected the intermediate-*P* upper units during the Early Ordovician (Abati et al., 1999), thereby pointing to their polymetamorphic and polyorogenic character. The igneous rocks and a good part of the metamorphic parageneses belong to the Early Ordovician event, but the exact extent of the Variscan metamorphic imprint is not yet fully established. Crosscutting relationships suggest that a few of the extensional detachments are pre-Variscan, possibly related to the Ordovician event (Abati, 2000; González Cuadra et al., 1999), whereas others, such as the Ponte Carreira detachment and the Corredoiras detachment, are early Variscan (Fig. 3).

Fornás unit. This unit crops out in the southwest of the Ordenes Complex, and consists of heterogeneous and well-foliated amphibolites. Van Zuuren (1969) described preserved igneous textures, indicating a gabbroic origin, and granulite facies relics. The dominant amphibolite facies foliation developed subsequently, during the exhumation, in a shear zone that carried them into contact with the overlying mesozonal schists and paragneisses. This shear zone is a subtractive detachment, and was subsequently affected by recumbent folds with east vergence, repeated by a thrust and back-folded (Plate 3, section 7; Fig. 3).

Monte Castelo unit. A massive two-pyroxene gabbro crops out in the west of the complex. It has a tholeiitic character, similar in composition to modern island-arc basalts (see Andonaegui et al., this volume). Textures vary from granular to intergranular and ophitic. The presence of olivine and the common ophitic textures point to a relatively shallow emplacement. U-Pb analyses on zircons from the gabbro are concordant and overlap, giving a precise age of 499 ± 2 Ma (Abati et al., 1999), interpreted as the age of crystallization of the protolith (Fig. 4). Near its bottom, the gabbro is cut by granulite facies prograde shear zones, which produced a complete recrystallization of the igneous paragenesis. Monazite inclusions within biotite in a granulite facies rock of pelitic composition inside the gabbro yielded an age of 498 ± 2 Ma (Fig. 4), whereas rutiles in the same sample gave ages of 380–390 Ma, ~100 m.y. younger than the ages yielded by the monazites (Abati et al., 1999).

Separating the gabbro from the overlying amphibolite facies paragneisses of the O Pino unit, a subtractive detachment occurs, marked by narrow granitic bands 50–100 m thick (Plate 3, section 9), and by the phyllonitization of the rocks (Abati, 2000). Kinematic criteria show a consistent top-to-the-south or southwest motion. The granites appear deformed and show abundant restites of paragneisses from the surrounding metasediments. Zircons and monazites from the granites give an age of 500 ± 2 Ma (Abati et al., 1999).

Corredoiras unit. Most of the unit is a coarse-grained metagranodiorite that crops out in the eastern part of the Ordenes Complex (Plates 2 and 3). U-Pb dating on zircons gave 500 ± 2 Ma (Abati et al., 1999), considered the age of intrusion (a similar age was obtained by Kuijper, 1980). The massif is variably gneissified, and kilometer-scale xenoliths of hornfelses and migmatites occur inside the orthogneisses. The xenoliths, hornfelsic in origin as indicated by preserved relics, were migmatized during a subsequent high-grade metamorphic event that also affected the orthogneisses. This event has been dated in two migmatite samples, by monazites included in biotite from the high-grade association, giving ages of 493 ± 2 and 484 ± 2 Ma (Abati et al., 1999). The temperature necessary for migmatization was above the closing temperature of the U-Pb system in monazite, ensuring that the ages date the high-grade metamorphic fabric and that monazites are not contact-metamorphism relics. The Barrovian character of this high-grade metamorphism points to an Early Ordovician compressional event.

Following crustal thickening, an extensional detachment brought into contact metasediments of the garnet zone above, with the high-grade orthogneisses below (Figs. 3 and 5). Subsequently, both the orthogneisses and the detachment were folded and overprinted by ductile thrusts developed under a new Barrovian metamorphic gradient, in staurolite-zone conditions (González Cuadra et al., 1999). These thrusts were overprinted by the mylonitic zone associated with the Corredoiras detachment (see Plate 3, section 1).

O Pino unit. A thick pile of mesozonal monotonous schists and paragneisses overlies the high-grade metaigneous massifs. This is a classical Barrovian pile, with metamorphic zones ranging from the sillimanite below, to the almandine above (Castiñeiras et al., 2000). Monazites from semipelitic paragneisses of the sillimanite zone have yielded ages of 493 ± 1.3 and 496 ± 3 Ma (Abati et al., 1999).

The unit appears as two large lens-shaped massifs separated from the underlying higher grade units by old extensional detachments, and from the overlying low-grade unit by another extensional low-dipping fault, the Ponte Carreira detachment. This fault is probably Variscan, and is marked by a zone of phyllonitization with abundant quartz veins and kinematic criteria indicating top-to-the-north motion (Díaz García, 2000).

Betanzos unit. This unit occupies the uppermost structural position and consists of 2000–3000 m of pelites and graywackes that can be separated into two formations (Díaz García, 2000). The lower formation is composed of metapelites, often carbonaceous, with alternations of gray to black quartzites. The upper formation has a flyschoid appearance and consists of metagraywackes and slates, with minor conglomerates. The facies and sedimentary structures indicate a turbiditic character (Matte and Capdevila, 1978; Gutiérrez Alonso et al., 2000). Most of the sandstones are feldspathic graywackes with a framework of quartz and fresh plagioclase. Rock fragments with vitric and microgranular textures are common in the polymictic conglomerates and coarse-grained graywackes, together with slates, cherts, and clasts of bipyramidal quartz. The abundance of volcanic components, when considered together with the chemistry of the Monte Castelo gabbro, suggests a volcanic arc environment for the intermediate-*P* upper units. However, the lack of truly volcanic rocks points to a dissected magmatic arc, the Betanzos unit representing a forearc or backarc basin fill.

The sediments were affected by greenschist facies metamorphism and by early recumbent and late steep folds (Matte and Capdevila, 1978). Two whole-rock $^{40}Ar/^{39}Ar$ analyses carried out in the greenschist facies metapelites yielded 333.3 ± 0.3 and 294.7 ± 0.2 Ma (Dallmeyer et al., 1997), indicating that thermal activity took place during most of Carboniferous time.

COMPARING THE AUTOCHTHON AND ALLOCHTHON STRUCTURAL HISTORIES

The structural evolution of the autochthon (and the lower allochthon or schistose domain, see Fig. 2) is relatively simple; structures correspond to three main deformation events that developed during convergence following collision (Pérez-Estaún et al., 1991).

The first event (D_1) produced large-scale recumbent folds with east vergence and overturned limbs attaining 30 km in length (Matte, 1968; Bastida et al., 1986). These folds have an axial planar cleavage, which is the oldest penetrative fabric recognized. The recumbent folds were transported toward the east in the hanging wall of second-phase (D_2) ductile shear zones and thrusts (Marcos, 1973; Bastida et al., 1986). These ductile shear zones are associated with a second cleavage development and recumbent folding, but these folds are smaller, rarely exceeding a few meters. Other structures in the authochthon are extensional detachments, synkinematic to late kinematic with the thrusting phase, and late steep folds and transcurrent shear zones, usually ascribed to the third deformation phase (D_3).

The allochthonous complexes are more complicated; for example, the basal units were probably part of the Gondwana margin like the autochthon. However, the large recumbent anticline identified in the Lalín unit (Plate 3, section 5) folds a foliation (S_2) that overprints an earlier tectonic foliation (S_1), which is preserved as microinclusions in porphyroblasts. This recumbent fold may be comparable to D_1 structures in the autochthon, but for the basal units, is at least a third-phase structure. The Lalín-Forcarei thrust may be considered equivalent to the second-phase thrusts of the autochthon (D_2), but here the foliation developed along its shear zone is at least S_3.

There are also differences in the ophiolites. The structural history of the Vila de Cruces unit might, in principle, be compared to that of the autochthon: a first phase of recumbent folding followed by a second one with smaller structures, and by a thrust system. However, the abundance of serpentinite lenses intercalated between metapelites and greenstones and involved in the first recumbent folding suggests an early imbrication, prior to the recumbent folding.

Two different thrust systems have been identified in the ophiolitic Careón unit. The older system was initiated in relatively high metamorphic conditions (625–680 °C and 11–12 kbar) and was responsible for the stacking of the three main thrust sheets identified. Top-to-the-east kinematics characterize this mantle-rooted thrust system, which represents the imbrication of oceanic slivers in the early Variscan accretionary wedge (Díaz García et al., 1999a). Subsequently, retrogressive greenschist facies metamorphism concentrated close to the thrust surfaces indicates continued motion of the thrusts sheets into higher structural levels. This former thrust system was overprinted by a younger one, which was entirely developed in low-grade conditions, and shows top-to-the-southeast motion. The later system affected mostly the upper and lower boundaries of the Careón unit, and is related to the emplacement of the ophiolitic and upper units into their present position.

The major differences may be expected in the upper units, because of their polyorogenic evolution. Structural analysis and radiometric ages permit the definition of two complete cycles of burial followed by exhumation. Ordovician compression is deduced from the metamorphic parageneses and isotopic dating, but no unequivocal associated structures (other than relic foliations and granulite facies prograde shear zones in the Monte Castelo gabbro) have been identified. Conversely, extensional structures associated with the pre-Variscan exhumation have been found. The detachments that separate the massifs of Fornás, Corredoiras, and Monte Castelo from the O Pino unit (Plate 3; Fig. 3) are examples of extensional structures that seem to belong to this category. The first two structures were overprinted by recumbent folds and thrusts with typical Variscan attitude and kinematics (González Cuadra et al., 1999). The third example has granitoids of Early Ordovician age marking the detachment (Abati, 2000).

TECTONIC INTERPRETATION

Pre-Variscan tectogenesis

Little is known about the Early Ordovician event, except that it involved shortening and crustal thickening (intermediate-to high-*P* metamorphism) and was related to coeval magmatism of arc affinity. Both characteristics, together with the volcanic components so abundant in the low-grade Betanzos unit, imply an active continental margin or island-arc environment (Abati et al., 1999; Abati, 2000; Díaz García, 2000; Andonaegui et al., this volume).

The Early Ordovician plutonic and metamorphic ages of the upper units correlate with the Penobscottian event in the Appalachians and contemporaneous events (Finnmarkian, Grampian) in other parts of the Caledonian belt. However, these early Taconian episodes are commonly ascribed to convergence related to the closure of the Iapetus ocean, and occurred along the northern margin of Avalon (Hossack and Cooper, 1986; Soper, 1988; Hutton, 1989; Keppie, 1989; Neuman and Max, 1989; Ohta et al., 1989; Cawood et al., 1995; Hibbard and Samson, 1995; Winchester and Van Staal, 1995). Given the position of Iberia close to the junction of the Appalachians, the Caledonides and the Variscides (Fig. 1), it could be possible that a piece of lithosphere detached from the early Taconian belt was thrust on top of the Iberian autochthon. The large displacements associated with the thrust systems described in the following could, in principle, support such an interpretation.

However, Early Ordovician metamorphic ages also have been found in areas of the Variscan belt far from the junction of the three orogenic belts, such as the Alps (Von Raumer and Neubauer, 1993), particularly in the Tauern Window (Frisch and Neubauer, 1989), the Black Forest, the Bohemian Massif, and the Holy Cross Mountains (Franke, 1989), that support a relationship with peri-Gondwanan activity. Convergence inside the Rheic ocean may explain the Early Ordovician metamorphism and magmatism without the intervention of a volcanic arc extraneous to the Gondwana realm.

Early Variscan structures: Building the orogenic wedge

The Variscan history can be better established, thanks to the wealth of structural and petrologic data available and to the isotopic ages of the main fabrics. The Early Devonian protolith age obtained in the Careón unit provides evidence for oceanic crust generation and delimits the upper limit for the closure of the intervening ocean and the lower limit for the beginning of intracontinental deformation.

Diachronism of deformation. The ages of the retrogressive amphibolite facies foliation in the high-*P*–high-*T* upper units (405–380 Ma) mark the onset of Variscan deformation in the allochthonous units. They are followed by those on the upper ophiolites (390–375 Ma), which are prograde, and by the greenschist facies foliation in the lower ophiolites (370–365 Ma) and the posteclogitic, exhumation foliation in the basal units (375–365 Ma). These age intervals (Fig. 7, A and B) show younging of the main Variscan foliations from the top to the bottom of the nappe pile. This younging continued: the age of the main foliation in the autochthon, 360–335 Ma for the first cleavage, is associated with D_1 recumbent folding (Dallmeyer at al., 1997). The oldest age corresponds to the more internal areas and to the units occupying a relatively higher structural position within the autochthon, in the vicinity of the allochthonous complexes (Fig. 7C). The youngest age was obtained at the contact with the foreland thrust belt, which was structurally lower in the orogenic wedge (see cross section in Fig. 2).

The diachronism shown by the structures in both the allochthon and the autochthon, with ages older in the more internal and structurally higher units, can be extrapolated farther to the west. Because northwestern Iberia was close to the Grand Banks of Newfoundland at the end of the Variscan cycle (Lefort, 1989), the extrapolation may be prolonged into Atlantic Canada. The first cleavage in the Meguma terrane, the outermost accreted terrane of Canadian Appalachians, is dated as 415–390 Ma (Dallmeyer and Keppie, 1987). The accretion of Avalon to Laurentia occurred during the Late Ordovician to Early Silurian, along the late Taconian and Salinian deformations, between 460 and 430 Ma (Cawood et al., 1995; Murphy et al., 1995). Surface geology and deep seismic reflection profiles show that Avalon was underthrust westward beneath the Laurentian plate, perhaps as much as 200 km (Hatcher, 1989; Keen et al., 1989; Quinlan et al., 1992).

Accretionary sequence. The deformation in the Iberian allochthonous complexes and, in particular, the subduction of some of their units, probably reflects the continued understacking of continental, arc, and oceanic fragments to the west (Fig. 7A). Underthrusting followed the accretion of Avalon to Laurentia, and resulted from the convergence between Laurentia and Gondwana (Martínez Catalán et al., 1997, 1999). The 405–390 Ma ages would reflect the underthrusting of the upper units. These units underwent decompression during their emplacement onto the ophiolitic units. Synchronous amphibolite facies metamorphism, dated as 390–380 Ma, was retrogressive in the high-*P*–high-*T* upper units, and prograde in the underlying ophiolites, where it reflects the westward understacking of the oceanic lithosphere related to the closure of the intervening ocean (Díaz García et al., 1999a). In turn, this closure was immediately followed by the subduction of the outer continental edge of Gondwana (Fig. 7B). The *P-T* conditions reached by the basal units during this event vary according to the position inside the crustal subduction complex. Although affected by subsequent recumbent folding, a prefold reconstruction indicates that higher pressure assemblages occurred to the west: therefore, a westward polarity is deduced for the subduction.

After having registered the peak pressures, the metamorphic evolution in most of the units accreted to the orogenic wedge was governed by a strong decompression. Normally, the development of the regional foliation was synkinematic with decompression, which was attained via thrusting and unroofing of the thrust sheets. Thrusts associated with the initial Variscan accretion are probably preserved, but their identification is by no means unequivocal. The most probable thrusts of this type have been identified in the Careón (Díaz García et al., 1999a) and Corredoiras units (Fig. 3). Removal of the overburden was partially accomplished by extensional detachments, such as the Corredoiras detachment and the Ponte Carreira detachment (Plate 3; Fig. 3), with roughly north-south motion, implying crustal flow normal to the direction of shortening.

Emplacement of the allochthonous units: Thrusts and detachments

The role of thrusting in the emplacement of the allochthonous complexes of northwestern Spain was put forward by Ries and Shackleton (1971) and Bayer and Matte (1979). In the Ordenes Complex, contributions were made by Hilgen (1971), Martínez Catalán et al. (1984), and Díaz García (1990).

Other than the early Variscan thrusts associated with the accretion of the allochthonous units to form the initial orogenic wedge, the more important thrusts identified so far are the Lalín-Forcarei thrust, separating the basal units from the lower allochthon, and the thrust system that carried the upper units on top of the ophiolites. These thrusts were cut by a low-dipping normal fault with top-to-the-northwest motion, the Pico Sacro detachment. This fault overprints all of the thrusts and many granite intrusions, and was affected by late upright folding. Furthermore, it was cut by steeply dipping faults with normal and strike-slip components in the east and west parts of the complex (Plates 2 and 3; Fig. 3).

Thrusting of the basal units. The Lalín-Forcarei thrust postdates the high-*P* event in the basal units: therefore, it must be younger than 365 Ma (Santos Zalduegui et al., 1995). Thrusting was preceded by the development of a huge recumbent anticline, and it seems probable that both the fold and the Lalín-Forcarei thrust were developed in a wide regional shear zone responsible for the exhumation of the basal units.

The thrust sheet was overlain by the Campo Marzo unit, a piece of hot, probably subophiolitic mantle, that induced a heating from above and gave rise to an inverted metamorphic gradient in the basal units during decompression (Fig. 5). Its lower limit is a slightly folded low-dipping fault (Campo Marco detachment) with kinematic criteria indicating top-to-the-south motion. In the Santiago unit, also heated by an overlying source, kinematic criteria in the foliation coeval with the inverted metamorphism also supply a top-to-the-south sense of shear. This means motion normal to the direction of shortening, and in the northwestern Iberian Massif, this is usually associated with synorogenic extension (Escuder Viruete et al., 1994; Díez Balda et al., 1995; Díaz García et al., 1999b).

The emplacement of the hot slab followed the recumbent folding of the basal units, and was contemporaneous with the development of a new regional foliation and of migmatization in the upper parts of the Forcarei, Lalín, and Santiago units. The new foliation (S_3) is roughly synchronous with the emplacement of the Lalín and Forcarei units (Martínez Catalán et al., 1996), although the isograds of the inverted metamorphism reached and overprinted the Lalín-Forcarei thrust once motion had ceased on it, due to the low velocity of heat transfer in rocks.

If the migmatization of the Agualada unit was caused by the emplacement of the same mantle unit preserved in Campo Marzo, the 346 Ma age obtained by Abati and Dunning (2002) in Agualada can be taken as a reference for the age of the Lalín-Forcarei thrust. This age is consistent with the regional cleavage developed in the underlying lower allochthon during the emplacement of the overlying basal units, where a cooling age of 340 ± 0.9 Ma was obtained by Dallmeyer et al. (1997; $^{40}Ar/^{39}Ar$ whole-rock analysis on phyllite). Furthermore, Dallmeyer et al. (1997) dated the second-phase foliation in the autochthon, related to thrust faults, as between 345 Ma, in the internal areas, and 320 Ma, at the limit with the foreland thrust belt (Fig. 7C).

Taking into account the uncertainties on the age of the high-*P* event and of the Lalín-Forcarei thrust, this important thrust, dated as between 365 and 340 Ma, seems to follow a complex but essentially normal sequence of burying and exhumation. Subsequently, thrusting propagated into the authochthon, prograding toward the external zones where the thrusts and associated ductile shear zones overprinted recumbent folds but no previous thrust faults.

Thrusting of the upper and ophiolitic units. The emplacement of the upper units and ophiolites took place along a system of out-of-sequence thrusts developed under low-*T* conditions.

One of the more important thrusts of this system separates the upper units from the Bazar, Careón, and Vila de Cruces units. In the southeast of the Ordenes Complex, it carried the high-*P*–high-*T* upper units of Sobrado, Melide, and Belmil, together with the overlying Corredoiras unit, over the ophiolites (Plate 3, section 2; Fig. 3). Furthermore, the fault cut across two different thrust sheets of the previously stacked Careón unit. The associated mylonites and ultramylonites show a stretching northwest-southeast lineation and top-to-the-southeast sense of motion.

The Sobrado and Belmil units crop out in northeast-southwest–striking antiforms that seem to be related to the thrusting. The first unit is a duplex, of antiformal stack type, where the imbricates are three high-*P*–high-*T* slices reflecting *P-T* conditions progressively higher downward in the stack. This may be explained by the thrusts cutting across a granulite-eclogite slab placed onto the ophiolites by an early thrust, and sampling it as shown in Figure 6A. The Belmil unit is the uppermost horse of an antiformal stack that also involves horses of the Careón and Vila de Cruces units (Plate 3; Figs. 3 and 6B).

In both cases, the mylonites of the Corredoiras detachment surround the high-*P*–high-*T* rocks, which crop out as tectonic windows. Mylonitization occurred under amphibolite facies conditions and has been dated as ca. 375 Ma by Dallmeyer et al. (1997). The Corredoiras detachment clearly predates the out-of-sequence thrusts, which proceeded under greenschist facies conditions. Other minor slices of high-*P*–high-*T* mafic rocks occur in the area (Plates 2 and 3) and are viewed as small isolated horses related to this major thrust system.

Southwest of Belmil, a thrust separates the intermediate-*P* upper units from the lower ophiolites (Vila de Cruces unit), although the intervening Careón unit is not totally absent and crops out in two isolated horses (Plates 2 and 3, section 4). In the hanging wall of the main thrust, dragging of the Corredoiras orthogneiss during eastward motion left a tail of orthogneiss more than 15 km long, and a dismembered orthogneiss horse on top of a high-*P*–high-*T* horse to the south of Touro. In the footwall, duplexes formed in the Vila de Cruces unit. They include the hinterland dipping and antiformal stack types, as well as a few foreland-dipping horses of Vila de Cruces and Careón units in the Belmil duplex.

In the west of the complex, the intermediate-*P* upper units were thrust over the ophiolitic Bazar unit, and this over the basal Santiago unit (Plate 3, section 9; Fig. 3). Northwest of Carballo, an imbricate thrust carried an ophiolitic horse on top of the O Pino unit (Plate 2). A comparable tectonic mixing is seen in the Belmil duplex, where a horse of high-*P*–high-*T* gneisses is intercalated between imbricates of the Careón unit (Plate 3, section 3).

Although thrust faults are usually additive, the out-of-sequence thrusts are characterized by subtractions such as the disappearance of the Careón unit southwest of Belmil, and of the high-*P*–high-*T* units along most of the contact between the intermediate-*P* upper units and the ophiolites. Other subtractions, such as the lack of Vila de Cruces unit to the west and of the basal units to the east, may be partially due to the late extensional Pico Sacro detachment, but the thrust system also played a role. There is no doubt that the faults are of the thrust type, because of the presence of contractional imbricates. The disappearance of units may be attributed to superposition of two thrust generations. This can be seen in the Careón unit, where young, low-temperature thrusts cut an older imbricate stack. Consequently, an out-of-sequence character is deduced for this thrust system.

The out-of-sequence thrusts include duplexes for which displacement equals (antiformal stacks) or exceeds (foreland-dipping duplexes) fault spacing (Mitra and Boyer, 1986), and the isolated horses involve still larger displacements. Figure 6 shows the evolution of the out-of-sequence thrusts in the southeast of the Ordenes Complex. A partial restoration has been attempted in the Vila de Cruces unit, obtaining more than 50 km of shortening only inside this unit (Fig. 6C).

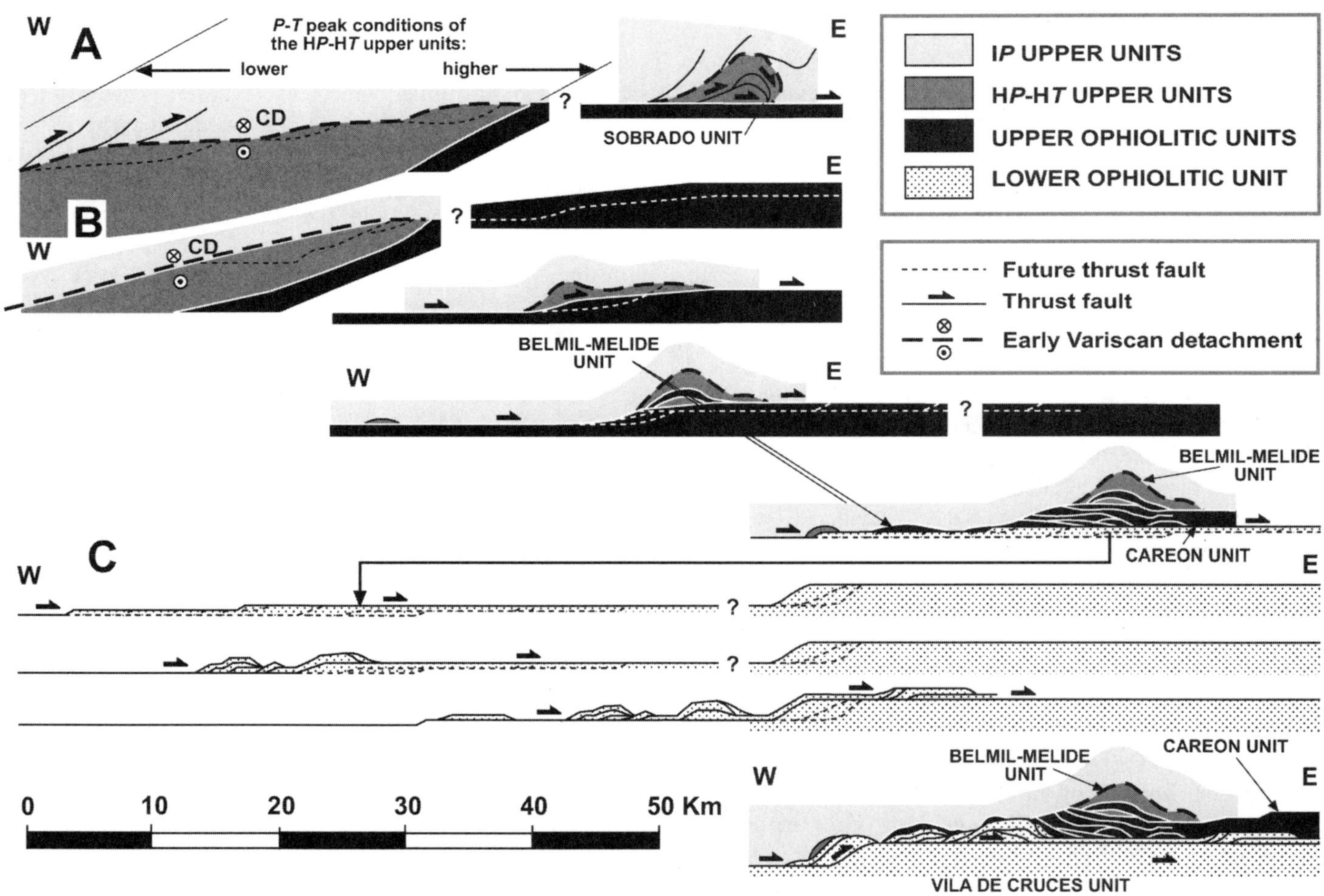

Figure 6. Development of structures associated with out-of-sequence thrusts. A: Development of Sobrado antiformal stack formed by three horses with metamorphic grade increasing downward. B: Development of antiformal stack involving Belmil unit, and horses of Careón and Vila de Cruces units. C: Partial restoration and possible sequence of thrusting of Vila de Cruces imbricates, showing more than 50 km of shortening. I*P*, intermediate-pressure; H*P*-H*T*, high-pressure–high-temperature.

The Corredoiras detachment had developed long before these late thrusts, overprinting early Variscan imbricate structures cutting across the Corredoiras unit (Plate 3, section 1; Figs. 3 and 6A). However, the Corredoiras detachment was reactivated during the out-of-sequence thrusting. Its mylonites, developed under amphibolite facies conditions and with top-to-the-north or northeast sense of shear, were involved in top-to-the-southeast thrusting and mylonitized in greenschist facies conditions to the east of Sobrado and Belmil (Plate 3). In a late pulse, brittle reverse faults dissected the duplexes (Plate 3, sections 1 and 3; Fig. 3).

The out-of-sequence-thrusts cut the Agualada unit, so that they are younger than 346 Ma, the age of the migmatization in this unit (Abati and Dunning, 2002). They are also younger than the Lalín-Forcarei thrust, structurally beneath and overprinted by amphibolite facies conditions, a metamorphism that never affected the out-of-sequence thrusts. However, they must be older than the late extensional Pico Sacro detachment, bracketed between 323 ± 11 and 314 ± 6 Ma (see following).

An age constraint is given by a $^{40}Ar/^{39}Ar$ cooling age of 325 Ma in a greenschist facies ultramylonitic gneiss of the upper slice of the Sobrado unit (Dallmeyer et al., 1997; see Fig. 4) that seems to mark a thrust contact. Dallmeyer et al. (1997) also recorded a 316 Ma age in a phyllonite at the base of the Cabo Ortegal Complex. These represent cooling ages of fine-grained white micas, and probably mark different episodes of out-of-sequence thrusting, because the emplacement of the upper and ophiolitic units was probably a protracted event. In any case, the out-of-sequence thrusts are between 40 and 70 m.y. younger than the amphibolite and greenschist facies foliations dated in the upper and ophiolitic units, and interpreted as related to the building of the orogenic wedge.

Figure 7D depicts the interpretation given to the out-of-sequence thrusts. A narrow band of low-dipping thrust faults cuts and displaces former boundaries between units stacked previously in a more or less sequential way. Thrusting directed to the southeast or east-southeast would have sampled the high-P–high-T upper units and the ophiolites, which appear sandwiched between the intermediate-P upper units and the basal units. According to this interpretation, dismembering of the units, particularly of the high-P–high-T upper units and of the ophiolites, should mostly be attributed to the out-of-sequence thrusts. The low inclination of the out-of-sequence thrusts accounts for the fact that they produced little exhumation of the units involved, which were relatively exhumed and cold.

Pico Sacro detachment system. The Pico Sacro detachment is the last and more continuous extensional detachment in Ordenes. Initially described by Van Zuuren (1969), it can be followed from the southeast near Palas de Rei, to Santiago in the southwest, and continues northward along the contact of the Ordenes Complex nearly up to the coast (Plates 2 and 3; Fig. 3). It shows a subtractive character, indicated by the omission of units and by a jump in the grade of the late significant metamorphism. South of Vila de Cruces, the fault cuts the basal units: the normal limb of the recumbent fold making the northern part of the Lalín unit occurs in the footwall of the Pico Sacro detachment and reappears, displaced to the west, as the Santiago unit at its hanging wall.

The detachment shows a consistent top-to-the-northwest motion. Along most of its length, there is a metamorphic contrast between the high-grade schists, migmatites, and granites of its footwall and the less metamorphic, often low-grade rocks of its hanging wall. Some granites were late enough to pierce the detachment (Plate 3). On petrological grounds, the vertical component of the fault is estimated as ~5 km (Marquínez García, 1984), and the horizontal component cannot exceed 20 km, because of the initial continuity between the orthogneisses in the north of Lalín and in the south of Santiago (Martínez Catalán et al., 1996).

Several associated normal faults occur in the hanging wall of the Pico Sacro detachment, in the south and east. The most important defines the western limit of the Corredoiras unit. This is a listric fault, downthrown to the west, that joins the Pico Sacro detachment to the southwest of Vila de Cruces, in an area characterized by a network of conjugated normal faults (Plates 2 and 3; Fig. 3). However, to the west of this network, several contractional structures seem to partially compensate the extension associated with the Pico Sacro detachment to the east. These structures include back folding of the Fornás unit and reverse faulting with top-to-the-west motion affecting the Monte Castelo unit (Díaz García, 1990). These structures seem to have followed local reactivation of an important out-of-sequence thrust (Plate 3; Fig. 3).

The Pico Sacro detachment cuts the granodiorites of Negreira in the west, and Palas de Rei in the east (Plates 2 and 3), dated as 350 ± 11 and 323 ± 11 Ma, respectively (Bellido et al., 1992). It also overprints a set of late, roughly northwest-southeast upright folds: the Lalín and Forcarei synforms and the intervening antiform (Plates 2 and 3; Fig. 3). However, it is in turn folded by a younger set of upright folds, striking north-south to northeast-southwest (Plate 3; Fig. 3), that are responsible for the present synformal structure of the Ordenes Complex (excluding the Lalín and Forcarei units). These folds are associated with transcurrent shear zones, of which the sinistral Palas de Rei shear zone in the east is the most important (Plate 2). The age of these late upright folds was estimated as 314 ± 6 Ma by Capdevila and Vialette (1970; corrected by Ries, 1979). This age, together with that of the Palas de Rei granodiorite, brackets the Pico Sacro detachment system between 323 ± 11 and 314 ± 6 Ma.

High-angle faults developed finally at the east and west margins of the Ordenes Complex. One of these faults, essentially normal, separates the complex from the Variscan granitoids west of Palas de Rei. Another occurs west of the Forcarei synform, separating the granites from the metasediments. This fault has a sinistral strike-slip component and a dip-slip movement, and can be followed discontinuously to the north up to the eastern limit of the Monte Castelo unit. The dip-slip movement, downthrown to the east, is probably the cause of the flexure

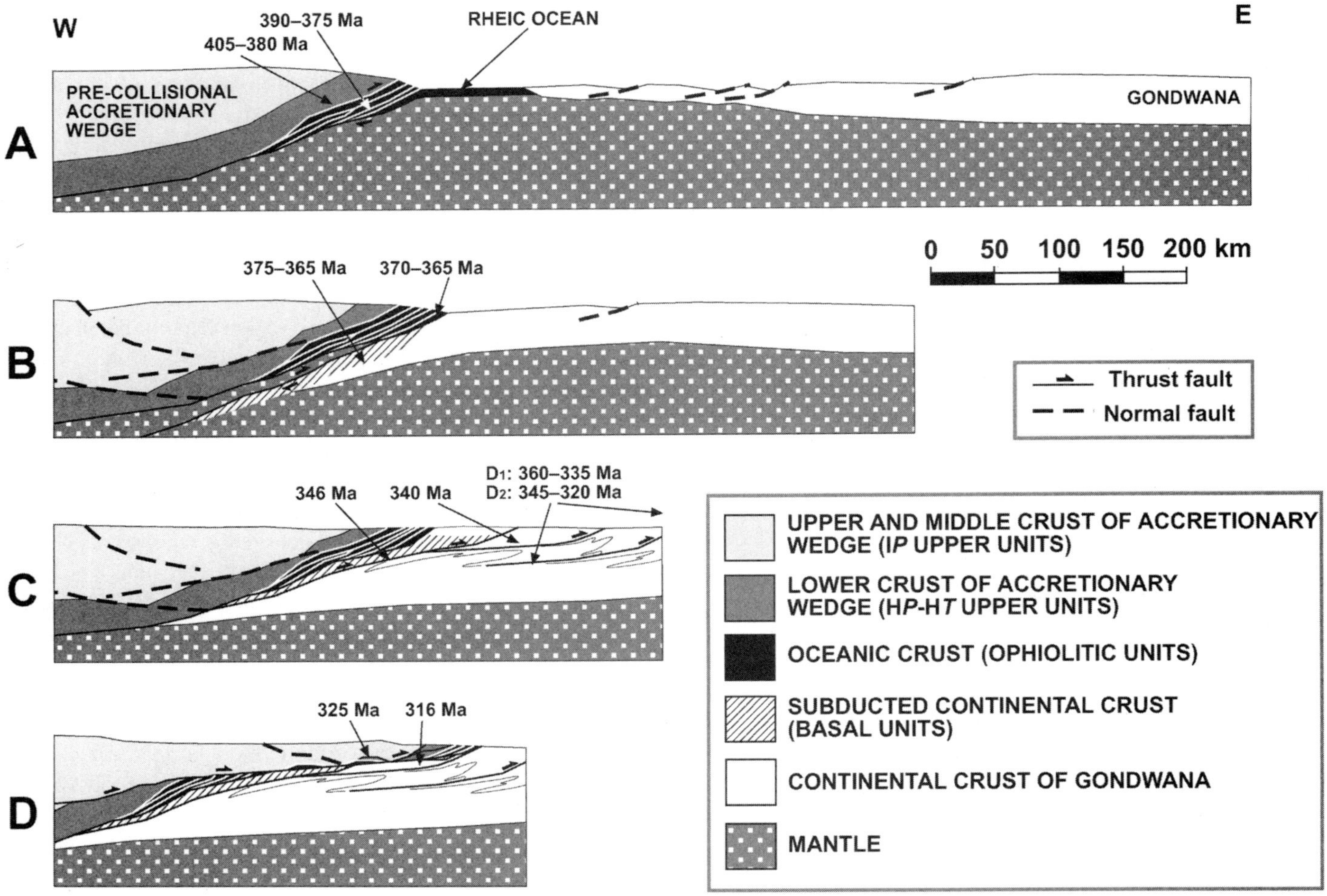

Figure 7. Proposed stages in emplacement of allochthonous complexes of northwestern Iberia. A: Building of accretionary wedge by progressive terrane accretion to Laurentia during Devonian. B: Subduction of outer edge of continental margin of Gondwana. C: Thrusting of basal units and lower allochthon, and thickening of Gondwana continental crust. D: Development of out-of-sequence thrusts and emplacement of upper and ophiolitic units to their present position. U-Pb and $^{40}Ar/^{39}Ar$ ages of regional fabrics after Dallmeyer et al. (1991, 1997), Schäfer et al. (1993), Ordóñez Casado et al. (2001), Santos Zalduegui et al. (1996), Valverde Vaquero and Fernández (1996), and Fernández-Suárez et al. (2002). *IP*, intermediate-pressure; *HP-HT*, high-pressure–high-temperature.

undergone by the overall subhorizontal structure in the steeply dipping corridor north of Santiago (Plate 3, section 8; Fig. 3).

IMPLICATIONS FOR THE VARISCAN-APPALACHIAN GEODYNAMICS

The Early Ordovician magmatic and tectonometamorphic event identified in the upper units reflects convergence in the peri-Gondwanan realm. This interpretation contrasts with the well-established fact that the Iberian autochthon was undergoing extension during the Early Ordovician (Ribeiro and Floor, 1987; Liñán and Quesada, 1990), related to peri-Gondwanan terrane dispersion (Fortey and Cocks, 1988; Soper, 1988). The apparent contradiction is solved by Von Raumer et al. (2001) and Stampfli et al. (this volume) by relating the drifting of European Variscan terranes to slab rollback of the ProtoTethys and Rheic oceanic lithospheres. Therefore, a short-lived orogenic event parallels, in this peri-Gondwanan terrane, the early Taconic convergence related to the closure of the Iapetus ocean in the Laurentian margin.

The upper units also contain the first evidences of Variscan convergence. The 405–390 Ma metamorphic ages reflect their accretion to an orogenic wedge that was being constructed on the Laurentian side. This time interval was preceded by deformation in Atlantic Canada (Dallmeyer and Keppie, 1987; Cawood et al., 1995; Murphy et al., 1995), and followed by accretion of oceanic lithosphere (ophiolitic units, Fig. 7A) and by limited subduction of the basal units. No evidence exists for separation of the Iberian autochthon from Gondwana during the Paleozoic. Therefore, the basal units must represent the outer edge

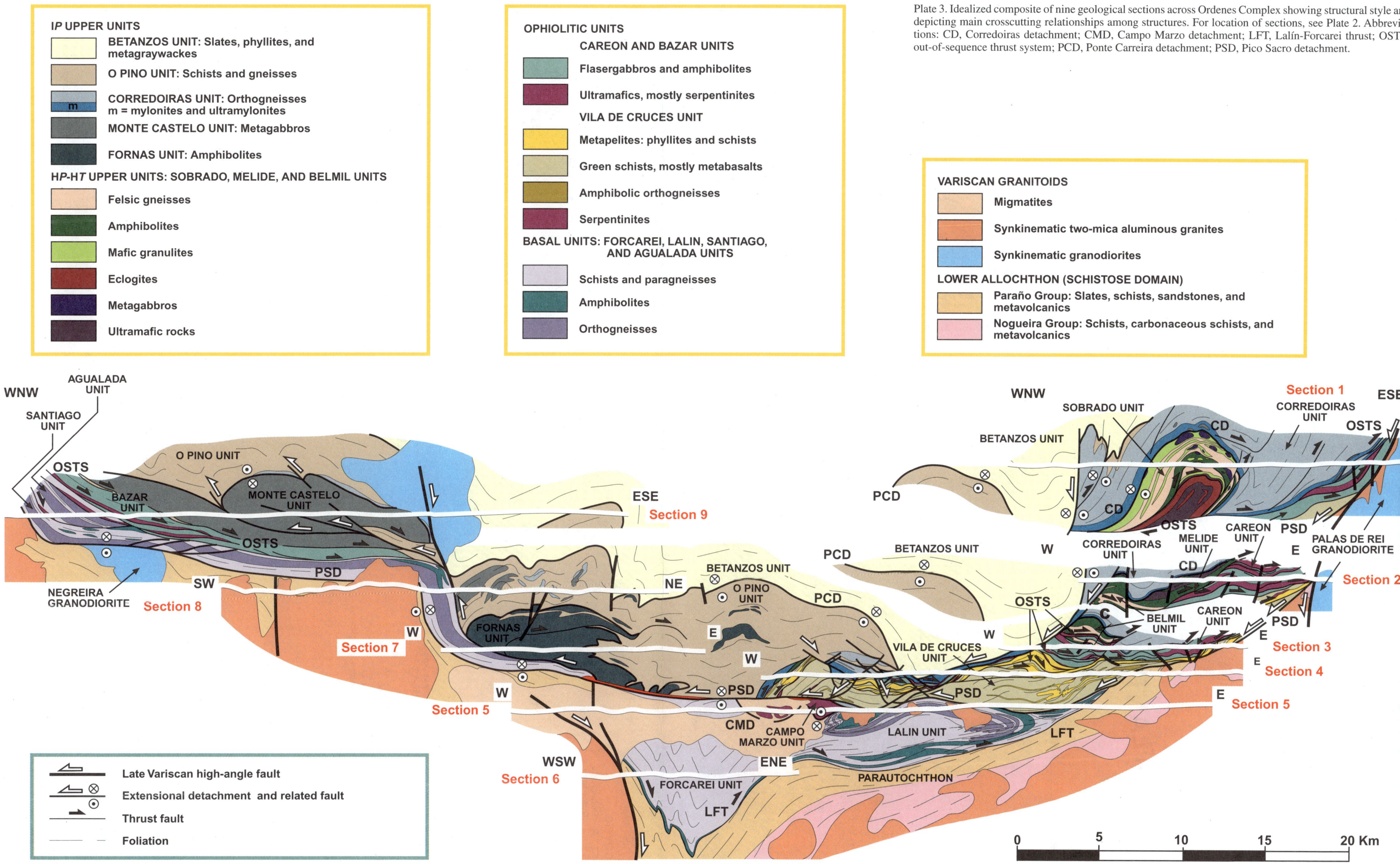

Plate 3. Idealized composite of nine geological sections across Ordenes Complex showing structural style an depicting main crosscutting relationships among structures. For location of sections, see Plate 2. Abbrevia tions: CD, Corredoiras detachment; CMD, Campo Marzo detachment; LFT, Lalín-Forcarei thrust; OSTS out-of-sequence thrust system; PCD, Ponte Carreira detachment; PSD, Pico Sacro detachment.

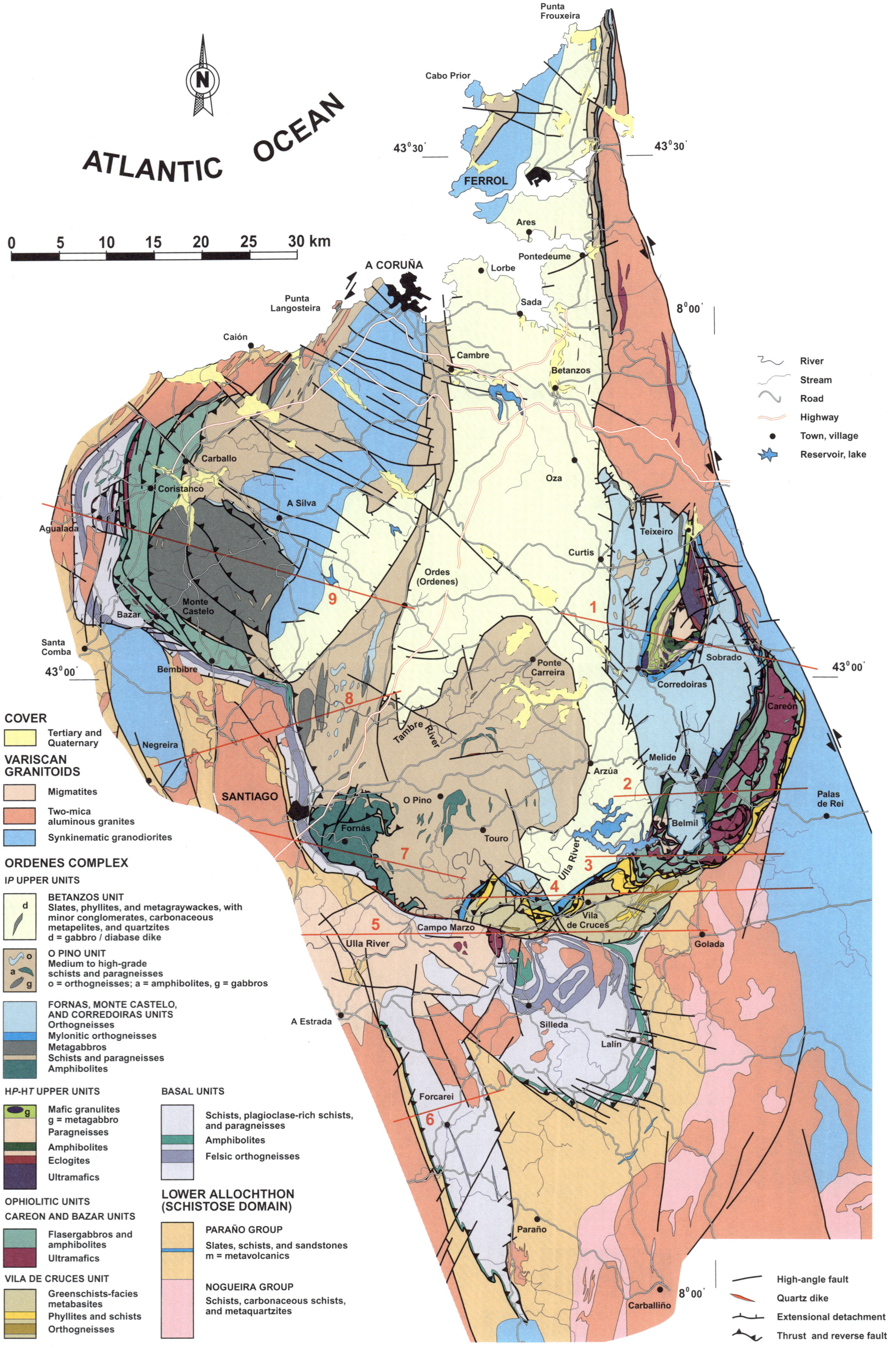

Plate 2. Geological map of Ordenes Complex (northwestern Spain).

of the Gondwana continental margin (Fig. 7B), and the end of subduction, dated as 374–365 Ma (Van Calsteren et al., 1979; Santos Zalduegui et al., 1995), marks the onset of deformation of the autochthonous cover and basement of Gondwana (Fig. 7C), which took place during the Early Carboniferous.

Accretion of the Iberian allochthonous units was contemporaneous with the Acadian orogeny in the Appalachians that occurred from the end of Silurian to the end of Devonian (Hatcher, 1989; Osberg et al., 1989). Our data indicate that seafloor spreading was still active 395 m.y. ago in the Rheic oceanic realm (Díaz García et al., 1999a), but Gondwana-Laurentia convergence caused final oceanic closure during the Late Devonian. Continent-continent collision was accomplished during the Acadian orogeny, between Iberia and the northern Appalachians. However, convergence did not cease. The lack of significant early Carboniferous deformation in the Appalachians was compensated for by shortening and thickening on the Iberian autochthon, which, together with the rest of the European Variscides, registered the main orogenic event during the early Carboniferous.

During the late Carboniferous, a system of low-dipping thrusts cut and displaced the units, stacked previously in a more or less sequential way (Fig. 7D). A large translation toward the foreland seems associated with this out-of-sequence thrust system, which could reflect a major geotectonic event. Timing of the out-of-sequence thrusts, loosely defined as between 325 and 315 Ma, coincides with the early Alleghanian orogeny (Hatcher, 1989, 2001), which initiated continent-continent collision in the central and southern Appalachians (Hatcher, this volume). A change in the relative motion of Laurentia and Gondwana has been suggested for the Alleghanian orogeny, even though there is not agreement on the sense of rotation of Gondwana relative to Laurentia (counterclockwise, Gates et al., 1986; clockwise, Hatcher, this volume). The change of thrust transport direction from east-west (building of the accretionary wedge and in-sequence thrusts) to northwest-southeast (out-of-sequence thrusts) would rather support a counterclockwise rotation of Iberia, which did not necessarily remain passively attached to Gondwana.

CONCLUSION

The structural evolution of the Iberian allochthonous complexes is being progressively unveiled. From what we currently know, the polyorogenic character of the upper units, deduced from isotopic data, is becoming evident. The pre-Variscan structural evolution is poorly characterized, but several macrostructures in the upper units seem to be pre-Variscan. These include granulite and eclogite facies shear zones and extensional detachments.

Conversely, our understanding of the Variscan evolution is far greater. The tectonothermal evolution and isotopic ages of the regional foliations in the different allochthonous units suggest a history of subduction and progressive accretion of lithospheric pieces to an orogenic wedge, in a context of convergence between Gondwana and Laurentia (Martínez Catalán et al., 1997, 1999). The continuous understacking of continental and oceanic fragments toward the west began with the upper units, or with more internal units not represented in the northwestern Iberian complexes (Fig. 7A), and ended with the basal units (Fig. 7B). The latter represent the outermost margin of Gondwana, and their subduction marked the closure of the intervening ocean and the transition from subduction to collision. Terrane accretion took place during the Devonian and, during the Late Devonian and the early Carboniferous, the deformation progressed toward inner realms of the Gondwana margin (Fig. 7C).

New thrust generation reflects renewed shortening in the internal parts of the orogenic wedge. Associated with continued Gondwana-Laurentia convergence and possibly to a change in plate dynamics, the small thickness of the new allochthon probably resulted from the impossibility of continued continental subduction once the thick parts of the Gondwana continental crust were involved in the orogenic wedge and were additionally thickened by folding and thrusting (Fig. 7D). The buoyancy inherent to continental crust may have triggered the rupture along low-dipping thrusts to accommodate continued convergence.

Contractional structures are common in a context of convergence. Less representative, but also common, is the presence of detachments showing subtractive relationships. They are interpreted as late orogenic to postorogenic structures related to stress relaxation, but also as synorogenic structures related to gravitational equilibration on the growing orogenic wedge. Because two orogenic cycles, Early Ordovician and Variscan, have been identified in the Ordenes Complex, and the youngest of them involved several episodes of burial and exhumation, overprinting of contractional and extensional structures should be expected.

Even though each allochthonous unit and group of units underwent a particular history, many of them share the following evolutionary stages.

1. The first stage was burial accompanied by ductile deformation related to subduction and/or understacking.
2. The second stage was main exhumation, primarily associated with complex ductile shearing and involving (1) development of regional fabrics (amphibolite to greenschist facies foliation); (2) recumbent folding (large folds identified in all but the Careón and Bazar units); and (3) in-sequence, mainly ductile thrusting or complex sequence of folding and thrusting (involving more or less coeval structural thinning and tectonic denudation, and exhumation of structurally higher units coeval with accretion of lower units below).
3. The third stage was out-of-sequence thrusting with large displacements but little exhumation.
4. The fourth stage developed late extensional detachments and associated normal faults.

5. The fifth stage was upright folding, strike-slip ductile shearing, and high-angle faulting.

ACKNOWLEDGMENTS

This study has been funded by the Spanish agencies Dirección General de Investigación Científica y Técnica and Dirección General de Enseñanza Superior e Investigación Científica, and includes results from projects PB88-0145-C02, PB91-0192-C02, PB94-1396-C02, and PB97-0234-C02. We thank J. Malavieille for a few clever comments that helped to improve the graphic presentation of data and J.B. Murphy for acute suggestions and exhaustive correction of English grammar that resulted in a considerable improvement of the text.

REFERENCES CITED

Abati, J., 2000, Petrología metamórfica y geocronología de la unidad culminante del Complejo de Ordenes en la región de Carballo (Galicia, NW del Macizo Ibérico) [Ph.D. thesis]: Madrid, Spain, Universidad Complutense, 200 p.

Abati, J., and Dunning, G.R., 2002, Edad U-Pb en monacitas y rutilos de los paragneises de la Unidad de Agualada (Complejo de Ordenes, NW del Macizo Ibérico: Geogaceta, v. 32 (in press).

Abati, J., Dunning, G.R., Arenas, R., Díaz García, F., González Cuadra, P., Martínez Catalán, J.R., and Andonaegui, P., 1999, Early Ordovician orogenic event in Galicia (NW Spain): Evidence from U-Pb ages in the uppermost unit of the Ordenes Complex: Earth and Planetary Science Letters, v. 165, p. 213–228.

Arenas, R., 1991, Opposite *P, T, t* paths of Hercynian metamorphism between the upper units of the Cabo Ortegal Complex and their substratum (northwest of the Iberian Massif): Tectonophysics, v. 191, p. 347–364.

Arenas, R., and Martínez Catalán, J.R., 1993, High-pressure and high-temperature metabasites from the Sobrado Antiform (Northwest of the Iberian Massif, Spain): A petrologic and field study in a granulite-eclogite transition zone [abs.]: Terra abstracts, Abstracts Supplement no. 4, Terra Nova, v. 5, p. 1.

Arenas, R., Abati, J., Martínez Catalán, J.R., Díaz García, F., and Rubio Pascual, F.J., 1997, *P-T* evolution of eclogites from the Agualada unit (Ordenes Complex, NW Iberian Massif, Spain): Implications for crustal subduction: Lithos, v. 40, p. 221–242.

Arenas, R., Gil Ibarguchi, J.I., González Lodeiro, F., Klein, E., Martínez Catalán, J.R., Ortega-Girones, E., Pablo Maciá, J.G. de, and Peinado, M., 1986, Tectonostratigraphic units in the complexes with mafic and related rocks of the NW of the Iberian Massif: Hercynica, v. 2, p. 87–110.

Arenas, R., Rubio Pascual, F.J., Díaz García, F., and Martínez Catalán, J.R., 1995, High-pressure micro-inclusions and development of an inverted metamorphic gradient in the Santiago Schists (Ordenes Complex, NW Iberian Massif, Spain): evidence of subduction and syn-collisional decompression: Journal of Metamorphic Geology, v. 13, p. 141–164.

Bard, J.P., Capdevila, R., and Matte, Ph., 1971, Structure de la chaîne hercynienne de la Meseta Ibérique: Comparaison avec les segments voisins, *in* Histoire structurale du Golfe de Gascogne, Editions Technip, Symposiums, Publications de l'Institut Français du Pétrole, v. 1.4, p. 1–68.

Bastida, F., Martínez Catalán, J.R., and Pulgar, J.A., 1986, Structural, metamorphic and magmatic history of the Mondoñedo nappe (Hercynian belt, NW Spain): Journal of Structural Geology, v. 8, p. 415–430.

Bayer, R., and Matte, Ph., 1979, Is the mafic/ultramafic massif of the Cabo Ortegal (NW Spain) a nappe emplaced during a Variscan obduction: A new gravity interpretation: Tectonophysics, v. 57, p. T9–T18.

Bellido, F., Brandle, J.L., Lasala, M., and Reyes, J., 1992, Consideraciones petrológicas y cronológicas sobre las rocas graníticas hercínicas de Galicia: Cuadernos do Laboratorio Xeolóxico de Laxe, v. 17, p. 241–261.

Bessoles, B., 1977, Géologie de l'Afrique: Le craton ouest-africain: Mémoires du Bureau de Recherches Géologiques et Minières 88, 402 p.

Bouchez, J.L., Lister, G.S., and Nicolas, A., 1983, Fabric asymmetry and shear sense in movement zones: Geologische Rundschau, v. 72, p. 401–419.

Caby, R., 1989, Precambrian terranes of Benin-Nigeria and northeast Brazil and the Late Proterozoic south Atlantic fit, *in* Dallmeyer, R.D., ed., Terranes in the Circum-Atlantic Paleozoic Orogens: Boulder, Colorado, Geological Society of America Special Paper 230, p. 145–158.

Capdevila, R., and Vialette, Y., 1970, Estimation radiométrique de l'âge de la deuxième phase tectonique hercynienne en Galice Moyenne (Nord-Ouest de l'Espagne): Comptes Rendus de l'Académie des Sciences, Paris, v. 270, p. 2527–2530.

Castiñeiras, P., Arenas, R., Martínez Catalán, J.R., and González del Tánago, J., 2000, Mineral assemblages and *P-T* constraints in Ky-St-Grt schists from the Arinteiro antiform (Ordenes Complex, NW Iberian Massif): Variscan-Appalachian dynamics: The building of the Upper Paleozoic basement, Basement Tectonics 15: A Coruña, Spain, Program and Abstracts, p. 81–84.

Cawood, P.A., Van Gool, J.A.M., and Dunning, G.R., 1995, Collisional tectonics along the Laurentian margin of the Newfoundland Appalachians, *in* Hibbard, J.P., Van Staal, C.R., and Cawood, P.A., eds., Current perspectives in the Appalachian-Caledonian Orogen: Geological Association of Canada Special Paper 41, p. 283–301.

Dallmeyer, R.D., and Gil Ibarguchi, J.I., 1990, Age of amphibolitic metamorphism in the ophiolitic unit of the Morais allochthon (Portugal): Implications for early Hercynian orogenesis in the Iberian Massif: Journal of the Geological Society of London, v. 147, p. 873–878.

Dallmeyer, R.D., and Keppie, J.D., 1987, Polyphase late Paleozoic tectonothermal evolution of the southwestern Meguma Terrane, Nova Scotia: Evidence from $^{40}Ar/^{39}Ar$ mineral ages: Canadian Journal of Earth Sciences, v. 24, p. 1242–1254.

Dallmeyer, R.D., Ribeiro, A., and Marques, F., 1991, Polyphase Variscan emplacement of exotic terranes (Morais and Bragança Massifs) onto Iberian successions: Evidence from $^{40}Ar/^{39}Ar$ mineral ages: Lithos, v. 27, p. 133–144.

Dallmeyer, R.D., Martínez Catalán, J.R., Arenas, R., Gil Ibarguchi, J.I., Gutiérrez Alonso, G., Farias, P., Aller, J., and Bastida, F., 1997, Diachronous Variscan tectonothermal activity in the NW Iberian Massif: Evidence from $^{40}Ar/^{39}Ar$ dating of regional fabrics: Tectonophysics, v. 277, p. 307–337.

Díaz García, F., 1990, La geología del sector occidental del Complejo de Ordenes (Cordillera Hercínica, NW de España): Coruña, Spain, Laboratorio Xeolóxico de Laxe, Nova Terra, v. 3, 230 p.

Díaz García, F., 1993, Análisis comparativo de la foliación regional y estructuras asociadas en el Dominio de Santiago y su autóctono relativo (Galicia, NW de España): Revista de la Sociedad Geológica de España, v. 6, p. 105–114.

Díaz García, F., 2000, Lithostratigraphy and structure of the A Coruña unit. Uppermost units of the Variscan Belt, NW Spain: Variscan-Appalachian dynamics: The building of the Upper Paleozoic basement, Basement Tectonics 15: A Coruña, Spain, Program and Abstracts, p. 92–94.

Díaz García, F., Arenas, R., Martínez Catalán, J.R., González del Tánago, J., and Dunning, G., 1999a, Tectonic evolution of the Careón ophiolite (Northwest Spain): A remnant of oceanic lithosphere in the Variscan belt: The Journal of Geology, v. 107, p. 587–605.

Díaz García, F., Martínez Catalán, J.R., Arenas, R., and González Cuadra, P., 1999b, Structural and kinematic analysis of the Corredoiras Detachment: Evidence for early variscan orogenic extension in the Ordenes Complex, NW Spain: International Journal of Earth Sciences, v. 88, p. 337–351.

Díez Balda, M.A., Martínez Catalán, J.R., and Ayarza, P., 1995, Syn-collisional extensional collapse parallel to the orogenic trend in a domain of steep tectonics: The Salamanca detachment zone (Central Iberian zone, Spain): Journal of Structural Geology, v. 17, p. 163–182.

Escuder Viruete, J., Arenas, R., and Martínez Catalán, J.R., 1994, Tectonothermal evolution associated with Variscan crustal extension in the Tormes Gneiss Dome (NW Salamanca, Iberian Massif, Spain): Tectonophysics, v. 238, p. 117–138.

Farias, P., Gallastegui, G., González-Lodeiro, F., Marquínez, J., Martín Parra, L.M., Martínez Catalán, J.R., Pablo Maciá, J.G. de, and Rodríguez Fernández, L.R., 1987, Aportaciones al conocimiento de la litoestratigrafía y estructura de Galicia Central: Memórias da Faculdade de Ciências, Universidade do Porto, v. 1, p. 411–431.

Fernández-Suárez, J., Corfu, F., Arenas, R., Marcos, A., Martínez Catalán, J.R., Díaz García, F., Abati, J., and Fernández, F.J., 2002, U-Pb evidence for a polyorogenic evolution of the *HP-HT* units of the NW Iberian Massif: Contributions to Mineralogy and Petrology, v. 143, p. 236–253.

Fortey, R.A., and Cocks, L.R.M., 1988, Arenig to Llandovery faunal distributions in the Caledonides, *in* Harris, A.L., and Fettes, D.J., eds., The Caledonian-Appalachian Orogen: London, Geological Society of London Special Publication 38, p. 233–246.

Franke, W., 1989, Tectonostratigraphic units in the Variscan belt of central Europe, *in* Dallmeyer, R.D., ed., Terranes in the Circum-Atlantic Paleozoic Orogens: Boulder, Colorado, Geological Society of America Special Paper 230, p. 67–90.

Frisch, W., and Neubauer, F., 1989, Pre-Alpine terranes and tectonic zoning in the eastern Alps, *in* Dallmeyer, R.D., ed., Terranes in the Circum-Atlantic Paleozoic Orogens: Boulder, Colorado, Geological Society of America Special Paper 230, p. 91–100.

García Garzón, J., Pablo Maciá, J.G. de, and Llamas Borrajo, J.F., 1981, Edades absolutas obtenidas mediante el método Rb/Sr en dos cuerpos de ortoneises en Galicia Occidental: Boletín Geológico y Minero de España, v. 92, p. 463–466.

Gates, A.E., Simpson, C., and Glover III, L., 1986, Appalachian Carboniferous dextral strike-slip faults: An example from Brookneal, Virginia: Tectonics, v. 5, p. 119–133.

Gebauer, D., 1993, Intra-grain zircon dating within the Iberian Massif: Ollo de Sapo augengneisses, bimodal gneisses from the Massif de Guilleries (Girona), graywacke of the Tentudía Group (Serie Negra, SW Spain) and the high-*P*/high-*T*-rock association at Cabo Ortegal (Galicia): Comunicações XII Reunião de Geologia do Oeste Peninsular, Universidade de Evora, v. 2, p. 41–46.

Gil Ibarguchi, J.I., Mendia, M., Girardeau, J., and Peucat, J.J., 1990, Petrology of eclogites and clynopiroxene-garnet metabasites from the Cabo Ortegal Complex (northwestern Spain): Lithos,v. 25, p. 133–162.

Girardeau, J., and Gil Ibarguchi, I., 1991, Pyroxenite-rich peridotites of the Cabo Ortegal Complex (Northwestern Spain): Evidence for large-scale upper-mantle heterogeneity: Journal of Petrology, v. 32 (Special Lherzolites Issue), p. 135–154.

González Cuadra, P., Martínez Catalán, J.R., Arenas, R., Díaz García, F., and Abati, J., 1999, Polyorogenic evolution of the Uppermost units of the Ordenes Complex (NW Spain): Journal of Conference Abstracts, European Union of Geosciences 10, p. 34.

Gutiérrez-Alonso, G., Barba, P., and Díaz García, F., 2000, Sedimentology and structure of the low grade uppermost unit of the Ordenes Complex at the coastal section. Superposition of structural trends and processes: Variscan-Appalachian dynamics: The building of the Upper Paleozoic basement, Basement Tectonics 15: A Coruña, Spain, Program and Abstracts, p. 234–235.

Hatcher, R.D., Jr., 1989, Tectonic synthesis of the U.S. Appalachians, *in* Hatcher, R.D., Thomas, W.A., and Viele, G.W., eds., The Appalachian-Ouachita Orogen in the United States: Boulder, Colorado, Geological Society of America, Geology of North America, v. F-2, p. 511–535.

Hatcher, R.D., Jr., 2001, Rheological partitioning during multiple reactivation of the Paleozoic Brevard fault zone, Southern Appalachians, USA, *in* Holdsworth, R.E., Strachan, R.A., Magloughlin, J.F., and Knipe, R.J., eds., The nature and tectonic significance of fault zone weakening: London, Geological Society of London Special Publication 186, p. 257–271.

Hibbard, J.P., and Samson, S.D., 1995, Orogenesis exotic to the Iapetan cycle in the southern Appalachians, *in* Hibbard, J.P., Van Staal, C.R., and Cawood, P.A., eds., Current perspectives in the Appalachian-Caledonian Orogen: Geological Association of Canada Special Paper 41, p. 191–205.

Hilgen, J.D., 1971, The Lalín unit: A new structural element in the Hercynian Orogen of Galicia (NW Spain): Koninklijke Nederlandse Akademie van Wetenschappen, Amsterdam, Series B, v. 74, p. 398–407.

Holdaway, M.J., 1971, Stability of andalusite and the aluminum silicate phase diagram: American Journal of Science, v. 271, p. 97–131.

Hossack, J.R., and Cooper, M.A., 1986, Collision tectonics in the Scandinavian Caledonides, *in* Coward, M.P., and Ries, A.C., eds., Collision tectonics: London, Geological Society of London Special Publication 19, p. 287–304.

Hubregtse, J.J. M.V., 1973, High-grade metamorphic rocks of the Mellid area, Galicia, N.W. Spain: Leidse Geologische Mededelingen, v. 49, p. 9–31.

Hutton, D.H. W., 1989, Pre-Alleghanian terrane tectonics in the British and Irish Caledonides, *in* Dallmeyer, R.D., ed., Terranes in the Circum-Atlantic Paleozoic Orogens: Boulder, Colorado, Geological Society of America Special Paper 230, p. 67–90.

Jamieson, R.A., 1986, *P-T* paths from high temperature shear zones beneath ophiolites: Journal of Metamorphic Geology, v. 4, p. 3–22.

Keen, C., Peddy, C., de Voogd, B., and Matthews, D., 1989, Conjugate margins of Canada and Europe: Results from deep reflection profiling: Geology, v. 17, p. 173–176.

Keppie, J.D., 1989, Northern Appalachian terranes and their accretionary history, *in* Dallmeyer, R.D., ed., Terranes in the Circum-Atlantic Paleozoic Orogens: Boulder, Colorado, Geological Society of America Special Paper 230, p. 159–192.

Kuijper, R.P., 1980, Precambrian U-Pb zircon ages from Western Galicia (NW Spain): Earth-Science Reviews, v. 16, p. 313–316.

Lancelot, J.R., Allegret, A., and Iglesias Ponce de León, M., 1985, Outline of Upper Precambrian and lower Paleozoic evolution of the Iberian Peninsula according to U-Pb dating of zircons: Earth and Planetary Science Letters, v. 74, p. 325–337.

Lefort, J.P., 1989, Basement correlation across the north Atlantic: Berlin, Germany, Springer-Verlag, 148 p.

Liñán, E., and Quesada, C., 1990, Ossa-Morena zone: Stratigraphy, rift phase (Cambrian), *in* Dallmeyer, R.D., and Martínez García, E., eds., Pre-Mesozoic geology of Iberia: Berlin, Germany, Springer-Verlag, p. 259–266.

Marcos, A., 1973, Las Series del Paleozoico Inferior y la estructura herciniana del occidente de Asturias (NW de España): Trabajos de Geología, Universidad de Oviedo, v. 6, p. 1–113.

Marcos, A., Marquínez, J., Pérez-Estaún, A., Pulgar, J.A., and Bastida, F., 1984, Nuevas aportaciones al conocimiento de la evolución tectonometamórfica del Complejo de Cabo Ortegal (NW de España): Cuadernos do Laboratorio Xeolóxico de Laxe, v. 7, p. 125–137.

Marquínez García, J.L., 1984, La geología del área esquistosa de Galicia Central (Cordillera Herciniana, NW de España): Memorias del Instituto Geológico y Minero de España, v. 100, 231 p.

Martínez Catalán, J.R., and Arenas, R., 1992, Deformación extensional de las unidades alóctonas superiores de la parte oriental del Complejo de Ordenes (Galicia): Geogaceta, v. 11, p. 108–111.

Martínez Catalán, J.R., Arenas, R., Díaz García, F., and Abati, J., 1997, Variscan accretionary complex of northwest Iberia: Terrane correlation and succession of tectonothermal events: Geology, v. 25, p. 1103–1106.

Martínez Catalán, J.R., Arenas, R., Díaz García, F., and Abati, J., 1999, Allochthonous units in the Variscan belt of NW Iberia. Terranes and accretionary history, *in* Sinha, A.K., ed., Basement Tectonics 13: Dordrecht, Netherlands, Kluwer, p. 65–84.

Martínez Catalán, J.R., Arenas, R., Díaz García, F., Rubio Pascual, F.J., Abati, J., and Marquínez, J., 1996, Variscan exhumation of a subducted Paleozoic continental margin: The basal units of the Ordenes Complex, Galicia, NW Spain: Tectonics, v. 15, p. 106–121.

Martínez Catalán, J.R., Klein, E., Pablo Maciá, J.G., and González Lodeiro, F., 1984, El Complejo de Ordenes: Subdivisión, descripción y discusión sobre su origen: Cuadernos do Laboratorio Xeolóxico de Laxe, v. 7, p. 139–210.

Matte, Ph., 1968, La structure de la virgation hercynienne de Galice (Espagne): Revue de Géologie Alpine, v. 44, p. 1–128.

Matte, Ph., and Capdevila, R., 1978, Tectonique en grands plis couchés et plissements superposés d'âge hercynien dans la série de Ordenes-Betanzos (Galice Occidentale): Coruña, Spain, Cuadernos del Seminario de Estudios Cerámicos de Sargadelos, v. 27, p. 193–201.

Mendia Aranguren, M.S., 1996, Petrología de la Unidad Eclogítica del Complejo de Cabo Ortegal (NW de España) [Ph.D. thesis]: Bilbao, Spain, Universidad del País Vasco, 463 p.

Mitra, G., and Boyer, S.E., 1986, Energy balance and deformation mechanisms of duplexes: Journal of Structural Geology, v. 8, p. 291–304.

Murphy, J.B., and Nance, R.D., 1991, Supercontinent model for the contrasting character of late Proterozoic orogenic belts: Geology, v. 19, p. 469–472.

Murphy, J.B., Nance, R.D., Keppie, J.D., Dostal, J., and Cousens, B.L., 1995, Odyssey of West Avalonia: Isotopic constraints for Late Proterozoic III–Early Silurian Paleogeography, *in* Hibbard, J.P., Van Staal, C.R., and Cawood P.A., eds., Current perspectives in the Appalachian-Caledonian orogen: Geological Association of Canada Special Paper 41, v. 227–237.

Neuman, R.B., and Max, M.D., 1989, Penobscottian-Grampian-Finnmarkian orogenies as indicators of terrane linkages, *in* Dallmeyer, R.D., ed., Terranes in the Circum-Atlantic Paleozoic Orogens: Boulder, Colorado, Geological Society of America Special Paper 230, p. 31–45.

Ochsner, A., 1993, U-Pb geochronology of the Upper Proterozoic-Lower Paleozoic geodynamic evolution in the Ossa-Morena zone (SW Iberia): Constraints on the timing of the Cadomian orogeny [Ph.D. thesis]: Zurich, Switzerland, Swiss Federal Institute of Technology, 249 p.

Ohta, Y., Dallmeyer, R.D., and Peucat, J.J., 1989, Caledonian terranes in Svalbard, *in* Dallmeyer, R.D., ed., Terranes in the Circum-Atlantic Paleozoic Orogens: Boulder, Colorado, Geological Society of America Special Paper 230, p. 1–15.

Ordóñez Casado, B., Gebauer, D., Schäfer, H.J., Gil Ibarguchi, J.I., and Peucat, J.J., 2001, A single Devonian subduction event for the *HP/HT* metamorphism of the Cabo Ortegal complex within the Iberian Massif: Tectonophysics, v. 332, p. 359–385.

Osberg, P.H., Tull, J.F., Robinson, P., Hon, R., and Butler, J.R., 1989, The Acadian Orogeny, *in* Hatcher, R.D., Thomas, W.A., and Viele, G.W., eds., The Appalachian-Ouachita Orogen in the United States: Boulder, Colorado, Geological Society of America, Geology of North America, v. F-2, p. 179–232.

Pérez-Estaún, A., Martínez Catalán, J.R., and Bastida, F., 1991, Crustal thickening and deformation sequence in the footwall to the suture of the Variscan Belt of northwest Spain: Tectonophysics, v. 191, p. 243–253.

Peucat, J.J., Bernard-Griffiths, J., Gil Ibarguchi, J.I., Dallmeyer, R.D., Menot, R.P., Cornichet, J., and Iglesias Ponce de León, M., 1990, Geochemical and geochronological cross section of the deep Variscan crust: The Cabo Ortegal high-pressure nappe (northwestern Spain): Tectonophysics, v. 177, p. 263–292.

Pin, C., Ortega Cuesta, L.A., and Gil Ibarguchi, J.I., 1992, Mantle-derived, early Paleozoic A-type metagranitoids from the NW Iberian Massif: Nd isotope and trace-element constraints: Bulletin de la Société Géologique de France, v. 163, p. 483–494.

Quinlan, G.M., Hall, J., Williams, H., Wright, J.A., Colman-Sadd, S.P., O'Brien, S.J., Stockmal, G.S., and Marillier, F., 1992, Lithoprobe onshore seismic reflection transects across the Newfoundland Appalachians: Canadian Journal of Earth Sciences, v. 29, p. 1865–1877.

Ribeiro, A., Dias, R., and Brandao Silva, J., 1995, Genesis of the Ibero-Armorican arc: Geodinamica Acta, v. 8, p. 173–184.

Ribeiro, A., Pereira, E., and Dias, R., 1990, Structure in the northwest of the Iberian Peninsula, *in* Dallmeyer, R.D., and Martínez García, E., eds., Pre-Mesozoic geology of Iberia: Berlin, Germany, Springer-Verlag, p. 220–236.

Ribeiro, M.L., and Floor, P., 1987, Magmatismo peralcalino no Maciço Hespérico: Sua distribução e significado geodinámico, *in* Bea, F., Carnicero, A., Gonzalo, J.C., López-Plaza, M., and Rodríguez Alonso, M.D., eds., Geología de los granitoides y rocas asociadas del Macizo Hespérico: Madrid, Spain, Rueda, p. 211–221.

Ries, A.C., 1979, Variscan metamorphism and K-Ar dates in the Variscan fold belt of south Brittany and NW Spain: Journal of the Geological Society of London, v. 136, p. 89–103.

Ries, A.C., and Shackleton, R.M., 1971, Catazonal complexes of northwest Spain and north Portugal remnants of a Hercynian thrust plate: Nature Physical Sciences, v. 234, p. 65–69.

Santos Zalduegui, J.F., Schärer, U., and Gil Ibarguchi, J.I., 1995, Isotope constraints on the age and origin of magmatism and metamorphism in the Malpica-Tuy allochthon, Galicia, NW-Spain: Chemical Geology, v. 121, p. 91–103.

Santos Zalduegui, J.F., Schärer, U., Gil Ibarguchi, J.I., and Girardeau, J., 1996, Origin and evolution of the Paleozoic Cabo Ortegal ultramafic-mafic complex (NW Spain): U-Pb, Rb-Sr and Pb-Pb isotope data: Chemical Geology, v. 129, p. 281–304.

Schäfer, H.J., Gebauer, D., Gil Ibarguchi, J.I., and Peucat, J.J., 1993, Ion-microprobe U-Pb zircon dating on the *HP/HT* Cabo Ortegal Complex (Galicia, NW Spain): Preliminary results: Terra Abstracts, v. 5.4, p. 22.

Soper, N.J., 1988, Timing and geometry of collision, terrane accretion and sinistral strike-slip events in the British Caledonides, *in* Harris, A.L., and Fettes, D.J., eds., The Caledonian-Appalachian Orogen: London, Geological Society of London Special Publication 38, p. 481–492.

Valverde Vaquero, P., and Fernández, F.J., 1996, Edad de enfriamiento U/Pb en rutilos del Gneis de Chímparra (Cabo Ortegal, NO de España): Geogaceta, v. 20, p. 475–478.

Van Calsteren, P.W.C., Boelrijk, N.A.I.M., Hebeda, E.H., Priem, H.N.A., den Tex, E., Verdurmen, E.A.T.H., and Verschure, R.H., 1979, Isotopic dating of older elements (including the Cabo Ortegal mafic-ultramafic complex) in the Hercynian Orogen of NW Spain: Manifestations of a presumed Early Paleozoic Mantle-plume: Chemical Geology, v. 24, p. 35–56.

Van Zuuren, A., 1969, Structural petrology of an area near Santiago de Compostela (NW Spain): Leidse Geologische Mededelingen, v. 45, p. 1–71.

Vogel, D.E., 1967, Petrology of an eclogite- and pyrigarnite-bearing polymetamorphic rock Complex at Cabo Ortegal, NW Spain: Leidse Geologische Mededelingen, v. 40, p. 121–213.

Von Raumer, J.F., and Neubauer, F., 1993, Late Precambrian and Paleozoic evolution of the Alpine basement—An overview, *in* Von Raumer, J.F., and Neubauer, F., eds., Pre-Mesozoic geology in the Alps: Berlin, Germany, Springer-Verlag, p. 625–639.

Von Raumer, J.F., Stampfli, G.M., Borel, G., and Bussy, F., 2001, The organization of pre-Variscan Basement areas at the Gondwana margin: International Journal of Earth Sciences, v. 91, p. 35–52.

Winchester, J.A., and Van Staal, C.R., 1995, Volcanic and sedimentary terrane correlation between Dunnage and Gander zones of the Canadian Appalachians and the British Caledonides reviewed, *in* Hibbard, J.P., Van Staal, C.R., and Cawood, P.A., eds., Current perspectives in the Appalachian-Caledonian Orogen: Geological Association of Canada Special Paper 41, p. 95–114.

Yardley, B.W.D., 1989, An introduction to metamorphic petrology: New York, USA, Longman Scientific and Technical, 248 p.

Manuscript Accepted by the Society February 8, 2002

Printed in the U.S.A.

Geological Society of America
Special Paper 364
2002

Cross section through the central part of the Iberian Pyrite Belt, South Portuguese zone (Spain)

Carles Soriano*
Institut de Ciències de la Terra Jaume Almera, Consell Superior d'Investigacions Científiques, C/ Lluis Solé Sabarís s/n 08028, Barcelona, Spain
Josep Maria Casas
Department de Geodinàmica i Geofísica, Universitat de Barcelona, c/ Martí i Franqués s/n, 08028 Barcelona, Spain

ABSTRACT

The Variscan structure of the Iberian Pyrite Belt is characterized by fold and thrust tectonics. Three roughly coaxial deformation phases can be distinguished that include the development of a pervasive D2 axial plane cleavage, D1 and D2 south-verging folds, D1 and D2 southward-propagating thrusts, and D3 open and subvertical folds. D2 thrusts developed in a forward-breaking sequence with respect to D1 thrusts. The basal décollement level of the D2 thrust system can be tentatively correlated with a high conductive level located at a depth of 8 km in the northern part of the Iberian Pyrite Belt. The style of fold and thrust displacement varies throughout the different structural units. D1 structures predominate in the northern part of the area and in the higher structural units whereas they are absent in the intermediate and lowermost units, which are mainly characterized by D2 structures. The style of folds evolves from D1 tight and recumbent, through D2 asymmetric and south verging, to D3 open and subvertical. This distribution suggests that deformation propagated southward and decreased in intensity from the D1 to D3 events. Sense of cleavage transection is not systematic in the eastern part of the Iberian Pyrite Belt and its relation to sinistral transpression is doubtful. Map-scale west-verging asymmetric folds are proposed as better markers of this transpression.

INTRODUCTION

The main structural feature of the Variscan orogen in western Europe is the Ibero-Armorican arc. The southern branch of the Ibero-Armorican arc is known as the Iberian Massif and crops out in the western part of the Iberian Peninsula (Fig. 1). The Iberian Massif comprises an autochthonous central part and exotic oceanic terranes located to the northwest and southwest of the Iberian Peninsula. The exotic terrane in the southwest is known as the South Portuguese zone and is bounded to the north by the Ossa Morena zone (Fig. 1). This boundary is interpreted as a major suture in the Variscan orogen corresponding to the former site of subduction of the South Portuguese plate beneath the Ossa Morena plate and subsequent continental collision (Munhá et al., 1986; Crespo-Blanc and Orozco, 1988; Ribeiro et al., 1990; Quesada, 1991; Dallmeyer et al., 1993).

The Variscan structure of the South Portuguese zone, except the suture zone with the Ossa Morena zone, is still poorly known, and most works are concerned with the Portuguese part (Ribeiro et al., 1983; Oliveira, 1990; Silva et al., 1990). The

*E-mail: csoriano@ija.csic.es.

Soriano, C., and Casas, J.M., 2002, Cross section through the central part of the Iberian Pyrite Belt, South Portuguese zone (Spain), *in* Martínez Catalán, J.R., Hatcher, R.D., Jr., Arenas, R., and Díaz García, F., eds., Variscan-Appalachian dynamics: The building of the late Paleozoic basement: Boulder, Colorado, Geological Society of America Special Paper 364, p. 183–197.

South Portuguese zone comprises a tectonostratigraphic domain that is noted for its great deposits of massive sulfide, and is known as the Iberian Pyrite Belt. Few works have been published that aim to describe in detail the overall structure of the central and eastern parts of Iberian Pyrite Belt (Simancas, 1986; Quesada, 1998). Understanding the Variscan structures and the history of Variscan deformation is crucial to predicting ore-deposit location at depth and to decipher the role of the Iberian Pyrite Belt in the process of continental collision between the South Portuguese plate and the Ossa Morena plate.

In this chapter we present a cross section through the central part of the Iberian Pyrite Belt. To illustrate the structure of this part of the belt, field sketches, maps, and detailed cross sections of areas close to the cross-section plane are provided. The structure of the three structural units intersected in the cross section plane is discussed. An evolution of the Variscan deformation is proposed in accordance with the structures and the deformation phases observed.

TECTONOSTRATIGRAPHIC DOMAINS OF THE SOUTH PORTUGUESE ZONE

The South Portuguese zone is composed of five tectonostratigraphic units that record the process of accretion to the Iberian autochthon during oblique plate convergence in late Paleozoic time (Oliveira, 1990; Quesada, 1991). Rock ages, intensity of Variscan deformation, and regional metamorphism increase from south to north in the South Portuguese zone (Munhá, 1990; Silva et al., 1990). Regional metamorphism is considered syntectonic, of Namurian to Westphalian age (Simancas, 1986; Munhá, 1990; Silva et al., 1990; Giese et al., 1994). Large intrusive bodies of granitic, gabbroic, tonalitic, and dioritic composition crop out at the northeastern part of the South Portuguese zone and the southern part of the Ossa Morena zone, and are thought to be coincident with this regional metamorphism (Dallmeyer et al., 1993; De La Rosa et al., 1993). The domains of the South Portuguese zone are listed in the following, from north to south.

Beja-Acebuches Ophiolitic Complex

This domain (Munhá et al., 1986) is a narrow band as wide as 1.5 km exposed discontinuously at the southern boundary of the Ossa Morena zone (Fig. 1). The stratigraphy of the Beja-Acebuches Ophiolitic Complex is composed, from top to bottom, of pillowed metabasalts, amphibolites that correspond to metamorphosed multiple sheeted dike intrusions, flaser metagabbros, and scattered serpentinite bodies (Munhá et al., 1986). The complex displays a polyphasic deformation history that comprises as many as three deformation events (Crespo-Blanc and Orozco, 1991; Fonseca and Ribeiro, 1993; Giese et al., 1994). Beds dip mainly to the north and polarity is right-way up, while the main structures are thrusts and cleavages that trend west-northwest–east-southeast and dip north-northeast (Crespo-Blanc and Orozco, 1991; Castro et al., 1996). The northern contact of the Beja-Acebuches Ophiolitic Complex with the Ossa Morena zone and southern contact of the complex with the Pulo do Lobo accretionary prism are two thrusts dipping to the north with a southward sense of motion and a component of sinistral shear (Crespo-Blanc and Orozco, 1991; Fonseca and Ribeiro, 1993; Giese et al., 1994; Quesada et al., 1994). The complex exhibits an inverted regional metamorphic gradient, from greenschist facies in the south to granulite facies in the north (Crespo-Blanc and Orozco, 1991; Fonseca and Ribeiro, 1993; Giese et al., 1994; Castro et al., 1996). A retrograde dynamic metamorphism has been related to a late sinistral shear movement in the southern boundary (Crespo-Blanc and Orozco, 1988). Regional metamorphism is low pressure and has been dated as 340 Ma according to $^{40}Ar/^{39}Ar$ dating of hornblende in the amphibolites (Dallmeyer et al., 1993). The age of the Beja-Acebuches Ophiolitic Complex rocks has not been determined, but an Early Devonian age has been proposed (Oliveira, 1990; Crespo-Blanc and Orozco, 1991). There is a general agreement in interpreting the Beja-Acebuches Ophiolitic Complex as the remnants of an oceanic lithosphere representing a major suture zone in the Variscan orogen of western Europe (Munhá et al., 1986; Crespo-Blanc and Orozco, 1991; Dallmeyer et al., 1993; Fonseca and Ribeiro, 1993; Giese et al., 1994; Quesada et al., 1994; Castro et al., 1996).

Pulo do Lobo accretionary prism

This domain (Quesada, 1991) is exposed south of the Beja-Acebuches Ophiolitic Complex. Due to the discontinuous exposure of the Beja-Acebuches Ophiolitic Complex and late Variscan northeast-southwest strike-slip faulting (Simancas, 1985) that displaces the boundary between the Ossa Morena zone and the South Portuguese zone, the Pulo do Lobo accretionary prism is also in contact with the Ossa Morena zone along a sinistral strike-slip fault (Fig. 1). The prism is composed of siliciclastic metasediments, mainly phyllites, quartzites, and graywackes (Oliveira et al., 1986; Giese et al., 1988; Oliveira, 1990; Giese et al., 1994). An oceanic melange that comprises metasediments and amphibolites and serpentinites with mid-ocean ridge (MORB) type geochemical affinities is present at the base of the stratigraphic succession (Fig. 1) (Munhá, 1983; Eden, 1991; Quesada et al., 1994). The overall structure of the Pulo do Lobo accretionary prism has been interpreted to be an antiformal stacking of thrust sheets (Quesada et al., 1994). As many as three deformation phases have been distinguished in the Pulo do Lobo accretionary prism depending on the structural level and the stratigraphic position (Silva et al., 1990; Crespo-Blanc and Orozco, 1991; Giese et al., 1994; Quesada et al., 1994). Thrusts, fold axes, and cleavage surfaces trend west-northwest–east-southeast and fold vergence is to the south-southwest (Silva et al., 1990; Giese et al., 1994). Regional metamorphism increases from low grade in the south to greenschist facies and amphibolite facies in the north of the accretionary prism (Munhá, 1990; Giese et al., 1994). The age of Pulo do Lobo accretionary prism rocks ranges from late Givetian–early Frasnian to late Famennian, according to palynomorphs, acritarchs, and spores (Oliveira et al., 1986; Giese et al., 1988; Eden, 1991). The Pulo do Lobo rocks have been interpreted in a broad sense to be accretionary prism deposits of a forearc tectonic setting (Quesada et al., 1994). Oceanic melanges are believed to incorporate the stratigraphy during the process of oceanic crust subduction and obduction (Quesada et al., 1994).

Iberian Pyrite Belt

This domain extends over 300 km in east-west and northwest-southeast directions south of the Pulo do Lobo accretionary prism, and it is in contact with the prism by a north-dipping thrust fault (Figs. 1 and 2). The main feature of the Iberian Pyrite Belt is the occurrence of polymetallic sulfide deposits associated with basic and silicic volcanic rocks, which are interbedded with Late Devonian to early Carboniferous turbiditic siliciclastic deposits. Sedimentation and volcanism are continuous through the stratigraphic sequence of the Iberian Pyrite Belt, and no internal unconformities have been observed (Soriano and Martí, 1999). Deposition of these rocks took place in a submarine (below wave base) continental platform environment (Oliveira, 1990; Soriano and Martí, 1999). The lithostratigraphy of the Iberian Pyrite Belt is described in the following. Variscan structure of the Iberian Pyrite Belt is characterized by folding and thrusting tectonics with southward vergence and piggyback propagation mode (Silva et al., 1990; Soriano, 1997). Regional metamorphism is low grade, ranging from zeolite facies in the south to greenschist in the north, and is superimposed upon hydrothermal alteration that occurred during sulfide deposition (Barriga, 1990; Munhá, 1990). The Iberian Pyrite Belt ore-bearing basin has been interpreted as a transtensional basin developed on the continental crust of the south Portuguese plate (Oliveira, 1990).

Baixo Alentejo Flysch

This domain is mainly composed of siliciclastic turbidite sediments, conglomerates, sandstones, and shales of the late Visean to Namurian Mértola and Mira Formations (Oliveira et al., 1979; Oliveira, 1983). Conformably underlying this flysch succession are the rocks of the Volcano-Sedimentary Complex of the Iberian Pyrite Belt, which crop out in the core of the Cercal antiform (Silva et al., 1990), near the Atlantic Ocean (Fig. 1). The northern boundary of the Baixo Alentejo Flysch with the Iberian Pyrite Belt is a north-dipping thrust (Figs. 1 and 2). Regional metamorphism ranges from zeolite facies of very low grade in the north of the Baixo Alentejo Flysch to nonmetamorphic conditions in the south (Munhá, 1990). Variscan deformation of the Baixo Alentejo Flysch is characteristic of an imbricate southward-directed fold and thrust belt. Silva et al. (1990) recognized one deformation phase that includes asymmetric south-verging folds and an axial plane cleavage of northwest-southeast orientation. Facies distribution suggests a southwest progradation of turbidite fan lobes, and paleocurrent data indicate a strong northwest-southeast control on the turbidite currents (Oliveira, 1990).

Southwest Portugal domain

This domain (Oliveira, 1983) is composed of siliciclastic subtidal sediments of late Famennian to early Tournaisian age overlain by a carbonate and evaporitic subtidal succession of late Tournaisian to late Namurian age (Oliveira, 1990). On top of the carbonate and evaporitic succession there are epiclastic graywackes and shales of middle Namurian to early Westphalian age that display paleocurrents and facies distribution characteristics of a flysch prograding from the north and northeast (Oliveira, 1990). The Southwest Portugal domain northern boundary with the Baixo Alentejo Flysch is a southward-directed thrust (Fig. 1). The Variscan structure of the Southwest Portugal domain comprises open antiforms and synforms at a kilometer scale and southward-verging minor folds with well-developed axial plane cleavage (Silva et al., 1990).

STRATIGRAPHY OF THE IBERIAN PYRITE BELT

The Iberian Pyrite Belt has been divided into three lithostratigraphic units, based on the occurrence of volcanic and volcaniclastic rock types and associated hydrothermal deposits (Strauss, 1970; Schermerhorn, 1971). From the base to the top of the stratigraphic succession these units are the Phyllite and Quartzite unit, the Volcano-Sedimentary Complex, and the Culm Group (Fig. 3).

Phyllite and Quartzite unit

This unit is composed of siliciclastic prevolcanic rocks deposited in a subtidal environment from fan delta and sand-bar systems on a shallow-marine continental platform (Moreno and Sáez, 1990; Oliveira, 1990; Moreno et al., 1996). The base of this unit is not exposed, but stratigraphic sections have minimum thicknesses of 300–400 m. The Phyllite and Quartzite unit is Frasnian to late Famennian in age according to conodont fauna located in rare limestone lenses (Boogaard and Schermerhorn, 1975a, 1980, 1981; Fantinet et al., 1976) and palynomorphs in slate (Lake et al., 1988).

Volcano-Sedimentary Complex

This synvolcanic unit hosts the sulfide ore bodies of the Iberian Pyrite Belt. It is composed of intrusions and lavas of mafic and silicic compositions, volcaniclastic and siliciclastic rocks, and hydrothermal deposits, including massive sulfides and jaspers. The Volcano-Sedimentary Complex conformably overlies the Phyllite and Quartzite unit, and measured thickness

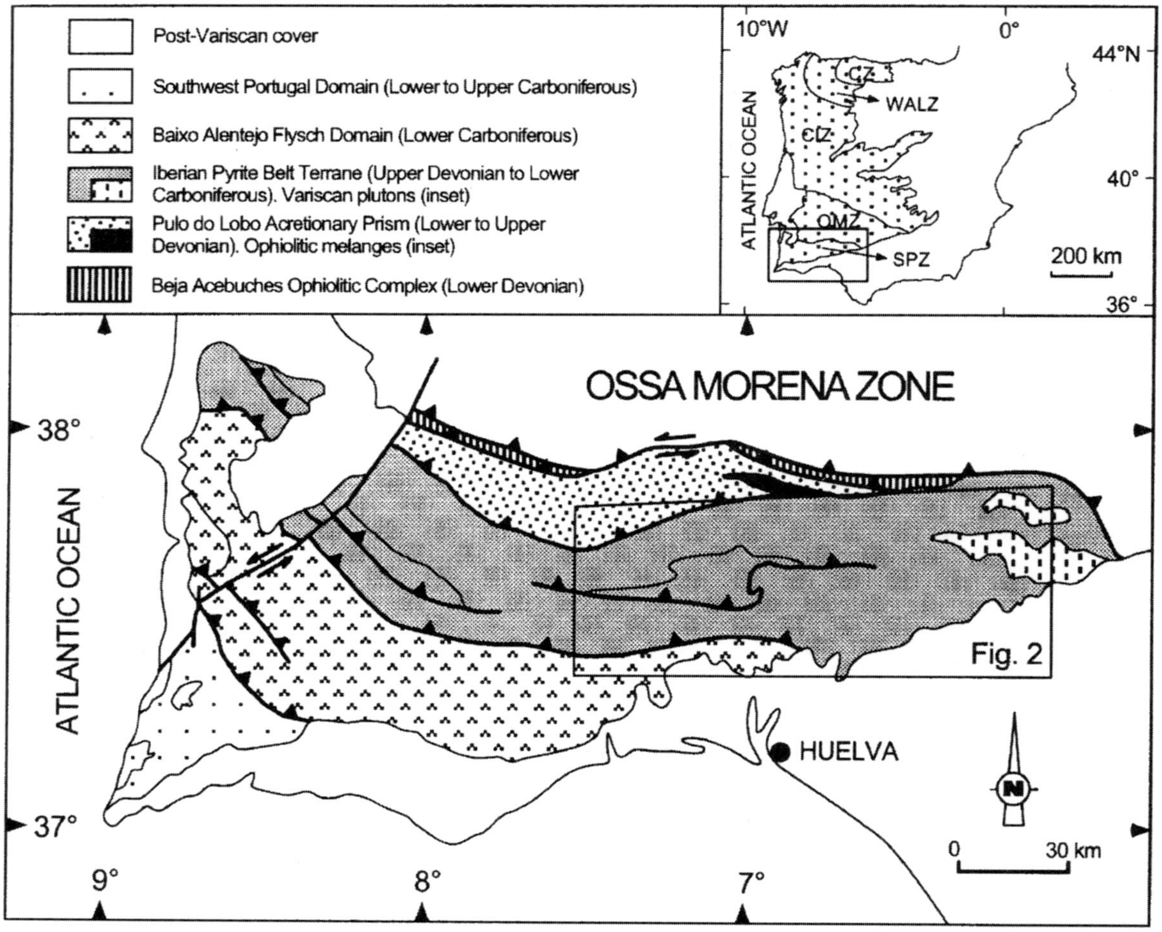

Figure 1. Tectonostratigraphic terrane map of South Portuguese zone (modified from Oliveira, 1990; Quesada, 1991). Inset: Variscan geological domains of Iberian Massif (shaded): CZ, Cantabrian zone; WALZ, West Asturian–Leonese zone; CIZ, Central Iberian zone; OMZ, Ossa Morena zone; SPZ, South Portuguese zone (after Julivert et al., 1974).

in stratigraphic sections varies from 300 to 800 m (Soriano and Martí, 1999). Rare limestone lenses bearing conodont fauna (Boogaard and Schermerhorn, 1975b; Oliveira, 1992), black shales bearing palynomorphs (Pereira et al., 1996), and zircons in stockwork zones (Nesbitt et al., 1999) provide a late Famennian to late Visean age.

The Volcano-Sedimentary Complex rocks show a marked east-west structural trend as well as an important change in the distribution of rocks from north to south (Figs. 2 and 3). Volcanic rocks of silicic and mafic compositions are more abundant in the northern and eastern parts of the basin, where they compose stratigraphic piles as thick as 500 m (Figs. 2 and 3). Silicic and mafic intrusions and lavas tend to be located in the lower parts of stratigraphic successions of the Volcano-Sedimentary Complex in the southern and western parts of the Iberian Pyrite Belt, whereas they occupy higher stratigraphic positions in the north and east (Figs. 2 and 3).

The volcanism of the Iberian Pyrite Belt was mainly effusive, although explosive volcanism may have occurred outside the ore-bearing basin. Facies analysis and facies distribution suggest that water depth increased to the east and north in the basin (Soriano and Martí, 1999). Volcanism occurred within a maximum time span of 30 Ma in the continental crust of the South Portuguese plate (Mitjavila et al., 1997). This volcanism is considered to occur during a longer, continuous, oblique plate convergence process between the South Portuguese and Ossa Morena plates (Silva et al., 1990; Fonseca and Ribeiro, 1993; Quesada et al., 1994). First, the oceanic crust of the South Portuguese plate was consumed beneath the continental crust of the Ossa Morena plate, then continental collision commenced and volcanism occurred in the Iberian Pyrite Belt, prior to the main compressional deformation and thrusting events (Silva et al., 1990; Quesada et al., 1994).

Volcanic rocks of the Iberian Pyrite Belt display compositional bimodality. Tholeiitic basalts and rhyolites predominate, although minor andesite and dacite have also been recognized (Munhá, 1983; Mitjavila et al., 1997; Thiéblemont et al., 1998). Relative abundances of rock types, their Sr and Nd isotope com-

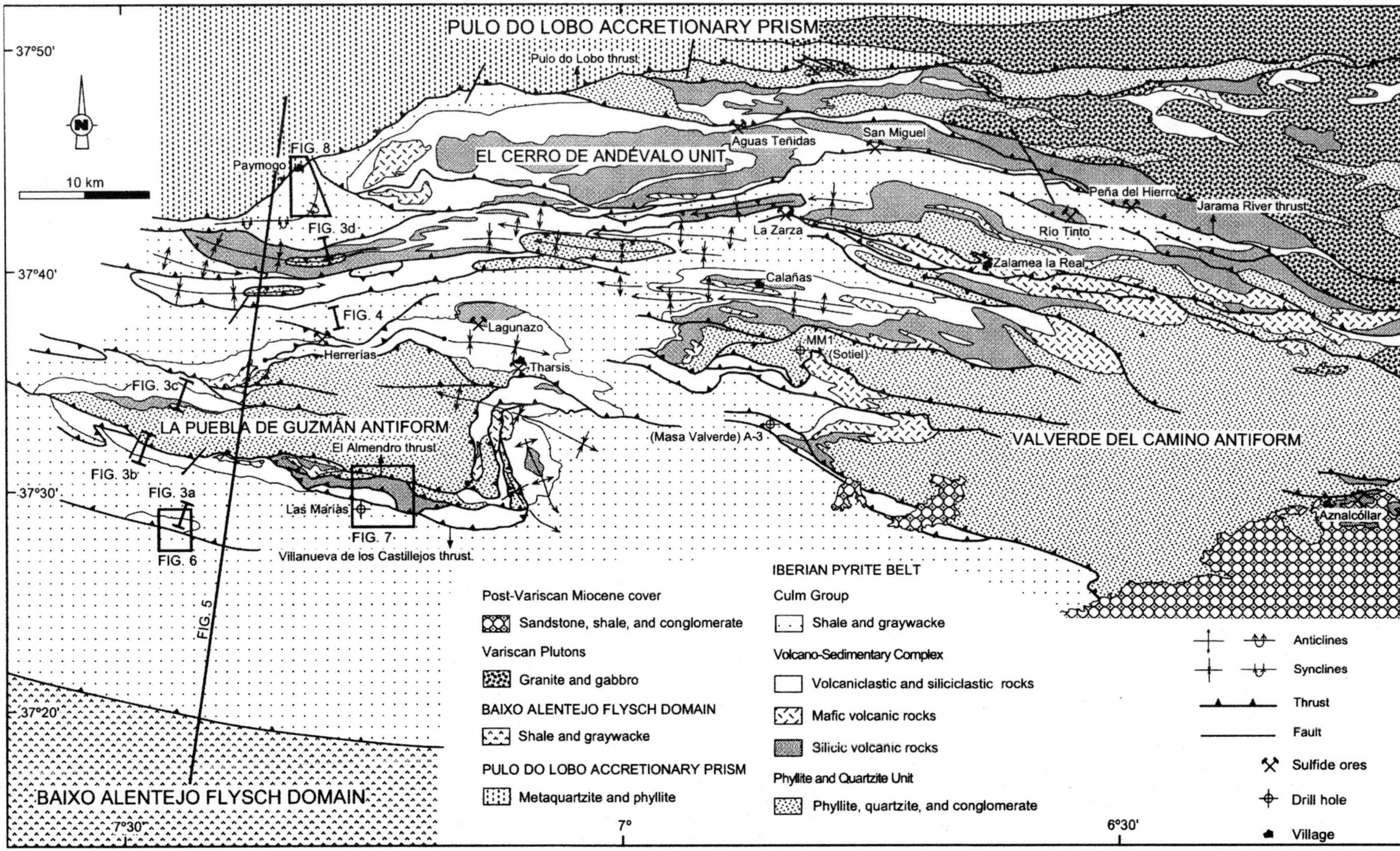

Figure 2. Geological map of eastern part of the Iberian Pyrite Belt (modified from Instituto Geológico y Minero de España, 1982).

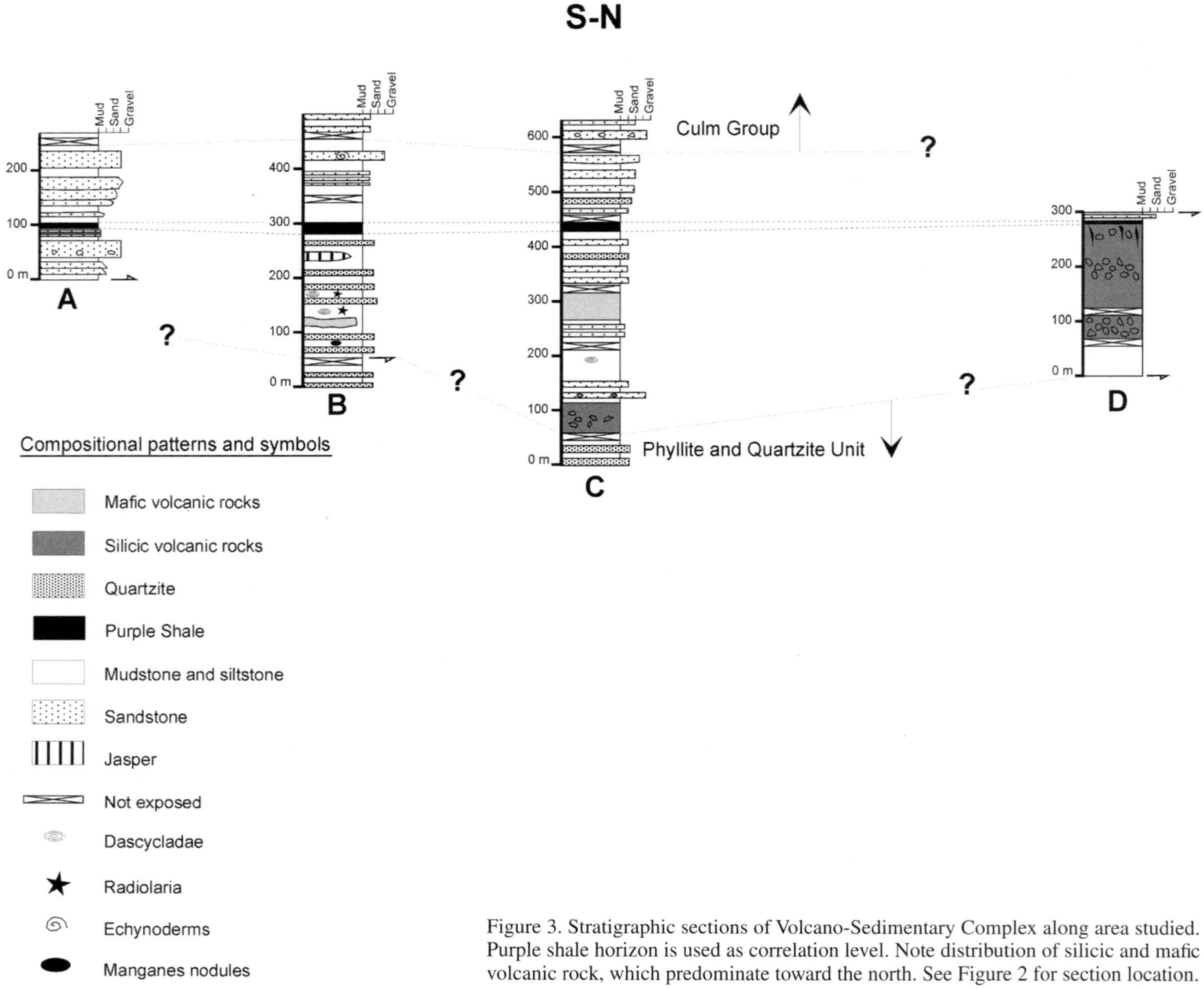

Figure 3. Stratigraphic sections of Volcano-Sedimentary Complex along area studied. Purple shale horizon is used as correlation level. Note distribution of silicic and mafic volcanic rock, which predominate toward the north. See Figure 2 for section location.

positions, and major trace and rare earth element geochemistry suggest that these four groups of rocks are not related by fractional crystallization processes (Mitjavila et al., 1997). The basalts are attributed to variable degrees of partial melting of an asthenospheric mantle source and were contaminated by crustal partial melts (Munhá, 1983; Mitjavila et al., 1997). The more evolved rock types were produced by partial melting of different parts of the continental crust due to the accumulation of these basaltic melts at the base of the crust (Munhá, 1983; Mitjavila et al., 1997).

Culm Group

This is a postvolcanic unit composed of epiclastic Variscan flysch turbiditic deposits, which conformably overlie the Volcano-Sedimentary Complex. The top of this unit is eroded, but the maximum measured thickness is 3000 m (Moreno, 1993). This unit was deposited by submarine fans prograding southward onto a shallow- to deep-water continental platform (Oliveira, 1990; Moreno, 1993). Goniatite and conodont fauna yield a late Visean to Westphalian age (Oliveira, 1983).

VARISCAN DEFORMATION PHASES

Variscan deformation in the Iberian Pyrite Belt is polyphasic. Most authors agree in distinguishing as many as three deformation phases that are regionally extended and involve the three lithostratigraphic units of the Iberian Pyrite Belt (Lécolle, 1977; Routhier et al., 1980; Instituto Geológico y Minero de España, 1982; Simancas, 1986; Silva et al., 1990; González-Clavijo et al., 1994; Soriano, 1997). Cleavages, parasitic folds, and thrusts characterize these three deformation phases: these structures display an east-west trend in the area studied (Fig. 2), whereas they trend northwest-southeast more to the west, near the Atlantic Ocean (Fig. 1).

D1 structures comprise a weakly developed millimeter to centimeter spaced cleavage (S1) dipping shallowly to moderately north-northeast. S1 is axial planar to decimeter-scale second-order folds located in the limbs of first-order folds. First-order F1 folds normally exceed outcrop scale, verge to the south-southeast, display fold axes trending east-west and may exhibit a recumbent style (Silva et al., 1990; Soriano, 1997). The occurrence of these folds has been interpreted from exposures of continuous section with no faults and monocline bedding, in which beds show opposite stratigraphic polarities (normal and reverse), as deduced from sedimentary criteria, but identical structural polarity with respect to S2 cleavage Fig. 4). This suggests the existence of folds developed before S2 cleavage. Thrusts corresponding to the D1 deformation phase have an east-west trend, propagate to the south, and limit D1 thrust nappes folded and thrusted by D2 structures (Figs. 4 and 5).

D2 structures are the most prominent in the Iberian Pyrite Belt and comprise a pervasive, pressure-solution, millimeter to centimeter spaced cleavage (S2) that locally transposes bedding and dips moderately to steeply to north-northeast (Fig. 5). S2 is axial planar to asymmetric, south-verging folds to three orders of magnitude that display subhorizontal to subvertical folds axes trending east-west. Most D2 folds display hinge thickening, and some disharmonic folds have been observed in slate beds that alternate with sandstones. D2 thrusts propagate to the south, trend east-west, dip moderately (normally less than S2) north-northeast, and commonly displace D2 folds (Fig. 5). S2 commonly transects D2 folds in many parts of the Iberian Pyrite Belt, although there is not a general agreement in the sense of this transection. Ribeiro et al. (1983) and Silva et al. (1990) described a clockwise transection in the western part of the Iberian Pyrite Belt, and interpreted it as an evidence for a sinistral shear component during transpressive deformation. Counterclockwise transection has been described in the eastern part and suggested as an effect of dextral movements parallel to strike (Andrews, 1989). However, in the eastern part of the belt, Simancas (1986), Soriano (1997), and Quesada (1998) described both clockwise and counterclockwise transections that can be related to oblique movements. Our observations concur

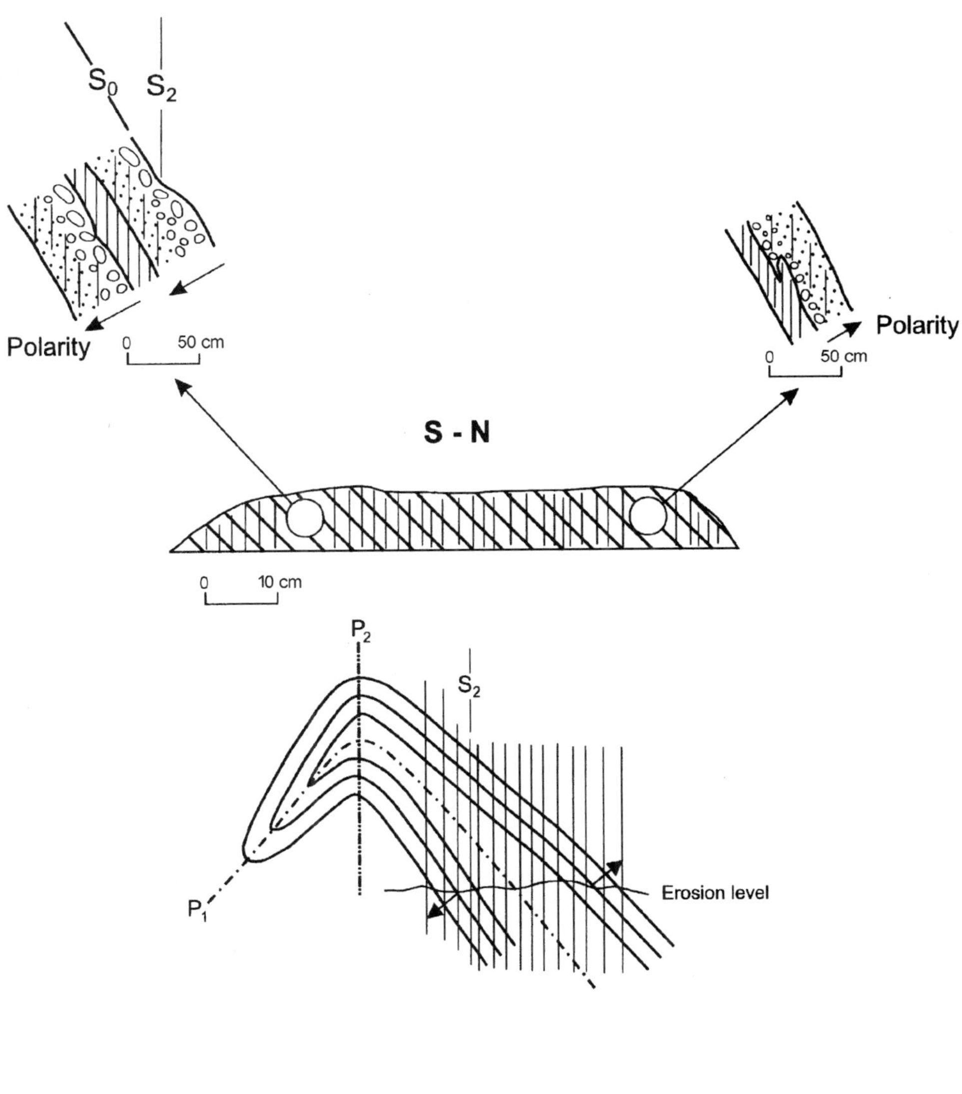

Figure 4. Occurrence of D1 fold in rocks of Culm Group: turbidite sandstone beds in southern part of outcrop show graded bedding, indicating reverse sedimentary polarity (top to south), whereas in northern part of outcrop the same beds show normal sedimentary polarity (top to north). Because no internal structural disruptions are observed, and bedding-S2 cleavage is constant, indicating normal structural polarity all along exposure, D1 fold has been interpreted. See Figure 2 for location.

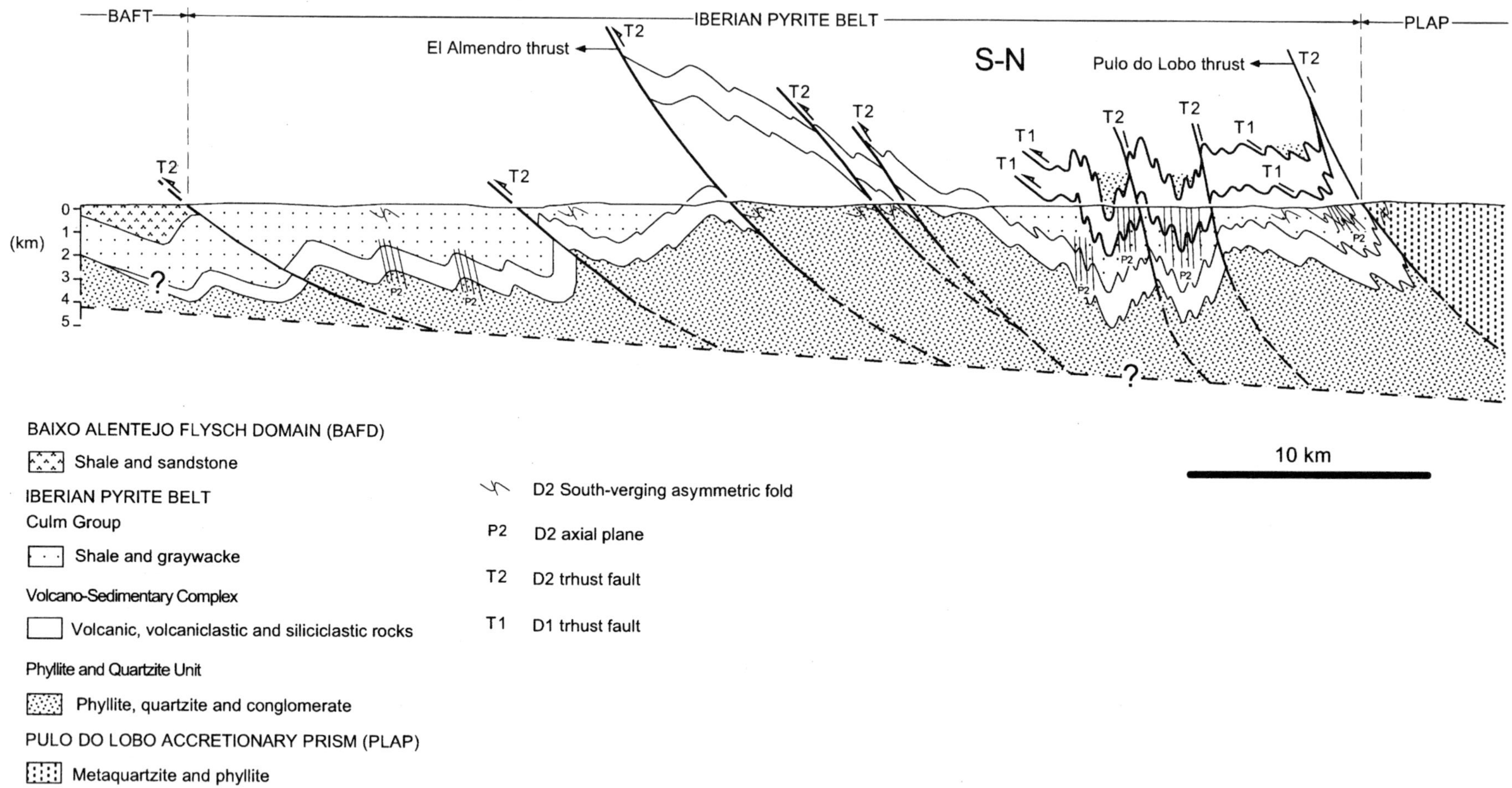

Figure 5. Cross-section through central part of the Iberian Pyrite Belt. See Figure 2 for section location.

with these works, i.e., that there is not a systematic sense of transection. We have observed clockwise and counterclockwise transections in a single outcrop, and S2 cleavage parallel to D2 fold axial surfaces.

D3 structures are locally developed and they do not interfere with D1 structures. D3 structures include a spaced, centimeter to decimeter cleavage (S3) that forms crenulations in S2 and dips steeply south-southwest. This cleavage is axial planar to open, east-west–trending folds located in the limbs of second-order D2 folds. D3 folds are of decimeter to meter scale, usually parallel, and display subhorizontal fold axes.

STRUCTURE OF THE CENTRAL PART OF THE IBERIAN PYRITE BELT

Based on the tectonic style, the structural position in the Iberian Pyrite Belt context, and the predominance of structures belonging to a particular deformation phase, three structural units have been distinguished in the eastern part of the Iberian Pyrite Belt (Soriano, 1997). From south to north they are the La Puebla de Guzmán antiform, Valverde del Camino antiform, and El Cerro de Andévalo unit (Fig. 2). La Puebla de Guzmán and Valverde del Camino antiforms are structural equivalents of the paraautochthonous units of Oliveira (1990) and Silva et al. (1990), and the El Cerro de Andévalo unit is equivalent to the allochthonous units of Oliveira (1990) and Silva et al. (1990).

The cross section in Figure 5 illustrates the structure of the central part of the Iberian Pyrite Belt, near the border between Portugal and Spain. From south to north it intersects the Sanlúcar de Guadiana subunit, La Puebla de Guzmán antiform, and El Cerro de Andévalo unit. The Valverde del Camino antiform is located farther east of the cross section (Fig. 2). This antiform is structurally located above the La Puebla de Guzmán antiform and beneath the El Cerro de Andévalo unit. The Valverde del Camino antiform displays macroscopic folds plunging west and deeper structural levels, represented by granites and gabbros, to the east (Fig. 2).

Sanlúcar de Guadiana subunit

The Sanlúcar de Guadiana subunit is a minor unit in the Iberian Pyrite Belt, extending as much as 7 km in an east-west direction, and composed of rock exposures of the Volcano-Sedimentary Complex and the Culm Group (Figs. 2 and 6). These rocks display a north-dipping cleavage that is axial planar to east-west trending folds, which correspond to the D2 deformation phase. Structural polarities are always consistent with the stratigraphic polarities shown by sedimentary structures. For example, in normal limbs, fold asymmetry and bedding-S2 relationships indicate normal polarity, which is also the polarity indicated by graded bedding. Reverse limbs are also indicated by both structural and sedimentary polarities. D1 structures have not been recognized in this unit. Two north-dipping, southward-propagating thrusts probably involve rocks of the Phyllite and Quartzite unit at depth (Fig. 6). The northernmost of these thrusts has a minimum displacement of 250 m and the southernmost has a displacement of almost 500 m. In the fault planes of these thrusts minor reverse shear zones characterized by S-C fabrics also dip to the north, and display displacements of as much as 10 cm to the south. These thrusts displace D2 folds and cleavage, but are interpreted as belonging to D2 deformation phase, because D3 thrust faults are rare in the Iberian Pyrite Belt and usually produce displacements <10 m.

The Sanlúcar de Guadiana subunit is the lowermost of the studied area, and is structurally located beneath the La Puebla de Guzmán antiform (Fig. 5). Macroscopic folds in this unit plunge toward the east at the eastern end of the unit and toward the west ant the western end (Figs. 2 and 6). Thus, the rocks of the Volcano-Sedimentary Complex illustrated in Figure 6 are intersected at an approximate depth of 2 km in the plane of the cross section of Figure 5.

La Puebla de Guzmán antiform

This D2 antiform extends 50 km in a west-northwest–east-southeast direction along the central part of the Iberian Pyrite Belt (Fig. 2). It is cored by rocks of the Phyllite and Quartzite unit and trends west-northwest–east-southeast, except in its southeastern part, where it bends to the north-northwest–south-southeast (Fig. 2).

Along the southern part of the antiform, the rocks of the Phyllite and Quartzite unit and the Volcano-Sedimentary Complex dip to the north (Fig. 7) and show stratigraphic (cross-lamination and ripple marks) and structural (bedding-S2 relationship) polarities that indicate normal bedding. The contact between these stratigraphic units is not exposed but, because the rocks of the Volcano-Sedimentary Complex are located beneath the rocks of the Phyllite and Quartzite unit, a thrust propagating to the south (the El Almendro thrust) has been postulated (Fig. 7). Minor reverse shear zones dipping north and rare fault planes dipping north with slickensides showing an oblique reverse sense of movement have been observed near the contact zone. Based on the consistent polarities shown by bedding-S2 relationships and sedimentary structures, another thrust (the Villanueva de los Castillejos thrust) located in a lower structural position is postulated at the contact between the Volcano-Sedimentary Complex and the Culm Group in the Villanueva de los Castillejos area (Fig. 7).

Two smaller thrusts have been postulated in a structural position intermediate between the El Almendro and Villanueva de los Castillejos thrusts. These thrusts develop in the hanging wall of the Villanueva de los Castillejos thrust, and form an imbricate thrust system (Fig. 7). In the thrust-fault zone of one of the former thrusts dip-slip slickensides displaying reverse sense of movement have been observed. S2 displays an asymptotic geometry that indicates a reverse shear sense of movement; S3 affects the fault-zone material (Fig. 7). This confirms that thrusting predates formation of S3 cleavage and belongs to the D2 deformation phase. A schematic cross-section restoration shows

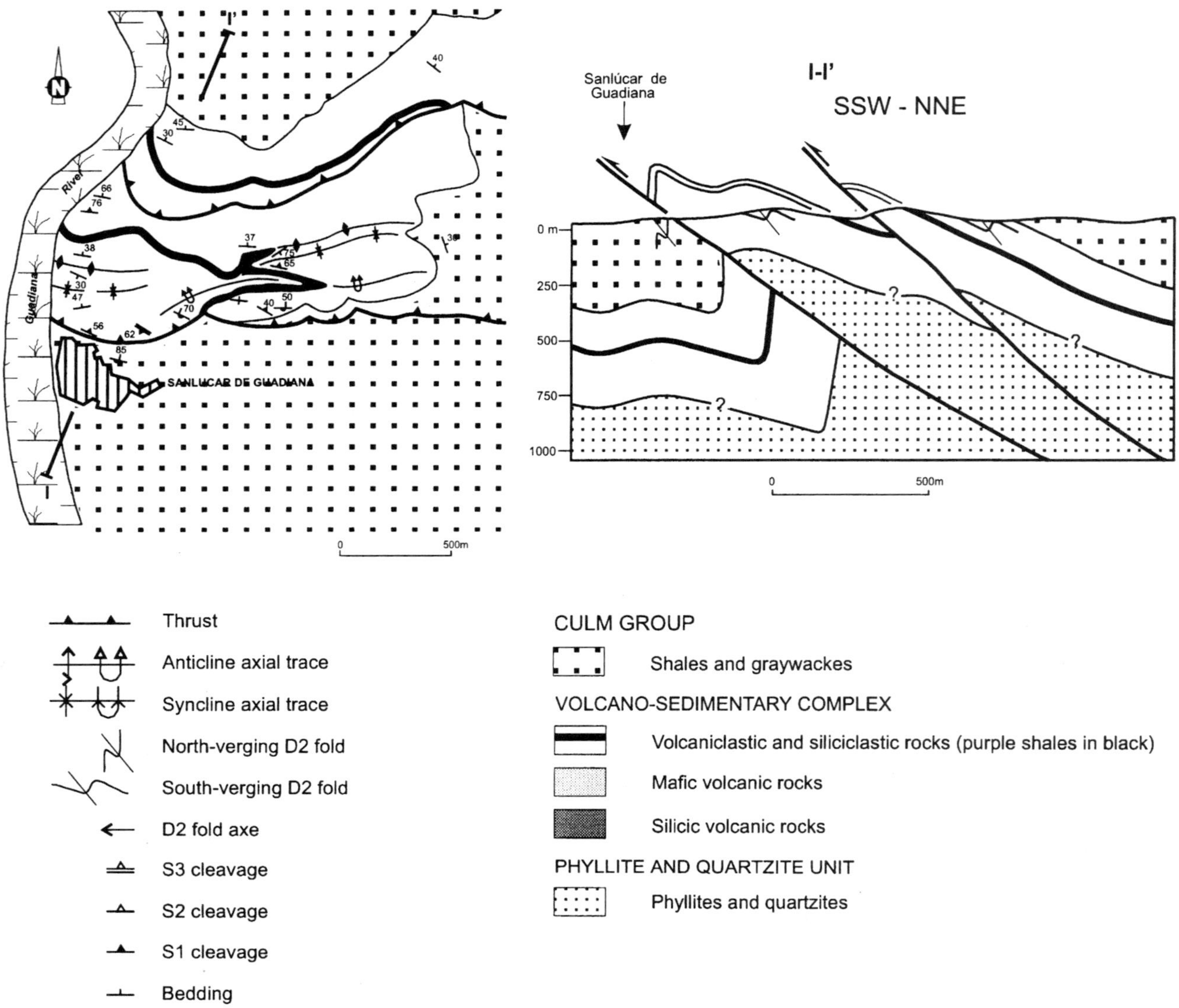

Figure 6. Map and cross section of Sanlúcar de Guadiana unit. See Figure 2 for location.

how thrusts cut higher stratigraphic levels in the sense of propagation, i.e., to the south (Fig. 7). The structurally higher El Almendro thrust involves rocks of the Phyllite and Quartzite unit and lower thrusts involve rocks of the Volcano-Sedimentary Complex and the Culm Group. Despite displacements of El Almendro thrust and the thrust immediately beneath not being well defined, mapping in the Villanueva de los Castillejos area suggests that the two décollement levels are located at the base and the top of a stratigraphic package composed of the stacking at least two cryptodomes of silicic lavas in the lower part of the Volcano-Sedimentary Complex (Fig. 7). The competence contrast between phyllites of the Phyllite and Quartzite unit and volcanic rocks of the Volcano-Sedimentary Complex favors the location of these décollements.

South of the Villanueva de los Castillejos thrust the Las Marías-1 drill hole intersects 538 m of Culm Group rocks dipping 45° to the north (Moreno and Vera, 1985). The base of the Culm Group has been tentatively located a few meters below the end of the drill hole, in the footwall of the Villanueva de los Castillejos thrust (Fig. 7). The branch line of the thrust system composed of the Villanueva de los Castillejos thrust and two more thrusts located structurally above it (Fig. 7) intersects the topographic surface toward the west (Fig. 2). Because no major topographic differences exist in the area between Figure 7 and the plane of cross section in Figure 5, it is suggested that this imbricate system plunges to the east.

In many areas of the northern part of La Puebla de Guzmán antiform beds dip to the north, and graded bedding on turbidite beds and ripple marks on sandstones of the three stratigraphic units of the Iberian Pyrite Belt indicate normal bedding. Although contacts between these stratigraphic units are usually not exposed, they are interpreted as conformable and dipping to the

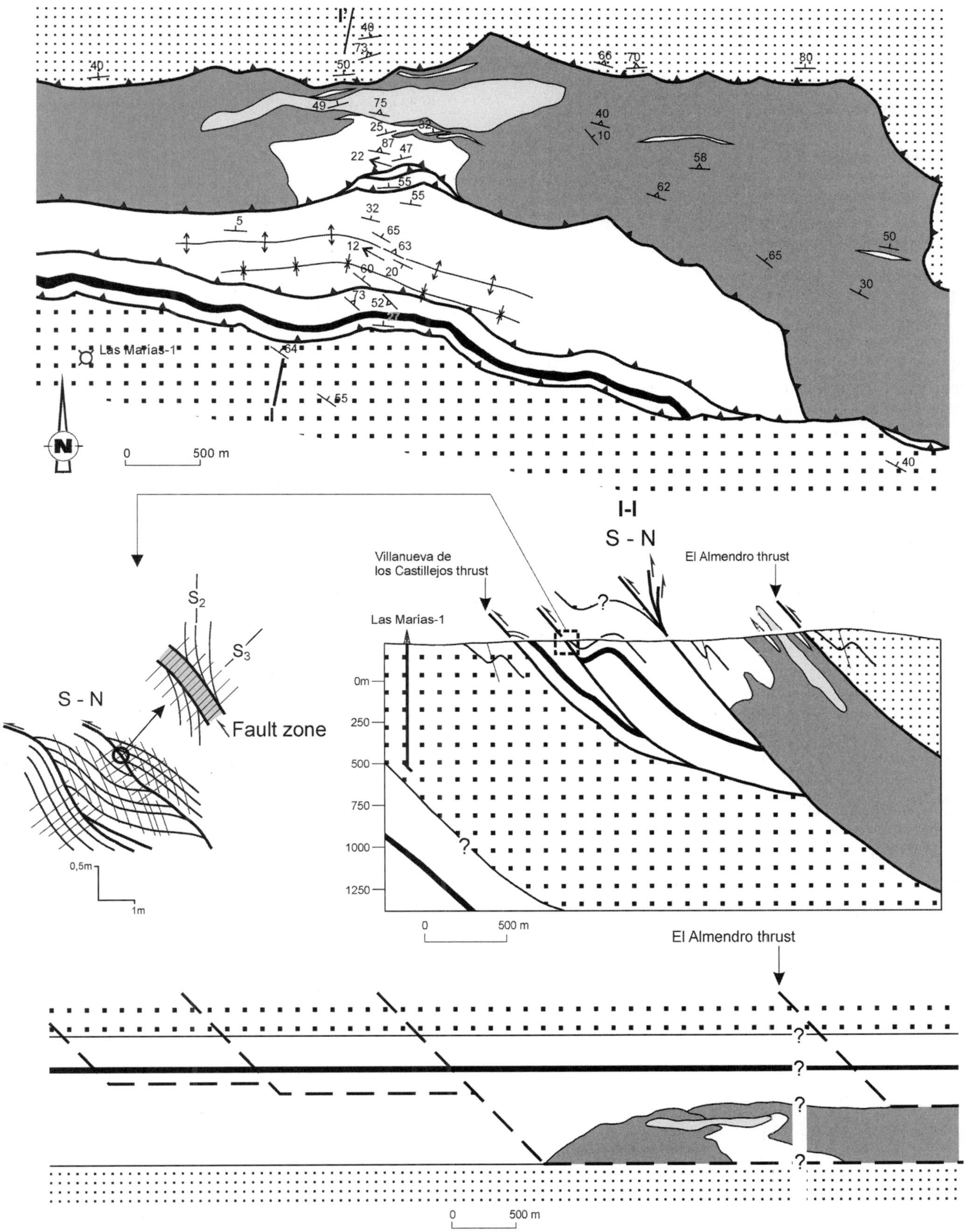

Figure 7. Map and cross-section of southern part of La Puebla de Guzmán antiform in Villanueva de los Castillejos area. Schematic restoration of D2 thrusts illustrates stratigraphy of Volcano-Sedimentary Complex and thrust-fault geometry. See Figure 2 for location and Figure 6 for legend.

north in most of the northern part of the antiform (Figs. 2, 3C, and 5). As a result, La Puebla de Guzmán antiform is here interpreted as a D2 thrust system composed of several thrusts that propagated toward the south-southwest (Figs. 2 and 5).

El Cerro de Andévalo unit

This is the uppermost structural unit and extends east-west along the northern part of the Iberian Pyrite Belt (Fig. 2). It is located above the La Puebla de Guzmán antiform and is mainly characterized by the occurrence of D1 structures and D1 thrust nappes. These thrust nappes are tightly folded by D2 folds and faulted by D2 thrusts (Figs. 2 and 5). D1 thrusts show a flat geometry at the base and the top of the Volcano-Sedimentary Complex (Fig. 5). This flat geometry is probably due to the competence contrast between silicic volcanic rocks of the Volcano-Sedimentary Complex and phyllites of the Phyllite and Quartzite unit. For these reasons the base and top of the Volcano-Sedimentary Complex can be interpreted as décollement levels for D1 thrusts. Similar D1 thrust nappes, such as the Mértola and Galé-Cela nappes, have been postulated in the Portuguese part of the Iberian Pyrite Belt (Silva et al., 1990) and could be correlated to the D1 thrust nappes forming the El Cerro de Andévalo unit. D1 thrust nappes propagated to the south and carry rocks of the Phyllite and Quartzite unit. Therefore, the lower décollement level of D1 nappes might be within the Phyllite and Quartzite unit or in lower stratigraphic units of the Pulo do Lobo accretionary prism, and located north of the contact between the Iberian Pyrite Belt and the Pulo do Lobo accretionary prism.

Contact of the Iberian Pyrite Belt with the Pulo do Lobo accretionary prism

The Iberian Pyrite Belt is overridden to the north by the Pulo do Lobo accretionary prism (Figs. 2, 5, and 8) on a northward-dipping thrust (Pulo do Lobo thrust). In the Paymogo area (Fig. 8), the Pulo do Lobo thrust is characterized by a 25-m-thick shear zone developed in shales and graywackes of the Culm Group in the footwall, and metaquartzites and phyllites of the Pulo do Lobo accretionary prism in the hanging wall. This shear zone mainly consists of oblique reverse shear planes dipping as much as 30° to the north. Beds of the Pulo do Lobo accretionary prism display D2 south-verging asymmetric mesoscopic folds and S2 features dip to the north (Fig. 8, A and B).

South of this shear zone (Fig. 8) several exposures of rocks of the Culm Group exhibit structures that belong to the three deformation phases and allow a complete reconstruction on the structural evolution. Beds of the Culm Group show D2 south-verging asymmetric folds (Fig. 8B). The reverse limb of the southernmost D2 syncline-anticline pair folds a south-verging D1 fold (Fig. 8B). Farther north, in the normal limb of the D2 fold, mesoscopic south-verging and almost recumbent D2 folds are refolded by open D3 folds with subvertical axial planes (Fig. 8B). In map view, D2 and D3 fold axial traces, and S1, S2, and S3 cleavages, are subparallel (Fig. 8A). This suggests a rough coaxiality between D1, D2, and D3 structures giving rise to type 3 fold interference patterns (Ramsay, 1967).

In the Paymogo area, D2 axial fold traces bend from an east-west direction toward a north-northeast–south-southwest direction (Fig. 8A). D2 thrusts and D2 axial fold traces located in the southeastern part of La Puebla de Guzmán antiform exhibit a similar bending (Fig. 2). In this part of the antiform, D2 structures describe a kilometre-scale west-verging fold (Fig. 2). We suggest that these map-scale folds and bends could be formed as a result of a sinistral transpressive tectonics, and that they can reflect this tectonic regime more effectively than transected folds.

Sequence of deformation

The lower décollement level of D1 thrust nappes can be located within the Pulo do Lobo accretionary prism. This suggests kilometer displacements on D1 thrust faults (Fig. 5). In the Iberian Pyrite Belt, D2 thrust-fault displacements, although kilometer scale, appear to be less than D1 (Fig. 5). An exception to this is the Pulo do Lobo thrust, which places Lower to Middle Devonian rocks of the Pulo do Lobo accretionary prism over early Carboniferous rocks of the Culm Group, and might have a great displacement (Figs. 2 and 5). D2 thrusts developed in a piggyback sequence as suggested by the Pulo do Lobo thrust and two lower D2 thrusts cutting D1 nappes (Fig. 5), and by the southward migration of the synorogenic deposits of the Culm Group (Oliveira, 1990). Basal décollement for the regional D2 thrust system can be tentatively correlated to the décollement of the South Portuguese zone (Fig. 5) that, according to Ribeiro et al. (1983), dips ~8° to the north. Thrusts bounding some of the tectonostratigraphic domains in the South Portuguese zone, such as the contact between the Baixo Alentejo Flysch and the Iberian Pyrite Belt (Quesada, 1991) and the contact between the Iberian Pyrite Belt and Pulo do Lobo accretionary prism (Pulo do Lobo thrust), may correspond to D2 thrusts that branch this basal décollement (Fig. 5). Based on magnetotelluric data, Monteiro et al. (1999) identified, in the northern part of the Iberian Pyrite Belt, a highly conductive horizon at 8 km depth. This horizon may correspond to the former basal décollement. Monteiro et al. (1999) suggested that this horizon is the base of the late Paleozoic, and corresponds to the brittle-ductile transition in the upper crust beneath the South Portuguese zone.

Figure 8. Structure of Culm Group beds in footwall of Pulo do Lobo thrust south of Paymogo. A: Map of Paymogo area showing axial fold traces. B: Detailed cross section showing interference between D1, D2, and D3 folds. C: Detail of D1 tight asymmetric fold located on overturned limb of asymmetric south-verging mesoscale D2 fold. Note north-vergent asymmetry of D2 minor folds in beds with opposite polarity and S1 cleavage folded by D2 folds. D: Subvertical D3 fold and S3 axial-planar cleavage superimposed on D2 fold. In opposite limbs of this D3 fold, S2 axial-planar cleavage and D2 fold asymmetry are consistent with south-verging D2 folds.

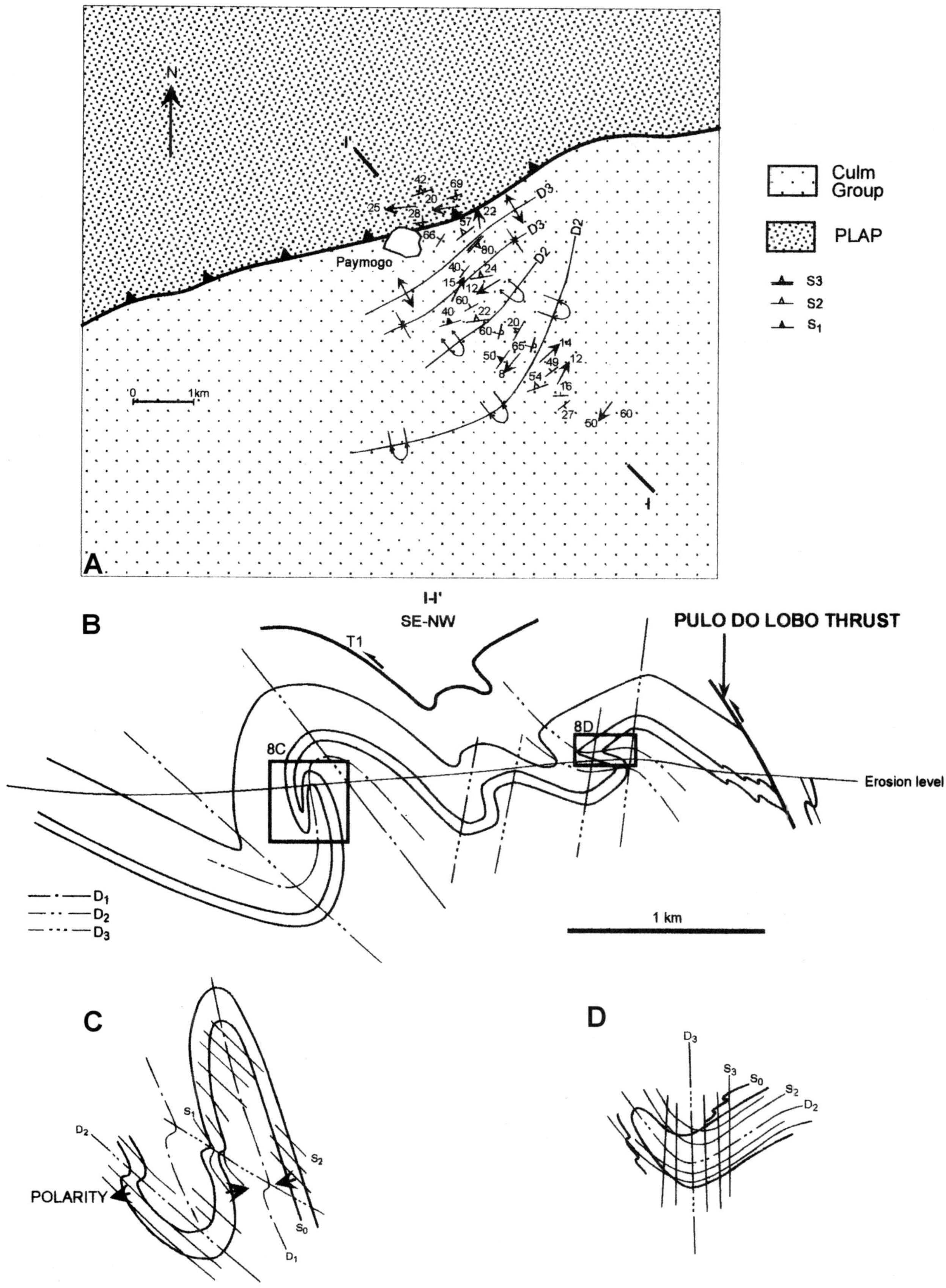

N
I'
Paymogo
D3
D2
0
1km
I
A
Culm Group
PLAP
S3
S2
S1
I-I'
SE-NW
B
T1
PULO DO LOBO THRUST
8C
8D
Erosion level
D1
D2
D3
1 km
C
POLARITY
S1
S2
S0
D2
D1
D
D3
S3
S0
S2
D2

D1 structures predominate in the northern part of the studied area and in the higher structural units whereas they are absent in La Puebla de Guzmán antiform and the Sanlúcar de Guadiana unit, which are mainly characterized by D2 structures (Figs. 2 and 5). Moreover, D2 folds are tighter to the north and D1 and D2 folds are predominantly south verging (Fig. 5). These observations and the southward propagation of thrusts suggest that deformation propagates from north to the south. From D1 to D3, folds are less tight and dip of the fold axial surfaces evolves from subhorizontal to subvertical (Fig. 8). In addition, displacement on thrust faults seems to be greater for D1 than for D2 thrusts. Due to coaxiality between deformation phases, the pervasive S2 cleavage observed may be a composite cleavage as a result of transposition of a previous S1 cleavage. Taking these observations into account, it is suggested that deformation was of maximum intensity during D1 and progressively decreased to the D3 deformation event.

CONCLUSIONS

Variscan deformation in the Iberian Pyrite Belt is characterized by fold and thrust tectonics developed over three roughly coaxial deformation phases. As a result, the Variscan structure is characterized by interference of folds of different orders of magnitude. A pervasive S2 cleavage affects all lithologies. The distribution of style of folds and displacements on thrusts faults during D1, D2, and D3 and throughout the structural units of the studied area suggests that deformation propagated southward and decreased in intensity from D1 to D3 deformation events. D2 thrusts developed in a piggyback sequence and branch to a basal décollement that dips to the north and can be tentatively correlated to a high conductive level located at a depth of 8 km in the northern part of the Iberian Pyrite Belt. D2 thrusts developed in the footwall of D1 thrusts nappes and some of them (i.e., Pulo do Lobo thrust) break previous D1 thrusts. S2 cleavage transecting folds is a common feature in the Iberian Pyrite Belt and has been interpreted as a result of transpressive tectonics. However, in the studied area no systematic sense of transection has been observed, preventing us from characterizing the sense of transpression in this part of the belt by means of transecting cleavage. Alternatively, map-scale west-verging folds that involve D2 structures may reflect a sinistral tranpressive tectonics better than transecting cleavage.

ACKNOWLEDGMENTS

Soriano benefited from a Ministerio de Educación y Cultura postdoctoral fellowship during this research. Insightful comments of Manu Gómez improved a first version of the manuscript. We thank J.R. Andrews and C. Quesada for careful reviews.

REFERENCES CITED

Andrews, J.R., 1989, The Iberian Pyrite Belt, an oblique-slip mobile zone: Notes et Mémoires du Service Géologique du Maroc, v. 335, p. 361–366.

Barriga, F.J.A.S., 1990, Metallogenesis in the Iberian Pyrite Belt, *in* Dallmeyer, R.D., and Martinez García, E., eds., Pre-Mesozoic Geology of Iberia: Berlin, Springer-Verlag, p. 369–379.

Boogaard, M.V.D., and Schermerhorn, L.J.G., 1975a, Conodont faunas from Portugal and southwestern Spain, Part 2: A Famennian conodont fauna at Cabezas de Pasto: Scripta Geologica, v. 28, p. 1–18.

Boogaard, M.V.D., and Schermerhorn, L.J.G., 1975b, Conodont faunas from Portugal and southwestern Spain, Part 3: Carboniferous conodonts at Sotiel Coronada: Scripta Geologica, v. 28, p. 37–43.

Boogaard, M.V.D., and Schermerhorn, L.J.G., 1980, Conodont faunas from Portugal and southwestern Spain, Part 4: A Famennian conodont fauna near Nerva (Rio Tinto): Scripta Geologica, v. 56, p. 1–14.

Boogaard, M.V.D., and Schermerhorn, L.J.G., 1981, Conodont faunas from Portugal and southwestern Spain, Part 6: A Lower Fammenian conodont fauna at Monte do Forno da Cal (South Portugal): Scripta Geologica, v. 63, p. 1–16.

Castro, A., Fernández, C., De la Rosa, J.D., Moreno, I., and Rogers, G., 1996, Significance of MORB-derived amphibolites from the Aracena metamorphic belt, Southwest Spain: Journal of Petrology, v. 37, p. 235–260.

Crespo-Blanc, A., and Orozco, M., 1988, The Southern Iberian shear zone: A major boundary in the Hercynian folded belt: Tectonophysics, v. 148, p. 221–227.

Crespo-Blanc, A., and Orozco, M., 1991, The boundary between the Ossa-Morena and Southportuguese zones (Southern Iberian Massif): A major suture in the European Hercynian Chain: Geologische Rundschau, v. 80, p. 691–702.

Dallmeyer, R.D., Fonseca, P.E., Quesada, C., and Ribeiro, A. 1993, $^{40}Ar/^{39}Ar$ mineral age constraints for the tectonothermal evolution of a Variscan suture in southwest Iberia: Tectonophysics, v. 222, p. 177–194.

De la Rosa, J.D., Rogers, G., and Castro, A., 1993, Relaciones $^{87}Sr/^{86}Sr$ de rocas básicas y granitoides del batolito de la Sierra Norte de Sevilla: Revista de la Sociedad Geológica de España, v. 6, p. 141–149.

Eden, C.P., 1991, Tectonostratigraphic analysis of the northern extent of the oceanic exotic terrane, Northwestern Huelva province, Spain [Ph.D. Thesis]: Southampton, UK, University of Southampton, 281 p.

Fantinet, D., Dreesen, R., Dusar, M., and Termier, G., 1976, Faunes Famenniennes de certains horizons calcaires dans la Formation Quartzitophylladique aux environs de Mértola (Portugal méridional): Comunicações dos Serviços Geológicos de Portugal, v. LX, p. 121–137.

Fonseca, P., and Ribeiro, A., 1993, Tectonics of the Beja-Acebuches ophiolite: A major suture in the Iberian Variscan foldbelt: Geologische Rundschau, v. 82, p. 440–447.

Giese, U., Reitz, E., and Walter, R., 1988, Contributions to the stratigraphy of the Pulo do Lobo Succession in Southwest Spain: Comunicações dos Serviços Geológicos de Portugal, v. 74, p. 79–84.

Giese, U., von Hoegen, R., Hoymann, K.-H., Kramm, U., and Walter, R., 1994, The Paleozoic evolution of the Ossa Morena zone and its boundary to the South Portuguese zone in SW Spain: Geological constraints and geodynamic interpretation of a suture in the Iberian Variscan orogen: Neues Jahrbuch Geologie Paläontologie Abhandlungen, v. 192, p. 383–412.

González-Clavijo E., Oswin W.M., Ferrer A., and Fernández-Gianotti J., 1994, Esquema estructural de un sector de la zona Surportuguesa al este de las minas de Rio Tinto: Boletín Geológico y Minero, v. 105, p. 225–235.

Instituto Geológico y Minero de España, 1982, Síntesis geológica de la Faja Pirítica del SO de España: Colección Memorias del Instituto Geológico y Minero de España, v. 98, p. 1–106.

Julivert, M., Fontboté, J.M., Ribeiro, A., and Conde, L., 1974, Mapa tectónico de la Península Ibérica y Baleares: Instituto Geológico y Minero de España, Escala 1:1,000,000, p. 1–101.

Lake, P.A., Oswin, W.M., and Marshall, J.E.A., 1988, A palynological approach to terrane analysis in the South Portuguese zone: Trabajos de Geología, Universidad de Oviedo, v. 17, p. 125–131.

Lécolle, M., 1977, La ceinture sud-ibérique: Un exemple de province à amas sulfurés volcano-sédimentaires (tectonique, métamorphisme, stratigraphie, volcanisme, paléogéographie et métallogenie) [Thèse d'Etat]: Paris, Université Pierre et Marie Curie, 609 p.

Mitjavila, J., Martí, J., and Soriano, C., 1997, Magmatic evolution and tectonic setting of the Pyrite Belt volcanism: Journal of Petrology, v. 38-6, p. 727–755.

Monteiro, F.A., Pous, J., Almeida, E.P., Queralt, P., Marcuello, A., Matias, H., and Mendes, L.A., 1999, Magnetotelluric survey of electrical conductivity of the crust across the Ossa Morena and South Portuguese zone suture: Tectonophysics, v. 313, p. 449–462.

Moreno, C., 1993, Postvolcanic Paleozoic of the Iberian Pyrite Belt: An example of basin morphologic control on sediment distribution in a turbidite basin: Journal of Sedimentary Petrology, v. 63, p. 1118–1128.

Moreno, C, and Vera, J. A., 1985, El Culm al sur del anticlinorio de La Puebla de Guzmán (Huelva, España): Boletín Geológico y Minero, v. XCVI-II, p. 141–151.

Moreno, C., and Sáez, R., 1990, Sedimentacion marina somera en el Devónico del anticlinorio de Puebla de Guzmán, Faja Pirítica Ibérica: Geogaceta, v. 8, p. 62–64.

Moreno, C., Sierra, S., and Sáez, R., 1996, Evidence for catastrophism at the Famennian-Dinantian boundary in the Iberian Pyrite Belt, *in* Strogen, P.S., and Jones, G.Ll., eds., Recent advances in Lower Carboniferous geology: Geological Society of London Special Publication 107, p. 153–162.

Munhá, J., 1983, Hercynian magmatism in the Iberian Pyrite Belt, *in* Sousa, M.J.L, and Oliveira, J.T., eds., The Carboniferous of Portugal: Memorias Serviços Geológicos Portugal, v. 29, p. 39–81.

Munhá, J., 1990, Metamorphic evolution of the South Portuguese/Pulo do Lobo zone, *in* Dallmeyer, R.D., and Martínez García E. eds., Pre-Mesozoic geology of Iberia: Berlin, Springer-Verlag, p. 363–369.

Munhá, J.M., Oliveira, J.T., Ribeiro, A., Oliveira, V., Quesada, C., and Kerrich, R., 1986, Beja-Acebuches ophiolite: Characterization and geodynamic significance [abs.]: Maleo, v. 13, p. 1–31.

Nesbitt, R.W., Pascual, E., Fanning, C.M., Toscano, M., Sáez, R., and Almodóvar, G.R., 1999, U-Pb dating of stockwork zircons from the eastern Iberian Pyrite Belt: Journal of the Geological Society of London, v. 156, p. 7–10.

Oliveira, J.T., Horn, M., and Paproth, E., 1979, Preliminary note on the stratigraphy of the Baixo Alentejo Flysch Group, Carboniferous of Portugal, and on the paleogeographic development compared to corresponding units in northwest Germany: Comunicações dos Serviços Geologicos de Portugal, v. 65, p. 151–168.

Oliveira, J.T., 1983, The marine Carboniferous of Portugal: A stratigraphic and sedimentological approach, *in* Sousa, M.J.L. and Oliveira, J.T. eds., The Carboniferous of Portugal: Memorias Serviços Geológicos Portugal, v. 29, p. 3–37.

Oliveira, J.T., Cunha, T., Streel, M., and Vanguestaine, M., 1986, Dating the Horta da Torre Formation, a new lithostratigraphic unit of the Ferreira-Ficalho Group: Comunicações dos Serviços Geológicos de Portugal, v. 72, p. 129–137.

Oliveira, J.T., 1990, Stratigraphy and synsedimentary tectonism: South Portuguese zone, *in* Dallmeyer, R.D., and Martínez García, E. eds., Pre-Mesozoic geology of Iberia: Berlin, Springer-Verlag, p. 334–347.

Oliveira, J.T., Coord., 1992, , Carta geológica de Portugal: Serviços Geológicos de Portugal, scale 1:200,000, sheet 8.

Pereira, Z., Sáez, R., Pons, J.M., Oliveira, J.T., and Moreno, C., 1996, Edad devónica (Struniense) de las mineralizaciones de Aznalcóllar (Faja Pirítica Ibérica) en base a palinología: Geogaceta, v. 20, p. 1609–1612.

Quesada, C., 1991, Geological constraints on the Paleozoic tectonic evolution of the tectonostratigraphic terranes in the Iberian Massif: Tectonophysics, v. 185, p. 225–245.

Quesada, C., Fonseca, P.E., Munhá, J., Oliveira, J.T., and Ribeiro, A., 1994, The Beja-Acebuches Ophiolite (Southern Iberia Variscan fold belt): Geological characterization and geodynamic significance: Boletín Geológico y Minero, v. 105-1, p. 3–49.

Quesada, C., 1998, A reappraisal of the structure of the Spanish segment of the Iberian Pyrite Belt: Mineralium Deposita, v. 33, p. 31–44.

Ramsay, J.G., 1967, Folding and fracturing of rocks: New York, McGraw-Hill, p. 560.

Ribeiro, A., Oliveira, J.T., and Silva, J.B., 1983, La estructura de la Zona Sur Portuguesa, *in* Comba, J.A., coord., Geología de España: Madrid Instituto Geológico y Minero de España, v. 1, p. 504–511.

Ribeiro, A., Quesada, C., and Dallmeyer, R.D., 1990, Geodynamic evolution of the Iberian Massif, *in* Dallmeyer, R.D., and Martínez García, E. eds., Pre-Mesozoic geology of Iberia: Berlin, Springer-Verlag, p. 399–409.

Routhier, P., Aye, F., Boyer, C., Lécolle, M., Molière, P., Roger, G., and Picot, P., 1980, La ceinture sud-ibérique à amas sulfurés dans sa partie espagnole mediane: Bureau Recherches Géologiques Minières, Mémoires 94, p. 1–265.

Schermerhorn, J.L.G., 1971, An outline stratigraphy of the Pyrite Belt: Boletín Geológico y Minero, v. 82, p. 239–268.

Silva, J.B., Oliveira, J.T., and Ribeiro, A., 1990, Structural outline: South Portuguese zone, *in* Dallmeyer, R.D., and Martínez García, E., eds., Pre-Mesozoic geology of Iberia: Berlin, Springer-Verlag, p. 348–363.

Simancas, J.F., 1985, Una zona de cizalla frágil tardía en el límite entre Ossa Morena y la Zona Sur Portuguesa (Macizo Ibérico): Cuadernos de Geología, v., 12, p. 105–112.

Simancas, J.F., 1986, La deformación en el sector oriental de la zona Sudportuguesa: Boletín Geológico y Minero, v. 97, p. 148–159.

Soriano, C., 1997, Evolució geodinàmica de la Faja Pirítica Ibèrica, Zona Sud Portuguesa [Ph. D. thesis]: Barcelona, Spain, Universitat de Barcelona, 265 p.

Soriano, C., and Martí, J., 1999, Facies analysis of volcano-sedimentary successions hosting massive sulfide deposits in the Iberian Pyrite Belt, Spain: Economic Geology, v. 94, p. 867–883.

Strauss, G.K., 1970, Sobre la geología de la provincia piritífera del Suroeste de la Península Ibérica y sus yacimientos, en especial sobre la mina de pirita de Lousal (Portugal): Memorias Instituto Geológico Minero España, v. 77, p. 1–266.

Thiéblemont, D., Pascual, E., and Stein, G., 1998, Magmatism in the Iberian Pyrite Belt: Petrological constraints on a metallogenetic model: Mineralium Deposita, v. 33, p. 98–110.

Manuscript Accepted by the Society February 8, 2002

Printed in the U.S.A.

Geological Society of America
Special Paper 364
2002

Alleghanian (Appalachian) orogeny, a product of zipper tectonics: Rotational transpressive continent-continent collision and closing of ancient oceans along irregular margins

Robert D. Hatcher Jr.
Department of Geological Sciences, University of Tennessee, Knoxville, Tennessee 37996-1410, USA

ABSTRACT

This chapter modifies the transpressional-oblique collision model for closing of oceans by suggesting that rotation of colliding masses and zippered closing of the ocean occurs following initial oblique collision. Zipper tectonics involves closing of an ocean from one end to the other as colliding elements touch and subduction eliminates the intervening ocean; rotation and transpressional collision are required. While this mechanism probably has operated in numerous other orogenics, evidence from the Carboniferous-Permian (320–260 Ma) Alleghanian orogen suggests that the mechanism operated here. The Appalachian orogen formed by events that began in the early Paleozoic and terminated with the Alleghanian orogeny, driven by continent-continent collision and final closing of the Theic ocean. The diachronous Alleghanian orogen is thrust dominated south of New England, but both New England and the southern parts contain an internal dextral strike-slip domain. Rotational oblique collision of the irregular continental margin of Africa with North America resulted in diachronous deformation, metamorphism, and plutonism along the continental margin, then crustal detachment and formation of the Blue Ridge–Piedmont megathrust sheet that drove foreland deformation in front of it, and then normal faulting. Previously plate models had assumed that the Alleghanian was the result of head-on collision, but they could not explain the diachroneity of orogeny, the abundance of only dextral strike-slip faulting throughout the Appalachians, and the origin of step-over basins in the northern Appalachians. Zipper closing of the Theic ocean by rotational oblique collision (northeast to southwest, present geographic frame), produced internal dextral strike-slip deformation in the southern and central Appalachians, and formed a dextral strike-slip system (initially sinistral) and stepover basins in New England and the Canadian Maritimes.

INTRODUCTION

The Appalachian orogen was constructed via several lithospheric convergence events throughout much of the Paleozoic. The Late Proterozoic–early Paleozoic rifted Laurentian margin was destroyed by Ordovician arc collision and obduction of oceanic crust onto the margin that produced the Taconic orogeny. During the middle Paleozoic (Late Devonian–early Mississippian) the Acadian orogeny produced deformation, metamorphism, and plutonism in the internal parts of the orogen and a clastic wedge that spread sediment onto the foreland from New York to Virginia. The climactic event that affected the Appalachians and ended the tectonic cycle was the late Mississippian to Permian Alleghanian (Appalachian) orogeny (Fig. 1).

Hatcher, R.D., Jr., 2002, Alleghanian (Appalachian) orogeny, a product of zipper tectonics: Rotational transpressive continent-continent collision and closing of ancient oceans along irregular margins, *in* Martínez Catalán, J.R., Hatcher, R.D., Jr., Arenas, R., and Díaz García, F., eds., Variscan-Appalachian dynamics: The building of the late Paleozoic basement: Boulder, Colorado, Geological Society of America Special Paper 364, p. 199–208.

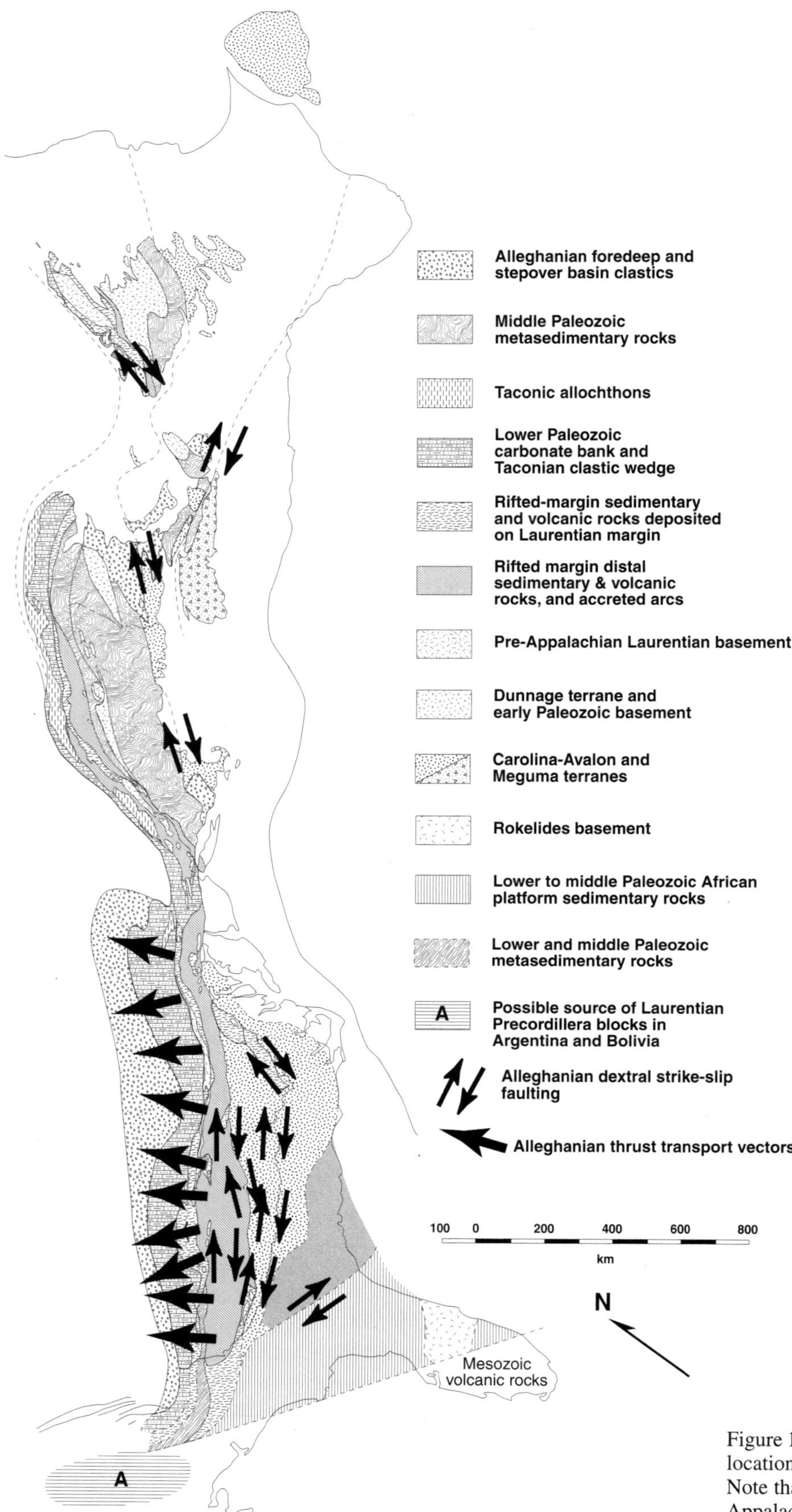

Figure 1. Simplified tectonic map of Appalachians showing locations of Alleghanian thrust and strike-slip domains. Note that there is no Alleghanian thrusting in the northern Appalachians, but that there is both dextral strike-slip faulting and thrusting in southern and central Appalachians. Thrusting appears to be younger than strike-slip faulting.

This event is the product of the final closing of the Theic ocean by collision of Africa with Laurentia. The event produced dextral strike-slip faulting and plutons throughout the orogen, stepover basins in the north, and dominance of thrust faulting in the southern and central Appalachians. Extensive Alleghanian metamorphism has long been known in eastern New England (Rodgers, 1970; Mosher, 1983; Hatcher et al., 1989); a linear belt of Alleghanian amphibolite facies metamorphism and deformation also has been identified in the eastern Piedmont of the southern Appalachians, traceable from Virginia to Alabama (Hatcher et al., 1990).

The purpose of this chapter is to introduce the zipper tectonics mechanism for closing of the Theic ocean and to suggest that this mechanism can be used to explain numerous attributes of the Appalachian orogen that earlier plate models fail to explain. Zipper tectonics involves closing of an ocean that proceeds from one end of the closing ocean to the other as colliding land masses touch and subduction eliminates the intervening ocean; transpressional collision accompanied by rotation is required for this to occur. The process of rotational oblique collision and progressive, zipper closing of oceans, using the Late Carboniferous–Permian (320–260 Ma) Alleghanian collision of western Africa with eastern North America, is discussed as the primary example. This process has been alluded to in past syntheses of the Appalachians (e.g., Rodgers, 1970; Hatcher, 1971, 1972) and other chains (e.g., Şengör, 1987; Şengör et al., 1993), but has not been analyzed in detail. This study reveals that zippered closing may occur as a partly head-on, partly rotational process, and the results consist of a complex interaction of transpressive and head-on collisional elements, producing a spectrum of structures with displacement vectors ranging from orogen parallel to oblique to orogen normal.

It has long been accepted that the Appalachian mountain chain was formed by the collision of Africa with North America (e.g., Wilson, 1966; Rodgers, 1970). The Alleghanian event, however, affected different parts of the orogen in different ways. The northern half of the Appalachians from New York northward contains only strike-slip faults, stepover basins, a metamorphic core, and a number of plutons that formed during the Alleghanian event. From New York southward, the entire aspect of the orogen changes such that the elegant Alleghanian foreland appears to have been formed as the great crystalline Blue Ridge–Piedmont megathrust sheet moved toward the interior of the continent out of the collision zone. Previously, most plate models had assumed that the Alleghanian was the result of head-on collision (see Hatcher, 1972, 1978, 1987; Lefort, 1984). These models, however, could not explain the diachroneity of orogeny in different parts of the orogen, the abundance of dextral and only dextral strike-slip faulting throughout the Appalachians, and the origin of step-over basins in the northern Appalachians.

Mountain chains are one of the main products of continent-continent collision, a fact recognized by some of the earliest contributors to development of continental drift theory, including Ampherer (1906) and Wegener (1912), and plate tectonics theory, such as Argand (1925), Hess (1938, 1962), Isacks et al. (1968), and Dewey and Bird (1970). Harland (1971) suggested that collision may not always be head on, and offered the terms transpression and transtension to describe oblique collision and extension processes. Burke and Dewey (1973) and Tapponier and Molnar (1977) recognized the irregular nature of collision in the convergence zone in southern Asia, and the importance of a rigid, irregularly shaped indenter (India). Both near- and far-field effects resulted throughout much of China, Afghanistan, Pakistan, and southeastern Asia, a process subsequently modeled by Tapponier et al. (1982). "Escape tectonics" has been described as the escape of blocks from a collision zone where oblique collision or the irregular shape of the indenter provides space into which escape may occur. Escape of large continental blocks probably also occurred in the Appalachians, where irregular colliding margins produced promontories that became indentors, causing large blocks to move laterally and large strike-slip faults to transport blocks away from the locus of collision in the internal parts of the mountain chain (Lefort, 1984; Hatcher, 1989).

COLLISION MODES AND PRODUCTS OF COLLISION

Several pure modes of collision have been recognized previously: head on, orogen parallel, and orogen oblique. Rotational collision, based primarily on paleomagnetic evidence, has also been suggested. Not all of these mechanisms have been demonstrated to be involved in continent-continent collision, perhaps because of the mechanical problems of wholesale rotation of large continental blocks, and possibly also because two-dimensional analysis used by most geologists has hindered analysis of rotational, oblique, and zippered modes. Rotation and orogen-parallel motion of small blocks (terranes) is easier to demonstrate and account for geometrically (see Beck, 1980; Cox, 1980; Irving et al., 1980), but paleomagnetic and paleoclimatic data indicate that larger rotations of continental blocks have occurred (e.g., see Van der Voo, 1988; Ziegler, 1990). Scotese and McKerrow (1990) also made a reconstruction for the late Paleozoic Carboniferous and Permian that suggests virtually the same as the analysis herein, that the continents around the Theic ocean closed from north to south, providing independent evidence that zipper closing may be the mechanism for closing of the Theic ocean. Their reconstruction was based on paleomagnetic and biogeographic data.

Head-on collision has been documented in a number of mountain chains, indicated by the continent-directed overthrust sheets present in most orogens. Despite this, a number of thrust belts contain orogen-parallel or oblique-slip elements, but thrust belts still may be the best indicators of head-on collision. Thrust belts, containing both thin-skinned and crustal thrusts in the southern and central Appalachians (Hatcher et al., 1989), Canadian Cordillera (Price and Mountjoy, 1970; Price, 1981), British (Elliott and Johnson, 1980; McClay and Coward, 1981) and

Scandinavian Caledonides (Roberts and Gee, 1985), central and eastern Alps (Trümpy, 1960; Laubscher, 1971, 1988, 1992), Himalayas (Brunel and Kienast, 1986), and Urals (Hamilton, 1970; Şengör et al., 1993), while containing evidence for oblique motion, all yield direct evidence for head-on collision. Closer examination of all these orogens reveals complications that do not negate this conclusion, but instead provide evidence for a more complex history of superposition of thrusts on earlier structures that may have had different kinematic histories, or vice versa. In addition to thrust belts, high pressure-temperature (*P-T*) (amphibolite facies or higher) regional metamorphism and the presence of S-type granitoids may provide additional evidence of head-on collision (Dewey and Bird, 1970; O'Brien et al., 2001), although oblique collision may provide some of the same results (e.g., in the Zagros Mountains; Şengör, 1984, 1987). This is no less true of the southern and central Appalachians, where the more external foreland fold-thrust belt appears to have been driven in front of the Blue Ridge–Piedmont crystalline megathrust sheet. The Blue Ridge–Piedmont sheet, however, contains a complex array of Alleghanian dextral strike-slip faults, plutons, and a Carboniferous-Permian metamorphic core that appears to have been transported continentward in the megathrust sheet (Hatcher, 1989; Hatcher et al., 1989).

Oblique (transpressional) collision has been suggested for parts of several orogenic belts, including the Appalachians (Bird and Dewey, 1970; Hatcher, 1971, 1972, 1989; Rodgers, 1987; Stockmal et al., 1987), Alps-Carpathians (Dewey et al., 1973; Laubscher, 1992), Caledonides in Spitzbergen (Harland, 1971), and Variscides (Arthaud and Matte, 1977; Weber and Behr, 1983). Supporting evidence for this mechanism is commonly large strike-slip faults. A domain of Alleghanian dextral strike-slip faulting exists in the internal parts of the Appalachians (Hatcher, 1989), and in the southern Appalachians the western boundary of the domain consists of the multiply reactivated Brevard fault (Hatcher, 2001). Several other major dextral faults exist southeast of the Brevard (Ocmulgee, central Piedmont, Modoc, Nutbush Creek, Hylas, Roanoke Rapids), and the displacements of all of them remain unknown. A complex array of Alleghanian east-vergent thrust and dextral strike-slip faults (Clinton-Newbury, Bloody Bluff, Lake Char, Hope Valley, Honey Hill) also exists in southeastern New England (Gromet, 1989; Snoke and Mosher, in Hatcher et al., 1989).

Transcurrent collisional orogens are perhaps best exemplified by the North American Cordillera, where outboard terranes have been transported northward relative to North America during the late Mesozoic and Cenozoic (Coney et al., 1980; Irving et al., 1980). Blocks ranging in area from a few square kilometers to many thousands of square kilometers have been accreted either singly or as composite superterranes to western North America outboard of the miogeocline (Monger et al., 1982). The terrane-accretion process is commonly thought to have affected deformation and metamorphism, although A-type subduction of the miogeocline here should necessarily be disconnected from the northward-directed terrane translation and docking process. Davis et al. (1978) and Monger and Price (1979) concluded the Cordillera formed by collision of the North American miogeocline with outboard terranes. Several terranes were accreted as two large composite terranes collided with Laurentia during Middle Jurassic to Early Cretaceous and Late Cretaceous to Paleocene time (Monger et al., 1982). Terrane accretion thus resulted in numerous small- to medium-size linear composite blocks distributed along the length of the orogen, with crustal thrusts and/or small blocks at bends. Both S- and I-type granites were emplaced, and high *P-T* metamorphism occurred during both accretionary events. In the Appalachians the Carolina-Avalon superterrane was accreted to Laurentia before or early in the Alleghanian, and other smaller terranes may have formed and been accreted even earlier (Horton et al., 1989). The Alleghanian appears to have been a continent-continent collision event, except in Europe, where numerous small terranes were swept up into a collision zone that spanned the area from Spain to eastern Europe (Franke, 1989).

Block rotation has been documented by geologic and paleomagnetic data in several mountain chains. Some of the best-documented small- to medium-size rotated blocks are in the Cordillera of western North America, where Cenozoic rotations of more than 50° have been estimated (Cox, 1980). Both Beck (1980) and Cox (1980) provided mechanisms (all involving simple shear) whereby both clockwise and counterclockwise rotation of blocks can occur in a dextral (or sinistral) transcurrent plate boundary system. Dewey et al. (1973) suggested rotation of small and large blocks in the Alps. The model experiments of Tapponier et al. (1982) suggested that Asia rotated ~25° during the Asia-India collision and escape of Indochina. Similar block rotations have not been identified in the Appalachians.

Oblique-rotational collision is the mechanism employed here to explain the zippered closing of the Theic ocean during the Permian Alleghanian-Variscan orogeny in eastern North America and western Eurafrica. This mechanism may have also been appropriate for closing of parts of the Tethys to form parts of the Alps (Dewey et al., 1973; Laubscher, 1992), the Altaid realm in Asia (Şengör et al., 1993), and Himalayas (Tapponier et al., 1982), as well as parts of the intervening Tethyan realm (Şengör, 1984, 1987).

ZIPPER CLOSING OF THE THEIC OCEAN AND THE ALLEGHANIAN OROGENY

The Alleghanian orogeny was a diachronous event that lasted from 325 to 260 Ma (Secor et al., 1986). It produced a clastic wedge that began shedding sediment into the Appalachian basin in the late Early Carboniferous. Sediment spread from the central to the southern Appalachians, providing evidence for uplift of the outer margin. Plutons, both S- and I-type, were intruded in the New England and southern Appalachians, with the greatest number and variety of compositions occurring in the southern Appalachians. Sinha and Zietz (1982) suggested that a magmatic arc existed in the southern Appalachians at this time, because of the arcuate distribution of Alleghanian granitoid I-type plutons. The occurrence of an extensive suite of Alleghanian plutons in the Appalachian orogen suggests that westward subduction occurred beneath North America at this time (Hatcher, 1972, 1978, 1987; Sinha and Zietz, 1982). Although Alleghanian metamorphism is restricted to southeastern New England and the easternmost exposed parts of the southern Appalachians in South Carolina and Georgia, it yields a Barrovian sequence of isograds and reached upper amphibolite facies (sillimanite grade) conditions in both segments (see Snoke and Mosher, in Hatcher et al., 1989). Nothing in the metamorphic or plutonic patterns provides any clear evidence favoring a zipper mechanism for closing of the Theic ocean, but there is also no contrary evidence from these data sets.

The Alleghanian orogeny completed the Appalachian Wilson cycle and produced structures—mostly thrusts and continent-directed folds—that dominate the southern and central Appalachian orogen. The purpose of this section is to present the pertinent elements of Alleghanian history to demonstrate the mechanism of zipper tectonics using the Alleghanian orogen as an example. Lengthy summaries of Appalachian tectonics are readily available (e.g., Rodgers, 1970; Rankin, 1976; Hall and Robinson, 1982; Williams and Hatcher, 1983; Hatcher, 1978, 1987, 1989; Hatcher et al., 1989; Horton et al., 1989). Primary reference documents for this discussion that show localities and structures in the U.S. and Canadian Appalachians are the tectonic maps by Keppie (1985) and Hatcher et al. (1990) and Figure 1.

Appalachians prior to the Alleghanian orogeny

The Appalachian orogen, prior to the Alleghanian orogeny, consisted of an assemblage of terranes of different sizes that were accreted during the Taconian (Ordovician) and possibly the Acadian (Late Devonian–Early Carboniferous) events (Gromet, 1989; Hatcher, 1987, 1989; Horton et al., 1989). The central and southern Appalachian foreland remained laterally extensive and only locally deformed, whereas the New England and Canadian Appalachian foreland had already been deformed and shortened during the Taconian event (Drake et al., 1989). Ordovician-Silurian and Devonian clastic wedges that were deposited on the foreland attest to uplift or overthrusting on the continental margin. The Appalachians prior to the Alleghanian thus consisted of consolidated basement now part of North America (Rodgers, 1987; Hatcher, 1988, 1999; Hatcher et al., 1989) that had formed during the earlier Paleozoic orogenies, along with recycled crustal elements mostly from the Middle Proterozoic Grenvillian orogeny, and the exotic Avalon-Carolina (Armorica) terrane. The Theic ocean (see Van der Voo, 1983, for discussion of the extent and definitions of the Iapetus, Rheic, and Theic oceans) remained open prior to the Alleghanian, but was closed by north to south collision during the Alleghanian-Variscan event (see also Scotese and McKerrow, 1990).

Alleghanian deformation sequence

Preliminary phase. Early Alleghanian (late Early Carboniferous) dextral strike slip in the internides, whose initial motion was sinistral strike slip in New England, produced pull-apart stepover basins in New England, the Canadian Maritimes, and Newfoundland (Bradley, 1982; Mosher, 1983). Reactivation of earlier dip- and strike-slip faults (e.g., Brevard) in dextral strike slip occurred in the southern and central Appalachian internides.

Lackawanna phase. Formation of early thrusts (late Early Carboniferous Lackawanna phase of Geiser and Engelder, 1983) in the central Appalachian foreland was possibly coeval with strike-slip motion in the internides, and may indicate arrival of an indenter (the Reguibat promontory of Lefort, 1984) and initiation of head-on collision (Geiser and Engelder, 1983; Hatcher, 1989; Hatcher, in Hatcher et al., 1989). The overprinting of early Alleghanian (Lackawanna?) ductile dextral strike-slip (Reed and Bryant, 1964; Bobyarchick, 1984; Gates et al., 1986; Bobyarchick et al., 1988; Hatcher, 1989) to late Alleghanian (Early Permian) brittle dip-slip motion (Hatcher, 1971, 1989; Bobyarchick et al., 1988) is readily documented in the Brevard fault zone in the southern Appalachians (Hatcher, in Hatcher et al., 1989; Hatcher, 2001). Similar reactivation has not been documented on other Alleghanian faults in the Appalachians. Vauchez et al. (1987) attempted to model the collision of the Reguibat indenter as a mirror image of the Tapponier et al. (1982) extrusion tectonics model for the Himalayan region, but did not take into account the rotational component or the general southwestward transport of all blocks, both north and south of the indenter.

Main phase. Westward-directed Late Carboniferous–Permian thrusts formed in the Appalachian foreland by faults propagating westward from the ductile-brittle transition in the internides cutting through basement formed during earlier (Paleozoic and Precambrian) orogenies as the Blue Ridge–Piedmont megathrust (Hatcher, 1999), then ramped into the cooler parts of the orogen into the Lower Cambrian clastic sequence just above the basement, then into detachments in the Upper Devonian to Carboniferous clastic rocks in the western part of the orogen where the Cambrian detachment pinches. Thrusts reached maximum displacement in the southern Appalachians where the Blue Ridge–Piedmont and foreland thrusts have a cumulative minimum displacement of 400 km.

A variety of weak units was available within the stratigraphic succession on the platform into which the crustal thrusts of the southern and central Appalachians could propagate. Exceptionally thin stratigraphic sequences on promontories, like the Reading Prong in New Jersey, southeastern New York, and northeastern Pennsylvania, and in the parts of the southern Appalachians in Georgia, Alabama, and the westernmost foreland of Tennessee and Virginia, limited displacement on Alleghanian thrusts. In contrast, the exceptionally thick sequences along either the Late Proterozoic rifted margin in Tennessee and western North Carolina or in Ordovician foredeeps in Tennessee and Pennsylvania permitted much greater displacement on thrusts that formed in embayments, particularly in Tennessee.

Possible collision sequence

The possible zippered collision sequence in the Appalachians (Fig. 2) can be summarized as follows.

1. Initial contact of Gondwana with Laurentia (Fig. 2A)—beginning the closing of the Theic ocean—probably occurred in the New England and Canadian Maritimes Appalachians during the Early Carboniferous, indicated by the early strike-slip deformation (initially sinistral, but quickly reversed to dextral) and formation of stepover basins (Bradley, 1982; Mosher, 1983). Prior to contact, Gondwana was far south of Laurentia (Van der Voo, 1982, 1983, 1988). After initial contact, Africa (Gondwana) began to rotate clockwise relative to North America, allowing more of the continental margins farther south to come into contact.

2. The Reguibat promontory (Lefort, 1984) appears to have been the first more southerly element along the African margin to contact the central Appalachian margin (Fig. 2, A and B). This probably occurred during the late Early Carboniferous (late Mississippian), because easterly derived clastic sediment began arriving on the central Appalachian continental margin in Pennsylvania and New York, and spread southward from there (see Thomas, in Hatcher et al., 1989). Paralleling this was the first deformation of the nearby foreland by the Lackawanna phase in the central Appalachians (early Late Carboniferous; i.e., early Pennsylvanian) deformation, dextral strike-slip deformation in the southern and central Appalachian internides, and formation of dextral pull-apart basins in New England.

3. As the Reguibat promontory continued to impinge on the North American margin and Africa rotated southward, it pro-

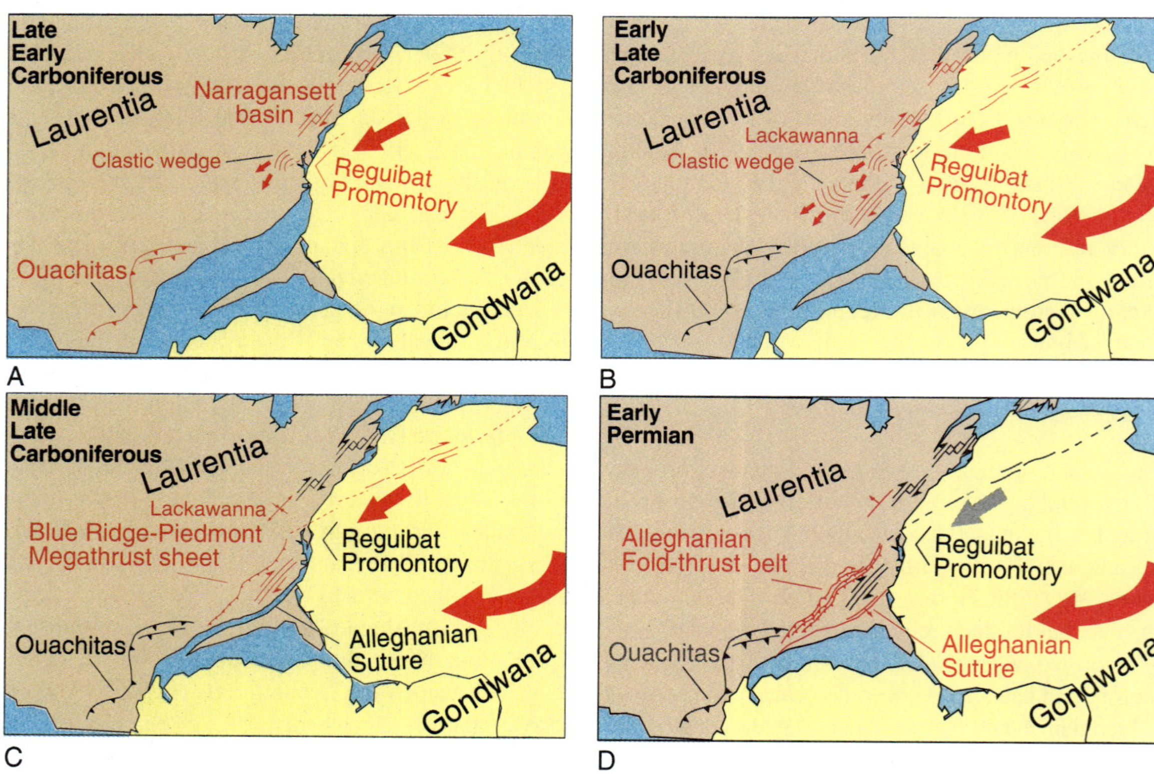

Figure 2. Zipper closing of Theic ocean to form the Alleghanian orogen (continents shown on Robinson projection; reconstruction modified from that in Ziegler, 1990). Red lines and symbols indicate feature is active in interval shown. A: Initial contact between Gondwana and Laurentia occurred in late Early Carboniferous (late Mississippian) producing initially sinistral faulting in New England followed immediately by dextral motion and pull-apart basins, first shedding of clastic sediments onto continent, and Lackawanna phase deformation. B: Southward movement and rotation of Gondwana with respect to Laurentia in early Late Carboniferous (early Pennsylvanian) produced dextral motion throughout orogen, waning of Lackawanna phase deformation, and greater dispersal of sediments onto Laurentian foreland. C: Continued clockwise rotation of Gondwana with respect to Laurentia during Late Carboniferous closed Theic ocean southward, bringing Gondwana into head-on collision with Laurentia, producing first elements of Blue Ridge–Piedmont megathrust sheet. D: Early Permian head-on collision of Gondwana with Laurentia produced major transport on Blue Ridge–Piedmont megathrust sheet and drove foreland fold-thrust belt (Valley and Ridge and Plateau) ahead of it.

duced dextral motion on a number of faults (Hylas, Brookneal, Nutbush Creek, Modoc) and reactivated others (Brevard, Dean Creek) in a dextral mode (Fig. 2C). This probably occurred during the Late Carboniferous to Early Permian (Secor et al., 1986) and may be a continuation of the second event.

4. Head-on collision of Africa with southeastern North America formed, propagated, and drove the Blue Ridge–Piedmont megathrust sheet onto the continental margin, deforming the foreland and the clastic wedge that formed ahead of the advancing sheet (Fig. 2D). This was the culmination of the Alleghanian orogeny that probably occurred during the Early Permian.

5. Vestiges of relaxation and extensional collapse are not widespread in the Appalachians, but the Lake Char fault in southeastern New England (Goldstein, 1982) and the Augusta fault along the Georgia–South Carolina border (Maher, 1987) both contain kinematic indicators that show the faults moved in a hanging-wall-down sense, and that these motions appear to record some of the youngest Paleozoic movement in the Appalachians.

ZIPPER TECTONICS—DISCUSSION

The sequence of deformation described assumes, based on data available principally from the Appalachians, that collision of North America occurred with Gondwana and that it occurred zipper fashion from north to south (clockwise), closing the Theic ocean. In contrast, Sacks and Secor (1990) concluded that the Laurentia-Gondwana collision occurred by counterclockwise rotation of Gondwana following initial collision of South America with southern North America to form the Ouachitas. While most agree that stratigraphic data confirm the Ouachita deformation occurred and ended in the Carboniferous (see Morris, 1989; Viele and Thomas, 1989), and that the Alleghanian is a Carboniferous-Permian event (see Secor et al., 1986), most do not consider the Ouachita orogen to be the product of continent-continent collision, but the product of A-type subduction of Laurentia beneath an arc-accretionary complex located seaward from the Carboniferous southern margin of Laurentia, and one that contrasts markedly with collisional orogens like the Appalachians (Keller and Cebull, 1973; Viele, 1979a, 1979b; Viele and Thomas, 1989). More likely, the ocean never really totally closed south of Laurentia after it closed along the eastern margin (Keller and Cebull, 1973), consistent with the Ouachitas not being a collisional orogen. It is also difficult to simultaneously produce southward (dextral) transport and cratonward thrusting along the eastern Laurentian margin in a collision setting where the two continents approach from the northwest and southeast. Hence the zipper model proposed here for the north to south collision of Laurentia and Gondwana appears more likely.

The zipper closing mechanism should be documentable in other orogenic belts. Oblique, rotational collision could partly be responsible for the Asia-India collision that produced the Himalayas and extrusion of Indochina, as indicated by the model experiments of Tapponier *et al.* (1982), along with Afghanistan and Pakistan. Combined transpressive collision and the irregular continental margins of late Mesozoic–Tertiary Europe and Africa may have produced zipper closing of the Tethys ocean and resulted in the complexly arcuate Alpine system. Dextral transpressive continent-continent collision between Africa and Europe that culminated in the late or early Eocene closing of the Tethys ocean probably involved a zipper mechanism. This event rotated the Penninic realm counterclockwise, and produced the Silvretta (Ötztal) nappe that obducted African basement onto Europe (Laubscher, 1988, 1992). Later in the Tertiary (late Oligocene to early Miocene), continued westward dextral motion of the Adria indenter with respect to Europe, along with some northward transport, and uplift, produced the "Ligurian knot" of the western Alps and permitted westward unroofing of the Alpine system (Laubscher, 1992).

CONCLUSIONS

1. Zipper tectonics involves closing of an ocean that proceeds from one end to the other as colliding land masses touch and subduction eliminates the intervening ocean; rotation and transpressional collision are required for this to occur. Zipper closing of both small and large oceans, involving orogen-parallel transport, oblique collision, and rotation to produce head-on collision, should be considered a viable mechanism where evidence of nearly coeval superposed or incremental deformation with differences in vergence can be demonstrated in the same orogen.

2. The closing of the Paleozoic Theic ocean occurred by progressive north-south zippered collision of Africa with North America to produce the climactic event, the Alleghanian orogeny, which produced the Appalachians. Such a mechanism and north-south polarity of closing best explains the dominance of early Alleghanian strike-slip faulting in the northern Appalachians and dominance of later Alleghanian thrusting in the southern Appalachians. Similar zipper closing of larger and smaller oceans may have occurred in formation of the Himalayas and Alps.

ACKNOWLEDGMENTS

I am grateful to A.G. Goldstein for inviting me to present an earlier version of this paper as part of a symposium on Alleghanian tectonics at the 1990 Northeastern Section Meeting of the Geological Society of America, and to H.P. Laubscher for inviting me to present this paper in the symposium, "Subduction Processes, Strike-slip Faulting and Continental Collision in the Tethyan Orogenic Assembly," at the 1991 European Union of Geoscientists Meeting in Strasbourg, France. I thank the Geologisches Vereinigung for providing me with travel funds to participate in the Strasbourg symposium. Comments on the technical content of an earlier draft of this paper by A.M.C. Şengör,

D.T. Secor Jr., and P.J. Lemiszki are very much appreciated. Reviewer comments by S. Marshak and the late N. Rast resulted in considerable improvement of the manuscript and are very much appreciated. Support for work on Alleghanian structures in the Appalachians has been provided by grants from the Tectonics and Continental Dynamics programs of the U.S. National Science Foundation, the U.S. Nuclear Regulatory Commission, the U.S. Department of Energy National Energy Technology Laboratory, and the University of Tennessee Science Alliance Center of Excellence Distinguished Scientist Stipend.

REFERENCES CITED

Ampherer, O., 1906, Über das Bewegungsbild von Faltengebirgen: Jahrbuch der Kaiserlich Königlichen Geologischen Reichsanstalt, v. 56, p. 529–622.

Argand, E., 1925 (trans. by A.V. Carozzi, 1977), The tectonics of Asia: New York, Hafner, 218 p.

Arthaud, F., and Matte, P., 1977, Late Paleozoic strike-slip faulting in southern Europe and northern Africa: Result of a right-lateral shear zone between the Appalachians and the Urals: Geological Society of America Bulletin, v. 88, p. 1305–1320.

Beck, M.E., Jr., 1980, Paleomagnetic record of plate-margin tectonic processes along the western edge of North America: Journal of Geophysical Research, v. 85, p. 7115–7131.

Bird, J.M., and Dewey, J.F., 1970, Lithosphere plate-continental margin tectonics and the evolution of the Appalachian orogen: Geological Society of America Bulletin, v. 81, p. 1031–1060.

Bobyarchick, A.R., 1984, A late Paleozoic component of strike-slip in the Brevard zone, southern Appalachians: Geological Society of America Abstracts with Programs, v. 19, no. 3, p. 126.

Bobyarchick, A.R., Edelman, S.H., and Horton, J.W., Jr., 1988, The role of dextral strike-slip in the displacement history of the Brevard zone, *in* Secor, D.T., Jr., ed., Southeastern geological excursions: Columbia, South Carolina, Geological Society of America Southeastern Section Field Trip Guidebook, p. 53–154.

Bradley, D.C., 1982, Subsidence in late Paleozoic basins in the northern Appalachians: Tectonics, v. 1, p. 107–123.

Brunel, M., and Kienast, J.-R., 1986, Étude pétro-structurale des chevauchements ductiles himalayens sur la traversale de l'Everest-Makalu (Népal oriental): Canadian Journal of Earth Sciences, v. 23, p. 1117–1137.

Burke, K., and Dewey, J.F., 1973, Plume generated triple junctions: Key indicators in applying plate tectonics to old rocks: Journal of Geology, v. 81, p. 406–433.

Coney, P.J., Jones, D.L., and Monger, J.W.H., 1980, Cordilleran suspect terranes: Nature, v. 288, p. 329–333.

Cox, A., 1980, Rotation of microplates in western North America, *in* Strangway, D.W., ed., The continental crust and its mineral deposits: Geological Association of Canada Special Paper 20, p. 305–321.

Davis, G.A., Monger, J.W.H., and Price, R.A., 1978, Mesozoic construction of the Cordilleran "collage," central British Columbia to central California, *in* Howell, D.A., and McDougall, K.A., eds., Mesozoic paleogeography of the western United States: Society of Economic Paleontologists and Mineralogists, Pacific Section, v. 2, p. 1–32.

Dewey, J.F., and Bird, J.M., 1970, Mountain belts and the new global tectonics: Journal of Geophysical Research, v. 75, p. 2626–2647.

Dewey, J.F., Pittman, W.C., III, Ryan, W.B.F., and Bonnin, J., 1973, Plate tectonics and the evolution of the Alpine system: Geological Society of America Bulletin, v. 84, p. 3137–3180.

Drake, A.A., Jr., Sinha, A.K., Laird, J., and Guy, R.E., 1989, The Taconic orogen, *in* Hatcher, R.D., Jr., Thomas, W.A., and Viele, G.W., eds., The Appalachian-Ouachita orogen in the United States: Boulder, Colorado, Geological Society of America, The Geology of North America, v. F-2, p. 101–177.

Elliott, D., and Johnson, M.R.W., 1980, Structural evolution in the northern part of the Moine thrust belt, N.W. Scotland: Transactions of the Royal Society of Edinburgh, v. 71, p. 69–96.

Franke, W., 1989, Tectonostratigraphic units of the Variscan belt of central Europe, *in* Dallmeyer, R.D., ed., Terranes in the Circum-Atlantic Paleozoic orogens: Boulder, Colorado, Geological Society of America Special Paper 230, p. 67–90.

Gates, A.E., Simpson, C., and Glover, L., III, 1986, Appalachian Carboniferous dextral strike-slip faults: An example from Brookneal, Virginia: Tectonics, v. 5, p. 119–33.

Geiser, P.A., and Engelder, T., 1983, The distribution of layer parallel shortening fabrics in the Appalachian foreland of New York and Pennsylvania: Evidence for two non-coaxial phases of the Alleghanian orogeny, *in* Hatcher, R.D., Jr., Williams, H. and Zeitz, I. eds., Contributions to the tectonics and geophysics of mountain chains: Boulder, Colorado, Geological Society of America Memoir 158, p. 161–175.

Goldstein, A.G., 1982, Geometry and kinematics of ductile faulting in a portion of the Lake Char mylonite zone, Massachusetts and Connecticut: American Journal of Science, v. 282, p. 1378–1405.

Gromet, L.P., 1989, Avalonian terranes and late Paleozoic tectonism in southeastern New England; constraints and problems, *in* Dallmeyer, R.D., ed., Terranes in the Circum-Atlantic Paleozoic orogens: Boulder, Colorado, Geological Society of America Special Paper 230, p. 193–211.

Hall, L.M., and Robinson, P., 1982, Stratigraphic-tectonic subdivisions of southern New England, *in* St. Julien, P., and Beland, J., eds., Major structural zones and faults in the northern Appalachians: Geological Association of Canada Special Paper 24, p. 15–42.

Hamilton, W.B., 1970, The Uralides and the motion of the Russian and Siberian platforms: Geological Society of America Bulletin, v. 81, p. 2553–2576.

Harland, W.B., 1971, Tectonic transpression in Caledonian Spitzbergen: Geological Magazine, v. 108, p. 27–42.

Hatcher, R.D., Jr., 1971, Structural, petrologic and stratigraphic evidence favoring a thrust solution to the Brevard problem: American Journal of Science, v. 270, p. 177–202.

Hatcher, R.D., Jr., 1972, Developmental model for the southern Appalachians: Geological Society of America Bulletin, v. 83, p. 2735–2760.

Hatcher, R.D., Jr., 1978, Tectonics of the western Piedmont and Blue Ridge: Review and speculation: American Journal of Science, v. 278, p. 276–304.

Hatcher, R.D., Jr., 1987, Tectonics of the southern and central Appalachian internides: Annual Review of Earth and Planetary Sciences, v. 15, p. 337–362.

Hatcher, R.D., Jr., 1988, Basement-cover relationships in the Appalachian-Caledonian-Variscan orogen: Mid-Devonian (end of Acadian orogeny) to end of Permian, *in* Harris, A.L., and Fettes, D.J., eds, The Caledonian-Appalachian orogen: Geological Society of London Special Publication 38, p. 507–514.

Hatcher, R.D., Jr., 1989, U.S. Appalachians synthesis, *in* Hatcher, R.D., Jr., Thomas, W.A., and Viele, G.W., eds., The Appalachian-Ouachita orogen in the United States: Boulder, Colorado, Geological Society of America, The Geology of North America, v. F-2, p. 511–535.

Hatcher, R.D., Jr., 1999, Crust-forming processes, *in* Sinha, A.K. ed., Basement tectonics 13: Dordrecht, Netherlands, Kluwer Academic Publishers, p. 99–118.

Hatcher, R.D., Jr., 2001, Rheological partitioning during multiple reactivation of the Palaeozoic Brevard fault zone, southern Appalachians, USA, *in* Holdsworth, R.E., Strachan, R.A., Macloughlin, J.F., and Knipe, R.J., eds., The nature and significance of fault zone weakening: Geological Society of London Special Publication 186, p. 255–269.

Hatcher, R.D., Jr., Thomas, W.A., Geiser, P.A., Snoke, A.W., Mosher, S., and Wiltschko, D.V., 1989, Alleghanian orogen, *in* Hatcher, R.D., Jr., Thomas, W.A., and Viele, G.W., eds., The Appalachian-Ouachita orogen in the

United States: Boulder, Colorado, Geological Society of America, The Geology of North America, v. F-2, p. 233–318.

Hatcher, R.D., Jr., Osberg, P.H., Drake, A.A., Jr., Robinson, P., and Thomas, W.A., 1990, Tectonic map of the U.S. Appalachians, *in* Hatcher, R.D., Jr., Thomas, W.A., and Viele, G.W., eds., The Appalachian-Ouachita orogen in the United States: Boulder, Colorado, Geological Society of America, The Geology of North America, v. F-2, Plate 1, scale 1:2,500,000.

Hess, H.H., 1938, Gravity anomalies and island arc structure: Proceedings of the American Philosophical Society, v. 79, p. 71–95.

Hess, H.H., 1962, History of ocean basins, *in* Engel, A. E. J., James, H. L., and Leonard, B. F., eds, Petrologic studies: A volume in honor of A.F. Buddington: Boulder, Colorado, Geological Society of America, p. 599–620.

Horton, J.W., Jr., Drake, A.A., Jr., and Rankin, D.W., 1989, Tectonostratigraphic terranes and their Paleozoic boundaries in the central and southern Appalachians, *in* Dallmeyer, R.D., ed., Terranes in the Circum-Atlantic Paleozoic orogens: Boulder, Colorado, Geological Society of America Special Paper 230, p. 213–245.

Irving, E., Monger, J.W.H., and Yole, R.W., 1980, New paleomagnetic evidence for displaced terranes in British Columbia, *in* Strangway, D.W., ed., The continental crust and its mineral deposits: Geological Association of Canada Special Paper 20, p. 441–456.

Isacks, B., Oliver, J., and Sykes, L.R., 1968, Seismology and the new global tectonics: Journal of Geophysical Research, v. 72, p. 5855–5900.

Keller, G.R., and Cebull, S.E., 1973, Plate tectonics and the Ouachita system in Texas, Oklahoma, and Arkansas: Geological Society of America Bulletin, v. 84, p. 1659–1666.

Keppie, J.D., 1985, The Appalachian collage, *in* Gee, D.G., and Sturt, B.A., eds, The Caledonide orogen—Scandinavia and related areas: London, John Wiley & Sons, p. 1217–1226.

Laubscher, H.P., 1971, The large-scale kinematics of the western Alps and northern Appenines and its palinspastic implications: American Journal of Science, v. 271, p. 193–226.

Laubscher, H.P., 1988, Material balance in Alpine orogeny: Geological Society of America Bulletin, v. 100, p. 1313–1328.

Laubscher, H.P., 1992, The arc of the western Alps today: Eclogae geologicae Helvetiae, v. 84, p. 631–659.

Lefort, J.-P., 1984, Mise en évidence d'une virgation carbonifère induite par la dorsale Reguibat (Mauritanie) dans les Appalaches du Sud (U.S.A.): Arguments géophysiques: Bulletin de la Société Géologique de France, v. 26, p. 1293–1303.

Maher, H.D., Jr., 1987, Kinematic history of mylonitic rocks from the Augusta fault zone, South Carolina and Georgia: American Journal of Science, v. 287, p. 795–816.

McClay, K.R., and Coward, M.P., 1981, The Moine thrust zone: An overview, *in* McClay, K.R., and Price, N.J., eds., Thrust and nappe tectonics: Geological Society of London Special Publication 9, p. 241–260.

Monger, J.W.H., and Price, R.A., 1979, Geodynamic evolution of the Canadian Cordillera—Progress and problems: Canadian Journal of Earth Science, v. 16, p. 770–791.

Monger, J.W.H., Price, R.A., and Templeman-Kluitt, D.J., 1982, Tectonic accretion and the origin of the two major metamorphic and plutonic welts in the Canadian Cordillera: Geology, v. 10, p. 70–75.

Morris, R.C., 1989, Stratigraphy and sedimentary history of post-Arkansas Novaculite Carboniferous rocks of the Ouachita Mountains, *in* Hatcher, R.D., Jr., Thomas, W.A., and Viele, G.W., eds., The Appalachian-Ouachita orogen in the United States: Boulder, Colorado, Geological Society of America, The Geology of North America, v. F-2, p. 591–602.

Mosher, S., 1983, Kinematic history of the Narragansett basin, Massachusetts and Rhode Island: Constraints on late Paleozoic plate reconstructions: Tectonics, v. 2, p. 327–344.

O'Brien, P.J., Zotov, N., Law, R., Khan, M.J., and Jan, M. Q, 2001, Coesite in Himalayan eclogite and implications for models of India-Asia collision: Geology, v. 29, p. 435–438.

Price, R.A., 1981, The Cordilleran foreland thrust and fold belt in the southern Canadian Rocky Mountains, *in* McClay, K.R., and Price, N.J., eds., Thrust and nappe tectonics: Geological Society of London Special Publication 9, p. 427–448.

Price, R.A., and Mountjoy, E.W., 1970, Geologic structure of the Canadian Rocky Mountains between Bow and Athabasca Rivers—A progress report: Geological Association of Canada Special Paper 6, p. 7–25.

Rankin, D.W., 1976, Appalachian salients and recesses: Late Precambrian continental breakup and the opening of the Iapetus Ocean: Journal of Geophysical Research, v. 81, p. 5605–5619.

Reed, J.C., and Bryant, Bruce, 1964, Evidence for strike-slip faulting along the Brevard zone in North Carolina: Geological Society of America Bulletin, v. 75, p. 1177–1196.

Roberts, D., and Gee, D.G., 1985, An introduction to the structure of the Scandinavian Caledonides, *in* Gee, D.G., and Sturt, B.A., eds, The Caledonide orogen—Scandinavia and related areas: London, John Wiley & Sons, p. 55–68.

Rodgers, J., 1970, Tectonics of the Appalachians: New York, John Wiley & Sons, 271 p.

Rodgers, J., 1987, Chains of basement uplifts within cratons marginal to orogenic belts: American Journal of Science, v. 287, p. 661–692.

Sacks, P.E., and Secor, D.T., Jr., 1990, Kinematics of late Paleozoic continental collision between Laurentia and Gondwana: Science, v. 250, p. 1702–1705.

Scotese, C.R., and McKerrow, W.S., 1990, Revised world maps and introduction, *in* McKerrow, W.S., and Scotese, C.R., eds., Palaeozoic palaeogeography and biogeography: London, Geological Society of London Memoir 12, p. 1–21.

Secor, D.T., Jr., Snoke, A.W., and Dallmeyer, R.D., 1986, Character of the Alleghanian orogeny in the southern Appalachians: Part III: Regional tectonic relations: Geological Society of America Bulletin, v. 97, p. 1345–1353.

Şengör, A.M.C., 1984, The Cimmeride orogenic system and the tectonics of Eurasia: Boulder, Colorado, Geological Society of America Special Paper 195, 82 p.

Şengör, A.M.C., 1987, Tectonic subdivisions and evolution of Asia: Bulletin of the Technical University of Istanbul, v. 40, p. 355–435.

Şengör, A.M.C., Natal'in, B.A., and Burtman, V.S., 1993, Evolution of the Altaid tectonic collage and Palaeozoic crustal growth in Eurasia: Nature, v. 364, p. 299–307.

Sinha, A.K., and Zietz, I, 1982, Geophysical and geochemical evidence for a Hercynian magmatic arc, Maryland to Georgia: Geology, v. 10, p. 593–596.

Stockmal, G.S., Colman-Sadd, S.P., Keen, C.E., O'Brien, S.J., and Quinlan, G., 1987, Collision along an irregular margin: A regional plate tectonic interpretation of the Canadian Appalachians: Canadian Journal of Earth Science, v. 24, p. 1098–1107.

Tapponier, P., and Molnar, P., 1977, Active faulting and Cenozoic tectonics of China: Journal of Geophysical Research, v. 82, p. 2905–2930.

Tapponier, P., Pelzer, G., Le Dain, A.Y., Armijo, R., Cobbold, P., 1982, Propagating extrusion tectonics in Asia: New insights from simple experiments with plasticene: Geology, v. 10, p. 611–616.

Trümpy, R., 1960, Paleotectonic evolution of the central and western Alps: Geological Society of America Bulletin, v. 71, p. 843–908.

Van der Voo, R., 1982, Pre-Mesozoic paleomagnetism and plate tectonics: Annual Reviews of Earth and Planetary Science, v. 10, p. 191–220.

Van der Voo, R., 1983, A plate-tectonics model for the Paleozoic assembly of Pangaea based on Paleomagnetic data, *in* Hatcher, R.D., Jr., Williams, H., and Zietz, I., eds., Contributions to the tectonics and geophysics of mountain chains: Boulder, Colorado, Geological Society of America Memoir 158, p. 19–23.

Van der Voo, R., 1988, Paleozoic paleogeography of North America, Gondwana, and intervening displaced terranes: Comparisons of paleomagnetism with paleoclimatology and biogeographical patterns: Geological Society of America Bulletin, v. 100, p. 311–324.

Vauchez, A., Kessler, S.F., Lécorché, J.-P., and Villeneuve, M., 1987, Southward extrusion tectonics during the Carboniferous Africa—North America collision: Tectonophysics, v. 142, p. 317–322.

Viele, G.W., 1979a, Geological map and cross section, eastern Ouachita Mountains, Arkansas: Boulder, Colorado, Geological Society of America Map and Chart Series MC-28F.

Viele, G.W., 1979b, Geological map and cross section, eastern Ouachita Mountains, Arkansas: Map summary: Geological Society of America Bulletin, v. 90, p. 1096–1099.

Viele, G.W., and Thomas, W.A., 1989, Tectonic synthesis of the Ouachita orogenic belt, *in* Hatcher, R.D., Jr., Thomas, W.A., and Viele, G.W., eds., The Appalachian-Ouachita orogen in the United States: Boulder, Colorado, Geological Society of America, The Geology of North America, v. F-2, p. 695–728.

Weber, K., and Behr, H.-J., 1983, Geodynamic interpretation of the mid-European Variscides, *in* Martin, H., and Eder, F.W., eds, Intracontinental foldbelts: Berlin, Springer-Verlag, p. 427–469.

Wegener, A., 1912, Die Entstehung der Kontinente: Geologische Rundschau, v. 3, p. 276–292.

Williams, H., and Hatcher, R.D., Jr., 1983, Appalachian suspect terranes, *in* Hatcher, R.D., Jr., Williams, H., and Zietz, I., eds., Contributions to the tectonics and geophysics of mountain chains: Geological Society of America Memoir 158, p. 33–53.

Wilson, J.T., 1966, Did the Atlantic close and then reopen?: Nature, v. 211, p. 676–681.

Ziegler, P.A., 1990, Geological atlas of western and central Europe, 2nd ed.: Avon, England, Geological Society Publishing House, Shell International Petroleum Maatschappij, 238 p.

Manuscript Accepted by the Society February 8, 2002

Printed in the U.S.A.

Geological Society of America
Special Paper 364
2002

Megadisplacements and the Hercynian orogen of Gondwanan France and Iberia

David Shelley*
Department of Geological Sciences, University of Canterbury,
Private Bag 4800, Christchurch, New Zealand
Gérard Bossière*
Université de Nantes, Laboratoire de Planètologie et Géodynamique (UMR CNRS-6112),
BP 92208, 44322 Nantes Cedex 3, France

ABSTRACT

A model is presented for the formation of the principal Iberian and Armorican fault-bound terranes during the Hercynian orogeny, and their displacement and shuffling along an anastomosing complex of dextral shears along the northwest African boundary of Gondwana. Displacements are thousands of kilometers, and relate to movement of Laurentia around Gondwana and formation of Pangea. The terranes in Brittany seem reversed with respect to Gondwana, with the most proximal Paleozoic shelf sediments in the north and oceanic terranes in the south (though Gondwana lay to the south). This may be explained by movements around extensional and contractional bends on the dextral faults. The terranes of Brittany may have been sheared around Iberia along the Porto-Tomar shear zone with Iberia a promontory on the broad Gondwanan shelf, southwest of its present position. The Badajoz-Córdoba shear zone, a Precambrian suture, formed the southwestern edge of proto-Iberia, oblique to the Hercynian dextral shears, and it became a zone of sinistral bookshelf-type faulting, underthrusting, and high-pressure metamorphism as displacements were transferred from one dextral shear to another. Similarly, underthrusting and sinistral shearing occurred along the Léon terrane–Northern terrane boundary in France. The final Hercynian development of Iberia-Armorica is related to a 90° Permian swing in the shortening direction when the final bending of the Ibero-Armorican arc was imposed together with dextral shearing along the northern and southern coasts of present-day Iberia. The principal purpose of this paper is to show that such a model is reasonably possible.

INTRODUCTION

We suggested that the development of the Hercynian orogen in Gondwanan France and Iberia involved dextral displacements of 4000–5000 km, caused by the clockwise movement of Laurentia around northern Gondwana immediately prior to the formation of Pangea (Fig. 1; Shelley and Bossière, 2000). We proposed that the movement was represented by dextral faults such as the Porto-Tomar shear zone, the North Armorican shear zone, and the southern and northern branches of the South Armorican shear zone, and that movements along these faults were responsible for the shuffling and translation that led to the col-

*E-mails: d.shelley@geol.canterbury.ac.nz; Gerard.Bossiere@chimie.univ-nantes.fr.

Shelley, D., and Bossière, G., 2002, Megadisplacements and the Hercynian orogen of Gondwanan France and Iberia, *in* Martínez Catalán, J.R., Hatcher, R.D., Jr., Arenas, R., and Díaz García, F., eds., Variscan-Appalachian dynamics: The building of the late Paleozoic basement: Boulder, Colorado, Geological Society of America Special Paper 364, p. 209–222.

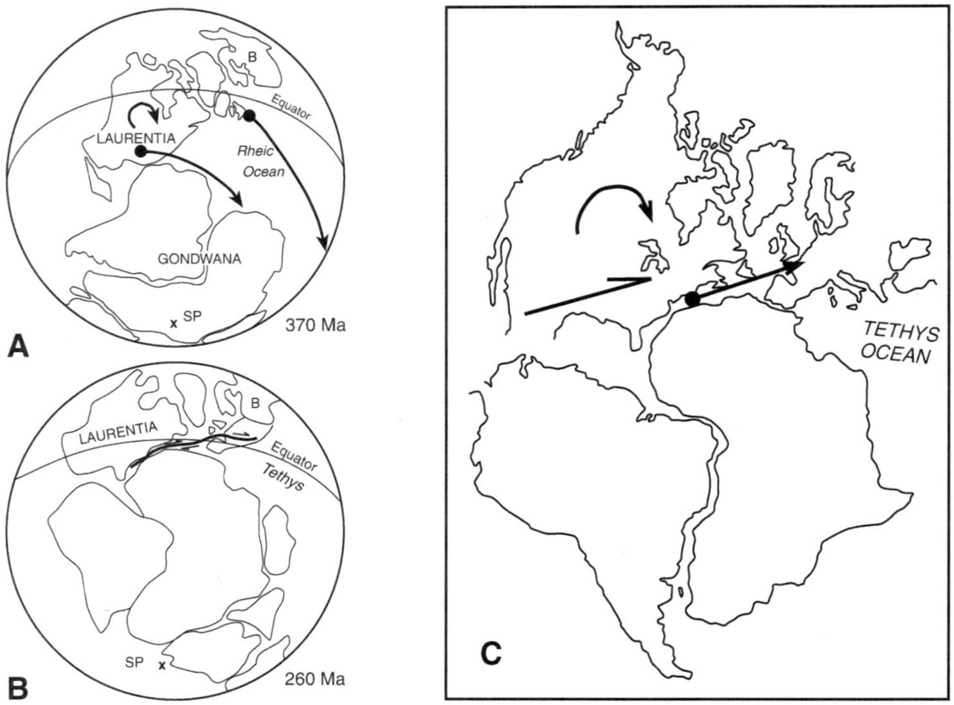

Figure 1. Laurentia and its dextral shearing along northern boundary of Gondwana to form Pangea and Hercynian orogen (reprinted from Shelley and Bossière, 2000, Fig. 3, copyright 2000, with permission from Elsevier Science). SP is South Pole, B is Baltica. Hercynian events took place between 370 and 260 Ma. A: Arrows show several thousand kilometers of dextral movement and clockwise motion necessary to form Pangea. B: Zone of dextral transpression we suggest was responsible for development of Hercynian orogen. C: Filled circle and arrow show how North Brittany may have been translated dextrally from northwest Africa to its present site in response to dextral translation of Laurentia. Necessary clockwise rotation of Laurentia and Baltica that provided convergent component of transpression is also indicated. Plate reconstructions in A are after Dalziel et al. (1994); those in B are after Dalziel (1995); those in C are after Bullard et al. (1965).

lage of disparate fault-bound terranes currently found in the orogen, particularly in France (Fig. 2). Specifically we suggested that the Northern terrane (Fig. 2) was displaced 2000 km from northwest Africa. Attention was also drawn to the South Armorican front or suture. The South Armorican front is, according to Lefort et al. (1997, their Fig. 12), another dextral fault, and we discuss herein whether it could possibly be a master fault in the proposed dextral displacements. However, as Rolet (1997) noted, the nature of this and other possible "sutures" remains uncertain due to the lack of deep seismic studies.

In our model we explained the oceanic terranes with Eohercynian structures as slivers of the Rheic ocean that formerly bordered Gondwana to the north. Within that ocean we suggested that there were volcanic arcs and subduction zones at a high angle to the Gondwana margin within which were caught up various thinned fragments of Gondwana that were rifted into the ocean during Cambrian-Ordovician extension. Previous explanations had been that the oceanic terranes in France represent a South Armorican ocean that opened and closed by orthogonal collision during the Paleozoic (Dubuisson et al., 1989); Lefort et al. (1997, their Fig. 10) portrayed the South Armorican front as a possible fossil plate boundary and the site of closure of this South Armorican ocean.

Herein we develop our previous model further. In particular, we discuss the relative ages of the various shear zones and the possible sequence of emplacement of the disparate tectonostratigraphic terranes that now make up Armorica and Iberia.

MAIN ELEMENTS OF THE HERCYNIAN OROGEN

Before discussing the sequence of emplacement of the tectonostratigraphic terranes it is necessary to discuss the various parts of the Hercynian orogen and their possible locations in a plate tectonic setting.

Plate tectonic setting

In our (Shelley and Bossière, 2000) model, the megadextral displacements are driven by the clockwise motion of Laurentia and Baltica about Gondwana (Fig. 1; Dalziel, 1995). However, there is no general agreement on the relative positions of Laurentia-Baltica and Gondwana. Kent and Van der Voo (1990), Dalziel et al. (1994), Dalziel (1995), and Torsvik (1998) suggested that Laurentia was tucked in to the west of South America in the Devonian, a position that requires massive dextral displacement in the Hercynian orogeny. Scotese and McKerrow (1990), Tait et al. (1997), and Stampfli (2000) placed Laurentia facing North Africa, from which position the amalgamation to form Pangea would have involved mainly orthogonal collision.

Therefore, the megadisplacements of terranes that we suggested (Shelley and Bossière, 2000) can only be supported if the Dalziel et al. (1994) and Dalziel (1995) type of plate reconstruction is valid; and, if geological evidence provides strong support for our (Shelley and Bossière, 2000) model, that is strong support for the Dalziel et al. (1994) and Dalziel (1995) model.

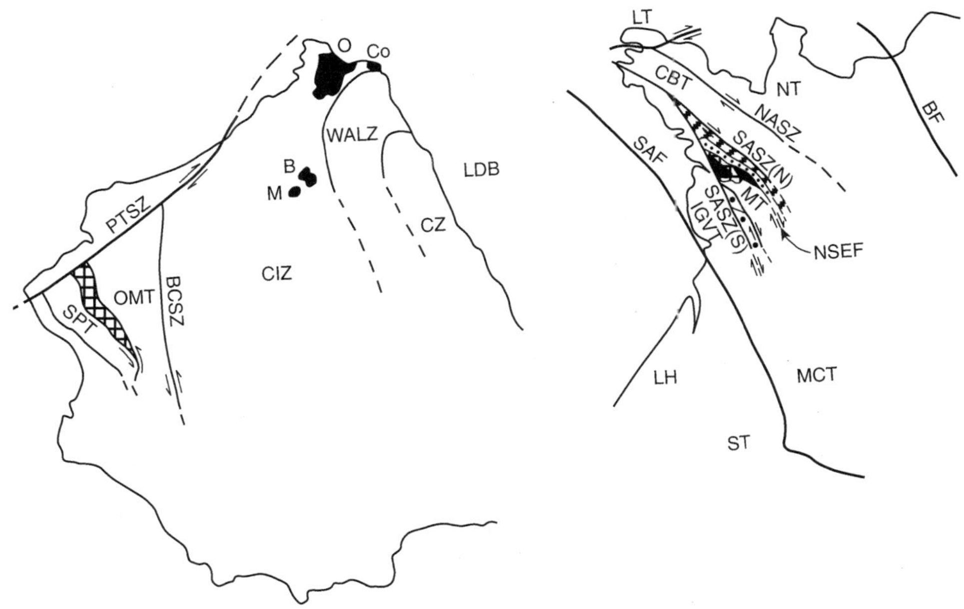

Figure 2. Tectonostratigraphic terranes of Gondwanan France and Iberia (reprinted from Shelley and Bossière, 2000, Fig. 1, copyright 2000, with permission from Elsevier Science). Relative positions of Iberia and France, at end of Paleozoic and before opening of Atlantic Ocean, are after Lefort (1989). In France: BF—Bray fault and possible suture of Rheic ocean; LT—Léon terrane; NT—Northern terrane; NASZ—North Armorican shear zone; CBT—Central Brittany terrane; SASZ(N)—northern branch of South Armorican shear zone; SASZ(S)—southern branch of South Armorican shear zone; thick lines—Lanvaux terrane; small dots—Saint Georges-sur-Loire terrane; NSEF—Nort-sur-Erdre fault; black area bounded to north by NSEF—Ancenis terrane; black area cut by SASZ(S)—Champtoceaux terrane; cross-hachures—Mauves-sur-Loire terrane; MT—Mauges terrane; large dots—Mortagne terrane; IGVT—Ile de Groix-Vendée terrane; SAF—South Armorican front; MCT—Massif Central terrane; LH—Landes High; ST—Southern terrane. To east of Armorican shear zones, and between them and MCT, Hercynian structures are buried under post-Hercynian Paris Basin sediments. Much of ST is covered by post-Hercynian Aquitaine Basin sediments. In Iberia: Co—Cabo Ortegal, O—Ordenes, B—Brangança, M—Morais, all representing klippe with ophiolite and other complexes; LDB—Le Danois Bank; CIZ (Central Iberian zone), WALZ (West Asturian–Leonese zone) and CZ (Cantabrian zone) represent Paleozoic sedimentary zones within Iberian terrane; PTSZ—Porto-Tomar shear zone; BCSZ—Badajoz-Córdoba shear zone; OMT—Ossa Morena terrane; cross-hachured area—Pulo do Lobo oceanic terrane; SPT—South Portuguese terrane. Eastern part of Iberia is mainly covered with post-Hercynian deposits, although inliers of Proterozoic and Paleozoic material with uncertain affinities occur throughout. Area west of PTSZ exposes post-Hercynian basinal sediments.

Location of Armorica

The Northern terrane of Armorica (Fig. 2) exposes proximal Ordovician-Devonian Gondwana shelf sediments on 2 Ga gneisses that match the northwest African craton (Nance and Murphy, 1996), and we (Shelley and Bossière, 2000) argued that the Northern terrane remained adjacent to the northwest African craton until the beginning of the Hercynian orogeny. An alternative is that the Northern terrane was displaced to its present position during the Pan-African (Cadomian) deformations, but this is unlikely given the dominantly sinistral Pan-African shearing.

Another possibility is that Armorica, including the Northern terrane, drifted away from Gondwana during Cambrian-Ordovician rifting and extension. Tait (1999) used paleomagnetic data to support this idea, and suggested that Armorica amalgamated with Laurentia-Baltica after drifting away, then rejoined Gondwana during the formation of Pangea. Tait's interpretation is based on the plate reconstruction of Bachtadse and Briden (1991), shown in her Figure 9. Tait regarded this interpretation as the most conservative. However, if an alternative plate reconstruction is used, based on the data of Hargraves et al. (1987), also shown in Figure 9 of Tait (1999), it is completely unnecessary to move Armorica from Gondwana. We argue that this, actually, is the more conservative interpretation. Two indications that Armorica did not drift away from Gondwana are (1) rifting and extension would have been accompanied by thinning of the Armorican crust, and consequently there would have been either distal or little sedimentation, not the proximal sedimentation observed; (2) drifting is not supported by the faunal assemblages (Robardet et al., 1994a, 2001; Robardet, 2000), and Robardet et al. (2001, page 13,309) pointed out "it is thus impossible that any wide ocean has separated Gondwana from Armorica during the Early Devonian."

Stampfli (2000) provided an interesting new perspective on the Hercynian orogeny in a paper dealing mainly with the development of Tethys. He showed Armorica adjacent to northwest Africa, as in our (Shelley and Bossière, 2000) model, prior to the orogeny, as part of the Hun superterrane. However, Stampfli adopted a plate setting in which Laurentia-Baltica faces North Gondwana prior to the orogeny, and this seems incompatible with the need to displace Armorica and other terranes such as Moesia dextrally by many thousands of kilometers. Part of his proposed plate setting is a subduction zone along the northern boundary of the Hun superterrane, dipping under Armorica since the Early Silurian. The problem is that the stable shelf sediments of Armorica provide no evidence of such subduction.

Location of Iberia

The Alpine orogeny has obscured much of the evidence for the relative positions of Iberia and Gondwana. The main part of the peninsula is usually described as the Iberian autochthon, and in the Matte and Ribeiro (1975) model for the Hercynian orogen, the Iberian autochthon forms an indentor that causes the curvature of the Ibero-Armorican arc. Unlike Armorica, there seems no doubt that Iberia remained part of Gondwana prior to the Hercynian orogeny; however, the autochthonous nature of Iberia should be discussed. If it was an indentor, was it always a promontory of northern Gondwana, close to its present position? If it were a promontory, could it have been to the west of its present position, and displaced dextrally along with Armorica? This possibility was suggested by Matte (1986, his Fig. 17), who showed the Hercynian belt, including Iberia, displaced en masse along the northwest edge of Africa. The main difference between our model and that of Matte is that Matte has few relative displacements within the Hercynian orogen, whereas we (Shelley and Bossière, 2000) proposed major reshufflings. Another difference is that Matte (1986) proposed the oceanic material in South Armorica and the Massif Central to represent oceanic sutures, formed more or less in situ, whereas we (Shelley and Bossière, 2000) explained the oceanic material as slices of Rheic ocean shuffled within displaced segments of the Gondwanan shelf.

Ibero-Armorican arc

This arcuate shape was ascribed by Matte and Ribeiro (1975), Matte (1986), and Dias and Ribeiro (1994) to the indentation of the Hercynian orogen by an autochthonous Iberian promontory. The sinistral shears of Iberia, the dextral shears of Armorica, and the thrusts associated with these shears are all seen as having formed during that indentation. As proposed, the indentation model assumes that the various main parts of the Hercynian orogen were in their present relative positions immediately prior to the formation of the arc and the main structures within it.

In contrast, we (Shelley and Bossière, 2000) proposed that the main structures within the arc represent the slicing and shuffling of various parts of the orogen during megadextral displacements. It was suggested that the arc represents bending of the displaced terranes around an autochthonous Iberian promontory, and that the sinistral faults of Iberia are bookshelf-type structures in response to dextral shearing along the Porto-Tomar shear zone. In our (Shelley and Bossière, 2000) model, the degree of bending of the arc is much less than that proposed in the indentor model.

Kollmeier et al. (2000), using calcite twin data from the central area of the Ibero-Armorica arc in Cantabria-Asturias (northern Iberia), and Weil et al. (2000), using paleomagnetic data from the same area, suggested that the maximum principal compressive stress during the late Carboniferous was east-west (in the geographic coordinates of present-day Iberia), that the various sedimentary belts extended more or less north-south, and that the bending of the arc relates to a switch of the compressive stress direction to north-south in the Permian. The model of Kollmeier et al. (2000) and Weill et al. (2000) does not seem to require an indentor, and they proposed that the sedimentary belts in the Cantabria-Asturias part of the arc were barely curved at the start of the Hercynian orogeny.

We are not totally convinced that the calcite twin data of Kollmeier et al. (2000) necessarily relate to regional stresses rather than localized strain fields, and in any case, results from the central part of the Ibero-Armorican arc cannot necessarily be extended to the entire arcuate structure, from France to southern Iberia. Most previous workers have suggested that possibly half the curvature of the Ibero-Armorican arc is primary, with folding only responsible for a counterclockwise rotation of 40°–50° in Iberia and a clockwise rotation of 40°–50° in Armorica (Ries et al., 1980; Perroud and Bonhommet, 1981; Perroud, 1986; Edel, 1987). Furthermore, if Dias and Ribeiro (1993, 1995) are correct in their observations that the dextral Porto-Tomar shear zone and the sinistral Badajoz-Córdoba shear zone were both active in the Late Devonian and Carboniferous, then the crosscutting relationship between the Porto-Tomar shear zone and the Iberian part of the arc makes it very difficult to ascribe all arc bending to folding in the latest Carboniferous and Permian. If that had been the case, the Porto-Tomar shear zone would have been folded so that its continuation would be more or less parallel to the northern coast of Spain. This would create enormous new problems in that the continuations of the Cantabrian, West Austurian–Leonese, and Central Iberian zones (Fig. 2) should be found under the Aquitaine Basin, or farther east and south of the Massif Central if there is a very large displacement along the Porto-Tomar shear zone. There is no evidence for this. The crosscutting relationship between the Devonian-Carboniferous Porto-Tomar shear zone and the Ibero-Armorican arc is therefore compelling evidence for a significant primary curvature to the Ibero-Armorican arc.

Edel (2001) proposed that all the Hercynian belt from Iberia to Bohemia drifted south by ~2200 km between 340 and 328 Ma, and that this was followed by a rapid clockwise rotation of 70°. Then, between 320 and 305 Ma, there was a second clockwise rotation of 45°. As Edel (2001) noted, the results indicate a much more mobilistic evolution of the Hercynides than is generally admitted. We discuss Edel's work in the following. Our main concern is that Edel extrapolated his paleomagnetic data to northwest France and Iberia from regions to the east. We do not think that this is justified. For example, it seems unlikely that northwest France was involved in a rotation of 115° when one considers the age and orientation of the dextral faults. Thus, the northern branch of the South Armorican shear zone, the tectonic St. Julien de Vouvantes Basin, and the Nort-sur-Erdre fault that bounds the Ancenis Culm Basin were all active from at least 340 Ma, and all were dextral structures (Fig. 6 in Shelley and Bossière, 2001). The Nort-sur-Erdre fault continued its dextral movement until at least 320 Ma, and the southern branch of the South Armorican shear zone was active from at least 320 Ma. Given that the Armorican shear zones are large-scale lithospheric discontinuities that extend to depths >100 km (Granet, 1999; Judenherc et al., 1999; Granet et al., 2000), we argued (Shelley and Bossière, 2002) that they cannot be the product of small displacements, nor could they be the conduits for granites unless they were already well developed. It seems likely that substantial displacement had already occurred on these faults before the syntectonic granites were emplaced, and shear-zone activity is probably over a longer period than generally acknowledged. The question in relation to Edel (2001) is how could these more or less parallel faults have continued to be active with a consistent sense of shear at the same time the entire orogenic belt was being rotated 115°. We do not think it is likely, and we suggest instead that those parts of the belt east of northwest France were rotated independently. We are not familiar with these eastern areas, but we wonder whether some of the rotation could be ascribed to block rotations during sinistral bookshelf deformation between the major dextral shears that Edel and Weber (1995, their Fig. 8) showed extending parallel to the Armorican shear zones, and that include the Bray fault, the Franconian transfer fault zone, and the Elbe fault. We do not know if such sinistral shears have been identified, but such structures exist in Iberia and North Brittany, and the accompanying clockwise rotations are discussed in the following. Edel and Wickert (1991) suggested the possibility that the first 70° of rotation was counterclockwise, not clockwise, and this requires further consideration; Edel (2001, p. 89) indicated that it is possible, although in his view improbable.

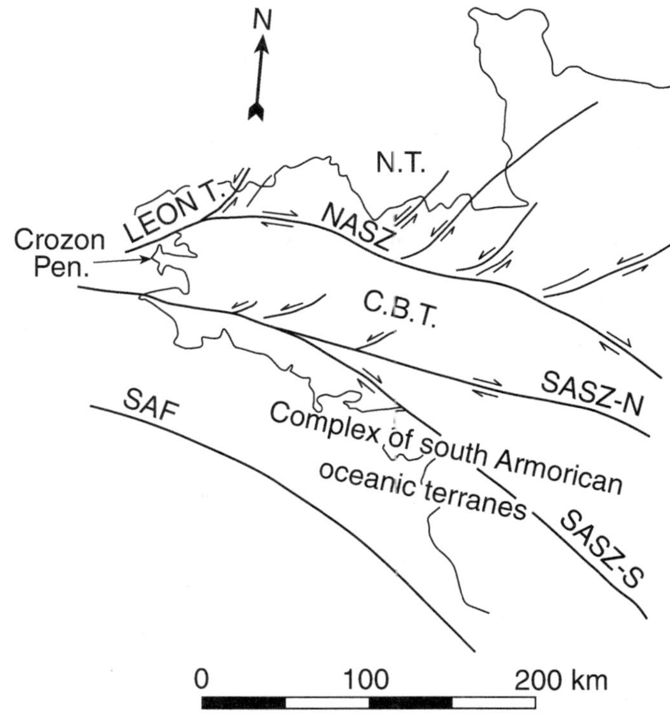

Figure 3. Map with positions of North Armorican and South Armorican shear zones (NASZ, SASZ; -S, -N are southern and northern branches, respectively) and terranes they delineate, including Northern and Central Brittany terranes (NT and CBT). As discussed in text, NT exposes west African 2 Ga cratonic rocks, and slightly more proximal Paleozoic shelf sediments than CBT. Complex of oceanic terranes crops out between SASZ (northern branch) and South Armorican front (SAF). Léon terrane (LEON T.) underthrust NT to southeast, but that boundary later became sinistral fault. Series of sinistral faults cut NT, and all are reactivated Cadomian structures. They merge with NASZ rather than crosscut it, indicating bookshelf-type movements related to dextral shearing. Fault positions are from Rolet et al. (1994). Crozon Pen.—Crozon Peninsula.

Iberia before the Hercynian orogeny

Primary arcuate sedimentary belts are consistent with Iberia as a promontory on the northwest African coast. The most likely initial orientation of the Porto-Tomar shear zone, which truncates that promontory, is parallel to the Moroccan–Western Saharan coast of Africa and its Hercynian dextral shears (such as the Western Meseta shear zone; Piqué et al., 1993). The currently curved Porto-Tomar shear zone may represent a contractional bend around the Iberian promontory, and we develop that idea in the following.

The Badajoz-Córdoba shear zone (Fig. 2) coincides, more or less, with an important Pan-African suture within Iberia (Abalos, 1992; Eguíluz et al., 2000). The Ossa-Morena terrane (Fig. 2) contains oceanic material of Cadomian (Pan-African) age, and its covering Paleozoic strata are more distal than those of Central Iberia. The Paleozoic Pulo do Lobo oceanic terrane is south of the Ossa-Morena terrane.

Oblique movements along the northern boundary of Gondwana in the Cadomian orogeny were characterized by sinistral shearing (Balé and Brun, 1983; Strachan et al., 1996). The sinistral shears in northern Brittany extend northeast-southwest, oblique to the North Armorican shear zone (Fig. 3). According to the paleomagnetic evidence (Edel, 1987, 2001), northern Brittany was rotated 40°–50° clockwise after 320 Ma (see preceding discussion of the Ibero-Armorican arc). In the Shelley and Bossière model, the Armorican shear zones were originally parallel to the northwest African dextral shears, prior to the 40°–50° clockwise rotation. According to this, the Cadomian shears might have been originally north-south, relative to present-day Africa. However, when the additional effects of sinistral Hercynian bookshelf shearing are taken into account, the original orientation of Cadomian structures may possibly have been closer to northwest-southeast in present-day African geographic coordinates. This is discussed in the following.

On the basis of the preceding discussion, we hypothesize that the Badajoz-Córdoba shear zone was an important control on the shape of the Gondwanan coastline, extended northwest-southeast, and marked the boundary between more proximal and more distal Paleozoic shelf sediments. The nearby Pan-African structures of western Africa appear to be wrapped around the Early Proterozoic–Archean cratons (e.g., Dallmeyer, 1990, his Fig. 8.8). Thus, to the east of the West African craton, the Pan-African orogen exhibits north-south–trending structures (Bertrand and Caby, 1978), and these swing around to east-west near Agadir to the north of the craton (Leblanc and Lancelot, 1980). On the western side of the craton the trend is generally north-south, but it swings around toward the northwest (Dallmeyer, 1990) exactly where we propose the Northern terrane was attached to Africa. If this northwest trend continued offshore as the Badajoz-Córdoba shear zone, it may have merged with the east-west–trending belt of Morocco.

Figure 4 shows, therefore, the proto-Iberian promontory with a northwest-southeast lineament along its southwestern side, extending from Buçaco toward the African coast, a speculative proposal that is consistent with the Pan-African patterns of West Africa, and consistent with the Shelley and Bossière model that shows how the Badajoz-Córdoba shear zone may have been rotated clockwise and sinistrally sheared in response to Hercynian dextral shearing.

Africa and North America during the Hercynian orogeny

The Hercynian belt of Gondwanan Iberia and France formed between Africa and North America, and it is relevant to note that dextral faulting along northeast-southwest lines was important in both these areas during the Hercynian orogeny. Dextral shearing in the northwestern and northern coastal areas of Africa is represented by the northeast-southwest–trending Western Meseta shear zone (Piqué et al., 1993), the Tizi n'Test fault (Mattauer et al., 1972), shear zones in eastern Morocco dated as ca. 370 Ma (Clauer et al., 1980; Huon et al., 1988), and the east-northeast–west-southwest–trending mylonites in Algeria (Bossière and Peucat, 1985, 1986; Peucat and Bossière, 1991). Matte (1986) noted that late Hercynian dextral shearing took place more or less parallel to the north African coast. The Western Meseta shear zone was active throughout the middle-late Paleozoic and controlled sedimentary basins as well as later Hercynian folding (Piqué et al., 1980; Piqué and Michard, 1989). The Tizi n'Test fault is late Hercynian but pre-Permian (Mattauer et al., 1972), and its orientation (east-northeast–west-southwest) compared to the earlier Western Meseta shear zone (northeast-southwest) could possibly reflect a clockwise rotation of shortening direction in the late Carboniferous, perhaps recording a transition during the change from east-west to north-south shortening noted by Kollmeier et al. (2000) and Weil et al. (2000) in Iberia. The north African region was subject to the later Alpine orogeny, which tends to obscure and overprint Hercynian events. In the Appalachians of North America, Alleghanian faulting along northeast-southwest lines is mainly dextral (Lefort, 1989, his Fig. 51; Hatcher, 2000).

EMPLACEMENT OF THE IBERIAN AND ARMORICAN TERRANES

In this section we develop, in detail, a model for the emplacement of the Iberian and Armorican terranes considering the propositions that Laurentia moved clockwise about Gondwana by several thousand kilometers during the Hercynian orogeny, that the Ibero-Armorican arc is in part primary, that Iberia may have been a promontory on the northwest African coast, and that the Northern terrane (Fig. 2) was adjacent to northwest Africa at the start of the orogeny. We did not attempt this in our previous paper (Shelley and Bossière, 2000).

Tikoff et al. (1999) noted that translation is the most difficult part of a deformation to identify and quantify. Understanding this is central to the concept of suspect tectonostratigraphic terranes: fault-bound disparate terranes cannot be presumed to have originated in their present relative positions, and the amount of translation involved in bringing them together may be indeterminable. Present-day rates of relative plate movement range to 16 cm/yr (DeMets et al., 1990). Using a rate of 10 cm/yr, we (Shelley and Bossière, 2000) showed that (1) the North Armorican shear zone could represent 2600 km of displacement between the Devonian shelf sediments and the 329 Ma granite (Guillet et al., 1985) that stitches together the terranes either side of the North Armorican shear zone (that granite was displaced by 15 km); (2) the South Armorican shear zone (northern branch) could represent 1500 km displacement and was stitched at 340 Ma (Jegouzo and Rossello, 1988); (3) the South Armorican shear zone (south branch) may represent 4500 km of displacement, and was stitched by granites ca. 300–310 Ma (Bernard-Griffiths et al., 1985). Higher or lower rates of relative movement may change these possible amounts of displacement, and the possibility is not necessarily a reality. Nevertheless, the important point is that the Hercynian orogeny is long-lasting enough to accommodate easily the amounts of displacement we propose.

Dextral shears and the reversed terrane relationships of Armorica

Robardet et al. (1994a) noted on the regional scale the apparent reversal of the expected shelf sedimentary sequences within France, so that the Northern terrane (Fig. 2) exposes proximal shelf sediments, and the Southern terrane (Fig. 2) exposes relatively distal sediments. Given that Gondwana was to the south, the opposite might have been expected. Robardet et al. (1994a, their Fig. 15) proposed major relative movements of these sedimentary successions in order to explain the relationship, but they did not provide a tectonic model.

The same kind of reversed relationship occurs on a finer scale. Ballèvre et al. (1992) showed that the sediments in northern Brittany become more distal from north to south. Furthermore, Robardet et al. (1994b) showed that proximal sedimentation started in the Central Brittany terrane (Figs. 2 and 3), and that when the proximal sedimentation commenced later in the Northern terrane, the sediments became more distal in Central Brittany. They interpreted this in terms of an Arenig marine transgression, with the Central Brittany and Northern terranes in their present relative positions. The reversals continue, so that the oceanic terranes of South Armorica (Fig. 3) are south of the Central Brittany terrane, whereas in the simplest model of the Gondwanan shelf one might have expected the oceanic terranes to be north of northern Brittany.

We believe that dextral transcurrent faulting can explain these reversals. First, the boundaries between these terranes are major dextral transcurrent faults: the North Armorican shear zone separates the Central Brittany and Northern terranes and the South Armorican shear zone (northern branch) separates the Central Brittany terrane from the terranes of South Armorica. These transcurrent faults are major lithospheric discontinuities that extend to subcrustal depths (the South Armorican shear zone, southern branch, to depths >100 km), according to the seismic tomographic results of Granet (1999), Judenherc et al. (1999), and Granet et al. (2000), implying considerable movement. Second, transcurrent fault patterns often form complex anastomosing arrays with contractional and extensional bends, rather like large-scale examples of SC planes and extensional shear bands, and such structures, like thrusts in section view, may reverse normal sequences. We propose a possible sequence of events for creating the reversed terrane sequence of Armorica. The fault array proposed is less complex than the mapped fault pattern of Armorica (e.g., Arthaud and Matte, 1977, their Fig. 7; Matte and Hirn, 1988, their Plate 1), and less complex than arrays in present-day transpressional settings such as New Zealand (Lamarche and Lebrun, 2000, their Figs. 5 and 8).

Sinistral shears and associated subduction in Iberia and Armorica

Rolet et al. (1994, their Fig. 19[III]) showed a series of sinistral Hercynian shears in northern Brittany. They are reactivated Cadomian sinistral shears. The way the shears merge with (rather than crosscut) the dextral North Armorican shear zone (Fig. 3) indicates bookshelf-type structures. This means that they rotated clockwise from their original orientations. We proposed that the sinistral Badajoz-Córdoba shear zone, which merges with (rather than crosscuts) the contemporaneous dextral Porto-Tomar shear zone, was rotated clockwise by bookshelf-type movements during the Hercynian orogeny (Shelley and Bossière, 2000). The Badajoz-Córdoba shear zone is also a reactivated Cadomian structure.

Some reactivated Cadomian structures developed subduction zones or zones of underthrusting with high-pressure metamorphism. Thus, although much of the Hercynian metamorphism in the Ossa-Morena terrane is low pressure, Fonseca and Ribeiro (1993) suggested that oceanic material was subducted under the Ossa-Morena terrane in the Devonian and Carboniferous, and eclogite facies metamorphism of Carboniferous age (340 ± 13 Ma) has now been established by zircon SHRIMP dating (Ordóñez-Casado, 1998; reported in Pereira and Silva, 2001). Similarly, in northern Brittany, the Léon–Northern terrane boundary is oblique to the dextral shears, parallel to Cadomian structures, and a site of underthrusting as well as sinistral shearing (Fig. 3). If Cadomian structures formed steps between adjacent dextral shears, it is inevitable that underthrusting or subduction would have developed along them. Modern analogues are found between dextral shears off the coast of New Zealand (Lamarche and Lebrun, 2000).

Detailed model for the development of the Iberian-Armorican terranes

Figures 4–9 show the possible relationships of Africa, the Northern terrane, and Iberia prior to the Hercynian orogeny, and how the various parts may have been sliced up and displaced during Hercynian dextral shearing. Figure 4 represents the following (prior to the orogeny). (1) A broad shallow shelf for the deposition of the Gondwanan Ordovician to Devonian shelf sediments extended to the north and west of Africa. The shelf is similar, in principle, to that shown in Paris and Robardet (1990) and Robardet et al. (1994a). (2) The shelf sediments are divided schematically into proximal (open circles) and distal (dots). (3) Iberia is shown as a promontory southwest of its present position. (4) The Northern terrane was attached directly to where the northwest African cratonic rocks are currently exposed. (5) Crozon, now in Brittany, and Buçaco, now in Portugal, were in one sedimentary belt (filled circles), consistent with reports of the very close similarity in their Ordovician successions (Perroud and Bonhommet, 1981; Ballèvre et al., 1992). (6) The Rheic ocean is shown, with dextral shears and associated subduction zones and volcanic arcs related to movement of Laurentia around Gondwana and/or the opening and closing of the Rheic ocean (Fig. 6 in Shelley and Bossière, 2000). Thinned fragments of Gondwana, drifted away during Cambrian-Ordovician extension, would have become involved in subduction-zone metamorphism.

A possible sequence of events during the course of the orogeny is as follows.

1. The start of the Hercynian orogeny was marked by the encroachment of shear zones on the Gondwana shelf (Fig. 4).

2. Eventually a shear with an extensional bend developed so that oceanic material was brought into juxtaposition with the proximal rocks of Gondwana (Fig. 5). Movement around the extensional bend causes a counterclockwise rotation of the southwestern part of Iberia.

3. Dextral shears developed (Fig. 6), so that the Crozon Peninsula rocks were carried partly around a contractional bend along the South Armorican shear zone (northern branch). The distal sedimentary belt of what became the Ossa-Morena terrane

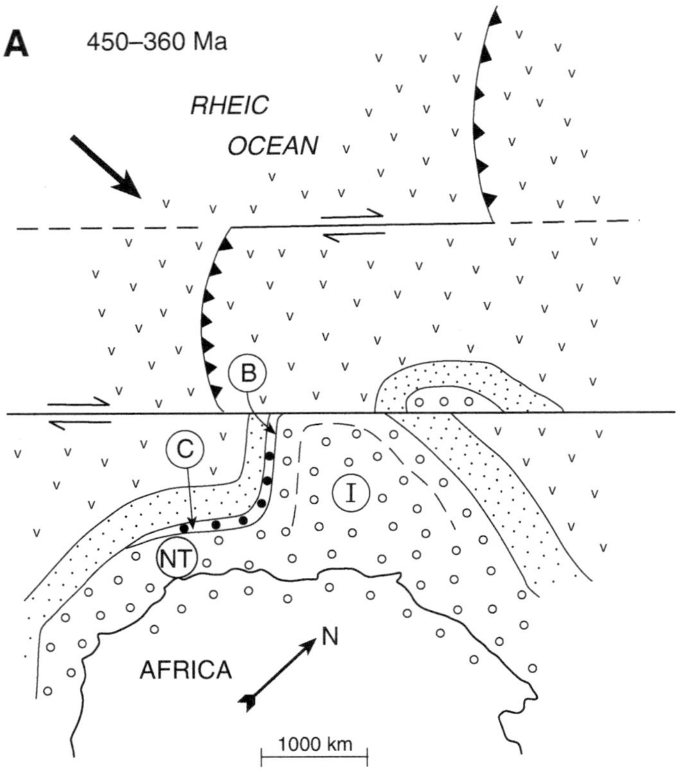

Figure 4. First stage in model for sequence of dextral translation of terranes to form Hercynian complex of Iberia and Armorica. Outline of present coastline of northwestern Africa is shown. Senses of displacement of faults and thrusts are only shown for times at which they were certainly active. In stage A of model, B—Buçaco (Portugal), C—Crozon Peninsula (Brittany), NT—Northern terrane of France, and I—Iberia. Arrow in Rheic ocean indicates shortening direction; "v" pattern indicates oceanic material, with or without oceanic sediments and with possible thinned drifted fragments of continental crust caught up in subduction zones; dots indicate distal shelf sediments; open circles represent proximal shelf sediments; infilled circles represent Buçaco-Crozon Peninsula sedimentary facies.

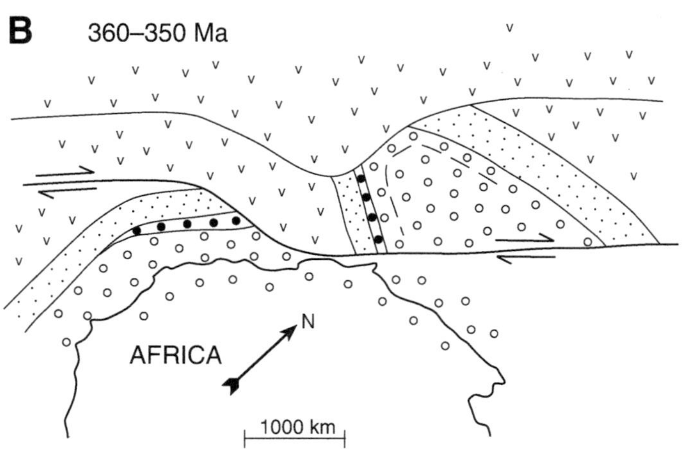

Figure 5. Stage B of model for sequence of dextral translation of terranes to form Hercynian complex of Iberia and Armorica. Outline of present coastline of northwestern Africa is shown. See caption to Figure 4 for details of symbols. See text for discussion.

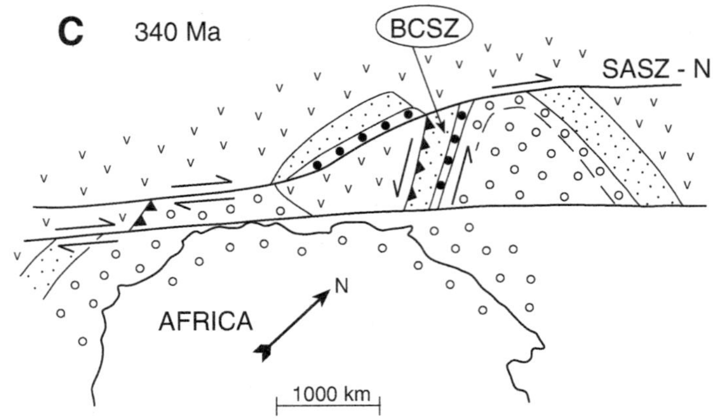

Figure 6. Stage C of model for sequence of dextral translation of terranes to form Hercynian complex of Iberia and Armorica. Outline of present coastline of northwestern Africa is shown. BCSZ—Badajoz-Córdoba shear zone; SASZ-N—South Armorican shear zone, northern branch. See caption to Figure 4 for details of symbols. See text for discussion.

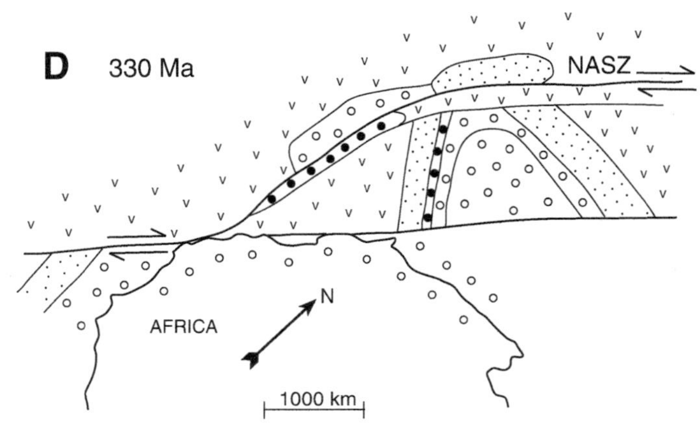

Figure 7. Stage D of model for sequence of dextral translation of terranes to form Hercynian complex of Iberia and Armorica. Outline of present coastline of northwestern Africa is shown. NASZ—North Armorican shear zone. See caption to Figure 4 for details of other symbols. See text for discussion.

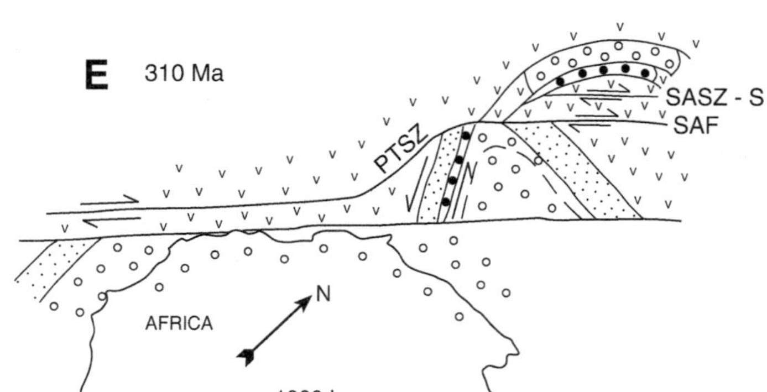

Figure 8. Stage E of model for sequence of dextral translation of terranes to form Hercynian complex of Iberia and Armorica. Outline of present coastline of northwestern Africa is shown. PTSZ—Porto-Tomar shear zone; SASZ-S—South Armorican shear zone, southern branch; SAF—South Armorican front. See caption to Figure 4 for details of other symbols. See text for discussion.

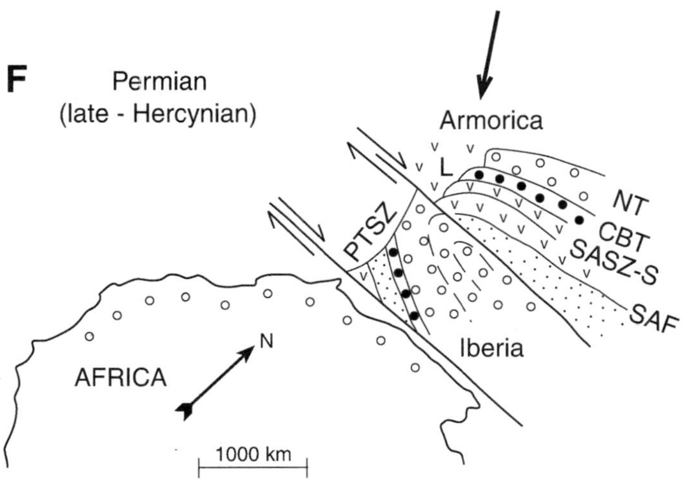

Figure 9. Stage F of model for sequence of dextral translation of terranes to form Hercynian complex of Iberia and Armorica. Outline of present coastline of northwestern Africa is shown. NT—Northern terrane; CBT—Central Brittany terrane; L—Léon terrane; PTSZ—Porto-Tomar shear zone; SASZ-S—South Armorican shear zone, southern branch; SAF—South Armorican front. See caption to Figure 4 for details of other symbols. See text for discussion.

was subjected to sinistral bookshelf shearing. In the southwestern part of the dextral shear system, a step from one shear to another caused subduction or underthrusting, and this later became the boundary between the Léon and Northern terranes of France. As noted herein, similar structures are forming today off the New Zealand coast (Lamarche and Lebrun, 2000).

4. One of the dextral shears in Figure 6 developed into a new contractional bend so that the Northern terrane rocks moved around those of Central Brittany (Crozon Peninsula) along what became the North Armorican shear zone (Fig. 7).

5. Another shear-zone system with contractional bends (Fig. 8) moves the oceanic rocks that were juxtaposed amongst the proximal sediments in Figure 5 along the Porto-Tomar shear zone, South Armorican shear zone (southern branch), and South Armorican front. Rocks of the Northern and Central Brittany terranes are carried along passively by this system.

6. In the late Hercynian (Permian), the shortening direction changed to near north-south, completing the formation of the Ibero-Armorican arc and rotating the rocks of Armorica clockwise (Fig. 9). At the same time, new dextral shears, the Chedabucto and Biscaye North Pyrenean faults (described by Arthaud and Matte, 1977, their Figs. 1 and 9) moved Iberia and Armorica eastward into their Pangea positions.

This sequence shows how the package of Armorican fault-bound terranes could have arrived in reverse sequence, and how the faults bounding them (North Armorican shear zone, South Armorican shear zone, northern and southern branches, and South Armorican front) may have originated. In this model, the South Armorican front, South Armorican shear zone (southern branch), and the Porto-Tomar shear zone are parts of the same fault system, and the Porto-Tomar shear zone may in part also represent an early position of the northern branch of the South Armorican shear zone (cf. Figs. 6 and 9). Some of the proposed rates of relative movement are very fast, but no faster than the fastest of present-day rates.

The sequence of events corresponds very well with the known sequence of stitching events along the faults. The northern branch of the South Armorican shear zone ceased substantial movement ca. 340 Ma, the North Armorican shear zone ceased movement ca. 329 Ma, and the southern branch of the South Armorican shear zone ceased movement ca. 300 Ma. These figures suggest that much or all of the dextral movement along the northern branch of the South Armorican shear zone and North Armorican shear zone was accomplished to the west of Iberia, and that the Porto-Tomar shear zone, southern branch of the South Armorican shear zone, and South Armorican front were the means of translating the entire package of Armorican terranes from the west of Iberia. The age of the sinistral shearing in northern Brittany is not well established, but Rolet et al. (1994, their Fig. 19[III]) suggested that it was taking place ca. 305 Ma, and in the Léon terrane it was taking place after this because it affects the Aber Ildut granite, dated as ca. 300 Ma (Deutsch and Chauris, 1965). J. Rolet (2000, personal commun.) noted that it is not possible to state it was not happening before this time. Because the North Armorican shear zone was mainly active before 329 Ma (Fig. 7), we suggest that some sinistral movement took place then, but the 15 km of movement on the North Armorican shear zone since 329 Ma (Guillet et al., 1985) may have continued until 300 Ma, accompanied by bookshelf sinistral shearing.

DISCUSSION

Other terrane relationships

Our model shows one possible way in which the present relationships between the principal tectonostratigraphic terranes of Gondwanan Iberia and France may have developed in a dextral translational regime. It may not be the only way, and many details remain for further discussion.

The Southern terrane of France (Fig. 2) includes relatively distal Gondwanan shelf sediments. We have not involved it in the terrane movements modeled here. There are two main alternatives: (1) the Southern terrane is essentially in situ; (2) it was displaced dextrally, in advance of the better known Armorican terranes to the north. Figures 4, 5, and 7 show that it is conceivable that packages of distal sediments were displaced along the dextral shears in advance of Iberia and/or the complex of Armorican terranes. However, it is probable that a belt of distal shelf sediments existed on the northern side of the Africa, before the Hercynian orogeny. Such a belt might have been relatively undisturbed and protected from the ensuing dextral displacements.

The Ancenis terrane (Fig. 2) provides a puzzle, being a small enclave of shelf sediments within the South Armorican oceanic terranes. We have suggested (Shelley and Bossière, 2001) that it represents an exotic duplex. It is not difficult to imagine a small package of shelf sediment being caught up within the complex anastomosing array of faults shown in Figures 2 and 4–9, but we are a long way from understanding this level of detail. We suggest that an analysis of the faunal affinities between the various parts of the Hercynian orogen will prove to be an important tool in resolving some of these tectonic questions.

Our discussion herein has been mainly about the terranes of Iberia and western France. We noted (Shelley and Bossière, 2000) that the Massif Central is a complex of oceanic material, essentially a continuation of the complex of South Armorican terranes, and the Hercynian orogen extends farther east into Germany, and southeast into Sardinia. We hesitate to suggest that all these areas were subject to the same degree of dextral transpression as northwest France and Iberia, because we are not familiar with them, but it is clear from Figure 8 in Edel and Weber (1995) that major dextral shears parallel to the Armorican shear zones exist. However, in common with Robardet (2000) and Robardet et al. (2001), we suggest that attention must be paid to the faunal assemblages before making structural and paleogeographical interpretations about the relationship of northwest France to areas farther east. For example, in the context of our discussion of Tait's (1999) suggestion that Armorica drifted away from Gondwana early in the Paleozoic, we noted that Robardet et al. (2001) viewed this suggestion as inconsistent with the faunal evidence, nor is it necessary from the paleomagnetic point of view. Turning to paleontology, we need to know what a "Bohemian fauna" means in tectonic terms. If megadisplacements characterize this orogen, where was "Bohemia" at the start of the orogeny? Robardet (2000) noted that it was always part of Gondwana; and Edel (2001) showed, from paleomagnetic data, that Bohemia drifted ~2000 km toward the south during the early stages of the orogeny. Is it possible, therefore, that Bohemia was originally a Gondwanan promontory northwest of Africa sheared dextrally into its present relative position? This would fit with the fact that the Ancenis terrane, also displaced dextrally to the south (Shelley and Bossière, 2001), displays Bohemian faunal characteristics (Henry, 1989; Cavet and Pillet, 1964). Clearly the integration of paleontological and tectonic models must be a major objective for future researchers.

We note that we have concentrated in this chapter on translational movements. The seismic tomographic work of Granet (1999), Judenherc et al. (1999), and Granet et al. (2000) shows that transcurrent faults dominate the structure of northwest France with dextral shears as major lithospheric discontinuities traced to >100 km depth. However, transpression is commonly partitioned into translation and thrusting (Dewey et al., 1998), and in Iberia, Aquitaine, and the Massif Central, thrusting becomes much more obviously important. As noted by Arenas et al. (2000, p. 108) in descriptions of the allochthonous terranes in northern Iberia (Fig. 2), thrust regimes are illustrated in cross section, and the amount of accompanying transcurrent movement tends not to be considered, even though it may be substantial. Some of the overthrust material in Iberia contains oceanic material derived from the Rheic ocean (Díaz García et al., 1999). In view of the model proposed here, a question to ask is whether some of this oceanic material represents slices of Rheic ocean from within the complex of terranes shown in Figures 6–9 rather than material derived directly from the main body of the ocean. Whether the allochthonous oceanic complexes of northwest Iberia came directly from the Rheic ocean or from slices bound by transcurrent faults, the east-west movement directions in the complexes (Arenas et al., 2000, Fig. 56; after Martínez Catalán et al., 1999) are consistent with the east-west shortening direction proposed in our model (Fig. 4).

Because the Hercynian orogeny involved a dextral transpression, the width of terranes would have narrowed northeastward during displacement, and if the overthrust terranes of Iberia, for example, represent slices of ocean where the Hercynian deformation was partitioned into overthrusting, some terrane slices may have been thinned and squeezed out altogether by this process. The clockwise rotation of Laurentia and Baltica relative to Gondwana in the reconstructions of Dalziel et al. (1994) and Dalziel (1995), as shown in Figure 1, might suggest that the amount of shortening associated with translation would increase progressively from west to east. However, work in many parts of the orogen has been aimed at quantifying the thrust movements, not the translations (e.g., Arenas et al., 2000, p. 108); here, we have emphasized the translations (Figs. 4–9). A complete model must integrate all the likely strains and movements, including those in Hercynian Africa.

Oceanic terranes: Fragments of the Rheic ocean or oceanic sutures?

Geologists have discussed the exotic terranes in terms of pieces of the Rheic ocean thrust over Gondwana during the collision (e.g., Díaz García et al., 1999). However, where the oceanic terranes are bound by transcurrent faults, as in France, they have more often been interpreted in terms of a smaller ocean that opened and closed. The word "suture" could reasonably be used for any major tectonostratigraphic terrane boundary, but it appears to be used in much of the literature on the Hercynian orogen for oceanic rocks interpreted to represent a preexisting ocean that disappeared through processes of subduction or obduction. This usage reflects a somewhat fixist approach to the Hercynian terranes, contrary to tectonostratigraphic modeling, which requires special evidence and pleading if one is to advocate in situ terrane relationships. The evidence from Cenozoic Southeast Asia (Hall, 1996), Cenozoic and Mesozoic New Zealand (Sutherland, 1999; Adams et al., 1998), and Mesozoic-Cenozoic northwest America (Andronicos et al., 1999) surely provides compelling reason to consider the Hercynian orogeny in terms of transcurrent movements that measure thousands of kilometers, and to view sutures in quite another way.

There is remarkably little evidence for the opening and closing of a South Armorican ocean. Thus, according to Robardet (2000), the faunal relationships between all the Gondwanan areas now represented by terranes in Armorica and southern France are too close for an ocean to have been between them. As we pointed out in Shelley and Bossière (2000), the volcanic arcs (Peucat et al., 1986), the thrusting on east-west lines (Brun and Burg, 1982), and the subduction to >70 km depth followed by rapid exhumation (Shelley and Bossière, 1999), are all evident in South Armorica, but it seems impossible to explain them in the general setting of stable platform sedimentation. East-west thrusting and extension are also difficult to reconcile with the opening and closing of an ocean at a high angle to that direction. Furthermore, there is no substantial evidence of the opening and closing of an ocean in the seismic tomography of Judenherc et al. (1999). What is now east-west thrusting and extension would originally have been northeast-southwest thrusting and extension, parallel to the megadisplacements of the Hercynian orogen, and consistent with structures developed in association with the subduction zones shown in Figure 4. The explanation of the oceanic terranes in terms of displaced slices of the Rheic ocean (or other associated oceanic areas) seems to us much more logical and in keeping with the facts.

CONCLUSIONS

We have shown that the Shelley and Bossière (2000) model for the Hercynian orogen of Gondwanan France and Iberia can be developed into a much more detailed model of displaced terranes. The model involves thousands of kilometers of movement within a complex anastomosing array of northeast-southwest–trending dextral transcurrent faults, and is broadly consistent with geological data from the constituent terranes. The model is speculative, but suspect tectonostratigraphic terrane theory requires speculative modeling as a preliminary to further testing. The commonly adopted position that the Iberian and French fault-bound Hercynian terranes evolved close to their present relative positions is not sustainable in tectonostratigraphic terrane theory without special evidence, which currently does not exist.

The main features of the detailed model shown in Figures 4–9 are as follows.

Prior to the Hercynian orogeny, Iberia formed a promontory on the northwest African coast as part of a broad early-middle Paleozoic shelf.

The important Iberian Precambrian suture, now the Badajoz-Córdoba shear zone, formed the southwestern margin of the Iberian promontory.

The Northern terrane of France originated southwest of the Iberian promontory, adjacent to what is now the Western Saharan coast of Africa.

Offshore from the Northern terrane was the Central Brittany terrane, which was part of a sedimentary belt continuous with the southwestern edge of the Iberian platform.

Before the orogeny, the neighboring Rheic ocean was the site of northeast-southwest dextral translational movements. Subduction zones extended northwest-southeast and were sinks for thinned continental crust rifted from Gondwana during Cambrian-Ordovician extension. The direction of extension and thrusting associated with subduction in the Rheic ocean was initially northeast-southwest.

The start of the orogeny proper ca. 360 Ma was marked by dextral shears cutting the shelf platform. An extensional bend on one shear juxtaposed Rheic oceanic material with proximal shelf sediments.

In a series of anastomosing dextral displacements, the Rheic oceanic material and the Northern and Central Brittany terranes were reversed in order and moved around Iberia.

These movements caused clockwise rotation and bookshelf sinistral shearing along the Badajoz-Córdoba shear zone. Similar movements occurred in the Northern terrane.

Steps from one shear to another resulted in underthrusting along the Badajoz-Córdoba shear zone and along the Léon–Northern terrane boundaries.

Slices of Rheic ocean carried the Eohercynian tectonic signatures of the Rheic ocean into the Hercynian orogen.

Permian shortening was from the north and caused the final bending of the Ibero-Armorican arc and the clockwise rotation of Armorica so that the Eohercynian thrusting and extension directions now extend more or less east-west.

Permian dextral shears define the southern and northern edges of Iberia.

Some Hercynian transpression was partitioned into thrusting and terrane shortening, but the model makes no attempt to quantify these effects.

ACKNOWLEDGMENTS

Shelley thanks the University of Canterbury for leave and the University of Nantes for a visiting professorship that facilitated further research on the Hercynian orogen and attendance at the "Basement Tectonics 15" conference in A Coruña, and Bossière thanks the University of Canterbury for a Visiting Erskine Fellowship. We thank J.B. Edel and an anonymous reviewer for their helpful and constructive comments.

REFERENCES CITED

Abalos, B., 1992, Variscan shear-zone deformation of late Precambrian basement in southwest Iberia: Implications for circum-Atlantic pre-Mesozoic tectonics: Journal of Structural Geology, v. 14, p. 807–823.

Adams, C.J., Barley, M.E., Fletcher, I.R., and Pickard, A.L., 1998, Evidence from U-Pb zircon and $^{40}Ar/^{39}Ar$ muscovite detrital mineral ages in metasandstones for movement of the Torlesse suspect terrane around the eastern margin of Gondwanaland: Terra Nova, v. 10, p. 183–189.

Andronicos, C.L., Hollister, L.S., Davidson, C., and Chardon, D., 1999, Kinematics and tectonic significance of transpressive structures within the Coast Plutonic Complex, British Columbia: Journal of Structural Geology, v. 21, p. 229–243.

Arenas, R., Díaz García, F., Martínez Catalán, J.R., Abati, J., González Cuadra, P., Andonaegui, P., González del Tánago, J., Rubio Pascual, F., Castiñeiras, P., and Gómez Barreiero, J. 2000, Structure and evolution of the Ordenes Complex: A Coruña, Spain, Basement Tectonics 15, Preconference Field Trip, 165 p.

Arthaud, F., and Matte, P., 1977, Late Paleozoic strike-slip faulting in southern Europe and northern Africa: Result of a right-lateral shear zone between the Appalachians and the Urals: Geological Society of America Bulletin, v. 88, p. 1305–1320.

Bachtadse, V., and Briden, J.C., 1991, Paleomagnetism of Devonian Ring Complexes from the Bayuda Desert, Sudan—New constraints on Gondwana's apparent polar wander path: Geophysical Journal International, v. 104, p. 635–646.

Balé, P., and Brun, J.P., 1983, Les chevauchements cadomiens de la Baie de Saint-Brieuc (Massif armoricain): Comptes Rendus de l'Académie des Sciences Paris, v. 297 (série II), p. 359–366.

Ballèvre, M., Paris, F., and Robardet, M., 1992, Corrélations ibéro-armoricaines au Paléozoïque: Une confrontation des données paléobiogéographiques et tectonométamorphiques: Comptes Rendus de l'Académie des Sciences Paris, v. 315 (série II), p. 1783–1789.

Bernard-Griffiths, J., Peucat, J.J., Sheppard, S., and Vidal, P., 1985, Petrogenesis of Hercynian leucogranites from the southern Armorican Massif: Contribution of REE and isotopic (Sr, Nd, Pb and O) geochemical data to the study of source rock characteristics and ages: Earth and Planetary Science Letters, v. 74, p. 235–250.

Bertrand, J.M.L., and Caby, R., 1978, Geodynamic evolution of the Pan-African orogenic belt: A new interpretation of the Hoggar Shield (Algerian Sahara): Geologische Rundschau, v. 67, p. 357–388.

Bossière, G., and Peucat, J.J., 1985, New geochronological information by Rb-Sr and U-Pb investigations from the pre-Alpine basement of Grande Kabylie (Algeria): Canadian Journal of Earth Sciences, v. 22, p. 675–685.

Bossière, G., and Peucat, J.J., 1986, Structural evidence and Rb-Sr, 39–40 Ar mica age relationships for the existence of an Hercynian deep crustal shear zone in Grande Kabylie (Algeria) and its Alpine reworking: Tectonophysics, v. 121, p. 277–294.

Brun, J.P., and Burg, J.P., 1982, Combined thrusting and wrenching in the Ibero-Armorican arc: A corner effect during continental collision: Earth and Planetary Science Letters, v. 61, p. 319–332.

Bullard, E., Everett, J.E., and Smith, A.G., 1965, The fit of the continents around the Atlantic: Philosophical Transactions of the Royal Society of London, v. 258 (series A), p. 41–51.

Cavet, P., and Pillet, J., 1964, Contribution à l'étude stratigraphique et paléontologique du synclinal d'Ancenis (Massif armoricain): Bulletin de la Société Géologique de France, v. 5, 7th series, p. 318–329.

Clauer, N., Jeannette, D., and Tisserant, D., 1980, Datation isotopique des cristallisations successives d'un socle cristallin et cristallophyllien (Haute Moulouya, Moyen Atlas): Geologische Rundschau, v. 69, p. 63–83.

Dallmeyer, R., 1990, The West African orogens and Circum-Atlantic correlatives, *in* Strachan, R.A., and Taylor, G.K., eds., Avalonian and Cadomian geology of the North Atlantic: Glasgow, Blackie, p. 134–165.

Dalziel, I.W.D., 1995, Earth before Pangea: Scientific American, v. 272, no. 1, p. 38–43.

Dalziel, I.W.D., Dalla Salda, L.H., and Gahagan, L.M., 1994, Paleozoic Laurentia-Gondwana interaction and the origin of the Appalachian-Andean mountain system: Geological Society of America Bulletin, v. 106, p. 243–252.

DeMets, C., Gordon, R.G., Argus, D.F., and Stein, S., 1990, Current plate motions: Geophysical Journal International, v. 101, p. 425–478.

Deutsch, S., and Chauris, L., 1965, Age de quelques formations cristallophylliennes et granitiques du Pays de Léon (Finistère): Comptes Rendus de l'Académie des Sciences Paris, v. 260, p. 615–617.

Dewey, J.F., Holdsworth, R.E., and Strachan, R.A., 1998, Transpression and transtension zones, *in* Holdsworth, R.E., et al., eds., Continental transpressional and transtensional tectonics: London, Geological Society of London Special Publication 135, p. 1–14.

Dias, R., and Ribeiro, A., 1993, Porto-Tomar shear zone, a major structure since the beginning of the Hercynian orogeny: Terra Abstracts, supplement number 6 to Terra Nova, v. 5, p. 11.

Dias, R., and Ribeiro, A., 1994, Constriction in a transpressive regime: An example in the Iberian branch of the Ibero-Armorican arc: Journal of Structural Geology, v. 16, p. 1543–1554.

Dias, R., and Ribeiro, A., 1995, The Ibero-Armorican arc: A collision effect against an irregular continent? Tectonophysics, v. 246, p. 113–128.

Díaz García, F., Arenas, R., Martínez Catalán, J.R., González del Tánago, J., and Dunning, G.R., 1999, Tectonic evolution of the Careón ophiolite (Northwest Spain): A remnant of oceanic lithosphere in the Variscan Belt: Journal of Geology, v. 107, p. 587–605.

Dubuisson, G., Mercier, J.C.C., Girardeau, J., and Frison, J.Y., 1989, Evidence for a lost ocean in Variscan terranes of the western Massif Central, France: Nature, v. 337, p. 729–732.

Edel, J.B., 1987, Paleopositions of the western Europe Hercynides during the late Carboniferous deduced from paleomagnetic data: Consequences for "stable Europe": Tectonophysics, v. 139, p. 31–41.

Edel, J.B., 2001, The rotations of the Variscides during the Carboniferous collision: Paleomagnetic constraints from the Vosges and the Massif Central (France): Tectonophysics, v. 332, p. 69–92.

Edel, J.B., and Weber, K., 1995, Cadomian terranes, wrench faulting and thrusting in the central Europe Variscides: Geophysical and geological evidence: Geologische Rundshau, v. 84, p. 412–432.

Edel, J.B., and Wickert, F., 1991, Paleopositions of the Saxothuringian (Northern Vosges, Pfalz, Odenwald, Spessart) in Variscan times: Paleomagnetic investigation: Earth and Planetary Science Letters, v. 103, p. 10–26.

Eguíluz, L., Gil Ibarguchi, J.I., Abalos, B., and Apraiz, A., 2000, Superposed Hercynian and Cadomian orogenic cycles in the Ossa-Morena zone and related areas of the Iberian Massif: Geological Society of America Bulletin, v. 112, p. 1398–1413.

Fonseca, P., and Ribeiro, A., 1993, Tectonics of the Beja-Acebuches ophiolite: A major suture in the Iberian Variscan foldbelt: Geologische Rundschau, v. 82, p. 440–447.

Granet, M., 1999, Des images du système lithosphère-asthénosphère sous la France: L'apport de la tomographie télésismique et de l'anisotropie sismique: Comité National Français de Géodésie et Géophysique Rapport Quadriennal 1995–1998, p. 71–88.

Granet, M., Judenherc, S., and Souriau, A., 2000, Des images du système lithosphère-asthénosphère sous la France et leurs implications géodynamiques: L'apport de la tomographie télésismique et de l'anisotropie sismique: Bulletin de la Société Géologique de France, v. 171, p. 149–167.

Guillet, P., Bouchez, J.L., and Vigneresse, J.L., 1985, Le complexe granitique de Plouaret (Bretagne): Mise en évidence structurale et gravimétrique de diapirs emboîtés: Bulletin de la Société Géologique de France, v. 1, 8th series, p. 503–513.

Hall, R., 1996, Reconstructing Cenozoic southeast Asia, *in* Hall, R., and Blundell, D., eds., Tectonic evolution of Southeast Asia: London, Geological Society of London Special Publication 106, p. 153–184.

Hargraves, R.B., Dawson, E.M., and Van Houten, F.B., 1987, Paleomagnetism and age of mid Paleozoic ring complexes in Niger, West Africa, and tectonic implications: Geophysical Journal of the Royal Astronomical Society, v. 90, p. 705–729.

Hatcher, R.D., Jr., 2000, Taconian, Acadian, and Alleghanian (Variscan) structural styles and accretionary processes in the southern and central Appalachians, *in* Díaz García, F., et al., eds., Variscan-Appalachian dynamics: The building of the Upper Paleozoic basement: A Coruña, Spain, Basement Tectonics 15, Program and Abstracts, p. 7.

Henry, J.L., 1989, Paléoenvironnements et dynamique de faunes de trilobites dans l'Ordovicien (Llanvirn supérieur–Caradoc basal) du Massif armoricain (France): Palaeogeography, Palaeoclimatology, Palaeoecology, v. 73, p. 139–153.

Huon, S., Piqué, A., and Clauer, N., 1988, Etude de l'orogenèse hercynienne au Maroc par la datation K-Ar de l'évolution métamorphique de schistes ardoisiers: Sciences Géologiques Bulletin, v. 40, p. 273–284.

Jegouzo, P., and Rossello, E.A., 1988, La branche nord du cisaillement sud-armoricain (France): Un essai d'évolution du déplacement par l'analyse des mylonites: Comptes Rendus de l'Académie des Sciences Paris, v. 307 (série II), p. 1825–1831.

Judenherc, S., Granet, M., Poupinet, G., Achauer, U., and Mocquet, A., 1999, Imaging the Hercynian suture in the French Armorican Massif: Eos (Transactions, American Geophysical Union), Fall meeting supplement, v. 79, no. 45, p. F592.

Kent, D.V., and Van der Voo, R., 1990, Paleozoic paleogeography from paleomagnetism of the Atlantic-bordering continents: Geological Society of London Memoir 12, p. 49–56.

Kollmeier, J.M., van der Pluijm, B.A., and Van der Voo, R., 2000, Analysis of Variscan dynamics; early bending of the Cantabria-Asturias arc, northern Spain: Earth and Planetary Science Letters, v. 181, p. 203–216.

Lamarche, G., and Lebrun, J.F., 2000, Transition from strike-slip faulting to oblique subduction: Active tectonics at the Puysegur Margin, South New Zealand: Tectonophysics, v. 316, p. 67–89.

Leblanc, M., and Lancelot, J.R., 1980, Interprétation géodynamique du domaine pan-africain (Précambrien terminal) de l'Anti-Atlas (Maroc) à partir de données géologiques et géochronologiques: Canadian Journal of Earth Sciences, v. 17, p. 142–155.

Lefort, J.P., 1989, Basement correlation across the North Atlantic: Berlin, Springer-Verlag, 148 p.

Lefort, J.P., Bois, C., Liewig, N., Peucat, J.J., Agarwal, B., and Gariel, O., 1997, Contribution of the ECORS Bay of Biscay deep seismic profile to the location of the southern Variscan front beneath the Aquitaine basin (France): Mémoire de la Société Géologique de France, v. 171, p. 79–96.

Martínez Catalán, J.R., Arenas, R., Díaz García, F., and Abati, J., 1999, Allochthonous units in the Variscan belt of northwest Iberia: terranes and accretionary history, *in* Sinha, A.K., ed., Basement Tectonics 13: Dordrecht, Netherlands, Kluwer, p. 65–84.

Mattauer, M., Proust, F., and Tapponier, P., 1972, Major strike-slip fault of Late Hercynian age in Morocco: Nature, v. 237, p. 160–162.

Matte, P., 1986, Tectonics and plate tectonics model for the Variscan Belt of Europe: Tectonophysics, v. 126, p. 329–374.

Matte, P., and Hirn, A., 1988, Seismic signature and tectonic cross-section of the Variscan crust in western France: Tectonics, v. 7, p. 141–155.

Matte, P., and Ribeiro, A., 1975, Forme et orientation de l'ellipsoide de déformation dans la virgation hercynienne de Galice: Relations avec le plissement et hypothèses sur la genèse de l'arc ibéro-armoricain: Comptes Rendus de l'Académie des Sciences Paris, v. 280 (series D), p. 2825–2828.

Nance, R.D., and Murphy, J.B., 1996, Basement isotopic signatures and Neoproterozoic paleogeography of Avalonian-Cadomian and related terranes in the circum-North Atlantic, *in* Nance, R.D., and Thompson, M.D., eds., Avalonian and related peri-Gondwanan terranes of the cirum-North Atlantic: Boulder, Colorado, Geological Society of America Special Paper 304, p. 333–346.

Ordóñez-Casado, B., 1998, Geochronological studies of the Pre-Mesozoic basement of the Iberian Massif: The Ossa Morena Zone and Allochthonous Complexes within the Central Iberian Zone [Ph.D. Thesis]: Zürich, Eidgenössische Technische Hochschule, v. 12940, 235 p.

Paris, F., and Robardet, M., 1990, Early Palaeozoic palaeobiogeography of the Variscan regions: Tectonophysics, v. 177, p. 193–213.

Pereira, M.F., and Silva, J.B., 2001, A new model for the Hercynian orogen of Gondwanan France and Iberia: Discussion: Journal of Structural Geology, v. 23, p. 835–838.

Perroud, H., 1986, Paleomagnetic evidence for tectonic rotations in the Variscan Mountain Belt: Tectonics, v. 5, p. 205–214.

Perroud, H., and Bonhommet, N., 1981, Paleomagnetism of the Ibero-Armorican arc and the Hercynian orogeny in Western Europe: Nature, v. 292, p. 445–448.

Peucat, J.J., and Bossière, G., 1991, Age U-Pb fini-hercynien de la ceinture mylonitique de haute pression-haute température en Grande Kabylie (Algérie): Comptes Rendus de l'Académie des Sciences Paris, v. 313 (série II), p. 1261–1267.

Peucat, J.J., Paris, F., and Chalet, M., 1986, U-Pb zircon dating of volcanic rocks close to the Silurian-Devonian boundary from Vendée (western France): Chemical Geology, v. 59, p. 133–142.

Piqué, A., Bossière, G., Bouillin, J.P., Chalouan, A., and Hoepffner, C., 1993, Southern margin of the Variscan belt: The northwestern Gondwana mobile zone (eastern Morocco and northern Algeria): Geologische Rundschau, v. 82, p. 432–439.

Piqué, A., Jeannette, D., and Michard, A., 1980, The Western Meseta shear zone, a major and permanent feature of the Hercynian belt in Morocco: Journal of Structural Geology, v. 2, p. 55–61.

Piqué, A., and Michard, A., 1989, Moroccan Hercynides: A synopsis. The Paleozoic sedimentary and tectonic evolution at the northern margin of West Africa: American Journal of Science, v. 289, p. 286–330.

Ries, A.C., Richardson, A., and Shackelton, R.M., 1980, Rotation of the Iberian arc: Palaeomagnetic results from north Spain: Earth and Planetary Science Letters, v. 50, p. 301–310.

Robardet, M., 2000, An alternative approach to consider the Variscan belt in southwest Europe: The pre-orogenic paleogeographical constraints, *in* Díaz García, F., et al., eds., Variscan-Appalachian dynamics: The building of the upper Paleozoic basement: A Coruña, Spain, Basement Tectonics 15, Program and abstracts, p. 23–26.

Robardet, M., Verniers, J., Feist, R., and Paris, F., 1994a, Le Paléozoïque anté-varisque de France, contexte paléogéographique et géodynamique: Géologie de la France, v. 3, p. 3–31.

Robardet, M., Bonjour, J.L., Paris, F., Morzadec, P., and Racheboeuf, P.R., 1994b, Ordovician, Silurian, and Devonian of the Medio-North-Armorican Domain, *in* Keppie, J.D., ed., Pre-Mesozoic Geology in France: Berlin, Springer-Verlag, p. 142–151.

Robardet, M., Paris, F., and Plusquellec, Y., 2001, Comment on "New Early Devonian paleomagnetic data from northwest France: Paleogeography and implications for the Armorican microplate hypothesis" by J. Tait: Journal of Geophysical Research, v. 106 (B7), p. 13,307–13,310.

Rolet, J., 1997, The concealed basement of Aquitaine: Mémoire de la Société Géologique de France, v. 171, p. 115–141.

Rolet, J., Gresselin, F., Jegouzo, P., Ledru, P., and Wyns, R., 1994, Intracontinental Hercynian events in the Armorican Massif, *in* Keppie, J.D., ed., Pre-Mesozoic geology in France: Berlin, Springer-Verlag, p. 195–219.

Scotese, C.R., and McKerrow, W.S., 1990, Revised world maps and introduction: Geological Society of London Memoir 12, p. 1–21.

Shelley, D., and Bossière, G., 1999, Ile de Groix: Retrogression and structural developments in an extensional régime: Journal of Structural Geology, v. 21, p. 1441–1455.

Shelley, D., and Bossière, G., 2000, A new model for the Hercynian orogen of Gondwanan France and Iberia: Journal of Structural Geology, v. 22, p. 757–776.

Shelley, D., and Bossière, G., 2001, The Ancenis terrane: An exotic duplex in the Hercynian belt of Armorica, western France: Journal of Structural Geology, v. 23, p. 1597–1614.

Shelley, D., and Bossière, G., 2002, A new model for the Hercynian orogen of Gondwanan France and Iberia: Reply (to Cartier et al.): Journal of Structural Geology, v. 24, p. 1665-1668.

Stampfli, G.M., 2000, Tethyan oceans, *in* Bozkurt, E., et al., eds., Tectonics and Magmatism in Turkey and the Surrounding Area: London, Geological Society of London Special Publications 173, p. 1–23.

Strachan, R.A., D'Lemos, R.S., and Dallmeyer, R.D., 1996, Neoproterozoic evolution of an active plate margin: North Armorican Massif, France, *in*

Nance, R.D., and Thompson, M.D., eds., Avalonian and related peri-Gondwanan terranes of the circum-North Atlantic: Geological Society of America Special Paper 304, p. 319–332.

Sutherland, R., 1999, Cenozoic bending of New Zealand basement terranes and Alpine fault displacement: A brief review: New Zealand Journal of Geology and Geophysics, v. 42, p. 295–301.

Tait, J., 1999, New Early Devonian paleomagnetic data from northwest France: Paleogeography and implications for the Armorican microplate hypothesis: Journal of Geophysical Research, v. 104, no. B2, p. 2831–2839.

Tait, J.A., Bachtadse, V., Franke, W., and Soffel, H.C., 1997, Geodynamic evolution of the European Variscan fold belt: Paleomagnetic and geological constraints: Geologische Rundschau, v. 86, p. 585–598.

Tikoff, B., de Saint Blanquat, M., and Teyssier, C., 1999, Translation and the resolution of the pluton space problem: Journal of Structural Geology, v. 21, p. 1109–1117.

Torsvik, T.H., 1998, Palaeozoic palaeogeography: A North Atlantic viewpoint: Journal of the Geological Society of Sweden (GFF), v. 120, p. 109–118.

Weil, A.B., Van der Voo, R., van der Pluijm, B.A., and Parés, J.M., 2000, The formation of an orocline by multiphase deformation: A paleomagnetic investigation of the Cantabria-Astruias arc (northern Spain): Journal of Structural Geology, v. 22, p. 735–756.

Manuscript Accepted by the Society February 8, 2002

Printed in the U.S.A.

Geological Society of America
Special Paper 364
2002

Transcurrent strain partitioning along a suture zone in the Maures massif (France): Result of eastern indenter tectonics in European Variscides?

J.-P. Bellot*
Le Bourg 40990, Gourbera, France
G. Bronner
Faculté des Sciences et Techniques de St.-Jérôme, Case 441, 13397 Marseille Cedex 20, France
C. Laverne
Lab. Pétrologie Magmatique, Faculté des Sciences et Techniques de St.-Jérôme, Case 441, 13397 Marseille Cedex 20, France

ABSTRACT

The ultramafic-rich shear zone from the Variscan Maures massif is a part of a linear suture zone that extends for 1000 km from Sardinia to the Bohemian massif. Multiscale analyses of ultramafic porphyroclasts and associated mantle structures, including analysis of ductile and associated ductile-brittle deformation, demonstrate the strong overprinting of this suture zone by wrench tectonics during continental collision. For 30 km along the ultramafic-rich shear zone, noncoaxial strain indicates at all scales a major sinistral transcurrent shearing with a heterogeneous, partitioned, and progressive strain. This transpressional stage occurs during the mid-Visean exhumation of crustal nappes. In this chapter we point out a model of oblique collision in this Variscan range resulting from eastern indenter tectonics in the European Variscides. The implicate microplates, as well as the irregular margin of continents, were strongly deformed during the convergence and diachronous collision between Baltica-Laurentia and Gondwana. The ultramafic-rich shear zone from the Variscan Maures massif is a well-defined example of an unrooted lithospheric suture zone in a collisional belt, overprinted by transcurrent tectonics after continental subduction and prior to late orogenic extension.

INTRODUCTION

In a collisional system, regional-scale fault zones spatially associated with ancient or present-day magmatic arcs provide a unique opportunity to address both tectonic and magmatic phenomena at the crustal scale. Spectacular natural examples, including the Alps (Mancktelow, 1992), Appalachians (Vauchez et al., 1993), Caledonides (Gilotti and Hull, 1991), Himalayas (Molnar and Tapponier, 1975), Tasminides (Glen, 1992), and Caucasus (Jackson, 1992), in addition to physical modeling (e.g., Pinet and Cobbold, 1992; Fossen et al., 1994; Pubellier and Cobbold, 1996), establish that oblique plate convergence often results in large-scale transpressional tectonics accompanied by major transcurrent faults.

In recent years significant advances have been made in the understanding of the Variscide volcanic-arc complexes, in par-

*E-mail: jpbellot@wanadoo.fr.

Bellot, J.-P., Bronner, G., and Laverne, C., 2002, Transcurrent strain partitioning along a suture zone in the Maures massif (France): Result of eastern indenter tectonics in European Variscides?, *in* Martínez Catalán, J.R., Hatcher, R.D., Jr., Arenas, R., and Díaz García, F., eds., Variscan-Appalachian dynamics: The building of the late Paleozoic basement: Boulder, Colorado, Geological Society of America Special Paper 364, p. 223–237.

ticular in the well-exposed southeastern Variscan belt where new petrological, geochemical, and structural data have been acquired (Laverne et al., 1997; Bellot et al., 1999, 2000a, 2000b; Bouloton et al., 1998). On the basis of these data, Bellot et al. (2000b) noted that the geodynamic setting of the Maures arc and the proto-Apulian plate collision has a number of similarities with arc-continent collisions currently taking place along major suture zones such as the Urals (Brown and Spadea, 1999) or the northern Variscan orogen (Brun and Burg, 1982; Burg et al., 1987). Nevertheless, a comprehensive model detailing the active geodynamic processes associated with the convergence and final continent-continent collision, and a comparison with those in currently active settings, has not been fully developed for the southern Variscan belt, particularly with respect to the deep levels of such orogens. Studies of the geometric and kinematic characteristics and modes of interaction among structures that have accommodated convergence within collisional systems are necessary to better understand their tectonic evolution.

The ultramafic-rich shear zone from the Variscan Maures massif is part of a linear suture zone that extends for 1000 km from Sardinia to the Bohemian massif. Its structural study provides an opportunity to examine the structural development at middle to deep crustal levels of a Paleozoic transpressional system, where an accretionary wedge is largely overprinted by intracontinental deformations. Based on multiscale analyses using mapping, field, and structural studies of ultramafic porphyroclasts and associated mantle structures, this chapter provides the first geometric and kinematic delimitations on this major suture zone in this part of the southern Variscan belt.

GEOLOGIC SETTING

Variscan belt of western Europe

The Maures massif belongs to the western European Variscan belt (Fig. 1), which resulted from the collision between the two major continents Laurentia-Baltica and Gondwana (e.g., Matte, 1986) and numerous microplates as Armorica and Avalonia (Matte, 2001). Oceanic areas present during Cambrian to Early Ordovician time (Ménot et al., 1988; Peucat et al., 1990; Pin, 1990) were closed by oceanic and subsequent continental subduction recorded by Silurian to Devonian high- to ultrahigh-pressure metamorphism (Pin and Peucat, 1986; Matte, 1998). Continent-continent collision was characterized by nappe tectonics under high- to medium-grade metamorphism (380–360 Ma; Matte, 1991), followed by transpressional deformation (360–340 Ma; Costa, 1992).

After continental subduction, combined thrusting and wrenching in the northern and the southern flank of the belt accommodated oblique collision between microplates, particularly along the Ibero-Armorican arc (Brun and Burg, 1982; Burg et al., 1981, 1987; Soper et al., 1992; Dias and Ribeiro, 1995). Well-developed ductile wrench faults with a dominant noncoaxial strain have been characterized, such as the dextral South Armorican shear zone and the sinistral Coimbra-Cordoba shear zone (Berthe et al., 1979; Burg et al., 1981; Bale and Brun, 1986). Combined structural and geochronological investigations on syntectonic granites along major shear zones allow the timing of deformation to be defined and demonstrate a progressive and partitioned noncoaxial strain produced during Silurian-Devonian arc-continent and subsequent continent-continent collision.

Maures massif

The study area, located in southeast France between the Alps and the Mediterranean Sea, contains exposures of metamorphic rocks that constitute part of the western European Variscan belt (Fig. 1A). The exposed Provencal basement is composed of two blocks (the Maures and the Tanneron massifs) of the same basement, separated by a west-east Permian graben. The Maures massif juxtaposes two well-contrasted blocks separated by the major Grimaud fault (Vauchez and Bufalo, 1985, 1988; Bronner, 1986). Variscan formations, including Stephanian deposits, are crosscut by regional left-lateral Permian wrench faults that strike 80°–90°.

The Eastern block is composed of migmatitic gneisses with a 20° striking vertical foliation, upright folding, and a horizontal lineation associated with a ductile sinistral movement. The Western block is composed of several metasedimentary formations (Gueirard, 1960; Seyler, 1986) metamorphosed under greenschist to amphibolite facies conditions, increasing in grade from west to east (Buscail and Leyreloup, 1999). They have an axial-plane foliation related to isoclinal folds locally overprinted by a secondary foliation, parallel to axial-planes of folds overturned to the east (Arthaud and Matte, 1966; Bronner et al., 1971; Seyler and Crévola, 1982). The foliation carries a 290°–310° striking stretching lineation, associated with top-to-the-northwest normal shearing interpreted as the result of extensional tectonics (Morillon et al., 2000) or as a sign of a polyphased nappe tectonics (Bellot and Bronner, 2000).

Ultramafic-rich suture zone

The ultramafic-rich suture zone is a heterogeneous mylonitic zone, interpreted as an accretionary wedge containing sheets of

Figure 1. General geological features of Maures massif with location of ultramafic-rich shear zone (URSZ) in relation to Hermittan granite and regional geologic features. A: Synthetic geological map after Gueirard (1960), Tempier (1978), Seyler (1986), and Buscail and Leyreloup (1999). Square inset shows area of Figure 2. B: Pre-Permian reconstituted URSZ. C: Synthetic cross section throughout Maures massif parallel to northwest-southeast stretching lineation. Upper left inset: Location in context of western European Variscan belt (after Matte, 2001). Suture zone in northern Sardinia is after Elter et al. (1990), Cappelli et al. (1992), and Carmignani et al. (1994). Variscan areas: G, Galicia; SB, South Brittany; MC, Massif central; V, Vosges massif; R, Rhenish massif; A, Ardennes; B, Bohemian massif; S, Sardinia; C: Corsica; M: Maures massif; VA, Variscan Alps.

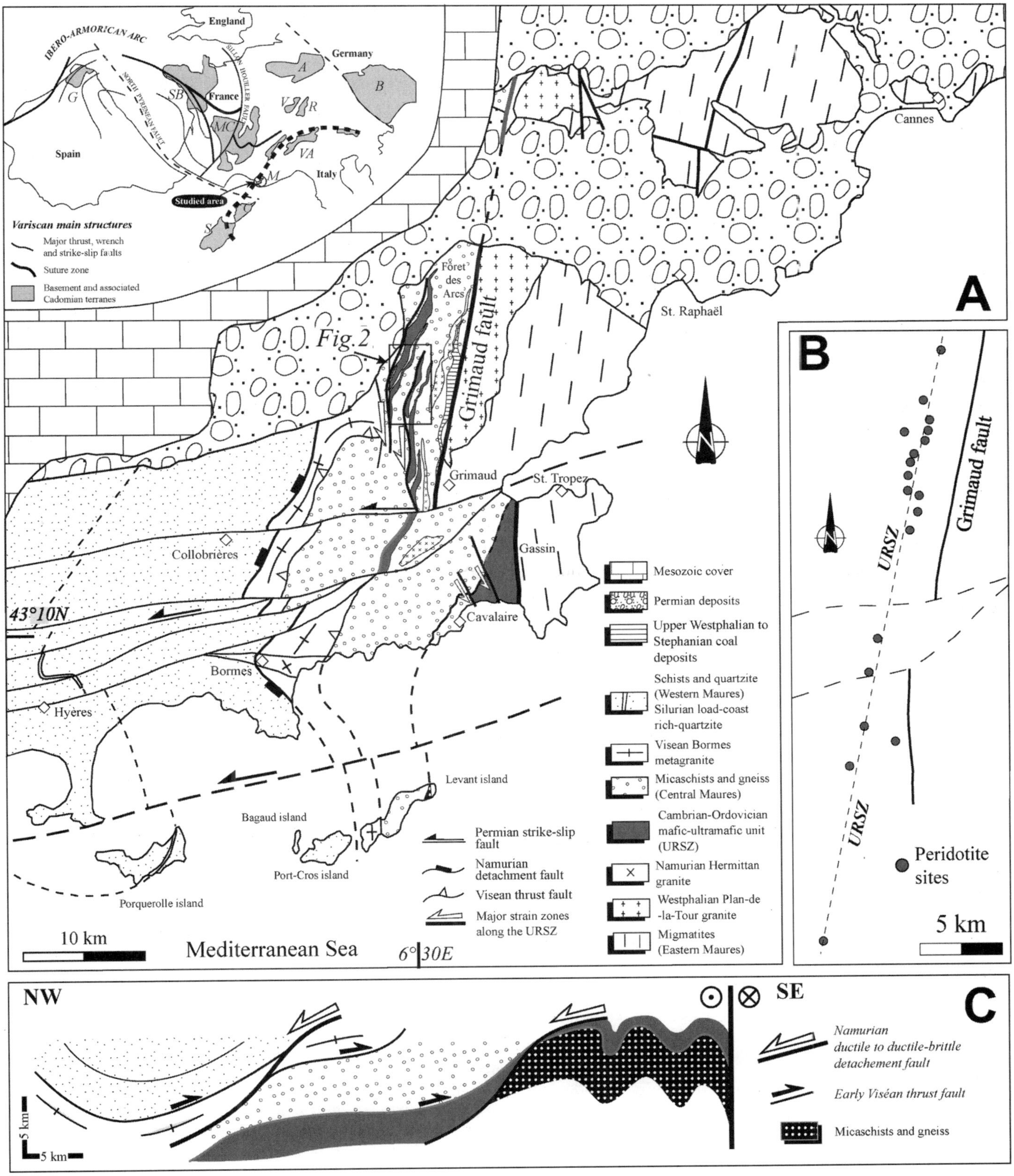

England
Germany
IBERO-ARMORICAN ARC
France
Spain
Italy
Studied area
Variscan main structures
Major thrust, wrench and strike-slip faults
Suture zone
Basement and associated Cadomian terranes
Cannes
St. Raphaël
Forêt des Arcs
Fig.2
Grimaud fault
Grimaud
St. Tropez
Gassin
Collobrières
43°10N
Cavalaire
Bormes
Hyères
Levant island
Bagaud island
Port-Cros island
Porquerolle island
10 km
Mediterranean Sea
6°30E
A
Mesozoic cover
Permian deposits
Upper Westphalian to Stephanian coal deposits
Schists and quartzite (Western Maures) Silurian load-coast rich-quartzite
Visean Bormes metagranite
Micaschists and gneiss (Central Maures)
Cambrian-Ordovician mafic-ultramafic unit (URSZ)
Namurian Hermittan granite
Westphalian Plan-de-la-Tour granite
Migmatites (Eastern Maures)
Permian strike-slip fault
Namurian detachment fault
Visean thrust fault
Major strain zones along the URSZ
B
URSZ
Grimaud fault
Peridotite sites
5 km
C
NW
SE
5 km
5 km
Namurian ductile to ductile-brittle detachement fault
Early Viséan thrust fault
Micaschists and gneiss

two contrasting relict lithospheric types (Bellot et al., 1999, 2000b): (1) lenses of boninitic-type ultramafic cumulates (Laverne et al., 1997; Bellot et al., 1999), alkaline and calc-alkaline orthogneiss (Seyler, 1986), ferro-gabbros and diorites of a supposed Lower Ordovician (500–460 Ma) arc, and (2) oceanic troctolites and gabbros (Bouloton et al., 1998) of supposed Late Cambrian ophiolitic members (ca. 500 Ma). Several types of metasedimentary rocks of unknown age are associated with the igneous rocks. Occurrences of garnet-peridotites, indicating pressure of 20–22 kbar, characterize the zone as a lithospheric-scale suture zone (Bellot et al., 2000b).

DUCTILE TRANSCURRENT FAULTING

Wrench shear zone

General features. Pre-Permian reconstitutions of the Maures massif reveal that the ultramafic-rich shear zone is a 50–250-m-thick, 50-km-long, linear shear zone (Fig. 1B) that is particularly well exposed in the northern part of the Maures massif. The ultramafic-rich shear zone is a tectonic melange of peridotites and metagabbro lenses dispersed in a layered matrix of gneiss and amphibolite. Anatectic sillimanite gneiss to the west and micaschists to the east are separated by a meter-thick high strain zone outlined by mafic and/or ultramafic lenses. The suture zone has a vertical to high, west-dipping foliation and a horizontal or gently north plunging mineral and stretching lineation (Figs. 2, 3, and 4), suggesting a predominantly strike-slip component. This subvertical shear zone contains well-preserved strain gradients, with relatively less deformed rocks grading into mylonites over a few meters or less. The metamorphic wall rocks have centimeter- to meter-wide high-strain zones with penetrative foliation and a poorly to moderately well defined stretching lineation. On approaching any of these shear zones, rocks grade from coarse-grained gneiss to finer grained straight and layered gneisses, amphibolites, annealed mylonitic metagabbros, and metaultramafic rocks. Mylonites zones range from a few meters to a few tens of meters thick.

S-L secondary fabric. The cores of ultramafic and mafic lenses within and adjacent to shear zones have a well-developed high-temperature S_1 foliation (Bellot et al., 2000b) oriented subparallel or oblique to the regional foliation (Figs. 2 and 4). This indicates that the regional foliation of gneiss and micaschists is clearly a secondary planar fabric (S_{1-2}) resulting from the progressive recrystallization during deformation of a previous foliation (S_1) preserved only in the lens cores. Amphibolites in strain shadows around ultramafic porphyroclasts present a composite S_{1-2} foliation, except in the contractional face of porphyroclasts, where the S_1 foliation is intensely reworked by disharmonic, decimeter- to meter-scale folds (as modeled by Mandal et al., 2000).

In the rims of lenses, macroscopic, vertical shear planes that are interpreted to be a secondary foliation plane largely overprint the S_1 foliation. In ultramafic mylonites, this S_2 foliation is defined by millimeter-scale compositional variations and the alignment of amphibole and chlorite planes, which wrap around olivine, pyroxene, and spinel porphyroclasts. In metagabbros, the S_2 foliation is characterized by the following features: (1) millimeter-scale bands rich in oriented amphibole ± biotite, (2) dynamically recrystallized quartz and plagioclase ribbons in flattened mineral aggregates, and (3) fractured pyroxene porphyroclasts.

In the field, mineral and stretching lineations associated with the S_2 foliation include: (1) the long axis of elongated serpentinite lenses, (2) meter to decameter-scale strain shadows of amphibolites around ultramafic porphyroclasts, (3) well-developed acicular centimeter-long amphibole, (4) millimeter-scale boudinaged pyroxene or amphibole crystals and elongated polygranular mineral aggregates of plagioclase in metagabbros, and (5) elongated quartz lenses or quartzofeldspathic aggregates in the most highly strained amphibole-rich orthogneiss.

Vertical and upright folding

In the cores of mafic and ultramafic lenses, two types of folds refolded the S_1 high-temperature foliation (Fig. 5). Asymmetric centimeter- to decimeter-scale Z-shaped folds, east-trending and verging to the east, present a north-south axis parallel to the stretching lineation, with a highly dipping to vertical axial-plane S_2 foliation. In addition, rare asymmetric, decimeter-scale, S-shaped folds with vertical axes are included in vertical boudins. An axial-plane foliation bears a subhorizontal lineation. All these post-S_{1-2} folds are preserved in the core of boudins with vertical XY planes. These have a common orientation of the finite strain ellipsoid, where the X and Z axes are horizontal and north-south and west-east oriented, respectively. This indicates a progressive and partitioned deformation that

Figure 2. A: Geologic map of northern part of studied shear zone. See Figure 1A for location. Foliation trajectories are drawn after general synthesis of our own data. Heavy white arrows indicate relative-transport direction along shear zone. Two major shear zones, north-northeast–south-southwest striking, are well identified. Eastern shear zone, 12 km long and 1 km thick, concentrated most of strike-slip deformation. It also includes numerous strain shadows of amphibolites around ultramafic porphyroclasts that defined asymmetrical mantle structures. Western shear zone appears as secondary branch connected to major Eastern shear zone. Amphibolites constitute large and regular bands, north-northeast–south-southwest striking, that did not favor formation of mantle structures during strike-slip shearing. However, along 6 km where Western fault zone is clearly identified, major asymmetrical structures are recognized. B: Equal-area, lower hemisphere, stereographic projections for structural data where contours are 1%, 2%, 4%, and 8% per 1% area. C: Tectonic model that favors a major strike-slip shearing is proposed in order to explain relationships between major and secondary shear zones, as well as foliation trajectories and asymmetrical macrostructures. D: Geological cross section throughout exposed shear zone. Right part crosscuts structures that are not represented in map in A, but drawn in Figure 1A. Strike-slip faulting, folding of S_{1-2} foliation, and granite emplacement during mid-Visean are coeval.

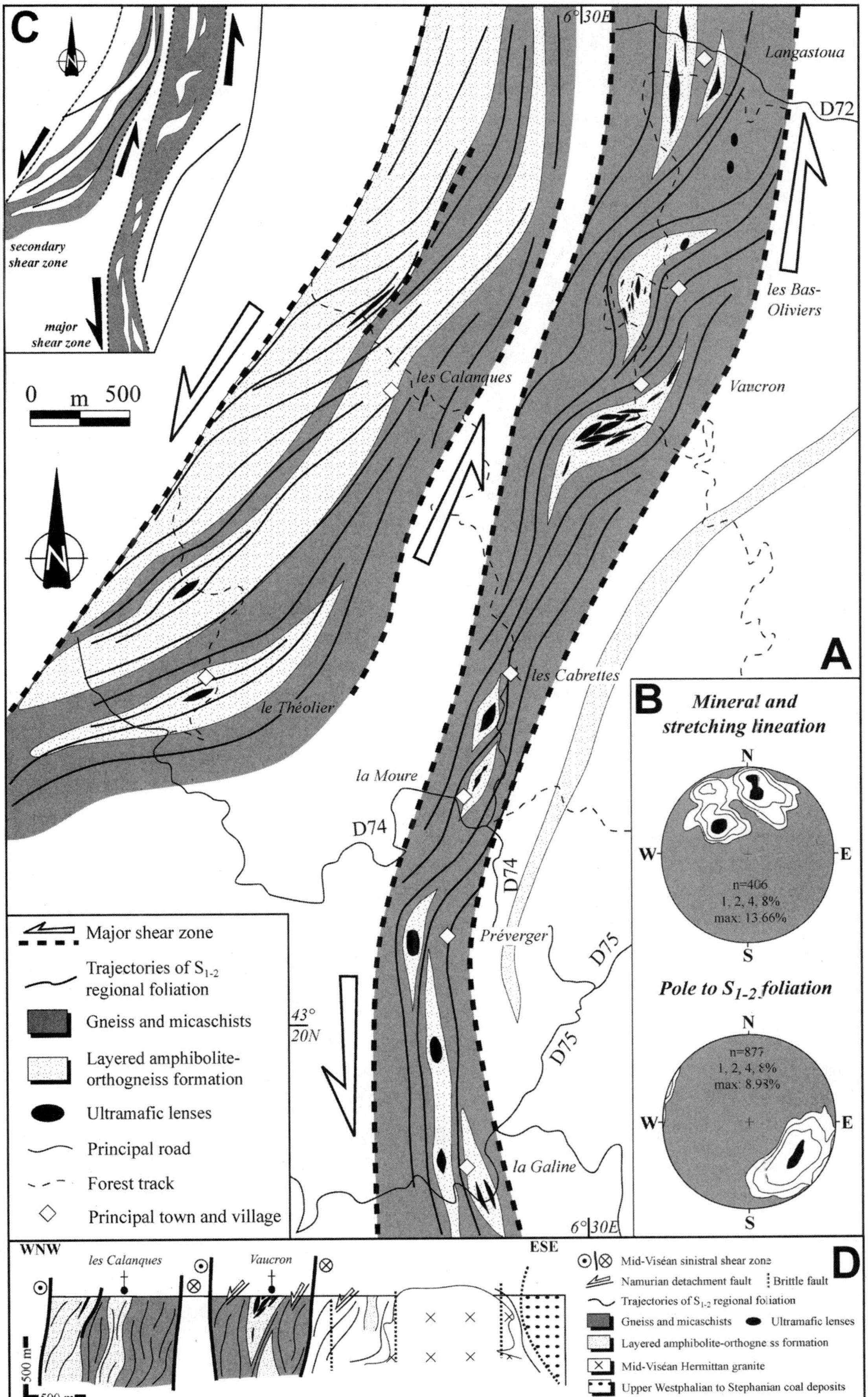

C
secondary shear zone
major shear zone
0 m 500
6°30E
Langastoua
D72
les Bas-Oliviers
Vaucron
les Calanques
les Cabrettes
le Théolier
la Moure
D74
D74
Préverger
D75
D75
43° 20N
la Galine
6°30E
A
B
Mineral and stretching lineation
N
W
E
S
n=406
1, 2, 4, 8%
max: 13.66%
Pole to S1-2 foliation
n=877
1, 2, 4, 8%
max: 8.93%
Major shear zone
Trajectories of S1-2 regional foliation
Gneiss and micaschists
Layered amphibolite-orthogneiss formation
Ultramafic lenses
Principal road
Forest track
Principal town and village
WNW
les Calanques
Vaucron
ESE
D
500 m
500 m
Mid-Viséan sinistral shear zone
Namurian detachment fault
Brittle fault
Trajectories of S1-2 regional foliation
Gneiss and micaschists
Ultramafic lenses
Layered amphibolite-orthogneiss formation
Mid-Viséan Hermittan granite
Upper Westphalian to Stephanian coal deposits

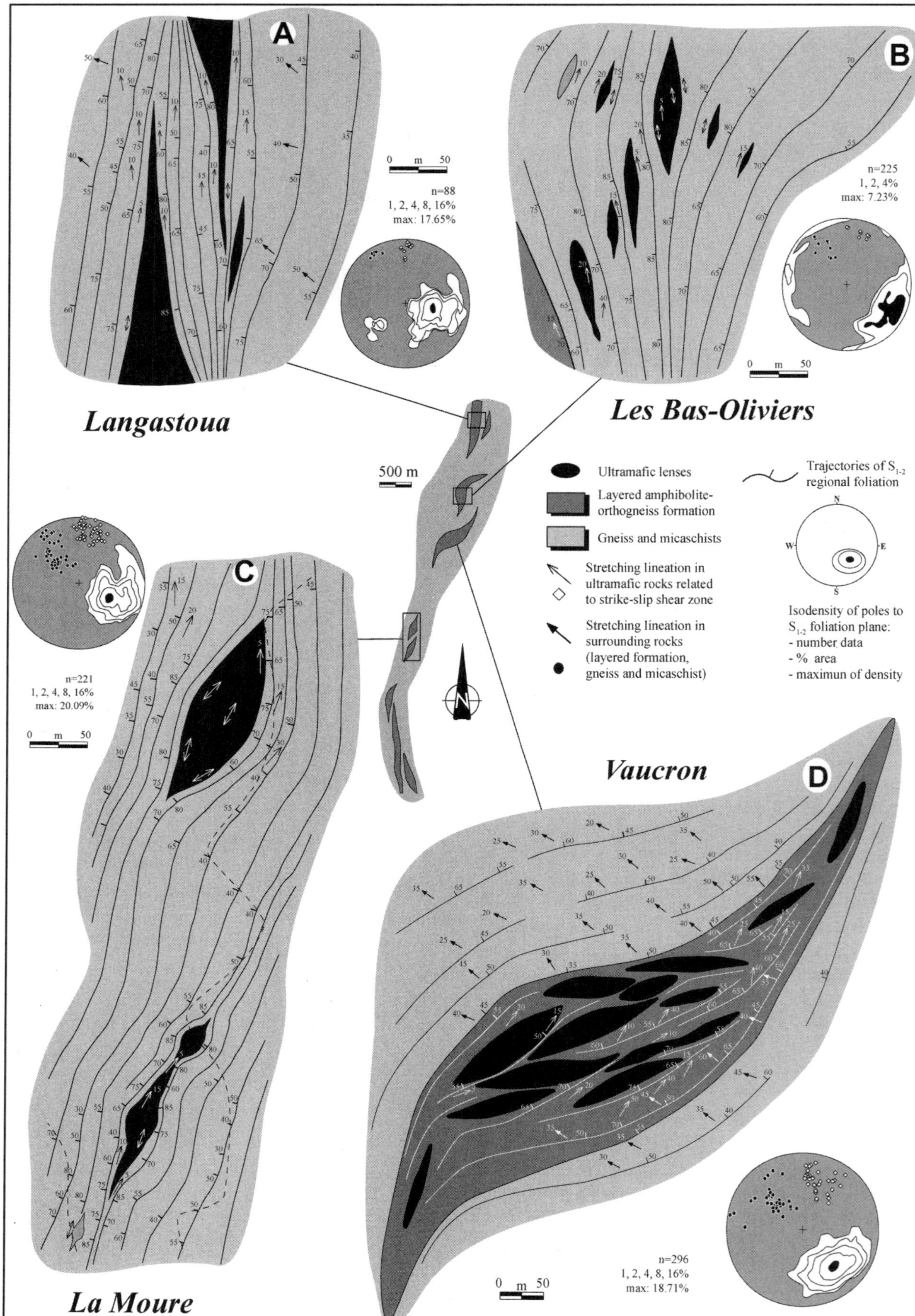
A
Langastoua
n=88
1, 2, 4, 8, 16%
max: 17.65%
0 m 50
B
Les Bas-Oliviers
n=225
1, 2, 4%
max: 7.23%
0 m 50
500 m
Ultramafic lenses
Layered amphibolite-orthogneiss formation
Gneiss and micaschists
Stretching lineation in ultramafic rocks related to strike-slip shear zone
Stretching lineation in surrounding rocks (layered formation, gneiss and micaschist)
Trajectories of S_{1-2} regional foliation
N
W
E
S
Isodensity of poles to S_{1-2} foliation plane:
- number data
- % area
- maximun of density
C
n=221
1, 2, 4, 8, 16%
max: 20.09%
0 m 50
La Moure
N
D
Vaucron
n=296
1, 2, 4, 8, 16%
max: 18.71%
0 m 50

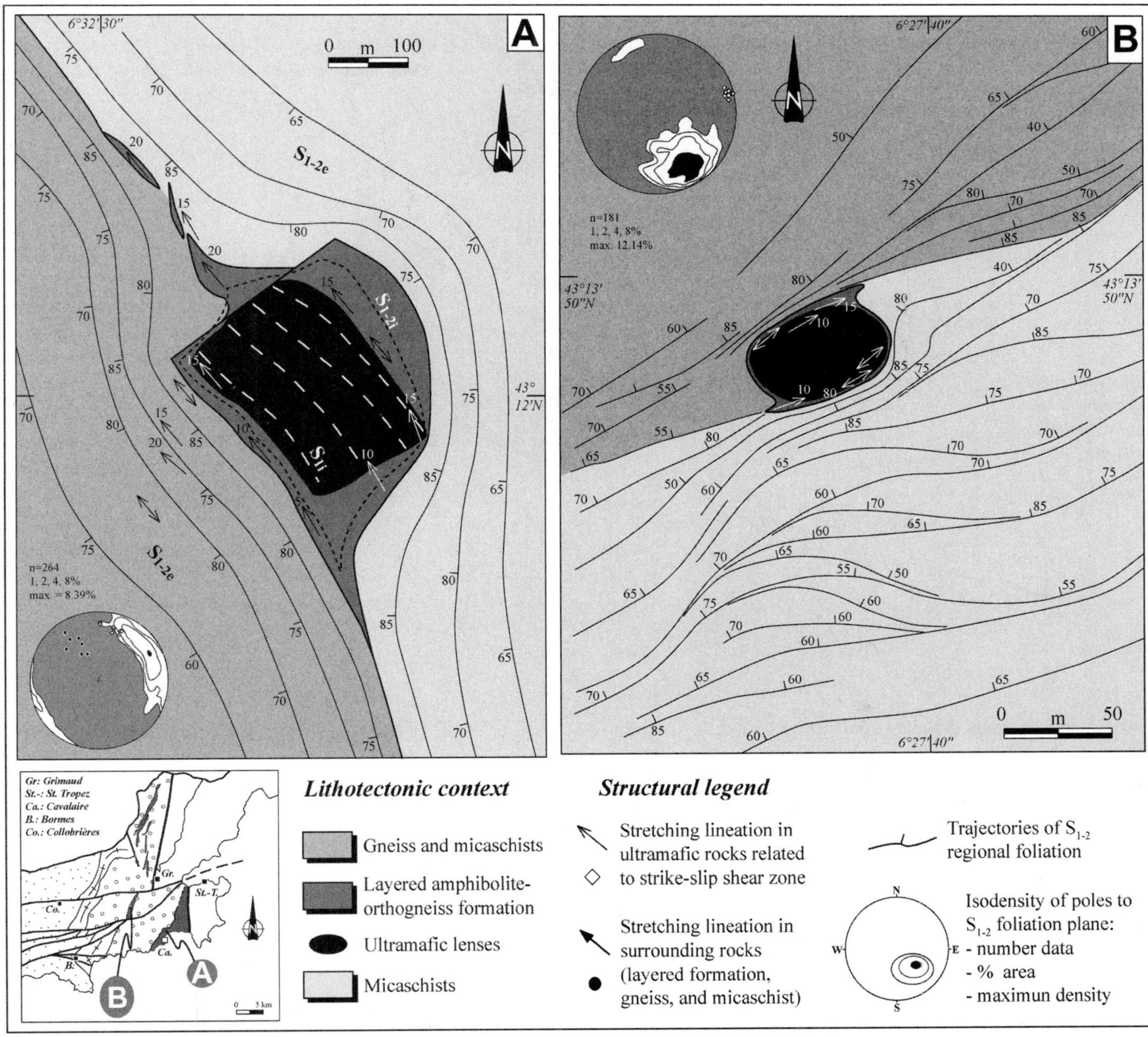

Figure 4. Geological map of two ultramafic megaporphyroclasts from (A) La Carrade and (B) Gorbières, and equal-area, lower hemisphere, stereographic projections for structural data. S_{1i} = internal high-temperature S_1 foliation in ultramafic rock; S_{1-2i} = internal composite S_{1-2} foliation in ultramafic rock; S_{1-2e} = external composite S_{1-2} foliation.

Figure 3. Detailed geological and structural maps of four ultramafic sites along peridotite-rich shear zone, from (A) Langastoua, (B) les Bas-Oliviers, (C) les Cabrettes, (D) Vaucron, and equal-area, lower hemisphere, stereographic projections for structural data. All detailed maps are located on general map simplified from Figure 2.

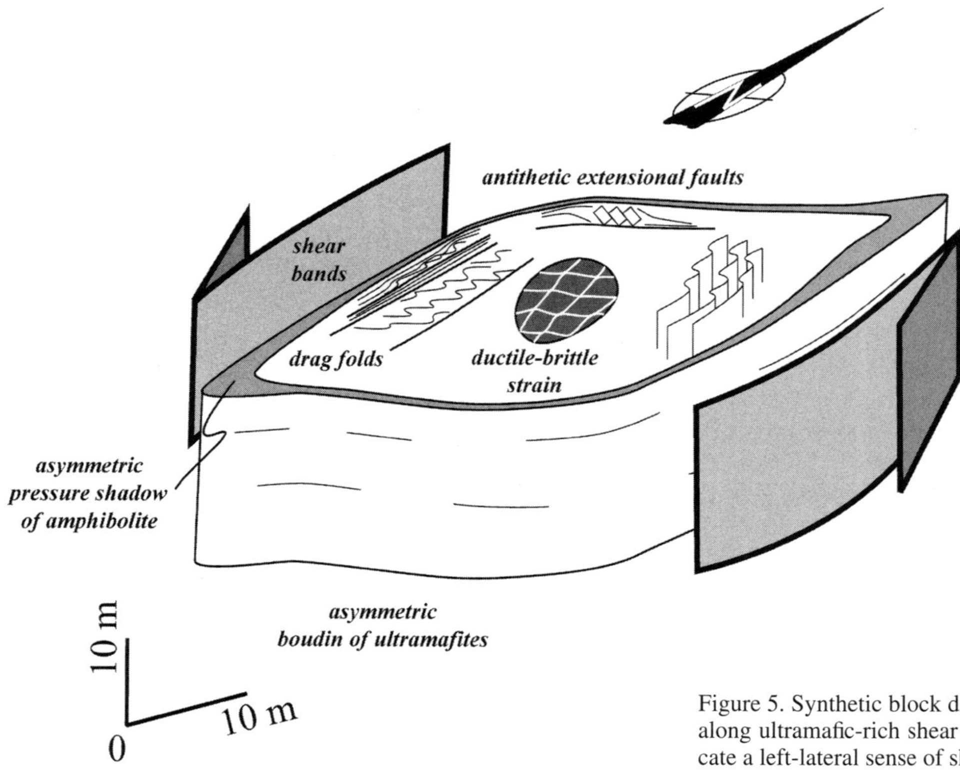

Figure 5. Synthetic block diagram of wrench deformations well defined along ultramafic-rich shear zone. Asymmetrical structures always indicate a left-lateral sense of shear.

preserved incremental stages in different samples of porphyroclasts during strike-slip movement. Due to deformation partitioning, several incremental stages of S_2 foliation formation may not have been preserved in individual samples.

Kinematic analyses

S-C fabric relationships (e.g., Berthé et al., 1979; Lister and Snoke, 1984) and/or asymmetric porphyroclasts (e.g., Simpson and Schmid, 1983) on surfaces oriented perpendicular to foliation and parallel to the stretching lineation were used to infer the relative-transport direction along the shear zone. Based on the gently north plunging orientation of lineation, a predominantly strike-slip sense of shear can be inferred. Kinematic analysis is approached by multiscale analyses of the studied structures.

Mapping analyses. The shear zone, with its west-dipping foliation and gently north plunging lineation, is ideally suited to carrying out a kinematic analysis in map view. At the map scale, the en echelon disposition of mafic-ultramafic lenses (Fig. 2) is typical of strike-slip faulting (i.e., Pubellier and Cobbold, 1996). This structure can be interpreted as regional-scale S/C structures associated with left-lateral kinematics. Ultramafic porphyroclasts surrounded by amphibolites form mantle structures (Passchier, 1994; Mandal et al., 2000): asymmetric strain shadows of syntectonic amphibolites at the La Carrade site (Fig. 4A) and asymmetric crystallization trails of dark and fine-grained amphibolites around megaporphyroclast and spectacular kilometer-scale S-C relationships or foliation fish features (Hanmer, 1986) at the Gorbière site (Fig. 4B) are consistent with left-lateral kinematics. The spectacular sigmoid from Vaucron is an asymmetric structure that reveals a noncoaxial strain associated with a sinistral sense of shear (Fig. 3D). Decameter-scale lenses characterize most of the ultramafic sites (Fig. 3, A and C): the regularly disposed lenses of Vaucron (Fig. 3D) contrast with the less regularly disposed lenses of the Bas-Oliviers area (Fig. 3B). Common symmetric strain shadows of amphibolites around more- or less-strongly elongated lenses, give evidence of a significant flattening component during deformation in the highest strain zones. However, rare asymmetric strain shadows of amphibolites around lenses, monoclinic boudins, and sigmoidal porphyroclasts indicate a noncoaxial deformation associated with a sinistral transcurrent movement (Fig. 5).

Mesoscopic analyses. On Levant Island (Fig. 1A), a mesoscopic kinematic analysis has been carried out on strongly deformed mafic and ultramafic tectonites. In the lens cores, the foliation of a highly sheared metagabbro, composed of white centimeter-scale layers of zoisite and decimeter-thick bands of amphibole, is folded by asymmetric, decimeter-scale, S-shaped folds with vertical axes (Fig. 5). In a highly strained zone, the foliation is crosscut by meter-long, centimeter-thick, extensional shear planes of synkinematic amphibolite and chlorite (Fig. 6A). These shear planes have a well-defined horizontal mineral lineation. Because the latter is not refolded, vertical post-S_{1-2} folds are interpreted as drag folds resulting from a

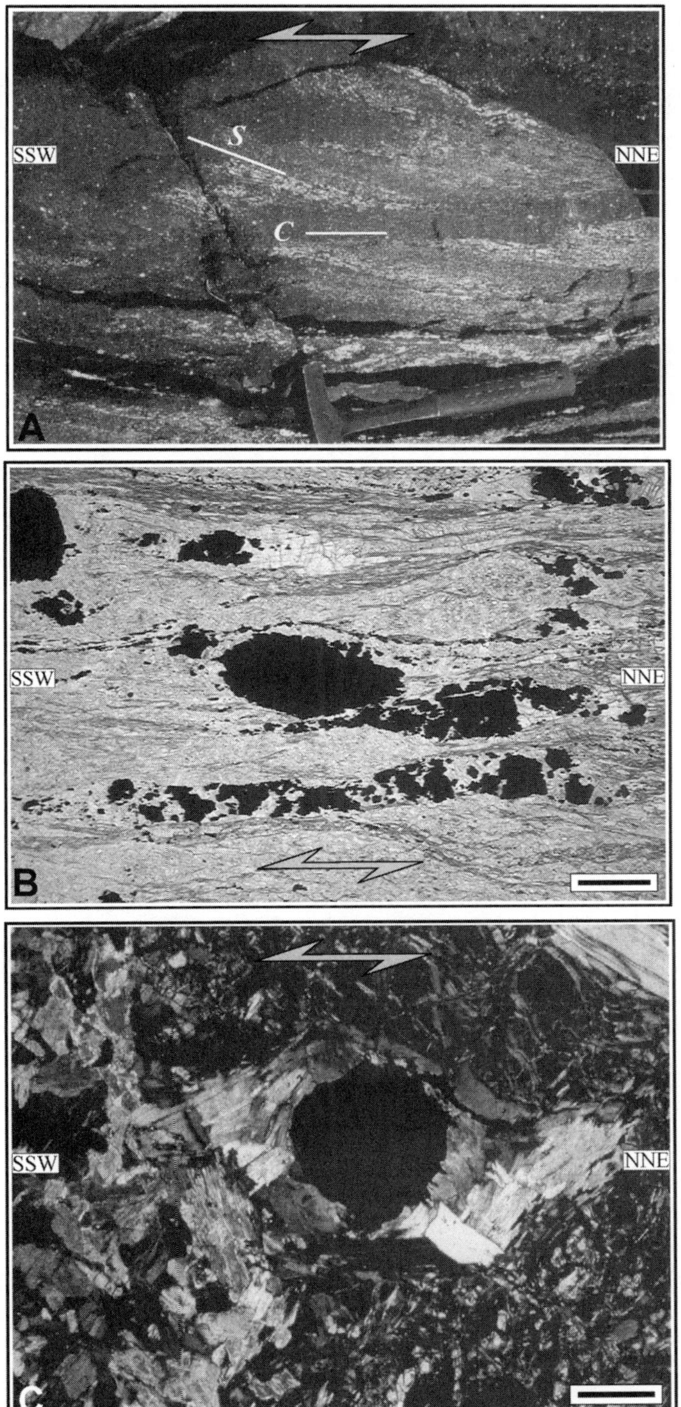

Figure 6. Kinematic criteria along ultramafic-rich shear zone, consistent with left-lateral movement. A: Meter-scale shear band crosscuts S_1 foliation of layered metagabbro (Levant Island). B: Asymmetric strain shadow of chlorite around chromite, chlorite-amphibole-magnetite shear bands, and sigmoidal fragmented and elongated chromite porphyroclast in mylonite. C: Asymmetric strain shadow of chlorite-amphibole around undeformed chromite in highly strained serpentinized metapyroxenite. Late amphibole-rich veins crosscuts foliation.

strike-slip fault. S-shaped folds are consistent with a sinistral sense of shear (Fig. 5). In white layers within metagabbros, intracrystal normal faults cut through 2-cm-long megacrysts of zoisite with asymmetric strain shadows of amphibole, also indicating a left-lateral kinematic sense (Fig. 5).

Microstructural analyses. The mineral recrystallizations during wrench deformation can be observed on XZ sections of the finite strain ellipsoid in mylonites. Common symmetric structures characterize an essentially coaxial strain: strain shadows of chlorite around chromian spinel and boudinaged olivine in dunites and pyroxenites, orthorhombic boudinaged chromite and symmetric strain shadows of amphibole around olivine in troctolite, and syntectonic ellipsoidal-shaped boudins of hornblendites. In the rim of sheared gabbroic lenses, layers of amphibole and plagioclase present an apparent symmetric fabric.

However, ultramafic tectonites also contain asymmetric microstructures consistent with a noncoaxial deformation associated with a sinistral component. Pyroxenites have rare highly strained zones with chlorite in strain shadows around chromite and a rare sigma-type system of chlorite around chromite (Fig. 6C). In dunite, microshear bands with fine layers of brown chlorite and magnetite crosscut the early foliation. Metatroctolites show edenitic-hornblende to hornblende and chlorite forming the mylonitic foliation, elongated and fragmented ferrochromite that occasionally forms an en echelon array, and chlorite in asymmetric strain shadows around pyroxene partly retrograded in hornblende (Bellot et al., 2000a) (Fig. 6B). In the high-strain zone, mylonites also attest to a noncoaxial strain with a sinistral shearing and show the following features: S-C relationships and penetrative shear planes are formed by centimeter-long amphibole (hornblende to actinolite) and ribbons of brown chlorite, fractured or elongated Cr-spinel, magnetite fine-grained levels, and boudinaged chromite with asymmetric strain shadows of chlorite (Fig. 6B).

Highly sheared gabbros contain a foliation formed by hornblende, biotite occurring as relict submillimeter crystals, sigmoidal hornblende, and late actinolite. Plagioclase crystals show a strong alignment. They are generally microfaulted, bent, and more commonly recrystallized to a fine-grained aggregate along the C′ plane. The preferred alignment of plagioclase could be of an early tectonic origin, overprinted later by subparallel dynamic solid-state recrystallization during regional deformation. In conclusion, syntectonic recrystallizations related to strike-slip shearing reveal a progressive and retrograde metamorphism from lower amphibolite to upper greenschist facies conditions.

DEFORMATION OF KILOMETER-SIZE PORPHYROCLASTS

Methodology

In shear zones, the various elements with their different rheology, dimensions, and compositions give rise to combine ductile and brittle deformation (Passchier and Trouw, 1996). A partitioned deformation is a common characteristic of ductile shear zones, and has been particularly well described along strike-slip fault zones. The detailed mapping of the ultramafic-rich shear zone reveals a heterogeneous composition, with kilometer-scale rigid bodies of dunites and pyroxenites in southern sites and decameter-scale, more plastic lenses of troctolites and plagioclase-rich gabbros in northern sites. All rigid clasts are included in a viscous matrix composed of layered gneiss and amphibolite. Depending on their composition and dimension, the noncoaxial strain in the shear zone produced a heterogeneous deformation: ultramafic porphyroclasts were fractured, whereas gabbroic lenses were transposed and dismembered into small lenses, and the layered matrix was intensely folded and boudinaged. We performed a kinematic analysis of ductile-brittle deformation of the largest ultramafic porphyroclast known in the ultramafic-rich shear zone (Fig. 4A). The ductile-brittle deformations were analyzed using the classical criteria (e.g., Petit, 1987).

Analysis and results

Detailed studies reveal that the D_1 fabric of dunite, pyroxenite, and gabbro is largely overprinted by an anastomosing network of faults occurring along well-defined north-trending lineaments. Most of the faults are vertical to subvertical, have prominent subhorizontal striae, and vary in length from meters to decameters. Striated fault planes form curved ductile-brittle shear surfaces that undulate and mantle ultramafic porphyroclasts. According to field analyses, a mesoscopic conjugate system of dominant north-northwest–trending left-separation and west-northwest–trending right-separation mesoscopic faults was identified within the mylonitic zone. This observation, together with a predominantly steeply dipping mylonitic foliation, is consistent with the shear zone using a sinistral fault zone. Using TectonicsFP software (Reiter and Acs, 2000), two distinct homogeneous populations of ductile-brittle faults were recognized. Each of them includes one family of faults with a predominant strike-slip component: the most abundant faults strike 155°–180° (average 65°–85°), but more rarely, well-defined faults strike between 105° and 120° (average 75°–90°). Both sets of faults display well-developed kinematic indicators on the fault surfaces, which indicate sinistral slip for the north-northwest–striking set and dextral slip for the west-northwest–striking set. In the field, ductile-brittle faults are crosscut by numerous brittle faults. A kinematic analysis of the mesoscopic

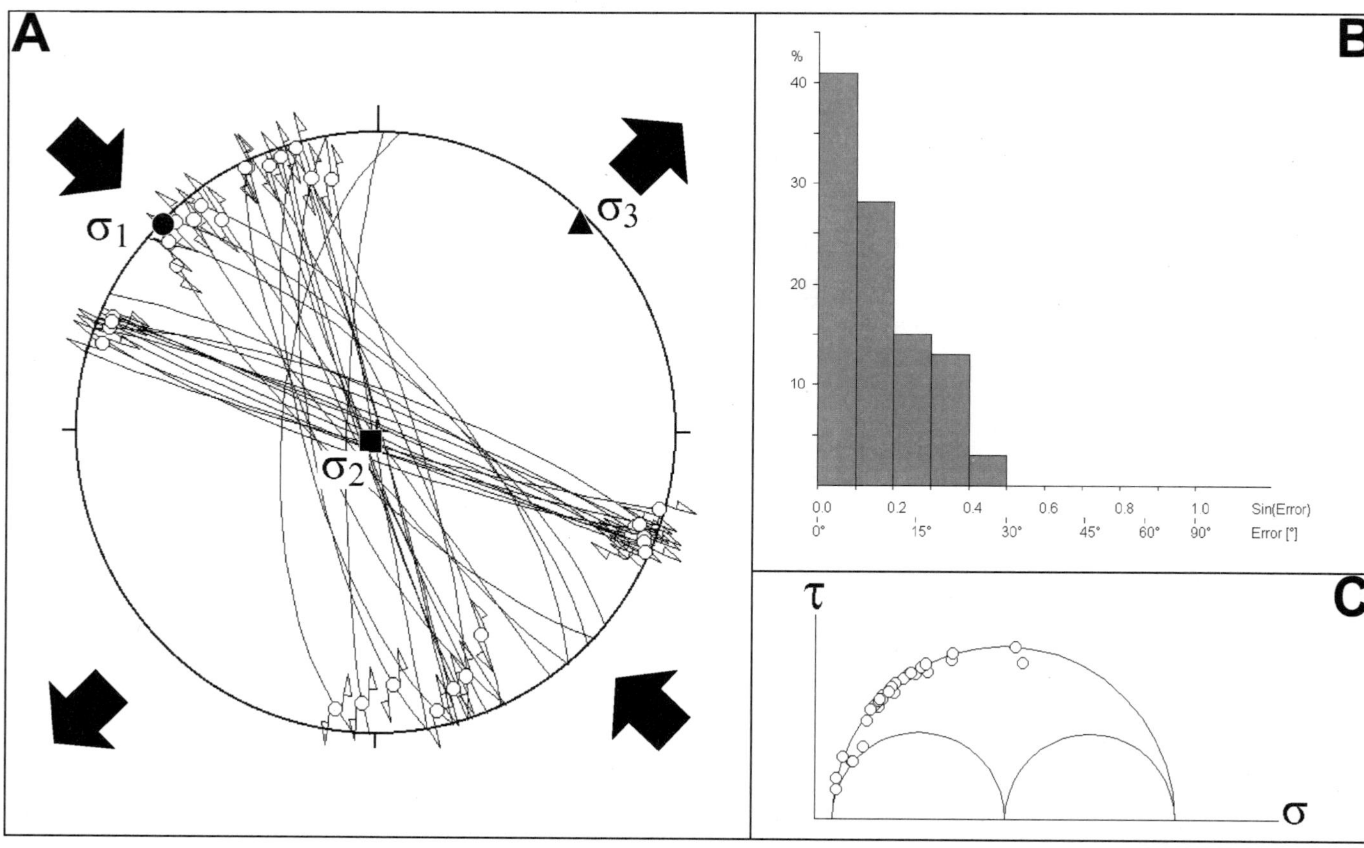

Figure 7. Orientation and kinematic data for ductile-brittle deformations analysis in ultramafic porphyroclasts at La Carrade and Gorbières, using TectonicsFP software (Reiter and Acs, 2000). A: Equal-area, lower hemisphere, stereographic projections for structural data. Arrows attached to fault traces correspond to measured slip vectors. Convergent large arrows give azimuths of computed maximum principal stress direction (σ1). B: Thick segments on fault traces show deviations between measured (s) and predicted (σ) slip vector on each plane. Corresponding histograms show number of faults (n) as function of difference in pitch between σ and s. C: Mohr diagram.

faults using TectonicsFP computer program is consistent with a north-northeast–trending sinistral strike-slip deformation zone for the first population of faults: σ_1 = 314°, 1°; σ_2 = 205°; 87°; σ_3 = 44°; 03° (Fig. 7). North-northwest–striking sinistral and west-northwest–striking dextral faults along with northeast-striking reverse faults are consistent with a horizontal maximum compressional stress direction trending northwest associated with an overall sinistral strike-slip tectonics.

DISCUSSION

Strain partitioning during retrograde metamorphism

We described a major sinistral wrench fault with a heterogeneous and partitioned strain during a retrograde medium-temperature–medium-pressure to low-temperature–low-pressure metamorphism for 30 km along the ultramafic-rich shear zone. Coexisting symmetric fabrics suggest that at least part of the deformation has a significant flattening component within most domains. The interpretation of symmetric structures, however, may not always be straightforward (e.g., Passchier and Trouw, 1996). Such deformation is typical of heterogeneous deformation along ductile shear zones with strain gradients (e.g., Brun and Burg, 1982; Burg et al., 1987). Ductile deformation almost characterizes metabasic and metapelitic rocks, whereas ductile-brittle deformation characterizes kilometer-size ultramafic porphyroclasts.

Timing of deformation

Structural analysis of the eastern part of the ultramafic-rich shear zone allows to define the timing of the regional deformation related to strike-slip shearing. In the southern part of the Central Maures, the Hermittan granite is a 20°-striking body located in the core of a 20°-striking, upright, regional-scale antiform that folds the regional S_1 foliation. U/Pb dating of zircon extracted from the syntectonic facies of the Hermittan granite has produced a date of 338 ± 6 Ma (Moussavou, 1998). This middle Visean age is interpreted as the age of crystallization of the granite. Because this antiform shows north-northeast–south-southwest longitudinal elongation and west-northwest–east-southeast shortening, and metapelites show a vertical axial-plane foliation of upright microfolds, it can be interpreted as the result of a west-northwest–east-southeast shortening during a compressive event.

The Hermittan antiform and the folds from the ultramafic-rich shear zone have several similarities: they (1) postdate early synmetamorphic events related to the lower Visean (ca. 345 Ma) by sedimentological analysis (Bronner, 1986; Bronner and Bellot, 2000); (2) have a similar finite strain ellipsoid characterized by a west-northwest horizontal shortening and a north-northeast horizontal elongation; and (3) predate ductile to ductile-brittle top-to-the-northwest normal shear bands related to the Westphalian (325–305 Ma) late orogenic extension (Morillon et al., 2000). Therefore, they both record the same middle Visean (340–330 Ma) transcurrent event.

In the Eastern Maures, a well-developed migmatization is dated as middle Visean by the U/Pb method on zircon extracted from the quartz-diorite from Reverdit (333 ± 4 Ma) and anatectic gneiss from Les Cavalières (334 ± 5 Ma) (Moussavou, 1998). Vauchez and Bufalo (1988) demonstrated that the migmatization is synchronous with a general left-lateral transcurrent tectonics associated with upright folding, which reworked an early foliation (Le Marrec, 1976). As a result, we interpret the middle Visean age in the Eastern Maures as the age of the sinistral transcurrent tectonics. By considering structural and geochronological data currently available on both sides of the Grimaud fault, the left-lateral transcurrent tectonics that largely affected the Maures units is related to the middle Visean.

Geodynamic significance of strike-slip shearing in the Maures massif

Structural data from the ultramafic-rich shear zone obtained in this study can be used to assess the influence of these parameters in the tectonics of (1) the Maures massif, (2) intracontinental deformation during a collisional system, and (3) the building of the western European Variscan belt.

Crustal-scale transcurrent tectonics in the Maures massif. The mid-Visean transcurrent tectonics occurs differently on both sides of the Grimaud fault: (1) dominant coaxial strain, with a major flattening increasing to the east, in the Western block, and (2) dominant noncoaxial strain with a constrictional strain ellipsoid in the Eastern block. A shift in the absorption of crustal shortening can be observed along the ultramafic-rich suture zone and the sinistral Grimaud fault that separates the two blocks. The Provencal basement is a 50-km-thick crustal block with a horizontally partitioned deformation resulting from a transpressional tectonics.

Intracontinental wrench fault overprinting an accretionary wedge. It is now established that intracontinental deformations, and particularly transcurrent shear zones, are typical results of incremental strain produced late during continent-continent oblique collision. Natural examples and analogue models reveal that thrusts and strike-slip faults are the characteristic features observed in terranes accreted during collision, as along the South Armorican shear zone (e.g., Pubellier and Cobbold, 1996; Mann et al., 1995; Stephens et al., 1996).

The exposed suture zone is interpreted as part of a large-scale thrust composed of a tectonic melange of sheets of island-arc and oceanic rock types (Bellot et al., 2000). The S_1 foliation and the northwest-southeast–trending lineation of ultramafites are interpreted as indicating stacking of the crustal nappes, with a probable northwest vergence (Arthaud and Matte, 1966; Bard and Caruba, 1981; Bellot and Bronner, 2000). The sinistral wrench fault partly overprinted the S_1 foliation. As a consequence, it can be interpreted as the reworking of the suture zone by a strike-slip shear zone during a collisional event. Despite differences in continental margin thickness, size, and geometry, some general phenomena seem to occur in continent-continent collision. En echelon and kilometer-scale boudins guided by large wrench faults within the accreted belt seem to form early in the collision zone as the accretion begins. According to the general migmatization synchronous with the left-lateral shearing, the major transpressional tectonics are related to the exhumation of crustal nappes during the late stage of the collision in this Variscan range. Because the lineation is not rigorously horizontal (north plunging) and the foliation not vertical (west dipping), the strike-slip shear zone shows a component of vertical movement, with a southeastward reversed sense. This inverted component of the sinistral transcurrent tectonics is the vertical movement that accommodates the exhumation of nappes.

Correlations with other Variscan areas. The position of the Maures massif is particular because it is a part of a north-south–trending segment in a general west-east elongated belt. Because the Maures massif is very small and isolated, it is rather difficult to establish its precise structural evolution. According to the most likely pre-Permian position based on geodynamic reconstitutions from paleomagnetic data (Edel, 2000), the Corsica-Sardinia belt trended north-south and probably continued into the Maures massif and the external crystalline massifs of the Western Alps (Arthaud and Matte, 1977). Based upon lithotectonic and tectonic correlations (Matte, 1991), paleogeographical reconstitutions (Tempier, 1978), and geochronological considerations, the Maures massif is correlated with the entire southern branch of the Paleozoic belt of western Europe, from the Corsica-Sardinia block to the Variscan Alps, and may be correlated with the Bohemian massif, as proposed by Matte (2001). The correlation between the provencal and the alpine basement is also suggested by recent gravimetric data (Masson et al., 1999) and supported by field observations (von Raumer and Neubauer, 1993; Matte, 1998). The comparison between northeastern Sardinia and the Maures massif, as discussed by Ricci and Sabatini (1978), Matte (1986), Vauchez and Bufalo (1988), Elter et al. (1990), and Onézime et al. (1999), allows us to emphasize the following similarities in structure and tectonic evolution: (1) the same metamorphic zonation, with a regional metamorphism grading from west to east from anchizonal to high-temperature amphibolite facies conditions increasing eastward (Arthaud and Matte, 1977; Buscail and Leyreloup, 1999); (2) the high-grade unit of the northernmost part of Sardinia, composed of granulitic gneisses, and mafic, ultramafic, and eclogitic rocks similar to the units of the Maures massif (Matte, 1991) and particularly to the ultramafic-rich shear zone; (3) the occurrence of relics of high-pressure granulite and eclogite facies metamorphism prior to regional metamorphism under intermediate pressure; (4) the chemical signature of oceanic gabbros (Ricci and Sabatini, 1978) and alkalic lavas (Seyler, 1986); (5) the good chronological correspondence of retrograde shear zones that have been active during the late Carboniferous (Elter et al., 1999); (6) S-type leucogranites, which are strictly associated with the main shear zone in the Variscan belt (Elter et al., 1999); and (7) major faults that tectonically controlled juxtaposition of domains with different metamorphic grades.

However, numerous problems appear with such a correlation because major facts are still unknown in eastern Variscides and particularly in the Maures massif. For example, the major vergence of nappes, the relations between tectonic processes and mineral recrystallizations during Carboniferous time, and the geodynamic evolution defined by geochronological data before the Namurian extension (Morillon et al., 2000) are still matters of debate in the Maures massif. As a consequence, the correlation between the ultramafic-rich shear zone from the Maures massif and an equivalent structure in northeastern Sardinia remains difficult to establish. Elter et al. (1990, 1999) pointed out the good relationship between the Posada-Asinara line in northern Sardinia (Elter et al., 1990; Cappelli et al., 1992; Carmignani et al., 1994) and the Grimaud strike-slip shear zone from the Maures massif (Vauchez and Bufalo, 1985, 1988; Onézime et al., 1999): they show ductile dextral kinematics, at middle to low temperatures, synchronous with syntectonic granite emplacement related to the late Carboniferous orogenic extension. Furthermore, they separate two contrasted metamorphic and lithological blocks, with high-pressure relics and migmatites to the east, and kyanite-biotite and/or staurolite-biotite micaschists to the west. However, the Grimaud fault has a polyphase tectonic history (Vauchez and Bufalo, 1988; Onézime et al., 1999), with mid-Visean left-lateral kinematics followed by Namurian right-lateral kinematics. It is surprising that such a sinistral transcurrent tectonics is unrecognized in Sardinia or in the Variscan Alps. Moreover, geometrical relationships suggest the rooting of the ultramafic-rich shear zone on the Grimaud fault zone. As a consequence, the ultramafic-rich shear zone of the Maures massif is correlated with the Posada-Asinara line in northern Sardinia.

Implication for the Variscan orogeny. Because the ultramafic-rich shear zone is probably part of a 1000-km-long suture zone, it is now possible to propose plate-scale geodynamic models involving an accretionary complex followed by continental collision. Major strike-slip ductile faults resulting from an oblique collision have been described in the Variscan belt, particularly around the Ibero-Armorican arc. The suture of the central part of the western European Variscan belt, partly overprinted by strike-slip fault, is a fundamental structure that should be included in the reconstitution of plate motions and the geodynamic history of the Variscan belt. Along the 20°-striking suture zone (Matte, 1991), early collision with a shortening in a west-northwest–east-southeast direction was accommodated initially by crustal thickening marked by westward thrust faults, whereas left-lateral transcurrent movement marked the waning stages of collision. Northwest-southeast thrusting and subsequent north-south–oriented sinistral wrenching may be well explained by eastern-southeast indenter tectonics that moved west-northwestward (Fig. 8A). The change of transverse lineation into longitudinal lineation was manifested by a plate-scale evolution from perpendicular to oblique collision during intracontinental deformation. Namurian dextral wrenching can be explained by a second phase of oblique collision characterized by southwestward indenter tectonics (Fig. 8B). Oblique collision may be due to a collisional effect on an irregular shape of the northern margin of the Gondwana block, as well as the implications of one or sev-

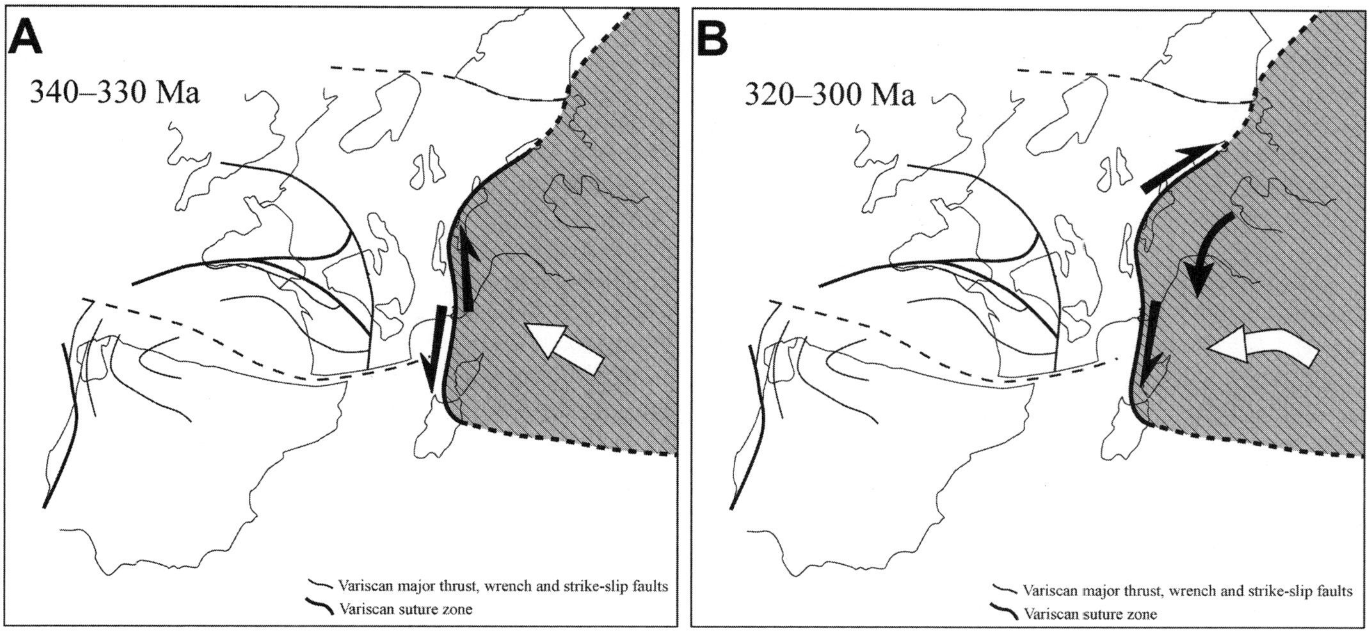

Figure 8. Plate-motion reconstruction of western European Variscides in order to illustrate oblique collision during possible eastern indenter tectonics, related to mid-Visean left-lateral shearing (A) and Namurian right-lateral shearing (B). Position of Sardinia-Corsica block is drawn after paleomagnetic data of Edel (2000), and orientation of main structures (Arthaud and Matte, 1977). Correlations between axial zone of northeastern Sardinia, Maures massif, Variscan Alps and Bohemian massif are explained in text.

eral microplates involved in the convergence of the Gondwana and Baltica-Laurentia blocks. Whatever the alternative hypothesis may be, the express changes in deformation in this part of the Variscan range occurred during the middle Carboniferous and almost certainly following the late Carboniferous.

CONCLUSION

We described a major transcurrent tectonics in the Maures massif as part of the southern European Variscan belt during middle Visean time. Signs of this major event are particularly well recorded along the ultramafic-rich shear zone, a good example of an unrooted lithospheric suture zone, overprinted by wrench tectonics after continental subduction and prior to late orogenic extension. In addition, we pointed out a model of oblique collision in this Variscan range, as the result of eastern indenter tectonics in the European Variscides. The general north-south convergence between the Baltica-Laurentia and Gondwana blocks (Matte, 2001) induces diachronous and oblique accretions and collisions, which favor crustal-scale transcurrent tectonics at the plate boundaries.

ACKNOWLEDGMENTS

This work has been supported by the Bureau de Recherches Géologiques et Minières through the financial support on a regional project on the Maures massif. We are particularly indebted to P. Rossi and C. Delor for their scientific support and their confidence about our work. We thank the staff of the Laboratoire de Pétrologie Magmatique (University of Marseille, France), where this work was done and thin sections studied over six months. We also thank Delphine Bellot for moral support during the field work; Rodolfo Carosi and an anonymous reviewer for very constructive comments and helpful suggestions on the manuscript; and David Vanko for helpful comments.

REFERENCES CITED

Arthaud, F., and Matte, P., 1966, Contribution à l'étude des tectoniques superposées dans la chaîne hercynienne: Étude microtectonique des séries métamorphiques du massif des Maures (Var): Paris, Comptes Rendus de l'Académie des Sciences, v. 262, p. 436–439.

Arthaud, F., and Matte, P., 1977, Synthèse provisoire sur l'évolution tectonique et les raccords entre les segments hercyniens situés autour du bassin Nord-Baléares (Sud de la France, Espagne, bloc Corso-Sarde): Rennes, France, Colloque International Centre National de la Recherche Scientifique, v. 243, p. 497–513.

Bale, P., and Brun, J.-P., 1986, Les complexes métamorphiques du Léon (NW Bretagne): un segment du domaine éo-hercynien sud armoricain translaté au Dévonien: Bulletin de la Société Géologique de France, v. 8, p. 471–477.

Bard, J.-P., and Caruba, C., 1981, Les séries leptyno-amphiboliques à éclogites relictuelles et serpentinites des Maures, marqueurs d'une paléosuture varisque affectant une croûte amincie?: Paris, Comptes Rendus de l'Académie des Sciences, v. 292, p. 611–614.

Bellot, J.-P., and Bronner, G., 2000, Are southeastward shearing in the Maures massif witness of syn-exhumation nappe tectonics?: Paris, Comptes Rendus de l'Académie des Sciences, v. 331, p. 659–665.

Bellot, J.-P., Laverne, C., and Bronner, G., 1999, Did ultrabasites from the hercynian Maures massif (France) originate in an arc context? Petrographical and geochemical evidences [abs.]: EUG10, Strasbourg, 28 Mars to 1 April, Journal of Conferences Abstract, Oxford, v. 4, p. 786.

Bellot, J.-P., Bronner, G., and Laverne, C., 2000a, Finite strain analysis and signification of ultramafic lenses from the Western Maures (SE France). Geodynamic implications: Paris, Comptes Rendus de l'Académie des Sciences, v. 331, p. 803–809.

Bellot, J.-P., Bronner, G., Triboulet, C., and Laverne, C., 2000b, Crust-mantle boundary connected with large crustal-scale shearing in the Variscan belt. Petrological, structural and metamorphic evidences from the Maures massif (SE France) [abs.]: Variscan-Appalachian dynamics: The building of the Upper Paleozoic basement: A Coruña, Spain, Basement Tectonics 15, Program and Abstract, p. 59–60.

Berthe, D., Choukroune, P., and Jegouzo, P., 1979, Orthogneiss, mylonite and noncoaxial deformation of granites: The example of the South Armorican Shear Zone: Journal of Structural Geology, v. 1, p. 31–42.

Bouloton, J., Goncalves, P., and Pin, C., 1998, Le pointement de péridotite à grenat-spinelle de La Croix-Valmer (Maures centrales): Un cumulat d'affinité océanique impliqué dans la subduction éohercynienne?: Paris, Comptes Rendus de l'Académie des Sciences, v. 326, p. 473–477.

Bronner, G., 1986, De schiste et d'eau, archipel d'Hyères: Cahier de Découverte-Parc National de Port-Cros, 84 p.

Bronner, G., and Bellot, J.-P., 2000, The inverted nodule-rich Bagaud-Malalongue Formation: An early Carboniferous synorogenic flysch in the Maures massif (SE France). Paleogeographic, structural and geodynamic implications [abs.]: Variscan-Appalachian dynamics: The building of the Upper Paleozoic basement: A Coruña, Spain, Basement Tectonics 15, Program and Abstract, p. 73–74.

Bronner, G., Lécorché, J.-P., and Orsini, J.-B., 1971, Un pli cônique kilométrique: L'île de Porquerolles, fragment méridional du Massif hercynien des Maures (Var, France): Paris, Comptes Rendus de l'Académie des Sciences, v. 272, p. 20–23.

Brown, D., and Spadea, P., 1999, Processes of forearc and accretionary complex formation during arc-continent collision in the southern Ural Mountains: Geology, v. 27, p. 649–652.

Brun, J.P., and Burg, J.-P., 1982, Combined thrusting and wrenching in the Ibero-Armorican arc: A corner effect during continental collision: Earth and Planetary Sciences Letter, v. 61, p. 319–332.

Burg, J.-P., Iglesias, M., Laurent, P., Matte, P., and Ribeiro, A., 1981, Variscan intra-continental deformation: the Coimbra-Cordoba shear zone (SW Iberian Peninsula): Tectonophysics, v. 78, p. 161–177.

Burg, J.-P., Bale, P., Brun, J.P., and Girardeau, J., 1987, Stretching lineation and transport direction in the Ibero-Armorican arc during the siluro-devonian collision: Geodinamica Acta, v. 1, p. 71–87.

Buscail, F., and Leyreloup, A.F., 1999, The collisional regional metamorphism in the Maures and Tanneron (south of France) area. A critical review [abs.]: EUG10, Strasbourg, 28 Mars to 1 April, Journal of Conferences Abstract, Oxford, v. 4, p. 790.

Cappelli, B., Carmignani, L., Castorina, F., Di Pisa, A., Oggiano, G., and Petrini, R., 1992, A hercynian suture zone in Sardinia: Geological and geochemical evidence: Geodinamica Acta, v. 7, p. 31–43.

Carmignani, L., Carosi, R., Di Pisa, A., Gattiglio, M., Musumeci, G., Oggiano, G., and Pertusati, P.C., 1994, The hercynian chain in Sardinia (Italy): Geodinamica Acta, v. 7, p. 31–47.

Costa, S., 1992, East-west diachronism of the collisional stage in the french Massif Central: Implications for the European Variscan Orogen: Geodinamica Acta, v. 5, p. 51–68.

Dias, R., and Ribeiro, A., 1995, The Ibero-Armorican Arc: A collision effect against an irregular continent?: Tectonophysics, v. 246, p. 113–128.

Edel, J.-B., 2000, Hypothesis for the late Variscan clockwise rotation of the Maures-Estérel-Corsica-Sardinia block. New paleomagnetic data from the Plan-de-la-Tour basin (Maures, Provence) and review of the previous data: Géologie de la France, no. 1, p. 3–19.

Elter, F.M., Musumeci, G., and Pertusati, P.C., 1990, Late hercynian shear zones in Sardinia: Tectonophysics, v. 176, p. 397–404.

Elter, F.M., Faure, M., Ghezzo, C., and Corsi, B., 1999, Late Hercynian shear zones in northeast Sardinia (Italy): Géologie de la France, v. 2, p. 3–16.

Fossen, H., Tikoff, T.B., and Teyssier, C.T., 1994, Strain modeling of transpressional and transtensional deformation: Norsk. geol. Tidsskr, v. 74, p. 134–145.

Gilotti, J.A., and Hull, J.M., 1991, A new tectonic model for the central Scandinavian Caledonides: Terra Nova, v. 3, p. 18.

Glen, R.A., 1992, Thrust, extensional, and strike-slip tectonics in an evolving Paleozoic orogen-a structural synthesis of the Lachlan orogen of the southeastern Autralia: Tectonophysics, v. 214, p. 341–380.

Gueirard, S., 1960, Description pétrographique et zonéographique des schistes cristallins des Maures (Var) [Thesis]: Marseille, France, University of Aix-Marseille, 194 p.

Hanmer, S., 1986, Asymmetrical pull-aparts and foliation fish as kinematic indicators: Journal of Structural Geology, v. 8, p. 112–122.

Jackson, J., 1992, Partitioning of strike-slip and convergent motion between Eurasia and Arabia in Eastern Turkey and the Caucasus: Journal of Geophysical Research, v. 97, p. 12 471–12 479.

Le Marrec, A., 1976, Reconnaissance pétrographique et structurale des formations cristallophylliennes catazonales du massif de Sainte-Maxime (quart NE du massif varisque des Maures, Var, France) [Thesis]: Marseille, France, University of Aix-Marseille, 126 p.

Laverne, C., Bronner, G., and Bellot, J.-P., 1997, Did the ultrabasites from the Hercynian Maures massif (Var), originate in a fore-arc basin? Petrological and mineralogical evidences: Paris, Comptes Rendus de l'Académie des Sciences, v. 325, p. 765–771.

Lister, G.S., and Snoke, A.W., 1984, S-C mylonites: Journal of Structural Geology, v. 6, p. 617–638.

Mancktelow, N.S., 1992, Neogene lateral extension during convergence in the Central Apls: Evidence from interrelated faulting and back-folding around the Simplon pass (Switzerland): Tectonophysics, v. 215, p. 295–317.

Mandal, N., Samanta, S.K., and Chakraborty, C., 2000, Progressive development of mantle structures around elongate porphyroclasts: Insights from numerical models: Journal of Structural Geology, v. 22, p. 993–1008.

Mann, P., Taylor, F.W., Lawrence Edwards, R., and Teh-Lung, K., 1995, Actively evolving microplate formation by oblique collision and sideways motion along strike-slip faults: An example from the northeastern Caribbean plate margin: Tectonophysics, v. 246, p. 1–69.

Masson, F., Verdun, J., Bayer, R., and Debeglia, N., 1999, Une nouvelle carte gravimétrique des Alpes et ses conséquences structurales et tectoniques: Paris, Comptes Rendus de l'Académie des Sciences, v. 329, p. 865–871.

Matte, P., 1986, Tectonics and plate tectonics model for the Variscan Belt of Europe: Tectonophysics, v. 126, p. 329–374.

Matte, P., 1991, Accretionary history and crustal evolution of the Variscan belt in Western Europe: Tectonophysics, v. 196, p. 309–337.

Matte, P., 1998, Continental subduction and exhumation of HP rocks in Paleozoic belts: Uralides and Variscides: Geological Society of Sweden, v. 120, p. 209–222.

Matte, P., 2001, The variscan collage and orogeny (480–290 Ma) and the tectonic definition of the Armorica microplate: A review: Terra Nova, v. 13, p. 122–128.

Ménot, R.P., Peucat, J.J., Scarenzi, D., and Piboule, M., 1988, 496 My age of plagiogranites in the Chamrousse ophiolite complex (external Crystalline massifs in the French Alps): Evidence of a lower Paleozoic oceanization: Earth and Planetary Sciences Letter, v. 88, p. 82–92.

Molnar, P., and Tapponier, P., 1975, Cenozoic tectonics of Asia: Effects of a continental collision: Science, v. 189, p. 419–426.

Morillon, A.-C., Féraud, G., Sosson, M., Ruffet, G., Crévola, G., and Lerouge, G., 2000, Diachronous cooling on both sides of a major strike-slip fault in the Variscan Maures massif (SE France), as deduced from a detailed $^{40}Ar/^{39}Ar$ study: Tectonophysics, v. 321, p. 103–126.

Moussavou, M., 1998, Contribution à l'histoire thermo-tectonique varisque du massif des Maures par la typologie des zircons et la géochronologie U/Pb sur minéraux accessoires [Ph.D. thesis]: Montpellier, France, University of Montpellier.

Onézime, J., Faure, M., and Crévola, G., 1999, Étude pétro-structurale du complexe granitique Rouet—Plan-de-la-Tour (massifs des Maures et du Tanneron occidental), Var: Paris, Comptes Rendus de l'Académie des Sciences, v. 328, p. 773–779.

Passchier, C.W., and Trouw, R.A.J., 1996, Microtectonics: Berlin, Springer-Verlag, 289 p.

Passchier, C.W., 1994, Mixing in flow perturbations: A model for development of mantle porphyroclasts in mylonites: Journal of Structural Geology, v. 16, p. 733–736.

Petit, J.P., 1987, Criteria for the sense of movement on fault surfaces in brittle rocks: Journal of Structural Geology, v. 9, p. 597–608.

Peucat, J.J., Bernard-Griffiths, J., Gil Ibarguchi, J.I., Dallmeyer, R.D., Menot, R.P., Cornichet, J., and Iglesias Ponce De Leone, M., 1990, Geochemical and geochronological cross section of the deep Variscan crust: The Cabo Ortegal high-pressure nappe (northwestern Spain): Tectonophysics, v. 177, p. 263–292.

Pin, C., and Peucat, J.J., 1986, Ages des épisodes de métamorphisme paléozoïques dans le Massif Central et le Massif armoricain: Bulletin de la Société Géologique de France, no. 3, 461–469.

Pin, C., 1990, Variscan oceans: Ages, origins and geodynamic implications inferred from geochemical and radiometric data: Tectonophysics, v. 177, p. 215–227.

Pinet, N., and Cobbold, P.R., 1992, Experimental insights into the partitioning of motion within zones of oblique subduction: Tectonophysics, v. 206, p. 371–388.

Pubellier, M., and Cobbold, P.R., 1996, Analogue models for the transpressional docking of volcanic arcs in the Western Pacific: Tectonophysics, v. 253, p. 33–52.

Reiter, F., and Acs, P., 2000, TectonicsFP Demo-Version 1.6, build 1.155 (http://go.to/TectonicsFP).

Ricci, C.A., and Sabatini, G., 1978, Petrogenetic affinity and geodynamic significance of metabasic rocks from Sardinia, Corsica and Provence: Neues Jahrb Mineral Mh, v. 1, p. 23–38.

Seyler, M., and Crévola, G., 1982, Mise au point sur la structure et l'évolution géodynamique de la partie centrale du Massif des Maures: Paris, Comptes Rendus de l'Académie des Sciences, v. 295, p. 243–246.

Seyler, M., 1986, Petrology and genesis of Hercynian alkaline orthogneiss from Provence, France: Journal of Petrology, v. 27, p. 1229–1251.

Simpson, C., and Schmid, S.M., 1983, An evaluation of criteria to deduce the sense of movement in sheared rocks: Geological Society of America Bulletin, v. 94, p. 1281–1288.

Soper, N.J., Strachan, R.A., Holdsworth, R.E., Gayer, R.A., and Greiling, R.O., 1992, Sinistral transpression and the Silurian closure of Iapetus: Journal of the Geological Society of London, v. 149, p. 871–880.

Stephens, M.B., Wahlgren, C.-H., Weijermars, R., and Cruden, A.R., 1996, Left-lateral transpressive deformation and its tectonic implications, Sveconorwegian orogen, Baltic Shield, southwestern Sweden: Precambrian Research, v. 79, p. 261–279.

Tempier, C., 1978, Les événements calédoniens dans les massifs varisques du Sud-Est de la France, Corse et Sardaigne: Ottawa, Ontario, Geological Survey of Canada Paper 22, p. 177–181.

Vauchez, A., and Bufalo, M., 1985, La limite Maures occidentals—Maures orientales (Var, France): Un décrochement ductile senestre majeur entre deux provinces structurales contrastées: Paris, Comptes Rendus de l'Académie des Sciences, v. 301, p. 1059–1062.

Vauchez, A., and Bufalo, M., 1988, Charriage crustal, anatexie et décrochements ductiles dans les Maures orientales (Var, France) au cours de l'orogenèse varisque: Geologische Rundschau, v. 77, p. 45–62.

Vauchez, A., Babaie, H.A., and Babaei, A., 1993, Orogen-parallel tangential motion in the Late Devonian-Early Carboniferous southern Appalachians internides: Canadian Journal of Earth Sciences, v. 30, p. 1297–1305.

von Raumer, J.F., and Neubauer, F., 1993, Late Precambrian and Palaeozoic evolution of the Alpine basement—An overview, *in* von Raumer, J.F., and Neubauer F., eds., The pre-Mesozoic geology in the Alps: Berlin, Springer-Verlag, p. 625–639.

Manuscript Accepted by the Society February 8, 2002

Printed in the U.S.A.

Geological Society of America
Special Paper 364
2002

Variscides between the Appalachians and the Urals: Similarities and differences between Paleozoic subduction and collision belts

Philippe Matte*
Laboratoire de Géophysique, Tectonique et Sédimentologie, UMR 5573, Centre National de la Recherche Scientifique, Université Montpellier 2, 34095 Montpellier, France

ABSTRACT

The Variscan Belt of western Europe is part of a large Paleozoic mountain belt, 1000 km wide and 8000 km long, that extended in Permian time from the Caucasus and Urals in Europe to the Appalachian and Ouachita Mountains in America. The tectonic evolution of these Paleozoic belts bordering Laurentia and Baltica is closely linked, but their geometry, tectonothermal history, and the amount of shortening are different.

The Appalachians and the Variscides are typical collision belts. They were built before the Triassic, between 480 and 250 Ma, and resulted from the diachronous collision of two major continents, Laurentia-Baltica to the northwest and Gondwana to the southeast. Between these two large continents, smaller intermediate continental plates (Avalonia, Armorica, Carolina) show a peculiar Late Proterozoic (650–550 Ma, Cadomian), soft, volcanic and sedimentary basement. These microplates detached from Gondwana during the Cambrian-Ordovician and were accreted to Baltica and Laurentia before the final Carboniferous collision of Gondwana with Baltica and Laurentia. The two belts show very large thrusts and recumbent folds evidencing low- to high-grade polymetamorphism. They were strongly eroded; the Moho is flat and between 30 and 40 km deep.

The Urals separating Europe from Asia were built at approximately the same time, but are not a typical collision belt between two large continents. The belt results from eastward subduction of continental crust beneath oceanic and/or island-arc lithosphere. This stage, contemporaneous with high-pressure metamorphism, was followed by westward subduction, accretion, and soft collision of island arcs and microcontinents. The amount of shortening is much less, as are the grade of metamorphism and the amount of erosion. A peculiar feature of the Urals is the preservation of an orogenic root corresponding to the mafic core of the orogen, the Moho deepening down to 55–60 km in the center of the orogen.

Differences in geometry, amount of shortening, grade of metamorphism, production of granites of these three belts depend on differences in collision processes and the nature of the preorogenic basement involved.

*E-mail: matte@dstu.univ-montp2.fr.

Matte, P., 2002, Variscides between the Appalachians and the Urals: Similarities and differences between Paleozoic subduction and collision belts, *in* Martínez Catalán, J.R., Hatcher, R.D., Jr., Arenas, R., and Díaz García, F., eds., Variscan-Appalachian dynamics: The building of the late Paleozoic basement: Boulder, Colorado, Geological Society of America Special Paper 364, p. 239–251.

INTRODUCTION

The Variscan Belt is part of a Paleozoic mountain system, now completely eroded, which shows great similarities with the Alpine belt. The following can be distinguished within the belt (Fig. 1).

1. Marginal belts of Andean type were built at the boundary between oceans and Precambrian cratons by subduction, during the Paleozoic, of oceanic lithosphere beneath the continent and accretion of volcanic arcs. In these belts the tectonometamorphic and plutonic events become younger from the continent toward the subduction zone.

2. Subduction-collision-subduction belts of Alpine-Himalayan type were built between large continents. In these belts tectonothermal events become generally younger from the suture-collision zone toward the external part of the belt and on both sides of the belt, where it is a fan-like belt generated from opposite subductions. Deformation propagated in time and space outward after collision; as in the Himalayas, subduction of continental lithosphere continued after oceanic subduction (Mattauer et al., 1999; Van der Voo et al., 1999). During this continental subduction the buoyant continental crust may be locally subducted at depth (200 km), but generally floats and forms an accretionary prism that is quickly eroded. These collision-subduction belts generally show significant shortening (to 500 km), and tectonothermal events were spread over 100 m.y. or more from the last oceanic stage to the last continental subduction stage, when orogeny reached its climax. It is thus better to define orogeny geographically than temporally (McKerrow et al., 2000). For example, the Caledonian orogeny lasted ~200 m.y. from northern Greenland–Scandinavia to the central Appalachians, and the Variscan-Appalachian orogeny lasted about 150 m.y. from Central Europe to the southern Appalachians.

3. Subduction-collision belts of Uralian type resulted from the subduction of a continent beneath oceanic lithosphere and were followed by opposite subduction and accretion of volcanic arcs and small continental arcs. These belts show large obduction of ophiolites as in Newfoundland (Dewey, 1976), but generally a modest shortening.

The aim of this chapter is to compare the Urals, the Variscan Belt of Europe, and the Appalachian-Mauritanides orogens in their tectonic evolution and to propose possible convergent histories and continental reconstructions between the Ordovician and the Permian.

URALS

The Urals are a linear belt extending along the 60°E meridian over 2500 km from the New Zemble Island to the Aral Sea; the maximum width of 550 km is in the southern part (Fig. 2). The Urals are unconformably overlain by continental Jurassic strata, which are overlain by shallow-marine Cretaceous sedimentary rocks. A linear suture, the Main Uralian fault, was reactivated as a normal fault. This suture extends from the polar to

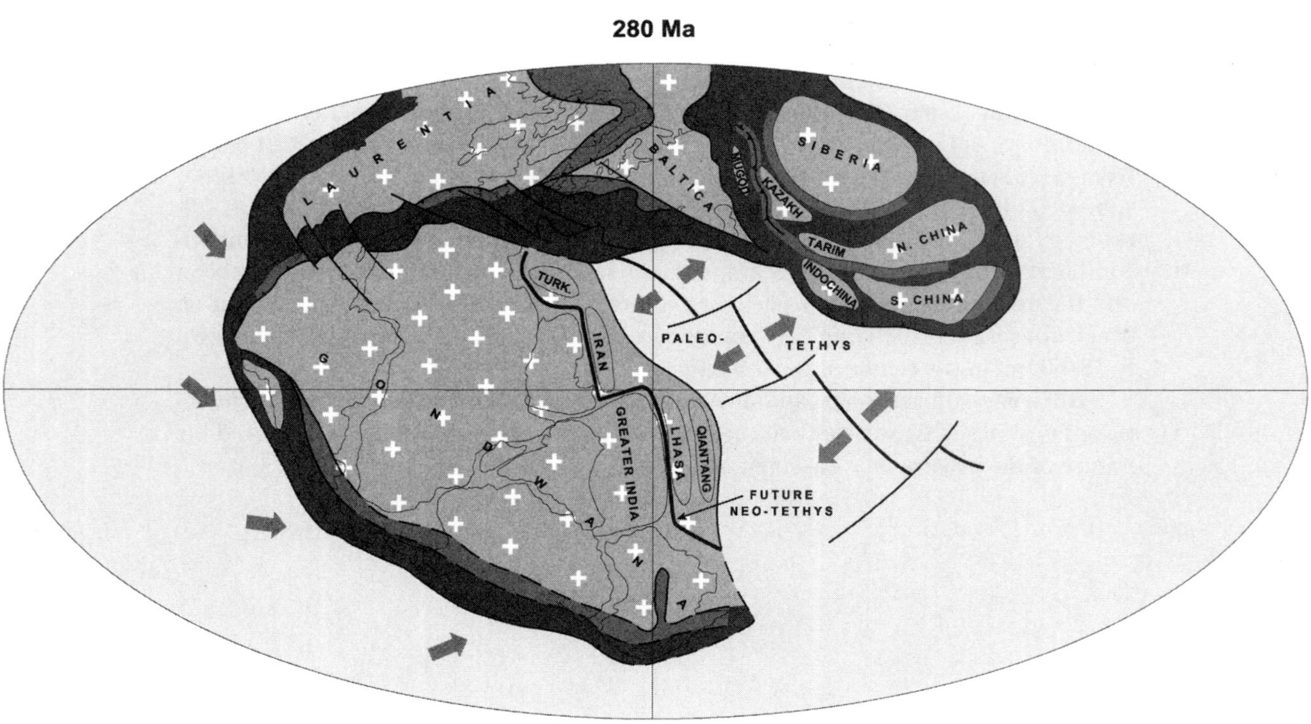

Figure 1. Relative (not paleolatitudinal) Permian (280 Ma) assembly of continents showing Paleozoic belts. Dark gray: 400–250 Ma, light gray: 450–400 Ma (best fit after J.L. Olivet, 2001, personal commun.).

the southern Urals, separating the European platform from an island arc over 2000 km.

The architecture of the belt is relatively well known in its southern part from surface geology (Puchkov, 1997), and an integrated seismic experiment (Berzin et al., 1996). The main characteristics are (Matte, 1995) as follows. (1) An orogenic root is due to the mafic ultramafic basement of the Magnitogorsk arc, which is in the center of the belt. (2) There is excellent preservation of ophiolites (Savelieva and Nesbitt, 1996) and of the island arc (Brown et al., 1998; Spadea et al., 1998). (3) Metamorphism is very low grade, except the high-pressure (*P*)–low-temperature (*T*) metamorphism developed in front of the suture. (4) Shortening is moderate compared to the Appalachians or the Variscides: slaty cleavage is not generally developed and there are no large areas with flat foliations and recumbent folds as in the Variscides of western Europe. (5) There are no crustal S-type granitoids in the continental European plate in front of the suture. Most of the granodiorites crop out to the east of the Main Uralian fault and are related to arc building. Some Permian peraluminous granites (Fig. 2) result from melting of the deep crust in the Mugodzhar block (Fershtater et al., 1997).

The fan-like shape of the orogen results from its plate tectonic evolution (Fig. 3). First the eastward oceanic subduction of the European plate built an island arc between 420 and 390 Ma. Then part of the continental European lithosphere was subducted beneath the island arc. The continental buoyant crust, in part metamorphosed in high-*P* conditions, was quickly exhumed ca. 380 Ma (Matte et al., 1993; Chemenda et al., 1997). Then oceanic subduction occurred westward beneath the island arc, accreting small continents like Mugodzhar and other island arcs (Valerianov) (Zonenshain et al., 1990) in middle-late Carboniferous time (Şengör et al., 1993). After the closing of all oceans and final collision with the Kazakhstan lower Paleozoic belts, the deformation continued in the western foreland, with thrusting in the Bashkirian anticlinal stack and the development of a late Carboniferous–Permian foredeep with blind duplexes, well known from oil prospecting, in Late Permian–early Mesozoic time (Sobornov, 1992).

EUROPEAN VARISCIDES

In contrast to linear belts like the Urals, the European Variscides, extending from Poland to southern Iberia (Fig. 4), show a complex, poorly understood sinuous pattern, with two main arcs, the Ibero-Armorican and the Bohemian. The mountains were completely eroded before the Permian-Triassic; the pre-Triassic reconstruction of the belt is difficult due to late Variscan strike-slip faulting (Arthaud and Matte, 1977; Bard, 1997), the formation of the Mesozoic-Cenozoic basins, the Alpine orogeny, and the opening of the Mediterranean Sea, leading to the dispersal of small continental blocks such as Kabylia, Corsica, Sardinia, and the various Alpine and Carpathian pre-Triassic massifs. The only good fit is between Iberia and France because the Cenozoic drift of Iberia relative to Europe is very

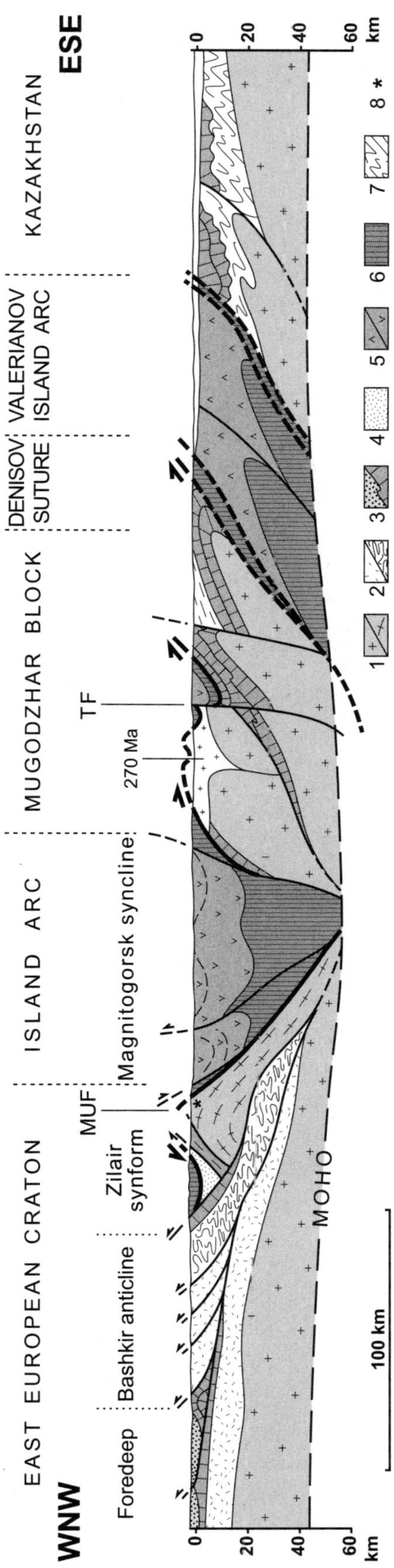

Figure 2. Schematic southern Urals section. 1, Precambrian basement; 2, Riphean sediments with or without late Vendian deformation and metamorphism; 3, shallow-water lower Paleozoic sediments and Permian molasse; 4, Zilair Upper Devonian chromite-bearing flysch; 5, island arcs; 6, oceanic mantle; 7, lower Paleozoic with pre-Devonian deformation in Kazakhstan; 8, high-pressure metamorphism (380 Ma). MUF is Main Uralian fault. TF: Troitsk fault.

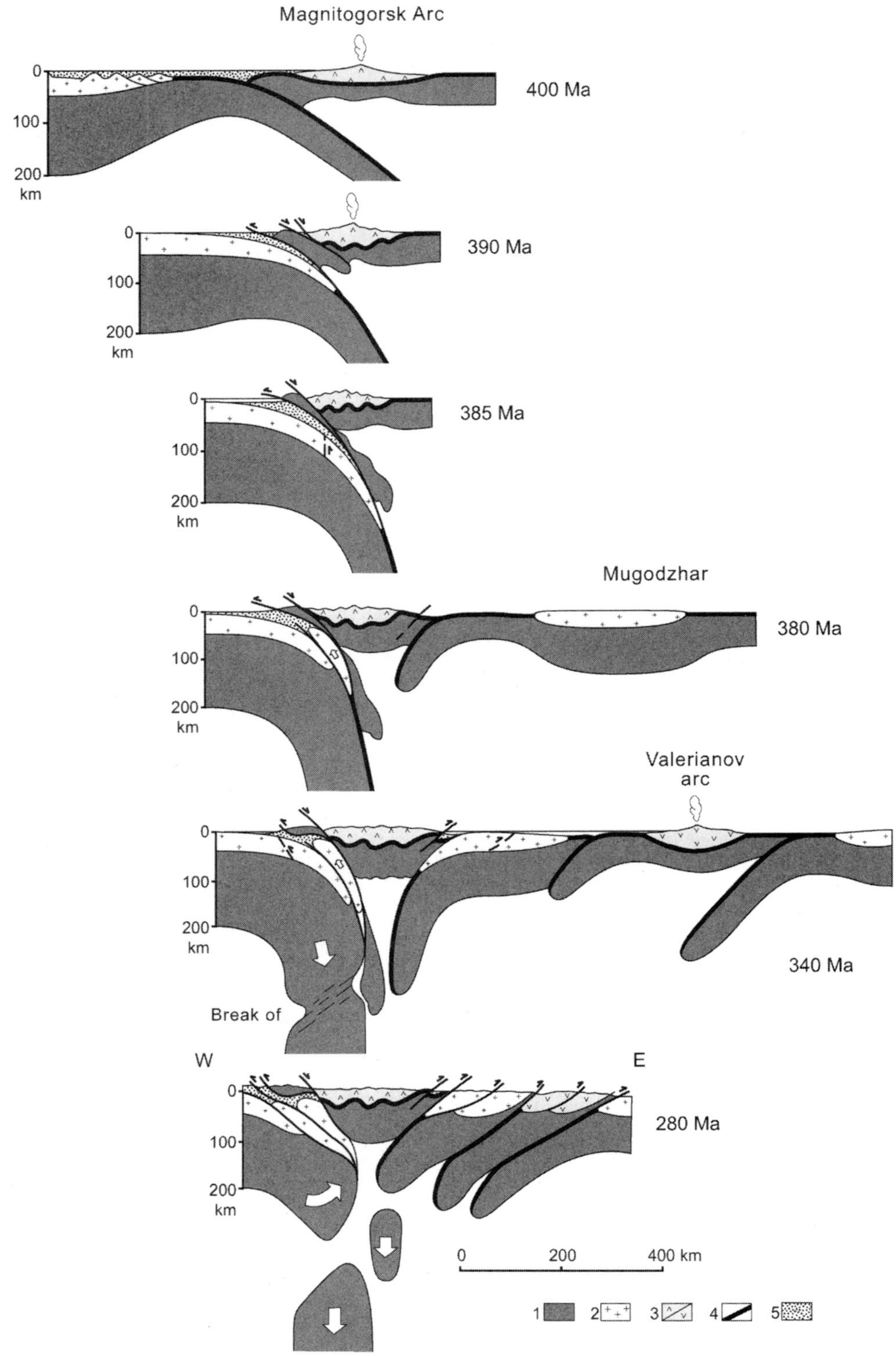

Figure 3. Possible plate tectonic evolution for southern Urals from Silurian to late Carboniferous. 1, mantle; 2, continental crust; 3, arc crust; 4, oceanic crust; 5, Paleozoic sediments in front of suture.

well defined (Olivet, 1996). Thus the two most complete sections through the stable Variscan basement are in the Iberian Peninsula and France (Fig. 5). These sections show the fan-like aspect of the Variscides with opposite vergences of nappes and recumbent folds migrating toward external Carboniferous basins.

Sutures are found on both sides of the belt; they are the roots of large nappes (reaching 200 km), containing 500–400 Ma ophiolitic rocks (Martínez Catalán et al., 1999) and high-*P* to ultrahigh-*P* rocks (Matte, 1998; Lardeaux et al., 2001). On the southern side of the belt, the Galicia–southern Brittany suture

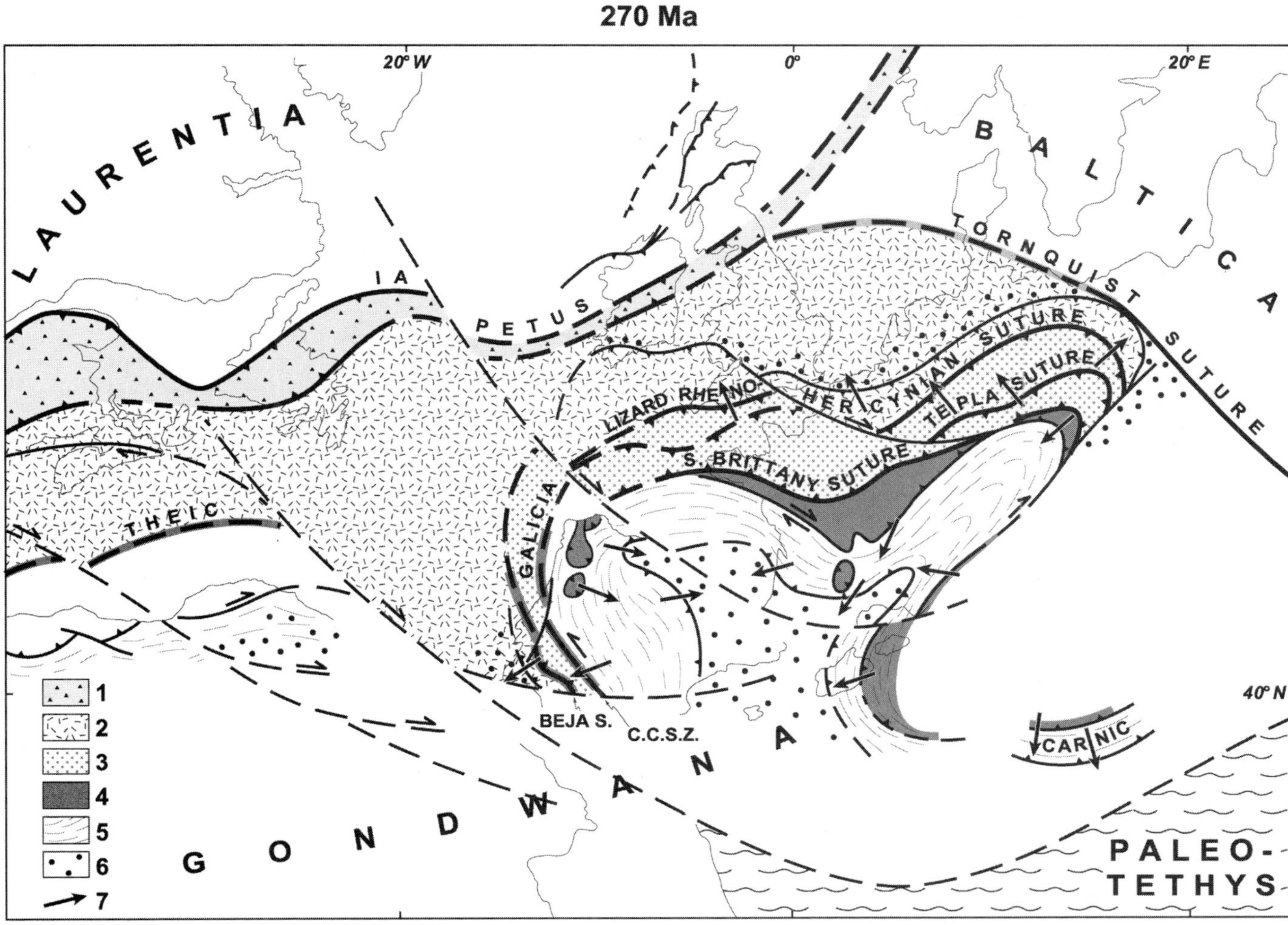

Figure 4. Possible configuration of Variscan Belt of western Europe and northern Africa in Permian time (270 Ma), showing main sutures and extension of Avalonia and Armorica continental microplates. 1, Iapetus Caledonian suture and Ordovician island arc; 2, Avalonia; 3, Armorica; 4, ophiolitic nappes rooted in Galicia South Brittany suture; 5, schistose nappes in southern Variscides; 6, Carboniferous (Visean to Westphalian) foredeep basins; 7, main vergence of nappes. C.C.S.Z. is Coïmbra-Córdoba shear zone. Beja S. is Beja suture.

extends from Central Iberia, where it is reactivated by a large-scale intracontinental shear zone (Coïmbra-Córdoba shear zone) to southern Brittany and the northern Massif Central. It must be considered as the root of the large (partly ophiolitic) nappes transported toward the core of the arc, to the east in Galicia and to the south in Brittany and the Massif Central. This suture is probably related through the Bohemian arc (Fig. 4) to a north-south–trending suture extending from the French External Alps, in which Paleozoic ophiolites have been described (Ménot et al., 1988), to Sardinia as the root of west-vergent nappes (Carmignani et al., 1994). On the northern side of the belt, the Lizard-Rheno-Hercynian and Tepla sutures (Franke, 2001) are relatively well defined from southern England, through Germany, to Poland. These sutures are the roots of northwest-transported nappes showing well-developed high-*P* and/or ultrahigh-*P* metamorphism in the ophiolitic rocks of the Tepla suture and its continental footwall (Erzgebirge). The oceanic rocks are dated as ca. 500 Ma and the high-*P* metamorphism as ca. 380 Ma, both in the Tepla suture and the Münchberg klippe, 100 km to the northwest (Beard et al., 1995). The Lizard-Rheno-Hercynian suture corresponds to a younger Lower Devonian oceanic basin that opened and closed later (Franke, 2001). The Beja suture in the Ossa Morena zone of southern Iberia (Castro et al, 1996) is characterized by southwest vergence, and could be the continuation of one of the German sutures.

The main characteristics of the European Variscides are the following. (1) The attitude of the Moho is flat, everywhere equilibrated at 30–35 km depth in the stable massifs (Cazes and Torreilles, 1988; Meissner and Bortfeld, 1990). (2) A deep level of erosion brings to the surface high-grade rocks unconformably overlain by Carboniferous to Permian-Triassic sediments. (3) There is significant shortening, with large recumbent folds and

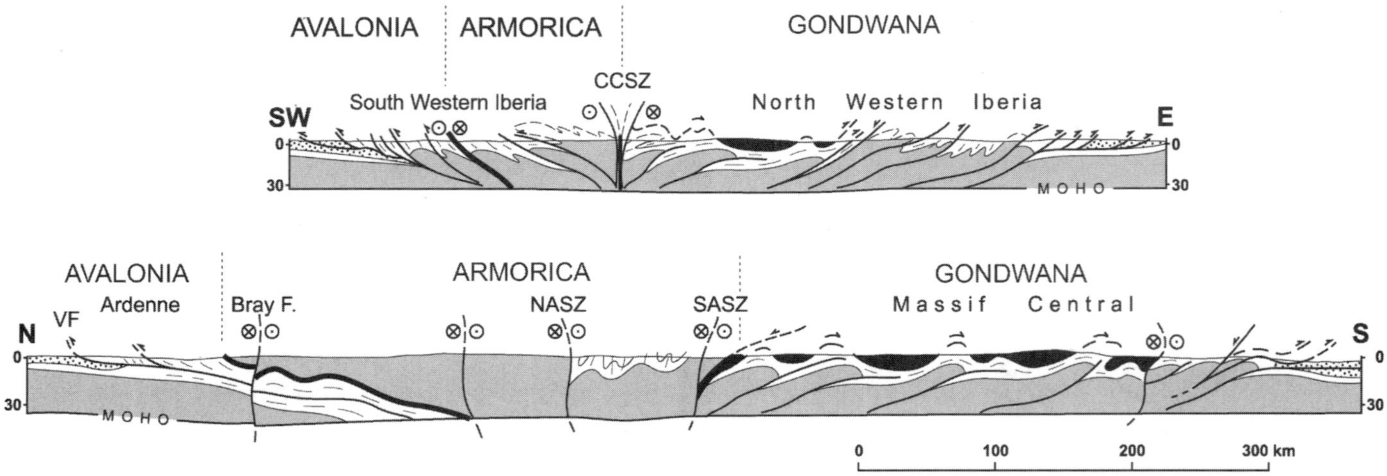

Figure 5. Two ultra-simplified cross sections at crustal scale through Variscan Belt in Iberian Peninsula and France, without granites, showing fan-like shape of the belt. Gray: pre-Paleozoic (essentially Late Proterozoic) low-grade basement. White: lower Paleozoic sediments. Black: Ophiolitic nappes and sutures. Dots: Carboniferous foredeeps. CCSZ: Coïmbra-Córdoba left-lateral shear zone. Bray F: Bray fault. NASZ and SASZ are North Armorican and South Armorican dextral shear zones. VF is Variscan front.

thrusts (Martínez Catalán et al., 1991; Matte, 1991). (4) A plurifacial metamorphism (high-*P*–low-*T* to high-*I*–low-*P*) occurred between 420 and 320 Ma, at times with inverted isograds, particularly in the French Massif Central (Burg et al., 1984; Schulz et al., 2001). (5) There is an enormous quantity of granitoids, especially S type produced by melting of the middle and lower crust, in front of the sutures, beneath the ophiolitic nappes (Lagarde et al., 1992).

The general fan-like shape of the belt is probably inherited from a double vergence subduction system (Matte, 1986), first oceanic, then continental (Fig. 6), corresponding to the consumption of two main oceans, the Rheic ocean to the northwest and the Galicia–South Brittany ocean to the southeast. The corresponding sutures extend around the Ibero-Armorican arc from the Bohemian Massif to southern Iberia (Fig. 4). The belt results from the collision of two large continents, Gondwana to the southeast and Laurussia to the northwest, and small intermediate continents, Avalonia to the northwest, between the Iapetus and the Rheic oceans, and Armorica to the southeast, between the Rheic and the Galicia–South Brittany oceans. Subduction and exhumation of deep continental crust along the sutures, on both sides of the belt, is attested to by the presence of high-*P*–ultra-high-*P* metamorphism between 400 and 360 Ma (Matte, 1998).

SOUTHERN APPALACHIANS–MAURITANIDES

The southern Appalachians and Mauritanides are two Paleozoic belts currently located on both sides of the Atlantic with a continentward vergence. With a classical Gondwana-Laurasia fit in Permian-Triassic time (Scotese and McKerrow, 1990; Scotese and Golonka, 1992), i.e., at the end of the orogeny, the southern Appalachians–Mauritanides appear as a very wide (1600 km) fan-like orogen, the greatest part of which (1000 km) is hidden beneath Mesozoic-Cenozoic coastal plain sediments in America and western Africa (Fig. 7).

The structure of the Appalachians is very well documented from extensive geological and geophysical studies (synthesized in Hatcher et al., 1989a). Like the Variscides, the Appalachians are a complex mountain belt in which the tectonothermal events propagated from the internal part toward the late Carboniferous foredeep, between 450 and 420 Ma (Taconic) 400 and 350 Ma (Acadian), and 330 and 270 Ma (Alleghanian). The main characteristics are (1) the flat attitude of the Moho around 40 km depth, and (2) a master decollement underlying the Appalachians from the surface in the external part to 20–30 km depth (Hatcher et al., 1989b) in the most internal part beneath the coastal plain. This decollement, well documented by COCORP profiling (Cook et al., 1979), was considered as essentially Alleghanian by Hatcher et al. (1989c) and Hatcher (1989) and would have transported part of the Precambrian Greenville basement and the internal metamorphic Appalachians (in part pre-Alleghanian), to the north onto the Laurentian basement and its sedimentary cover hundreds of kilometers. The minimum total Alleghanian shortening would be more than 300 km and the total shortening could reach 500 km. (3) Granites in front of the Acadian (possibly transported) Piedmont suture are scarce. Most of the granites are located south of the suture in the Carolina-Avalon terrane.

There are two main sutures. The Central Piedmont suture is between the Piedmont and the Carolina Avalon terrane, corresponding to the early (Taconic, Acadian?) collage of a Cambrian-Ordovician island arc and perhaps of the prolongation of the Avalon microcontinent, against Laurentia. The cryptic Alleghanian suture, underlined by the Brunswick magnetic

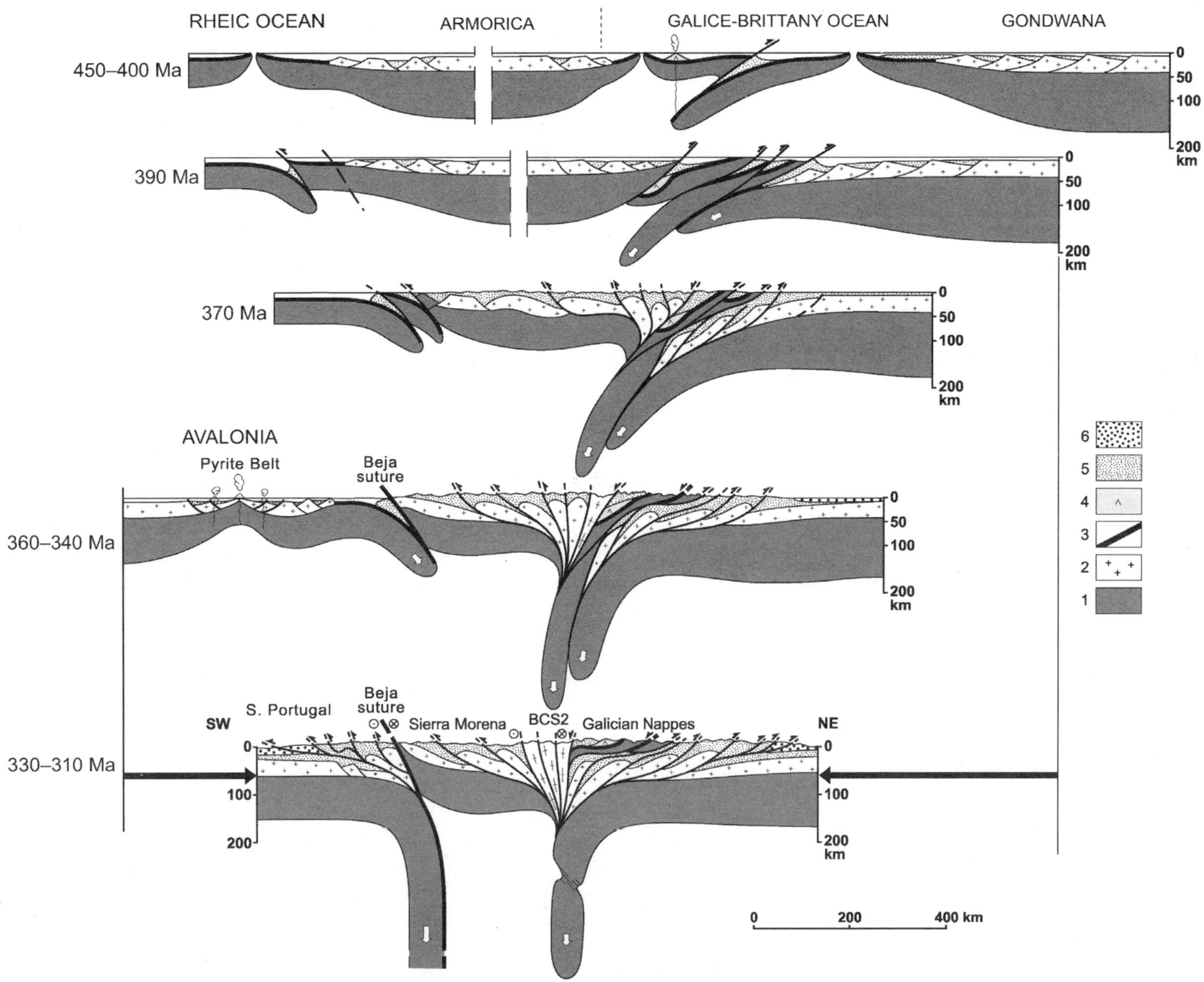

Figure 6. Possible Paleozoic convergent history of Iberian Variscides from Ordovician to Carboniferous time. 1, mantle; 2, continental crust; 3, oceanic crust; 4, possible immature island arc; 5, lower Paleozoic sediments; 6, Carboniferous.

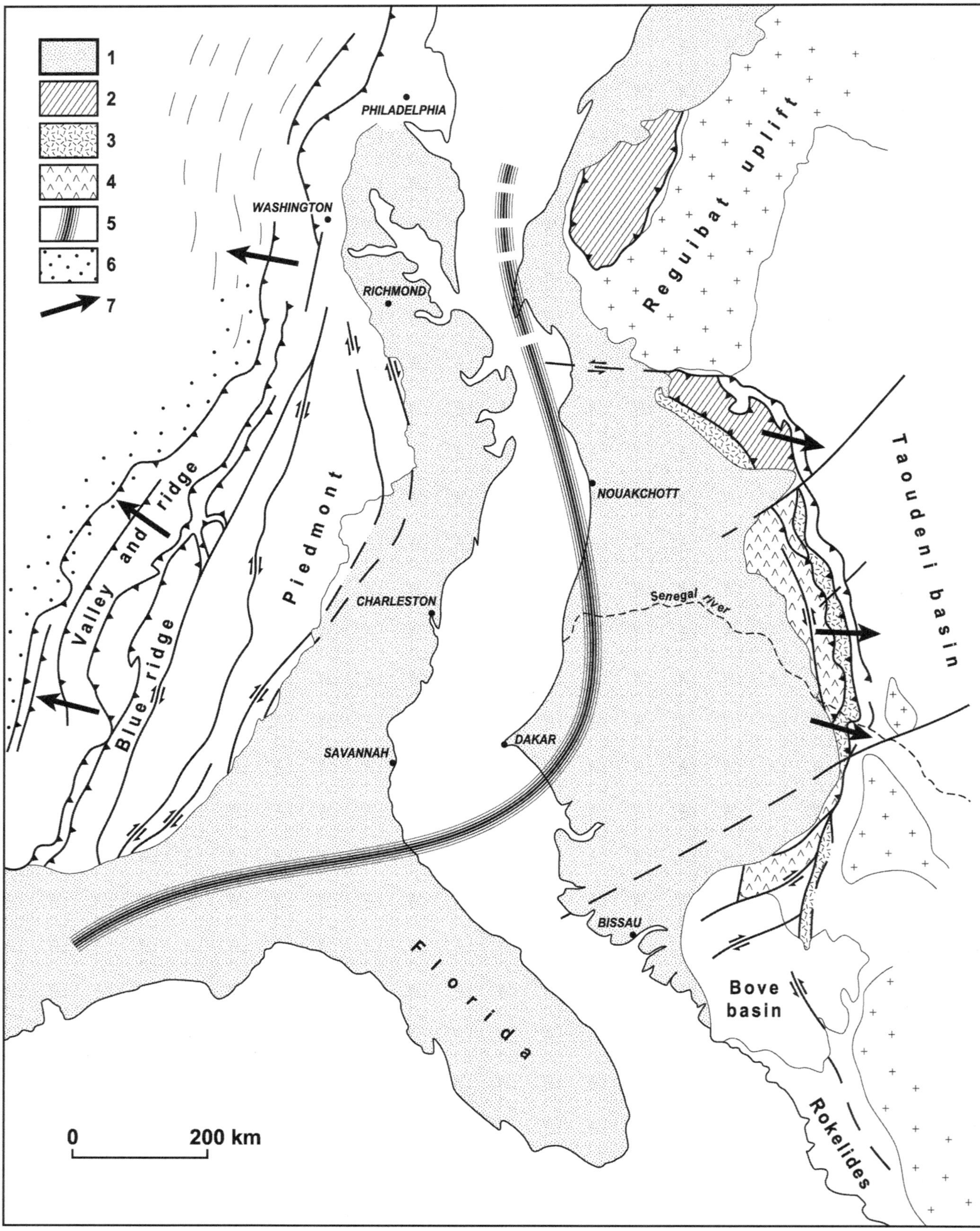

Figure 7. Simplified possible reconstruction for southern Appalachians–Mauritanides orogen in Permian time (modified after Vauchez et al., 1987). 1, coastal plain cover; 2, North Mauritanides klippes; 3, Mauritanian ophiolites; 4, Mauritanian volcanic arc; 5, possible Alleghanian-Mauritanian suture; 6, Appalachian foredeep; 7, nappe transport.

anomaly, is perhaps common with the Mauritanides (Nelson et al., 1985) and corresponds to the Laurentia-Africa final collision.

The Mauritanides form an arcuate thrust belt on the border of western Africa. The structure is poorly defined because only the very external part of the belt can be observed and because there is no vertical seismic line crossing the orogen. Another problem is the duality of Pan-African (680–600 Ma) versus late Variscan (320–280) tectonothermal events (Dallmeyer and Lécorché, 1990). The front visible part of the orogen in the central Mauritanides is characterized by the superposition of a complex allochthonous pile over the African old crystalline basement and its Proterozoic-Paleozoic (to Middle Devonian) nonmetamorphic cover (Lécorché et al., 1990).

Over a paraautochthonous unit made of Vendian to Devonian sediments, the allochthonous pile shows, from bottom to top, an undated ophiolitic low-grade series with serpentinites, and an overlying unit made of volcanic and magmatic rocks (680 Ma granodiorite). In the front part of the southern Mauritanides these units override the African basement over 50 km (thin-skinned tectonics; Burg et al., 1993). In the northern Mauritanides, other high-grade (to granulitic) undated allochthonous units overlie the ophiolites and the Precambrian autochthonous basement (Akjouit and western Sahara klippes) over 150 km. A high *P* metamorphism was found in the internal units overlying the ophiolites, where eclogites have been dated at 330 Ma (Le Goff et al., 2001).

There are two possible suture zones marked by gravity highs related to mafic and/or ultramafic rocks (Roussel and Lesquer, 1990): an internal suture borders the Atlantic coast, but this anomaly may correspond partly to the continent-ocean boundary; an external suture corresponds to the ophiolitic allochthonous unit. This suture may be a more or less rooted Pan-African suture derived from a Proterozoic ocean, as proposed by Lécorché et al. (1990). It may also correspond to rootless Paleozoic ophiolites largely transported to the east: in this case the total Permian-Carboniferous shortening of the Mauritanides could be very large, as in the southern Appalachians.

CONCLUSIONS

The Paleozoic mountain belts fringing the eastern and southern edges of Laurussia in the Permian were formed by convergent processes between 450 and 270 Ma. (Figs. 3, 6, 8, and 9). These belts resulted from a diachronous collision of volcanic arcs and small and/or large continents, e.g., Gondwana with Laurussia. All show evidence of oceanic and continental subduction, which plays an important role in mountain building. Their differences in architecture, shortening, metamorphism, granite production, and level of erosion reflect differences in collision processes and in the nature of the Precambrian basement involved.

In the Urals the eastward subduction of the East European continental passive margin beneath the Uralian oceanic lithosphere resulted in an Oman-type obduction and contemporaneous exhumation of high-*P* rocks (Chemenda et al., 1996), followed by the collage of the Magnitogorsk island arc. These events did not produce appreciable orogeny. Orogeny began later, during the middle-late Carboniferous, with the first foredeep molassic deposits on the European platform after closure of all oceanic basins and the final collision with Kazakhstan.

The low grade of metamorphism and the absence of Carboniferous granites in the Bashkirian foreland fold and thrust belt west of the suture (main Uralian fault) reflect a weak late Carboniferous–Permian crustal thickening and shortening (Brown et al., 1997).

The European Variscides show, like the Appalachians, a more complex history due to the collision of two large continents, Gondwana and Laurussia, and intermediate small continents (Avalonia and Armorica), detached from Gondwana during early Paleozoic time. These microplates have been essentially defined from paleomagnetism (Van der Voo, 1993; Torsvik, 1998), but the boundaries of these microplates must be also determined from geology, taking into account the oceanic sutures. Armorica is between the northern sutures of the belt (German sutures and Beja suture) and the southern suture (Galicia-Brittany-Sardinia suture). Avalonia is between the German–South Iberian sutures and the Iapetus suture.

Strong shortening (nearly 500 km on each side) between 370 Ma in the internides and 320 Ma in the externides, and corresponding continental subduction, produced large nappes (to 300 km). A strong crustal shortening and thickening resulted in high-grade polymetamorphism and melting, producing a great amount of granite of crustal origin between 360 and 290 Ma. Granite production was also enhanced by the low-grade water-rich Pan-African sedimentary and volcanic-sedimentary basement of Avalonia, Armorica, and the northern edge of Gondwana. Orogeny began from the early Carboniferous, when the first (locally wild) flyschs were deposited in the two opposite foredeeps overlain by late Carboniferous molasses with paralic coals. Denudation and strong erosion occurred as early as the early Carboniferous in the internal zones, bringing high-grade metamorphic rocks to the surface.

From the Middle Devonian to the early Carboniferous, the European Variscan Belt probably formed a unique orogen with the northern and central Appalachians. Gondwana then rotated clockwise, which resulted in a wider opening of the Paleo-Tethys and closure of the Theic ocean. In Europe, this stage is characterized by large strike-slip faulting (Arthaud and Matte, 1977; Bard, 1997). Final collision between Africa and America resulted in the building of the southern Appalachian–Mauritanian orogen.

One peculiarity of these two belts, like the Norvegian Caledonides, is the lack of Carboniferous granites in front of the suture. It is probably here the result of a very large thin-skinned decollement in the foreland, which does not necessarily imply a large crustal thickening, and of the dry character of the high-grade metamorphic Precambrian basement.

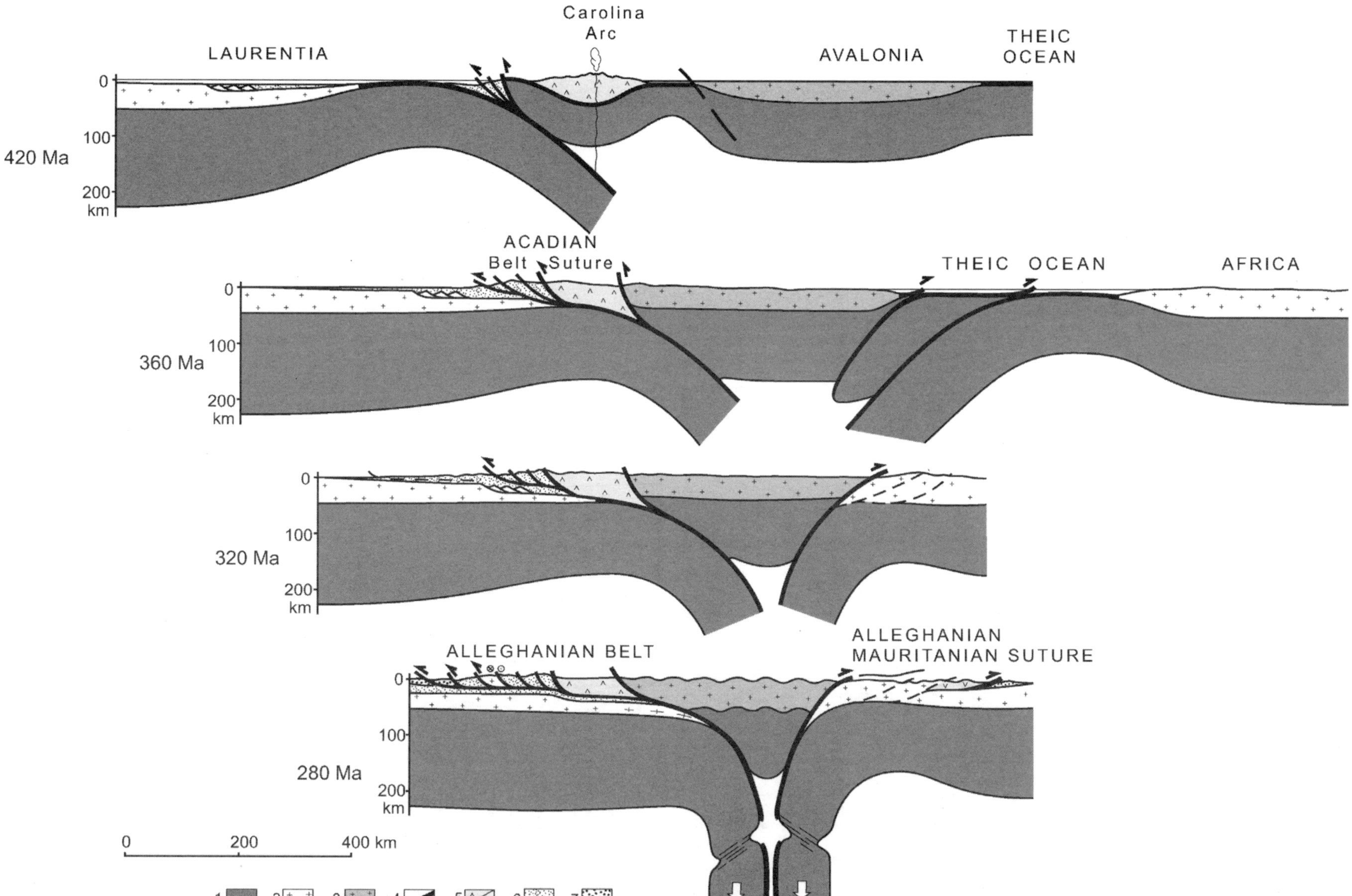

Figure 8. Possible Paleozoic convergent history of southern Appalachians (modified after Hatcher, 1989). 1, mantle; 2, continental crust of Laurentia and Africa; 3, continental and/or arc crust of Avalonia; 4, oceanic crust and sutures; 5, island arc crust; 6, lower Paleozoic sediments; 7, Carboniferous.

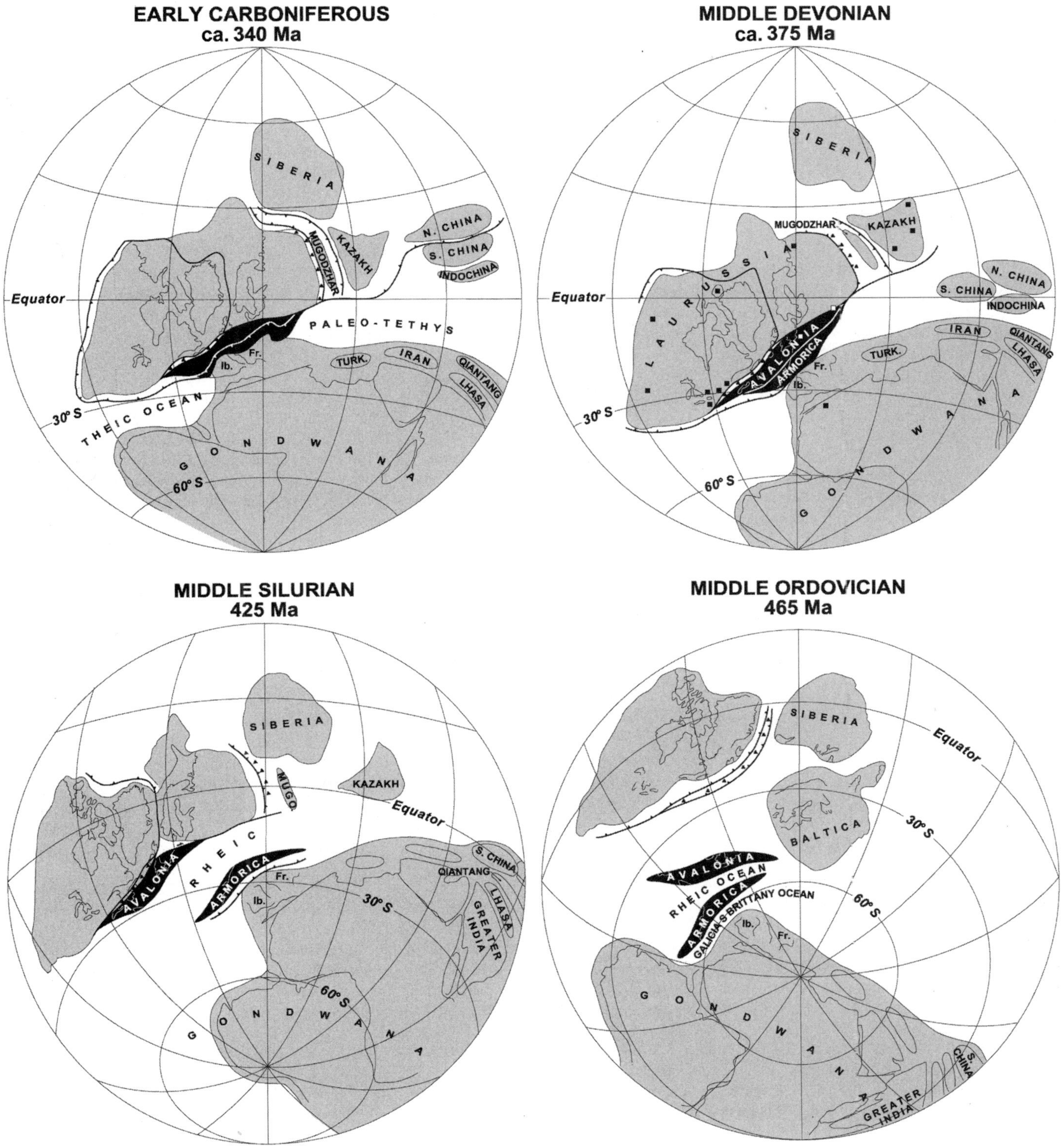

Figure 9. Tentative Paleozoic reconstructions from Middle Ordovician (465 Ma) to Lower Carboniferous (340 Ma) modified after Scotese and Golonka (1992) and Torsvik (1998), showing Armorica and Avalonia (black) continental microplates. Chevrons: island arcs. Black squares: distribution of Callixylon in Late Devonian (375 Ma) after Meyer-Berthaud et al. (1997).

ACKNOWLEDGMENTS

I thank José R. Martínez Catalán and Florentino Díaz García, who proposed that I write this paper, Christian Teyssier for his constructive review, and Anne Delplanque for the final version of the drawings.

REFERENCES CITED

Arthaud, F., and Matte, Ph., 1977, Late-Paleozoic strike-slip faulting in southern Europe and northern Africa: Result of a right lateral shear zone between the Appalachians and the Urals: Geological Society of America Bulletin, v. 88, p. 1305–1320.

Bard, J.P., 1997, Démembrement anté-mésozoïque de la chaîne varisque d'Europe occidentale et d'Afrique du Nord: Rôle essentiel des grands décrochements transpressifs dextres accompagnant la rotation-translation horaire de l'Afrique durant le Stéphanien: Paris, Comptes Rendus de l'Académie des Sciences, v. 324, p. 693–704.

Beard, B.L., Medaris, L.G., Johnson, C.M., Jelinek, E., Tonika, J., and Riciputi, L.R., 1995, Geochronology and geochemistry of eclogites from the Marianské Laznë complex, Czech Republic: Implications for Variscan orogenesis: Geologische Rundschau, v. 84, p. 552–567.

Berzin, R., Oncken, O., Knapp, J.H., Perez Estaún, A., Hismatulin, T., Yunusov, N., and Lipilin, A., 1996, Orogenic evolution of the Ural Mountains: Results from an Integrated Seismic Experiment: Science, v. 274, p. 220–221.

Brown, D., Alvarez-Marron, J., and Perez Estaún, A., 1997, Geometric and kinematic evolution of the foreland thrust and fold belt in the southern Urals: Tectonics, v. 16, p. 551–562.

Brown, D., Juhlin, C., Alvarez-Marron, J., Perez Estaún, and Oslianski, A., 1998, Crustal scale structure and evolution of an arc-continent collision zone in the southern Urals, Russia: Tectonics, v. 17, p. 158–171.

Burg, J.-P., Corsini, M., Diop, C., and Maurin J-C., 1993, Structure et cinématique du Sud de la chaîne des Mauritanides: Un système de nappes tégumentaires varisques: Paris, Comptes Rendus de l'Académie des Sciences, v. 317, p. 697–703.

Burg, J.-P., Leyreloup, A., Marchand, J., and Matte, Ph., 1984, Inverted metamorphic zonation and large scale thrusting in the Variscan Belt: An example in the French Massif Central, *in* Hutton, D.H.M., and Sanderson, P.J. eds., Variscan Tectonics of the North Atlantic region: London, Geological Society of London Special Publication 14, Blackwell Scientific Publishing, p. 47–61.

Carmignani, L., Carosi, R., Di Pisa, A., Gattiglio, M., Musumeci, G., Oggiano, G., and Pertusati, P.C., 1994, The hercynian chain in Sardinia (Italy): Geodynamica Acta, v. 7, p. 31–47.

Castro, A., Fernandez, C., De La Rosa, J.D., Moreno-Ventas, I., and Rogers, G., 1996, Significance of MORB-derived amphibolites from the Aracena metamorphic belt, Southwest Spain: Journal of Petrology, v. 37, p. 235–260.

Cazes, M., and Torreilles, G., 1988, Étude de la croûte terrestre par sismique profonde, Profil Nord de la France: Paris Institut Français du Pétrole, Institut National des Sciences de l'Univers-CNRS, Société Nationale Elf Aquitaine (Production), Éditions Technip (27 rue Ginoux, 75737 Paris cedex 15), 2 vols., Paris, 260 p., 15 planches H.T.

Chemenda, A., Mattauer, M., and Bokun, A.N., 1996, Continental subduction and a mechanism for exhumation of high-pressure metamorphic rocks: New modelling and field data from Oman: Earth and Planetary Science Letters v. 143, p. 173–182.

Chemenda, A., Matte, Ph., and Sokolov, V., 1997, A model of Paleozoic obduction and exhumation of high-pressure/low-temperature rocks in the southern Urals: Tectonophysics, v. 276, p. 217–227.

Cook, F.A., Albaugh, D.S., Brown, L.D., Kaufman, S., Oliver, J.E., and Hatcher, R.D., Jr., 1979, Thin-skinned tectonics in the crystalline southern Appalachians: COCORP seismic-reflection profiling of the Blue Ridge and Piedmont: Geology, v. 7, p. 563–567.

Dallmeyer, R.D., and Lécorché, J.P., 1990, $^{40}Ar/^{39}Ar$ polyorogenic mineral age record in the northern Mauritanides orogen, West Africa: Tectonophysics, v. 177, p. 81–107.

Dewey, J.F., 1976, Ophiolite obduction: Tectonophysics, v. 31, p. 93–120.

Fershtater, G.B., Montero, P., Borodina, N.S., Pusharev, E.V., Smirnov, V.N., and Bea, F., 1997, Uralian magmatism, an overview: Tectonophysics, v. 276, p. 87–102.

Franke, W, 2001, The mid-European segment of the Variscides: Tectonostratigraphic units, terranes boundaries and pate tectonic evolution: Journal of the Geological Society of London, v. 197, p. 35–61.

Hatcher, R.D., Jr, 1989, Tectonic synthesis of the U.S. Appalachians, *in* Hatcher, R.D., Jr., Thomas, W.A., and Viele, G.W., eds., The Appalachian-Ouachita orogen in the United States: Boulder, Colorado, Geological Society of America, The Geology of North America, v. F-2, p. 511–535.

Hatcher, R.D., Jr., Thomas, W.A., and Viele, G.W., editors, 1989a, The Appalachian-Ouachita orogen in the United States: Boulder, Colorado, Geological Society of America, The Geology of North America, v. F-2, 767 p., 12 plates.

Hatcher, R.D., Jr., Osberg, P.H., Drake, A.A., Jr., Robinson, P., and Thomas, W.A., 1989b, Tectonic map of the U.S. Appalachians, *in* Hatcher, R.D., Jr., Thomas, W.A., and Viele, G.W., eds., The Appalachian-Ouachita orogen in the United States: Boulder, Colorado, Geological Society of America, The Geology of North America, v. F-2, Plate 1, scale 1:2,500,000.

Hatcher, R.D., Jr., Thomas, W.A., Geiser, P.A., Snoke, A.W., Mosher, S., and Wiltschko, D.V., 1989c, Alleghanian orogen, *in* Hatcher, R.D., Jr., Thomas, W.A., and Viele, G.W., eds., The Appalachian-Ouachita orogen in the United States: Boulder, Colorado, Geological Society of America, The Geology of North America, v. F-2, p. 233–318.

Lagarde, J-L., Capdevila, R., and Fourcade, S., 1992, Granites et collision continentale: L'exemple des granitoïdes carbonifères dans la chaîne hercynienne ouest-européenne: Bulletin de la Société Géologique de France, v. 163, p. 597–610.

Lardeaux, J.M., Ledru, P., Daniel, I., and Duchene, S., 2001, The Variscan French Massif Central—A new addition to the ultra-high pressure metamorphic "club": Exhumation processes and geodynamic consequences: Tectonophysics, v. 332, p. 143–167.

Le Goff, E., Guerrot, C., Maurin, G., Johan, V., Tegyvey, M., and Ben Zerga, M., 2001, Découverte d'éclogites hercyniennes dans la chaîne des Mauritanides (Afrique de l'Ouest): Paris, Comptes Rendus de l'Académie des Sciences, v. 333, p. 711-718.

Lécorché, J.P., Bronner, G., Dallmeyer, R.D., Rocci, G., and Roussel, J., 1990, The Mauritanide Orogen and its northern extensions (Western Sahara and Zemmour), West Africa, *in* Dallmeyer, R.D., and Lécorché, J.P., eds, The West African Orogens and Circum-Atlantic correlatives: Berlin, Springer-Verlag, p. 187–227.

McKerrow, W.S., Mac Niocaill, C., and Dewey, J.F., 2000, The Caledonian orogeny redefined: Journal of the Geological Society of London, v. 157, p. 1149–1154

Martínez Catalán, J.R., Perez Estaún, A., Bastida, F., Pulgar, J.A., and Marcos, A., 1991, Structure of the West Asturian-Leonese Zone, *in* Dallmeyer, R.D., and Martínez García, E., eds., Pre-Mesozoic Geology of Iberia : Berlin, Springer-Verlag, p. 103–114.

Martínez Catalán, J.R., Arenas, R., Díaz García, F., and Abati, J., 1999, Allochthonous units in the Variscan Belt of NW Iberia, *in* Sinha, A.K., ed., Terranes and accretionary history, Basement Tectonics 13: Dordrecht, Netherlands, Kluwer Academic Publishers, p. 65–84.

Mattauer, M., Matte, Ph., and Olivet, J.L., 1999, A 3-D model of the India-Asia collision at plate scale: Paris, Comptes Rendus de l'Académie des Sciences, v. 328, p. 499–508.

Matte, Ph., 1986, La chaîne varisque parmi les chaînes paléozoïques péri-atlantiques, modèle d'évolution et position des grands blocs continentaux au Permian-Carbonifère: Bulletin de la Société Géologique de France, v. 1, p. 9–21.

Matte, P., 1991, Accretionary history and crustal evolution of the Variscan Belt in western Europe: Tectonophysics, v. 196, p. 309–339.

Matte P., 1995, Southern Uralides and Variscides: Comparison of their anatomies and evolutions: Geologie en Mijnbouw, v. 74, p. 151–166.

Matte, P., 1998, Continental subduction and exhumation of HP rocks in Paleozoic belts: Uralides and Variscides: Journal of the Geological Society of Sweden (GFF), (Special Issue: Tectonics and General History of Phanerozoic Orogens) v. 120, p. 209–222.

Matte, Ph., Maluski, H., Caby, R., Nicolas, A., Kepezhinskas, P., and Sobolev, S., 1993, Geodynamic model and $^{39}Ar/^{40}Ar$ dating for the generation and emplacement of the high pressure metamorphic rocks in SW Urals: Paris, Comptes Rendus de l'Académie des Sciences, v. 317, p. 1667–1674.

Meissner, R., and Bortfeld, K., 1990, DEKORP Atlas: Results of Deutsches Kontinentales Reflexionsseismisches Programm: Berlin, Springer-Verlag, 80 seismic sections and 5 further figures.

Ménot, R.-P., Peucat, J.J., Scarenzi, D., and Piboule, M., 1988, 496 M.y. age of plagiogranites in the Chamrousse ophiolite complex (external crystalline massifs in the French Alps): Evidence of lower Paleozoic oceanization: Earth and Planetary Science Letters, v. 88, p. 82, 92.

Meyer-Berthaud, B., Wendt, J., and Galtier, J., 1997, First record of a large Callixylon trunk from the late Devonian of Gondwana: Geological Magazine, v. 134, p. 847–853.

Nelson, K.D., Arnow, J.A., McBride, J.H., Willemin, J.H., Huang, J., Zheng, L., Oliver, J.E., Brown, L.D., and Kaufman, S., 1985, New COCORP profiling in the southeastern United States, part I and II: Geology, v. 13, p. 714–721.

Olivet, J.L, 1996, La cinématique de la plaque Ibérique: Bulletin du Centre de Recherche Exploration-Production, Elf Aquitaine, v. 20, no. 1, p. 131–195.

Puchkov, V.N., 1997, Structure and geodynamics of the Uralian orogen, *in* Burg, J.-P., and Ford, M., eds., Orogeny through time: London, Geological Society of London Special Publication 121, p. 201–236.

Roussel, J., and Lesquer, A., 1990, Geophysics and the crustal structure of west Africa, *in* Dallmeyer, R.D., and Lécorché, J.P., eds., The West African Orogens and Circum-Atlantic correlatives: Berlin, Springer-Verlag, p. 9–28.

Savelieva, G.N., and Nesbitt, R.W., 1996, A synthesis of the stratigraphic and tectonic setting of the Uralian ophiolites: Journal of the Geological Society of London, v. 153, p. 525–535.

Schulz, B., Triboulet, C., Audren, C., and Feybesse, J.-L., 2001, *P-T* paths from metapelite garnet zonations, and crustal stacking in the inverted metamorphic sequence of La Sioule, French massif central: Zeitschrift Deutschland: Geologische Gesellschaft, v. 152, p. 1–25.

Scotese, C.R., and McKerrow, W.S., 1990, Revised world maps and introduction, *in* McKerrow, W.S., and Scotese, C.R., eds., Palaeozoic palaeogeography and biogeography: Geological Society of London Memoir 12, p. 1–21.

Scotese, C.R., and Golonka, J., 1992, PALEOMAP Paleogeographic Atlas. PALEOMAP Progress Report 20: Arlington, Texas, Department of Geology, University of Texas.

Şengör, A.M.C., Natal'in, B.A., and Burtman, V.S., 1993, Evolution of the Altaid tectonic collage and Paleozoic crustal growth in Eurasia: Nature, v. 364, p.299–307.

Sobornov, K., 1992, Blind duplex structure of the north Urals thrust belt front: Journal of Geodynamics, v. 15, p. 1–11.

Spadea, P., Kabanova, L.Y., and Scarrow, J.H., 1998, Petrology, geochemistry, and geodynamic significance of Mid-Devonian boninitic rocks from the Baimak-Buribai area (Magnitogorsk zone, southern Urals): Ofioliti, v. 23, p. 17–36.

Torsvik, T.H., 1998, Palaeozoic palaeogeography: A North Atlantic viewpoint: Journal of the Geological Society of Sweden (GFF), v. 120, p. 109–118.

Van der Voo, R., 1993, Paleomagnetism of the Atlantic, Tethys and Iapetus oceans: Cambridge, Cambridge University Press, 411 p.

Van der Voo, R., Spakman, W., and Bijward, H., 1999, Tethyan subducted slabs under India: Earth and Planetary Science Letters, v. 171, p. 7–20.

Vauchez, A., Kessler, S.F., Lécorché, J.P., and Villeneuve, M., 1987, Southward extrusion tectonics during the Carboniferous Africa-North America collision: Tectonophysics, v. 142, p. 317–322.

Zonenshain, L.P., Kuzmin, M.I., and Natapov, L.M., 1990, Geology of the USSR: A plate-tectonic synthesis, *in* Page, B., ed., American Geophysical Union Geodynamics Series, v. 21, 242 p.

Manuscript Accepted by the Society February 8, 2002

Printed in the U.S.A.

Geological Society of America
Special Paper 364
2002

Opposite subduction polarities connected by transform faults in the Iberian Massif and western European Variscides

Fernando Simancas*
Francisco González Lodeiro
Inmaculada Expósito Ramos
Antonio Azor
David Martínez Poyatos
Departamento de Geodinámica, Universidad de Granada, Campus de Fuentenueva, 18002 Granada, Spain

ABSTRACT

The Iberian Massif has been divided into two branches of opposite vergence, thus obtaining the image of a cylindrical bivergent orogen. However, we show that the architecture of the Iberian Massif is not cylindrical, the two branches representing opposite subduction polarities connected by a transform fault. The northern branch has suture units in the allochthonous Galicia Trás-os-Montes zone; its Variscan evolution includes west-dipping subduction, obduction of oceanic-derived units and eastward propagation of the deformation. The southern branch shows southwest vergence and two sutures: (1) the boundary South Portuguese zone–Ossa-Morena zone, and (2) the boundary Ossa-Morena zone–Central Iberian zone. The Central Iberian zone is the footwall of the allochthonous terranes of the Galicia Trás-os-Montes zone in the northern branch, but it is the hanging wall of the suture-related rocks in the southern branch. The change in subduction polarity is explained by the existence of a transform fault connecting both domains, presently represented by the Porto-Tomar fault. The correlation of the Iberian sutures with other Variscan sutures in Europe has a main difficulty in the stratigraphic and paleontological similarities between the Armorican domain and the Central Iberian zone, which prevent the Massif Central ocean and the Galicia ocean to be considered as continuous. Instead, another transform would have connected these two oceanic domains. The closure of such oceanic domains bounded by transform faults, followed by rotation and escape tectonics in opposite directions, generated the Ibero-Armorican arc.

INTRODUCTION

The Iberian Massif is the largest Variscan outcrop in Europe, showing perhaps a nearly complete transect of this Paleozoic orogen (Fig. 1A). To the east and south, upper Proterozoic and Paleozoic rocks are bounded by Tertiary continental deposits, Mesozoic platform sediments, and ranges of the Alpine front (Pyrenees, Iberian Chain, and Betics). To the west, the boundary is the Porto-Tomar fault, which has a long history (Fig. 1B; Ribeiro et al., 1980). A broad structure, the Ibero-Armorican arc, connects the Iberian Variscides with the Variscan Massif of Central Europe (Fig. 1A; Bard et al., 1971).

*E-mail: simancas@ugr.es.

Simancas, F., González Lodeiro, F., Expósito Ramos, I., Azor, A., and Martínez Poyatos, D., 2002, Opposite subduction polarities connected by transform faults in the Iberian Massif and western European Variscides, *in* Martínez Catalán, J.R., Hatcher, R.D., Jr., Arenas, R., and Díaz García, F., eds., Variscan-Appalachian dynamics: The building of the late Paleozoic basement: Boulder, Colorado, Geological Society of America Special Paper 364, p. 253–262.

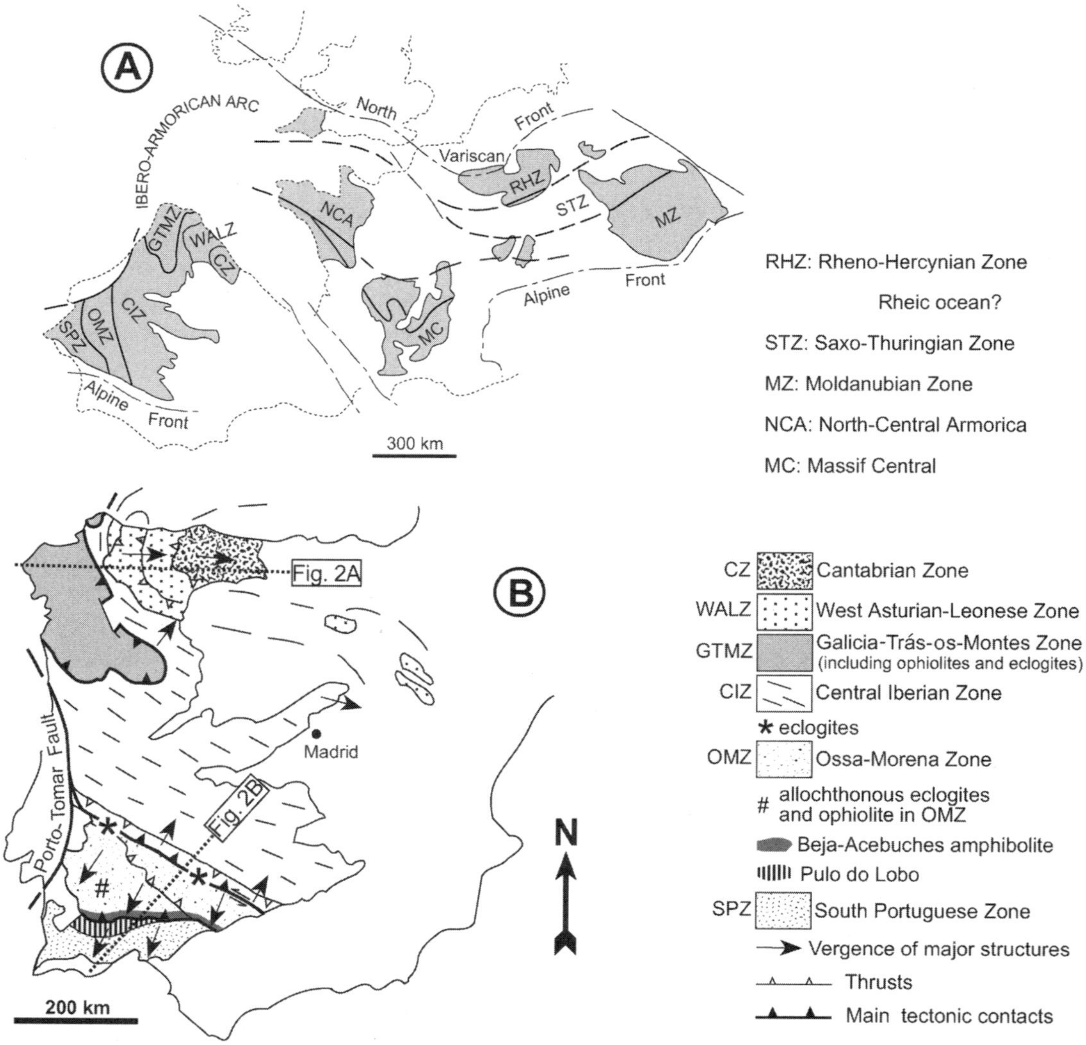

Figure 1. A: Simplified sketch of the Variscan belt showing the position of the main massifs (in grey pattern) and zones. B: Main zones, sutures and tectonic vergences in the Iberian Massif. The location of cross sections in Figure 2 is indicated.

The Iberian Massif is divided into a number of zones according to stratigraphic and tectonic criteria (Fig. 1B). The relationships between these zones are of very different significance. The Cantabrian, West Asturian–Leonese, and Central Iberian zones were part of the same Paleozoic continent (Bard et al., 1980; Matte, 1986; Martínez Catalán et al., 1997), which was Gondwana (Robardet et al., 1993) or an independent Armorican microcontinent (Tait et al., 1997). The paleogeographic significance of the Ossa-Morena zone is controversial. It may represent a fragment of that Paleozoic continent, which became an independent block during the Silurian and was amalgamated during the Devonian-Carboniferous continental collision of Gondwana or Armorica with Laurussia (Azor et al., 1994a; Simancas et al., 2001). The South Portuguese zone is thought to be a continental crust belonging to the other margin of the Rheic ocean, i.e., part of the Laurussia margin, in the same way as the Rheno-Hercynian zone of southern England and Central Europe (Fig. 1A; Crespo Blanc and Orozco, 1988; Fonseca and Ribeiro, 1993). A special case is the Galicia Trás-os-Montes zone (Fig. 1B), a complex allochthonous unit emplaced onto the Central Iberian zone that includes fragments of (1) the thinnest part of the Gondwana or Armorica margin, (2) an adjacent oceanic realm (ophiolitic units), and (3) continental crust of the other side (Arenas et al., 1986; Ribeiro et al., 1990; Martínez Catalán et al., 1996).

In this paper, we first summarize the main features of a west-east transect across Galicia Trás-os-Montes, Central Iberian, West Asturian–Leonese, and Cantabrian zones (northern branch), then a northeast-southwest transect across the Central Iberian, Ossa-Morena, and South Portuguese zones (southern branch); we argue that these two cross sections cannot be joined in order to build a single representative cross section of all the

Iberian Massif. Therefore, a noncylindrical tectonic scenario is envisaged for this transect of the Variscan orogen.

NORTHERN BRANCH

The northern Iberia cross section is well known and is briefly summarized here (Figs. 1B and 2A). It shows a clear west to east orogenic polarity, evidenced by the following features.

1. The main structures are recumbent folds and thrusts with east vergence (Pérez Estaún et al., 1991).

2. In the western part, allochthonous continental and ophiolitic terranes crop out (Galicia Trás-os-Montes zone), having reached granulitic and eclogitic metamorphic conditions in some of the units (Arenas et al., 1986; Martínez Catalán et al., 1996). To the east of these internally derived terranes, there is a medium- to low-grade schistose domain with east-vergent recumbent folds and thrusts (northern Central Iberian and West Asturian–Leonese zones). Farther east, the Cantabrian zone displays all the characteristics of an external zone: from the Cambrian to the early Carboniferous, it was a shoreline sedimentary environment (Truyols et al., 1990), then it received synorogenic deposits (Sánchez de Posada et al., 1990; Heredia et al., 1990), and finally it was transformed into a thin-skinned, unmetamorphosed, and nonschistose tectonic domain (Pérez Estaún et al., 1991).

3. The age of the first tectonometamorphic event ranges from Early Devonian in the Galicia Trás-os-Montes zone to Westphalian in the Cantabrian zone (Dallmeyer et al., 1997), thus providing evidence for the migration of the orogenic front. The features of the northern branch of the Iberian Massif lead us to infer the development of a complex orogenic wedge that initiated in the Early Devonian with the closure of an oceanic realm. During the Early and Middle Devonian, some units of the wedge underwent subduction, exhumation, and obduction (Martínez Catalán et al., 1996). The orogenic wedge (Fig. 2A) progressed eastward over the continental margin during Devonian and early Carboniferous time (Pérez Estaún et al., 1991; Pereira et al., 1999), until its arrest during the Stephanian (Heredia et al., 1990).

SOUTHERN BRANCH

The southern cross section of the Iberian Massif shows orogenic polarity from northeast to southwest (Fig. 1B). The polarity is marked by south-vergent recumbent folds and thrusts in the Ossa-Morena zone (Expósito, 2000) and south-vergent thin-skinned thrusts and overturned folds in the South Portuguese zone (Silva et al., 1990). A general cross section of the southern branch is shown in Figure 2B. The tectonic evolution of this branch is controversial, so its description is longer than that of the northern branch.

Boundary between the Central Iberian and the Ossa-Morena zones

One of the main points of discussion about the evolution of southwestern Iberia is the relationship between the Central Iberian zone and the Ossa-Morena zone, i.e., whether they were united during the Paleozoic Era. In this respect, several features must be emphasized: (1) the Middle Cambrian–Ordovician sequence of the Ossa-Morena zone, including abundant magmatism, indicates a continental rifting starting in Early-Middle Cambrian time (Liñán and Quesada, 1990; Mata and Munhá, 1990; Giese and Bühn, 1993); (2) the Silurian sequence in the Ossa-Morena zone has no clastic influx and is different from the

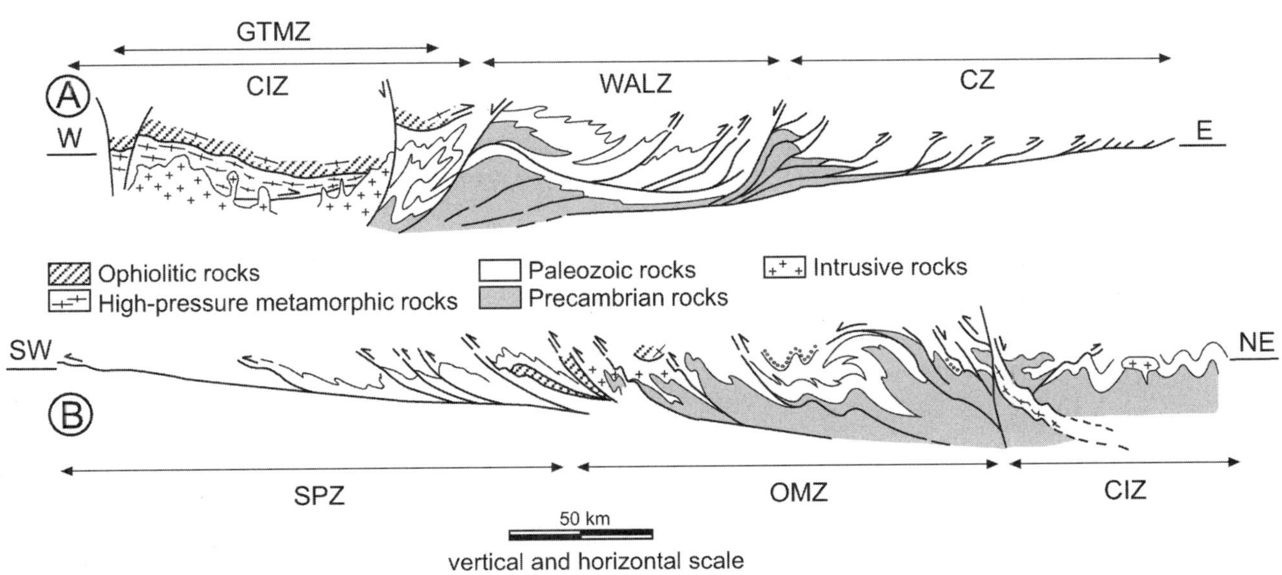

Figure 2. A: Cross-section of the northern branch of the Iberian Massif modified after Pérez Estaún et al. (1991). See location in Figure 1B. B: Cross-section of the southern branch of the Iberian Massif. See location in Figure 1B. Abbreviations as in Figure 1.

Silurian sequence of the Central Iberian zone (Robardet and Gutiérrez Marco, 1990; Gutiérrez Marco et al., 1998). Gutiérrez Marco et al. (1998) considered that the Ossa-Morena zone should occupy a very external position in the Gondwana or Armorica margin in order to prevent the arrival of terrigenous sediments, but for the same reason the Ossa-Morena zone may have been an independent terrane at that time. The investigation of the tectonometamorphic evolution of the Central Iberian–Ossa-Morena zone boundary has revealed it to be a major tectonic contact in which the large-scale recumbent folds and thrusts of the Ossa-Morena zone are footwall structures of the crustal-scale thrust of the Central Iberian zone onto the Ossa-Morena zone (Fig. 2B; Azor et al., 1994a; Expósito, 2000; Simancas et al., 2001). Eclogites crop out at this boundary (Abalos et al., 1991), for which a very young age of 340 ± 13 Ma has been proposed (single-grain U-Pb zircon dating; Ordóñez, 1998). Furthermore, magnetic regional anomalies in the southern Central Iberian zone have been modeled as a wedge of lower crustal-mantle rocks gently dipping under the Central Iberian zone and nearly emerging in the Central Iberian–Ossa-Morena zone boundary (Aller, 1996). All these data point toward the interpretation of the Ossa-Morena zone as a continental terrane independent from Gondwana (or Armorica) in Silurian time, so that the Central Iberian–Ossa-Morena zone boundary seems to be the suture of an oceanic realm.

Boundary between the Ossa-Morena and the South Portuguese zones

The Ossa-Morena–South Portuguese zone boundary has been interpreted as the suture of another oceanic realm, perhaps the Rheic ocean between Armorica (or Gondwana) and Laurussia, based on the oceanic signature of two basaltic-gabbroic units in this boundary (Figs. 1B and 2B): the Beja-Acebuches amphibolites and the Pulo do Lobo greenschists. On the basis of the mixed geochemical characteristics of the Beja-Acebuches amphibolites, a small ocean or backarc basin has been suggested (Dupuy et al., 1979; Munhá et al., 1989). The Pulo do Lobo, a schistose unit that crops out just to the south of the Beja-Acebuches amphibolites and contains metabasalts of oceanic affinity (Munhá, 1983), could represent a Devonian accretionary prism (Silva et al., 1990), although the corresponding volcanic arc is lacking.

An important issue concerning the evolution of the early collisional stage in southwest Iberia is the rooting and emplacement of small allochthonous outcrops in the southern part of the Ossa-Morena zone (Figs. 1B and 2B); the outcrops include eclogites from continental margin rocks and ophiolitic slices (Fonseca et al., 1999). As described by Fonseca et al. (1999), the geochemical signature of the ophiolitic slices is different from the Beja-Acebuches amphibolites, implying that they belong to a different oceanic realm. Geographic proximity and northward kinematic indicators favor the rooting of these allochthonous units in the Ossa-Morena–South Portuguese zone suture (Fonseca et al., 1999). However, the overall kinematic picture of the Ossa-Morena zone with southwestward tectonic transport, together with the need of a different oceanic domain, suggest that these allochthonous units may also have traveled from the Central Iberian–Ossa-Morena zone boundary.

Despite the existence of a number of unclear points concerning the precollision paleogeography and the collision tectonics of the Ossa-Morena–South Portuguese zone boundary, the presence of units with oceanic affinity is not debated, and consequently it is clear that there was some kind of intervening ocean. Because the South Portuguese zone correlates well with the Rheno-Hercynian zone of Central Europe (Oliveira et al., 1979), the latter obviously an external zone of the Variscan orogen that formed part of the Laurussia continental margin (Fig. 1A; Berthelsen, 1992; Franke, 1992), the same paleogeographic position is believed for the South Portuguese zone.

Tectonic evolution of the southern branch

It seems that every one of the three zones of the southern branch of the Iberian Massif (Central Iberian, Ossa-Morena, and South Portuguese zones) belonged to different continental masses in the Silurian. The collision between the Central Iberian zone and the Ossa-Morena zone would have occurred after closure of a small intervening oceanic domain. Collisional-type deformation in the Ossa-Morena zone, expressed as recumbent folds and thrusts with southwest vergence, is interpreted as the footwall deformation of the continental thrust of the Central Iberian zone onto the Ossa-Morena zone (Fig. 2B). A thin band in the southern Central Iberian zone is also affected by recumbent folds, but with opposite (northeast) vergence, representing a small domain of backfolding (Figs. 1B and 2B; Azor et al., 1994b). The age of this first Variscan deformation is Devonian, as suggested by synorogenic deposits dated as Early Devonian (Pereira et al., 1998; Piçarra et al., 1998), and by the unconformity at the base of the early Carboniferous deposits in the Ossa-Morena zone and the southernmost Central Iberian zone (Expósito, 2000; Martínez Poyatos et al., 1995). In this geological context, the age of 340 ± 13 Ma for the eclogites in the Central Iberian–Ossa-Morena zone boundary (Ordoñez, 1998) seems too young, probably representing the age of the high-temperature–intermediate-pressure metamorphism that postdates the high-pressure event.

The closure of the ocean between the Ossa-Morena zone and the South Portuguese zone must have also taken place before the Late Devonian because: (1) terrigenous deposits of this age postdate the emplacement of the ophiolitic Beja-Acebuches amphibolites (Oliveira, 1990), and (2) the age of the Beja gabbro, a nondeformed igneous body containing xenoliths of Acebuches amphibolites and other metamorphic rocks, is 350 Ma (U-Pb on zircon; Pin et al., 1999). According to Fonseca and Ribeiro (1993), the Beja-Acebuches ophiolite was obducted to the north, onto the continental margin of the Ossa-Morena zone. However, it mostly crops out under continental granulites of the

Ossa-Morena zone, as a north-dipping metagabbroic and metabasaltic slice in the bottom of which are metabasalts (Quesada, 1991). This attitude of the ophiolitic slice seems to be caused by tectonic overturning subsequent to its emplacement. In the same way, the ophiolitic slices and the high-pressure continental rocks that form allochthonous outcrops in the southern Ossa-Morena zone (Figs. 1B and 2B) would have been emplaced to the north, according to Fonseca et al. (1999). The probably coeval accretionary prism of Pulo do Lobo would rather witness a north-dipping subduction (Silva et al., 1990).

A turnover stage is manifested by the extension recorded in the South Portuguese zone during the Tournaisian and early Visean, which has given way to abundant rifting magmatism (Munhá, 1983) and the associated massive sulfide deposits constituting the famous Iberian Pyrite Belt. The tectonic scenario would have been transtensional, taking into account that in southwestern Iberia there are many kinematic indicators pointing toward a left-lateral component of displacement (Burg et al., 1981; Crespo Blanc, 1992; Fonseca and Ribeiro, 1993; Azor et al., 1994a; Expósito, 2000). Early Carboniferous transtension is well documented in the South Portuguese zone, but would probably have affected all of southwestern Iberia, because there are also early Carboniferous basins and bimodal volcanism in the Ossa-Morena and southern Central Iberian zones (Martínez Poyatos et al., 1998). The tectonic meaning of this extensional stage during collision is not yet well understood, although a tentative proposal is presented in the following.

Compression restarted in Namurian and Westphalian time, refolding previously folded rocks and producing tectonic inversion of the early Carboniferous basins. The southward migration of the middle Carboniferous deformation is well documented in the South Portuguese zone by the corresponding migration of the flysch deposits, which range in age from late Visean in the northern South Portuguese zone to early Westphalian in the southwest corner of Portugal (Oliveira, 1990). The very late stage of the continental collision is expressed as left-lateral strike-slip faulting of late Carboniferous age, in contrast with Central and eastern Europe, where right-lateral strike-slip faulting prevails. The presumed plate kinematics at this stage of advanced collision (Arthaud and Matte, 1977), and perhaps earlier (Badham, 1982; Shelley and Bossière, 2000), is consistent with a right-lateral pattern of faulting. The left-lateral component of movement recorded in the collisional history in southwest Iberia is due to the special orientation of this region as a part of the Ibero-Armorican arc (see following).

CENTRAL IBERIAN ZONE

The Central Iberian zone belongs to both branches of the Iberian Massif (Fig. 1B). However, it shows different characteristics in each transect.

In northwest Iberia, the Central Iberian zone crops out around the allochthonous complexes of the Galicia Trás-os-Montes zone as the relative autochthon (Fig. 1B). The deformation of the Central Iberian zone is expressed as east- to northeast-vergent overturned to recumbent folds. Synorogenic flyschoid units have been dated paleontologically as Frasnian (Pereira et al., 1999) and perhaps even Early Devonian (González Clavijo, 1997), implying that the onset of the deformation, i.e., the emplacement of the allochthonous Galicia Trás-os-Montes zone onto the autochthonous Central Iberian zone, is Devonian. Thus, the deformation history is claimed to be continuous in time from the stacking of the Galicia Trás-os-Montes zone units (Martínez Catalán et al., 1996) to the propagation of the deformation toward the foreland (Fig. 2A; Pérez Estaún et al., 1991).

The geometric picture and the timing of the deformation in the southern part of Central Iberian zone (Fig. 1B) are different. In a narrow band close to the Ossa-Morena zone, there is a sequence of Late Proterozoic and early Paleozoic rocks affected by northeast-vergent folds (Figs. 1B and 2B). The age of this deformation is Devonian because early Carboniferous deposits unconformably overlie the folded sequence (Martínez Poyatos et al., 1995). Despite the similarities both in style and time with the deformation encountered in the northwest part of the Central Iberian zone, Devonian deformation is restricted in the south to a 30-km-wide band. North of this band, no trace of Devonian deformation exists except for a generalized lacuna in the Middle Devonian; the first deformational structures in the central domain of the Central Iberian zone are upright folds of Namurian-Westphalian age (Fig. 2B; Díez Balda et al., 1990; Martínez Poyatos et al., 1995). Devonian collisional-type deformation is detected in the Central Iberian zone of northwest Iberia, but not in the central and southern parts, except in a restricted back-folding domain of the Central Iberian–Ossa-Morena zone boundary (Figs. 1B and 2B).

A single composite transect of the Iberian Massif, obtained from the combination of the northern and southern branches (Figs. 1B and 2), is the image of a bivergent orogen (Matte, 1986, 1991; Martínez Catalán et al., 1997). However, from the preceding discussion, it can be derived that the Central Iberian zone is not cylindrical, which prevents joining two laterally displaced cross sections (Fig. 1B). Instead, a noncylindrical tectonic scenario explaining the features of the Iberian Massif must be invoked.

TECTONIC SCENARIO FOR THE IBERIAN MASSIF

In the northern branch of the Iberian Massif subduction to the west (in present-day coordinates) can be deduced from (1) the thrusting of suture terrains of the Galicia Trás-os-Montes zone over the Central Iberian zone (Figs. 1B and 2A), (2) the east vergence of the structures, and (3) the eastern migration of the deformation. In the southern branch, the Central Iberian zone is a comparatively less deformed crustal domain that constitutes the hanging wall of the Central Iberian–Ossa-Morena zone suture (Fig. 2B), suggesting that subduction here had northeast polarity. The link between the two opposite subduction zones may have been a transform fault, and the Porto-Tomar shear zone (Fig. 1B; Ribeiro et al., 1980; Dias and Ribeiro, 1993) is tentatively taken as related to such a transform (Fig. 3).

We envisage for the time just prior to collision (the Silurian-Devonian boundary) a plate tectonic pattern in which the Cantabrian, West Asturian–Leonese, and Central Iberian zones belonged to the same continent, the Ossa-Morena zone being an isolated continental mass (Fig. 3A). In the oceanic domain between the Central Iberian zone and the Ossa-Morena zone, a transform fault connected two subduction zones of opposite polarity. The subduction polarity on one side of the transform allowed obduction onto the Central Iberian zone, piggybacking continental crust previously emplaced on it (Fig. 3B); such an orogenic wedge added to its base material of the underthrust continental edge so as to constitute the Galicia Trás-os-Montes zone (Fig. 3C; Martínez Catalán et al., 1996). On the other side of the transform fault, thrusting of the Central Iberian zone onto the Ossa-Morena zone was taking place at the same time (Azor et al., 1994a; Simancas et al., 2001). Subsequently, the deformation propagated toward the respective footwalls, i.e., toward the southwest in the Ossa-Morena zone and toward the east in the northern Central Iberian zone, both branches progressively rotating to form the Ibero-Armorican arc (Fig. 3, C and D). This arch-shaped structure was considered by Brun and Burg (1982) and Burg et al. (1987) to be essentially the result of a corner effect in a tectonic context reminiscent of that proposed here, although with a different organization. We consider that, although the arc probably started under the influence of such an effect, its main architecture can be attributed to progressive indentation of a continental salient (Matte and Ribeiro, 1975), as in the northwest Himalaya (Treloar and Coward, 1991). Most of the arc rotation should have been accomplished during the Carboniferous, although a significative tightening could have taken place in Early Permian time (Weil et al., 2000).

Impingement of an indentor provokes escape tectonics (Tapponier et al., 1982). In the Iberian case, there is lateral escape of a crustal wedge formed by Ossa-Morena and South Portuguese zones, bounded by the Central Iberian–Ossa-Morena zone contact and by the Porto-Tomar fault (Fig. 3C). The Porto-Tomar fault (Ribeiro et al., 1980) is a Carboniferous continental shear zone that we believe nucleated from, and connected with, the transform fault separating the two opposite subduction

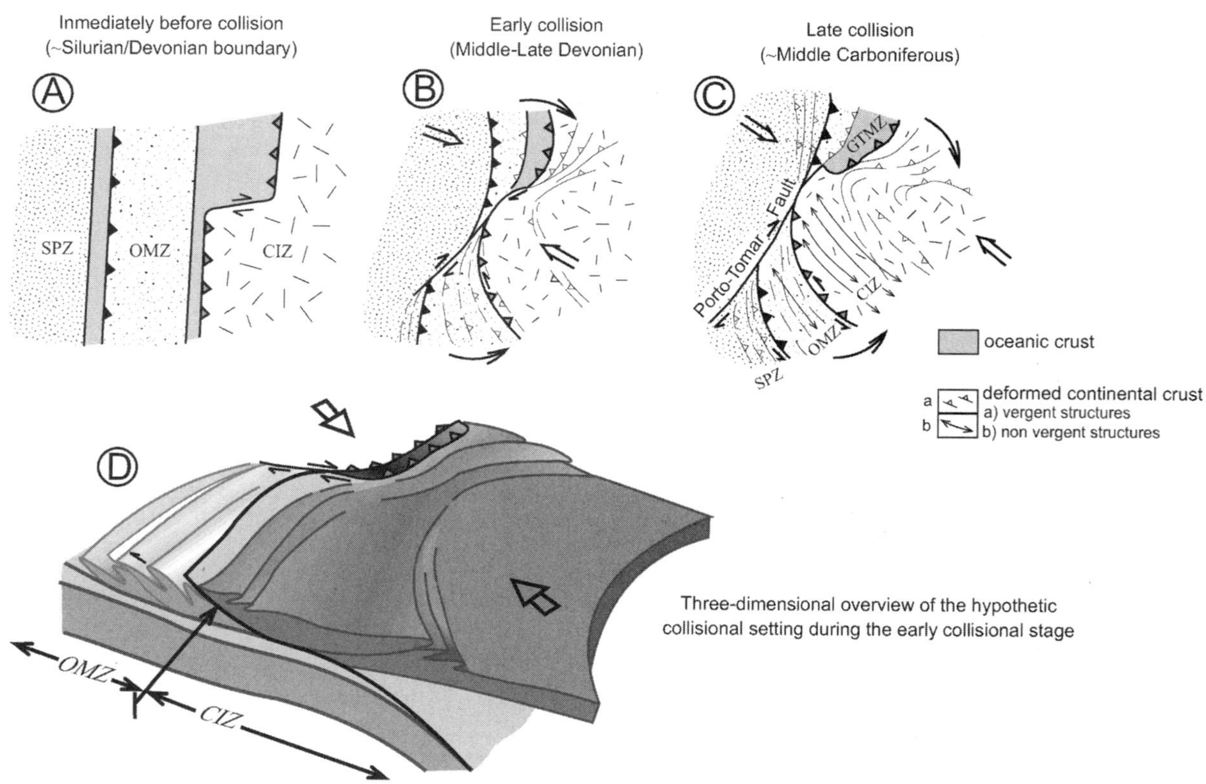

Figure 3. Sketches showing the Variscan evolution proposed for the Iberian Massif. A: Pre-collisional stage with the South Portuguese zone–Ossa-Morena zone and Ossa-Morena zone–Central Iberian zone oceanic domains being subducted. For the Ossa-Morena zone-Central Iberian zone suture, a transform fault connects two segments of opposite subduction polarity. B: Early collisional stage with the transform fault connecting the northern branch of east-vergence and the Southern one of southwest vergence (in present-day coordinates). C: Late collisional stage. D: Three dimensional sketch for the early collisional stage. See text for further explanations. Sketches are not to scale.

zones. The space problem created by indentation can also account for the middle Carboniferous shortening affecting the Iberian Massif (Fig. 3C). This shortening is manifested as upright folding in the central and southern Central Iberian zone and in the Ossa-Morena zone, whereas foreland-vergent structures formed in the external zones (Fig. 3C). In southwest Iberia, final indentation before blocking is reflected as semibrittle to brittle left-lateral strike-slip faulting of Stephanian age (Simancas, 1983; Crespo Blanc, 1992; Jackson and Sanderson, 1992). Left-lateral kinematics in southwest Iberia contrast with prominent right-lateral faulting of the same age in Central and eastern Europe, the latter being congruent with the general plate kinematics at the final stage of the Variscan collision (Arthaud and Matte, 1977). Undoubtely, the reason for such a different kinematics in southwest Iberia is its rotated orientation with respect to the rest of the European Variscides, due to development of the Ibero-Armorican arc (Fig. 1A; Ribeiro et al., 1995).

In the northern branch of the Iberian Massif, crustal compression can be deduced to be continuous from the start of the continental collision (Early Devonian) to its end in the Stephanian. Extensional structures formed at different times, but they have been interpreted as gravitational collapse of a thick and continuously evolving orogenic wedge (Martínez Catalán et al., 1996; Pérez Estaún et al., 1991; Díez Balda et al., 1995). The same time span for the collisional event is found in the southern branch, but here crustal compression was interrupted in the early Carboniferous by a transient tensional stage that originated basins and bimodal intraplate volcanism. In the Rheno-Hercynian zone of Central Europe (Fig. 1A), a prominent continental extension involving bimodal magmatism developed in Devonian–early Carboniferous time, representing an important problem of the Variscan orogeny (Franke, 1992). Some researchers have suggested extension in a suprasubduction domain to explain the Devonian-Carboniferous extensional events (Floyd, 1982; Martínez Catalán et al., 1997). However, the presumed polarity of the subduction near the Rheno-Hercynian (Franke, 1992) and in southwest Iberia (Fig. 3) is not consistent with this interpretation. Moreover, in southwest Iberia early Carboniferous extension is not confined to the South Portuguese zone, the equivalent of the Rheno-Hercynian zone, but also affects the Ossa-Morena zone and the southern part of the Central Iberian zone, implying that it cannot be taken as synkinematic extension in the upper plate of a subduction zone. The tectonic scenario proposed here suggests that a small clockwise rotation of the Central Iberian–West Asturian–Leonese–Cantabrian zone indentor may result in a change from transpressional to transtensional tectonic regime in southwest Iberia (Fig. 3C).

CORRELATIONS AROUND THE IBERO-ARMORICAN ARC

Because of the similar Carboniferous evolution and the outer position of both zones in the Ibero-Armorican arc, the South Portuguese zone (Iberia) and the Rheno-Hercynian zone (Central Europe) have been widely correlated, assuming that they could have formed part of a continuous continental domain (Figs. 1A and 4A). This implies, in turn, that the Ossa-Morena–South Portuguese zone suture is equivalent to the suture marked by the Lizard Complex in Cornwall (southwest England) and by the contact between the Rheno-Hercynian and the Saxo-Thuringian zones (Fig. 4, A and B). Other correlations between zones and sutures are controversial.

The suture located north of the French Massif Central (Fig. 1A) has been followed toward Iberia in the allochthonous terrains of the Galicia Trás-os-Montes zone and in the Central Iberian–Ossa-Morena zone boundary (Matte and Burg, 1981;

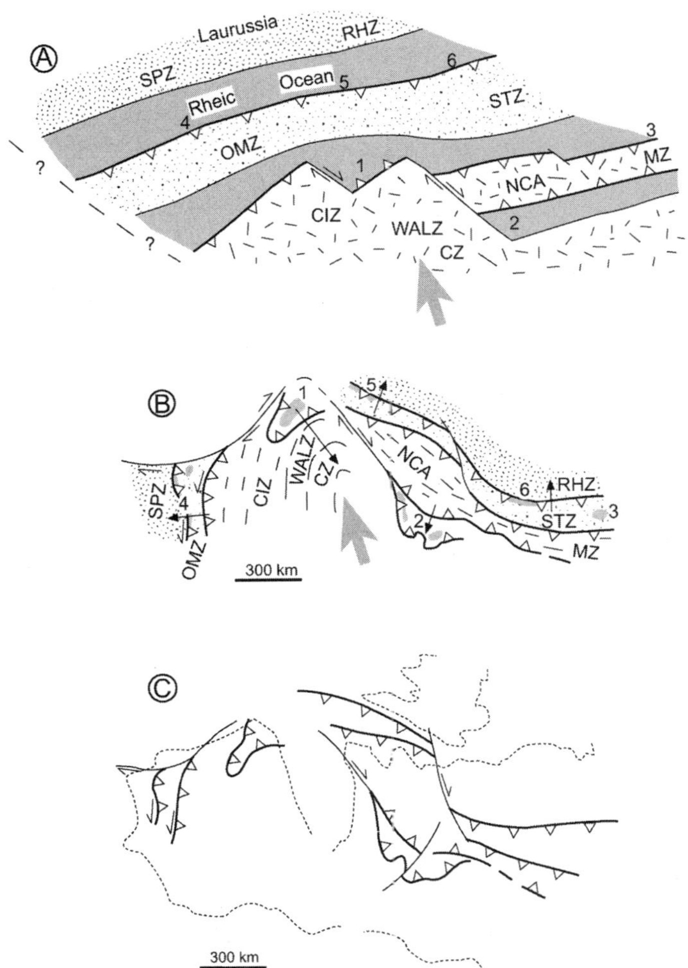

Figure 4. A: Hypothetical reconstruction of the Variscan belt prior to collision with the oceanic domains and the relative positions of the different zones. B: zones and sutures after closure of the oceanic domains and formation of the Ibero-Armorican arc. Abbreviations as in Figure 1. Grey pattern corresponds to ophiolitic outcrops, and numbers 1 to 6 in Figures 4A and 4B depict the oceanic domains and their corresponding ophiolites; 1—Galicia Trás-os-Montes, 2—Massif Central, 3—Münchberg, 4—Beja-Acebuches, 5—Lizard, 6—Giessen. C: Geographical location (dot line) of the Variscan sutures in western Europe (note, however, that the position of Iberia is prior to the opening of the Gulf of Biscay).

Burg et al., 1987; Matte, 1986, 1991). However, Ballèvre et al. (1992) have shown the sedimentary identity between North-Central Armorica (Fig. 1A) and the Central Iberian zone during the Ordovician, arguing that no suture can cross such a continuous paleogeographical domain. On this ground, we propose the plate configuration shown in Figure 4A, in which a transform fault connects the oceanic domain of the Massif Central with the oceanic domain between the Central Iberian zone and the Ossa-Morena zone. Thus, paleogeographical continuity is possible between North-Central Armorica and the Central Iberian zone, appearing the configuration of the indentor constituted by the Central Iberian zone, West Asturian–Leonese zone and Cantabrian zone. The collision that took place after closure of the oceanic domains gave way to the tectonic pattern presented in Figure 4B.

CONCLUSIONS

Despite the opposite orogenic polarities of the Northern and southern branches of the Iberian Massif, a detailed analysis of their tectonic evolution suggests an alternative to the usually accepted image of a cylindrical bivergent orogen. In the northern branch, the Variscan tectonic evolution started with a west-dipping subduction, followed by obduction of oceanic-derived units and eastward propagation of the deformation. The southern branch includes two oceanic sutures, both related to northern-dipping subduction. The tectonic evolution here seems more complex, with an initial Devonian compressional stage responsible for large-scale southwest-vergent recumbent folds and thrusts, followed by a generalized transtensional Early Carboniferous stage, and a second compressional stage at Middle-Late Carboniferous. The change in subduction polarity between the northern branch, where the Central Iberian zone is the footwall of the allochthonous terranes of the Galicia Trás-os-Montes zone , and the southern branch, where the Central Iberian zone is the hanging wall of the suture-related rocks, is explained by the existence of a transform fault. This transform fault would enter the continent during collision and produced the present-day Porto-Tomar fault.

Previous attempts to correlate the Iberian Massif with other Variscan European massifs through the Ibero-Armorican arc fail when suspect sutures, regional vergences and stratigraphic data are put together. The main difficulty arises from the stratigraphic and paleontological similarities between the Armorican domain and the Central Iberian zone, which prevent the Massif Central ocean and the Galicia ocean to be considered as continuous. We propose the existence of another transform fault connecting these two oceanic domains. The closure of such oceanic domains bounded by transform faults, followed by collisional deformation, rotation and escape tectonics in opposite directions would have generated the Ibero-Armorican arc.

ACKNOWLEDGMENTS

Financial suport was given by the CICYT (Spanish Ministry of Education) grant number PB96-1452-C03-01. We thank Francisco Gonzálvez García for improving our English text. Comments and suggestions by D. Roeder and J.P. Burg are appreciated.

REFERENCES CITED

Abalos, B., Gil Ibarguchi, J.I., and Eguiluz, L., 1991, Cadomian subduction/collision and Variscan transpression in the Badajoz-Córdoba shear belt, southwest Spain: Tectonophysics, v. 199, p. 51–72.

Aller, J., 1996, 2.5 modelling and structural implications of magnetic anomalies associated with the boundary between the Central-Iberian and Ossa-Morena zones in the Hornachos sector (Hercynian Belt, southwest Spain): Revista de la Sociedad Geológica de España, v. 9, p. 608–621.

Arenas, R., Gil Ibarguchi, J.I., González Lodeiro, F., Klein, E., Martínez Catalán, J.R., Ortega, E., Pablo Macía, J.G., and Peinado, M., 1986, Tectonostratigraphic units in the complexes with mafic and related rocks of the NW of the Iberian Massif: Hercynica, v. 2, p. 87–110.

Arthaud, F., and Matte, P., 1977, Late Paleozoic strike-slip faulting in southern Europe and northern Africa: Result of a right lateral shear zone between the Appalachians and the Urals: Geological Society of America Bulletin, v. 88, p. 1305–1320.

Azor, A., González Lodeiro, F., and Simancas, J.F., 1994a, Tectonic evolution of the boundary between the Central Iberian and Ossa-Morena zones (Variscan belt, southwest Spain): Tectonics, v. 13, p. 45–61.

Azor, A., González Lodeiro, F., Martínez Poyatos, D., and Simancas, J.F., 1994b, Regional significance of kilometric-scale NE-vergent recumbent folds associated with E- to SE-directed shear on the southern border of the Central Iberian zone (Hornachos-Oliva region, Variscan belt, Iberian Peninsula): Geologische Rundschau, v. 83, p. 377–387.

Badham, J.P.N., 1982, Strike-slip orogens, an explanation for the Hercynides: Journal of the Geological Society of London, v. 139, p. 493–504.

Ballèvre, M., Paris, F., and Robardet, M., 1992, Corrélations ibéro-armoricaines au Paléozoïque: Une confrontation des données paléobiogéographiques et tectono-métamorphiques: Comptes Rendus de l'Académie des Sciences de Paris, t. II, v. 315, p. 1783–1789.

Bard, J.P., Capdevila, R., and Matte, Ph., 1971, La structure de la chaîne Hercynienne de la Meseta Ibérique: Comparaison avec les segments voisins, *in* Histoire structurale du Golfe de Gascogne: Publications de l'Institut Français du Pétrole, v. 22, I, 4, p. 1–68.

Bard, J.P., Burg, J.-P., Matte, P., and Ribeiro, A., 1980, La chaîne hercynienne d'Europe occidental en terme de tectonique des plaques: Mémoire du Bureau de Recherche Géologique et Minière, v. 108, p. 233–246.

Berthelsen, A., 1992, Mobile Europe, *in* Blundell, D., Freeman, R., and Mueller, S., eds., A continent revealed: The European geotraverse: Cambridge, Cambridge University Press, p. 11–27.

Brun, J.P., and Burg, J.-P., 1982, Combined thrusting and wrenching in the Ibero-Armorican arc: A corner affect during continental collision: Earth and Planetary Science Letters, v. 61, p. 319–332.

Burg, J.-P., Iglesias, M., Laurent, P., Matte, P., and Riberio, A., 1981, Variscan intracontinental deformation: the Coimbra-Badajoz Shear zone (SW Iberian Peninsula): Tectonophysics, v. 78, p. 161–177.

Burg, J.-P., Bale, P., Brun, J.P., and Girardeau, J., 1987, Stretching lineation and transport direction in the Ibero-Armorican arc during the siluro-devonian collision: Geodinamica Acta, v. 1, p. 71–87.

Crespo Blanc, A., 1992, Structure and kinematics of a sinistral transpressive suture between the Ossa-Morena and the South Portuguese zones, South Iberian Massif: Journal of the Geological Society of London, v. 149, p. 401–411.

Crespo Blanc, A., and Orozco, M., 1988, The Southern Iberian shear zone: A major boundary in the Hercynian foldbelt: Tectonophysics, v. 148, p. 221–227.

Dallmeyer, R.D., Martínez Catalán, J.R., Arenas, R., Gil Ibarguchi, J.I., Gutiérrez Alonso, G., Farias, P., Bastida, F., and Aller, J., 1997, Diachronous tectonothermal activity in the NW Iberian Massif: Evidence from $^{40}Ar/^{39}Ar$ data of regional fabrics: Tectonophysics, v. 277, p. 307–337.

Dias, R., and Ribeiro, A., 1993, Porto-Tomar Shear zone, a major structure since the beginning of the Hercynian orogeny: Terra Abstracts, supplement number 6 to Terra Nova, v. 5, p. 11.

Díez Balda, M.A., Martínez Catalán, J.R., and Ayarza, P., 1995, Syn-collisional extensional collapse parallel to the orogenic trend in a domain of steep tectonics: The Salamanca Detachment zone (Central Iberian zone, Spain): Journal of Structural Geology, v. 17, p. 163–182.

Díez Balda, M.A., Vegas, R., and González Lodeiro, F., 1990, Part IV (Central-Iberian zone), Autochthonous sequences: Structure, *in* Dallmeyer, R.D., and Martínez García, E., eds., Pre-Mesozoic geology of Iberia: Berlin-Heidelberg, Springer-Verlag, p. 172–188.

Dupuy, C., Dostal, J., and Bard, J.P., 1979, Trace elements geochemistry of Paleozoic amphibolites from SW Spain: Tschermaks Mineralogische Petrographische Mitteilungen, v. 26, p. 87–93.

Expósito, I., 2000, Evolución estructural de la mitad septentrional de la Zona de Ossa-Morena y su relación con el límite Zona de Ossa-Morena/Zona Centroibérica [Ph.D. thesis]: Granada, Spain, Universidad de Granada, 295 p.

Floyd, P.A., 1982, Chemical variation in Hercynian basalts relative to plate tectonics: Journal of the Geological Society of London, v. 139, p. 505–520.

Fonseca, P, and Ribeiro, A., 1993, Tectonics of the Beja-Acebuches ophiolite: A major suture in the Iberian Variscan Foldbelt: Geologische Rundschau, v. 82, p. 440–447.

Fonseca, P., Munhá, J., Pedro, J., Rosas, F., Moita, P., Aráujo, A., and Leal, N., 1999, Variscan ophiolites and high-pressure metamorphism in Southern Iberia: Ofioliti, v. 24, p. 259–268.

Franke, W., 1992, Phanerozoic structures and events in Central Europe, *in* Blundell, D., Freeman, R., and Mueller, S., eds., A continent revealed: The European geotraverse: Cambridge, Cambridge University Press, p. 164–180.

Giese, U., and Bühn, B., 1993, Early Paleozoic rifting and bimodal volcanism in the Ossa-Morena zone of southwest Spain: Geologische Rundschau, v. 83, p. 143–160.

González Clavijo, E., 1997, La geología del sinforme de Alcañices, oeste de Zamora [Ph.D. thesis]: Salamanca, Spain, Universidad de Salamanca, 330 p.

Gutiérrez Marco, J.C., Robardet, M., and Piçarra, J.M., 1998, Silurian Stratigraphy and Paleogeography of the Iberian Peninsula (Spain and Portugal): Temas Geológico-Mineros, ITGE, Madrid, v. 23, p. 13–39.

Heredia, N., Rodríguez Fernández, L.R., and Wagner, R.H., 1990, Part II (Cantabrian and Palentian zones), Stratigraphy: Carboniferous of the Palentian zone, *in* Dallmeyer, R.D., and Martínez García, E., eds., Pre-Mesozoic geology of Iberia: Berlin-Heidelberg, Springer-Verlag, p. 34–38.

Jackson, P., and Sanderson, D.J., 1992, Scaling of faults displacements from the Badajoz-Córdoba shear zone, SW Spain: Tectonophysics, v. 210, p. 179–190.

Liñán, E., and Quesada, C., 1990, Part V (Ossa-Morena zone), stratigraphy: Rift phase (Cambrian), *in* Dallmeyer, R.D., and Martínez García, E., eds., Pre-Mesozoic geology of Iberia: Berlin-Heidelberg, Springer-Verlag, p. 259–266.

Martínez Catalán, J.R., Arenas, R., Díaz García, F., Rubio Pascual, F.J., Abati, J., and Marquínez, J., 1996, Variscan exhumation of a subducted Paleozoic continental margin: The basal unit of the Ordenes Complex, Galicia, NW Spain: Tectonics, v. 15, p. 106–121.

Martínez Catalán, J.R., Arenas, R., Díaz García, F., and Abati, J., 1997, Variscan accretionary complex of northwest Iberia: Terrane correlation and succession of tectonothermal events: Geology, v. 25, p. 1103–1106.

Martínez Poyatos, D., Azor, A., González Lodeiro, F., and Simancas, J.F., 1995, Timing of the Variscan structures on both sides of the Ossa-Morena/Central Iberian contact (southwest Iberian Massif): Comptes Rendus de l'Académie des Sciences de Paris, t. II, v. 321, p. 609–615.

Martinez Poyatos, D., Simancas, J.F., Azor, A., Gonzalez Lodeiro, F., and Expósito, I., 1998, Recumbent back-folding and extensional collapse in the hanging wall of the Ossa-Morena/Central Iberian suture (Iberian Peninsula): Acta Universitatis Carolinae, v. 42, p. 301–302.

Mata, J., and Munhá, J., 1990, Magmatogénese de Metavulcanitos Câmbricos do Nordeste Alentejano. Os estádios iniciais de "rifting" continental: Comunicaçoes dos Serviços Geologicos de Portugal, v. 76, p. 61–89.

Matte, Ph., 1986, Tectonics and plate tectonics model for the Variscan Belt of Europe: Tectonophysics, v. 126, p. 329–374.

Matte, Ph., 1991, Accretionary history and crustal evolution of the Variscan belt in Western Europe: Tectonophysics, v. 196, p. 309–337.

Matte, Ph., and Burg, J.-P., 1981, Sutures, thrusts and nappes in the Variscan arc of western Europe: Plate tectonics implications, *in* McClay, K., and Price, N.J., eds., Thrust and nappe tectonics: London, Geological Society of London Special Publication 9, p. 353–358.

Matte, Ph., and Ribeiro, A., 1975, Forme et orientation de la virgation hercynienne de Galice. Relations avec le plissement et hypothéses sur la genèse de l'arc Ibero-Armoricain: Comptes Rendus de l'Académie des Sciences de Paris, v. 280D, p. 2825–2828.

Munhá, J., 1983, Hercynian magmatism in the Iberian Pyrite Belt, *in* Sousa, A., and Oliveira, J.T., eds., The Carboniferous of Portugal: Memorias dos Serviços Geologicos de Portugal, v. 29, p. 39–81.

Munhá, J., Oliveira, J.T., Ribeiro, A., Quesada, C., Fonseca, P., and Castro, P., 1989, Accreted terrains in southern Iberia: the Beja-Acebuches ophiolite and related oceanic sequences: International Geological Congress, 28th, Washington, Abstracts, v. 2, p. 312–314.

Oliveira, J.T., Horn, M., and Paproth, E., 1979, Preliminary note on the stratigraphy of the Baixo Alentejo Flysch Group, Carboniferous of Southern Portugal and on the paleogeographic development, compared to corresponding units in Northwest Germany: Comunicaçoes dos Serviços Geologicos de Portugal, v. 65, p. 151–168.

Oliveira, J.T., 1990, Part VI (South Portuguese zone): Stratigraphy and synsedimentary tectonism, *in* Dallmeyer, R.D., and Martínez García, E., eds., Pre-Mesozoic geology of Iberia: Berlin-Heidelberg, Springer-Verlag, p. 334–347.

Ordóñez, B., 1998, Geochronological studies of the Pre-Mesozoic basement of the Iberian Massif: The Ossa-Morena zone and the allochthonous complexes within the Central Iberian zone [Ph.D. dissert]: Zürich, Switzerland, Eidgenössische Technische Hochschule, v. 12.940, 235 p.

Pereira, Z., Piçarra, J.M., and Oliveira, J.T., 1998, Palinomorfos do Devónico Inferior da regiao do Barrancos (Zona de Ossa-Morena): Comunicaçoes do V Congresso Nacional de Geologia, Portugal, v. 84, p. A18–20.

Pereira, Z., Meireles, C., and Pereira, E., 1999, Upper Devonian palynomorphs of NE sector of Trás-os-Montes (Central Iberian zone): XV Reunión de Geología del Oeste Peninsular (International Meeting on Cadomian orogens) and Annual Meeting of IGCP Project 376 (Laurentia-Gondwana connections before Pangea), Badajoz, Spain, Extended Abstracts, p. 201–206.

Pérez Estaún, A., Martínez Catalán, J.R., and Bastida, F., 1991, Crustal thickening and deformation sequence in the footwall to the suture of the Variscan belt of northwest Spain: Tectonophysics, v. 191, p. 243–253.

Piçarra, J.M., Pereira, Z., and Oliveira, J.T., 1998, Novos datos sobre a idade da sucessao Silúrico-Devónica do Sinclinal de Terena, na regiao de Barrancos. Implicaçoes geodinamicas: Comunicaçoes do V Congresso Nacional de Geologia, Portugal, v. 84, p. A15–16.

Pin, C., Paquette, J.L., and Fonseca, P., 1999, 350 Ma (U-Pb zircon) igneous emplacement age and Sr-Nd isotopic study of the Beja gabbroic complex (S Portugal): XV Reunión de Geología del Oeste Peninsular (International Meeting on Cadomian orogens) and Annual Meeting of IGCP Project 376 (Laurentia-Gondwana connections before Pangea), Badajoz, Spain, Extended Abstracts, p. 219–222.

Quesada, C., 1991, Introducción al significado geodinámico del límite entre las zonas de Ossa-Morena y Surportuguesa: XI Reunión de Geología del Oeste Peninsular, Aracena-Beja, Spain and Portugal, Field Trip Guide-Book, p. 2–24.

Ribeiro, A., Dias, R., and Brandao Silva, J., 1995, Genesis of the Ibero-Armorican arc: Geodinamica Acta, v. 8, p. 173–184.

Ribeiro, A., Pereira, E., and Severo, L., 1980, Análisis da deformaçao da zona de cizalhamento Porto-Tomar na transversal de Oliveira de Azemeis: Comunicaçoes Serviços Geologicos de Portugal, v. 66, p. 3–9.

Ribeiro, A., Pereira, E., and Dias, R., 1990, Part IV (Central Iberian zone), Allochthonous sequences: Structure in the northwest of the Iberian Peninsula, *in* Dallmeyer, R.D., and Martínez García, E., eds., Pre-Mesozoic geology of Iberia: Berlin-Heidelberg, Springer-Verlag, p. 220–236.

Robardet, M., and Gutiérrez Marco, J.C., 1990, Part VII: Sedimentary and faunal domains in the Iberian Peninsula during lower Paleozoic times, *in* Dallmeyer, R.D., and Martínez García, E., eds., Pre-Mesozoic geology of Iberia: Berlin-Heidelberg, Springer-Verlag, p. 383–395.

Robardet, M., Blaise, J., Bouix, E., Gourvennec, R., Lardeux, H., Le Herisse, A., Le Menn, J., Melou, M., Paris, F., Plusquellec, Y., Poncet, J., Regnault, S., Rioult, M., and Weyant, M., 1993, Paléogéographie de l'Europe occidentale de l'Ordovicien au Dévonien: Bulletin de la Société Géologique de France, v. 164, p. 683–695.

Sánchez de Posada, L.C., Martínez Chacón, M.L., Mendez Fernández, C., Menéndez Álvarez, J.R., Truyols, J., and Villa, E., 1990, Part II (Cantabrian and Palentian zones), Stratigraphy: Carboniferous Pre-Stephanian Rocks of the Asturian-Leonese Domain (Cantabrian zone), *in* Dallmeyer, R.D., and Martínez García, E., eds., Pre-Mesozoic geology of Iberia: Berlin-Heidelberg, Springer-Verlag, p. 24–33.

Sheley, D., and Bossière, G., 2000, A new model for the Hercynian orogen of Gondwanan France and Iberia: Journal of Structural Geology, v. 22, p. 757–776.

Silva, J.B., Oliveira, J.T., and Ribeiro, A., 1990, Part VI (South Portuguese zone): Structural outline, *in* Dallmeyer, R.D., and Martínez García, E., eds., Pre-Mesozoic geology of Iberia: Berlin-Heidelberg, Springer-Verlag, p. 348–362.

Simancas, J.F., 1983, Geología de la extremidad oriental de la Zona Sudportuguesa [Ph.D. thesis]: Granada, Spain, Universidad de Granada, 439 p.

Simancas, J.F., Martínez Poyatos, D., Expósito, I., Azor, A., and González Lodeiro, F., 2001, Structure of the SW Iberian Massif on both sides of a Variscan suture: The Ossa-Morena/Central Iberian contact: Tectonophysics, v. 332, p. 295–308.

Tait, J.A., Bachtadse, V., Franke, W., and Soffel, H.C., 1997, Geodynamic evolution of the European Variscan fold belt: Paleomagnetic and geological constraints: Geologische Rundschau, v. 86, p. 585–598.

Tapponier, P., Peltzer, G., Le Dain, A.Y., Armijo, R., and Cobbing, P., 1982, Propagating extrusion tectonics in Asia: New insights from simple experiments with plasticine: Geology, v. 10, p. 611–616.

Treloar, P.J., and Coward, M.P., 1991, Indian Plate motion and shape: Constraints on the geometry of the Himalayan orogen: Tectonophysics, v. 191, p. 189–198.

Truyols, J., Arbizu, M.A., García Alcalde, J.L., García López, S., Mendez Bedia, I., Soto, F., and Truyols Massoni, M., 1990, Part II (Cantabrian and Palentian zones), Stratigraphy: The Asturian-Leonese Domain (Cantabrian zone), *in* Dallmeyer, R.D., and Martínez García, E., eds., Pre-Mesozoic geology of Iberia: Berlin-Heidelberg, Springer-Verlag, p. 10–19.

Weil, A.B., Van der Voo, R., van der Pluijm, B.A., and Parés, J.M., 2000, The formation of an orocline by multiphase deformation: A paleomagnetic investigation of the Cantabria-Asturias Arc (northern Spain): Journal of Structural Geology, v. 22, p. 735–756.

Manuscript Accepted by the Society February 8, 2002

Printed in the U.S.A.

Geological Society of America
Special Paper 364
2002

Paleozoic evolution of pre-Variscan terranes: From Gondwana to the Variscan collision

Gérard M. Stampfli
Institut de Géologie et Paléontologie, Université de Lausanne, CH-1015 Lausanne, Switzerland
Jürgen F. von Raumer
Institut de Minéralogie et Pétrographie, Université de Fribourg, CH-1700 Fribourg, Switzerland
Gilles D. Borel
Institut de Géologie et Paléontologie, Université de Lausanne, CH-1015 Lausanne, Switzerland

ABSTRACT

The well-known Variscan basement areas of Europe contain relic terranes with a pre-Variscan evolution testifying to their peri-Gondwanan origin (e.g., relics of Neoproterozoic volcanic arcs, and subsequent stages of accretionary wedges, backarc rifting, and spreading). The evolution of these terranes was guided by the diachronous subduction of the proto-Tethys oceanic ridge under different segments of the Gondwana margin. This subduction triggered the emplacement of magmatic bodies and the formation of backarc rifts, some of which became major oceanic realms (Rheic, paleo-Tethys). Consequently, the drifting of Avalonia was followed, after the Silurian and a short Ordovician orogenic event, by the drifting of Armorica and Alpine domains, accompanied by the opening of the paleo-Tethys. The slab rollback of the Rheic ocean is viewed as the major mechanism for the drifting of the European Variscan terranes. This, in turn, generated a large slab pull force responsible for the opening of major rift zones within the passive Eurasian margin. Therefore, the first Middle Devonian Variscan orogenic event is viewed as the result of a collision between terranes detached from Gondwana (grouped as the Hun superterrane) and terranes detached from Eurasia. Subsequently, the amalgamated terranes collided with Eurasia in a second Variscan orogenic event in Visean time, accompanied by large-scale lateral escape of major parts of the accreted margin. Final collision of Gondwana with Laurussia did not take place before Late Carboniferous time and was responsible for the Alleghanian orogeny.

INTRODUCTION

Relics of the Variscan mountain chain are well known from many places in Europe (e.g., Iberia, Armorica, Moesia, the French Central Massif, the Saxo-Thuringian and Moldanubian domains, and Alpine pre-Mesozoic basement areas; Fig. 1), and modern reviews reveal their complex evolution since the Devonian (Franke, 1989, 1992; Dallmeyer and Martínez García, 1990; von Raumer and Neubauer, 1993, 1994; Keppie, 1994; Dallmeyer et al., 1995; Matte, 1998; Arenas et al., 2000; Franke et al., 2000). As consequences of Variscan and/or Alpine orogenic events, pre-Variscan elements in these areas mostly appear as polymetamorphic domains. Geotectonic nomenclature and zonation in these classical areas of Variscan evolution mirror the main Variscan tectonic structures (e.g., Suess, 1909; Kossmat, 1927; Stille, 1951), and evidently cannot be valid for the description of pre-Variscan elements. Relics of distinct geological periods from the Proterozoic to the Ordovician have been ob-

Stampfli, G.M., von Raumer, J.F., and Borel, G.D., 2002, Paleozoic evolution of pre-Variscan terranes: From Gondwana to the Variscan collision, *in* Martínez Catalán, J.R., Hatcher, R.D., Jr., Arenas, R., and Díaz García, F., eds., Variscan-Appalachian dynamics: The building of the late Paleozoic basement: Boulder, Colorado, Geological Society of America Special Paper 364, p. 263–280.

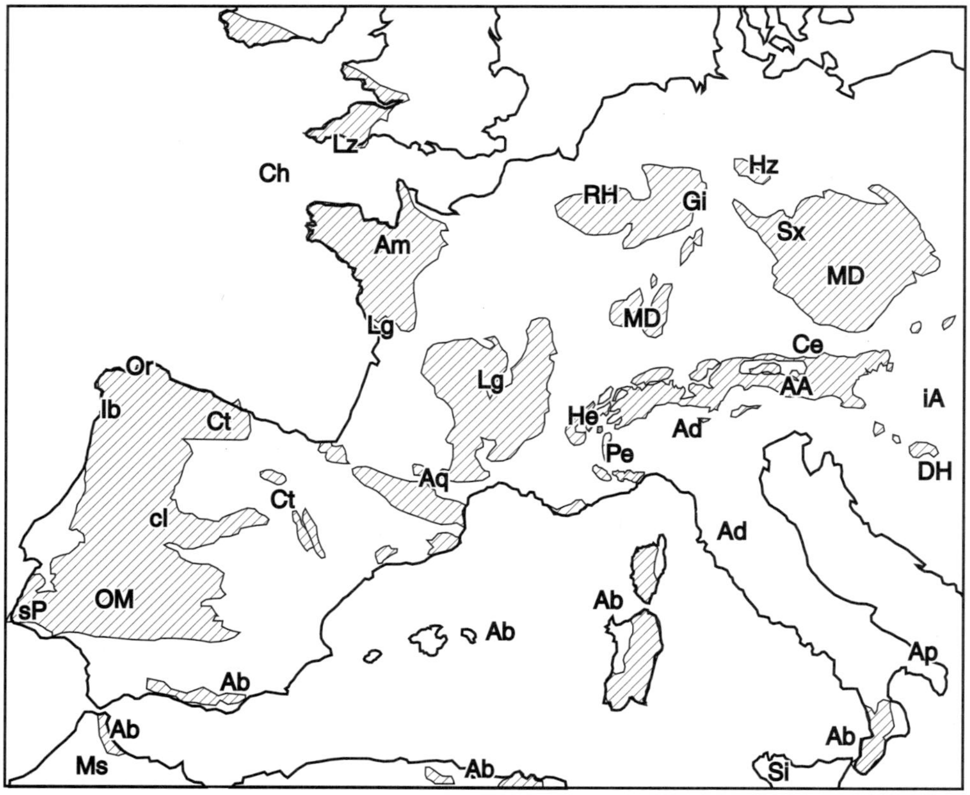

Figure 1. Present-day locations of terranes and blocks for western Europe. AA, Austroalpine; Ab, Alboran (Betic-Rif-Calabria-Kabbilies-Sardinia); Ad, Adria (Tuscan Paleozoic-Southern Alps); Am, Armorica; Ap, Apulia; Aq, Aquitaine (Montagne Noire-Pyrenees); Ce, Cetic; Ch, Channel; cI, Central Iberia; Ct, Cantabria-Asturia-Ebro; DH, Dinarides-Hellenides; Gi, Giesen; He, Helvetic; Hz, Harz; iA, intra-Alpine (Tizia-Transdanubian-Bükk); Ib, allochthonous units of northwestern Iberia; Lg, Ligeria (Massif Central–South Britanny); Lz, Lizzard; MD, Moldanubian; Ms, Meseta; OM, Ossa-Morena; Or, Ordenes ophiolites; Pe, Penninic; RH, Rheno-Hercynian; Si, Sicanian basin; sP, south-Portuguese; Sx, Saxo-Thuringian.

served in many of the basement units. The oldest elements were considered to be part of a Late Proterozoic supercontinent (e.g., Hoffmann, 1991; Unrug, 1997) and may have been detached from what is known as Gondwana or Laurentia-Baltica or Siberia. Examples for the Gondwana origin were given by Zwart and Dornsiepen (1978) and Ziegler (1984), and tectonic complications occurring in such polyorogenic basement massifs were illustrated by Hatcher (1983) for the Appalachians. It is the aim of this contribution to discuss the plate tectonic evolution of these European regions, from the Ordovician onward, in a larger context of global palinspastic reconstructions.

REVIEW OF PRE-VARISCAN EVOLUTION

Pre-Variscan relics include, besides Cadomian-type basement units, evidence for a sequence of Late Precambrian to early Paleozoic plate tectonic settings (e.g., successive stages of development of oceanic crust, volcanic arcs, active margins, and collision zones). Their corresponding evolution has to be discussed in the general framework of their peri-Gondwanan location. Alpine basement areas (Stampfli, 1996; von Raumer, 1998; von Raumer and Stampfli, 2000) as well as Avalonia have to be included in the discussion.

In von Raumer et al. (2002) we tried to compare the early Paleozoic plate tectonic evolution of Avalonia and of microcontinents formerly situated at its lateral eastern continuation along the Gondwana margin (e.g., Cadomia, and the Alpine terranes), and we proposed a similar evolution of all these terranes until the breakoff of Avalonia. Based on the presence of Late Cadomian (550–520 Ma) granitoids, comparable Neoproterozoic to Cambrian detrital sediments and volcanites, and Cambrian oceanic crust, we suggested that initial stages of the Rheic ocean should have been preserved in the microcontinents formerly located in the eastern prolongation of Avalonia at the Gondwana margin (Fig. 2). Using a model of continuous Gondwana-directed subduction since the Neoproterozoic and comparing time of rifting, breakoff, and emplacement of granitoids, we distinguished several steps of a plate tectonic evolution summarized as follows.

1. A Neoproterozoic active margin setting with formation of volcanic arcs is observed along the entire length of the future microcontinents at the Gondwanan border (e.g., Fernández Suárez et al., 2000; Schaltegger et al., 1997; Zulauf et al., 1999). Granites of Neoproterozoic age (ca. 550 Ma), common in many Gondwana-derived basement blocks, probably indicate slab breakoff at the end of the Cadomian orogeny. Zircons in these granites carry the evidence of peri-Gondwanan origin. Latest Proterozoic to Early Cambrian sedimentary troughs developed prior to the opening of the Rheic ocean, which resulted from continued oblique subduction and rifting in a backarc situation.

2. The drift of Avalonia and the opening of the Rheic ocean were enhanced after the subduction of the mid-oceanic ridge, under Gondwana, of what we called the proto-Tethys ocean (the former peri-Gondwanan ocean, Fig. 2). Large-scale magmatic

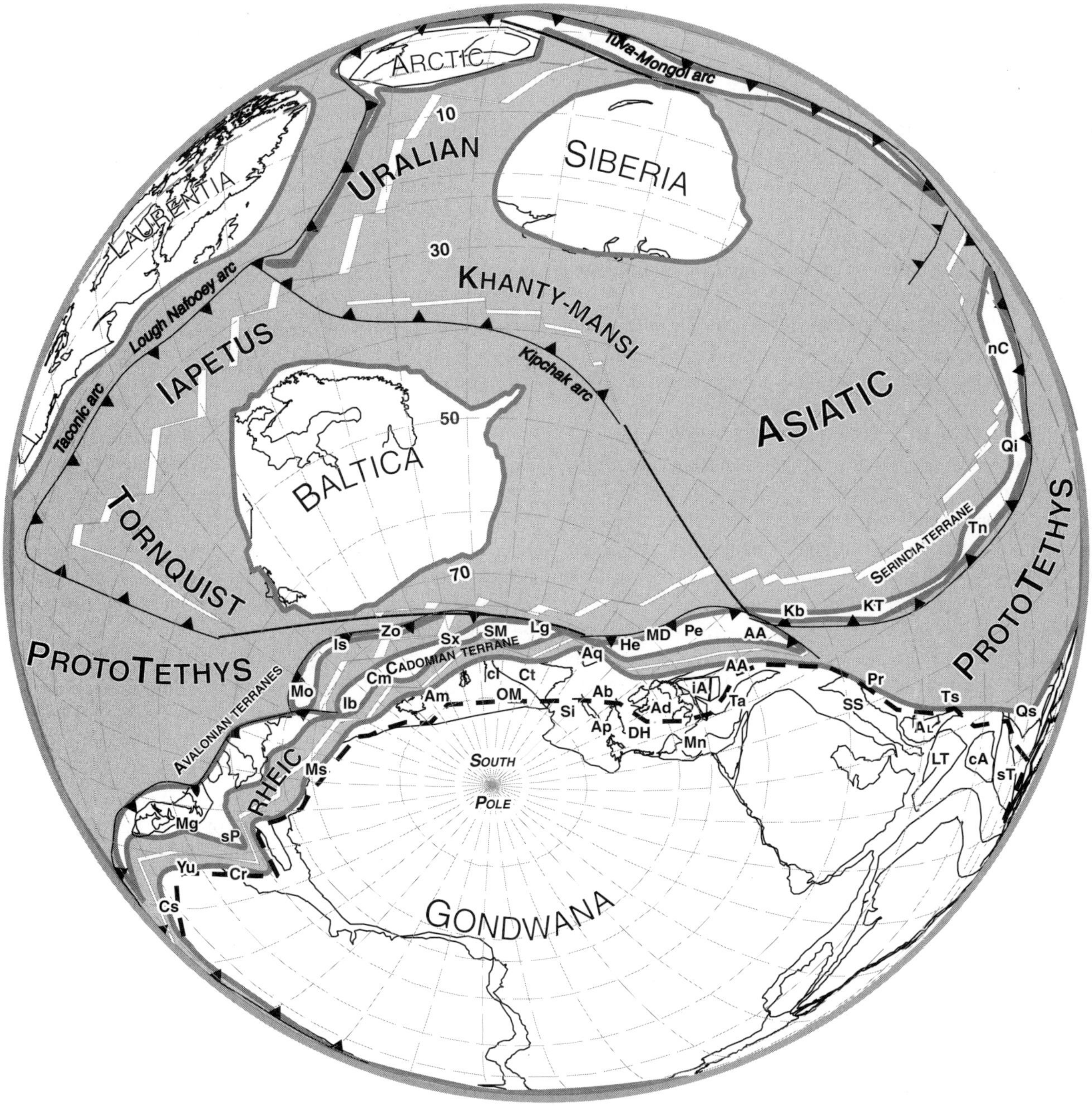

Figure 2. Location of pre-Variscan basement units at Gondwanan margin during Early Ordovician (490 Ma), modified from Stampfli (2000), showing early stages of Rheic ocean spreading. After short separation from Gondwana, Cadomia reaccreted to Gondwana in Middle Ordovician time. Thereafter Hun superterrane detached from Gondwana during opening of paleo-Tethys (dashed line along Gondwanan border). Avalonia: Is, Istanbul; Mg, Meguma; Mo, Moesia; sP, south Portuguese; Zo, Zonguldak. (Dean et al., 2000, proposed an Avalonian origin for the Istanbul Paleozoic; see also Seston et al., 2000, and Winchester, 2002). Cadomia: AA, Austro-Alpine; Cm, Cadomia; He, Helvetic; Ib, allochthons from northwestern Iberia; Lg, Ligerian; MD, Moldanubian; Pe, Penninic; SM, Serbo-Macedonian; Sx, Saxo-Thuringian. Serindia: Kb, Karaburun; KT, Karakum-Turan; nC, north China; Qi, Qilian; Tn, north Tarim. Gondwana: Ab, Alboran; Ad, Adria; Al, Alborz; Am, Armorica; Ap, Apulia; Aq, Aquitaine; cA, central Afghanistan; cI, central Iberia; Cr, Carolina; Cs, Chortis; Ct, Cantabria; DH, Dinarides-Hellenides; iA, intra-Alpine; LT, Lut-Tabas; Mn, Menderes; Ms, Meseta; OM, Ossa-Morena; Pr, Pamir; Qs, south Qinling; SS, Sanandaj-Sirjan; Si, Sicanian basin; sT, south Tibet; Ta, Taurus; Ts, south Tarim; Yu, Yucatan.

pulses of granites and/or gabbros ca. 500 Ma indicate this increased thermal activity (e.g., Abati et al., 1999). In the eastern continuation of Avalonia, only embryonic stages of the Rheic rifting may have existed (Fig. 2). Drifting was hampered by the still-existing mid-oceanic ridge of the proto-Tethys, the collision of which with the detaching terranes triggered the consumption of this embryonic eastern Rheic ocean. The amalgamation of volcanic arcs and continental ribbons with Gondwana occurred in a short-lived orogenic pulse. The resulting cordillera started to collapse during the Late Ordovician, leading to the opening of the paleo-Tethys rift. The chemical evolution of granitoids is the mirror of the general evolution from Cambrian-Ordovician rifting, to Cambrian-Ordovician active margin, and Ordovician amalgamation.

3. Mid-ocean ridge subduction during the Ordovician, in the former eastern prolongation of Avalonia, triggered not only the intrusion of many Ordovician granitoids, but also facilitated the opening of paleo-Tethys and the Late Silurian drift of the composite Hun superterrane (Stampfli, 2000). There is little evidence of this episode, neither sedimentation in a backarc setting (e.g., Saxo-Thuringian domain; Linnemann and Buschmann, 1995), nor Late Ordovician–Early Silurian active margin settings (e.g., Reischmann and Anthes, 1996) in many of the basement areas composing Cadomia (sensu lato; see following), except in the Alpine areas (Stampfli, 1996; von Raumer, 1998).

In the European pre-Variscan basement areas, hidden in the Variscan and Alpine mountain chains, a striking comparability of pre-Silurian evolutions shows that the pre-Variscan elements had similar related locations along the Gondwana margin. Many contain Cadomian basement with related evidence of Late Proterozoic detrital sedimentation and volcanic-arc development, relics of a Rheic ocean, Cambrian-Ordovician accretionary wedges, evidence of an Ordovician orogenic event with related granite intrusions, and subsequent volcanicity and sedimentation indicating the opening of paleo-Tethys. The occurrence of active margin settings during the Early Silurian supports a southward subduction of the Rheic and proto-Tethys oceans.

VARISCAN COMPLICATIONS—DISCUSSION

The pre-Variscan elements discussed herein have been interpreted from a Gondwana point of view (von Raumer et al., 2002), without regard to their post-Silurian evolution. Plate tectonic reconstructions of the Variscan history depend on paleomagnetic data and models of Variscan evolution. Independent of the model applied, the pre-Variscan elements mentioned herein were strongly transformed during the Variscan collision, and many of these relics appear today as polymetamorphic and migmatized domains, wherein much information has been lost. It is evident that size and contours of the many continental fragments have changed considerably. Nonetheless, in our reconstructions the original outlines are used to facilitate recognition of well-known basement areas. Evidently, after the Silurian, the Gondwana-derived continental blocks (Ziegler, 1984) started to be involved in the global Variscan orogenic cycle. This is not the place to discuss all the models currently available, and the reader is referred to the references given in the introduction, and to the new observations and data presented during the field trips at the Fifteenth Basement Tectonics Meeting in A Coruña, Spain (Arenas et al., 2000; Gil Ibarguchi et al., 2000). In northwestern Spain, large-scale nappes and their ramps and horses involved all lithospheric levels, from the upper mantle to the upper crust, thus redistributing the former lateral orogenic zonation. Comparable observations come from the mid-European Variscides (Matte et al., 1990; Schulmann et al., 1991; Mingram, 1998; Stipska et al., 1998). It is evident that a pre-Variscan orogenic zonation has been involved in the Variscan collisional events (e.g., Martínez Catalán et al., 1997; Arenas et al., 2000), and relics of oceanic domains appear as fragments within large divergent orogenic belts (Pin, 1990; Matte, 1991; Martínez Catalán et al., 1999).

Although we do not discuss details about the Variscan metamorphic evolution, we add some new points of view concerning the oceanic evolution. Such a discussion is needed in relation to the different oceanic realms of the Variscan domain (e.g., Iapetus, Rheic, Galicia–Massif Central oceans) that have already been identified. Although many occurrences of so-called amphibolites known from the Variscan mountain chain still need to be geochemically characterized and dated, a comparative approach (von Raumer et al., 2002, and references therein) may furnish additional arguments for comparing plate tectonic events across continental fragments derived from Gondwana. In the following summary of related arguments, based on the inferred former peri-Gondwanan location, we assume, instead of multioceanic models, the existence of one aborted Rheic ocean, contemporaneous with the drift of Avalonia, in all basement units derived from Gondwana (i.e., Cadomia sensu lato and the Alpine terranes). This model includes the Pulo do Lobo, Galicia, and Massif Central oceans (Robardet, 2000, and also localities from the Bohemian massif; Crowley et al., 2000), and fits the interpretation of the Cambrian events in northwestern Iberia (Abati et al., 1999). Pieces of this suture zone, in the basement assemblages of Cadomia (sensu lato) and the Alpine areas, were accreted or obducted to the Gondwana margin in the Ordovician, whereas Avalonia underwent Ordovician-Silurian collision with Laurentia-Baltica. The main point we develop is that the continuation of subduction of the Rheic–proto-Tethys oceans under the remaining peri-Gondwanan blocks triggered magmatic events (the subducting ridge being a heat source) and backarc spreading (with the formation of sedimentary basins and extrusion of volcanics) from the Middle Ordovician and, finally, the opening of paleo-Tethys, from the Silurian. Therefore, this new proposal considers that the Variscan collision in Europe took place between Gondwana-derived terranes and Laurussia and not between Laurussia and Gondwana (Stampfli et al., 2000).

GENERAL PRINCIPLES

In North America and western Europe, Variscide collisional processes are usually inferred to have ranged from the Early Devonian to the Late Carboniferous–Early Permian; and the Tethyan cycle (opening of the Alpine Tethys–Central Atlantic system) not to have started before Middle Triassic time. An apparent lack of major tectonic events during the Permian and Triassic is certainly responsible for the focus of attention on the Carboniferous history of the Variscides of Central Europe. However, the Variscan domain extends over the entire Alpine area and even further in the Dinarides and Hellenides. It also extends in time as deformations become younger, possibly grading into the eo-Cimmerian (Middle to Late Triassic) deformations, southward and eastward. This assumption is based on the fact that the paleo-Tethyan domain was not fully closed in southeastern Europe before the Late Permian. This is shown by Early Permian to Middle Triassic fully pelagic sequences found in Sicily (Catalano et al., 1988) and similar Carboniferous to Middle Triassic sequences in Crete (Krahl et al., 1986; Stampfli et al., 2002), located at the southern border of the Variscan domain. In the Hellenides and farther east, the final closure of this oceanic realm generally took place during the Carnian (Şengör, 1984).

Stampfli et al. (1991) and Stampfli (1996) discussed this diachronous closure of the large paleo-Tethys ocean, insisting on the development of backarc oceans or basins within the Permian-Triassic Eurasian margin (Ziegler and Stampfli, 2001) and the closure of paleo-Tethys between terranes drifting away from Eurasia (e.g., Pelagonia; Vavassis et al., 2000) and terranes drifting away from Gondwana (the Cimmerian blocks of Şengör, 1979; Stampfli, 2000; Stampfli et al., 2001a, 2001b).

Subsequently, the Atlantic-Alpine-Tethys system opened north of this eo-Cimmerian collisional zone, which thereafter was fully incorporated into the Alpine fold belt. This explains why the end member of the Variscide orogeny, the eo-Cimmerian event, is usually not taken into consideration by many. For most of those who study the Hercynian, the southern part of Variscan Europe (e.g., Spain, southern France) is usually regarded as stable Gondwana, which certainly it was in early Paleozoic time, whereas it was part of Gondwana-derived terranes accreted to Laurussia between the Late Devonian and Early Carboniferous. We formerly grouped these terranes as the Hun superterrane (von Raumer et al., 1998; Stampfli, 2000). In view of their relatively independent kinematic evolution (Fig. 3), we propose labeling its eastern components (Karakum-Turan, Tarim, north China, south China, north Tibet, and Indochina) the Asiatic Hunic terranes, whereas its western part is labeled the European Hunic terranes and comprises three major blocks: Armorica (sensu lato), Cantabria-Aquitaine-Ligeria-Moldanubia, and Alboran–Adria–intra-Alpine–Cetic (Fig. 4).

The main outcome of this proposal is that the Variscan collision must be polyphase and polymetamorphic. Initially it was made of the accretion of major terranes along the European segment of the passive margin of Laurussia (Avalonia), corresponding to the closure of the Rheic ocean in the Late Ordovician (Fig. 3). Thereafter, Gondwana collided with Laurussia, including previously accreted terranes, mainly along the Alleghanian segment of Laurussia; this last event was diachronous and younging eastward and corresponded to the closure of the paleo-Tethys.

This scheme implies that after the accretion of the European Hunic terranes to Laurussia (Avalonia), the ocean located to the south of them (paleo-Tethys) started subducting northward, the Laurussian margin becoming an active margin. Subsequent subduction of the mid-oceanic ridge of paleo-Tethys led, in Visean time, to a Variscan cordillera stage.

HUN SUPERTERRANE

The European Hunic terranes include all continental fragments accreted to Laurussia during the Variscan cycle and inferred to have previously been in lateral continuity with Avalonia along the Gondwana margin (Fig. 2). We place Armorica (sensu lato) (Ossa-Morena, Central Iberia, Brittany, Saxo-Thuringia) north of North Africa (Fig. 2) based on paleomagnetic data (e.g., Torsvik and Eide, 1998; Torsvik et al., 1992) and sedimentological and faunal data (e.g., Paris and Robardet, 1990; Robardet et al., 1994; Robardet, 1996). These data do not show a major separation of Armorica from Gondwana before the Early Devonian. We propose that all the other Hunic terranes were in lateral continuity to Armorica (sensu lato), forming a ribbon-like superterrane. Their drifting from Gondwana is delimited by paleomagnetic data from the eastern part of the European Hunic terrane, like the Noric-Bosnian block (Schätz et al., 1997) and the Bohemian block (Krs and Pruner, 1999).

The Asiatic Hunic terranes elements are represented by Turan and Pamir (the Kara-Kum–Tarim terrane of Khain, 1994; Zonenshain et al., 1990), together with Tarim and north China, which are inferred to have escaped from Gondwana in the Early Devonian. This escape followed the accretion to the northern parts of these regions of the Serindia terrane in the Late Silurian–Early Devonian (Meng and Zhang, 1999; Yin and Nie, 1996), as well as the accretion of island arcs in Vietnam (Findley, 1998) and east China (Hutchison, 1989) at the same time; a similar development is also found along the Australian margin (e.g., Foster and Gray, 2000).

Therefore, the Hun superterrane in Early Ordovician (Fig. 2) to Early Silurian reconstructions is spread over a relatively large paleolatitudinal area (from 60° south to the equator). Changes of facies can be expected between Armorica and terranes in the Alps (e.g., Carnic, Austroalpine, and intra-Alpine domains) located within the tropical zone, which present a Silurian to Carboniferous stratigraphic evolution very similar to that of the Gondwanan margin in Iran (Alborz) and Turkey (Taurus).

The prerift-synrift sequences of the Hun superterrane present a uniform sedimentary evolution equal to that found on the Gondwanan border. For example, in the Armorica (sensu lato)

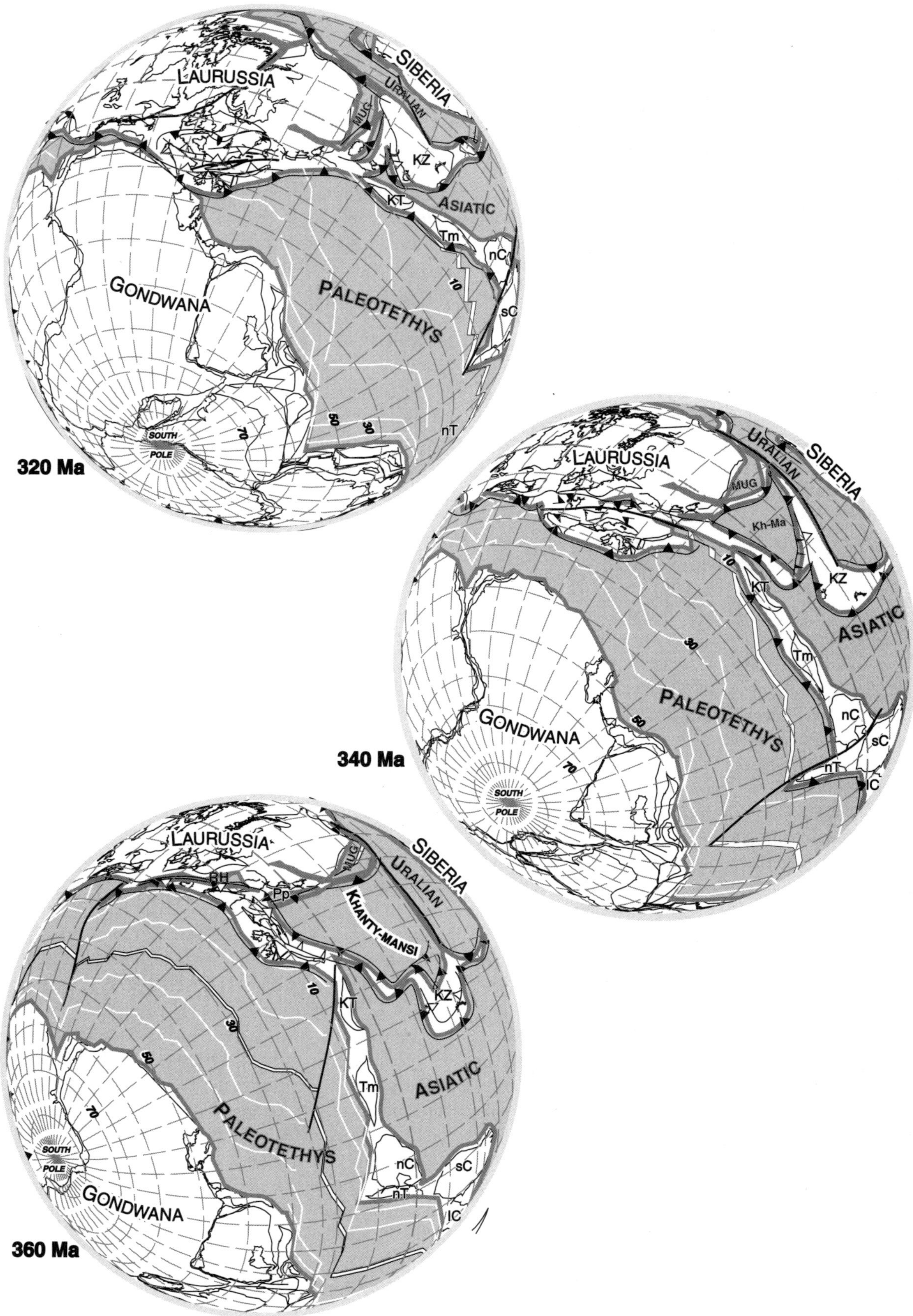

LAURUSSIA
SIBERIA
URALIAN
MUG
KZ
KT
ASIATIC
Tm
nC
GONDWANA
PALEOTETHYS
10
sC
SOUTH POLE
70
50
30
nT
320 Ma
LAURUSSIA
URALIAN
SIBERIA
MUG
Kh-Ma
10
KZ
KT
ASIATIC
30
Tm
PALEOTETHYS
nC
GONDWANA
50
sC
nT
IC
70
SOUTH POLE
340 Ma
LAURUSSIA
SIBERIA
MUG
URALIAN
RH
Pp
KHANTY-MANSI
10
KT
KZ
30
50
Tm
ASIATIC
70
PALEOTETHYS
SOUTH POLE
nC
sC
nT
GONDWANA
IC
360 Ma

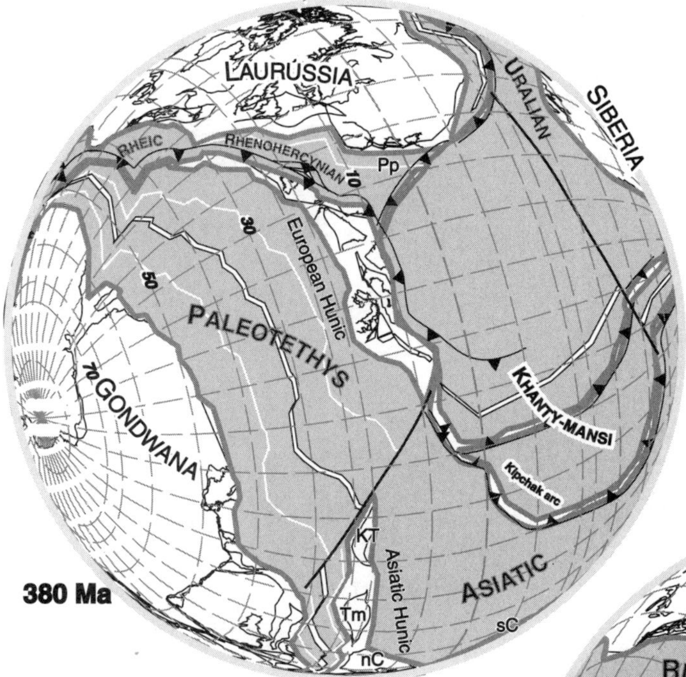

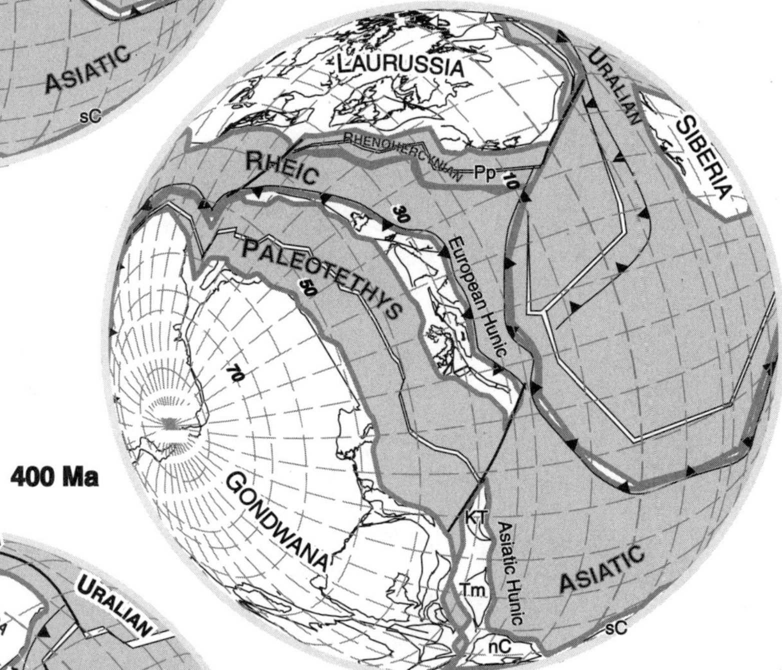

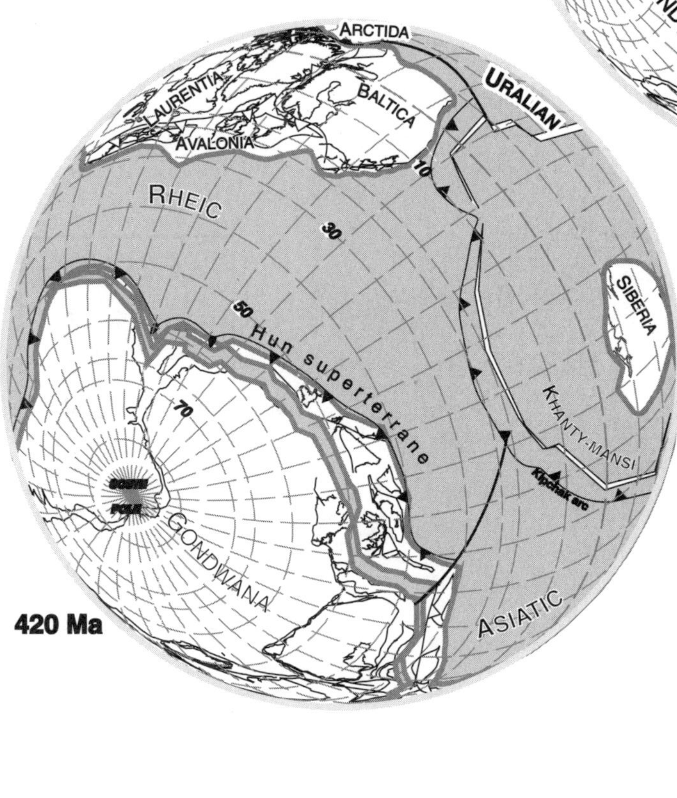

Figure 3 (on this and facing page). Drift history of Gondwana-derived basement areas between Late Silurian and late Carboniferous. 420 Ma projection is centered on present-day lat 10N, long 25E; 400 Ma and 380 Ma projections are centered on 10N, 20E; 360 Ma projection is centered on 05N, 25E; 340 Ma projection is centered on 10N, 25E; and 320 Ma projection is centered on 20N, 20E. These reconstructions were elaborated using GMAP program (Torsvik and Smethurst, 1994, 1999). Reference paleopoles are from Baltica (Torsvik and Smethurst, 1994, 1999). Position of Gondwana is constructed from paleomagnetic data (van der Voo, 1993; Klootwijk 1996) and admissible wander path (e.g., Tait et al., 2000; Stampfli and Borel, 2001). Mug, Mugdozar ocean; Pp, Paphlagonian ocean; KT, Karakum-Turan; Tm, Tarim; nC, north China; sC, south China; nT, north Tibet (Qiantang); IC, Indochina (Borneo included); KZ, Kazakhstan.

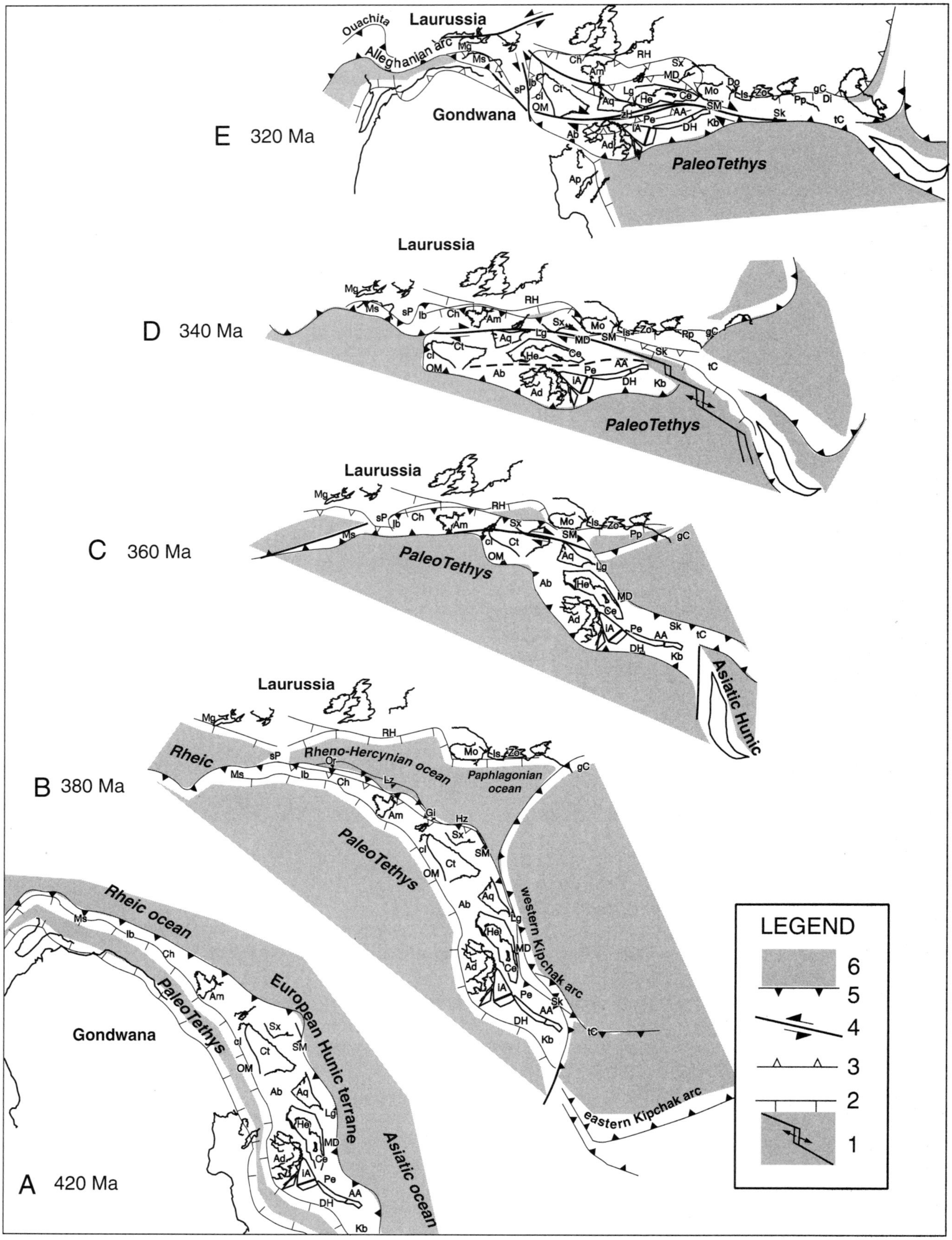

E 320 Ma
Ouachita
Laurussia
Alleghanian arc
Gondwana
PaleoTethys
D 340 Ma
Laurussia
PaleoTethys
C 360 Ma
Laurussia
PaleoTethys
Asiatic Hunic
B 380 Ma
Laurussia
Rheic
Rheno-Hercynian ocean
Paphlagonian ocean
PaleoTethys
western Kipchak arc
eastern Kipchak arc
A 420 Ma
Rheic ocean
Gondwana
PaleoTethys
European Hunic terrane
Asiatic ocean
LEGEND
6
5
4
3
2
1

segment (Robardet et al., 1994), Late Ordovician clastics (including minor volcanics) representing synrift formations are capped by Ashgillian glacial marine deposits, related to an ice cap that possibly developed on the nascent rift shoulders (e.g., Sardinia, Ghienne et al., 2000; Taurus, Monod et al., 2002). Early Silurian marine sediments containing cherts represent a southward deepening toward the rift zone and are dominated by black graptolite shales. These anoxic Silurian deposits characterize the widening of the rift zone, but also show that connections with major oceans were not yet realized. Silurian flood basalts have been reported in many areas, and can be regarded as contemporaneous with the onset of seafloor spreading.

The postrift evolution of the European Hunic terranes differed greatly depending on their final position in relation to the Rheic suture to the north or the paleo-Tethys suture to the south. The present juxtaposition of terranes cannot be used readily to understand their transformation during the Devonian and Carboniferous. Some terranes were deeply metamorphosed during the Middle Devonian, others were affected later in Visean time (340 Ma), and flysch development on some blocks did not start before the late Carboniferous (310 Ma). Thus it is obvious that some blocks were not involved in the first phase of metamorphism, which must have affected mainly the leading edge of the superterrane, being related to the subduction and suturing of the Rheic ocean. The European Hunic terranes were finally accreted individually to Laurussia, but not before the Visean, and were accompanied by widespread metamorphism and development of flysch. The former could be due to imbrication of the terranes and crustal thickening processes, or also to the subduction of a remnant mid-ocean ridge in the paleo-Tethys ocean (Fig. 3).

Large parts of the Hun superterrane kept Gondwanan faunal characteristics at least until Early Devonian time. The conclusion that a large ocean never separated Armorica and Gondwana (Robardet et al., 1990) could be explained by a connection between Gondwana and the Hun superterrane throughout its drifting history, being, in its westernmost sector, always attached to South America (Fig. 3). However, these Gondwanan faunal characteristics disappeared in Praguian time, when similar spore assemblages are found in North Africa and in the Rheno-Hercynian domain (Paris and Robardet, 1990). This can be explained if the Hun superterrane is viewed as a land bridge between the two domains. This also supports the idea that the collision of Armorica (sensu lato) with the Eurasian domain (or Eurasian outliers) occurred ca. 380 Ma. An earlier collision (Silurian) is not supported by these faunal data, or by the lack of Silurian synorogenic deposits on the Eurasian (Avalonian) margin (e.g., the Rheno-Hercynian domain) facing the Rheic ocean.

There seems to be a contradiction between the inference that Armorica should have docked with Laurussia in Middle Devonian time and the final welding in Namurian time, after the closure of the Rheno-Hercynian domain. To reconcile these different lines of evidence, we propose that Armorica collided in Middle Devonian time with blocks detached from the Laurussian (Avalonian) margin during an Early Devonian rifting event. The latter led to the opening of the Rheno-Hercynian basin in the Emsian, within the southern passive margin of Laurussia. The oceanic nature of this basin (the Rheno-Hercynian ocean) is proven by the ophiolitic and pelagic remnants found in the Lizard, Giessen, and Harz nappes.

There are other problems associated with the present juxtaposition of the different terranes. The Cantabrian-Aquitaine terrane (Fig. 4) developed a flysch sequence only in Moscovian time (ca. 310 Ma) and was finally juxtaposed with high-grade metamorphic nappes of the Galician zone, where the metamorphism is much older (Middle Devonian). The Ligerian-Moldanubian cordillera, extending from southern Brittany to central Europe, is also juxtaposed to areas to the north (Armorica sensu lato) and to the south (Aquitaine terrane, comprising the Montagne Noire and Pyrenees), where flysch development only started at the earliest in the late Visean or Namurian. However, large parts of the cordillera were strongly metamorphosed before that time. Relics of a major Middle Devonian metamorphic event are found in nearly all the metamorphic domains of the Variscan orogen. This has been conventionally interpreted as a collision involving major continents (i.e., Gondwana and Laurussia), but no syncollisional flysch has ever been described in neighboring regions and Gondwana was never close enough to Laurussia to generate a continent-continent collision (Tait et al., 2000; Stampfli and Borel, 2002).

Our model (Figs. 2 and 4A) places all the European Hunic terrane segments in continuity with each other in order to avoid these contradictions. The leading, northern border of the superterrane is regarded as an active margin, affected by metamorphism and plutonism, whereas its hinterland stayed away from this tectonothermal activity and developed a passive margin sedimentary succession. This situation prevailed until the Late Devonian (Fig. 4C), when major transcurrent events displaced the eastern segments westward and intraterrane collisions took place. The latter were accompanied by widespread flysch development and the building of a Visean cordillera, usually with diverging boundaries around the major blocks, giving rise to double-vergent cordilleras (e.g., Matte, 1991; Neubauer and Handler, 2000). The Rheic suture would have been located along the northern active margin of the superterrane, whereas the paleo-Tethys suture should be located along the southern border

Figure 4. Detailed plate tectonic evolution of pre-Variscan units preserved in European Variscan mountain chain (see text for explanation). AA, Austro-Alpine; Ab, Alboran; Ad, Adria–south Alpine; Al, Alborz; Am, Armorica; Ap, Apulia; Aq, Aquitaine; Ce, Cetic; Ch, Channel; cI, central Iberia; Ct, Cantabria; DH, Dinarides-Hellenides; Di, Dizi; Do, Dobrogea; gC, great Caucasus; Gi, Giessen; He, Helvetic; Hz, Harz; iA, intra-Alpine; Ib, northwestern Iberia allochthon; Is, Istanbul; Kb, Karaburun; KT, Karakum-Turan; Lg, Ligerian; Lz, lizard; MD, Moldadubian; Mg, Meguma; Mo, Moesia; Ms, Meseta; OM, Ossa-Morena; Or, Ordenes ophiolites; Pe, Penninic; Pp, Paphlagonian; RH, Rheno-Hercynian; Sk, Sakarya; SM, Serbo-Macedonian; sP, south Portuguese; Sx, Saxo-Thuringian; tC, trans-Caucasus; Zo, Zonguldak.

of the superterrane. However, in view of the major Carboniferous lateral displacements and rotation (Edel, 2000, 2001), suture duplication took place and led to present-day multiocean models.

The present southernmost portion of the European Hunic terrane (the Noric-Bosnian terrane of Frisch and Neubauer, 1989), comprising Alboran, Adria, the intra-Alpine domain (Carnic and Julian Alps, Tizia, Apuseni, Transdanubian, and Bükk units), and domains located in the Dinarides and Hellenides, together with north Sardinia and part of the southern Alps, was transported southwestward during the final Laurussia-Gondwana collision in the late Carboniferous (Fig. 4E). It collided with the Visean cordillera to form a double vergent orogen (Neubauer and Handler, 2000), following the fast northward drift of Gondwana (Fig. 3). Other metamorphic units found in the Alps (e.g., Cetic terrane) represent part of the Visean cordillera, possibly transported westward with the Noric-Bosnian block. In this context the Penninic domain would have been located formerly to the east of the Helvetic domain (Giorgis et al., 1999). This late Carboniferous collisional event was responsible for the final tectonic configuration of the Variscan orogen. These southernmost Variscan units are also characterised by the development of late Carboniferous magmatic arcs along their southern margins, dominated by calc-alkaline intrusions (Stampfli, 1996, and references therein), extending from the Alboran domain (e.g., Calabria with a transition from arc to rift between 295 and 275 Ma; Acquafreda et al., 1994) to the Hellenides (Vavassis et al., 2000) and possibly to the Pontides. The magmatic arcs were replaced in Permian time by major rift zones leading to the opening of backarc basins in Late Permian–Triassic time (Stampfli, 2000; Stampfli et al., 2001a, 2001b; Ziegler and Stampfli, 2001).

The Carboniferous lateral displacements and rotations imply the presence of major transcurrent faults and the opening of Gulf of California–type oceans within the European Hunic terrane (Fig. 4D). It is obvious that high-pressure rocks characterizing the Middle Devonian event had to reach the surface relatively rapidly, and the transcurrent movements were locally largely transtensive, creating intramountain basins, usually starting in the Late Devonian and found in the middle of the Ligerian-Moldanubian cordillera. Younger Carboniferous coal basins are also widespread. In this context the Zone Houillère (Cortesogno et al., 1993), extending over 100 km or more in the Penninic domain (Fig. 4E), is regarded as a late Carboniferous pull-apart basin along one of these major faults (Giorgis et al., 1999). It was accompanied by late Carboniferous–Early Permian granites and minor gabbros, also found elsewhere in the Alps (e.g., Capuzzo and Bussy, 2000) and emplaced in a scenario of cordillera construction and destruction, but in a general context of a still active margin.

Pelagic sequences of the Chios-Karaburun domain in the Aegean region (Stampfli et al., 2002), juxtaposed with Variscan metamorphic blocks (Pelagonia, Sakarya), could also be part of such gulf-like deep basin and/or of the paleo-Tethys accretionary prism (Kozur, 1997, 1998). Carboniferous to Permian mid-ocean ridge basalt (MORB) found in the Tavas unit of the Lycian nappe (Kozur, 1999; Kozur et al., 1998), and in eastern Iran (Ruttner, 1993) could be related to such Gulf of California–type oceans and/or to the paleo-Tethys (Fig. 4D).

These Visean lateral displacements would also involve major crustal thickening in transpressive areas and the buildup of high reliefs, leading to a cordillera stage, which lasted as long as a relatively buoyant part of the paleo-Tethyan slab was subducting under the Eurasian margin. The major geodynamic event at that time was the subduction of the paleo-Tethys mid-ocean ridge. Thereafter, from the Late Carboniferous onward, the increasing age of the subducting paleo-Tethyan slab generated important slab rollback and general extension affected the cordillera from the Early Permian.

MIDDLE DEVONIAN PHASE

The opening of the paleo-Tethys along the European Hunic segment and westward is viewed as backarc spreading related to Gondwana-directed (southward) subduction of the Rheic ocean (Fig. 3). The principle governing the drifting of the European Hunic terranes away from Gondwana is the roll-back toward Laurussia of the Rheic slab after Ordovician subduction of its mid-oceanic ridge. The strong pull force of the major subducting Rheic slab also triggered the opening of rifts in the subducting plate, leading to the opening of small oceanic basins (the Rheno-Hercynian ocean). This could explain the early collision (during the Devonian) of Gondwana-derived European Hunic terranes with Laurussia-derived terranes, whereas major collision and closure of the Rheno-Hercynian basin only took place in Late Carboniferous time. We review next the evolution of the intervening elements, the Rheno-Hercynian basin, the European Hunic active margin in Armorica, the Ligerian cordillera, the composite Middle Devonian event, and the Appalachians.

Rheno-Hercynian basin

The Rheno-Hercynian basin is characterized by important volcanism from the Early Devonian (e.g., Walliser, 1981; Ziegler, 1988) that extended through most of the Devonian period. Geochemical characterization of this volcanism (Floyd, 1995) has shown the purely ensialic extensional nature of this basin, and the absence of any subduction-related signatures. MORBs have been found in many places (Lizard, Giessen, Harz) and point to seafloor spreading, most likely starting in the Emsian. From the Namurian onward, this basin became a flexural basin in the foreland of the advancing Variscan nappes; the sedimentary records do not show evidence of any tectonic event before this.

The prerift sequence is locally composed of relatively complete Ordovician to Silurian sequences (e.g., east of the Rhine; Franke, 1995), or a Silurian sequence with a gap between the Silurian and Devonian (e.g., Moravo-Silesian region; Dvorak, 1995). Therefore, there is no indication of any middle Paleozoic

(Caledonian) event in what we regard as the hinterland part, i.e., the southern passive margin of the Avalonia terranes, whereas by contrast, in its front part (e.g., the present northern Variscan foreland), the Ordovician-Silurian sequence is clearly deformed due to the suturing of Avalonia to Baltica. The rift shoulder uplift occurred in the Early Devonian, marked by clastic input derived from the south or by a so-called Caledonian unconformity in the Rhenish Massif. Thereafter, sedimentation graded from Early Devonian synrift deposits to Middle Devonian to early Carboniferous pelagic deposits (Franke, 1995).

European Hunic active margin in Armorica (sensu lato)

The accretionary prism to the south of the Rheno-Hercynian ocean is composed of the Giessen-Harz nappes and the northern Phyllite zone. The mid-German Crystalline High played the role of backstop; it is characterized by volcanic-arc activity in Silurian time (e.g., Reischmann and Anthes, 1996; Anthes and Reischmann, 2001). Pelagic sediments, extending from the Silurian to the Early Carboniferous, are found in melange or slivers in the accretionary sequences together with MORB and other basalts of intraplate affinity (seamounts; e.g., Flick et al., 1988; Nesbor et al., 1993). The Lizard ophiolite of Cornwall is considered to be a Devonian ophiolite; its emplacement in the accretionary prism was dated as Famennian (365–370 Ma) by Sandeman et al. (1995) and could correspond to the collision of the Rheno-Hercynian mid-oceanic ridge with the prism.

The Rheno-Hercynian prism evolved from an older accretionary belt developed during the southward subduction of the Rheic ocean in Silurian time. The Rheic prism incorporated a detached Eurasian block, located south of the Rheno-Hercynian ocean; detached from the already thinned Avalonian passive margin, this block could have been easily subducted. Its underplating and the subduction of the buoyant young Rheno-Hercynian oceanic lithosphere provided the necessary condition for Devonian high-pressure rocks to be exhumed. The absence of pelagic material older than Emsian in the Giessen-Harz nappe makes it difficult to place the Rheic prism in this domain; therefore, the Rheic suture should be placed in the Northern Phyllite zone (Franke, 2000). In the Wippra area of the Phyllite zone, MORBs are supposed to be partly of Ordovician and Silurian age (Meisl, 1995), thus representing the Rheic ocean floor. Ordovician and Silurian fauna are described in this zone, some with tropical or even boreal affinities, which can be taken as a Rheic signature.

A Silurian high-pressure event, recorded in the Leon domain in northern Brittany (Le Corre et al., 1991), could be regarded as a western continuation of this Rheic suture. In Galicia (Marcos et al., 2000; Arenas et al., 2000) high-pressure metamorphism in the allochthonous nappe is in the range 390–380 Ma. This Middle Devonian metamorphic complex comprises several types of Ordovician ophiolitic fragments transformed into eclogites, protolith ages ranging from 490 to 460 Ma and therefore possibly pertaining to the Rheic ocean. It also contains other ultramafic rocks (e.g., in the middle of the Ordenes complex) with younger ages (390–380 Ma) (Ordoñez Casado, 1998; Díaz García et al., 1999; Pin et al., 2000), most likely representing the extension of the Rheno-Hercynian ocean in that region (Fig. 4B). Thus the Middle Devonian event would have sutured the Rheic ocean and created a new accretionary wedge, including ophiolites of the Rheno-Hercynian ocean. The second metamorphic event in the Cabo Ortegal sequence, dated as ca. 345–350 Ma (Ordoñez Casado, 1998), would correspond to the subduction of the paleo-Tethyan mid-oceanic ridge and the buildup of the Visean cordillera. Younger ages (330–345 Ma) were found in more external domains in the Ossa-Morena zone (Ordoñez Casado, 1998) and mark the final suturing of Armorica (sensu lato) with the South Portuguese promontory. The Beja-Acebuches ophiolitic complex, separating the Ossa-Morena zone from the South Portuguese zone (Oliveira and Quesada, 1998; Eguiluz et al., 2000), would then correspond again to the Rheic suture. The Pulo do Lobo accretionary complex, located in a more external position, comprises normal-MORB remnants, unconformably overlain by Late Devonian flysch. Therefore, it can be regarded as the Rheic accretionary prism, comprising a fragment of the Rheno-Hercynian ocean.

The Late Devonian–Carboniferous development of the South Portuguese zone would be directly related to the onset of paleo-Tethys northward subduction, after the Middle Devonian event. The flysch of the South Portuguese zone was derived from backarc type basic rocks that cannot be of Rheno-Hercynian origin, but could be paleo-Tethyan. Thereafter, there is a widespread development of a volcanic-sedimentary complex (the Pyrite Belt) of Late Famennian–Visean age and bimodal signature, where felsic volcanics predominate (Thiéblemont et al., 1994). This belt may represent a forearc-type basin to the newly established active margin of Laurussia. Such Late Devonian basins, where extension predominates, are also known in the Meseta and Meguma-Avalon domain (e.g., Piqué and Skehan, 1992). These basins characterize the subduction and rollback of the nonbuoyant northern part of the paleo-Tethyan slab. As subduction proceeded, the mid-ocean ridge finally collided with the margin, probably in Visean time, generating numerous granitic intrusions and closing these basins. Oroclinal bending of the cordillera then took place and deformed the originally linear features of the active margin (Weill et al., 2001). Such an oroclinal bending of a large terrane was recently proposed by Johnston (2001) for the Great Alaskan Terrane (SAYBIA).

Ligerian cordillera

In our model, suturing of the Rheic ocean took place all along the outer border of the western part of the European Hunic terranes during a Middle Devonian accretionary phase (Fig. 4B). We have already extended the situation described herein from Armorica (sensu lato) to the Iberian allochthonous units of Galicia, and we infer that it extended eastward to the Münchberg nappe and to the western Sudetes. However, this Middle Devonian event also affected areas located south of the Armorica (sensu lato) domain. The Middle Devonian eo-Variscan metamorphic event affected the Massif Central and other northern European Variscan units (Faure et al., 1997) and was accompanied by Givetian-Frasnian high-pressure events dated as 380–370 Ma. This event, affecting simultaneously areas now imbricated in the entire Variscan orogen, could not be a major continent-continent collision, because Gondwana was still far away to the south at that time (e.g., Tait et al., 2000; Stampfli and Borel, 2002), and the Rheno-Hercynian ocean far from closing (Fig. 3). To avoid multiplying the number of oceans and terranes, we infer that all the areas affected by this Ligerian phase were formerly located on the leading edge of the eastern part of the European Hunic terranes. An example of this is the Middle Devonian suture between the Saxo-Thuringian and Tepla-Barrandian domains (Franke et al., 1995), where the Saxo-Thuringian domain, being of clear European Hunic affinity, cannot represent a block detached from Laurussia during the opening of the Rheno-Hercynian ocean. However, the sedimentary sequences in the Saxo-Thuringian basin did not record the Middle Devonian collisional event; pelagic conditions predominate from Silurian to Early Carboniferous time (Falk et al., 1995). Therefore, as is the case for many other parts of this Middle Devonian Rheic suture, it must have been laterally displaced, and the present relationship between the two domains remains ambiguous.

The same reasoning could be applied to other potential Rheic suture zones like the Moldanubian and the Massif Central domains. It is not yet clear if the Moldanubian zone underwent this Middle Devonian event (Vràna et al., 1995), but it seems that the major cordillera-building processes affecting this zone occurred in the Early Carboniferous, most likely due to intra-Hunic collisional events. However, the high-pressure rocks dated as Middle Silurian in Bavaria (427 ± 5 Ma; von Quadt and Gebauer, 1993) point to an active margin setting of the Moldanubian zone at that time (like the German Crystalline zone and the Leon domain), but not necessarily to collision. Therefore, the Moldanubian domain could also represent the same leading accretionary edge of the European Hunic terranes at that time and should have been located eastward of the Sudetes in prolongation of the Armorica (sensu lato) terrane (Fig. 4).

Similarly, the Ligerian cordillera of central France (Quenardel et al., 1991), extending up to the South Armorican domain (Le Corre et al., 1991), represents a large part of this Devonian Rheic suture zone and would also have been located east of the Sudetes. The metamorphism of the Massif Central was between 400 and 360 Ma, whereas poorly dated older high-pressure events could be related to Silurian subduction of the Rheic ocean. Lardeaux et al. (2001) proposed an oblique continent-arc collision for this region, a scenario similar to our Figure 4B. The high-pressure event of South Armorica (Champtoceaux complex) is dated as ca. 360 Ma (Ballèvre et al., 2000), including continent-derived protoliths of Early Ordovician age. These units were exhumed in the Visean and largely imbricated by thrusting and shearing during dextral movements in the Late Carboniferous. We have here the juxtaposition of two different domains, the Ligerian and Armorican, along a continental suture. Subduction-related granitic activity in Armorica was mainly Carboniferous and may correspond to the subduction of paleo-Tethys after the Middle Devonian event.

Composite Middle Devonian event

As shown in Figures 3 and 4, the Middle Devonian event would have been created in a different geodynamic context along the northern border of the Hun superterrane. From Portugal-Galicia to the Sudetes, this event corresponds to the consumption of a Rheno-Hercynian intervening terrane and mid-oceanic ridge by the Rheic accretionary prism. This situation can be extrapolated eastward in view of a potential continuation of the Rheno-Hercynian ocean toward the Black Sea and Caucasus. In northern Turkey, Kozur et al. (1999) and Kozur (1999) found remnants of a Carboniferous to Permian pelagic domain (Paphlagonia) formerly located just south of the Istanbul zone, which shows a development very similar to the Rheno-Hercynian domain (Göncüoglu and Kozur, 1998; Kozur et al., 2000). This Paphlagonian ocean could be extended eastward to the Dizi area of the western Great Caucasus (Adamia and Kutelia, 1987). The difference between the situation east and west of the Moesian promontory is that the final closure of these eastern pelagic realms did not take place before the Permian.

In the oriental part of the European Hunic terranes, we propose that the Ligerian-Moldanubian domain collided with an island-arc system derived from the subduction of the Asiatic ocean (Zonenshain et al., 1985), a time equivalent of the Rheic ocean. It was a large ocean connected to Panthalassa and developed numerous island arcs. It was proposed that the amalgamation of such island arcs (the Kipchak arc system) gave birth to the Kazakhstan plate (Şengör and Natal'in, 1996). In view of the convergence between Gondwana and the future continents composing Laurasia, a collision between an Asiatic island-arc system and the European Hunic terranes was probably unavoidable. We tentatively place domains such as the Sakarya zone of northern Turkey and part of the trans-Caucasus in this western Kipchak island-arc system. Elements of this arc would also be found in the Ligerian-Moldanubian cordillera, but most of this cordillera would most likely belong to the European Hunic terranes.

Appalachians and Meguma-Meseta dilemma

In the Appalachians the situation is less clear, although potentially simpler. It is less clear because a large part of the hinterland was lost in the opening of the Atlantic Ocean; it is apparently simpler because the mountain-building processes there appear to be a continuum of deformation (Keppie, 1989; Piqué and Skehan, 1992). Some correlations are proposed between, for example, the Meguma belt and the Ligerian cordillera of central France (Rast and Skehan, 1993), which would somewhat complicate the story. One of the problems to discuss here is the concept of tectonic phases, and, more precisely, the Acadian. This concept can only be used if it is supported by a consistent plate tectonic scheme; otherwise it is meaningless. The Acadian phase proper should be restricted to the docking of Avalonia to Laurentia, a docking generally accepted to be finished by Late Silurian time for east Avalonia but later for west Avalonia (Early Devonian; Friend et al., 2000). Therefore, younger events could correspond to the docking of Hun-like terranes to Laurentia, followed by the onset of subduction of paleo-Tethys under this continent. The docking of such terranes to Laurentia would be more or less simultaneous to the Middle Devonian event further west. Dallmeyer and Keppie (1987) have shown that the Meguma terrane was affected by an Early to Middle Devonian tectonometamorphic event, which could confirm the accretion of terranes at that time. The Meguma terrane was later intruded by Late Devonian granites, showing that subduction of paleo-Tethys under North America had already started at that time. Was the Meguma block accreted to North America in Middle Devonian time (Hun origin), or was it already part of Avalonia and detached from Avalonia and accreted again? These are main questions that are also relevant for the Moroccan Meseta, which is often considered to be part of the Meguma terrane. The absence of Silurian metamorphism in the Meguma and Meseta terranes is not a proof of their separation from Avalonia, because only the leading edge of Avalonia should have undergone such metamorphism. The leading northern edge of Meguma was affected by the Acadian orogeny (slaty cleavage dated as 405 Ma; Keppie, 1989), which would exclude a Hun origin, whereas its southern border remained a passive margin attached, in that case, to the Rheic ocean. Then it became an active margin, documented by Emsian deformation, extending southward to the Carolina region, where a tectonometamorphic event was dated as 340–360 Ma (Dallmeyer et al., 1986). A younger event, dated as ca. 268–315 Ma, marked the final collision with Gondwana.

A magmatic arc development is also known in the Meseta starting in the Tournaisian (Aouli granitoids, Oukemeni and Bourne, 1993) and postcollisional granites were emplaced until the Early Permian (Amenzou and Badra, 1996). However, the Meseta (e.g., Piqué, 1989) is characterized on its southeastern border by the development of a south-facing passive margin, with a prominent Early Devonian rift shoulder in the Rehamna region, passing to the southeast to Early to Middle Devonian deep-water deposits, therefore representing potential northern paleo-Tethyan marginal series characteristic of Hun-like terranes (see following). This margin was deformed during a Late Devonian tectonometamorphic phase (Huon et al., 1993), followed by the installation of an arc in Visean time, clearly marking the change from a passive to active margin setting. Late Visean (ca. 325 Ma) wild-flysch-type deposits, also present in the High Atlas (Jenny, 1988), mark the onset of deformation during the collision with Gondwana. Therefore, we can question the Meguma-Meseta connection, the Meguma being of Avalonian origin and the Meseta of Hun origin in our point of view. The limit between the two terranes could be located just offshore Morocco, west of the El Jadida escarpment, where low-grade metamorphic rock have been drilled (Kreuser et al., 1984).

We propose extending the Hun superterrane to Ecuador in order to include terranes, found now around the Gulf of Mexico area, that were detached from Gondwana in the Paleozoic (Dallmeyer, 1989; Keppie et al., 1996). Consequently, Hun-like terranes must have collided with North America. Potential candidates could be represented by the Carolina terrane (Horton et al., 1989) and related blocks, apparently separated from Laurentia by oceanic elements (Bel Air–Rising Sun terrane). Terranes implicated in the Alleghanian-Ouachita orogen are other potential candidates, like the Sabine block (Keller and Hatcher, 1999), as well as the Yucatan, Chortis, Mexican (Oaxaca, Arequipa-Antofalla), and other Central American terranes (Keppie et al., 1996).

PALEO-TETHYS EVOLUTION

The paleo-Tethys is more or less completely ignored by those who follow the classic Hercynian ideas; therefore it is important to present here the main lines of its geodynamic evolution. The opening of paleo-Tethys is relatively well defined on an Iranian transect (Alborz Range, north Iran; Stampfli, 1978; Stampfli et al., 1991, 2001a, 2001b) representing the southern Gondwanan margin of the eastern branch of the ocean. Late Ordovician to Early Devonian flood basalts, rift shoulder uplift in the Silurian, followed by the onset of thermal subsidence in the Devonian, point to a Late Ordovician–Silurian rifting phase. Seafloor spreading took place in the Late Silurian or Early Devonian and the rift shoulders were completely flooded in the Late Devonian, following regional thermal subsidence of the passive margin. From Late Devonian until Middle Triassic time a carbonate-dominated passive margin developed. A similar evolution is found in the Cimmerian part of Turkey (for details see references in Stampfli 1996; Göncüoglu and Kozur, 1998).

Toward the west, along the African domain, there are few data concerning the detachment of the European Hunic terranes from Gondwana. In the High Atlas of Morocco (e.g., Destombes, 1971), the Silurian unconformably overlies the Ordovician and presents locally, at its top, a conglomeratic sequence; the overlying Emsian-Eifelian sequence is locally very condensed and represented by open marine carbonates. This starvation event represents the onset of important thermal subsidence, which can be related to the drifting of either Armorica or the Meseta from Africa. On the basis of current information, it is unclear whether the opening took place simultaneously all along the Gondwana margin. Our preference is that the opening was earlier along the western branch of the ocean (Fig. 3). Farther west, along the Gondwana border, a major Carboniferous sedimentary wedge developed directly on the Precambrian basement along the Amazonian craton in Ecuador (Litherland et al., 1994); the absence of any sequence that could represent the early Paleozoic active or passive margin of this craton allows the opening of paleo-

Tethys to be extended to the northern part of South America (as discussed herein about the Mexican terranes origin; Fig. 3).

The northern Hunic margin of the paleo-Tethys ocean is well represented in the middle part of the European Hunic terranes in the Carnic Alps (Schönlaub and Histon, 1999: Laufer et al., 2001), Tuscany, Sardinia, and the Alboran fragments (cf. Stampfli, 1996); it is also characterized by a Late Ordovician–Early Silurian clastic and often volcanic synrift sequence (Silurian flood basalts are also known in Sardinia and the Rif; Piqué, 1989). Rift-related thermal uplift, erosion, and tilting took place in Silurian time and are often (wrongly) related to the Taconic event (Tollmann, 1985). Open marine conditions started in the Silurian, being represented by a graptolites facies; a more general flooding took place in the Early Devonian and marked the onset of widespread thermal subsidence related to seafloor spreading.

The Saxo-Thuringian domain was part of the northern margin of paleo-Tethys before the lateral displacement of the Moldanubian zone to the south of it. Its autochthonous sequence (Falk et al., 1995) is marked by basin deepening in Silurian time, accompanied by lavas and tuffs in the Ludlow representing the synrift event, whereas pelagic Gedinnian to Givetian sediments represent the drift sequence.

On the northern margin, the Visean usually marks the onset of widespread flysch deposition, often accompanied by volcanic activity. We regard this major change as marking the general aggregation of the different terranes to Eurasia to form the Variscan cordillera; it also marks the onset of paleo-Tethys subduction and the transformation of the margin from passive to active, the flysch troughs usually representing forearc basins. Accretionary sequences related to this subduction are little known, most likely because important subduction erosion took place during the cordillera stage, as observed now along the South American active margin. Potential paleo-Tethyan accretionary sequences are located in the southern part of the Variscan orogen, and in all cases were metamorphosed and intruded by subsequent Late Carboniferous granites; in addition, they were usually involved in eo-Cimmerian and Alpine deformation. However, pelagic Devonian-Carboniferous to Early Triassic pelagic sediments of paleo-Tethyan origin are found in Sicily, the Dinarides, Hellenides, and Turkey (Karaburun) (Stampfli et al., 2002).

CONCLUSIONS

Deciphering the evolution of former circum-Gondwana terranes is a feasible enterprise; similar geological evolutions are found in these terranes. The explanation of the present complexity should not be sought in complex plate tectonic scenarios involving numerous oceanic realms; on the contrary, and in view of the similarities, a simple model is preferable. A single terrane model is also proposed for the accretion of all the Alaskan terranes (Johnston, 2001).

We propose continuous southward subduction of oceanic realms under the Gondwanan border, starting in the Late Neoproterozoic,which triggered the detachment of three main terranes. First Avalonia in the Early Ordovician, then the European Hunic terranes in the Late Silurian, promptly followed by the eastern Hunic terranes in the Early Devonian.

The accretion of these terranes to Laurussia was complex. Whereas the Avalonia superterrane had a relatively simple evolution with a classical collision of an active and passive margin, a more complex scenario is necessary to explain the Variscan collage. In order to take into account similarities in the different parts of the European Hunic terranes, we propose that areas affected by the Middle Devonian high-pressure phase were located on the leading accretionary edge of these terranes, whereas areas not affected by this major eo-Variscan event were located on the paleo-Tethys margin of the terrane. This Middle Devonian eo-Variscan event is inferred to be related to accretion of buoyant material derived from Laurussia and subduction of a peri-Laurussian ocean, whereas farther east the event is related to collision with an island-arc system.

To explain the subsequent large-scale mixture of active and passive margins, important lateral displacements and rotations must be invoked: most who study Hercynian ideas would agree with this proposition (e.g., Matte et al., 1990; Edel, 2000, 2001), but the majority would place these translations in a context of continent-continent collision. We propose that these took place during the displacement of terranes along a still-active margin, the translations being accompanied by transtensional and tranpressional events leading to the opening of Gulf of California–type oceans and in other places to the buildup of cordilleras.

During the growth of the late Carboniferous cordillera, two types of geological evolution developed. The westward one is toward a continent-continent collision where the accreted terranes got squeezed between Laurussia and Gondwana (this is the prevailing scenario for the Alleghanian regions), whereas eastward, subduction continued with a general rollback of the paleo-Tethyan slab. This, in turn, generated the opening of numerous backarc basins and oceans, starting in the Early Permian and lasting until the Middle Triassic closure of the paleo-Tethys oceanic domain.

Postcollisional Permian-Carboniferous granites, for example found in Morocco (e.g., Amenzou and Badra, 1996), should be related to slab detachment when major crustal attenuation through generalized extension is not documented. In other western European regions, postcollisional Permian-Carboniferous granites should be related to slab detachment and/or the collapse of the cordillera, but not to postcollisional processes, the final collision being far distant in time and space. The final closure of paleo-Tethys from Sicily to the Caucasus took place during the eo-Cimmerian cycle, and the closure of backarc oceans issued from the paleo-Tethys slab rollback took place only in Cretaceous time (Stampfli et al., 2001b).

We hope that our provocative suggestions will trigger a new round of discussion for the coming years; more field data should be gathered for a better approach of Variscan history, and paleoreconstructions on a larger scale should be included.

ACKNOWLEDGMENTS

We thank the conveners of the Basement Tectonics 15 Galicia 2000 congress (R. Arenas, Madrid; F. Diaz Garcia, Oviedo; J.R. Martinez-Catalan, Salamanca) for providing an ideal environment for stimulating discussion, encouraging us to publish this paper, and for encouraging remarks. We also thank J. Mosar, with whom these reconstructions were initiated, and H. Kozur for sharing key information on Paleozoic paleogeography. Our warm thanks to Dave Gee (Uppsala) for his engaged criticism and a readable English version, and we thank an anonymous reviewer for constructive suggestions.

REFERENCES CITED

Abati, J., Dunning, G.R., Arenas, R., Díaz García, F., Gonzáles Cuadra, P., Martínez Catalán, J.R., and Andonaegui, P., 1999, Early Ordovician orogenic event in Galicia (NW Spain): Evidence from U-Pb ages in the uppermost unit of the Ordenes Complex: Earth and Planetary Science Letters 165, p. 213–228.

Acquafredda, P., Lorenzoni, S., and Zanettin Lorenzoni, E., 1994, Paleozoic sequences and evolution of the Calabrian-Peloritan Arc (Southern Italy): Terra Nova, v. 6, p. 582–594.

Adamia, S., and Kutelia, Z., 1987, paleo-Tethys-Tethys: Continuous development: IGCP, v. 5, newsletter 7, p. 109–115.

Amenzou, M., and Badra, L., 1996, Les granites d'Azegour et de Brikiine (Maroc): Implications génétiques d'après la typologie des zircons: Comptes Rendus de l'Académie des Sciences de Paris, Série II, v. 323, no. 3, p. 213–220.

Anthes, G., and Reischmann, T., 2001, Timing of granitoid magmatism in the eastern mid-German crystalline rise: Journal of Geodynamics, v. 31, p. 119–143.

Arenas, R., Díaz García, F., Martínez Catalán, J.R., Abati, J., González Cuadra, P., Andonaegui, P., Gonzáles del Tánago, J., Rubio Pascual, F., Castiñeiras, P., and Gomez Barreiro, J., 2000, Structure and evolution of the Ordenes Complex: A Coruña, Spain, Basement Tectonics 15, Pre-Conference Field Trip.

Ballèvre, M., Bosse, V., Capdevilla, R., de Jong, K., Féraud, G., Guerrot, C., Peucat, J.J., and Ruffet, G., 2000, Structure and thermo chronology of the Champtoceaux complex (France): A key for Ibero-Armorican correlation, *in* Basement tectonics 15: A Coruña, Spain, p. 52–55.

Capuzzo, N., and Bussy, F., 2000, High precision dating and origin of synsedimentary volcanism in the late Carboniferous Salvan-Dorénaz basin (Aiguilles-Rouges massif, western Alps: Schweizerische Mineralogische und Petrographische Mitteilungen, v. 80, p. 147–167.

Catalano, R., Distefano, P., and Kozür, H., 1988, New results in the Permian and Triassic stratigraphy of western Sicily with special reference to the section at Torrente San Calogero, SW of Pietra di Salome (Sosio valley): Società Geologica Italiana, v. Atti del 74 Congresso Nazionale, v. A, p. 119–125.

Cortesogno, L., Dallagiovanna, G., Gaggero, L., and Vanossi, M., 1993, Elements of the Paleozoic history of the Ligurian Alps, *in* von Raumer, J.F., and Neubauer, F., eds., Pre-Mesozoic geology of the Alps: Berlin, Springer-Verlag, p. 257–277.

Crowley, Q.G., Floyd, P.A., Stodra, V., Winchester, J.A., and Holland, J.G., 2000, The Marianske-Lazne complex, remnants of a subducted early Paleozoic seaway: Basement Tectonics 15, A Coruña, Spain, Program and Abstracts, p. 86–89

Dallmeyer, R.D., 1989, Contrasting accreted terranes in the southern Appalachian orogen and Atlantic-Gulf Coastal Plains and their correlations with West African sequences, *in* Dallmeyer, R.D., ed., terranes in the circum-Atlantic Paleozoic orogens: Boulder, Colorado, Geological Society of America Special Paper 230, p. 247–267.

Dallmeyer, R.D., and Keppie, J.D., 1987, Polyphase late Paleozoic tectonothermal evolution of the southwestern Meguma terrane, Nova Scotia: Evidence from $^{40}Ar/^{39}Ar$ mineral ages: Canadian Journal of Earth Science, v. 24, p. 1242–1254.

Dallmeyer, R.D., and Martínez García, E., 1990, Pre-Mesozoic geology of Iberia: Berlin, Springer-Verlag, 416 p.

Dallmeyer, R.D., Franke, W., and Weber, K., 1995, Pre-Permian geology of central and eastern Europe: Berlin, Springer-Verlag, 604 p.

Dallmeyer, R.D., Wright, J.E., Secor Jr, D.T., and Snoke, A.W., 1986, Character of the Alleghanian orogeny in the southern Appalachians: II: Geochronological constraints on the tectonothermal evolution of the eastern Piedmont in South Carolina: Geological Society of America Bulletin, v. 97, p. 1329–1344.

Dean W.T., Monod, O., Rickards, R.B., Demir, O., and Bultynck, P., 2000, Lower Paleozoic stratigraphy and palaeontology, Karadere-Zirze area, Pontu Mountains, northern Turkey: Geological Magazine, 137, p. 555–582.

Destombes, J., 1971, L'Ordovicien du Maroc: Essai de synthèse stratigraphique: Mémoires du Bureau de Recherches Géologiques et Minières Orléans, v. 73, p. 237–263.

Díaz García, F., Arenas, R., Martínez Catalán, J.R., González del Tánago, J., and Dunning, G.R., 1999, Tetctonic evolution of the careon ophiolite (northwest Spain): A remnant of the oceanic lithosphere in the Variscan belt: Journal of Geology, v. 107, p. 587–605.

Dvorak, J., 1995, ChapterVIII.B.1 Stratigraphy, *in* Dallmeyer, R.D., Franke, W., and Weber, K., eds., Pre-Permian geology of Central and Eastern Europe: Berlin Heidelberg New York, Springer-Verlag, p. 478–490.

Edel, J.B., 2000, Hypothèse d'une ample rotation horaire tardi-varisque en bloc Maures-Estérel-Corse-Sardaigne: Géologie de la France, v. 1, p. 3–19.

Edel, J.B., 2001, The rotations of the Variscides during the Carboniferous collision: Paleomagnetic constraints from the Vosges and the Massif Central (France): Tectonnophysics, v. 332, p. 69–92.

Eguiluz, L., Gil Ibarguchi, J.I., Abalos, B., and Apraiz, A., 2000, Superposed Hercynian and Cadomian orogenic cycles in the Ossa-Morena zone and related areas of the Iberian Massif: Geological Society of America Bulletin, v. 112, p. 1398–1413.

Falk, F., Franke, W., and Kurze, M., 1995, Chapter V.B.1 Stratigraphie, *in* Dallmeyer, R.D., Franke, W., and Weber, K., eds., Pre-Permian geology of Central and Eastern Europe: Berlin Heidelberg New York, Springer-Verlag, p. 221–234.

Faure, M., Leloix, Ch., Roig, J.-Y., 1997, L'évolution polycyclique de la chaîne hercynienne: Bulletin de la Société Géologique de France, 168, p. 695–705.

Fernández-Suárez, J., Gutiérrez-Alonso, G., Jenner, G.A., Tubrett, M.N., 2000, New ideas on the Proterozoic–early Paleozoic evolution of NW Iberia: Insights from U-Pb detrital zircon ages: Precambrian Research 102, p.185–206.

Findley, R.H., 1998, The Song Ma anticlinorium, northern Vietnam: The structure of an allochthonous terrane containing an early Paleozoic island arc sequence: Journal of Asian Earth Sciences, v. 15, no. 6, p. 453–464.

Flick, H., Pfefferkorn, H.W., and Schmidt, J., 1988, Presence of land-plants on a volcanic island in a Devonian sea (Rhenish mountains, West Germany): Review of Paleobotany and Palynology, v. 56, p. 177–181.

Floyd, P.A., 1995, Chapter III.B.3: Igneous activity, *in* Dallmeyer, R.D., Franke, W., and Weber, K., eds., Pre-Permian geology of Central and Eastern Europe: Berlin Heidelberg New York, Springer-Verlag, p. 59–81.

Flügel, H.W., 1990 Das voralpine Basement im Alpin-Mediterranen Belt—Überblick und Problematik: Jahrbuch der Geologischen Bundesanstalt, v. 133, p. 181–221.

Foster, D.A., and Gray, D.R., 2000, Evolution and structure of the lachlan fold belt (orogen) of Eastern Australia: Annual Review of Earth and Planetary Sciences, v. 28, p. 47–80.

Franke, W., 1989, Variscan Plate tectonics in Central Europe—Current ideas and open questions: Tectonophysics, 169, p.221–228.

Franke, W., 1992, Phanerozoic structures and events in Central Europe, *in* Blundell, D., Freeman, R., and Mueller, St., eds, A continent revealed: Cambridge, Cambridge University Press 1992, pp 164–180.

Franke, W., 1995, Chapter III.B.1: Stratigraphy, *in* Dallmeyer, R.D., Franke, W., and Weber, K., eds., Pre-Permian geology of Central and Eastern Europe: Berlin Heidelberg New York, Springer-Verlag, p. 33–49.

Franke, W., 2000, The mid-European segment of the Variscides: tectonostratigraphic units, terrane boundaries and plate tectonic evolution, *in* Franke, W., Haak, V., Oncken, O., and Tanner, D., eds., Orogenic processes: Quantification and modelling in the Variscan belt: Geological Society of London Special Publication 179, p. 35–62.

Franke, W., Kreuzer, H., Okrusch, M., Schüssler, U., and Seidel, E., 1995, ChapterV.C.1: Stratigraphy, structure, and igneous activity, *in* Dallmeyer, R.D., Franke, W., and Weber, K., eds., Pre-Permian geology of Central and Eastern Europe: Berlin Heidelberg New York, Springer-Verlag, p. 277–294.

Franke, W., Haak, V., Oncken, O., Tanner, D., 2000, Orogenic processes: Quantification and modelling in the Variscan belt: Geological Society Special Publication 179, 459p.

Friend, P.F., Williams, B.P.J., Ford, M., and Williams, E.A., 2000, Kinematics and dynamics of Old Red Sandstone basins, *in* Friend, P.F., and Williams, B.P.J., eds., New perspectives on the Old Red Sandstone: Geological Society of London Special Publication 180, p. 29–60.

Frisch, W., and Neubauer, F., 1989, Pre-Alpine terranes and tectonic zoning in the eastern Alps, *in* Dallmeyer, R.D., ed., Terranes in the circum-Atlantic Paleozoic orogens: Boulder, Colorado, Geological Society of America Special Paper 230, p. 91–114.

Ghienne, J.F., Bartier, D., Leone, F., and Loi, A., 2000, Caractérisation des horizons manganésifères de l'Ordovicien supérieur de Sardaigne: Relation avec la glaciation fini-ordovicienne: Comptes Rendus de l'Académie des Sciences de Paris, v. 331, p. 257–264.

Gil Ibarguchi, J.I., Ábalos, B., Azcárraga, J., Mendia, M., Puelles, P., 2000, A petrological and structural excursion through the high-grade/high-pressure allochthonous units of the Cabo Ortegal complex (NW Spain): Basement Tectonics 15, mid-Conference field trip, A Coruña, Spain.

Giorgis, D., Thélin, P., Stampfli, G.M., and Bussy, F., 1999, The Mont-Mort metapelites: Variscan metamorphism and geodynamic context (Briançonnais basement, western Alps, Switzerland): Schweizerische Mineralogische und Petrographische Mitteilungen, v. 79, p. 381–398.

Göncüoglu, C., and Kozur, H., 1998, Remarks to the pre-Variscan development in Turkey: Schr. Staatl. Mus. Min. Geol. Dresden, v. 9, p. 137–138.

Hatcher, R.D., 1983, Basement massifs in the Appalachians: Their role in deformation during the Appalachian orogenies: Geological Journal, v. 18, p. 255–265.

Heinisch, H., Schätz, M., Bachtadse, V., and Tait, J., 2000, Paleomagnetic and sedimentological constraints on the position of Northern Graywacke Zone, Eastern Alps within the Paleozoic terrane assemblage: Schriften der Alfred-Wegener-Stiftung Terra Nostra, v. 2000, no. 1, p. 54.

Hoffman, P.F., 1991, Did the breakout of Laurentia turn Gondwanaland inside out?: Science, v. 252, p. 1409–1412.

Horton, J.W., Drake, A.A., and Rankin, D.W., 1989, Tectonostratigraphic terranes and their Paleozoic boundaries in the central and southern Appalachians, *in* Dallmeyer, D., ed., Terranes in the circum-Atlantic Paleozoic orogens: Boulder, Colorado, Geological Society of America Special Paper 230, p. 213–245.

Huon, S., Cornée, J.J., Piqué, A., Rais, N., Clauer, N., Liewig, N., and Zayane, R., 1993, Mise en évidence d'un métamorphisme statique d'âge triassico-liasique lié à l'ouverture de l'Atlantique: Bulletin de la Société Géologique de France, v. 164, no. 2, p. 165–176.

Hutchison, C.S., 1989, Geological evolution of South-East Asia: Clarendon press Oxford, Oxford Monographs on Geology and Geophysics, 368 p.

Jenny, J., 1988, Carte géologique du Maroc au 1:100,000, feuille Azilal, Mémoire explicatif: Notes et Mémoires du Service géologique du Maroc, v. 399bis, 104 p.

Johnston, S.T., 2001, The Great Alaskan Terrane wreak: Reconciliation of paleomagnetic and geological data in the northern Cordillera: Earth and Planetary Science Letters, v. 193, p. 259–272.

Keller, G.R., and Hatcher, R.D., 1999, Some comparisons of the structure and evolution of the southern Appalachian-Ouachita orogen and portions of the Trans-European Suture Zone region: Tectonophysics, v. 314, p. 43–68.

Keppie, J.D., 1989, Northern Appalachian terranes and their accretionary history, *in* Dallmeyer, D., ed., Terranes in the circum-Atlantic Paleozoic orogens: Boulder, Colorado, Geological Society of America Special Paper 230, p. 159–192.

Keppie, J.D., 1994, Pre-Mesozoic geology in France: Berlin-Heidelberg-New York, Springer-Verlag, 514 p.

Keppie, J.D., Dostal, J., Murphy, J.B., and Nance, R.D., 1996, terrane transfer between Laurentia and western Gondwana in the early Paleozoic: Constraints on global reconstructions, *in* Nance, R.D., and Thompson, M.D., eds., Avalonian and related peri-Gondwana terranes of the circum-Atlantic: Boulder, Colorado, Geological Society of America Special Paper 304, p. 369–380.

Khain, V.E., 1994, Geology of Northern Eurasia (Ex-USSR): Beiträge zur regionalen Geologie der Erde, v. 24, p. 346–385.

Klootwijk, C., 1996, Phanerozoic configurations of Greater Australia : Evolution of the North West Shelf: Australian Geological Survey Organisation, Record 1996/51–1996/53.

Kossmat, F., 1927, Gliederung des varistischen Gebirgsbaues: Abhandlungen der sächsischen geologischen Landesanstalt 1, p. 1–39.

Kozur, H., 1997, First discovery of Muellerisphaerida (inc. sedis) and Eoalbaillella (Radiolaria) in Turkey and the age of the siliciclastic sequence (clastic series) in Karaburun peninsula: Freiberger Forschungsheft, v. C 466, p. 33–59.

Kozur, H., 1998, The age of the siliciclastic series ("Karareis formation") of the western Karaburun peninsula, western Turkey: Paleontologia Polonica, v. 58, p. 171–189.

Kozur, H., 1999, Permian development in the western Tethys, *in* Shallow Tethys 5, Chiang Mai: Department of Geological Sciences, University of Chiang Mai, p. 101–135.

Kozur, H.W., Senel, M., and Tekin, K., 1998, First evidence of Hercynian Lower carboniferous deep-water sediments in the Lycian nappes, SW Turkey: Geologia Croatica, v. 51, no. 1, p. 15–22.

Kozur, H., Aydin, M., Demir, O., and Yakar, H., 1999, Geological evolution of middle Pontide area (northern Turkey) and eastward continuation of the Meliata-Hallstatt ocean: Geologica Carpathica, v. 50, p. 154–155.

Kozur, H., Aydin, M., Demir, O., Yakar, H., Göncüoglu, M.C., and Kuru, F., 2000, New stratigraphic and palaeogeographic results from the Palaeozoic and Early Mesozoic of the Middle Pontides (Northern Turkey) in the Azdavay, Devrekani, Küre and Inebolu areas: Implications for the Carboniferous-Early Cretaceous geodynamic evolution and some related remarks to the Karakaya oceanic rift basin: Geologia Croatica, v. 53/2, p. 209–268.

Krahl, J., Kauffmann, G., Richter, D., Kozur, H., Möller, I., Förster, O., Heinritzi, F., and Dornsiepen, U., 1986, Neue Fossilfunde in der Phyllit-Gruppe Ostkretas (Griechenland): Zeitschrift der deutschen geologischen Gesellschaft, v. 137, p. 523–536.

Kreuser, H., Muller, P., Reinecke, T., and Wissmann, G., 1984, Petrography and K-Ar dating of the Mazagan granodiorite, *in* Hinz, K., Winterer, E.L., and others, eds., Initial Reports of the Deep Sea Drilling Project, Leg 9, holes 554A and B: Washington, D.C., U.S. Government Printing Office, p. 543–549.

Krs, M., and Pruner, P., 1999, To the paleomagnetic investigations of paleogeography of the Barrandian terrane, Bohemian massif: Acta Universitatis Carolinae–Geologica, v. 43, p. 519–522.

Lardeaux, J.M., Ledru, P., Daniel, I., and Duchene, S., 2001, The Variscan French Massif Central—A new addition to the ultrahigh pressure metamorphic "club": Exhumation processes and geodynamic consequences: Tectonophysics, v. 332, p. 143–167.

Laüfer, A.L., Hubich, D., and Loeschke, J., 2001, Variscan geodynamic evolution of the Carnia Alps (Austria/Italy): International Journal of Earth Sciences, v. 90, p. 855–870.

Le Corre, C., Auvray, B., Ballevre, M., and Robardet, M., 1991, Le Massif Armoricain, *in* Piqué, A., ed., Les massifs anciens de France: Sciences Géologiques (Strasbourg) Bulletin, p. 31–104.

Linnemann, U., Buschmann, B., 1995, Die cadomische Diskordanz im Saxothuringikum (oberkambrisch-tremadocische overlap-Sequenzen): Zeitschrift für geologische Wissenschaften 23, p. 707–727.

Litherland, M., Aspden, J.A., and Jemielita, R.A., 1994, The metamorphic belts of Ecuador: Overseas Memoir of the British Geological Survey, 147 p.

Marcos, A., Fernandez Rodriguez, F.J., and Llana Fúnez, S., 2000, Structure of the Cabo Ortegal Nappe, *in* Basement tectonics 15, A Coruña, p. 1–59.

Martínez Catalán, J.R., Arenas, R., Díaz García, F., Abati, J., 1997, Variscan accretionary complexes of northwest Iberia: Terrane correlation and succession of tectonothermal events: Geology, 25, p. 1103–1106.

Martínez Catalán, J.R., Arenas, R., Díaz García, F., Abati, J., 1999, Allochthonous units in the Variscan belt of NW Iberia, *in* Sinha, A.K., ed., Basement Tectonics 13: Dordrecht, Netherlands, Kluver Academic Publishers, p. 65–84.

Matte, P., 1991, Accretionary history and crustal evolution of the Variscan belt in Western Europe: Tectonophysics 180, p. 309–338.

Matte, P., 1998, Continental subduction and exhumation of HP rocks in Paleozoic orogenic belts: Uralides and Variscides: Geologiska Forhandlinger a Föreninger 120, 209–222.

Matte, P., Maluski, H., Rajlich, P., Franke, W., 1990, Terrane boundaries in the Bohemian massif: Result of large-scale Variscan shearing: Tectonophysics 177, p.151–170.

Meisl, S., 1995, Chapter III.C.3: Igneous activity, *in* Dallmeyer, R.D., Franke, W., and Weber, K., eds., Pre-Permian geology of Central and Eastern Europe: Berlin Heidelberg New York, Springer-Verlag, p. 118–137.

Meng, Q.-R., and Zhang, G.-W., 1999, Timing of collision of the north and south China blocks: Controversy and reconcialation: Geology, v. 27, no. 2, p. 123–126.

Mingram, B., 1998, The Erzgebirge, Germany—A subducted part of northern Gondwana: Geochemical evidence for repetition of early Paleozoic metasedimentary sequences in metamorphic thrust units: Geological Magazine 135, p. 785–801.

Monod, O., Ghienne, J.-F., Dean, W.T., Kozlu, H., Günay, Y., Paris, F., Le Hérissé, A., and Rickards, R.B., 2002, Ordovician glaciation in southern Turkey, in 1st international symposium of the Faculty of Mines (ITü) on Earth Science and Engineering, Istanbul, p. 82.

Nesbor, H.D., Buggisch, W., Flick, H., Horn, M., and Lippert, H.J., 1993, Vulkanismus im Devon des Rhenohercynikums: Geologische Abhandlungen Hessen, v. 98, p. 3–87.

Neubauer, F., and Handler, R., 2000, Variscan orogeny in the Eastern Alps and Bohemian Massif: How do these units correlate: Mitteilungen der Österreichischen Geologischen Gesellschaft, v. 92, p. 35–59

Oliveira, J.T., and Quesada, C., 1998, A comparison of stratigraphy, structure and paleogeography of the South-Portuguese zone and south-west England: Geoscience in South-West England, v. 9, p. 141–150.

Ordoñez Casado, B., 1998, Geochronological studies of the pre-Mesozoic basement of the Iberian Massif: The Ossa Morena zone and the allochthonous complexes within the Central Iberian zone: Eidgenössische Technische Hochschule Zürich, n. 12,940, 235 p.

Oukemeni, D., and Bourne, J.H., 1993, Études géochimiques des granitoides du pluton d'Aouli, Haute Moulouya, Maroc: Journal of African Earth Sciences, v. 17, p. 429–443.

Paris, F., and Robardet, M., 1990, Early Paleozoic paleogeography of the Variscan regions: Tectonophysics, v. 177, p. 193–213.

Pin, Ch., 1990, Variscan oceans: Ages, origins and geodynamic implications inferred from geochemical and radiometric data: Tectonophysics 177, p. 215–227.

Pin, C., Paquette, J.L., Gil Ibarguchi, J.I., Santos Zalduegui, J.F., Rodríguez Aller, J., and Ortega Cuesta, L.A., 2000, Geochronological and geochemical constraints on the origin of ophiolitic units from the northwestern Iberian massif, *in* Variscan Appalachian Dynamics: The building of the upper Paleozoic basement. Basement tectonics 15, La Coruña, p. 146–147.

Piqué, A., 1989, Variscan terranes in Morocco, *in* Dallmeyer, R.D., ed., Terranes in the circum-Atlantic Paleozoic orogens: Boulder, Colorado, Geological Society of America Special Paper 230, p. 115–129.

Piqué, A.,and Skehan, J.W., 1992, Late Paleozoic orogenesis in western Africa and eastern north America: Closure of the Theic ocean: Tectonics, v. 11, no. 2, p. 392–404.

Quenardel, J., Santallier, D., Burg, J., Bril, H., Cathelineau, M., and Marignac, C., 1991, Le Massif Central, *in* Piqué, A., ed., Les massifs anciens de France: Sciences Géologiques (Strasbourg) Bulletin, p. 105–206.

Rast, N., and Skehan, J.W., 1993, Mid-Paleozoic orogenesis in the North Atlantic: The Acadian orogeny: Boulder, Colorado, Geological Society of America Special Paper 275, p. 1–25.

Reischmann, T., and Anthes, G., 1996, Geochronology of the mid-German crystalline rise west of the River Rhine: Geologische Rundschau, v. 85, p. 761–774.

Robardet, M., 1996, Early Paleozoic palaeogeography of North-Gondwanan Europe, *in* Baldis, B., and Aceñolaza F.G., eds, Early Paleozoic evolution in NW Gondwana: Serie correlación geologica 12, p. 167–180.

Robardet, M., 2000, An alternative approach to consider the Variscan Belt in SW Europe: The pre-orogenic palaeogeographical constraints: Basement Tectonics 15, A Coruña, Spain, Program and Abstracts, p. 23–26.

Robardet, M., Paris, F., and Racheboeuf, P.R., 1990, Paleogeographic evolution of southwest Europe during early Paleozoic time, *in* McKerrow, W.S., and Scotese, C.R., eds., Paleozoic paleogeography and biogeography: Boulder, Colorado, Geological Society of America Memoir 12, p. 411–419.

Robardet, M., Verniers, J., Feist, R., and Paris, F., 1994, Le Paléozoïque antévarisque de France, contexte paléogéographique et géodynamique: Géologie de la France, v. 3, p. 3–31.

Ruttner, A.W., 1993, Southern borderland of Triassic Laurasia in NE Iran: Geologische Rundschau, v. 82, p. 110–120.

Sandeman, H.A., Chen, Y., Clark, A.H., and Farrar, E., 1995, Constraints on the *P-T* conditions and age of emplacement of the Lizard ophiolite Cornwal: Amphibole-plagioclase thermobarometry and Ar/Ar geochronology of basal amphibolites: Canadian Journal of Earth Science, v. 32, p. 261–272.

Schaltegger, U., Nägler, Th. N., Corfu, F., Maggetti, M., Galetti, G., and Stosch, H.G., 1997, A Cambrian island arc in the Silvretta nappe: Constraints from geochemistry and geochronology: Schweizerische Mineralogische Petrographische Mitteilungen, 77, p. 351–359.

Schätz, M., Bachtadse, V., Tait, J., and Soffel, H., 1997, Paleomagnetic results from lower Devonian sediments of the southern Alps: Schriften der Alfred-Wegener-Stiftung Terra Nostra, v. 97/5, p. 152–155.

Schönlaub, H.P., and Histon, K., 1999, The Paleozoic evolution of the southern Alps: Mitteilungen der Österreichischen Geologischen Gesellschaft, v. 92, p. 15–34.

Schulmann, K., Ledru, P., Autran, A., Melka, R., Lardeux, J.M., Urban, M., Lobkowicz, M., 1991, Evolution of nappes in the eastern margin of the Bohemian Massif: A kinematic interpretation: Geologische Rundschau 80, p. 73–92.

Şengör, A.M.C., 1979, Mid-Mezozoic closure of Permo-Triassic Tethys and its implications: Nature, v. 279, p. 590–593.

Şengör, A.M.C., 1984, The Cimmeride orogenic system and the tectonics of Eurasia: Boulder, Colorado, Geological Society of America Special Paper 195, p. 82.

Şengör, A.M.C., and Natal'in, B.A., 1996, Paleotectonics of Asia: Fragments of a synthesis, *in* Yin, A., and Harrison, M., eds., The tectonic evolution of Asia: Cambridge, Cambridge University press, p. 443–486.

Seston, R., Winchester, J.A., Piasecki, M.A.J., Crowley, Q.G., and Floyd, P.A., 2000, A structural model for the western-central Sudetes; a deformed stack of Variscan thrust sheets: Journal of the Geological Society of London, v. 157 Part 6, p. 1155–1167.

Stampfli, G.M., 1978, Étude géologique generale de l'Elbourz oriental au sud de Gonbad-e-Qabus (Iran NE) [PhD. Thesis]: University Genève, no. 1868, 329 p.

Stampfli, G.M., 1996, The Intra-Alpine terrain: A paleo-Tethyan remnant in the Alpine Variscides: Eclogae Geologicae Helvetiae, v. 89, no. 1, p. 13–42.

Stampfli, G.M., 2000, Tethyan oceans, *in* Bozkurt, E., Winchester, J.A., and Piper, J.D.A., eds., Tectonics and magmatism in Turkey and surrounding area: London, Geological Society of London Special Publication 173, p. 163–185.

Stampfli, G.M., and Borel, G.D., 2002, A plate tectonic model for the Paleozoic and Mesozoic: Earth and planetary Science Letters, v. 196, p. 17–33.

Stampfli, G.M., Marcoux, J., and Baud, A., 1991, Tethyan margins in space and time, *in* Channell, J.E.T., Winterer, E.L., and Jansa, L.F., eds., Paleogeography and paleoceanography of Tethys: Palaeogeography, Palaeoclimatology, Palaeoecology, p. 373–410.

Stampfli, G.M., Borel, G., and von Raumer, J., 2000, terrane accretion in the Variscan domain: Basement Tectonics 15, A Coruña, Spain, Program and Abstracts, p. 167–169.

Stampfli, G.M., Mosar, J., Favre, P., Pillevuit, A., and Vannay, J.-C., 2001a, Permo-Mesozoic evolution of the western Tethyan realm: The Neotethys/East-Mediterranean connection, *in* Ziegler, P.A., Cavazza, W., Robertson, A.H.F.R., and Crasquin-Soleau, S., eds., Peri-Tethys Memoir 6: Peritethyan rift/wrench basins and passive margins, Mémoire Musée National Histoire Naturelle, v. 186, p. 51-108.

Stampfli, G.M., Borel, G., Cavazza, W., Mosar, J., and Ziegler, P.A., 2001b, The paleotectonic atlas of the Peritethyan domain: European Geophysical Society (CD ROM).

Stampfli, G.M., Vavassis, I., De Bono, A., Rosselet, F., Matti, B., and Bellini, M., 2002, Remnants of the Paleotethys oceanic suture-zone in the western Tethyan area, *in* Cassinis, G., and Decandia, F.A., eds., Stratigraphic and structural evolution on the Late Carboniferous to Triassic continental and marine successions in Tuscany (Italy): Regional Reports and General Correlation, Bolletino della Società Geologica Italiana, special issue (in press).

Stille, H., 1951, Das mitteleuropäische Variszische Grundgebirge im Bilde des gesamteuropäischen: Geologisches Jahrbuch Beihefte, v. 2, p. 1–138.

Stipska, P., Schulmann, K., Kroener, A., 1998, Role of Cambrian-Ordovician rifting in Variscan collision at the NE margin of the Bohemian Massif: Acta Universitatis Carolinae Geologica, v. 42, p. 343–344.

Suess, E., 1909, Das Antlitz der Erde: Braumüller Wien.

Tait, J., Schätz, M., Bachtadse, V., Soffel, H., 2000, Palaeomagnetism and Paleozoic palaeogeography of Gondwana and European terranes, *in* Franke, W., Haak, V., Oncken, O., Tanner, D., eds., Orogenic processes: Quantification and modelling in the Variscan belt: London, Geological Society of London Special Publication 179, p. 21–34.

Thiéblemont, D., Marcoux, E., Tegyey, M., and Leistel, J.-M., 1994, Genèse de la province pyriteuse sud-ibérique dans un paléo-prisme d'accrétion? Arguments pétrologiques: Bulletin de la Société Géologique de France, v. 165, p. 407–423.

Tollmann, A., 1985, Geologie von Österreich, Band II: Wien, Franz Deuticke, 710 p.

Torsvik, T.H., and Smethurst, M.A., 1994, Geographic mapping and paleoreconstruction package (GMAP): Software description and examples of paleoreconstructions: Norges geologiske undersøkelse (NGU).

Torsvik, T.H., and Eide, E.A., 1998, Phanerozoic paleogeography and geodynamics with Atlantic details: Norges geologiske undersøkelse (NGU) report, v. 98.001, 82 p.

Torsvik, T.H., and Smethurst, M.A., 1999, Plate tectonic modelling: Virtual reality with GMAP: Computers & Geosciences, v. 25, p. 395–402.

Torsvik, T.H., Smethurst, M.A., Van der Voo, R., Trench, A., Abrahamsen, N., and Halvorsen, E., 1992, Baltica. A synopsis of Vendian-Permian paleomagnetic data and their paleotectonic implications: Earth-Science Reviews, v. 33, p. 133–152.

Unrug, R., 1997, Rodinia to Gondwana: The geodynamic map of Gondwana Supercontinent assembly: GSA Today v. 7, no. 1, p. 1–6

Van der Voo, R., 1993, Paleomagnetism of the Atlantic, Tethys, and Iapetus oceans: Cambridge, Cambridge University Press, 411 p

Vavassis, I., De Bono, A., Stampfli, G.M., Giorgis, D., Valloton, A., and Amelin, Y., 2000, U-Pb and Art-Ar geochronological data from the Pelagonian basement in Evia (Greece): Geodynamic implications for the evolution of paleo-Tethys: Schweizerische Mineralogische und Petrographische Mitteilungen, v. 80, p. 21–43.

von Quadt, A., and Gebauer, D., 1993, Sm-Nd and U-Pb dating of eclogites and granulites from the Oberpfalz, NE Bavaria, Germany: Chemical Geology, v. 109, p. 317–339.

von Raumer, J.F., 1998, The Paleozoic evolution in the Alps—From Gondwana to Pangea: Geologische Rundschau 87, p. 407–435.

von Raumer, J.F., and Neubauer, F., 1993, The Pre-Mesozoic geology in the Alps: Springer, Heidelberg, 671 p.

von Raumer, J.F., and Neubauer, F., 1994, The Paleozoic evolution of the Alps. Schweizerische Mineralogische Petrographische Mitteilungen v. 74, p. 459–467.

von Raumer, J.F., and Stampfli, G.M., 2000, Comparing the peri-Gondwanan evolution of pre-Variscan domains: Basement Tectonics 15, A Coruña, Spain, Program and Abstracts, p. 18–20.

von Raumer, J., Stampfli, G.M., and Mosar, J., 1998, From Gondwana to Pangea—An Alpine point of view: Terra Nostra, v. Schriften der A. Wegener Stiftung 98/2, no. Giessen, p. 154–156.

von Raumer, J.F., Stampfli, G.M., Borel, G., and Bussy, F., 2002, The Organization of pre-Variscan basement areas at the Gondwana margin: International Journal of Earth Sciences, v. 91, p. 35–52.

Vràna, P., Blümel, P., and Petrakakis, K., 1995, Chapter VII.C.4: Metamorphic evolution, *in* Dallmeyer, R.D., Franke, W., and Weber, K., eds., Pre-Permian geology of Central and Eastern Europe: Berlin Heidelberg New York, Springer-Verlag, p. 453–466.

Walliser, O., 1981, The geosynclinal development of the Rheinische Schiefergebirge (Rheno-Hercynian Zone of the Variscides; Germany), *in* Zwart, H.J., and Dornsiepen, U.F., eds., The Variscan Orogen in Europe, Geologie en Mijnbouw, p. 89–96.

Weill, A.B., Van der Voo, R., and van der Pluijm, B.A., 2001, Oroclinal bending and evidence against the Pangea megashear: The Cantabria-Asturias arc (northern Spain): Geology, v. 29, no. 11, p. 991–994.

Winchester, J.A., 2002, Micro-continent-promontory interactions: Affinities and possible emplacement mechanism for Proterozoic basement in northern Turkey, *in* 1st International Symposium of the Faculty of Mines (ITÜ) on Earth Science and Engineering, Istanbul: Istanbul, ITÜ, p. 76.

Yin, A., and Nie, S., 1996, A Phanerozoic palinspastic reconstruction of China and its neighboring regions, *in* Yin, A., and Harrison, M., eds., The tectonic evolution of Asia: Cambridge, Cambridge University Press, p. 442–485.

Ziegler, P.A., 1984, Caledonian and Hercynian crustal consolidation of Western and Central Europe—A working hypothesis: Geologie en Mijnbouw 63, p. 93–108.

Ziegler, P.A., 1988, Evolution of the Arctic-North Atlantic and the Western Tethys: American Association of Petroleum Geologists Memoir 43, 198 p.

Ziegler, P.A., and Stampfli, G.M., 2001, Late Paleozoic Early Mesozoic plate boundary reorganisation: Collapse of the Variscan orogen and opening of Neotethys, *in* Cassinis, R., ed., The continental Permian of the Southern Alps and Sardinia (Italy) regional reports and general correlations: Brescia, Annali Museo Civico Science Naturali, Brescia. v.25, p. 17–34.

Zonenshain, L.P., Kuzmin, M.I., and Kononov, M.V., 1985, Absolute reconstructions of the Paleozoic oceans: Earth and Planetary Science Letters, v. 74, p. 103–116.

Zonenshain, L.P., Kuzmin, M.I., and Natapov, L.M., 1990, Geology of the USSR: A plate tectonic synthesis: Washington, D.C., American Geophysical Union Geodynamics Series Monograph 21, p. 241.

Zulauf, G., Schitter, F., Riegler, G., Finger, F., Fiala, J., Vejnar, Z., 1999, Age constraints on the Cadomian evolution of the Tepla Barrandian unit (Bohemian Massif) through electron microprobe dating of metamorphic monazite: Zeitschrift der deutschen geologischen Gesellschaft, 150, p. 627–639.

Zwart, H.J. and Dornsiepen, U.F., 1978, The tectonic framework of Central and Western Europe: Geologie en Mijnbouw 57, p. 627–654

Manuscript Accepted by the Society February 8, 2002

Printed in the U.S.A.

Geological Society of America
Special Paper 364
2002

Review of seismic reflector signatures of crustal deformation in the Appalachian-Caledonide orogen with reference to the Spanish Variscides and the Uralides

John H. McBride
Department of Geology, Brigham Young University, P.O. Box 24606, Provo, Utah 84602, USA
James H. Knapp
Department of Geological Sciences, University of South Carolina, 700 Sumter Street, Columbia, South Carolina 29208, USA

ABSTRACT

The availability of thousands of kilometers of deep seismic reflection profiles from the Paleozoic orogenic belts around the North Atlantic Ocean provides an unparalleled opportunity to synthesize the contrasts and similarities of crustal deformation fabrics of compressional orogens. Dominating reflector fabrics are (1) shallow crustal sequences of subhorizontal reflections (décollement levels) that abruptly merge with complex dipping zones (sutures or root zones), (2) broad antiformal reflector patterns, in places underlain by a dipping planar reflector and interpreted as asymmetric folding and shearing in the deep crust, (3) complex, subhorizontal reflectors in the middle to lower crust that can be explained in places as intense ductile shearing related to terrane accretion and continental convergence, and (4) zones of deeply penetrating and mostly planar dipping features, interpreted as reverse-sense brittle faults or ductile shear zones. The structure of reflectors nearer the Moho discontinuity is more complex and variable relative to planar reflectors higher in the crust. Reflectors that curve into the Moho seem to typify orogens where the synorogenic lowermost crust is preserved and imply that the Moho acted as a décollement. In contrast, cases where dipping planar reflectors are abruptly cut off by a flat Moho suggest a postorogenic process that has destroyed and reequilibrated lowermost crustal structure. Other variations between the diverse reflector fabrics can in part be understood (and predicted) in terms of variations between thin- and thick-skinned deformation regimes, which are usually not viewed simultaneously on any one profile or group of profiles. More recently acquired deep seismic profiles from the Spanish Variscides and the Uralides provide examples of convergent belts where most of the full crustal orogenic structure has been preserved intact. These may serve as analogs for less well preserved or evolving orogenic belts.

INTRODUCTION

After the first images of the crustal structure of the southern Appalachians appeared on seismic reflection profiles in the 1970s (Cook et al., 1979), thousands of kilometers of deep seismic data were acquired over the greater Appalachian-Caledonide-Variscan orogen on both sides of the Atlantic Ocean (e.g., Klemperer and Hobbs, 1991; Meissner and Bortfeld, 1990; Pérez-Estaún et al., 1994). The interpretation of the North Atlantic Paleozoic orogens is better defined than any other, espe-

McBride, J.H., and Knapp, J.H., Review of seismic reflector signatures of crustal deformation in the Appalachian-Caledonide orogen with reference to the Spanish Variscides and the Uralides, *in* Martínez Catalán, J.R., Hatcher, R.D., Jr., Arenas, R., and Díaz García, F., eds., Variscan-Appalachian dynamics: The building of the late Paleozoic basement: Boulder, Colorado, Geological Society of America Special Paper 364, p. 281–300.

cially in terms of available geophysical data. The use of deep reflection data from this region can serve as an instructive model for the application of this technique to other compressional orogens. Many of the crustal framework structures around the North Atlantic (e.g., Appalachian décollement, central Piedmont suture, Moine thrust, and Iapetus suture) are among the best-studied Paleozoic orogenic features in the world. However, significant questions remain, especially on how orogen-scale décollements in the shallow crust are linked to crustal-penetrating reflection patterns interpreted as suture or root zones and the meaning of reflectivity variations within such root zones.

The purpose of this study is to provide a synoptic review of key seismic reflector signatures associated with Paleozoic orogenesis around the North Atlantic in order to indicate the dominant structural styles and deformation modes in the crust. For this review, previously published examples, chosen from both sides of the Appalachian-Caledonide orogen (Fig. 1, A and B), exemplify the diversity of deformation fabrics. We integrate these observations with comparisons from the Spanish Variscides and the Urals, which have a comparatively limited amount of deep reflection data but are well studied geologically. Several syntheses of geophysical data from the North Atlantic rim pertain to Paleozoic compressional tectonics (Brewer and Smythe, 1984; Cook et al., 1988; Quinlan et al., 1993; Hall and Quinlan, 1994; Keller and Hatcher, 1999). In particular, Keller and Hatcher (1999) provided a synthesis of the paleogeographic and tectonic divisions in the North American Appalachians and their European counterparts.

COMPARATIVE OROGENIC ANATOMY AROUND THE NORTH ATLANTIC OCEAN

North American Appalachians

Southern United States Appalachians. Initial observations from deep reflection profiles in the southern Appalachians indicated that the igneous-metamorphic core of the orogen has been transported continentward above a regional décollement a distance of at least 260 km over relatively undeformed lower Paleozoic shelf sedimentary strata (Appalachian décollement; Cook et al., 1980, 1981) (Fig. 1A). The inner core of the southern Appalachians, beginning on the west, consists of (1) the Blue Ridge belt, (2) the Piedmont terrane, which is a complex assemblage of Late Proterozoic to middle Paleozoic igneous and metamorphic rocks, roughly equivalent to the Dunnage zone in the Newfoundland Appalachians (Williams, 1979; Keller and Hatcher, 1999), and (3) metasedimentary and metavolcanic rocks of the Carolina-Avalon terranes (including the slate belts), which were later accreted mainly as volcanic arcs to the developing orogen during the middle and late Paleozoic (Hatcher, 1987). The Central Piedmont suture separates Carolina-Avalon exotic terranes, which arrived from the east, from Piedmont and Blue Ridge rocks within the core of the orogen (Fig. 1A). The terranes exotic to North America are postulated to have been emplaced during multiple episodes of Wilson cycle ocean-closure events such as the eastward subduction of the Piedmont terrane beneath the Carolina-Avalon terranes (Hatcher, 1987). The thin-skinned interpretation (i.e., horizontal transport on shallow crustal faults, originally meaning without basement involvement) of the Appalachian core engendered by the early deep seismic reflection observations (Cook et al., 1979) implied that prominent Paleozoic tectonic boundaries mapped at the surface were at least partially transported above a master décollement. Additional deep seismic profiles cross some of these boundaries, such as the Acadian (Middle Devonian to early Carboniferous) Central Piedmont and Alleghanian (late Carboniferous, equivalent to Variscan) sutures (Fig. 1A), which function as root zones for the décollement. These profiles show a pattern of dipping reflections that dip down from the level of décollement and extend into the middle and lower crust (Fig. 2A). The prominent dipping reflections are interpreted to represent brittle faults or, at deeper levels of original deformation, discrete ductile shear zones (McBride et al., 1996).

One of the suture zones best documented from reflection profiles in the Appalachians is the Alleghanian suture buried beneath the coastal plain in Georgia that separates a fragment of west African Gondwana (i.e., African platform sedimentary rocks; Chowns and Williams, 1983; Rankin et al., 1993; Thomas, 1993) from Precambrian Laurentian crust and exotic Paleozoic accreted terranes (Fig. 1B) (Nelson et al., 1985). Two reflection transects show crustal-penetrating south-dipping reflectors that reach down to the Moho discontinuity without crossing into the upper mantle (e.g., Fig. 2, A–C). This pattern includes prominent diffractions within the middle crust that suggest strong deformation and/or igneous intrusion (Fig. 2B). Sadowiak and Wever (1990) recognized such close associations of dipping reflections and diffractions as typical of suture zones. Along a key deep reflection transect in western Georgia (Figs. 1A and 2A), the Acadian and Alleghanian sutures are probably so close (Rankin et al., 1993) that they either share the same root zone, or the latter obliterated the former. Migrating part of the stacked data from western Georgia (Arnow, 1987) resolves some of the dipping reflections and diffractions into geologically precise images (Fig. 2D). These images imply north-verging antiformal thrust patterns consistent with a compressional suture zone. Such antiformal patterns are characteristic of this region outboard of the Central Piedmont suture (Cook et al., 1981), where the Appalachian décollement emerges from its root zone (i.e., Alleghanian suture) beneath the coastal plain (Fig. 2E). The pattern of antiformal reflectors underlain by quasi-planar reflectors is similar to geological patterns mapped at the surface in compressional orogens, especially suture zones (Sadowiak and Wever, 1990; McBride and Nelson, 1991; Blundell, 1993; McBride and England, 1999). In the southern Appalachians, this pattern is overwhelmingly univergent (i.e., verging northwest, toward the continent). The nearly flat Moho against which south-dipping crustal reflections are abruptly truncated (Fig. 2C) is thought to be superimposed either by Mesozoic

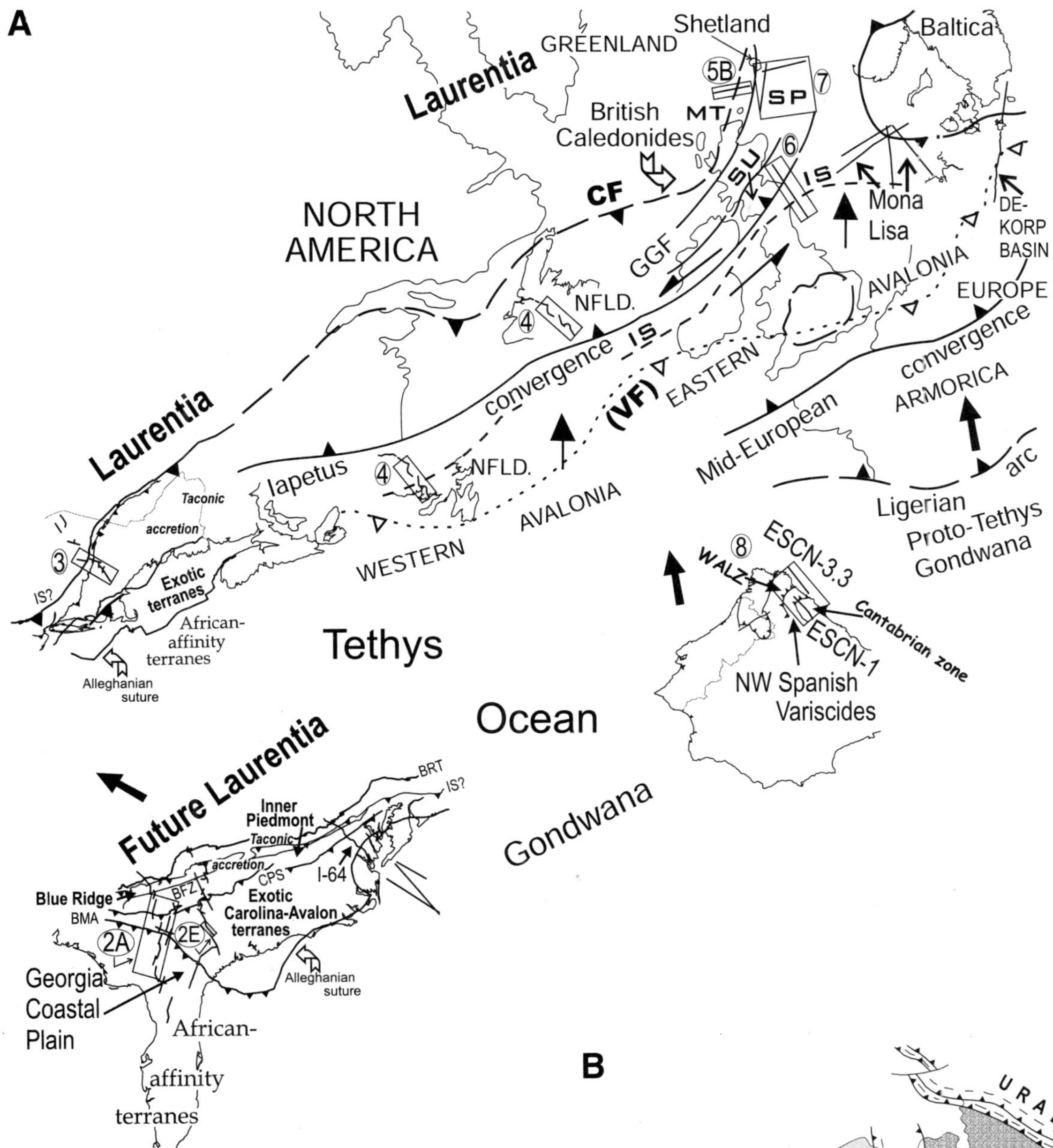

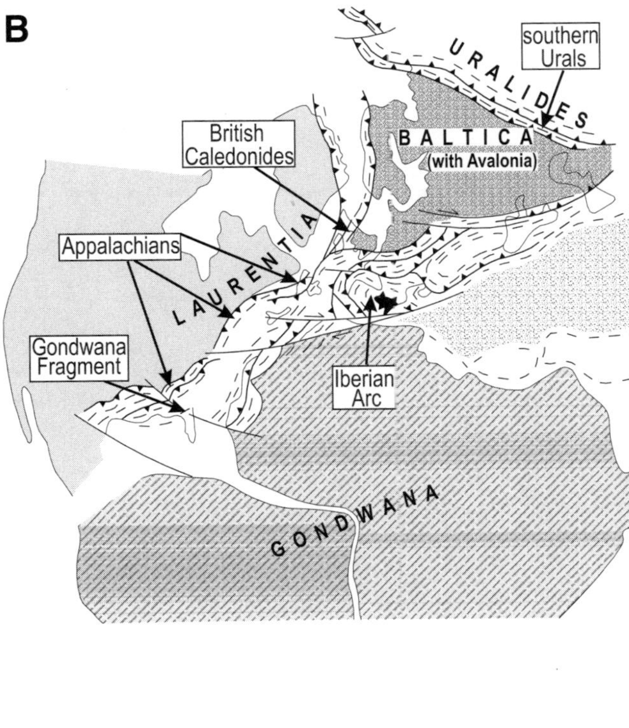

Figure 1. A: Location map of major orogenic structures of Appalachian-Caledonian orogen shown as generalized paleogeographic reconstruction during Silurian-Devonian time (after Soper, 1988; PALEOMAP Project, C.R. Scotese, University of Texas at Arlington). Circled numbers refer to figures showing interpreted seismic reflection profiles. Abbreviations: BMA, Brunswick magnetic anomaly; BFZ, Brevard thrust zone; BRT, Blue Ridge thrust; CF, Caledonian front; CPS, Central Piedmont suture; GGF, Great Glen fault; IS, Iapetus suture; MT, Moine thrust; NFLD, Newfoundland; SP, Shetland platform; SU, Southern Uplands; VF, Variscan front; WALZ, West Asturian Leonese zone. For more detailed profile locations and geological settings, see references cited in captions of subsequent figures. B: 270 Ma reconstruction of orogen corresponding to Alleghanian-Variscan deformation (from Gee and Zeyen, 1996). Areas studied in this report are indicated.

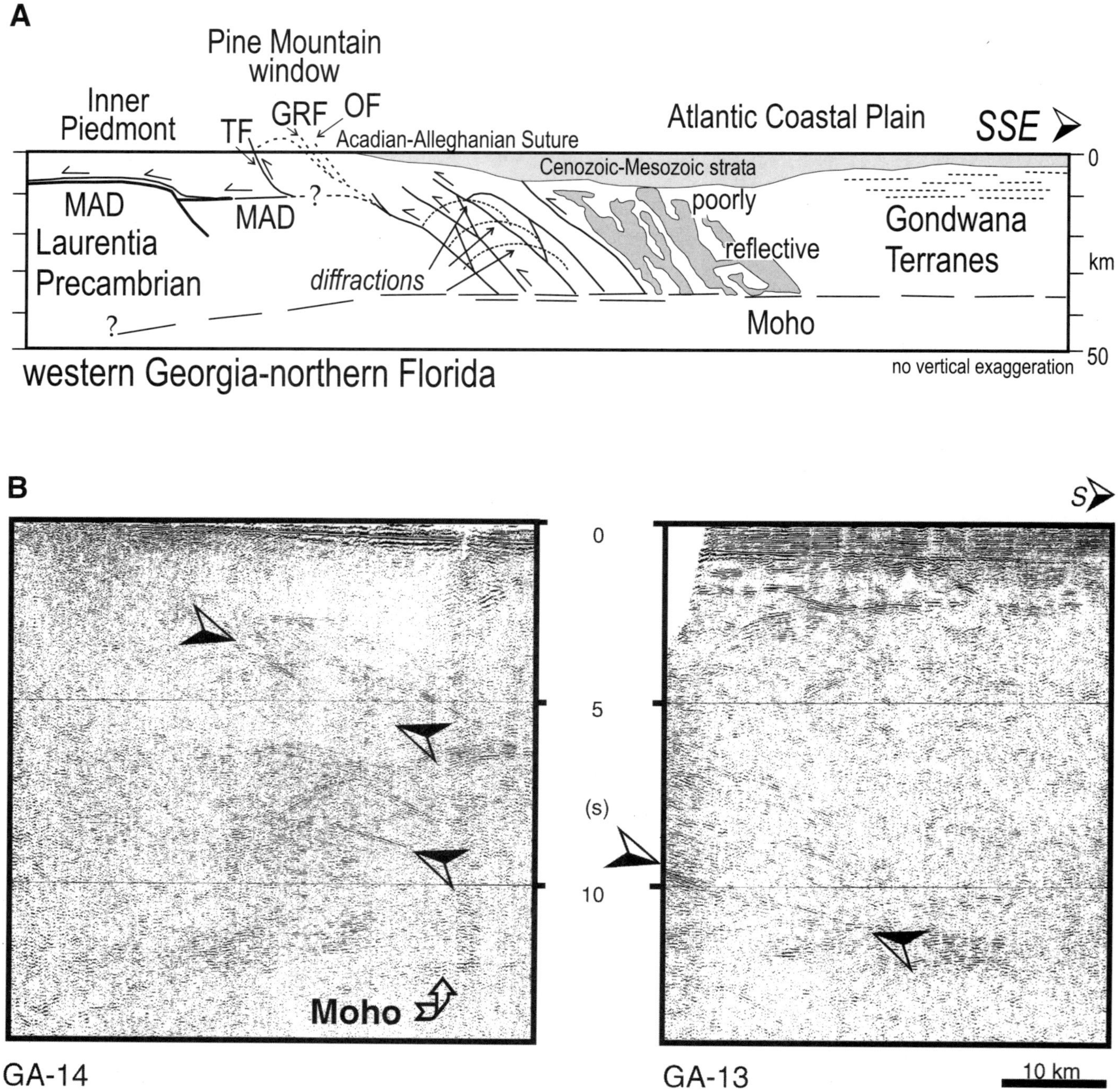

Figure 2. A: Simplified interpretation through southern Appalachians in western Georgia based on COCORP (Consortium for Continental Reflection Profiling Syndicate) deep seismic reflection profiles (see Fig. 1A) (modified from McBride and Nelson, 1991; McBride et al., 2001). MAD is master Appalachian décollement, TF is Towaliga fault, GRF is Goat Rock fault, and OF is Ocmulgee fault. B: Excerpt of COCORP Georgia deep seismic profiles 13 and 14 (unmigrated) (Nelson et al., 1985; Nelson, 1988), which corresponds to Acadian-Alleghanian suture in Figure 2A. Arrows indicate dipping reflections referred to in text. C: Migrated (constant velocity of 4.5 km/s) excerpt of COCORP Georgia profile 13. Arrows indicate dipping reflections referred to in text. D: Detailed migrated (constant velocity of 6.0 km/s) section of COCORP Georgia profile 14 and interpretive line drawing showing prominent antiformal structure (Arnow, 1987; McBride and Nelson, 1991). E: Interpretive line drawings of excerpts of unmigrated COCORP Georgia profiles 5 and 8 (modified from Cook et al., 1981). Open arrows indicate décollement reflections and solid arrows indicate antiformal patterns.

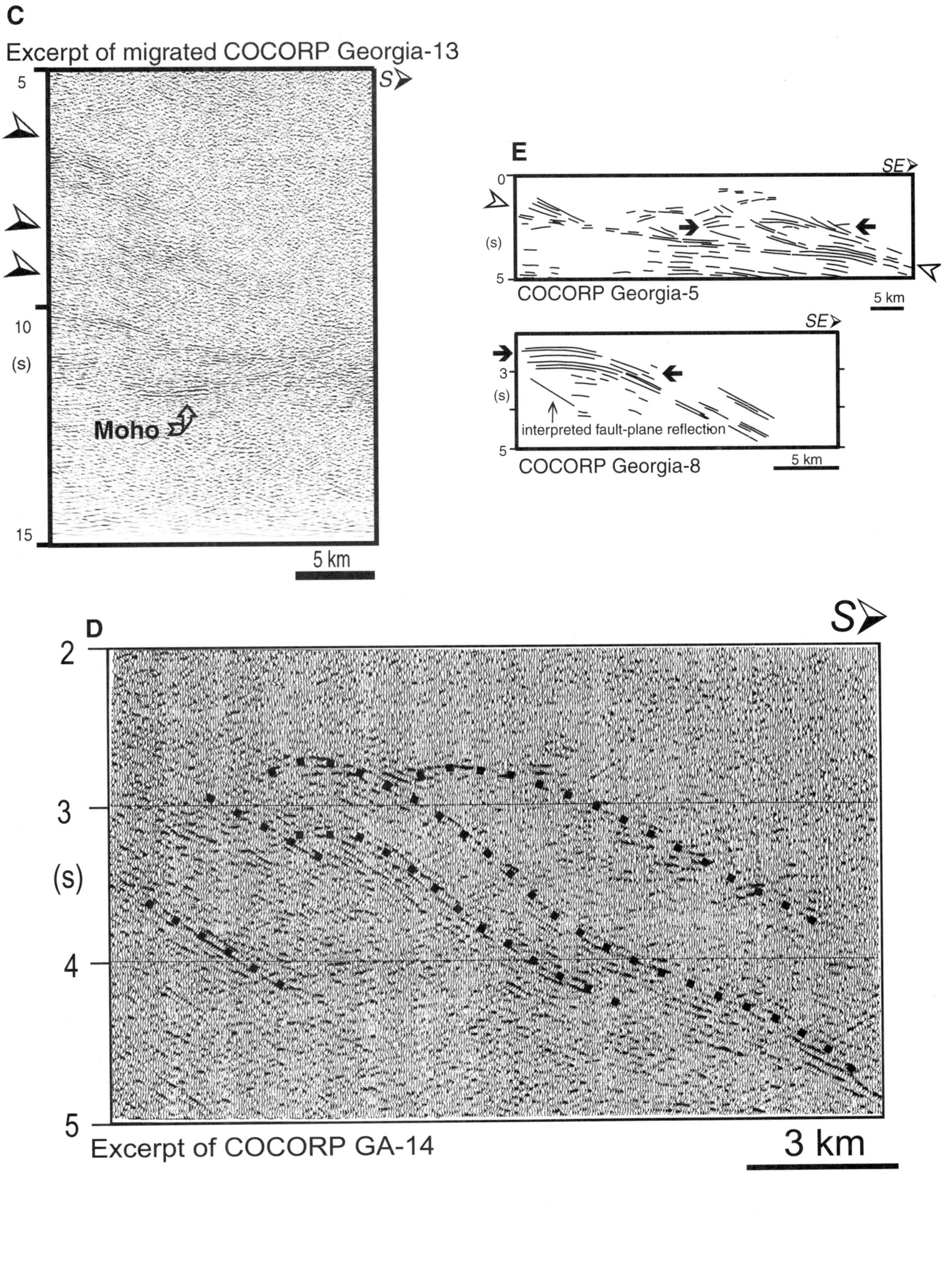
C
Excerpt of migrated COCORP Georgia-13
S
5
10
(s)
15
Moho
5 km
E
SE
0
(s)
5
COCORP Georgia-5
5 km
SE
3
(s)
5
interpreted fault-plane reflection
COCORP Georgia-8
5 km
D
S
2
3
(s)
4
5
Excerpt of COCORP GA-14
3 km

mantle heating and consequent magmatism associated with the breakup of Pangea or by postorogenic delamination of the lithosphere (Nelson, 1992).

Northeastern United States Appalachians. A seismic reflection traverse over the New England Appalachians (Fig. 1A) provides a structural cross section, beginning on the west in New York, through autochthonous Laurentian basement (Adirondack Mountains), the Taconic foreland allochthons, and the Merrimack synclinorium in New Hampshire (Ando et al., 1984). Beneath the Merrimack synclinorium, a belt of metamorphic rocks roughly correlative with the Gander zone of the Newfoundland Appalachians (Williams, 1979), the reflection profile shows an east-dipping fabric that merges upward with a shallow horizontal reflection sequence below the Taconic allochthons (Fig. 3A). In a broad sense, the structural picture of the New England transect is similar to that of the southern Appalachians (Fig. 2A) (and central Virginia "I-64" profile [Fig. 1A; Çoruh et al., 1988]). A thin-skinned deformation regime in the foreland merges abruptly with a thick-skinned dipping reflector zone interpreted as planar thrusts that are part of a root zone for the Taconic allochthons. However, beneath the Merrimack synclinorium, the broad dipping reflector zone in places appears to merge with a subhorizontal lower crustal reflective layer at or just above the Moho (Phinney and Roy-Chowdhury, 1989) (Fig. 3B), rather than being abruptly cut off by the Moho as in the southern Appalachians (Fig. 2, B and C). Phinney and Roy-Chowdhury (1989) chose a single reflector that penetrates much of the crust as the suture between Laurentian and later accreted terranes (Fig. 3B), whereas Ando et al. (1984) interpreted all of the dipping reflectors as composing a broad root zone (Fig. 3A). Part of this inferred root zone includes asymmetric antiformal reflector geometries, suggesting a west (continentward) vergence in one direction only. A characteristic common to nearly all of the Appalachian deep reflection profiles in the United States is that the crust (and upper mantle) beneath the shallow foreland décollement is virtually nonreflective. This lack of foreland reflectivity has been shown not to be an artifact of poor signal penetration (Arnow, 1987), and may indicate a lack of coherent post-Grenvillian deformation in the crust beneath the foreland décollement. The Uralian foreland, by contrast, exhibits both a reflective crust and well-defined reflective Moho (Knapp et al., 1996; Diaconescu et al., 1998).

Newfoundland Appalachians. The Newfoundland Appalachians are divisible into a distinct series of narrow, approximately parallel accretionary belts in close proximity with the Proterozoic Grenvillian orogen (Williams et al., 1999). Several thousand kilometers of deep seismic reflection profiles, acquired both onland and offshore, show pronounced reflectivity throughout the crust and across many parts of the orogen (Quinlan et al., 1992). The reflection data from central Newfoundland provide probably the highest quality profiles across the Appalachians in North America (Hall and Quinlan, 1994; Hall et al., 1998).

One of the most distinctive reflection profiles is the eastern Meelpaeg transect line 9 (Figs. 1A and 4A), which crosses two of the major orogenic zones, the Gander and Dunnage zones (Williams, 1979). The Dunnage zone consists of Late Cambrian to Early Ordovician volcanic rocks, ophiolites, and melanges, whereas the Gander zone is dominated by polydeformed sedimentary and metamorphic rocks of roughly the same age (Williams et al., 1999). Despite the fact that Newfoundland is a classic example of tectonic lithofacies zonation (Williams, 1979), the reflectivity pattern within the crust as a whole does not easily lend itself to zonation. In particular, the relation between the Gander and Dunnage zones as mapped at the surface and deeper crustal reflection fabrics is not readily apparent (Fig. 4A). The upper crustal (0 to 6 or 7 s) portion of the Meelpaeg transect includes highly coherent northwest- and southeast-dipping reflections that are usually not collinear with deeper reflections in the middle and lower crust (Fig. 4, A and B). Quinlan et al. (1992) interpreted the combination of dipping and subhorizontal reflectors in the upper crust as representing the ramp and flat structure typical of foreland fold belts. Dipping reflections in the lower crust appear to flatten gradually toward the prominent horizontal Moho reflection (Fig. 4B). Quinlan et al. (1992) suggested that the northwest-dipping fabric developed during a Middle Silurian (i.e., Taconian) orogeny, when Gondwana crust on the east (correlative with northwestern Spain) was thrust beneath Laurentian crust to the west by as much as 200 km. Much of the northwest-dipping reflection geometries result from this latest event, although an earlier Ordovician-Silurian subduction episode (southeast directed) was postulated by Quinlan et al. (1992) to have produced the shallower southeast-dipping reflections, which have subsequently been truncated by Laurentia overriding Gondwana (Fig. 4A). The final thrust boundary (or suture) between Laurentian and Gondwanan crust is defined not so much by a discrete reflection, but by an interpreted change in reflectivity pattern (Fig. 4A). This boundary was interpreted by Quinlan et al. (1992, their Fig. 10) to dip through most of the crust over a distance of ~300 km (i.e., it does not define a discrete subhorizontal décollement).

British Caledonides

The Caledonides of the United Kingdom and Ireland represent the earliest-studied compressional orogen, whose classic deformational structures have served as benchmark geologic features. Almost no regional deep seismic reflection profiles are available onland across the Caledonides or other geologic provinces in the United Kingdom or Ireland. Virtually all deep reflection information comes from the shallow continental shelves bordering the North Sea and the eastern Atlantic margin, which host the world's largest concentration of offshore deep reflection profiles (Klemperer and Hobbs, 1991; Snyder and Hobbs, 1999).

Correlation of tectonic lithofacies belts and their individual structural boundaries between the North American Appalachians and the western European Caledonides remains an enduring problem. The best-documented correlations are those between

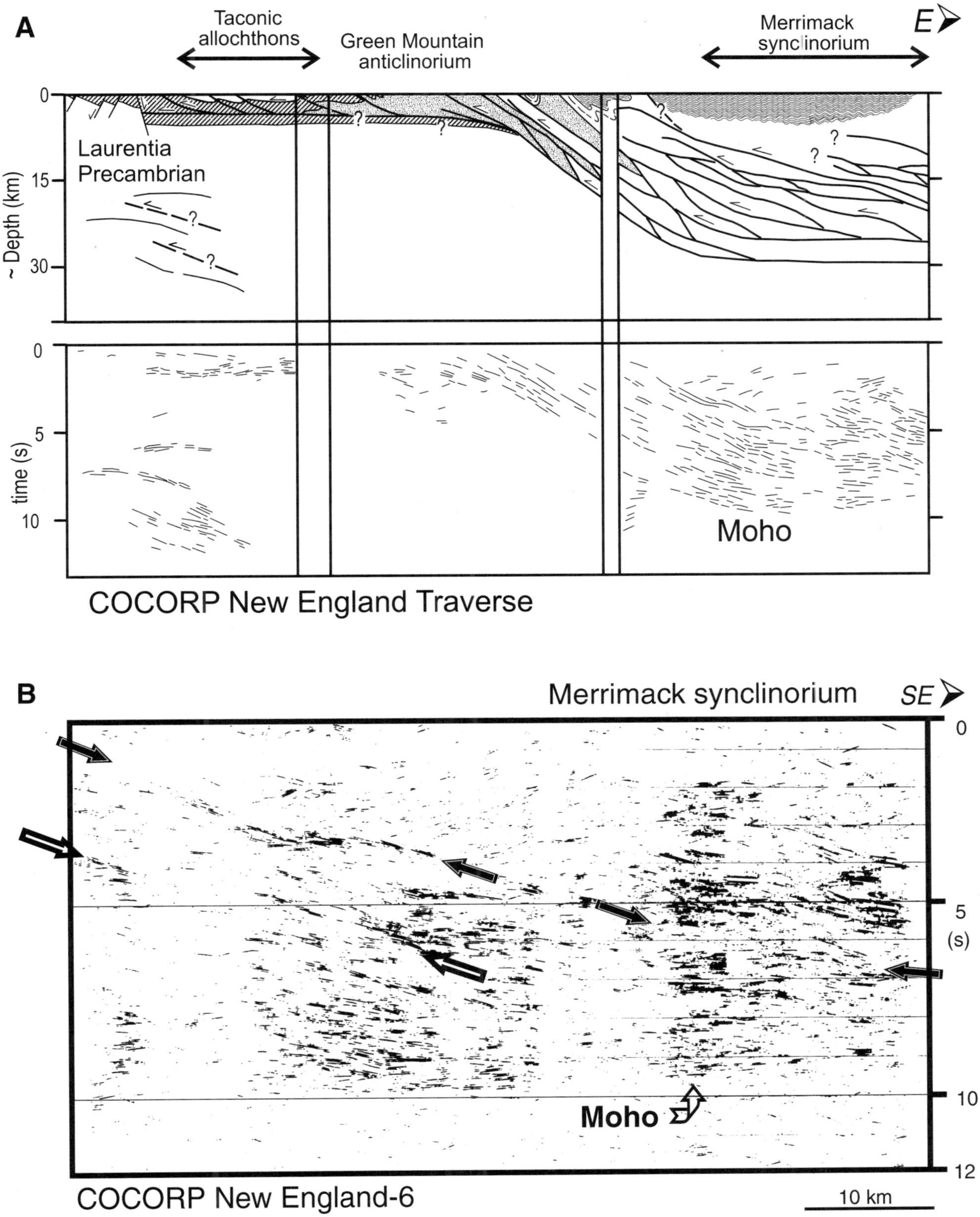

Figure 3. A: Simplified drawing and structural interpretation through New England Appalachians in New York, Vermont, and New Hampshire, based on COCORP (Consortium for Continental Reflection Profiling Syndicate) deep seismic reflection profiles (Ando et al., 1983) (see Fig. 1A). Shaded pattern is Precambrian basement, diagonal ruled pattern is mostly lower Paleozoic shelf and miogeoclinal rocks, blank pattern is transported Taconic rocks in western part of section and thrust-imbricated basinal facies and basement rocks east of Green Mountain anticlinorium, and wavy line pattern is Silurian and younger rocks (Ando et al., 1983). B: Migrated (constant velocity of 4.5 km/s) portion of COCORP New England line 6 corresponding to right panel in A (Phinney and Roy-Chowdhury, 1989) (see Fig. 1A). Arrows refer to dipping reflectors discussed in text. Open arrows indicate eastern extension of Laurentian basement (i.e., suture) as interpreted by Phinney and Roy-Chowdhury (1989). Moho interpretation is that of Phinney and Roy-Chowdhury (1989).

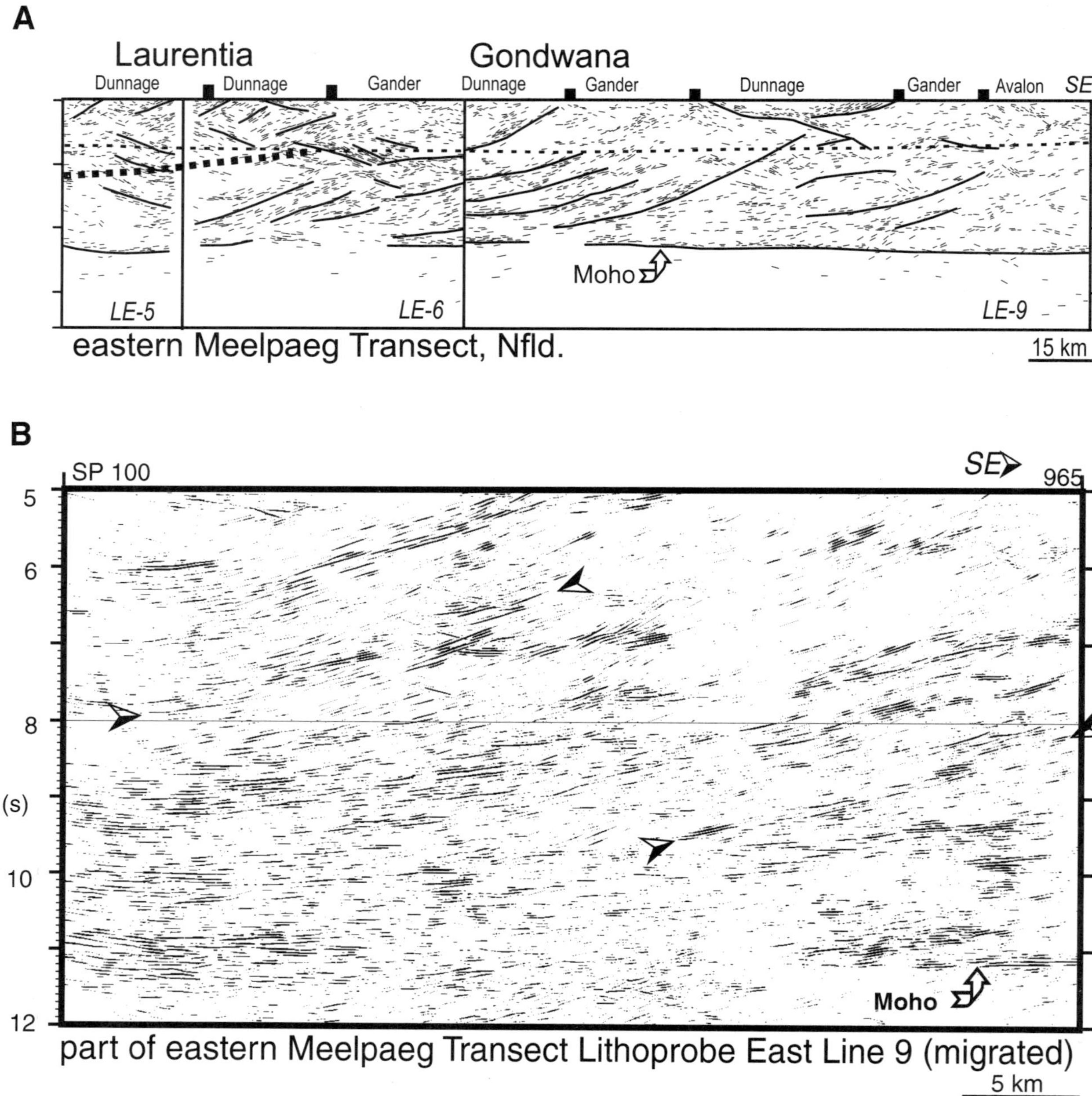

Figure 4. A: Simplified interpretive drawing through eastern Newfoundland Appalachians based on Lithoprobe deep seismic reflection profiles (Quinlan et al., 1992) (see Fig. 1A). Thick dashed line indicates boundary (suture) between Laurentia and Gondwana and thin dashed line indicates schematic lower limit of tectonic zones (e.g., Gander) (Quinlan et al., 1992). Note that possible discontinuity might exist at roughly 4–5 s based on changing dipping reflector fabrics (Hall et al., 1998). B: Eastern portion of migrated section of Lithoprobe profile LE-9 in eastern Newfoundland (see Fig. 1A) (Quinlan et al., 1992; courtesy of Lithoprobe). Arrows indicate dipping and curved reflections referred to in text.

Newfoundland and the British Isles involving Laurentian terranes in Scotland and Avalonian components in Wales and England (Winchester and van Staal, 1995; McKerrow et al., 2000; Armstrong and Owen, 2001). The Iapetus suture in England, as recognized from paleontological data, is basically the same boundary as the Acadian Central Piedmont suture in the southern Appalachians that separates Avalonia and North American Laurentia (Soper, 1988; Keller and Hatcher, 1999) (Fig. 1, A and B). As pointed out by Soper (1988), however, during Silurian-Devonian time, the Iapetus suture in the British Isles was undergoing more of a transpressive strike-slip deformation (Fig. 1A). The timing of the main compressional orogeny in the Scottish Caledonides and the throughgoing Moine thrust zone (Middle Ordovician to Middle Silurian) (Soper, 1988) is more correlative with the older Taconian orogeny in the Appalachians (Hatcher, 1987).

Moine thrust system. The Moine thrust is one of the classic structures of the Caledonide-Appalachian system, and may structurally be analogous to, although older than, the Central Piedmont suture (Fig. 1A). Where observed in northwestern Scotland, the thrust floors a thick east-dipping mylonite zone separating metamorphic rocks of the thrust belt on the east from the less deformed Precambrian foreland and cover sequence on the west (Barr et al., 1986). This entire system has been interpreted as a foreland-propagating thrust sequence that constitutes a progression beginning in the east of older, more ductile thrusting giving way to younger, more brittle thrusting westward (Barr et al. 1986; Butler 1986) (Fig. 5A). Although estimates of total crustal shortening across this system are debatable, Elliott and Johnston (1980), Butler and Coward (1984), Barr et al. (1986), and others have reasoned that it must be 50–100 km. Deep reflection profiles over the offshore projection of the Moine thrust between Shetland and the Scottish mainland (Fig. 1A) reveal a thick wedge of east-dipping reflections in the middle to lower crust (Fig. 5, B and C). This wedge is limited above by a highly coherent reflection (Fig. 5B; M–M′) that projects up to the offshore projection of the Moine thrust (McBride and England, 1994). The thrust separates an unreflective upper crust from a highly reflective crust below and can be projected to near the base of the crust on the basis of weaker collinear reflection segments and by the distinct change in the dipping reflector fabric (Fig. 5C). The most remarkable aspect of this pattern is that the distinct planar boundary seems to act as a leading edge of highly reflective (deformed by compression?) Precambrian foreland crust (i.e., the footwall of the suture zone) that continues down to near the Moho (Fig. 5B). This pattern persists along strike across the platform south of the Shetland Islands (McBride and England, 1995), which is less affected by sedimentary basin formation (Kirton and Hitchen, 1987), but is distorted beneath the extended areas farther south nearer the mainland (Snyder, 1990).

The uppermost mantle beneath the Caledonian orogen north and west of the Scottish mainland also contains prominent dipping reflectors, as documented in numerous studies (McGeary and Warner, 1985; Snyder and Flack, 1990; McBride et al., 1995; Warner et al., 1996; Snyder et al., 1997; Morgan et al., 2000). These mantle reflectors do not generally connect across the Moho discontinuity with the Caledonian reflection patterns in the upper to middle crust and thus seem kinematically distinct (McBride et al., 1996). Snyder et al. (1997) reviewed the broad range of proposed origins for mantle reflectivity and emphasized that many mantle reflectors beneath the British-Irish Caledonides track the position of the Neoproterozoic Laurentian margin and thus may be related to the opening (900–600 Ma) or closing (500–400 Ma) of the Iapetus ocean. The fact that the mantle reflectors beneath the Paleozoic orogen cannot be uniquely correlated spatially with a particular geologic feature observed at the surface or in the upper to middle crust (McBride et al., 1995) precludes a unique solution for their origin.

Iapetus suture. The Iapetus suture represents one arm of the convergent Caledonian triple junction between the Laurentia, Baltica, and Gondwanan terranes (Fig. 1, A and B) (Meissner et al., 1994). In England, the suture separates the Southern Uplands terrane on the north from Gondwanan Eastern Avalonia on the south (Figs. 1A and 6A) and so represents closure of the Iapetus ocean. The Southern Uplands is the most outboard of the Caledonian terranes accreted to Laurentia before the final closure of the Iapetus ocean (Soper, 1988). Suturing is inferred to have been associated with northward subduction during the Middle Silurian (Soper et al., 1992). Although the suture is a paleogeographically important boundary, its geological expression on land is not clear; however, geophysically it projects to an extremely prominent boundary offshore (Fig. 6, A and B). As seen for the Moine thrust between Shetland and mainland Scotland, deep seismic profiles (Soper et al., 1992) across the Iapetus suture reveal thick wedges of dipping reflections in the middle to lower crust, interpreted as thrust or ductile shear-zone fabrics. The upper and northernmost reflector in the wedge represents a boundary between reflective and unreflective crust (I–I′, Fig. 6A). The seismic expression of the suture is thus interpreted as the leading edge of Avalon crust (i.e., the footwall of the suture zone as observed for the Moine thrust; Fig. 5B). Both southeast- and northwest-dipping reflectors are present, but the northwest-dipping pattern clearly dominates the crust and may continue slightly into the uppermost mantle, where it appears to either deform and/or merge with a locally northwest-dipping Moho surface. Subhorizontal reflectivity in the lower crust just north of I–I′ (Fig. 6, A and B) may represent the thrust-imbricated leading edge of deep Avalon crust thrust under and accreted to Caledonian crust.

Shetland platform. The shallow platforms around Shetland (Fig. 1A) are among the few areas of the continental shelf of northwestern Europe that were deformed by the Caledonian orogeny but remained relatively insulated from subsequent rifting (Klemperer and Hurich, 1990), thus providing an acoustic window into the deep structure of the orogen. Interpretation of a large database of deep reflection data indicates that the crust beneath the platform retains much of its original Caledonian and older architecture; however, the unique tectonic character of this part of the orogen is emphasized by the variation in reflector trends (McGeary, 1987). The most striking tectonic feature beneath the platform is the combination of subhorizontal, dipping, and antiformal elements that appear on a north-south profile across the East Shetland platform (Fig. 7A). The antiforms are underlain by bright north-dipping reflections within the lowermost crust interpreted as thrust faults (or shear zones) with a southward vergence (Fig. 7A). Patterns of antiformal structures mapped in the central Highlands of Scotland (Treagus and King, 1978) may be analogous to the observed reflector patterns. The antiformal pattern correlates with two levels of uneven and diffractive lower crustal reflectivity on numerous approximately east-west (i.e., strike) profiles (Fig. 7B) (McBride and England, 1999). These profiles indicate that dipping upper crustal reflectors, interpreted to be related to Caledonian deformation, sole into the upper surface

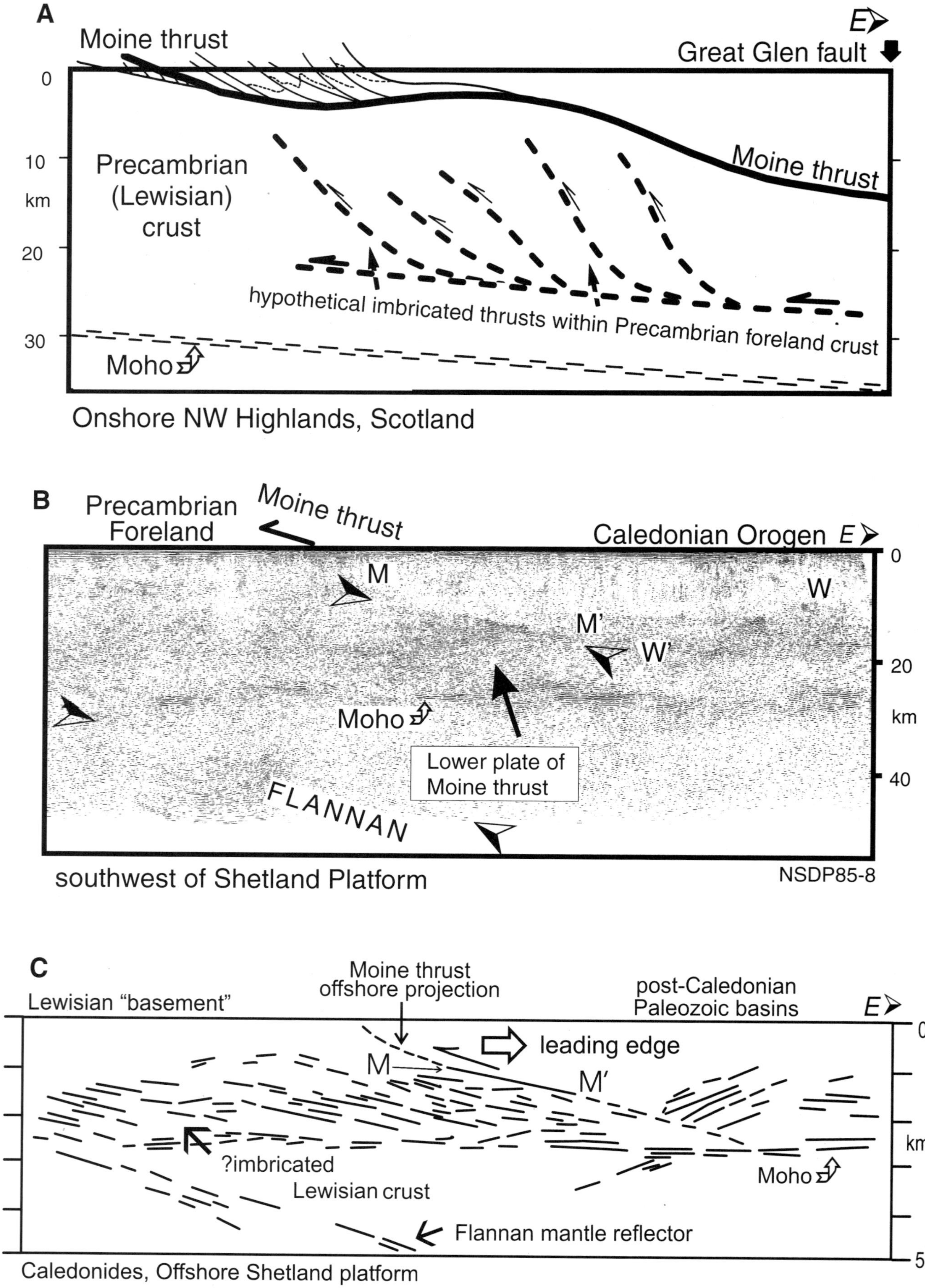

A
Moine thrust
E
Great Glen fault
0
10
km
20
30
Precambrian (Lewisian) crust
Moine thrust
hypothetical imbricated thrusts within Precambrian foreland crust
Moho
Onshore NW Highlands, Scotland
B
Precambrian Foreland
Moine thrust
Caledonian Orogen
E
0
M
W
M'
W'
20
km
40
Moho
Lower plate of Moine thrust
FLANNAN
southwest of Shetland Platform
NSDP85-8
C
Lewisian "basement"
Moine thrust offshore projection
post-Caledonian Paleozoic basins
E
0
leading edge
M
M'
km
?imbricated Lewisian crust
Moho
Flannan mantle reflector
50
Caledonides, Offshore Shetland platform

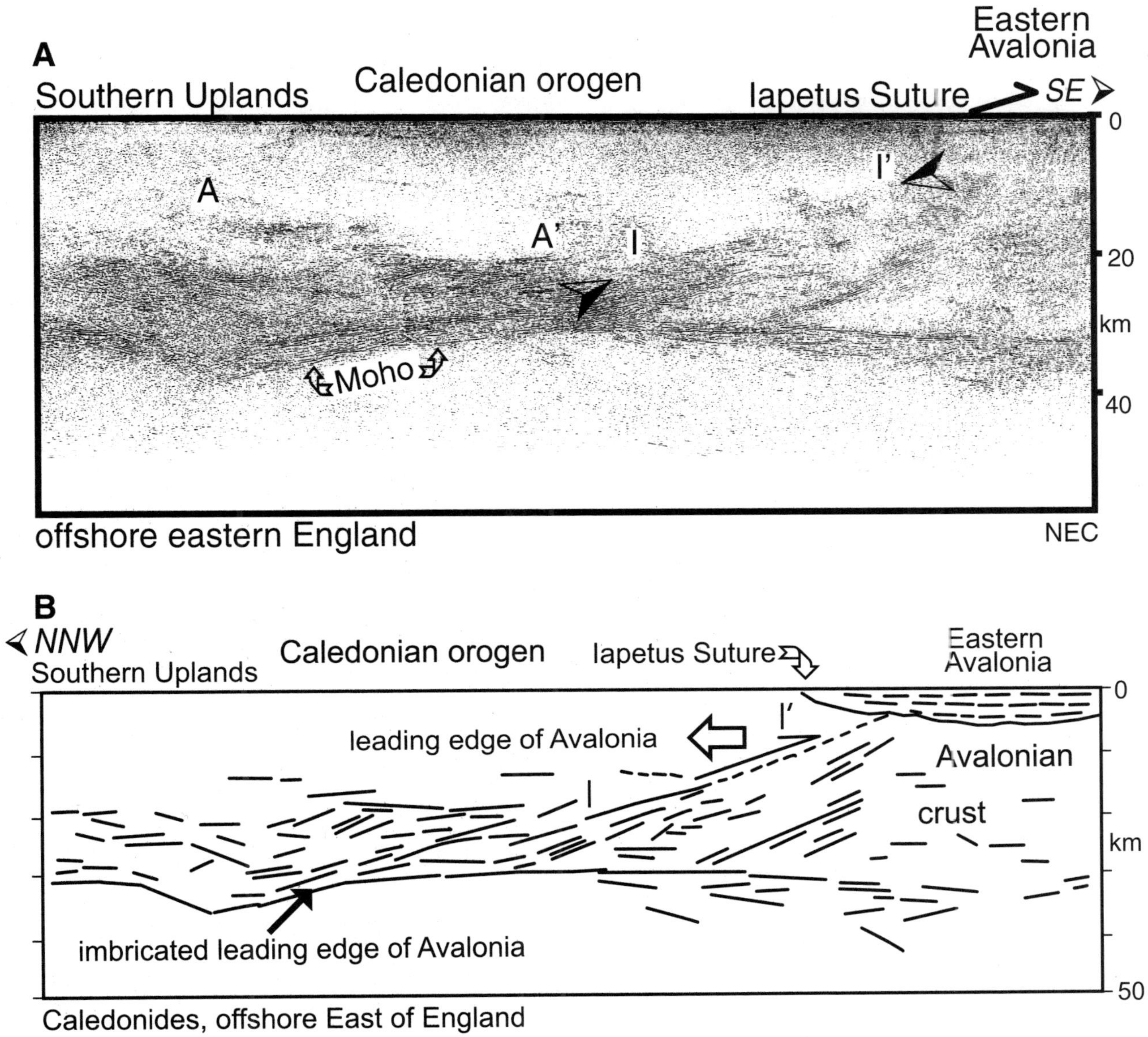

Figure 6. A: Migration and depth conversion of British Institutions Reflection Profiling Syndicate (BIRPS) NEC (North-East Coast) seismic reflection profile (Soper et al., 1992; McBride et al., 1996) (Fig. 1A). I–I′ is dipping reflector within Iapetus suture. No vertical exaggeration. Possible reversal of reflector dip appears in hanging wall of suture zone (A–A′). B: Simplified schematic drawing of seismic reflection profile in A showing relative position with respect to surface trace of Iapetus suture. No vertical exaggeration.

of layered lower crust. This surface may thus indicate a décollement level (Fig. 7B). The interpretation that the compressional thrust structure correlates with the two-tiered lower crustal reflectivity as shown in Figure 7B implies that what would typically be described as layered lower crust has a thrust-belt origin. Thus, the compressional explanation of the layered lower crust observed below both the Shetland platform and the Iapetus suture contradicts conventional interpretations that such reflectivity is in general caused by mantle-derived igneous intrusion or by deformation fabrics associated with stretching in response to continental rifting (see Warner, 1990, for review).

Figure 5. A: Geologic interpretation for Moine nappes beneath Northwest Highlands of Scotland (Butler and Coward, 1984) (Fig. 1A). No vertical exaggeration. B: Migration and depth conversion of Geco-Prakla NSDP85-8 seismic reflection profile (McBride and England, 1994) (Fig. 1A). M-M' is interpreted Moine thrust reflector. W–W′ is oppositely dipping reflector that appears to be cut off by M–M′. Flannan refers to dipping mantle reflector (McBride et al., 1995, 1996). No vertical exaggeration. C: Simplified schematic drawing of seismic reflection profile in B showing relative position with respect to surface trace of Caledonian orogenic front (locally Moine thrust). No vertical exaggeration.

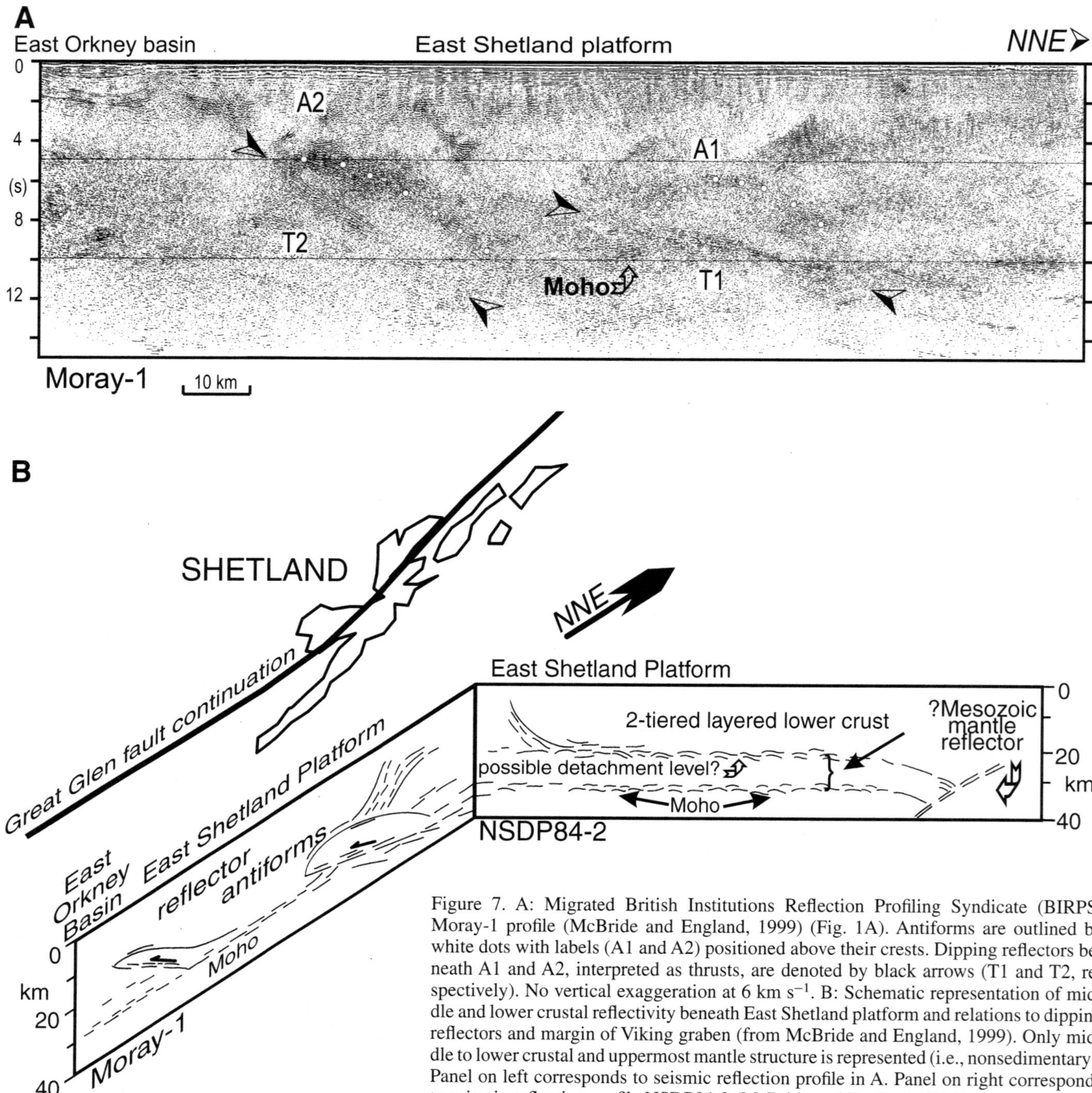

Figure 7. A: Migrated British Institutions Reflection Profiling Syndicate (BIRPS) Moray-1 profile (McBride and England, 1999) (Fig. 1A). Antiforms are outlined by white dots with labels (A1 and A2) positioned above their crests. Dipping reflectors beneath A1 and A2, interpreted as thrusts, are denoted by black arrows (T1 and T2, respectively). No vertical exaggeration at 6 km s^{-1}. B: Schematic representation of middle and lower crustal reflectivity beneath East Shetland platform and relations to dipping reflectors and margin of Viking graben (from McBride and England, 1999). Only middle to lower crustal and uppermost mantle structure is represented (i.e., nonsedimentary). Panel on left corresponds to seismic reflection profile in A. Panel on right corresponds to seismic reflection profile NSDP84-2 (McBride and England, 1999).

Variscides of northwestern Spain

The exposed Variscan terranes (Carboniferous, equivalent to early Alleghanian in the Appalachians) of northwestern Spain represent a well-studied window into late Paleozoic deformation within a greater area of the Alpine mountain belt (Fig. 1, A and B). The Spanish Variscides, representing the westernmost extent of late Paleozoic deformation of the European continent where it converged with the area of present-day Newfoundland, are today preserved as a horseshoe-shaped orocline possibly affected locally by an indenter of Gondwanan origin (Lefort, 1989; Matte, 1991) (Fig. 1B). While acknowledging the voluminous literature on geophysical studies of Variscan tectonics, especially in Germany and France (see Meissner and Sadowiak [1992] and Matte and Hirn [1988] for reviews), the Spanish Variscides have special relevance to the Appalachians because of the striking similarities in deep reflection signatures. The Variscides in northwestern Spain consist of a well-developed east-vergent foreland fold and thrust belt to the east (Cantabrian zone) that is closely associated with a hinterland zone to the west, the West Asturian–Leonese zone (Pérez-Estaún et al., 1994) (Fig. 1A). An east-west deep reflection transect shows a nearly horizontal band of strong reflections that extends for a distance of ~100 km corresponding to the Cantabrian zone décollement (Fig. 8, A and B) (Pérez-Estaún et al., 1994; Ayarza et al., 1998). This band dips down and merges to the west with a zone of west-dipping reflectors beneath the hinterland Narcea antiform, which at the surface consists of a series of stacked thrusts (Pérez-Estaún et al., 1991). Pérez-Estaún et al. (1994) interpreted this complex dipping reflector zone as marking a Variscan zone of shearing and mid-crustal indentation that marks the transition between thin-skinned deformation to the east and a thick-skinned regime to the west (Fig. 8B). Farther west beneath the West Asturian–Leonese zone, a thick, complexly reflective crust involves interpreted thrust imbrication at multiple levels in the crust, including possible imbrication of the lowermost crust and adjacent mantle (Fig. 8A) (Ayarza et al., 1998). Ayarza et al. (1998) suggested that the deepest imbrication could represent either a Variscan crustal root zone or the superimposed effects of Alpine-age underthrusting. The well-expressed thin-skinned regime beneath the Cantabrian zone emerges from beneath the Narcea antiform from what can be considered to be a localized root zone underlain by the original Paleozoic Moho with the preserved Variscan deformation. Such a picture is in contrast with offshore British Caledonian thrust systems, in which thin-skinned deformation is not manifest on seismic profiles, but similar to that observed for the Alleghanian suture in the southern Appalachians (Fig. 2A). As observed for much of the U.S. Appalachian foreland, the Precambrian crust beneath the Cantabrian zone is transparent except for infrequent Moho reflections (Pérez-Estaún et al., 1994).

Uralian orogen

The Ural Mountains, which divide Europe and Asia, are the remnant of another major zone of continental convergence that, together with the Appalachian, Caledonian, and Variscan orogens, contributed to the construction of the Pangea supercontinent during late Paleozoic time (Fig. 1B). The development of the Uralide orogen was analogous to that of the southern Appalachians (Knapp et al., 1993) and culminated during late Carboniferous–Permian time (equivalent to Variscan or Alleghanian). The convergence of an intricate collage of accreted island-arc–rifted continental fragments with the eastern European craton on the west was driven by the westward movement of the Siberian craton on the east (Şengör et al., 1993; Puchkov, 1997). The development of the southern Uralide orogen differs from most parts of the Appalachian-Caledonian system in having undergone little or no postorogenic extension or magmatism (Berzin et al., 1996). Knapp et al. (1998) argued that the middle and northern Urals underwent postorogenic extension related to the formation of the West Siberian basin, but that the southern Urals remain locked within the Eurasian continent and the extension did not ultimately lead to ocean basin formation. During 1995, an integrated seismic reflection–wide-angle transect was surveyed across the entire orogen (URSEIS) beginning on the west from the Eastern European craton, crossing the Main Uralian fault (suture), and extending over the collage of microcontinental fragments (e.g., trans-Uralian zone) (Fig. 9A) (Berzin et al., 1996; Knapp et al., 1996; Echtler et al., 1996; Carbonell et al., 1996).

The first-order result from the reflection profile (Fig. 9B) is the presence of thickened Uralian crust beneath the core of the bivergent orogen, with crust-penetrating reflectors in both the foreland and hinterland (Fig. 9A). The most distinctive reflection signature appears beneath the trans-Uralian and East Uralian zones (i.e., part of the collage of microcontinental fragments), where the upper crust is cut by discontinuous east-dipping reflections; however, the dominating fabric beneath the accreted terranes is a broad zone of west-dipping reflections that penetrates the entire crust and that projects up to shear zones mapped at the surface within the Paleozoic collage. These reflections also continue down to and appear to merge with a highly reflective, west-dipping layered sequence that correlates with the Moho discontinuity (Fig. 9B). The apparent soling of dipping crustal shear zones into the Moho would imply coeval deformation on the Moho, which acted as a structural décollement (Echtler et al., 1996; Steer et al, 1998). The west-dipping reflectivity is limited on the west by the Troitsk fault, which is interpreted as a major suture between the trans-Uralian and East Uralian zones (Fig. 9B). Within the basement and cover sequence of the East European craton, immediately west of the Main Uralian suture, the crust is cut by gently east-dipping reflections that flatten toward the middle crust. These reflectors correlate updip to west-vergent thrusts in the Paleozoic platform sequence of the craton and project eastward down into a zone of diffuse reflectivity in the root of the orogen (Fig. 9B). The east-dipping reflections describe an imbrication of the crust deformed by antiformal structures in the footwall of the Main Uralian suture, which corresponds to the leading edge of the autochthonous Precambrian craton. The leading edge projects down to the center of the crustal root (or keel) beneath the core of the orogen.

SYNTHESIS

Relation between dipping and horizontal reflectivity

Prominent dipping reflections are the dominant reflector fabric throughout most of the Appalachian-Caledonian-Variscan orogen, with the foremost exception of the East Shetland platform, which shows prominent middle and lower crustal horizontal layering. Confirmed reflections from known inclined fault or tectonic zones in the middle to lower crust have been recorded from several surveys, for example, the Wind River

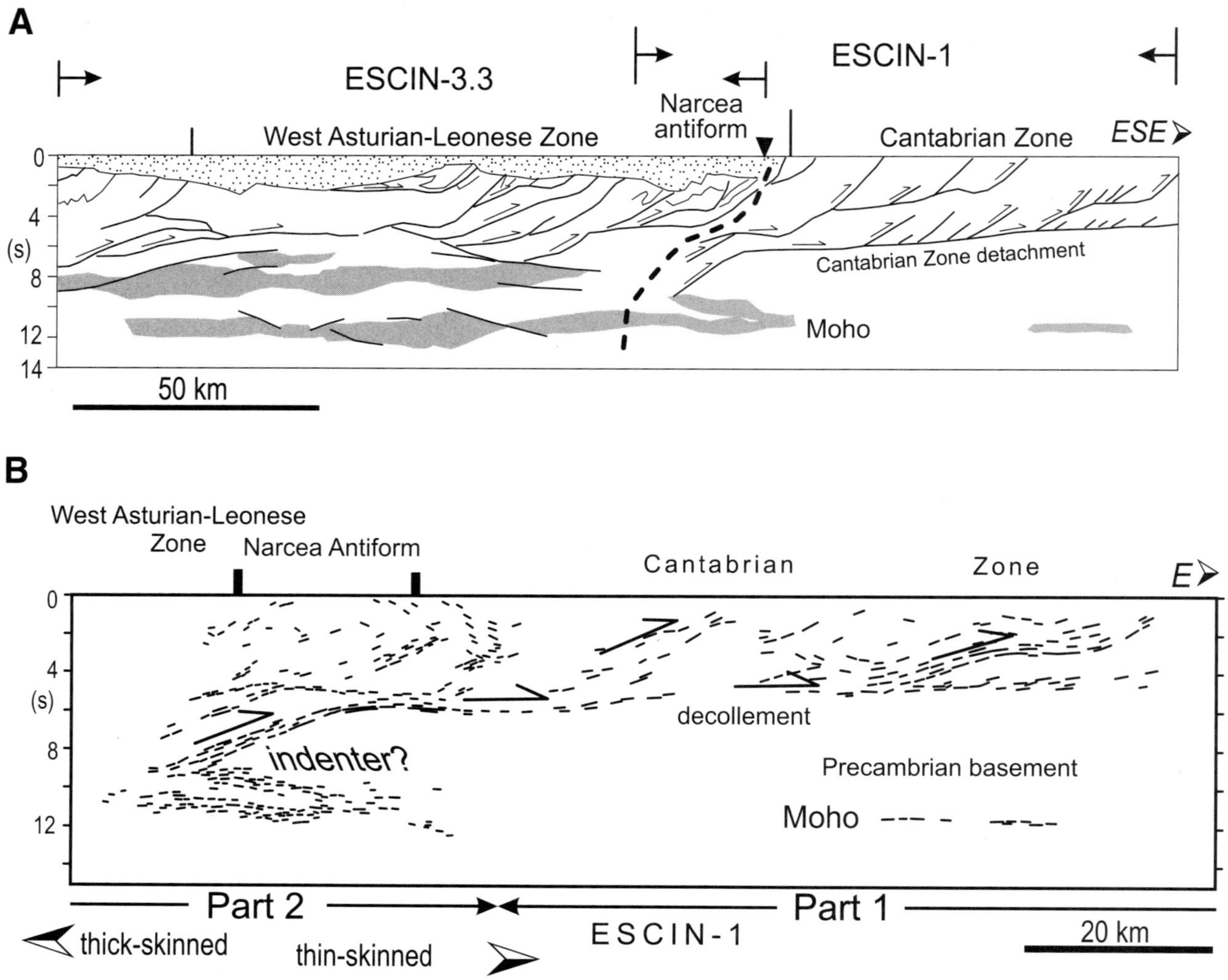

Figure 8. A: Interpretation of deep seismic profiles Estudio Sísmico de la Corteza Ibérica Norte (ESCIN) lines 1 and 3.3 from Spanish Variscides (Ayarza et al., 1998) (Fig. 1A). Coarse stipple is Mesozoic to Holocene sediments, fine shading is reflective lower crust, dashed line is interpreted suture (Ayarza et al., 1998). B: Detailed line drawing interpretation of ESCIN-1 over Cantabrian zone (Pérez-Estaún et al., 1994), which corresponds approximately to right half of A.

thrust in Wyoming (Smithson et al., 1979) and the Grenville front in Ontario and Ohio (Green et al., 1988; Culotta et al., 1990). Fracturing, fault gouge, mylonitization, and enhanced pore pressure are ways in which the reflection coefficient can be increased enough to produce a bright reflection (Hurich et al., 1985). Our interpretations, and those of previous investigators, of the seismic reflection profiles are influenced by the geological context of a compressional orogen. Thus, reflector fabrics, especially dipping reflectors, are seen as interpretable as deformation; however, given the high relative amplitude and coherency of many dipping crustal reflectors, it seems likely that igneous intrusion (e.g., dikes and sills) may have enhanced, or in some cases complicated, original deformation-related boundaries. For example, reflectivity within the Alleghanian suture zone beneath the Coastal Plain in eastern Georgia may have been enhanced by local tabular intrusion of mafic igneous rocks (Pratt et al., 1991). In parts of the Scandinavian Caledonide orogen, strong dipping and subhorizontal crustal reflections are interpreted as mafic igneous sill complexes (Juhojuntti et al., 2001), even though preliminary interpretations were in terms of deformation fabrics. Shear zones in contractional and/or transpressive orogens, such as the Appalachians and the Alps, may channel the ascent of igneous melts in the crust (Brown and Solar, 1998; Rosenberg, 2001). The crystallization of such magmas along a fault plane could markedly enhance its reflectivity.

Dipping reflections that penetrate the middle and lower crust typically approach the Moho discontinuity as planar or almost planar reflections, as in the Alleghanian suture zone beneath the coastal plain in Georgia (Fig. 2C) or Moine thrust offshore Shetland (Fig. 5C). Above the lowermost crust, dipping reflections commonly occur in subparallel sets and retain nearly constant dip on traveltime sections without showing obvious

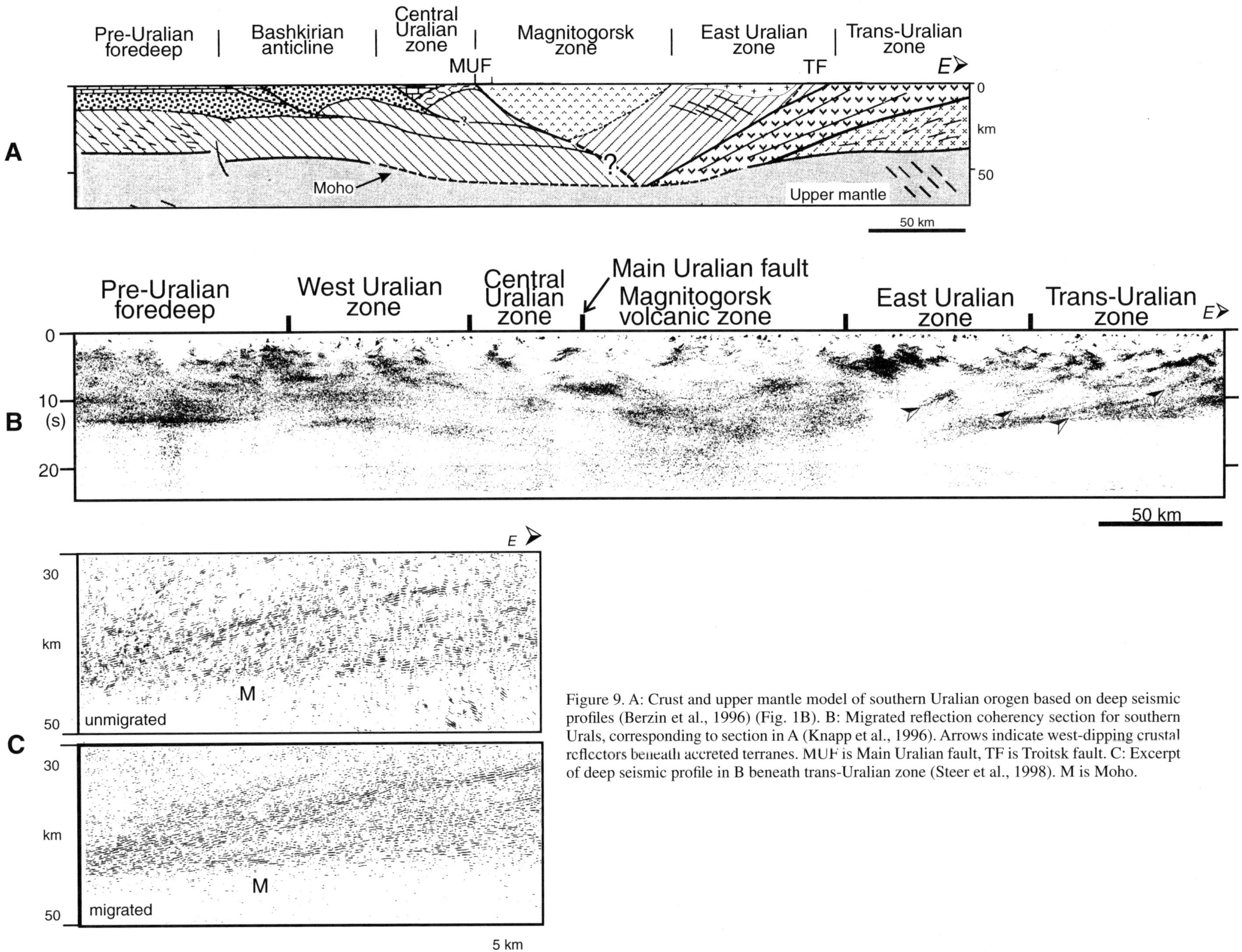

Figure 9. A: Crust and upper mantle model of southern Uralian orogen based on deep seismic profiles (Berzin et al., 1996) (Fig. 1B). B: Migrated reflection coherency section for southern Urals, corresponding to section in A (Knapp et al., 1996). Arrows indicate west-dipping crustal reflectors beneath accreted terranes. MUF is Main Uralian fault, TF is Troitsk fault. C: Excerpt of deep seismic profile in B beneath trans-Uralian zone (Steer et al., 1998). M is Moho.

ramp and flat or curved reflector surfaces. Broad zones of straight crustal-penetrating reflectors suggest large-scale imbrication of most of the crust. In some places, most notably in the Newfoundland Appalachians (Fig. 4, A and B), the lower lengths of otherwise planar reflections decrease in dip above the Moho discontinuity. In general, Cook and Varsek (1994) interpreted the curved lowermost sections of some reflector patterns at or just above the Moho as indicating a mechanically weak zone that developed coeval with the main compressional deformation in the orogen (i.e., the Moho is a décollement). Such an explanation can also be advanced for the crust beneath the accreted terranes of the eastern Uralian orogen (Fig. 9A). Alternative explanations for Moho development are that: (1) the thin zone of planar lower crustal reflectivity just at the Moho may represent a relict delamination surface created by the peeling away of the original upper mantle lithosphere (and lower crust?) just after or late during the compressional deformation due to density instabilities caused by subducted lithosphere directly beneath the orogen (Nelson, 1992; Willett et al., 1993; Meissner and Mooney, 1998); or (2) the flat reflective Moho and coincident thinned crust, such as observed oceanward of the Central Piedmont suture (Fig. 2A), are the result of Mesozoic mantle-derived mafic intrusion of the lower crust associated with Atlantic or North Sea rifting and crustal extension (see Warner, 1990, for a review).

For example, the central and southern Appalachians oceanward of the Central Piedmont suture (Fig. 2A) have undergone significant postorogenic magmatism related to the initial rifting of the Atlantic as shown by the presence of Triassic-Jurassic sedimentary basins and associated tholeiitic basaltic dikes, flows, and sills (Costain et al., 1989; McBride et al., 1989; McBride and Nelson, 1991). The redevelopment of the Moho discontinuity, whether by annealing or delamination, could have destroyed the original Paleozoic compressional structure in the lowermost crust. In the Newfoundland Appalachians, any Mesozoic thermal event related to opening of the Atlantic Ocean was apparently relatively subdued, as evidenced by Triassic and Jurassic basin sediments and volcanic rocks of Atlantic Canada now being mostly confined to the Bay of Fundy in Nova Scotia (Williams et al., 1999). Thus, a curved reflector pattern like that preserved above the Moho in eastern Newfoundland (Fig. 4A), as well as observed in the relatively unextended eastern Avalonian-Caledonian crust divided by the Iapetus suture (Fig. 6B), could have been obliterated in the lowermost crust of the southern Appalachians (Fig. 2A), where postorogenic modification of the lower crust was more pronounced. The Iapetus suture profile (Fig. 6B) provides a rare example of a dipping reflector pattern, associated with a crustal suture, that disrupts the Moho at the point of intersection, suggesting that the original lower crustal structure has been preserved and was unaffected by the subsequent extension and magmatism in the nearby North Sea. Similarly, the projection of the Main Uralian suture down to the center of the crustal root beneath the orogen in the southern Urals implies preservation of original, synorogenic crustal structure (Fig. 9A).

The dual levels of uneven and diffractive (or corrugated) reflectivity in the lower crust beneath the East Shetland platform correlate on an orthogonal profile with south-vergent folded and imbricated thrust structure in the lower crust (Fig. 7B). This provides one of the clearest examples of lower crustal layering that correlates with compressional structure and thus represents the original synorogenic structure and not a postconvergent process. Watts et al. (1990) reasoned that the prominent layered lower crust appearing on deep reflection profiles in the Valencia trough, western Mediterranean Sea, was inherited from an earlier deformation unrelated to the rifting of the trough.

Thin-skinned versus thick-skinned deformation regimes

Perhaps the most important orogenic variation around the North Atlantic Ocean involves the differential development and/or preservation of thin-skinned versus thick-skinned deformation. Cook and Varsek (1994) argued that compressional orogens at a lithospheric scale describe only a thin-skinned deformation regime by assuming that all material is transported laterally along décollements at multiple levels. Therefore, the distinction between thin- and thick-skinned regimes as defined by Cook and Varsek (1994) would be superficial. Nevertheless, we suggest that this concept aids in understanding variations between crusts dominated by shallow horizontal versus deep-dipping reflector patterns. The variation in regimes is exemplified in the complete transects through the southern Appalachians (Fig. 2A) and the northwest Spanish Variscides (Fig. 8A), which are among the few that show the geometrical relationship between thin-skinned deformation involving a well-formed décollement and thick-skinned crustal-penetrating deformation emerging from a suture or root zone.

As observed from reflection profiles in the southern Appalachians, thin-skinned deformation is expressed as a shallow crustal décollement beneath the foreland that is closely coupled with a narrow zone of thick-skinned deformation beneath the hinterland expressed as dipping reflectors that usually include antiformal underthrusted duplex patterns (Fig. 2, A, D, and E). The seismic transect through the Spanish Variscides images a foreland décollement sequence (Cantabrian zone) in close proximity to a possible Variscan shear or suture zone (Narcea antiform, Fig. 8B). The near juxtaposition of thin- and thick-skinned crustal deformation here is similar to the relationship between the Appalachian décollement and the Alleghanian-Acadian suture in the southeastern United States (Fig. 2A) and implies a common deformation process for these two parts of the orogen. In contrast, in the offshore British Caledonides (Figs. 5C and 6B) and the Uralian orogen (Fig. 9A), reflector patterns are dominated by crust-penetrating dipping surfaces, implying local thick-skinned deformation or large-scale crustal imbrication. For the Caledonides, this is in spite of onland geological inferences of a regional décollement (thin-skinned regime) beneath the core of the Scottish Caledonides (Butler and Coward, 1984) (Fig. 5A). The lack of an observable shallow regional décollement in this part of the Caledonides may imply a very restricted foreland development (or preservation) relative to the U.S. Appalachians. This may also be due to deeper levels of erosion, compared to the U.S. Appalachians, and/or less development of subsequent Alleghanian (Variscan) deformation spread across the Caledonides. For example, the higher grade of metamorphism in Shetland rocks implies a deeper level of erosion (Fettes et al., 1985), which might explain the apparent discrepancy between mainland Scotland (Fig. 5A) and the area nearer Shetland (Fig. 5C). The crustal structure of Newfoundland (Fig. 4A) and the adjacent offshore seems anomalous compared to the U.S. Appalachians because no décollement is observed, but only steep reflector fabrics representing Taconian (Quinlan et al., 1992) deformation. Hall et al. (1998) suggested that an interpreted discontinuity between upper and lower crustal dipping reflectors implies a mid-crustal rheological detachment; however, no horizontal reflection sequence analogous to the Appalachian décollement is observed.

Suture and compressional zone structure

Among the most common reflector signatures preserved in suture and compressional zones are the antiformal patterns such as seen below the southern Appalachians (Fig. 2, D and E), the New England Appalachians (Fig. 3B), the Shetland platform (Fig. 7A), the Narcea antiform (Fig. 8B), and west of the Main Uralian suture (Fig. 9A). Such antiformal patterns can be interpreted as asymmetric thrust horses (i.e., large slices or slivers separated from their original mass) that root at a level much deeper in the crust (e.g., St. Julien et al., 1983). A buried antiform may be difficult to image seismically because a convex-upward surface tends to defocus acoustic energy; however, such patterns can be imaged at great depth, notably in the Shetland platform (McBride and England, 1999), and provide important kinematic evidence for compressional deformation and the vergence direction. Blundell (1993) documented underthrusted antiformal patterns from the southern British Caledonides and interpreted them as thrust deformation associated with the Caledonian orogeny. From a field structural geology point of view, these structures appear to be scaled-up versions of fault-propagation folds, which are characteristic of fold and thrust belts.

The most remarkable contrast between reflector structure in the North American Appalachians and the British Caledonides is in the singular nature of the boundary of reflective wedges beneath the Caledonides. A collisional orogenic suture is usually conceived geologically as a complex zone of broad deformation, rather than a single easily recognizable line, especially at deep structural levels in older, more eroded orogens (Dewey, 1977). This concept appears to be partly in contrast with observations of a single bright planar reflection that projects down from an orogenic structure mapped at the surface into the deep crust (e.g., Figs. 5C and 6B). This singular boundary is inferred to correspond to a deformation front or leading edge of terrane accretion (e.g., Figs. 5C and 6B). These wedges are capped by a highly coherent reflection (e.g., Fig. 5B). The inference that these capping reflections constitute a significant boundary or discrete suture (i.e., not merely one of a sequence of thrusts) is supported by their separating an unreflective upper crust from a highly reflective crust below (McBride et al., 1996). Such a prominent reflection boundary also appears beneath the Troitsk fault suture in the eastern Uralides (Fig. 9A) dividing a reflective crust to the east from a less reflective one immediately to the west. The occurrence of a wedge of dipping reflectors in the footwalls or lower plates of the Moine thrust system and of both the Main Uralian and Troitsk fault sutures can be interpreted as imbrication of crustal packages beneath the major sutures of the orogenic belt (e.g., Fig. 5A).

The unusually distinct expression of suture boundaries beneath the Moine thrust and Iapetus suture (Figs. 5C and 6B) points to a strong contrast in physical properties across what must otherwise be a complex deformation zone (Soper et al., 1992). However, high-quality reflection profiles located over the central North Sea basin and the North German basin of the southern Baltic Sea and northern Germany (Fig. 1A) indicate a strikingly different picture. Reflector images of the early Caledonian (Late Ordovician to Silurian) convergent boundary between Baltica and eastern Avalonia (Fig. 1B; Caledonian deformation front of Meissner et al., 1994) show only a very thin zone of relatively weak south-dipping reflections that steeply penetrate the crust (DEKORP-BASIN Research Group, 1999; Abramovitz and Thybo, 2000). The strong contrast across reflector boundaries beneath the British Caledonides could be explained by (1) a lack of modification by post-Devonian basin formation and crustal extension, as occurred beneath the central North Sea and southern Baltic Sea or by (2) an absence of superimposed late Paleozoic convergence as occurred in the southern Appalachians (Keller and Hatcher, 1999). Repeated convergent reactivation and modification of preexisting structures could conceivably create a complexly and broadened distribution of deformation expressed as subdued reflectivity.

The central convergent zones beneath the Iapetus suture, the West Asturian Leonese zone, and the Main Uralian fault all share the common characteristic of a crustal root or narrow keel into which the suture zone projects down from the foreland part of the orogen. Although this relationship between a suture zone and a crustal root or keel is not commonly observed on deep seismic reflection profiles, it may represent the original whole-crustal structure before postorogenic processes destroyed deeper parts of the orogen.

SUMMARY

Four dominating reflector fabrics can be recognized within the Appalachian-Caledonide orogen.

Shallow crustal sequences of nearly horizontal reflections (décollement levels) abruptly merge with complex dipping zones (sutures or root zones) (e.g., U.S. Appalachians; Spanish Variscides).

Broad antiformal reflector patterns, in places underlain by a dipping planar reflector, are interpreted as asymmetric folding and shearing in the deep crust (e.g., U.S. Appalachians; Shetland platform).

When observed in a compressional orogenic context, complex, subhorizontal reflector patterns in the lower crust can be explained as layers of intense ductile shearing related to terrane accretion and continental convergence (e.g., Shetland platform; Iapetus suture).

Series of deeply penetrating mostly planar dipping reflectors are interpreted as reverse-sense brittle faults or ductile shear zones (e.g., Moine thrust; North American Appalachians; Uralides).

Many deep seismic profiles are dominated by a prominent dipping reflector that correlates with a metamorphic and/or terrane boundary and that divides the crust into a reflective lower section and a poorly reflective upper section. The planar continuation of the reflective wedge from the upper to middle crust toward the Moho suggests a local thick-skinned model involving imbricated thrust slices. The dipping reflector commonly marks the leading edge of highly reflective crust that may be more coherently deformed than poorly reflective crust (e.g., British Caledonides). A similarly distinct division is less typical for the North American Appalachians, where such reflector-defined boundaries appear to be more diffuse, and may indicate a difference in the degree of postorogenic reactivation or modification and/or difference in the lithospheric strength (e.g., a weaker crust deforms more readily, giving rise to complex and less discrete reflector patterns). In the U.S. Appalachians and the Spanish Variscides, a shallow subhorizontal décollement, neatly separating upper and lower plates in a thin-skinned geometry, dips downward into and merges with a relatively chaotic root zone of imbricate thrust and shear zones deeper in the crust. In most places along the orogen, the structure of intracrustal planar reflectors is more complex and variable nearer the Moho discontinuity: (1) examples of reflectors curving into the Moho seem to typify orogens where the synorogenic lowermost crust is preserved and imply that the Moho acted as a décollement, while (2) places where planar reflectors are abruptly cut off by a flat Moho suggest a postorogenic process that has destroyed and reequilibrated lowermost crustal structure. Igneous intrusive processes that occur after or in conjunction with brittle or ductile shearing could enhance or alter original deformation structure, for example, by reactivating weak zones. The layered reflection sequences in the lowermost crust beneath regions of known rifting and related magmatism thus could also be an expression of mafic sill intrusion.

ACKNOWLEDGMENTS

This work was supported in part by the Earthquake Engineering Research Centers Program of the National Science Foundation under Award EEC-9701785 (McBride). We thank G.R. Keller, W.A. Thomas, and an anonymous referee for reviews of a previous version that substantially improved the final manuscript.

REFERENCES CITED

Abramovitz, T., and Thybo, H., 2000, Seismic images of Caledonian, lithosphere-scale collision structures in the southeastern North Sea along Mona Lisa Profile 2: Tectonophysics, v. 317, p. 27–54.

Ando, C.J., Czuchra, B.L., Klemperer, S.L., Brown, L.D., Cheadle, M.J., Cook, F.A., Oliver, J.E., Kaufman, S., Walsh, T., Thompson, J.B., Jr., Lyons, J.B., and Rosenfeld, J.L., 1984, Crustal profile of mountain belt: COCORP deep seismic reflection profiling in New England Appalachians and implications for architecture of convergent mountain chains: AAPG Bulletin, v. 68, p. 819–837.

Ando, C.J., Cook, F.A., Oliver, J.E., Brown, L.D., and Kaufman, S., 1983, Crustal geometry of the Appalachian orogen from seismic reflection studies, *in* Hatcher, R.D., Jr., Williams, H., and Zietz, I, Contributions to the tectonics and geophysics of mountain chains: Boulder, Colorado, Geological Society of America Memoir 158, p. 83–101.

Armstrong, H.A., and Owen, A.W., 2001, Terrane evolution of the paratectonic Caledonides of northern Britain: Journal of the Geological Society, London, v. 158, p. 475–486.

Arnow, J.A., 1987, Deep structure from continental collision: COCORP deep seismic reflection profiles across the southern Appalachians [M.S. thesis]: Ithaca, New York, Cornell University, 82 p.

Ayarza, P., Martínez Catalán, J.R., Gallart, J., Pulgar, J., and Dañobeitia, J.J., 1998, Estudio Sísmico de la Corteza Ibérica Norte 3.3: A seismic image of the Variscan crust in the hinterland of the NW Iberian Massif: Tectonics, v. 17, p. 171–186.

Barr, D., Holdsworth, R.E., and Roberts, A.M., 1986, Caledonian ductile thrusting in a Precambrian metamorphic complex: The Moine of northwestern Scotland: Geological Society of America Bulletin, v. 97, p. 754–764.

Berzin, R., Oncken, O., Knapp, J.H., Pérez-Estaún, A., Hismatulin, T., Yunusov, N., and Lipilin, A., 1996, Orogenic evolution of the Ural Mountains: Results from an integrated seismic experiment: Science, v. 274, p. 220–221.

Blundell, D.J., 1993, Deep structure of the Anglo-Brabant massif revealed by seismic profiling: Geological Magazine, v. 130, p. 563–567.

Brewer, J.A., and Smythe, D.K., 1984, MOIST and the continuity of crustal reflector geometry along the Caledonian-Appalachian orogeny: Journal of the Geological Society, London, v. 141, p. 105–120.

Brown, M., and Solar, G.S., 1998, Shear-zone systems and melts: Feedback relations and self-organization in orogenic belts: Journal of Structural Geology, v. 20, p. 211–227.

Butler, R.W.H., 1986, Structural evolution in the Moine of Northwest Scotland: A Caledonian linked thrust system?: Geological Magazine, v. 123, p. 1–11.

Butler, R.W.H., and Coward, M.P. 1984. Geological constraints, structural evolution, and deep geology of the northwest Scottish Caledonides: Tectonics, v. 3, p. 347–365.

Carbonell, R., Pérez-Estaún, A., Gallart, J., Diaz, J., Kashubin, S., Mechie, J., Stadtlander, R., Schulze, A., Knapp, J.H., and Morozov, A., 1996, Crustal root beneath the Urals: Wide-angle seismic evidence: Science, v. 274, p. 222–224.

Chowns, T.M., and Williams, C.T., 1983, Pre-Cretaceous rocks beneath the Georgia Coastal Plain: Regional implications, *in* Gohn, G.S., ed., Studies related to the Charleston, South Carolina, earthquake of 1886: tectonics and seismicity, U.S. Geological Survey Professional Paper, p. L1–L42.

Cook, F.A., and Varsek, J.L., 1994, Orogen-scale décollements: Reviews of Geophysics, v. 32, p. 37–60.

Cook, F.A., Albaugh, D.S., Brown, L.D., Kaufman, S., Oliver, J.E., and Hatcher, R.D., Jr., 1979, Thin-skinned tectonics in the crystalline southern Appalachians: COCORP seismic-reflection profiling of the Blue Ridge and Piedmont: Geology, v. 7, p. 563–567.

Cook, F.A., Brown, L.D., and Oliver, J.E., 1980, The southern Appalachians and the growth of continents: Scientific American, v. 243, p. 156–168.

Cook, F.A., Brown, L.D., Kaufman, S., Oliver, J.E., and Petersen, T.A., 1981, COCORP seismic profiling of the Appalachian Orogen beneath the coastal plain of Georgia: Geological Society of America Bulletin, v. 92, p. 738–748.

Cook, F.A., Matthews, D.H., and Jacobs, A.W.B., 1988, Crustal and upper mantle structure of the Appalachian-Caledonide orogen from seismic results: Geological Society of London Special Publication 38, p. 21–33.

Çoruh, C., G.A. Bollinger, and J.K. Costain, 1988. Seismogenic structures in the central Virginia seismic zone: Geology v. 16, 748–751.

Costain, J.K., Hatcher, R.D., Jr., Çoruh, C., Pratt, T.L., Taylor, S.R., Litehiser, J.J., and Zietz, I., 1989, Geophysical characteristics of the Appalachian crust, in Hatcher, R.D., Jr., Thomas, W.A., and Viele, G.W., eds., The Appalachian-Ouachita Orogen in the United States: Boulder, Colorado, Geological Society of America, The Geology of North America, v. F-2, p. 385–416.

Culotta, R.C., Pratt, T., and Oliver, J., 1990, A tale of two sutures: COCORP's deep seismic surveys of the Grenville province in the eastern U.S. midcontinent: Geology, v. 18, p. 646–649.

DEKORP-BASIN Research Group, 1999, Deep crustal structure of the Northeast German basin: New DEKORP-BASIN '96 deep-profiling results: Geology, v. 27, p. 55–58.

Dewey, J.F., 1977, Suture zone complexities: a review: Tectonophysics, v. 40, p. 53–67.

Diaconescu, C.C., Knapp, J.H., Brown, L.D., Steer, D.N., and Stiller, M., 1998, Precambrian Moho offset and tectonic stability of the East European platform from the URSEIS deep seismic profile: Geology, v. 26, p. 211–214.

Echtler, H.P., Stiller, M., Steinhoff, F., Krawczyk, C., Suleimanov, A., Spiridonov, V., Knapp, J.H., Menshikov, Y., Alvarez-Marrón, J., and Yunusov, N., 1996, Preserved collisional crustal structure of the southern Urals revealed by vibroseis profiling: Science, v. 274, p. 224–226.

Elliott, D., and Johnston, M.R., 1980, Structural evolution of the northern part of the Moine thrust belt, NW Scotland: Transactions of the Royal Society of Edinburgh, v. 71, p. 69–96.

Fettes, D.J., Long, C.B., Bevins, R.E., Max, M.D., Oliver, G.J.H., Primmer, T.J., Thomas, L.J., and Yardley, B.W.D., 1985, Grade and time of metamorphism in the Caledonide orogen of Britain and Ireland, *in* Harris, A.L., ed., The nature and timing of orogenic activity in the Caledonian rocks of the British Isles: London, Geological Society of London Memoir 9, p. 41–53.

Gee, D.G., and Zeyen, H.J., editors, 1996, EUROPROBE 1996-Lithosphere Dynamics: Origin and Evolution of Continents: EUROPROBE Secretariat, Uppsala University, 138 p.

Green, A.G., Milkereit, B., Davidson, A., Spencer, C., Hutchinson, D.R., Cannon, W.F., Lee, M.W., Agena, W.F., Behrendt, J.C., and Hinze, W.J., 1988, Crustal structure of the Grenville Front and adjacent terranes: Geology, v. 16, p. 788–792.

Hall, J., Marillier, F., and Dehler, S., 1998, Geophysical studies of the structure of the Appalachian orogen in the Atlantic borderlands of Canada: Canadian Journal of Earth Sciences, v. 35, p. 1205–1221.

Hall, J., and Quinlan, G., 1994, A collisional crustal fabric pattern recognized from seismic reflection profiles of the Appalachian/Caledonide orogen: Tectonophysics, v. 232, p. 31–42.

Hatcher, R.D., Jr., 1987, Tectonics of the southern and central Appalachian Internides: Annual Reviews Earth and Planetary Sciences, v. 15, p. 337–362.

Hurich, C.A., Smithson, S.B., Fountain, D.M., and Humphreys, M.C., 1985, Seismic evidence of mylonite reflectivity and deep structure in the Kettle Dome metamorphic core complex, Washington: Geology, v. 13, p. 577–580.

Juhojuntti, N., Juhlin, C., and Dyrelius, D., 2001, Crustal reflectivity beneath the Central Scandinavian Caledonides: Tectonophysics, v. 334, p. 191–210.

Keller, G.R., and Hatcher, R.D., Jr., 1999, Some comparisons of the structure and evolution of the southern Appalachian-Ouachita orogen and portions of the Trans-European Suture Zone region: Tectonophysics, v. 314, p. 43–68.

Kirton, S.R., and Hitchen, K., 1987, Timing and style of crustal extension N of the Scottish mainland, *in* Coward, M.P., Dewey, J.F., and Hancock, P.L., Continental extensional tectonics: London, Geological Society of London Special Publication 28, p. 501–510.

Klemperer, S., and R. Hobbs, 1991, The BIRPS Atlas Deep seismic reflection profiles around the British Isles: Cambridge, Cambridge University Press, 128 p, 100 fold-outs.

Klemperer, S.L., and Hurich, C.A., 1990, Lithospheric structure of the North Sea from deep seismic reflection profiling, *in* Blundell, D.J., and Gibbs, A.D., eds., Tectonic evolution of the North Sea rifts: Oxford, Clarendon Press, p. 37–63.

Knapp, J.H., Sokolov, V.B., Hatcher, R.D., Jr., and Brown, L.D., 1993, Comparative crustal structure of the Urals and Appalachians: Geological Society of America Abstracts with Programs, v.25, no. 6, p. A-179.

Knapp, J.H., Steer, D.N., Brown, L.D., Berzin, R., Suleimanov, A., Stiller, M., Lüschen, E., Brown, D.L., Bulgakov, R., Kashubin, and Rybalka, A.V., 1996, Lithosphere-scale seismic image of the southern Urals from explosion-source reflection profiling: Science, v. 274, p. 226–228.

Knapp, J.H., Diaconescu, C.C., Bader, M.A., Kashubin, S.N., Rybalka, A., and Sokolov, V.B., 1998, Seismic reflection fabrics of continental collision and post-orogenic extension in the Middle Urals, Russia: Tectonophysics, v. 288, p. 115–126.

LeFort, J.-P., 1989, Basement correlation across the North Atlantic: Berlin, Springer-Verlag, 148 p.

Matte, P., 1991, Accretionary history and crustal evolution of the Variscan belt in western Europe: Tectonophysics, v. 196, p. 309–337.

Matte, P., and Hirn, A., 1988, Seismic signature and tectonic cross section of the Variscan crust in Western France: Tectonics, v. 7, p. 141–155.

McBride, J.H., and England, R.W., 1994. Deep seismic reflection structure of the Caledonian orogenic front west of Shetland: Journal of the Geological Society, London, v. 151, p. 9–16.

McBride, J.H., and England, R.W., 1995, Problems interpreting deep crustal reflectors beneath the Moine thrust between Shetland and the Scottish Highlands. Geological Magazine, v. 132, p. 465–472.

McBride, J.H., and England, R.W., 1999, Window into the Caledonian Orogen: Structure of the crust beneath the East Shetland Platform, United Kingdom: Geological Society of America Bulletin, v. 111, p. 1030–1041.

McBride, J.H., and Nelson, K.D., 1991. Deep seismic reflection constraints on Palaeozoic crustal structure and definition of the Moho in the buried Southern Appalachian orogen, *in* Meissner, R., Brown, L., Dürbaum, H.-J., Franke, W., Fuchs, K., and Seifert, F., eds., Continental lithosphere: Deep seismic reflections: Washington, D.C., American Geophysical Union Geodynamics Series 22, p. 9–20.

McBride, J.H., Nelson, K.D., and Brown, L.D., 1989, Evidence and implications of an extensive early Mesozoic rift basin and basalt/diabase sequence beneath the Southeast Coastal Plain: Geological Society of America Bulletin, v. 101, p. 512–520.

McBride, J.H., Snyder, D.B., Tate, M.P., England, R.W., and Hobbs, R.W., 1995, Upper mantle reflector structure and origin beneath the Scottish Caledonides: Tectonics, v. 14, p. 1351–1367.

McBride, J.H., Snyder, D.B., England, R.W., and Hobbs, R.W., 1996, Dipping reflectors beneath old orogens: A perspective from the British Caledonides: GSA Today v. 6, no. 4, p. 1–6.

McBride, J.H., Hatcher, R.D., Jr., and Stephenson, W.J., 2001, Toward a reconciliation of seismic reflection and field geologic mapping from the Pine Mountain belt: Preliminary results: Geological Society of America Abstracts with Programs, v. 33, no. 2, p. A-74.

McGeary, S., 1987, Nontypical BIRPS on the margin of the northern North Sea: The SHET survey: Geophysical Journal of the Royal Astronomical Society, v. 89, p. 231–238.

McGeary, S., and Warner, M.R., 1985, Seismic profiling the continental lithosphere: Nature, v. 317, p. 795–797.

McKerrow, W.S., Mac Niocaill, C., and Dewey, J.F., 2000, The Caledonian Orogen redefined: Journal of the Geological Society of London, v. 157, p. 1149–1154.

Meissner, R., and Bortfeld, R.K., 1990, The DEKORP Atlas: Berlin, Springer, 19 p., 80 seismic sections.

Meissner, R., and Mooney, W.D., 1998, Weakness of the lower continental crust: A condition for delamination, uplift, and escape: Tectonophysics, v. 296, p. 47–60.

Meissner, R., and Sadowiak, P., 1992, The terrane concept and its manifestation by deep reflection studies in the Variscides: Terra Nova, v. 4, p. 598–607.

Meissner, R., Sadowiak, P., Thomas, S.A., and BABEL Working Group, 1994, East Avalonia, the third partner in the Caledonian collisions: Evidence from deep seismic reflection data: Geologische Rundschau, v. 83, p. 186–196.

Morgan, R.P.L., Barton, P.J., Warner, M., Morgan, J., Price, C., and Jones, K., 2000, Lithospheric structure north of Scotland; I, P-wave modelling, deep reflection profiles, and gravity: Geophysical Journal International, v. 142, p. 716–736.

Nelson, K.D., 1988, The COCORP Atlas, Volume 1: Ithaca, New York, Institute for the Study of the Continents, Cornell University.

Nelson, K.D., 1992, Are crustal thickness variations in old mountain belts like the Appalachians a consequence of lithospheric delamination?: Geology, v. 20, p. 498–502.

Nelson, K.D., Arnow, J.A., McBride, J.H., Willemin, J.E., Huang, J., Zheng, L., Oliver, J.E., Brown, L.D., and Kaufman, S., 1985, New COCORP profiling on the southeastern United States. Part I: Late Paleozoic suture and Mesozoic rift basin: Geology, v. 13, p. 714–718.

Pérez-Estaún, A., Martínez-Catalán, J.R., and Bastida, F., 1991, Crustal thickening and deformation sequence in the footwall to the suture of the Variscan belt of northwest Spain: Tectonophysics, v. 191, p. 243–253.

Pérez-Estaún, A., Pulgar, J.A., Banda, E., Alvarez-Marrón, J., and ESCI-N Research Group, 1994, Crustal structure of the external variscides in northwest Spain from deep seismic reflection profiling: Tectonophysics, v. 232, p. 91–118.

Phinney, R.A., and Roy-Chowdhury, K., 1989, Reflection seismic studies of crustal structure in the eastern United States, *in* Pakiser, L.C., and Mooney, W.D., Geophysical framework of the continental United States: Boulder, Colorado, Geological Society of America Memoir 172, p. 613–653.

Pratt, T.L., Hauser, E.C., Hearn, T.M., and Reston, T.J., 1991, Reflection polarity of the midcrustal Surrency bright spot beneath southeastern Georgia; testing the fluid hypothesis: Journal of Geophysical Research, v. 96, p. 10,145–10,158.

Puchkov, V.N., 1997, Structure and geodynamics of the Uralian orogen, *in* Burg, J.-P., and Ford, M. eds., Orogeny through time: Geological Society of London Special Publication 121, p. 201–236.

Quinlan, G., Beaumont, C., and Hall, J., 1993, Tectonic model for crustal seismic reflectivity patterns in compressional orogens: Geology, v. 21, p. 663–666.

Quinlan, G.M., Hall, J., Williams, H., Wright, J.A., Colman-Sadd, S.P., O'Brian, S.J., Stockmal, G.S., and Marillier, F., 1992, Lithoprobe onshore seismic reflection transects across the Newfoundland Appalachians: Canadian Journal of Earth Science, v. 29, p. 1865–1877.

Rankin, D.W., Drake, A.A., Jr., Goldsmith, R., Hinze, W.J., Horton, J.W., Jr., Lidiak, E.G., McLelland, J.M., and Secor, D.T., Jr., 1993, Map showing subdivisions of the Grenville orogen and younger tectonostratigraphic terranes in eastern United States and adjacent Canada that may contain Precambrian rocks, Precambrian (Proterozoic) rocks east and southeast of the Grenville front, *in* Rankin, D.W., Reed, J.C., Jr., Bickford, M.E., Sims, P.K., Link, P.K., Houston, R.S., and Van Schmus, W.R., eds., Precambrian: Conterminous U.S.: Boulder, Colorado, Geological Society of America, Geology of North America, v. C-2, plate 5.

Rosenberg, C.L., 2001, Coupling of shear zones and magma ascent: A review of the Oligocene magmatism in the Alps: Geological Society of America Abstracts with Programs, v. 33, no. 6, p. A-212.

Sadowiak, P., and Wever, T., 1990, Reflection-diffraction seismic pattern at crustal suture zones: Tectonics: v. 9, p. 1495–1513.

Şengör, A.M.C., Natal'in, B.A., and Burtman, V.S., 1993, Evolution of the Altaid tectonic collage and Palaeozoic crustal growth in Eurasia: Nature v. 364, p. 299–307.

Smithson, S.B., Brewer, J.A., Kaufman, S., Oliver, J.E., and Hurich, C.A., 1979, Structure of the Laramide Wind River Uplift, Wyoming, from COCORP deep reflection data and from gravity data: Journal of Geophysical Research, v. 84, p. 5955–5972.

Snyder, D., and Hobbs, R., 1999, BIRPS Atlas II: A Second Decade of Deep Seismic Reflection Profiling: Geological Society Miscellaneous.

Snyder, D.B., 1990. The Moine Thrust in the BIRPS data set. Journal of the Geological Society of London, v. 147, p. 81–86.

Snyder, D.B., England, R.W., and McBride, J.H., 1997, Linkage between mantle and crustal structures and its bearing on inherited structures in northwestern Scotland: Journal of the Geological Society of London v. 154, p. 79–84.

Snyder, D.B., and Flack, C.A., 1990, A Caledonian age for reflections within the mantle lithosphere north and west of Scotland: Tectonics, v. 9, p. 903–922.

Soper, N.J., 1988, Timing and geometry of collision, terrane accretion and sinistral strike-slip events in the British Caledonides, *in* Harris, A.L., and Fettes, D.J. eds., The Caledonide-Appalachian Orogen: London, Geological Society of London, Special Publication 38, p. 481–492.

Soper, N.J., England, R.W., Snyder, D.B., and Ryan, P.D., 1992, The Iapetus suture zone in England, Scotland, and eastern Ireland: A reconciliation of geological and deep seismic data: Journal of the Geological Society of London, v. 149, p. 697–700.

Steer, D.N., Knapp, J.H., Brown, L.D., Berzin, R., Echtler, H., and Brown, D., 1998, Deep structure of the continental lithosphere in an unextended orogen: An explosive-source seismic reflection profile in the Urals (Urals Seismic Experiment and Integrated Studies [URSEIS 1995]): Tectonics, v. 17, p. 143–157.

St.-Julien, P., Slivitsky, A., and Feininger, T., 1983, A deep structural profile across the Appalachians of southern Quebec, *in* Hatcher, R.D., Jr., Williams, H., and Zietz, I., eds., Contributions to the tectonics and geophysics of mountain chains: Boulder, Colorado, Geological Society of America Memoir 158, p. 103–111.

Thomas, W.A., compiler, 1993, Pre-Mesozoic paleogeographic map of the Appalachian-Ouachita orogen beneath Atlantic and Gulf Coastal Plains: *in* Hatcher, R.D., Jr., Thomas, W.A., and Viele, G.W., eds., The Appalachian-Ouachita orogen in the United States: Boulder, Colorado, Geological Society of America, Geology of North America, v. F-2, plate 6.

Treagus, J.E., and King, G., 1978, A complete lower Dalradian succession in the Schiehallion District, central Perthshire: Scottish Journal of Geology, v. 14, p. 157–166.

Warner, M.R., 1990, Basalts, water, or shear zones in the lower continental crust?: Tectonophysics, v. 173, p. 163–174.

Warner, M., Morgan, J., Barton, P., Morgan, P., Price, C., and Jones, K., 1996, Seismic reflections from the mantle represent relict subduction zones within the continental lithosphere: Geology, v. 24, p. 39–42.

Watts, A.B., Torne, M., Buhl, P., Mauffret, A., Pascal, G., and Pinet, B., 1990, Evidence for reflectors in the lower continental crust before rifting in the Valencia trough: Nature, v. 348, p. 631–635.

Willett, S., Beaumont, C., and Fullsack, P., 1993, Mechanical model for the tectonics of doubly vergent compressional orogens: Geology, v. 21, p. 371–374.

Williams, H., 1979, Appalachian Orogen in Canada: Canadian Journal of Earth Sciences, v. 16, p. 792–807.

Williams, H., Dehler, S.A., Grant, A.C., and Oakey, G.N., 1999, Tectonics of Atlantic Canada: Geoscience Canada, v. 26, p. 51–70.

Winchester, J.A., and van Staal, C.R., 1995, Volcanic and sedimentary terrane correlation between the Dunnage and Gander zones of the Canadian Appalachians and the British Caledonides reviewed, *in* Hibbard, J.P., van Staal, C.R., and Cawood, P.A., eds., Current Perspectives in the Appalachian-Caledonian Orogen: Geological Association of Canada Special Paper 41, p. 95–114.

Manuscript Accepted by the Society February 8, 2002

Printed in the U.S.A.

Index